Nervous Systems
in Invertebrates

NATO ASI Series

Advanced Science Institutes Series

A series presenting the results of activities sponsored by the NATO Science Committee, which aims at the dissemination of advanced scientific and technological knowledge, with a view to strengthening links between scientific communities.

The series is published by an international board of publishers in conjunction with the NATO Scientific Affairs Division

A	**Life Sciences**	Plenum Publishing Corporation
B	**Physics**	New York and London
C	**Mathematical and Physical Sciences**	D. Reidel Publishing Company Dordrecht, Boston, and Lancaster
D	**Behavioral and Social Sciences**	Martinus Nijhoff Publishers
E	**Engineering and Materials Sciences**	The Hague, Boston, Dordrecht, and Lancaster
F	**Computer and Systems Sciences**	Springer-Verlag
G	**Ecological Sciences**	Berlin, Heidelberg, New York, London,
H	**Cell Biology**	Paris, and Tokyo

Recent Volumes in this Series

Volume 135—Signal Transduction and Protein Phosphorylation
edited by L. M. G. Heilmeyer

Volume 136—The Molecular Basis of Viral Replication
edited by R. Perez Bercoff

Volume 137—DNA—Ligand Interactions: From Drugs to Proteins
edited by Wilhelm Guschlbauer and Wolfram Saenger

Volume 138—Chaos in Biological Systems
edited by H. Degn, A. V. Holden, and L. F. Olsen

Volume 139—Lipid Mediators in the Immunology of Shock
edited by M. Paubert-Braquet

Volume 140—Plant Molecular Biology
edited by Diter von Wettstein and Nam-Hai Chua

Volume 141—Nervous Systems in Invertebrates
edited by M. A. Ali

Volume 142—A Multidisciplinary Approach to Myelin Diseases
edited by G. Serlupi Crescenzi

Series A: Life Sciences

Nervous Systems in Invertebrates

Edited by
M. A. Ali
University of Montreal
Montreal, Quebec, Canada

Plenum Press
New York and London
Published in cooperation with NATO Scientific Affairs Division

Proceedings of a NATO Advanced Study Institute on
Nervous Systems in Invertebrates,
held July 20-August 2, 1986,
at Bishop's University, Lennoxville, Quebec, Canada

Library of Congress Cataloging in Publication Data

NATO Advanced Study Institute on Nervous Systems in Invertebrates (1986:
 Bishop's University)
 Nervous systems in invertebrates / edited by M. A. Ali.
  p. cm.—(NATO ASI series. Series A, Life sciences: v. 141)
 "Proceedings of a NATO Advanced Study Institute on Nervous Systems
in Invertebrates, held July 20-August 2, 1986, at Bishop's University, Len-
noxville, Quebec, Canada"—T.p. verso.
 Includes bibliographies and index.
  ISBN-13: 978-1-4612-9084-1   e-ISBN-13: 978-1-4613-1955-9
  DOI: 10.1007/ 978-1-4613-1955-9
 1. Nervous system—Invertebrates—Congresses. I. Ali, M. A. (Mohamed
Ather), 1932-   . II. Title. III. Series. [DNLM: 1. Invertebrates—congresses.
2. Nervous System—congresses. QL 935 N279n]
 QL364.N38 1986
 592'.0188—dc19   87-25898
                                                              CIP

© 1987 Plenum Press, New York
Softcover reprint of the hardcover 1st edition 1987
A Division of Plenum Publishing Corporation
233 Spring Street, New York, N.Y. 10013

All rights reserved

No part of this book may be reproduced, stored in a retrieval system, or transmitted
in any form or by any means, electronic, mechanical, photocopying, microfilming,
recording, or otherwise, without written permission from the Publisher

PREFACE

The idea of holding an Advanced Study Institute (ASI) and getting a volume out, on the Nervous Systems in Invertebrates first cropped up in the summer of 1977 at the ASI on Sensory Ecology. I had prepared a review of the nervous systems in coelomates and noticed how much we depended on Bullock and Horridge's treatise on the one hand and how much new material and requirements has cropped up since 1965, when this classical work was published. Interest in the concerted study of pollution and environmental toxicology was growing in geometrical proportions and the use of invertebrates as indices was growing. As a teacher of a course on the biology of invertebrates since the beginning of my career I had also noticed how the interest of the students and the content of my course was shifting gradually and steadily from the traditional morphology-taxonomy type to the physiology-ecology-embryology orientation. Students were demanding to know the relevency of what they had to learn. Thus, after the ASI on Photoreception and Vision in Invertebrates held in 1982 the question of one on nervous systems was raised by a number of colleagues. It appeared then that the consensus was that the time was ripe to hold one and that it will be worthwhile. Therefore, as usual arrangements had to begin at least two years in advance. Most of the persons I contacted to lecture and write chapters on selected topics agreed enthusiastically. As is usual in the case of most ASIs, the programme had to be structured with the tutorial nature of the gathering and the ensuing volume in mind. This called for the selection of topics which were often imposed on the lecturers-authors. Also, as a NATO-ASI the choice of lecturers had to be made with as wide a national distribution as possible in mind. Of course, the reputation of the lecturer-author, his or her ability to present an interesting lecture and chapter and, his or her ability to get along with a heterogenous group over a two-week period had also to be taken into consideration. As the organiser, I was extremely lucky to gather a group of people who satisfied all these conditions as evidenced by the smooth way the ASI functioned. As I usually do, I asked the authors-lecturers to be as provocative and speculative as possible, especially in their oral presentations at the ASI. Most were so as evidenced by the lively discussions that ensued. At a meeting of the authors we ironed out the details and established general standards. Apart from the criticism the presentations received at the ASI, the finished products were also reviewed critically by the editor and at least one other competent person. As the organiser I attended every session and as editor read every chapter and learned a great deal about the matter and I hope that the users of this volume would find it of some use. The authors and I have tried to present the situation, as much as possible, as it reflects the actual state of affairs in this field. The concluding chapter, based on the rapporteur presentations and ensuing discussions which took place on the last day of the ASI tries to bring out as many perspectives as possible. We wanted to put in a glossary of terms but the constraints of time made this most impossible and I regret that it had to be so.

I am very grateful to my colleague Mary Ann Klyne for the help she gave in the organisation of the meeting and the editing of the volume. I thank Catherine Joron of Jacmar Informatique Inc. for the preparation of the typescript. Françoise Simard and Miss Margaret Pertwee helped with the various aspects of the organisation. I am also very appreciative of the help that Nick Strausfeld and Michel Anctil gave in the choice of lecturers-authors. Michel Anctil also kindly helped with the preparation of the introductory chapter.

Financial assistance was provided to a large extent by the Scientific Affairs Division of NATO and I thank the director of the ASI programme, Craig Sinclair, for his encouragement throughout. Other financial help came from the Natural Sciences and Engineering Research Council of Canada, FCAR du Québec and the Université de Montréal. I thank Jean-Luc Grégoire, vice-principal and Marcia Boisvert, coordinator of events at Bishop's University for their help. The director of my department, Roch Carbonneau, extended the numerous facilities of the department to facilitate the organisation of the ASI. My editor at Plenum Press, Patricia Vann has been patient, understanding and helpful and I am thankful to her for that.

Montréal, May 1987   M.A. ALI

CONTENTS

Introduction 1
    M.A. Ali

Ultrastructure of invertebrate synapses 3
    Jane A. Westfall

Synapse formation between identified invertebrate neurones
in vitro 29
    Jonathan P. Bacon

Identified neurons and cellular homologies 41
    Roger P. Croll

Functions of invertebrate glia 61
    V.W. Pentreath

Neuropeptides in invertebrates 105
    C.J.P. Grimmelikhuijzen, D. Graff, A. Groeger and
    I.D. McFarlane

Purification, characterisation and cellular distribution of
insect neuropeptides with special emphasis on their relationship to biologically active peptides of vertebrates 133
    Alan Thorpe and Hanne Duve

Neuroactive substances in the insect CNS 171
    Dick R. Nässel

Organization of conducting systems in "simple" invertebrates:
Porifera, Cnidaria and Stenophora 213
    Richard A. Satterlie and Andrew N. Spencer

Organisation and development of the peripheral nervous system
in annelids 265
    Susanna E. Blackshaw

Ontogenèse du système nerveux central des Chélicérates et sa
signification éco-éthologique 303
    Arturo Muñoz-Cuevas et Yves Coineau

The nervous system of the Crustacea with special reference
to the organisation of the sensory system 323
    M.S. Laverack

Aspects of the functional and chemical anatomy of the insect
brain 353
    Dick R. Nässel

Insect neurons: synaptic interactions, circuits and the control of behavior.. 393
    R.M. Robertson

Ontogenesis of the nervous system in Cephalopods.............. 443
    H.-J. Marthy

Nervous system in Chaetognatha................................ 461
    T. Goto and M. Yoshida

Neurobiology of the Echinodermata............................. 483
    J.L.S. Cobb

Tunicates... 527
    Q. Bone

Nervous mechanisms of spawning in regular echinoids........... 559
    M. Yoshida, H. Nogi and Y. Tani

Neural control mechanisms in bioluminescence.................. 573
    M. Anctil

Acoustic communication in crickets: Behavioral and neuronal mechanisms of song recognition and localization.............. 603
    Klaus Schildberger

A model for decision making in the insect nervous system   621
    J.S. Altman and J. Kien

General conclusions... 645
    M.A. Ali

Species Index... 651

Subject Index... 655

INTRODUCTION

M.A. ALI

Département de biologie

Université de Montréal

Montréal, Québec, Canada H3C 3J7

The invertebrates represent such a large chunk of the animal kingdom that their nervous systems simply cannot be ignored, would it be just to understand fundamental mechanisms of neuronal activity. This was understood decades ago by Hodgkin and Huxley, and Bullock working on the squid giant axon and synapses, by Kandel and his colleagues on the cellular neurobiology of learning and memory in *Aplysia*, etc. These efforts pioneered the model-oriented approach to the study of the invertebrate nervous systems.

The early realisation of the expository power of invertebrate neurobiological preparations led to the emergence of Bullock and Horridge's now classic monograph on the nervous systems of invertebrates. One had to take stock of what one knew of these nervous systems, their organisation and the behaviours they elicited and sustained, with an eye on disentangling from this mass of information new models most appropriate to shed light on neurobiological questions popping out by observing vertebrate, especially mammalian brains.

Although a few invertebrate model systems are exemplified in some of the contributions of this book, the main thrust of the latter is more in the tradition of Bullock and Horridge's approach. Its intent is to provide, on a reduced and somewhat more modest scale, a survey of the kinds of nervous systems that the invertebrates use to relate to their worlds, in order to get a feeling for the levels of understanding we have reached, and to highlight the riddles and puzzles and roadblocks which still succeed in preventing us from gaining a holistic understanding of the subject.

The first four chapters, in the vein of the first section of Bullock and Horridge's reference work, attempt to formulate general organisational principles regarding synaptic morphology (Westfall), synaptogenesis in cell culture (Bacon), cellular homologies as probes of the genealogy of identified neurones (Croll) and contemporary views on the role of glial cells (Pentreath). These chapters include issues that were largely of current interest for Bullock and Horridge in 1965, and yet remain so today with the advantage of having gained a deeper understanding of these topics in the meantime.

Bullock and Horridge had largely and deliberately ignored chemical neurotransmission in their monograph. Understandably so for reasons of space and because of the poor state of knowledge on the subject at the time. They had, however, included a substantial chapter on neurosecretion. The field has bounced back to haunt them in the 80s, especially due to the emergence of neuropeptides as major players of chemical communication within the nervous system of invertebrates. Several of the following chapters deal with neurotransmitters and neurotransmitter-specific pathways in invertebrate nervous systems. General aspects of the neurochemistry and distribution of invertebrate neuropeptides are introduced by Grimmelikhuijzen, Graff, Groeger and McFarlane. Insects provide good examples of neuropeptidergic systems and these are examined by Thorpe and Duve, and Nässel in two chapters on the neurochemistry and cellular localisation of neuropeptides in intensively investigated insect species. In addition, Nässel's chapter examines the distribution of classical neurotransmitters such as monoamines and amino acids in the insect CNS.

The next 10 chapters are loosely modelled after the systematic accounts of the invertebrate groups in Bullock and Horridge's monograph. However, space limitations and the extraordinary growth of knowledge of these nervous systems since 1965 have forced us to be very topical and very selective in the treatment of anatomical and physiological aspects of the nervous system of only the major invertebrate taxa. The taxa covered are the Porifera, Cnidaria and Platyhelminthes (Satterlie and Spencer), Annelida (Blackshaw), Chelicerata (Munoz-Cuevas and Coineau), Crustacea (Laverack), Insecta (Nässel, Robertson), Cephalopoda (Marthy), Chaetognatha (Goto and Yoshida), Echinodermata (Cobb) and Tunicata (Bone). Major themes raised by these authors are the evolutionary emergence of centralisation in the nervous system, the neurophysiological analysis of circuits and the control of behaviour, the cellular basis of integration and chemical transmission, and developmental issues as related to ecological-ethological problems.

The last 4 chapters cover miscellaneous topics relevant to specific, neurally controlled activities of some invertebrate taxa. Yoshida, Nogi and Tani examine how the gonads of sea urchins function as neurally controlled effector systems. The light-emitting effectors of various invertebrate groups are reviewed by Anctil from the point of view of their nervous control. An insect sensory function of great behavioural import, the acoustical communication system of crickets, is presented by Schildberger. The book concludes with the exposition of a model of neuronal integration in the insect nervous system by Altman and Kien.

ULTRASTRUCTURE OF INVERTEBRATE SYNAPSES

JANE A. WESTFALL

Department of Anatomy and Physiology

College of Veterinary Medicine

Kansas State University

Manhattan, KS 66505 USA

ABSTRACT

Invertebrate chemical synapses are characterized by a diversity of presynaptic vesicles and membrane-associated structures. They have in common with classical chemical synapses of vertebrates a pair of parallel densified membranes with a uniformly wide intercellular cleft containing intracleft material, a presynaptic aggregation of clear or dense-cored vesicles, and usually one or more mitochondria with nearby microtubules in the synaptic terminal or axonal varicosity. At these conventional synaptic foci some vesicles have thin filamentous connections to the presynaptic membrane. Invertebrate neuromuscular junctions often appear morphologically similar to interneuronal synapses because they lack the postsynaptic infoldings of vertebrate muscles. Electrical synapses presumably appear in all metazoa as morphologically identifiable gap junctions in which there is cytoplasmic continuity between two cells separated by a 2-3-nm-wide intercellular gap. In addition to these conventional synapses there are dyads, spine synapses, neuro-secretory-motor junctions, neuromuscular junctions with presynaptic dense bars, and gap junctions with vesicles in various invertebrate groups.

1. INTRODUCTION

Synapses are sites of rapid and precise information transfer between cells and are characterized ultrastructurally by parallel, close apposition of a pair of membranes. Chemical synapses presumably are present in all animal plyla with a nervous system and, in general, are characterized by vesicle-associated, paramembranous densities separated by a 15 to 30-nm-wide intercellular cleft. Information transfer occurs at these synapses as a result of release of a chemical by one neuron onto the surface of another neuron or effector cell.

The ultrastructure of invertebrate synapses is poorly understood compared to that of vertebrate synapses. Invertebrate synaptic foci or active zones often lack the striking presynaptic dense projections and postsynaptic densities characteristic of typical central synapses in vertebrates. Moreover, there may be only a few large irregular vesicles, often with dense cores, at invertebrate synaptic foci instead of the large aggregations of small, clear vesicles that are clustered at active zones in the vertebrate brain. Also, in invertebrate nervous systems, neurosecretory neurons can form synaptic contacts on other neurons and effector cells in addition to synaptoid contacts on noncellular lamellae.

Electrical synapses presumably appear in all metozoa as morphologically identifiable gap junctions in which there is cytoplasmic continuity between two cells separated by a 2 to 3-nm-wide intercellular gap. Gap junctions typically lack synaptic vesicles, except at septal synapses between giant axons of earthworms and crayfish. Electrical information transfer occurs between cells at these morphologically specialized junctions.

In 1978, Cobb and Pentreath analyzed the comparative morphology of invertebrate and vertebrate synapses and concluded that specialized chemical synapses are the exception rather than the rule in invertebrates. The present phylogenetic survey of synaptic morphology in invertebrates suggests that specialized chemical synapses are the rule in invertebrates, but that we need more investigations at a detailed ultrastructural level. Improved techniques of fixation and higher magnifications of serial sections through active synaptic foci in a variety of neural regions, both central and peripheral, will add greatly to our current knowledge of invertebrate synaptic structure.

2. CHARACTERISTICS OF INVERTEBRATE SYNAPSES

Invertebrate chemical synapses are characterized by a diversity of presynaptic vesicles and membrane-associated structures. They have in common with classical chemical synapses of vertebrates a pair of parallel, densified membranes with a uniformly wide, intercellular cleft containing intracleft material, a presynaptic aggregation of clear or dense-cored vesicles, and usually one or more mitochondria with nearby microtubules in the synaptic terminal or axonal varicosity. The paired synaptic membranes and intervening cleft are recognized by some increase in electron density owing to associated fine filaments, which in the cleft often appear as periodic striations (Figs. 1-4). Such junctional densities, when associated with a linear or stacked array of vesicles, represent sites of active synaptic foci. The triadic densifications of pre- and postsynaptic membranes and intracleft material are equal in length and constitute the synaptic membrane complex. The symmetry or asymmetry of the paramembranous densities is not a feature that can be discussed in invertebrate chemical synapses where our knowledge is limited and the synaptic foci vary greatly in their morphology. Some synaptic membrane complexes have thick paramembranous densities, whereas others have only a thin densification of the paired membranes (Figs. 1-2). Also, some synaptic membrane complexes are long and continuous, whereas others have one or more short interruptions along their length (Figs. 2-3). Many are very short and extremely difficult to locate at low mangifications with the electron microscope (Figs. 1, 4). In some animals, such as jellyfish, two-way or symmetrical synapses are common (Fig. 5). Several groups of animals have dyadic type synapses with elaborate presynaptic bodies in association with two postsynaptic cells (Figs. 6-7). Other variations include a slightly widened cleft with an intermediate periodic line and mixed clear and dense-cored vesicles. In the polychaete annelid, there is a presynaptic dense body ringed with clear vesicles (Fig. 8). In cephalopods spine synapses are present in which a halo of clear vesicles surrounds a postsynaptic invagination (Fig. 9). Sometimes, dense-cored

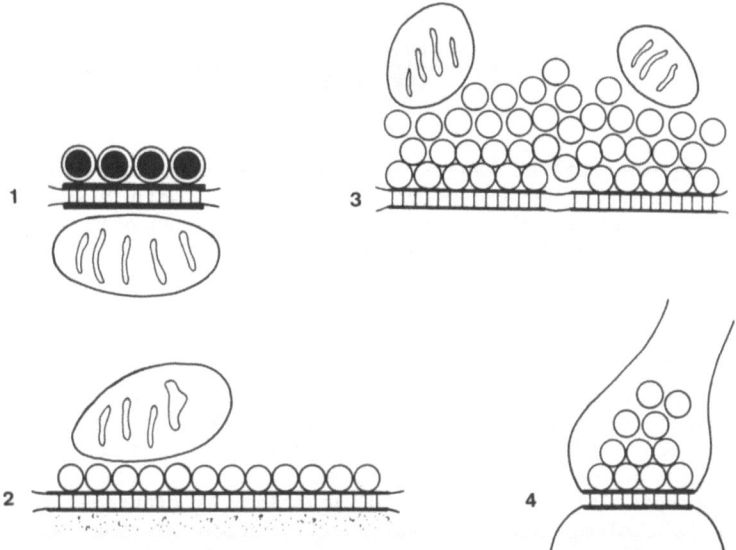

FIG. 1: Polarized, interneuronal synapse of Hydra (Hydrozoa, Cnidaria). A short, single row of dense-cored vesicles lies in contact with a presynaptic membrane density, which parallels the postsynaptic density and region of intracleft filaments. Such synapses occur en passant between axons and usually have a mitochondrion nearby (Westfall, original).

FIG. 2: Polarized, interneuronal synapses with mitochondria adjacent to a long, single row of clear vesicles paralleling a pair of thin, paramembranous densities with intracleft transverse filaments as observed in hydromedusae and jellyfish (Westfall, original).

FIG. 3: Polarized, interneuronal synapse with mitochondria adjacent to tiers of synaptic vesicles at a long, bipartite, synaptic membrane complex as seen in a larval mussel (Bivalvia, Mollusa). Note regular arrangement of initial row of vesicles at paired, synaptic membranes except for loss of continuity at interruption. After Zs-Nagy and Lábos (1969).

FIG. 4: Nerve terminal with tiered clear vesicles at a short, synaptic membrane complex in a sea urchin (Echinoidea, Echinodermata). After Cobb and Laverack (1966a).

vesicles are present in a synapse with predominently clear vesicles (Fig. 10). In other cases, neurosecretory-type granules predominate (Fig. 11). Occasionally, dense-cored vesicles are present at the presynaptic contact of a neurosecretory ending (Fig. 12). Neurosecretory endings may form true synaptic contacts or end in synaptoid contacts with small clear vesicles at an extracellular lamina (Fig. 13).

Invertebrate neuromuscular junctions often appear morphologically similar to interneuronal synapses because they lack the postsynaptic infoldings of vertebrate striated muscles and have a tendency to contact the granular cytoplasm of the underlying muscle cell. In some cnidarians, dense-cored vesicles may be present at the neuromuscular synapse.

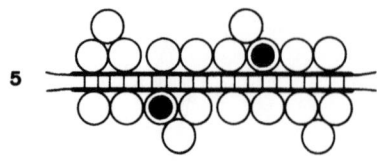

FIG. 5: Symmetrical or two-way interneuronal chemical synapse of the jellyfish *Cyanea* (Scyphozoa, Cnidaria). Note mixed clear and dense-cored vesicles on both sides of the synaptic membrane complex. After Horridge and Mackay (1962).

Usually the synaptic vesicles are clear and lie either in a single row closely apposed to the presynaptic membrane density or as tiers of vesicles (Figs. 14-15). In jellyfish, there is often a subsynaptic cisterna present at neuromuscular synapses (Fig. 16). Ctenophores have a unique, presynaptic triad of a row of vesicles, a flattened cisterna of endoplasmic reticulum, and a large mitochondrion at the neuromuscular synapses (Fig. 17). At some mollusc neuromuscular synapses, there are large aggregations of clear vesicles (Fig. 18). Synaptic vesicles generally are few in number at echinoderm neuromuscular junctions, where clear vesicles predominate at the presynaptic membrane (Fig. 19). In arthropods, the synaptic vesicles congregate at one or more hour glass-shaped, presynaptic dense bodies (Figs. 20-21). The neuromuscular cleft, usually of similar width to the interneuronal cleft, may be bisected by an intermediate periodic line (Figs. 18, 20-22) and/or contain faint cross filaments (Figs. 14-19). In ctenophore, earthworm, moth, lobster and crayfish neuromuscular synapses, a row of periodic filaments has been observed on the extracellular surface of the postsynaptic membrane (Figs. 17, 20, 23). In both crayfish and lobster, excitatory neuromuscular junctions can be distinguished from inhibitory neuromuscular junctions on the basis of their ultrastructure. Excitatory junctions have an abundance of clear, round vesicles, whereas inhibitory junctions contain fewer and less regular vesicles (Figs. 23-23).

Electrical synapses vary from typical gap junctions with a 2-3-nm-wide gap between a pair of parallel, closely apposed membranes in coelenterates (Fig. 24) to a somewhat wider gap with ribbed membranes and associated vesicles between giant axons of crayfish (Fig. 25). Although the junctions between giant axons sometimes appear to have the morphology of chemical synapses, there is electrophysiological evidence that they are low resistance junctions.

This brief ultrastructural survey of invertebrate synapses indicates that they have in common a uniformly constant apposition of paired membranes for each specialized synaptic contact but that there is great morphological diversity among synaptic components. Conventional synapses with paired densified membranes and associated vesicles constitute the majority of synapses found in all animal phyla with a nervous system, whereas unusual configurations such as dyads and synaptic spines are usually associated with sensory receptor cells. Large aggregations of neurosecretory granules and dense-cored vesicles generally occur in neuroendocrine organs.

The morphology and occurrence of different types of synapses in the primitive nerve net of cnidarians will be discussed next, followed by a selected review of synaptic variations described in several higher invertebrate groups.

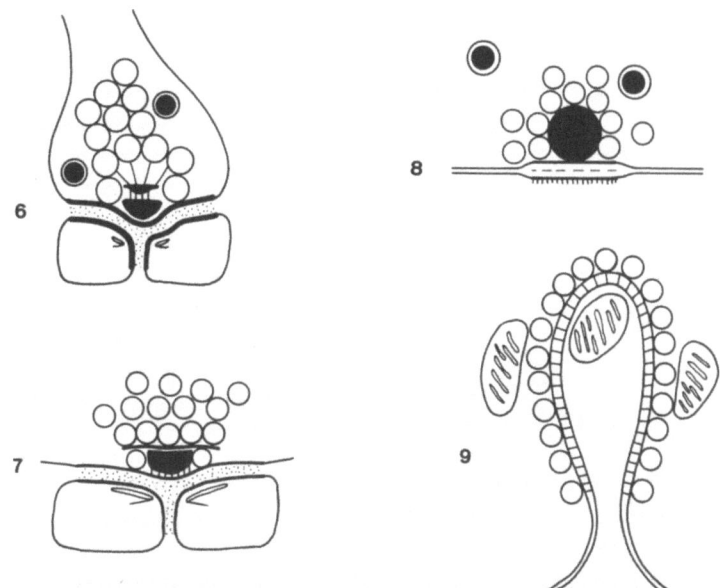

FIG. 6: Dyad with presynaptic clear and dense-core vesicles and pair of postsynaptic neurites with small cisternae of endoplasmic reticulum in the flatworm Gastrocotyle (Monogenea, Platyhelminthes). Note halo of clear synaptic vesicles with filamentous connections to a bipartite, presynaptic dense bar, intracleft density, and paired membrane thickenings. After Shaw (1981).

FIG. 7: Dyad with presynaptic clear vesicles and pair of postsynaptic elongate cisternae with medial whiskers in the fly eye (Insecta, Arthropoda). Note row of vesicles along top plate of synaptic bar. After Burkhardt and Braitenberg (1976).

FIG. 8: Presynaptic dense body surrounded by clear vesicles with slightly larger, dense-cored vesicles nearby in Nereis (Polychaeta, Annelida). Note intermediate, periodic line in widened cleft and postsynaptic specialization. After Dhainaut-Courtois and Warembourg (1969); Fisher and Tabor (1977).

FIG. 9: Postsynaptic spine with associated mitochondria in the octopus statocyst (Cephalopoda, Mollusca). Note arrangement of presynaptic clear vesicles along invaginated, synaptic cleft with transverse filaments. After Budelmann and Thies (1977).

3. CNIDARIA

Synapses in the cnidarians, the most primitive group of animals with a recognizable nervous system, range from short foci of parallel electron dense membranes with one to three or four dense-cored or clear vesicles in Hydra (Fig. 1) to long foci of many vesicles in various jellyfish (Fig. 2). Symmetrical synapses were reported first in the marginal ganglia of the jellyfish Cyanea (Horridge et al. 1962; Horridge and Mackay 1962) and were thought to transmit bidirectionally, similar to electrical synapses between giant fibers in earthworms and crayfish. This was the first

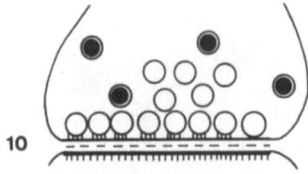
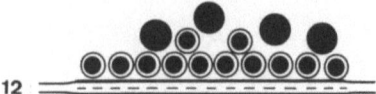
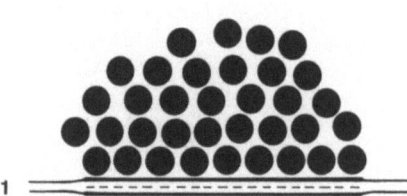
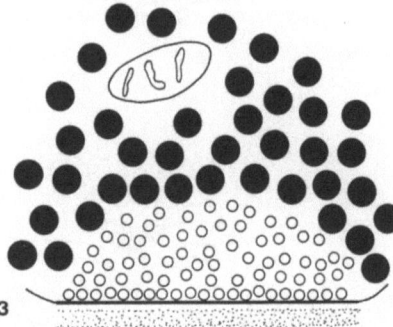

FIG. 10: Conventional interneuronal synapse with mixed clear and dense-cored vesicles in a gastropod mollusc. Note filamentous connections between initial row of clear vesicles and presynaptic membrane, an intracleft intermediate periodic line, and cytoplasmic densities on the postsynaptic membrane. After Coggeshall (1967).

FIG. 11: Interneuronal synapse with neurosecretory granules in Aplysia (Gastropoda, Mollusca). Note association of initial row of granules with synaptic membrane complex and intermediate, periodic line in slightly widened cleft. After Tremblay et al. (1979).

FIG. 12: Interneuronal synapse with dense-cored vesicles at the presynaptic membrane of a neurosecretory neuron in Aplysia. After Tremblay et al. (1979).

FIG. 13: Diagrammatic representation of a neurosecretory cell synaptoid contact on a noncellular lamella in crayfish (Crustacea, Arthropoda). Note mitochondrion among neurosecretory granules and small clear vesicles indicative of site of neuroendocrine release and vesicle recycling. After Bunt (1969).

report of vesicles on both sides of a 20-nm-wide synaptic cleft with parallel electron-dense membranes resembling vertebrate chemical synapses. The vesicles, 50-100 nm in diameter with an occasional electron-dense core, were closely apposed to the electron-dense membranes (Fig. 5). Recently, Anderson (1985) demonstrated physiologically that such synapses in the motor nerve net of Cyanea are bidirectional chemical synapses.

Polarized or unidirectional synapses in cnidarians were demonstrated ultrastructurally by Jha and Mackie (1967) in the marginal nerve ring of the hydromedusan Sarsia. Small tiers of clear and dense-cored vesicles (100-150 nm in diameter) were aggregated at membrane densities with a

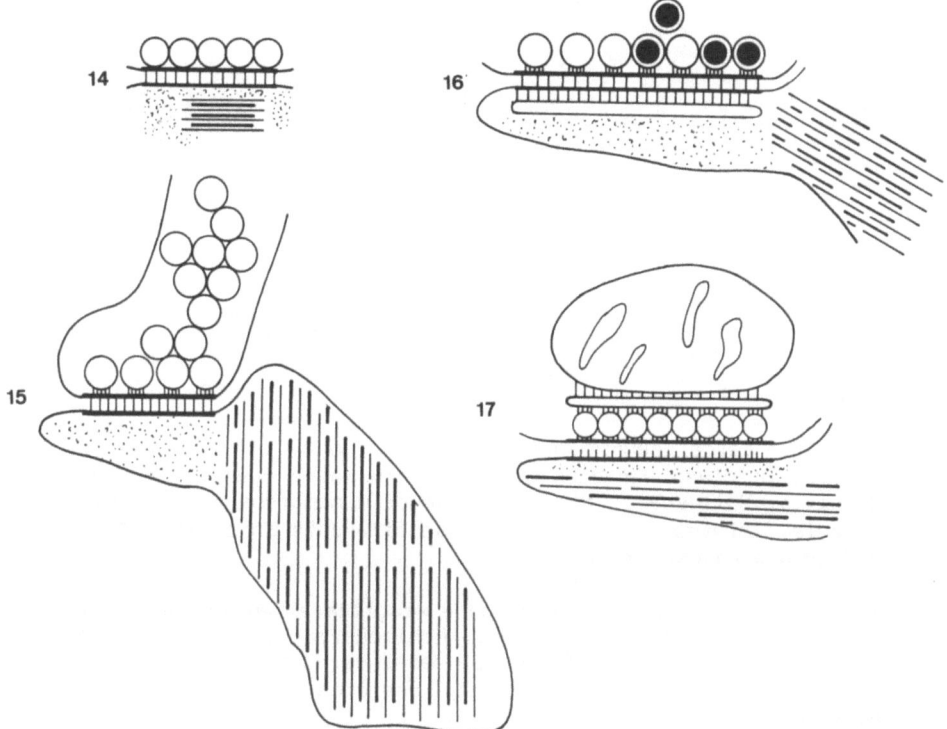

FIG. 14: En passant neuromuscular synapse with myonemes near a straight, postsynaptic membrane in the hydromedusan Aglantha (Hydrozoa, Cnidaria). Note short, single row of presynaptic clear vesicles paralleling paramembranous densities and intracleft filaments. After Singla (1978a).

FIG. 15: Nerve terminal synapse on a muscle cell process in the sea anemone Metridium (Anthozoa, Cnidaria). Note tiered arrangement of presynaptic clear vesicles and postsynaptic granular cytoplasm at contact site with longitudinal muscle. After Westfall (1970b).

FIG. 16: Neuromuscular synapse with mixed vesicles and a subsynaptic cisterna of endoplasmic reticulum observed in the jellyfishes Aurelia, Chrysaora, and Haliclystus (Scyphozoa, Cnidaria). Westfall, original.

FIG. 17: Ctenophore neuromuscular synapse with a presynaptic triad of mitochondrion, flattened cisterna of endoplasmic reticulum, and single row of clear vesicles at the presynaptic membrane. Note filamentous connections between presynaptic elements; postsynaptic membrane densification is periodic within the cleft. After Hernández-Nicaise (1968; 1973a).

20-nm-wide cleft between neurites. Buisson and Franc (1969) observed a single row of 100 to 150-nm-diameter vesicles at en passant synapses between neurites in the anthozoan Veretillum. Although Hydra was said to be the exception to the rule that cnidarians have synapses (Bullock and Horridge 1965), Westfall et al. (1970a, 1971) found ultrastructural evidence of synaptic foci with vesicles on one or both sides of paired,

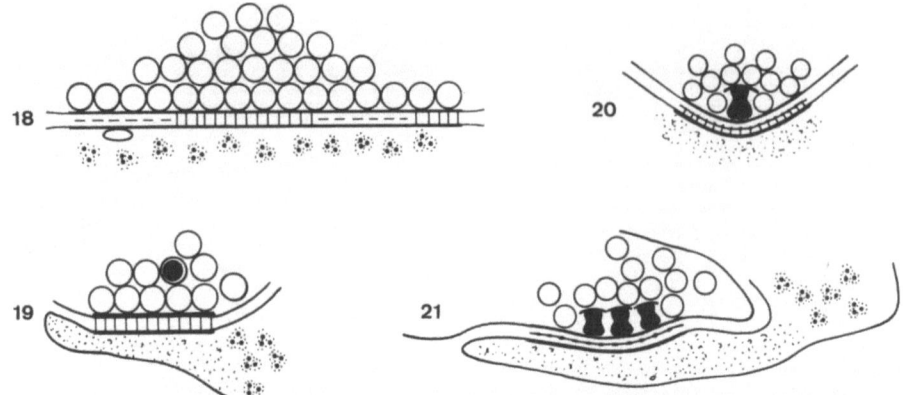

FIG. 18: Large neuromuscular synapse with tiered mass of presynaptic clear vesicles in the fresh water mussel, Anodonta (Bivalvia, Mollusca). Note long synaptic membrane complex with variable, intermediate line or cross filaments in cleft. After Zs-Nagy and Lábos (1969).

FIG. 19: Neuromuscular synapse on a winglike extension of muscle cell in a sea urchin (Echinoidea, Echinodermata). Note initial row of clear vesicles and single, dense-cored vesicle in second row of short tier of presynaptic vesicles. After Cobb and Laverack (1967).

FIG. 20: Presynaptic dense bar observed at annelid, moth, and lobster neuromuscular synapses. Note clear vesicles surrounding hour-glass-shaped presynaptic density, intermediate line in cleft, and periodic filaments on extracellular surface of postsynaptic membrane. After Rosenbluth (1972); Rheuben and Reese (1978); Govind and DeRosa (1983).

FIG. 21: Neuromuscular synapse on granular, cytoplasmic extension of muscle cell in a lobster (Crustacea, Arthropoda). Note clear vesicles associated with row of presynaptic, hour-glass-shaped densities paralleling intracleft intermediate line and postsynaptic density. After Govind and Pearce (1982).

parallel, electron-dense membranes between both neuronal soma and axons in this simple hydrozoan polyp. Thus, all cnidarians presumably have chemical synapses, but not all types of synapses have been well studied to date. Two difficulties are apparent in surveying the ultrastructural literature on cnidarian synapses. First, many investigators do not publish high magnification micrographs of their synapses, so morphological criteria cannot be well defined. Secondly, preservation of cnidarian tissues for electron microscopy is difficult, so that measurements of synaptic vesicles often are variable owing to vesicular swelling or other shape changes. In spite of these difficulties, I believe there is good ultrastructural evidence for interneuronal, neuromuscular, and neuronematocyte synapses in addition to neurosecretory endings and gap junctions in the cnidarian nervous system.

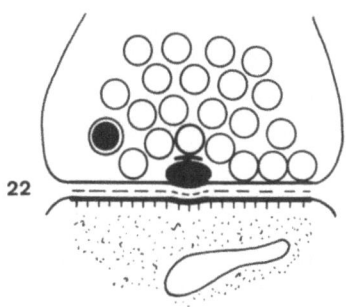 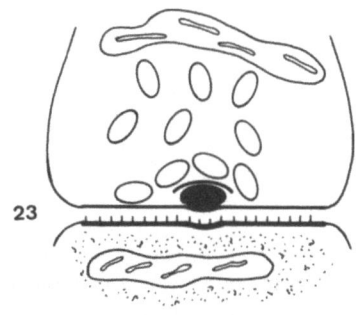

FIG. 22-23: Excitatory and inhibitory neuromuscular synapses found in crayfish and lobster (Crustacea, Arthropoda). Fig. 22. Note numerous, spherical, clear vesicles associated with presynaptic density of excitatory neuromuscular synapse. The synaptic membrane complex is slightly invaginated at the presynaptic density and contains an intermediate, periodic line and cytoplasmic, postsynaptic densities. Fig. 23. Note small number of scattered, flattened vesicles associated with presynaptic density of inhibitory neuromuscular synapse. Elongate, pre- and postsynaptic mitochondria and intracleft, periodic densities on the postsynaptic membrane are present. After Jahromi and Atwood (1974); King and Govind (1980).

3.1 Interneuronal synapses

Ultrastructural evidence for chemical synapses between neurons has been found in representatives of all three classes of the Cnidaria (Westfall et al. 1970b). In the Hydrozoa, interneuronal synapses have been demonstrated ultrastructurally in the following hydromedusae: Sarsia (Jha and Mackie 1967), Gonionemus (Westfall 1969, 1970a), Stomotoca (Mackie and Singla 1975), Polyorchis (Spencer 1979), Spirocodon (Toh et al. 1979), and Aglantha (Weber et al. 1982). They have also been observed in the simple fresh water polyp Hydra (Westfall et al. 1970a, 1971; Westfall 1973a; Westfall and Kinnamon 1978, 1984; Kinnamon and Westfall 1982) and in the hydroid polyps of Hydractinia (Stokes 1974) and Podocoryne (Pavans de Ceccatty 1979). In the Scyphozoa, they have been observed in the jellyfish Cyanea (Horridge et al. 1962; Horridge and Mackay 1962; Anderson and Schwab 1981), in the stauromedusan Haliclystus (Westfall 1973b; Singla 1976), and in the cubomedusan Tamoya (Yamasu and Yoshida 1976). In the ocelli of the cubomedusan Tamoya, processes of the second order neurons are invaginated into axons of the photoreceptor cells to form spine synapses similar to those observed in the squid photoreceptor, except that the 80-nm-diameter clear vesicles are fewer and larger in the more primitive system. In the Anthozoa, chemical synapses were first observed in the sea pen Veretillum (Franc 1968; Buisson and Franc 1969; Buisson 1970) and later in the sea anemones Metridium (Westfall 1979b), Ceriantheopsis (Peteya 1973a,b) and Actinia (Quaglia 1976), and octocorals Clavularia (Satterlie and Case 1980) and Virgularia (Satterlie et al. 1980). In an ultrastructural survey of types of synapses present in tentacles of 18 species of cnidarians, it was found that polarized or one-way interneuronal synapses were more common than symmetrical or two-way synapses (Westfall et al. 1970b).

FIG. 24: Typical interneuronal gap junction present in the coelenterate Hydra and in many other invertebrates (Westfall, original).

FIG. 25: Diagrammatic representation of a low resistance junction between septate lateral giant axons in the crayfish (Crustacea, Arthropoda). Note beading of intramembranous particles along a 4-5-nm-wide gap and clear vesicles paralleling both sides of junction. After Peracchia and Dulhunty (1976).

3.2. Neuromuscular synapses

Neuromuscular synapses in cnidarians have not been as thoroughly investigated at the ultrastructural level as interneuronal synapses. Where reported, however, their synaptic membrane complex does not differ significantly from that of interneuronal synapses. Vesicle-associated, densified, paired membranes are present both en passant (Fig. 14) and at nerve terminals (Fig. 15) along smooth and striated muscles. In the Hydrozoa, neuromuscular synapses in polyps tend to have shorter foci than those of hydromedusae. They have been demonstrated in the polyps Hydra (Westfall 1973a, b; Westfall and Kinnamon 1978, 1984; Kinnamon and Westfall 1982) and Podocoryne (Pavans de Ceccatty 1979); in the hydromedusans Gonionemus (Westfall 1979a, c), Aglantha (Singla 1978a), Polyorchis (Singla 1978b; Spencer 1979), and Aequorea (Satterlie and Spencer 1983); and in the siphonophore Nanomia (Mackie 1976). In the Scyphozoa, a subsynaptic cisterna of endoplasmic reticulum has been observed at neuromuscular synapses (Fig. 16). These have been described in the scyphistoma of Aurelia (Westfall 1973b) and in the scyphomedusae Haliclystis (Westfall 1973b), Chrysaora (Westfall 1973b), and Cyanea (Anderson and Schwab 1981). In the Anthozoa, tiers of clear vesicles have been observed at nerve terminals on muscle cells (Fig. 15). However, neuromuscular synapses have not been well described in this group as a whole. To date they have been demonstrated in the coral Astrangia (Westfall 1973b), in the sea anemones Metridium (Westfall 1970b) and Ceriantheopsis (Peteya 1973a), and in the octocoral Muricea (Satterlie and Case 1978).

3.3. Neuronematocyte synapses

Neuronematocyte synapses, which are specific to the Cnidaria, have only been demonstrated ultrastructurally in Gonionemus (Westfall 1969, 1970a, c), Hydra (Westfall et al. 1970a, b, 1971; Westfall 1973a, b; Westfall and Kinnamon 1978, 1984), and the octocoral Muricea (Satterlie and Case 1978). In Hydra they are characterized by small foci with only two or three clear or dense-cored vesicles but in Gonionemus the neuronematocyte synapse has longer foci with well-defined, paramembranous densities and intracleft material. In Muricea, clear vesicles are present at the neuronematocyte synapse.

3.4. Neurosecretory endings

Neurosecretory vesicles resemble dense-cored vesicles at synaptic foci in Hydra and some other cnidarian neurons. Neurosecretory endings with a large aggregation of granular vesicles have been reported in Hydra (Lentz 1965) and in the sea pen Veretillum (Buisson 1970). Such endings have also been associated with neuromuscular junctions in Hydra (Westfall 1973a) and in Cyanea (Anderson and Schwab 1981). Neurons with such junctions have been termed neurosecretory motorneurons in higher organisms (Osborne et al. 1971; Anwyl and Finlayson 1973). Westfall (1973a) observed that many ganglion cells in Hydra had the morphological features of a neurosecretory-sensory-motor-interneuron, suggesting that in this primitive animal phylum, neurons originated as multifunctional cells, which later evolved into the specialized sensory, motor, interneuronal, and neurosecretory types of cells characteristic of higher animal phyla.

3.5. Gap junctions

Interneuronal and neuromuscular gap junctions (Fig. 24) have been demonstrated ultrastructurally in the epidermis of Hydra (Westfall et al. 1980), and interneuronal gap junctions are present in the hydromedusans Polyorchis (Singla 1978b), Aglantha (Weber et al. 1982), and Aequorea (Satterlie and Spencer 1983). However, no gap junctions have been found between neurons in scyphomedusae or anthozoa. Gap junctions, although rare in the nervous system, are common between epitheliomuscular cells of the Hydrozoa.

4. CTENOPHORA

Synapses in ctenophores are unique in having a presynaptic triad of a row of synaptic vesicles with filamentous connections to an elongate sac of smooth endoplasmic reticulum (ER), which, in turn, is connected to a large mitochondrion (Hernandez-Nicaise 1968, 1973a,b, 1974a,b). Although Horridge was the first to characterize the ultrastructure of ctenophore synapses at comb plates (Horridge and Mackay 1964) and neuromuscular junctions (Horridge 1965), he failed to notice the flattened sac between mitochondrion and row of synaptic vesicles. This presynaptic triad appears to be characteristic of all types of chemical interneuronal and neuroeffector junctions in ctenophores (Fig. 17). The vesicles are closely aligned in a single row in intimate contact with the presynaptic membrane. Faint presynaptic projections attach the vesicles to the plasma membrane, and similar projections on the opposite side of the vesicles appear to connect them to the elongate sac. The sac of ER is also connected by dense projections to the outer mitochondrial membrane. The mitochondrion generally lacks well-defined cristae and matrix material. The synaptic vesicles average about 70 nm in diameter and occasionally contain an eccentric granule. Horridge observed 30 to 50-nm-wide clear vesicles at interneuronal and neuromuscular junctions; at interneuronal synapses, the cleft was slightly narrower (12 to 15-nm-wide) than at neuromuscular junctions (15 to 20-nm-wide). A cluster of vesicles may accumulate around the mitochondria in a nerve ending; also, synaptic foci with a row of three mitochondria may occur at a neuromuscular junction (Hernandez-Nicaise 1974a).

No distinct differences between interneuronal and neuromuscular intracleft material have been reported. However, I noticed a definite extracellular fringe on the outer surface of the postsynaptic membrane in one micrograph (Fig. 19, Hernandez-Nicaise 1973a) of a neuromuscular junction where the author reported a strong thickening of the muscle membrane. It would be interesting to see if a better fixation and a higher magnification reveal a series of extracellular membrane projections similar to that observed in type I junctions on earthworm muscle fibers (Rosenbluth 1972). No specialized subsynaptic features have been observed in ctenophores other that a thin, approximately 7-nm-wide, internal coating of the postsynaptic membrane.

5. PLATYHELMINTHES

Dyadic types of synapses, where one axon contacts two postsynaptic neurites, first appear in the Platyhelminthes (Webb 1977; Reuter 1981; Shaw 1981; Ferrero et al. 1985) and become well-documented in sensory systems of arthropods (Figs. 6-7). In a monogenean flatworm, Gastrocotyle trachuri, the presynaptic membrane has a 40 to 50-nm-high and 55 to 65-nm-wide dense body surrounded by a cloud of 25 to 50-nm-diameter clear vesicles and lies in a depression opposite the postsynaptic dyads (Shaw 1981). In longitudinal section, it appears two-parted with a thin top plate connected to the bar by a central row of cross filaments. A row of clear vesicles is connected to the top plate by fine filaments (Fig. 6). A few 50 to 90-nm-diameter, dense-cored vesicles are present in the presynaptic profile along with one or more mitochondria. The intercellular gap is 10-18 nm and contains some filamentous material near the presynaptic density. A pair of subsynaptic cisternae resemble the "bags" seen in arthropod dyads. A similar synapse with postsynaptic cytoplasmic filaments instead of cisternae was reported in the turbellarian flatworm Microstomum lineare (Reuter 1981) and in the cestode Hymenolepis microstoma (Webb 1977). In addition to these specialized synapses, there are conventional symmetrical interneuronal synapses (Chien and Koopovitz 1977) and neurosecretory endings with synaptoid release sites containing an aggregation of clear vesicles near a densification of the terminal membrane (Webb 1977). Neuroeffector synapses, which resemble conventional interneuronal synapses, occur on muscle cells (Chien and Koopowitz 1972) and rhabdite-secreting cells (Chien and Koopowitz 1977). In the sarconeural junctions of the free living flatworm Notoplana, the synapses are between long sarcoplasmic extensions and the nerve cells (Chien and Koopowitz 1972).

6. ANNELIDA

Dyadic synapses have not been reported very often in the annelids; however, poorly described examples of such synapses have been seen in the polychaete Myxicola infundibulum (Wells et al. 1972) and in the leech Hirudo medicinalis (Purves and McMahan 1972; Muller and McMahan 1976; Muller and Carbonetto 1979). They are characterized by the usual presynaptic density invaginated between two postsynaptic neurites with dense material in the cleft. In the tubiculous polychaete Myxicola, there are 100-nm-diameter, dense-cored vesicles, associated with the presynaptic depression, whereas in the leech Hirudo, there are small clear vesicles 50 nm in diameter near the tuft of dense material. This presynaptic specialization is portrayed as a long bar residing in a groove and surrounded by two longitudinal rows of vesicles. Distal to the small clear vesicles are dense-cored vesicles and mitochondria. The cleft is 30 nm in width with dense material associated both with the cleft and the postsynaptic neurites (Muller and McMahan 1976). In Nereis, there are large, round, dense bodies ringed by small clear synaptic vesicles (Dhainaut-Courtois and Warembourg 1969; Fisher and Tabor 1977). These bodies appear at conventional, monosynaptic sites where either an intracleft line without cytoplasmic densities (Dhainaut-Courtois and Warembourg 1969) or dense material and associated cytoplasmic densities (Fisher and Tabor 1977) are present (Fig. 8).

A variety of synaptic structures have been demonstrated in the earthworm in association with the dorsal giant fiber system in the ventral nerve cord. Some electrical synapses have a parallel row of 50 to 70-nm-diameter clear or dense-cored vesicles on both sides of a 6.5 to 7.5-nm gap junction (Oesterle and Barth 1973). Other junctions have rows of clear vesicles associated with only one side of a 20 to 30-nm-wide cleft and intracleft and postsynaptic dense material; these are considered chemical in nature (Günther and Schürmann 1973). Efferent synapses appear to have a large number of clear vesicles associated with a uniform band of presynaptic dense material, whereas afferent synapses have mixed clear and dense-cored vesicles in endings with little membrane specialization (Günther and Schürmann 1973). In the ventral cirrus of the polychaete Harmothoë, a typical interneural junction has an aggregation of 30 to 60-nm-diameter clear vesicles closely apposed to small densities on the presynaptic membrane (Lawry 1967). A parallel densification is present on the postsynaptic membrane, along with electron-dense material in the cleft.

Neurosecretory endings are found in the annelid brain with a synaptoid arrangement of granular vesicles and smaller electron lucent vesicles (Baskin 1974).

Annelid myoneural junctions are best described in the earthworm, in which either clear vesicles 50 nm in diameter or dense-cored vesicles appear at the synapse (Rosenbluth 1972). An unusual feature is the postjunctional membrane specialization, in which a prominent row of 20-nm long projections occur at 14-nm intervals on the outer dense lamina of the postjunctional membrane. A similar intracleft specialization is seen in the nematode Ascaris (Rosenbluth 1965) and in the crab Grapsis (Govind et al. 1973).

7. ARTHROPODA

Dyadic types of synapses with elaborate, presynaptic densities lying in a groove between two postsynaptic neurites are prevalent in the arthropods (Fig. 7). The first observations of T-shaped, dense, synaptic ribbons in the presynaptic profiles were made in fly eyes (Trujillo-Cenóz 1965, 1969). Several studies on the eyes of the house fly Musca domestica indicate that the T-shaped ribbon is, in fact, a bar with an overlying plate (Boschek 1971; Burkhardt and Braitenberg 1976; Fröhlich and Meinertzhagen 1982; Nichol and Meinertzhagen 1982; Saint Marie and Carlson 1982). In one study, the plate was obviously separated from the bar and connected to it by a row of cross filaments; another row of similar cross filaments connected the bar to the presynaptic membrane (Saint Marie and Carlson 1982). Vesicles 30-40 nm in diameter are regularly arranged near the presynaptic bar. Cross filaments are present within the cleft and on the cytoplasmic surface of the two postsynaptic neurites. A pair of postsynaptic bags and whiskers also have been described (Burkhardt and Braitenberg 1976). Dyadic synapses are present in the brain and eye of the bee (Schürmann 1971; Ribi 1981) and in the cockroach (Wood et al. 1977). In the antennal lobes of the moth Manduca sexta, typical dyads show parallel rows of clear vesicles on either side of synaptic bars cut tangentially (Tolbert and Hildebrand 1981). No synaptic vesicles were observed in a longtitudinal section of a presynaptic density in the dorsal, unpaired, medial neurons of the locust metathoracic ganglion (Watson 1984). However, in the locust ocelli, synaptic vesicles formed a halo around the dyadic bar synapses (Goodman et al. 1979).

In crustacean eyes, the synaptic ribbon is similar to that observed in insects. In the lobster optic lamina, a synaptic ribbon 110-150 nm by

500 nm lies at right angles to the presynaptic membrane and is covered on each side by a row of 30 to 60-nm-diameter synaptic vesicles (Hámori and Horridge 1966). In the lobster stomatogastric ganglion, the presynaptic tuft is associated with either irregular or round, clear vesicles (King 1976). Also, clear vesicles are present near the tuft and dense-cored vesicles distal to the tuft. Presynaptic dense bodies have been seen in the photoreceptors of the barnacle Balanus (Hudspeth and Stuart 1977), the crayfish Procambarus (Hafner 1974), and lobster Homarus (Hámori and Horridge 1966), in giant fiber synapses in the crayfish Astacus (Stirling 1972), and in the motor neurons and fourth abdominal ganglion of the crayfish Procambarus (Atwood and Pomerantz 1974; Skinner 1985). They have also been reported in the brain of the horseshoe crab Limulus (Fahrenbach 1979) and in the peripheral nervous system of the whip spider Zygilla (Foelix and Troyer 1980).

The crustacean neuromuscular synapse is best described in the lobster and is characterized by a presynaptic dense body with a broad, oblong base and small T-shaped top attached to it (King 1976; Meiss and Govind 1980; Govind and DeRosa 1983). It resembles the dense body of dyadic synapses by its tendency to be found in a V-shaped depression of the presynaptic membrane and to be surrounded by a cloud of small clear vesicles around 40 nm in diameter (Fig. 22). A tangential cut through the presynaptic dense body, variously referred to as a tuft (King 1976), a presynaptic dense body (Govind and Chiang 1979), or a dense bar (Meiss and Govind 1980) with an hour-glass shape (Govind and DeRosa 1983), reveals a halo of vesicles connected to the body by radiating short filaments (King 1976). The cleft is 20 nm wide with an intermediate electron-dense line and lies at a region of granular sarcoplasm extending from the muscle (Govind and Chiang 1979; Meiss and Govind 1980; Govind et al. 1980). In lobsters, a postsynaptic densification is apparent (King 1976; Meiss and Govind 1980; Govind and DeRosa 1983). In crayfish, an excitatory neuromuscular synapse (Fig. 22) is characterized by round, clear vesicles and an inhibitory synapse (Fig. 23) by irregular-shaped vesicles (Jahromi and Atwood 1974).

The neuromuscular synapse on striated insect muscle appears as a series of dense bars with filaments extending to one, long, thin lamella or plate to which the first row of vesicles are attached by fine filaments (Lane 1985). In the moth Manduca, there is a prominent intermediate line in the cleft (Rheuben and Reese 1978) (Fig. 20). Excitatory contacts are short with narrow clefts and thickened postsynaptic densities, whereas inhibitory contacts may be longer (Aizu 1982) or have wider clefts with a less distinct postsynaptic density (Titmus 1981).

Crustacean giant-fiber systems typically have large electrical junctions with or without associated vesicles (Peracchia and Dulhunty 1976). In crayfish lateral giant axons, the septate junction is characterized by a pair of beaded membranes separated by a 4-5 nm gap and bounded by two cytoplasmic rows of vesicles 40-70 nm in diameter (Fig. 25). Other giant axon synapses appear chemical in nature (Stirling 1972); hence, more structural-functional correlations need to be made to clarify this variation in morphological characteristics of synapses associated with low resistance junctions. In fly eyes, gap junctions with associated filaments are present between rhabdomeres but vesicles occur only at a distance (Ribi 1978).

8. MOLLUSCA

Spine synapses have been reported in several molluscs (Fig. 9). These synapses, consisting of a postsynaptic spine invaginating the presynaptic element, have been observed in the sea hare Aplysia (Bailey and Thompson 1979; Bailey et al. 1979), the snail Planorbis (Pentreath et al. 1975), and in the statocyst and brain of the octopus (Gray and Young 1964; Budelmann and Thies 1977). Small, clear vesicles are aligned along the length of the spine wherever electron dense material is observed in the cleft (Bailey and Thompson 1979). In the photoreceptors of the squid Loligo, some invaginationed spines are also associated with a narrow 2 to 4-nm-gap accompanied by a pair of flat cisterns and are considered to be electrotonic junctions (Cohen 1973).

Conventional synapses with 40 to 70-nm-diameter, clear vesicles associated with a linear presynaptic density are common in all three classes of the Mollusca (Fig. 3). The cleft appears to widen somewhat in many synapses and usually a postsynaptic density is present (Hama 1962; Gray and Young 1964; Nicaise et al. 1968; Zs-Nagy and Sakharov 1969; Froesch and Martin 1972; Gillette and Pomeranz 1975; Pentreath et al. 1975; Prior and Lipton 1977; Tremblay et al. 1979; Roubos and Moorer-van Delft 1979). Only in Aplysia did there appear to be an example of an invaginated, presynaptic tuft as in the dyadic synapses of arthropods (Bailey et al. 1981). The use of ethanolic phosphotungstic acid (EPTA) as a stain has demonstrated that many Aplysia active zones resemble vertebrate synapses in having 30 nm high by 40 nm wide isolated or interconnected linear presynaptic densities in parallel with a continuous intracleft densification and a periodic postcytoplasmic membrane densification (Bailey et al. 1981). Subsynaptic cisterns were present in the bivalve Glycimeris (Elekes 1978).

Neurosecretory-type endings are characteristic of the central nervous system of molluscs (Figs. 11-13). A variety of types of neurosecretory synapses have been described in the abdominal ganglion of Aplysia, ranging from presynaptic profiles with large dense-cored vesicles or large dense granules to a mixture of dense-cored vesicles and neurosecretory or clear vesicles (Tremblay et al. 1979). Other neurosecretory endings have release sites associated with an aggregation of small, clear vesicles at the presynaptic membrane (Wendelaar Bonga 1970, 1971; Colmers 1977) (Fig. 13). Paramembranous densities and intracleft material are present at some synapses but many have a sparcity of paramembranous material. In the giant fiber system of cephalopods, a single row of 50-nm-diameter clear vesicles is present on either side of a 10 nm gap in the squid Sepia and multiple rows of clear vesicles appear on either side of a 15 nm cleft in the squid Illex (Martin 1969).

Neuroglial synapses have been demonstrated in the gastropod Aplysia and Lymnaea (Colonnier et al. 1979; Roubos and Moorer-van Delft 1979; Schwartz and Shkolnick 1981). The presynaptic profile reveals clear vesicles in contact with linear densities at the membrane but there is little densification of the postsynaptic membrane.

At gastropod neuromuscular junctions, clear vesicles predominate with 40 to 50-nm-diameter vesicles crowded against the membrane and larger clear or dense-cored vesicles scattered throughout the axon profile (Rogers 1968; Nicaise et al. 1968; Kiss and Elekes 1972; Heyer et al. 1973; Orkand and Orkand 1975) (Fig. 18). Neuroglandular synapses have been described in Octopus vulgaris and contain a mixture of small, clear vesicles 30-60 nm in diameter and both small 50 to 60-nm-diameter and large 90-nm-diameter dense-cored vesicles (Ducros 1972).

9. ECHINODERMATA

Echinoderm synaptic foci have gone unrecognized for many years because they lack substantial membrane densification (see Pentreath and

Cobb 1972; Cobb and Pentreath 1978). In retrospect, however, it appears that interneuronal and neuromuscular synapses were demonstrated early on by Cobb and Laverack (1966a, b, 1967). In echinoderm synapses, densification of the paired membranes at both interneuronal (Fig. 4) and neuromuscular contacts (Fig. 19) is generally less than that seen in the Cnidaria. Thus, many echinoderm synapses have been identified solely by the vesicles aligned at paired membranes (Cobb 1970; Cobb and Pentreath 1977). Both polarized and nonpolarized synapses with clear vesicles ranging in size from 50-100 nm in diameter have been reported (Cobb 1970). A few, dense-cored vesicles are present distal to the closely aligned membranes of the synaptic complex. Dense material in the cleft may appear as faint cross filaments. Neuromuscular synapses contact a wing-like process of the muscle cell and resemble interneuronal synaptic membrane complexes (Cobb and Laverack 1966b). Neurosecretory endings packed with dense granules are present in ganglia (Cobb and Laverack 1966a).

There is evidence that chemical transmission occurs across the basal lamina separating ectoneural nerve endings from postsynaptic muscle cells of hyponeural motor neurons (Cobb 1985a). These endings are varicose, vesicle-filled profiles without membrane densification. Recently, synapses showing marked pre- and postsynaptic densities and material within the cleft have been described between hyponeural motor nerves and juxtaligamental cells (Cobb 1985b). These synapses are described in the chapter by Cobb (1987) in this volume.

10. PHYLOGENETIC COMPARISONS

10.1. Interneuronal synapses

Polarized synapses with single or multiple rows of clear or dense-cored vesicles at focal membrane specializations are present in every major phylum from coelenterates to echinoderms. Nonpolarized, chemical synapses with a symmetrical arrangement of vesicles on both sides of the synaptic contact are common in jellyfish but rare in higher organisms. Presynaptic dense bodies or specialized T-bars associated with postsynaptic dyads originate in flatworms and become highly developed in arthropod sense organs. A presynaptic dense body with round, clear vesicles at a conventional synapse has been observed only in polychaete annelids. A postsynaptic spine invaginating the presynaptic element has been found in a cnidarian photoreceptor and sensory organs of cephalopod molluscs. Mixed synapses containing (a) clear and dense-cored vesicles, (b) dense-cored vesicles and neurosecretory granules, (c) neurosecretory granules with a postsynaptic element, and (d) neurosecretory granules in a synaptoid contact with a basal lamina are prominent in the gastropod central nervous system.

10.2. Neuromuscular synapses

Neuromuscular synapses vary from highly specialized as in ctenophores, in which there is a presynaptic triad of mitochondrion, cisterna of endoplasmic reticulum, and row of synaptic vesicles, to conventional, generalized structures resembling interneuronal synapses as in cnidarians, molluscs, and echinoderms. Neuromuscular synapses with one or more presynaptic bars, a periodic, intermediate line in the cleft, and periodic filaments on the extracellular surface of the postsynaptic membrane are typical of many arthropods. Synaptic vesicle differences between excitatory and inhibitory neuromuscular synapses have been reported in both crayfish and lobster.

10.3. Electrical synapses

Typical gap junctions with a 3-nm-wide gap between a pair of closely apposed membranes occur in Hydra and other cnidarians. Septal junctions present between dorsal giant fibers in the earthworm have a 6.5 to 7.5-nm-wide gap with cytoplasmic vesicles on both sides, whereas in crayfish the gap is only 4-5 nm. In some cases, the presumed low resistance junctions resemble chemical synapses.

11. EVOLUTION

The morphological diversity of invertebrate synapses offers a challenge to electron microscopy, which led to their first structural elucidation. However, comparisons of invertebrate synapses to date show no overall evolutionary trend, except perhaps the evolution of synaptic dyads coincident with the development of a brain in the flatworms. Considering the great number and diversity of invertebrate species, such a result is not surprising. Obviously, more data are required on many more species, tissues, and phyla before any evolutionary trends in synaptic structure become apparent. Future ultrastructural research on invertebrate synapses needs to be oriented towards serial reconstructions at high magnifications, and structural interpretations need to be correlated with biochemical and electrophysiological studies.

12. ACKNOWLEDGMENTS

This is contribution No.87-130-B from the Kansas Agricultural Experiment Station, supported in part by United States PHS grant NS-10264, awarded by the National Institute of Neurological and Communicative Disorders and Stroke. I thank Mallory R. Hoover for the artwork.

13. REFERENCES

Aizu S (1982) Morphological differences between excitatory and inhibitory nerve terminals in cockroach coxal muscles. Tissue Cell 14: 329-339.

Anderson PAV (1985) Physiology of a bidirectional, excitatory, chemical synapse. J Neurophysiol 53: 821-835.

Anderson PAV, Schwab WE (1981) The organization and structure of nerve and muscle in the jellyfish Cyanea capillata (Coelenterata; Scyphozoa). J Morphol 170: 383-399.

Anwyl R, Finlayson LH (1973) The ultrastructure of neurons with both a motor and a neurosecretory function in the insect, Rhodnius prolixus. Z Zellforsch Mikrosk Anat 146: 367-374.

Atwood HL, Pomeranz B (1974) Crustacean motor neuron connections traced by backfilling for electron microscopy. J Cell Biol 63: 329-334.

Bailey CH, Thompson EB (1979) Indented synapses in Aplysia. Brain Res 173: 13-20.

Bailey CH, Thompson EB, Castellucci VF, Kandel ER (1979) Ultrastructure of the synapses of sensory neurons that mediate the gill-withdrawal relfex in Aplysia. J Neurocytol 8: 415-444.

Bailey CH, Kandel P, Chen M (1981) Active zones at Aplysia synapses: organization of presynaptic dense projections. J Neurophysiol 46: 356-368.

Baskin DG (1974) Further observations on the fine structure and development of the infracerebral complex ("infracerebral gland") of Nereis limnicola (Annelida, Polychaeta). Cell Tissue Res 154: 519-531.

Boschek CB (1971) On the fine structure of the peripheral retina and lamina ganglionaris of the fly Musca domestica. Z Zellforsch Mikrosk Anat 118: 369-409.

Budelmann BU, Thies G (1977) Secondary sensory cells in the gravity receptor system of the statocyst of Octopus vulgaris. Cell Tissue Res 182: 93-98.

Buisson B (1970) Les supports morphologiques de l'intégration dans la colonie de Veretillum cynomorium Pall. (Cnidaria, Pennatularia) Z Morphol Tiere 68: 1-36.

Buisson B, Franc S (1969) Structure et ultrastructure des cellules mésenchymateuses et nerveuses intramésogléennes de Veretillum cynomorium Pall. (Cnidaire, Pennatulidae) Vie Milieu 20: 279-292.

Bullock TH, Horridge GA (1965) Structure and function in the nervous systems of invertebrates. WH Freeman and Company, San Franciso and London. Vol. 1.

Bunt AH (1969) Formation of coated and "synaptic" vesicles within neurosecretory axon terminals of the crustacean sinus gland. J Ultrastruct Res 28: 411-421.

Burkhardt W, Braitenberg V (1976) Some peculiar synaptic complexes in the first visual ganglion of the fly, Musca domestica. Cell Tissue Res 173: 287-308.

Chien P, Koopowitz H (1972) The ultrastructure of neuromuscular systems in Notoplana acticola, a free-living polyclad flatworm. Z Zellforsch Mikrosk Anat 133: 277-288.

Chien PK, Koopowitz H (1977) Ultrastructure of nerve plexus in flatworms. III The infra-epithelial nervous system. Cell Tissue Res 176: 335-347.

Cobb JLS (1970) The significance of the radial nerve cords in asteroids and echinoids. Z Zellforsch Mikrosk Anat 108: 457-474.

Cobb JLS (1985a) The neurobiology of the ectoneural/hyponeural synaptic connection in an echinoderm. Biol Bull 168: 432-446.

Cobb JLS (1985b) The motor innervation of the oral plate ligament in the brittle star Ophiura ophiura (L.) Cell Tissue Res 242: 685-688.

Cobb JLS (1987) Neurobiology of the Echinodermata. (This volume).

Cobb JLS, Laverack MS (1966a) The lantern of Echinus esculentus (L.) II. Fine structure of hyponeural tissue and its connexions. Proc Soc Lond Ser B 164: 641-650.

Cobb JLS, Laverack MS (1966b) The lantern of *Echinus esculentus* (L.) III The fine structure of the lantern retractor muscle and its innervation. Proc Soc Lond Ser B 164: 651-658.

Cobb JLS, Laverack MS (1967) Neuromuscular systems in echinoderms. Symp Zool Soc Lond 20: 25-51.

Cobb JLS Pentreath VW (1977) Anatomical studies of simple invertebrate synapses utilizing stage rotation electron microscopy and densitometry. Tissue Cell 9: 125-135.

Cobb JLS, Pentreath VW (1978) Comparison of the morphology of synapses in invertebrate and vertebrate nervous systems: analysis of the significance of the anatomical differences and interpretations of the morphological specifications. Prog Neurobiol 10: 231-252.

Coggeshall RE (1967) A light and electron microscope study of the abdominal ganglion of *Aplysia californica*. J Neurophysiol 30: 1263-1287.

Cohen AI (1973) An ultrastructural analysis of the photoreceptors of the squid and their synaptic connections. III. Photoreceptor terminations in the optic lobes. J Comp Neurol 147: 399-426.

Colmers WF (1977) Neuronal and synaptic organization in the gravity receptor system of the statocyst of *Octopus vulgaris*. Cell Tissue Res 185: 491-503.

Colonnier M, Tremblay JP, McLennan H (1979) Synaptic contacts on glial cells in the abdominal ganglion of *Aplysia californica*. J Comp Neurol 188: 391-400.

Dhainaut-Courtois N, Warembourg M (1969) Etude ultrastructurale des neurones de la chaîne nerveuse de *Nereis pelagica* L. (Annélide Polychète). Z Zellforsch Mikrosk Anat 97: 260-273.

Ducros C (1972) Etude ultrastructurale de l'innervation des glandes salivaires postérieures chez *Octopus vulgaris*. III. L'innervation des tubules de la glande. Z Zellforsh Mikrosk Anat 132: 67-78.

Elekes K (1978) Ultrastructure of synapses in the central nervous system of lamellibranch molluscs. Acta Biol Acad Sci Hung 29: 139-154.

Fahrenbach WH (1979) The brain of the horseshoe crab (*Limulus polyphemus*). III. Cellular and synaptic organization of the corpora pedunculata. Tissue Cell 11: 163-200.

Ferrero EA, Lanfranchi A, Bedini C (1985) An ultrastructural account of otoplanid turbellaria neuroanatomy. I. The cerebral ganglion and the peripheral nerve net. Acta Zoologica (Stockh) 66: 63-74.

Fisher SK, Tabor GA (1977) Unusual presynaptic inclusions in the CNS of the marine polychaete, *Nereis virens*. Cell Tissue Res 177: 123-126.

Foelix RF, Troyer D (1980) Giant neurons and associated synapses in the peripheral nervous system of whip spiders. J Neurocytol 9: 517-535.

Franc S (1968) Les évolutions cellulaires et les rapports entre les tissus mésenchymateux et nerveux au cours de la régénération du pédoncule de *Veretillum cynomorium* pall. (Cnidaire "Anthozoaire"). Thèse, Université de Lyon.

Froesch D, Martin R (1972) Heterogeneity of synaptic vesicles in the squid giant fibre system. Brain Res 43: 573-579.

Fröhlich A, Meinertzhagen IA (1982) Synaptogenesis in the first optic neuropile of the fly's visual system J. Neurocytol 11: 159-180.

Gillette R, Pomeranz B (1975) Ultrastructural correlates of interneuronal function in the abdominal ganglion of Aplysia californica. J Neurobiol 6: 463-474.

Goodman LJ, Mobbs PG, Kirkham JB (1979) The fine structure of the ocelli of Schistocerca gregaria. The neural organization of the synaptic plexus. Cell Tissue Res 196: 487-510.

Govind CK, Chiang RG (1979) Correlation between presynaptic dense bodies and transmitter output at lobster neuromuscular terminals by serial section electron microscopy. Brain Res 161: 377-388.

Govind CK, DeRosa RA (1983) Fine structure of comparable synapses in a mature and larval lobster muscle. Tissue Cell 15: 97-106.

Govind CK, Pearce J (1982) Proliferation and relocation of developing lobster neuromuscular synapses. Dev Biol 90: 67-78.

Govind CK, Atwood HL, Land F (1973) Synaptic differentiation in a regenerating crab-limb muscle. Proc Nat Acad Sci USA 70: 822-826.

Govind CK, DeRosa RA, Pearce J (1980) Presynaptic dense bars at neuromuscular synapses of the lobster, Homarus americanus. Cell Tissue Res 207: 81-88.

Gray EG, Young JZ (1964) Electron microscopy of synaptic structure of octopus brain. J Cell Biol 21: 87-103.

Günther J, Schürmann FW (1973) Zur Feinstruktur des dorsalen Riesenfasersystems im Bauchmark des Regenwurms. II. Synaptische Beziehungen der proximalen Riesenfaserkollateralen. Z Zellforsh Mikrosk Anat 139: 369-396.

Hafner GS (1974) The ultrastructure of retinula cell endings in the compound eye of the crayfish. J Neurocytol 3: 295-311.

Hama K (1962) Some observations on the fine structure of the giant synapse in the stellate ganglion of the squid, Doryteuphis bleekeri. Z Zellforsch Mikrosk Anat 56: 437-444.

Hámori J, Horridge GA (1966) The lobster optic lamina. II. Types of synapse. J Cell Sci 1: 257-270.

Hernandez-Nicaise M-L (1968) Specialized connexions between nerve cells and mesenchymal cells in ctenophores. Nature 217: 1075-1076.

Hernandez-Nicaise M-L (1973a) Le système nerveux des Cténaires. I. Structure et ultrastructure des réseaux epithéliaux. Z Zellforsch Mikrosk Anat 137: 223-250.

Hernandez-Nicaise M-L (1973b) The nervous system of ctenophores. III. Ultrastructure of synapses. J Neurocytol 2: 249-263.

Hernandez-Nicaise M-L (1974a) Système nerveux et intégration chez les ctenaires. Etude ultrastructurale et comportementale. Thèse, Université Claude Bernard (Lyon I).

Hernandez-Nicaise M-L (1974b) Ultrastructural evidence for a sensory-motor neuron in Ctenophora. Tissue Cell 6: 43-47.

Heyer CB, Kater SB, Karlsson UL (1973) Neuromuscular systems in molluscs. Amer Zool 13: 247-270.

Horridge GA (1965) Non-motile sensory cilia and neuromuscular junctions in a ctenophore independent effector organ. Proc Soc Lond Ser B 162: 333-350.

Horridge GA, Mackay B (1962) Naked axons and symmetrical synapses in coelenterates. Q J Microsc Sci 103: 531-541.

Horridge GA, Mackay B (1964) Neurociliary synapses in Pleurobrachia (Ctenophora). Q J Microsc Sci 105: 165-174.

Horridge GA, Chapman DM, Mackay B (1962) Naked axons and symmetrical synapses in an elementary nervous system. Nature 193: 899-900.

Hudspeth AJ, Stuart AE (1977) Morphology and responses to light of the somata, axons and terminal regions of individual photoreceptors of the giant barnacle. J Physiol 272: 1-23.

Jahromi SS, Atwood HL (1974) Three-dimensional ultastructure of the crayfish neuromuscular apparatus. J Cell Biol 63: 599-613.

Jha RK, Mackie GO (1967) The recognition, distribution and ultrastructure of hydrozoan nerve elements. J Morphol 123: 43-62.

King DG (1976) Organization of crustacean neuropil. I. Patterns of synaptic connections in lobster stomatogastric ganglion. J Neurocytol 5: 207-237.

King JA, Govind CK (1980) Development of excitatory innervation in the lobster claw closer muscle. J Comp Neurol 194: 57-70.

Kinnamon JC, Westfall JA (1982) Type of neurons and synaptic connections at hypostome-tentacle junctions in Hydra. J Morphol 173: 119-128.

Kiss T, Elekes K (1972) Myo-neural junctions in the ventricle of the snail Helix pomatia L. Acta Biol Acad Sci Hung 23: 207-209

Lane NJ (1985) Structure of components of the nervous system. In: Kerkut GA, Gilbert LI (eds) Comprehensive insect physiology, biochemistry and pharmacology, vol 5. Permagon Press, Oxford, pp 1-47.

Lawry JV (1967) Structure and function of the parapodial cirri of the polynoid polychaete, Harmothoë. Z Zellforsch Mikrosk Anat 137: 223-250.

Lentz TL (1965) Fine structural changes in the nervous system in the regenerating hydra. J Exp Zool 159: 181-194.

Mackie GO (1976) The control of fast and slow muscle contractions in the siphonophore stem. In: Mackie GO (ed) Coelenterate ecology and behavior. Plenum Press, New York, pp 647-659.

Mackie GO, Singla CL (1975) Neurobiology of Stomatoca. I. Action systems. J Neurobiol 6: 339-356.

Martin R (1969) The structural organization of the intracerebral giant fiber system of cephalopods. Z Zellforsh Mikrosk Anat 97: 50-68.

Meiss DE, Govind CK (1980) Heterogeneity of excitatory synapses at the ends of single muscle fibers in lobster, Homarus americanus. J Neurobiol 11: 381-395.

Muller KJ, Carbonetto S (1979) The morphological and physiological properties of a regenerating synapse in the CNS of the leech. J Comp Neurol 185: 485-516.

Muller KJ, McMahan UJ (1976) The shapes of sensory and motor neurones and the distribution of their synapses in ganglia of the leech: a study using intracellular injection of horseradish peroxidase. Proc R Soc Lond B 194: 481-499.

Nicaise G, DeCeccatty MP, Baleydier C (1968) Ultrastructure des connexions entre cellules nerveuses, musculaires et glio-interstitielles chez Glossodoris. Z Zellforsh Mikrosk Anat 88: 470-486.

Nichol D, Meinertzhagen IA (1982) An anlysis of the number and composition of the synaptic populations formed by photoreceptors of the fly. J Comp Neurol 207: 29-44.

Oesterle D, Barth FG (1973) Zur Feinstruktur einer electrischen Synapse. Die Septen der dorsalen Riesenfasern von Regenwürmern (Lumbricus terrestris, Eisenia foetida). Z Zellforsch Mikrosk Anat 136: 139-152.

Orkand PM, Orkand RK (1975) Neuromuscular junctions in the buccal mass of Aplysia: fine structure and electrophysiology of excitatory transmission. J Neurobiol 6: 531-548.

Osborne MP, Finlayson LH, Rice MJ (1971) Neurosecretory endings associated with striated muscles in three insects (Schistocerca, Carausius, and Phormia) and a frog (Rana). Z Zellforsch Mikrosk Anat 166: 391-404.

Pavans de Ceccatty M (1979) Physiological and ultrastructural bases for a study of the ontogenesis of integration in Podocoryne carnea cultures. In: Tardent P, Tardent R (eds) Developmental and cellular biology of coelenterates. Proc 4th Intl Coelenterate Conference. Elsevier/North-Holland Biomedical Press, Amsterdam, pp 459-464.

Pentreath VW, Cobb, JLS (1972) Neurobiology of echinodermata. Biol Rev 47: 363-392.

Pentreath VW, Berry MS, Cobb JLS (1975) Nerve-ending specializations in the central ganglia of Planorbis corneus. Cell Tissue Res 163: 99-110.

Peracchia C, Dulhunty AF (1976) Low resistance junctions in crayfish. Structural changes with functional uncoupling. J Cell Biol 70: 419-439.

Peteya DJ (1973a) A light and electron microscope study of the nervous system of Ceriantheopsis americanus (Cnidaria, Ceriantharia) Z Zellforsh Mikrosk Anat 141: 301-317.

Peteya DJ (1973b) A possible proprioceptor in *Ceriantheopsis americanus* (Cnidaria, Ceriantharia) Z Zellforsh Mikrosk Anat 144: 1-10.

Prior DJ, Lipton BH (1977) An ultrastructural study of peripheral neurons and associated non-neural structures in the bivalve mollusc, *Spisula solidissima*. Tissue Cell 9: 223-240.

Purves D, McMahan UJ (1972) The distribution of synapses on a physiologically identified motor neuron in the central nervous system of the leech. An electron microscope study after the injection of the fluorescent dye procion yellow. J Cell Biol 55: 205-220.

Quaglia A (1976) Osservazioni sul sistema nervosa digli Antozoii. Boll Zool 43: 397-398.

Reuter M (1981) The nervous system of *Microstomum lineare* (Turbellaria, Macrostomida). II. The ultrastructure of synapses and neurosecretory release sites. Cell Tissue Res 218: 375-387.

Rheuben MB, Reese TB (1978) Three-dimensional structure and membrane specializations of moth excitatory neuromuscular synapse. J Ultrastruct Res 65: 95-111.

Ribi WA (1978) Gap junctions coupling photoreceptor axons in the first optic ganglion of the fly. Cell Tissue Res 195: 299-308.

Ribi WA (1981) The first optic ganglion of the bee. IV. Synaptic fine structure and connectivity patterns of receptor cell axons and first order interneurones. Cell Tissue Res 215: 443-464.

Rogers DC (1968) Fine structure of smooth muscle and neuromuscular junctions in optic tentacles of *Helix aspersa* and *Limas flavus*. Z Zellforsch Mikrosk Anat 89: 80-94.

Rosenbluth J (1965) Ultrastructure of somatic muscle cells in *Ascaris lumbricoides*. II. Intermuscular junctions, neuromuscular junctions, and glycogen stores. J Cell Biol 26: 579-591.

Rosenbluth J (1972) Myoneural junctions of two ultrastructurally distinct types of earthworm body wall muscle. J Cell Biol 54: 566-574.

Roubos EW, Moorer-van Delft CM (1979) Synaptology of the central nervous system of the freshwater snail *Lymnaea stagnalis* (L.), with particular reference to neurosecretion. Cell Tissue Res 198: 217-235.

Saint Marie RL, Carlson SD (1982) Synaptic vesicle activity in stimulated and unstimulated photoreceptor axons in the housefly. A freeze-fracture study. J Neurocytol 11: 747-761.

Satterlie RA, Case JF (1978) Neurobiology of the gorgonian coelenterates, *Muricea californica* and *Lophogorgia chilensis*. II. Morphology. Cell Tissue Res 187: 379-396.

Satterlie RA, Case JF (1980) Neurobiology of the stoloniferan octocoral *Clavularia* sp. J Exp Zool 212: 87-99.

Satterlie RA, Spencer AN (1983) Neuronal control of locomotion in hydrozoan medusae. A comparative study. J Comp Physiol 150: 195-206.

Satterlie RA, Anderson PAV, Case JF (1980) Colonial coordination in anthozoans: pennatulacea. Mar Behav Physiol 7: 25-46.

Schürmann FW (1971) Synaptic contacts of association fibres in the brain of the bee. Brain Res 26: 169-176.

Schwartz JH, Shkolnik LJ (1981) The giant sterotonergic neuron of Aplysia: a multi-targeted nerve cell. J Neurosci 1: 606-619.

Shaw MK (1981) The ultrastructure of synapses in the brain of Gastrocotyle trachuri (Monogenea, Platyhelminthes). Cell Tissue Res 220: 181-189.

Singla CL (1976) Ultrastructure and attachment of the basal disk of Haliclystus. In: Mackie GO (ed) Coelenterate ecology and behavior. Plenum Press, New York, pp 533-540.

Singla CL (1978a) Locomotion and neuromuscular system of Aglantha digitale. Cell Tissue Res 188: 317-327.

Singla CL (1978b) Fine structure of the neuromuscular system of Polyorchis penicillatus (Hydromedusae, Cnidaria). Cell Tissue Res 193: 163-174.

Skinner K (1985) The structure of the fourth abdominal ganglion of the crayfish Procambarus clarki (Girard). II. Synaptic neuropils. J Comp Neurol 234: 182-191.

Spencer AN (1979) Neurobiology of Polyorchis. II. Structure of effector systems. J Neurobiol 10: 95-117.

Stirling CA (1972) The ultrastructure of giant fiber and serial synapses in crayfish. Z Zellforsch Mikrosk Anat 131: 31-45.

Stokes DR (1974) Morphological substrates of conduction in the colonial hydroid Hydractinia echinata. I. An ectodermal nerve net. J Exp Zool 190: 19-46.

Titmus MJ (1981) Ultastructure of identified fast excitatory, slow excitatory and inhibitory neuromuscular junctions in the locusts. J Neurocytol 10: 363-385.

Toh Y, Yoshida M, Tateda H (1979) Fine structure of the ocellus of the hydromedusan, Spirocodon saltatrix. I. Receptor cells. J Ultrastruct Res 68: 341-352.

Tolbert LP, Hildebrand JG (1981) Organization and synaptic ultrastructure of glomeruli in the antennal lobes of the moth Manduca sexta: a study using thin sections and freeze-fracture. Proc R Soc Lond B 213: 279-301.

Tremblay JP, Colonnier M, McLennan H (1979) An electron microscope study of synaptic contacts in the abominal ganglion of Aplysia californica. J Comp Neurol 188: 367-390.

Trujillo-Cenóz O (1965) Some apsects of the structural organization of the intermediate retina of dipterans. J Ultrastruct Res 13: 1-33.

Trujillo-Cenóz O (1969) Some aspects of the structural organization of the medulla in muscoid flies. J Ultrastruct Res 27: 533-553.

Watson AHD (1984) The dorsal unpaired median neurons of the locust metathoracic ganglion: neuronal structure and diversity, and synapse distribution. J Neurocytol 13: 303-327.

Webb RA (1977) Evidence for neurosecretory cells in the cestode *Hymenolepis microstoma*. Can J Zool 55: 1726-1733.

Weber C, Singla CL, Kerfoot PAH (1982) Microanatomy of the subumbrellar motor innervation in *Aglantha digitale* (Hydromedusae: Trachylina). Cell Tissue Res 223: 305-312.

Wells J, Besso JA Jr, Boldosser WG, Parsons RL (1972) The fine structure of the nerve cord of *Myxicola infundibulum* (Annelida, Polychaeta). Z Zellforsh Mikrosk Anat 131: 141-148.

Wendelaar Bonga SE (1970) Ultrastructure and histochemistry of neurosecretory cells and neurohaemal areas in the pond snail, *Lymnea stagnalis* (L.) Z Zellforsch Mikrosk Anat 108: 190-224.

Wendelaar Bonga SE (1971) Formation, storage, and release of neurosecretory material studied by quantitative electron microscopy in the fresh water snail *Lymnaea stagnalis* (L.). Z Zellforsch Mikrosk Anat 113: 490-517.

Westfall JA (1969) Nervous control of nematocyst discharge: chemical synapses. Amer Zool 9: 1141.

Westfall JA (1970a) Ultrastructure of synapses in a primitive coelenterate. J Ultrastruct Res 32: 237-246.

Westfall JA (1970b) Synapses in a sea anemone, *Metridium* (Anthozoa). Electron Microsc Proc Int Congr 7th, Société Française de Microscopie Electronique, Paris 3: 717-718.

Westfall JA (1970c) The nematocyte complex in a hydromedusan, *Gonionemus vertens*. Z Zellforsch Mikrosk Anat 110: 457-470.

Westfall JA (1973a) Ultrastructural evidence for a granule-containing sensory-motor-interneuron in *Hydra littoralis*. J Ultrastruct Res 42: 268-282.

Westfall JA (1973b) Ultrastructural evidence for neuromuscular systems in coelenterates. Amer Zool 13: 237-246.

Westfall JA, Kinnamon JC (1978) A second sensory-motor-interneuron with neurosecretory granules in *Hydra*. J Neurocytol 7: 365-379.

Westfall JA, Kinnamon JC (1984) Perioral synaptic connections and their possible role in the feeding behavior of *Hydra*. Tissue Cell 16: 355-365.

Westfall JA, Yamataka S, Enos PD (1970a) Ultrastructure of synapses in *Hydra*. J Cell Biol 47: 226a.

Westfall JA, Yamataka S, Enos PD (1970b) An ultrastructural survey of synapses in tentacles of coelenterates. Amer Zool 10: 545.

Westfall JA, Yamataka S, Enos PD (1971) Ultrastructural evidence of polarized synapses in the nerve net of *Hydra*. J Cell Biol 51: 318-323.

Westfall JA, Kinnamon JC, Sims DE (1980) Neuro-epitheliomuscular cell and neuro-neuronal gap junctions in *Hydra*. J Neurocytol 9: 725-732.

Wood MR, Pfenninger KH, Cohen MJ (1977) Two types of presynaptic configurations in insect central synapses: an ultrastructural analysis. Brain Res 130: 25-45.

Yamasu T, Yoshida M (1976) Fine structure of complex ocelli of a cubomedusan, Tamoya bursaria Haeckel. Cell Tissue Res 170: 325-339.

Zs-Nagy I, Labós E (1969) Light and electron microscopical investigations on the adductor muscle and nervous elements in the larva of Anodonta cygnea L. Ann Biol Tihany 36: 123-133.

Zs-Nagy I, Sakharov DA (1969) Axo-somatic synapses in procerebrum of Gastropoda. Experientia 25: 258-259.

SYNAPSE FORMATION BETWEEN IDENTIFIED INVERTEBRATE NEURONES IN VITRO

JONATHAN P. BACON

School of Biological Sciences

University of Sussex

Falmer

Brighton BN1 9QG, United Kingdom

ABSTRACT

Identified neurones from the leech, snail and <u>Aplysia</u> can be removed from the CNS and maintained in culture. The cells grow fasciculating neurites and form specific electrical and chemical synaptic connexions. The accessibility of these cultured neurones to physiological and pharmacological investigation facilitates study of the many factors that promote synaptogenesis.

1. INTRODUCTION

Invertebrates provide excellent model systems for solving fundamental problems in developmental neurobiology. Perhaps the biggest advantage they offer over the vertebrates is that many of their neurones are identified. We have an almost complete curriculum vitae of many of these identified cells: time and place of birth from its precursor cells, the progression of growth and differentiation of the neurone, the function of the cell and its synaptic partners in the postembryonic nervous system, and the time and place of its death. The ease with which many identified cells can be stained during their development encourages one to consider them as developing in splendid isolation. This, of course, is not the case; their development progresses in the context of a complex orchestration of spatial and temporal cues, provided largely by other neurones.

One attempt to understand at least some developmental processes is to simplify the environment as much as possible, to pluck the cells out of the constraints of the CNS and grow them in culture where many of the parameters of the cell's environment can be controlled. This essay will discuss this approach on identified invertebrate neurones. Their removal from the postembryonic nervous system largely circumvents the processes of neurogenesis, differentiation and pathfinding; this allows one to focus on the problem of synapse formation.

I am deliberately restricting this review to the culture of identified invertebrate neurones. In doing so, I inevitably exclude much excellent work; an example is provided by the work of Beadle's group on dissociated insect neuronal cultures (Beadle and Hicks 1985). However not to restrict my attention in this way, would be tantamount to ignoring the tremendous advantages of studying identified cells. Work on three main preparations, the leech, the snail and the sea slug, Aplysia, will be described. This does, to my knowledge, encompass almost all of the work that has been done on synaptogenesis between identified invertebrate neurones in vitro.

2. THE LEECH

The medicinal leech, Hirudo medicinalis, with its primitive segmented body and 23 ventral ganglia, lends itself to this kind of experimentation since many of its neurones are identifiable under the dissecting microscope by their size and unique position within the ganglia. This was exploited by Ready and Nicholls (1979) who used fine microfilament lassos to remove cells from the ganglion. The cells were maintained in culture for several weeks and exhibited their normal (in vivo) patterns of action potentials and membrane properties. After 5-7 days, the neurones became multipolar, neurites growing out from the cell bodies in all directions. Fasciculation was seen in the dish and specific electrical synapses were formed when neurones were placed in pairs. Retzius (R) cells (these cause mucus secretion in the animal) formed electrical connexions with other R cells but not with sensory Pressure (P) cells. P cells did not form electrical connexions with each other. The finding that this pattern of in vitro connectivity resembled the situation in the ganglion was an early encouraging omen for future work.

Fuchs et al. (1981) extended this analysis by adding the sensory neurones Touch (T) and Nociceptive (N) and the (L) motorneurone to the R and P cells, increasing the permutations of possible identified synaptic partners. In addition to those described in the previous study, a number of electical connexions seen in the animal were also produced in the dish, notably non-rectifying L-L connexions and rectifying P-L and N-L connexions. "Novel" connexions (meaning synaptic connexions not found in the ganglion) did form; R and L cells became electrically coupled and in those cases where R cells failed to make electrical synapses with each other, chemically mediated inhibitory potentials were seen.

By application of leech blood to the culture medium and placing the cells very close together, Fuchs et al. (1982) were able to promote synapse formation between R and P cells. Transmission between them appeared chemical since each action potential in the R cell was followed by a hyperpolarising synaptic potential in the P cell after a constant delay, the transmission was unidirectional, the potentials exhibited facilitation and depression, were reversed by hyperpolarisation, and reduced Ca^{++} and raised Mg^{++} in the bath blocked transmission. The potential results from a decrease in membrane resistance, at least partly due to chloride conductance; high intra- cellular Cl^- in P reversed and amplified the potential. These experiments prompted a search for this connexion in the ganglion and one was found that was similar in many respects to the in vitro synapse.

Further work on this synapse in vitro has made it something of a classic. Henderson (1983) went through many of the usual criteria to establish that 5-hydroxytryptamine (5HT) was the transmitter used at this synapse. These criteria are synthesis (R cells in culture were shown to synthesise 5HT from radio-labelled precursors and did not synthesise other transmitters from their precursors), storage (Neutral Red and histofluorescence showed the presence of monoamines in R cells), release (R cells in vitro released 5HT when depolarised), and uptake (R cells in vitro accumulated 100 times more 5HT than did non-serotonergic cells);

chlorimipramine (a blocker of 5HT uptake) interfered with uptake. In addition, P cells in vitro responded to focally applied, pressure ejected, 5HT.

A problem with studying "chemical synapses" in culture could be that since neurones become so intertwined in the dish, release of transmitter by one cell could cause synaptic potentials in adjacent cells without there being any kind of synaptic specialisation. To refute the notion that transmitter is merely sloshing around indiscriminately between the neurones, one needs both EM evidence on synaptic structure and physiological evidence on the quantal nature of transmitter release.

Henderson et al. (1983) have attempted to provide this evidence on the R-P in vitro synapse. The two cells, which can be distinguished with the EM by the appearance of their cytoplasm, grew intertwining processes approaching within 20-25 nm of each other. Structures resembling synapses between the processes of R cells formed after 4 days in culture. These comprised vesicle clusters adjacent to a widened, straight intercellular cleft containing dense-staining material. R to P connexions looked less "classical" but vesicle aggregations in R at points where R and P were separated by only a narrow cleft were observed.

In conditions designed to reduce synaptic release (high Mg^{++} or low Ca^{++} in the culture medium), repeated stimulation of the R cell revealed quantal fluctuation, and failures of the post-synaptic potentials. Their amplitudes were distributed in accordance with the Poisson equation which produced values of m (the mean number of quanta liberated per trial) and a unitary potential size close to that measured from spontaneous events (Henderson et al. 1983).

A further study of this synapse used voltage clamp techniques to investigate the release of 5HT from the R cell. Sudden displacements of R's membrane potential caused slower and smaller changes in P's membrane potential by altering the tonic release of transmitter. These changes of R cell holding potential also altered the voltage noise recorded in the P cell. Noise analysis showed that these changes could be accounted for by quantal events with an amplitude similar to that measured in the previous experiments (approximately 0.15 mV). Two-shock facilitation was also demonstrated in this study (Dietzel et al. 1986).

The sensitivity of the isolated, growing P cell to locally applied, pulses of both 5HT and ACh is highest at the base of its growing neurites. This could be a common feature of cultured neurones since the anterior Pagoda cell shows a similar distribution of its ACh receptors (Pellegrino and Simonneau 1984).

3. APLYSIA

Molluscan neurones are easily accessible and often "colour coded". Alving (1968) was able to tie fine threads around the neurites of individual neurones but Chen et al. (1971) were the first to dissect single somata (with a small attached neurite) away from the Aplysia nervous system. These unidentified cells survived for at least 24 hours in sea water and some showed spontaneous activity. Not surprisingly, these cells showed no neuritic outgrowth. However in 1979, Kaczmarek et al. removed Aplysia bag cells (which are thought to be an homogeneous population of cells) and placed them in culture. Just as in the CNS, they showed electrical coupling and Lucifer-Yellow dye transfer in the dish with some neurite outgrowth.

Dagan and Levitan (1981) were successful in growing large Aplysia neurons in culture by using plasma clots and methylcellulose to partially immobilise the cells. Electrical, non-rectifying synapses formed between some cells of the buccal ganglion. Their ability to form synapses was not correlated with proximity, however, indicating possible specificity in the dish. This is in keeping with the fact that these cells are thought not to comprise an homogeneous population in vivo (Gardner 1971).

In 1983, Schacher and Proshansky added haemolymph to their culture medium which resulted in vigorous growth of identified neurons from Aplysia abdominal ganglion. Camardo et al. (1983) found that neuron L10 is able to form chemical inhibitory synapses with the left upper quadrant (LUQ) cells L2-L6; a burst of activity in L10 produced an IPSP in the LUQ cells. The L10-LUQ synapse in vivo is cholinergic (Giller and Schwartz 1971) and LUQ cells in vitro respond to iontophoresis of ACh. More importantly, these workers were unable to make the right upper quadrant (RUQ) cells, which reside within the ganglion and also have receptors for ACh, receive synaptic input from L10 in vitro. This shows that transmitter-receptor compatibility of two neurones is not a sufficient criterion for in vitro synapse formation (it is, of course, a necessary criterion).

A further example of in vitro synaptic specificity of Aplysia neurones was observed in co-cultures of neurosecretory bag cells and buccal cells identified only at the level of belonging to a particular population (bag cells are white and buccal neurons are orange or brown pigmented; Bodmer et al. 1984). Bag cells formed electrical synapses between themselves, as did buccal cells, but no electrical synapses formed between the two types. Novel unidirectional chemical synapses were found; a burst of spikes in a buccal cell sometimes caused a slow hyperpolarisation in an adjacent bag cell. The potential was due to a decrease in input resistance in the bag cell and disappeared in Ca^{++} free Ringer.

Bodmer and Levitan (1984) applied local iontophoretic pulses of ACh and 5HT to a number of identified Aplysia neurones in vitro. Forty-seven % of the neurones they tested were sensitive to cholinergic agonists, 14% to 5HT and 9% responded to both; this localised application of transmitter revealed that receptors are situated on the cell body and on the neurites. These neurones in culture generally exhibited the same neurotransmitter receptor-ion channel complexes as they did in vivo.

One of the largest neurons in the Aplysia nervous system is the serotonergic metacerebral cell (MCC) of the cerebral ganglion. Schacher (1985) succeeded in removing the cell from the ganglion complete with its neurite and two major branches. In vivo, the smaller-diameter branch enters the posterior lip nerve and innervates the lip musculature while the stouter branch enters the cerebro-buccal connective and synapses with motorneurons B1 and B2 in the buccal ganglion. The two branches have very different axon diameters and so remain distinguishable in vitro. By placing buccal cells B1 or B2 next to the two branches, Schacher (1985) found that the larger branch of MCC made synapses with buccal cells in 90% of the cases where neurites overlapped, whereas the smaller branch formed synapses in only 20% of cases. Outgrowth was generally more vigorous from the stouter branch. With targets nearer the big branch, reduced outgrowth from the other branch was observed. However, the differential ability of the two branches to form synapses with B1 and B2 appears not to be simply a growth-related phenomenon since in cases where the big branch grew poorly in culture, it still formed synapses with the buccal neurones.

Cell R2, the giant cholinergic cell of the abdominal ganglion, is capable of making a variety of synaptic connexions in culture when challenged with a number of other abdominal cells, L2-6, L11 and R15, as possible synaptic targets (Schacher et al. 1985). The synaptic connexion established between R2 and L2-6 was a two-component hyperpolarisation that appears to be mediated by ACh. The R2 to L11 connexion was more complicated with a fast, cholinergic, inhibitory component and a slow excitatory component mediated by some other transmitter. In contrast, the R2 to R15 connexion was a slow hyperpolarisation which is definitely not mediated by ACh since ACh iontophoresis caused a depolarisation of R15's membrane. It is interesting that despite the fact that R2 uses ACh as a transmitter and R15 has receptors for it, the synapses formed by these two cells used some other transmitter. FMRFamide is a possible candidate.

Since R2 is so large (up to several hundred microns in diameter in these studies), it has been feasible to study the molecular events accompanying its synaptogenesis with other neurons in the dish (Ambron et al. 1985). Exposure of the cell to [^{35}S] methionine revealed that R2 synthesises more than 300 polypeptides in vitro and that, rather undramatically, only 2 proteins (68-kd and 72-kd) showed enhanced production after synaptogenesis. It would appear that many of the proteins required for synaptogenesis are present in R2 prior to it contacting a target cell.

Aplysia's major neurobiological impact has been in the investigation of learning at the cellular level (Kandel 1979). For example, its gill withdrawal reflex shows two kinds of nonassociative learning, habituation and sensitisation, as well as classical conditioning. The advances made in the culture of Aplysia neurones, have stimulated an attempt to construct the circuitry of this reflex in the dish (Rayport and Schacher 1986). Two components of the circuit, the LE mechanosensory cells and gill motorneurone L7 from the abdominal ganglion were cultured with the MCC cell from the cerebral ganglion. This latter serotonergic cell was used to simulate the facilitatory interneurones in the system. The LE cell formed synapses with L7 which showed homosynaptic depression on repeated stimulation. Stimulation of the MCC cell at this point caused an immediate increase in the PSP size, replicating heterosynaptic facilitation.

4. THE SNAIL

Like the leech and Aplysia, snail neurones can be extracted relatively simply from the CNS. The first attempts to isolate single cells were made by Oomura and Maeno (1963) on the marine pulmonate mollusc, Onchidium verruculatum. They managed to tie fine silk thread around the neurites of individual neurones, and showed that the soma was still capable of generating action potentials.

A Russian group, working on Lymnea stagnalis, was the first to remove unidentified snail neurones from the CNS and maintain them in vitro (Kostenko et al. 1974). The cells maintained their physiological characteristics, responded to transmitter substances and showed very limited neuritic growth. Since then, this group has become interested in the factors necessary to promote neurite outgrowth; low intracellular Ca^{++} and pH seem to be the optimum conditions (Kostenko et al. 1983).

Kater's group in the USA took a different approach to discovering the optimum conditions for the growth of isolated, unidentified Helisoma trivolvis neurones in vitro. The most vigorous neurite outgrowth was achieved when either the cells were co-cultured with whole ganglia or were cultured in ganglia-conditioned media (Wong et al. 1981). The factor(s) that enhanced neuritic growth appeared tightly bound to the substrate and seemed to be unimportant for the physiology of neurones because cells grown in unconditioned defined medium (showing no neurite outgrowth) displayed normal physiological characteristics.

Conditioning factors that promote outgrowth of neurones from the snails, <u>Lymnea</u> and <u>Biomphalaria</u>, and the sea slug, <u>Aplysia</u>, are also produced by their respective brains (Wong et al. 1983). The growth promoting activity appears to be highly conserved and cross species reactive for the closely-related <u>Helisoma</u> and <u>Biomphalaria</u>. <u>Lymnea</u> and especially <u>Aplysia</u> are phylogenetically more distantly related and showed a concomitant lack of cross-species efficacy of their conditioning factor(s). The factor(s) must be proteinaceous since the activity is abolished by proteases such as chymotrypsin and trypsin on heating to 100°C but is stable to DNase and RNase treatment. Application of the protein synthesis inhibitor, anisomycin, to brains in culture medium showed that about 65% of the activity resides in some stored pool whereas the remainder is synthesised de novo (Wong et al. 1984).

Having established the optimum conditions necessary for the maintenance and sprouting of <u>Helisoma</u> neurones in culture, Hadley et al. (1983), were able to grow identified cells in vitro and to test the hypothesis that formation of new electrical synapses requires spatial and temporal co-ordination of neurite outgrowth. Work in vivo had shown that axotomy of neurone 5 of the buccal ganglion caused it to sprout within the ganglion. The sprouting caused neurone 5 to overlap its contra- lateral homologue but the two cells would only form "novel" electrical connexions if they sprouted simultaneously (Hadley and Kater 1983).

The clearest way to confirm the idea that simultaneous neuritic growth is a prerequisite for electrical synapse formation is to perform the experiments in the dish. By exploiting the fact that 2 days after plating the cells are still growing but at 7 days they reach a stable morphological state, Hadley et al. (1983) demonstrated that neurone 5 was incapable of forming electrical connexions with networks of older neurone 5's which had already formed connexions between themselves. This confirms that electrical synapse formation does indeed depend on interaction of mutually growing neurites.

By careful observation of these identified snail neurones in culture, Kater (1985) noticed that some cells in vitro appeared to be attracted to one another whereas others appeared to be repelled from one another. Obvious culprits for this attraction-repulsion phenomenon are neurotransmitters since they are known to be released from the growth cones of growing neurones in culture (Hume et al. 1983; Young and Poo 1983) and it was indeed found that bath application of 5HT stopped the advance of growth cones of neurone 19 from the buccal ganglion but had no effect on neurone 5 (Haydon et al. 1984). Isolating the growth cone to make it behave like a separate autonomous organelle demonstrated that this was the site of 5HT action. In keeping with the fact that 5HT stopped growth of neurone 19 but not 5, bath application of the transmitter had no effect on the formation of electrical synapses between neurones 5 but did render neurone 19 incapable of electrical synapse formation.

The growth cones of neurone 5 and 19 are different. Neurone 19 has fewer and longer filopodia with a greater interfilopodial distance compared to neurone 5. In addition, neurone 19 growth cones advance over the substrate more slowly and respond to 5HT by retracting their filopodia whereas neurone 5 growth cones are unaffected. Growth cones act autonomously since focal application of 5HT affected only the exposed growth cones, leaving other growth cones on the same cell to carry on growing (Haydon et al. 1985).

As mentioned previously, growth-cone movement in the dish also stops when neurones reach a stable state over a period of 4-7 days. Their morphology changes from a flattened, dark structure with many filopodia to a club-shaped, phase-bright structure with fewer filopodia. Since action potentials in the cultured neurones show a broadening over this same period, the ionic properties of the growth-cone membrane were investigated as it changed from the growing to the stable state (Cohan et al. 1985). Patch recording on the growth cones revealed an active channel of conductance 70pS. This channel was inactive in the stable growth cones but became active when the patch of membrane was pulled away from the cell. This suggests that some intracellular factor (which appears not to be Ca) is responsible for controlling this channel's activity.

The transmission of action potentials within cultured neurones can also affect their morphology. Cohan and Kater (1986) were able to arrest the growth of buccal neurone 19 by initiating action potentials in the cell using extracellular patch pipettes. The stimulus frequency of 4Hz was within the physiological range of the normal firing pattern of the neurone. When growth ceased in neurone 19 as a result of electrical stimulation, its growth cone rounded up and had fewer filopodia; this was similar to the response to 5HT (Haydon et al. 1984). Action potential activity curtailing growth appears to be the converse of the experiment where complete blocking of the action potentials of tadpole retinal ganglion cells using TTX caused an increase in growth of their terminal arborisation on the tectum (Reh and Constantine-Paton 1985).

Cohan and Kater (1986) speculate that electrical interaction between neurons may prevent further neurite outgrowth seeking additional synaptic partners and may therefore influence the consolidation of functional circuitry. The mechanism of this phenomenon remains unclear but their preliminary results do show that action potentials cause a Ca^{++} influx at the growth cone.

5. CONCLUSIONS

From the common starting point of being able to maintain identified neurones in culture and promote synapse formation, it is interesting that the work on these three different animals has diverged to this extent. This demarcation reflects both the interests of the various teams of investigators and, of course, the experiments that prove successful with a particular organism.

The leech work has concentrated on a thorough examination of the chemical in vitro synapse between the R and P cells. This synapse, which bears many hallmarks of a conventional in vivo synapse, is so accessible to pharmacological and physiological manipulations that it ought to continue to provide important data on synaptic function.

The _Aplysia_ work has tended to focus on the factors that allow synapse formation. Important results to come out of this work are the fact that transmitter-receptor compatibility is not a sufficient criterion for synapse formation though it is, of course, obviously necessary. Another important result is the fact that different regions of neurones are specialized for synapse formation with different targets. A recent exciting development is the investigation of the molecular events associated with synaptogenesis between identified cells.

Studies on snail neurons in culture have centred on the factors that promote electrical synapse formation. A number of routine neuronal functions such as transmitter release, transmission of action potentials and the simultaneous growth of both synaptic partners are shown to have dramatic effects on the establishment of connectivity.

A question that remains for all these preparations is how well do these identified neurones maintain their in vivo phenotype in the dish? It is arguable that this is not a central issue for those studying synaptic physiology in vitro but is, of course, for students of the specificity of synapse formation.

The first general point is that the alarm bells should not necessarily ring if a "novel" synaptic connexion forms in culture. Since these cells are freed from the constraints of the CNS by being placed in culture, they may be put into new anatomical relationships with one another, allowing "novel" synaptic relationships to develop. In addition, it is often very hard to prove that two cells are not, in fact, making a synaptic connexion in the animal; this is usually easier to attest in vitro.

A more alarming situation is when a synapse that forms in the animal fails to do so in the dish. The culture situation must lack the subtle spatial and temporal cues on which some synapse formation may rely. But generally when a particular synaptic connexion fails to form in the dish, the first thing to check would be the culture conditions. For example, the leech the R to P chemical synapse was only demonstrated when leech blood was added to the medium, the cells were placed closely together and other neurones were placed in the culture dish (Fuchs et al. 1982).

It would appear, however, that these optimum conditions have largely been established. Much of that pioneering work must have been tedious and time consuming but now the maintenance of neurones and the promotion of synapse formation in vitro is routine. Who would have thought, not 10 years ago when the first synaptic contacts were established between identified invertebrate neurones in vitro, that quantal analysis, the cellular basis of learning and fundamental developmental processes could be studied in small plastic dishes? This is an exciting time to be working on identified neurones in vitro.

6. ACKNOWLEDGEMENTS

I thank Will Fuller and Kevin Thompson for their help in the preparation of this essay.

7. REFERENCES

Alving BO (1968) Spontaneous activity in isolated somata of *Aplysia* pacemaker neurons. J Gen Physiol 51: 29.

Ambron TA, Den H, Schacher S (1985) Synaptogenesis by single identified neurons *in vitro*: Contribution of rapidly transported and newly synthesised proteins. J Neurosci 5: 2857-2865.

Beadle DJ, Hicks D (1985) Insect nerve culture. In: Kerkut GA, Gilbert LI (eds) Comprehensive insect physiology, biochemisty and pharmacology. Vol 5. Pergamon Press, Oxford, pp 181-211.

Bodmer R, Levitan IB (1984) Sensitivity of Aplysia neurons in primary culture to putative neurotransmitters. J Neurobiol 15: 429-440.

Bodmer R, Dagan D, Levitan IB (1984) Chemical and electronic connections between Aplysia neurons in primary culture. J Neurosci 4: 228-233.

Camardo J, Proshansky E, Schacher S (1983) Identified Aplysia neurons form specific chemical synapses in culture. J Neurosci 4: 2614-2620.

Chen CF, Von Baumgarten R, Takeda R (1971) Pacemaker properties of completely isolated neurons in Aplysia californica. Nature 233: 27-29.

Cohan CS, Kater SB (1986) Supression of neurite elongation and growth cone motility by electrical activity. Science 232- 1638-1640.

Cohan CS, Haydon PG, Kater BS (1985) Single channel activity differs in growing and nongrowing growth cones of isolated identified neurons of Helisoma. J Neurosci Res 13: 285-300.

Dagan D, Levitan IB (1981) Isolated identified Aplysia neurons in cell culture. J Neurosci 1: 736-740.

Dietzel ID, Drapeau P, Nicholls JG (1986) Voltage depedence of 5-hydroxy-tryptamine release at a synapse between identified leech neurones in culture. J Physiol 372: 191-205.

Fuchs PA, Nicholls JG, Ready D (1981) Membrane properties and selective connexions of identified leech neurons in culture. J Physiol 316: 203-223.

Fuchs PA, Henderson LP, Nicholls JG (1982) Chemical transmission between individual retzius and sensory neurones of the leech in culture. J Physiol 323: 195-210.

Gardner D (1971) Bilateral symmetry and interneuronal organization in the buccal ganglia of Aplysia. Science 173: 550-553.

Giller E, Schwartz JH (1971) Choline acetyltransferase in identified neurons of abdominal ganglion of Aplysia californica. J Neurophysiol 34: 93-107.

Hadley RD, Kater SB (1983) Competence to form electrical connections is restricted to growing neurites in the snail, Helisoma. J Neurosci 3: 924-932.

Hadley RD, Kater SB, Cohan CS (1983) Electrical synapse formation depends on interaction of mutually growing neurites. Science 221: 466-468.

Haydon PG, McCobb DP, Kater SB (1984) Serotonin selectively inhibits growth cone motility and synaptogenesis of specific identified neurons. Science 226: 561-564.

Haydon PG, Cohan CS, McCobb DP, Miller Hr, Kater SB (1985) Neuron-specific growth cone properties as seen in identified neurons of Helisoma. J Neurosci Res 13: 135-147.

Henderson LP (1983) The role of 5-hydroxytryptamine as a transmitter between identified leech neurones in culture. J Physiol 339: 309-324.

Henderson LP, Kuffler DP, Nicholls J, Zhang R (1983) Structural and functional analysis of synaptic transmission between identified leech neurones in culture. J Physiol 340: 347-358.

Hume RI, Role LW, Fishbach GD (1983) Acetylcholine release from growth cones detected with patches of acetylcholine receptor-rich membranes. Nature 305: 632-634.

Kaczmarek LK, Finbow M, Revel JP, Strumwasser F (1979) The morphology and coupling of Aplysia bag cells within the abdominal ganglion and in cell culture. J Neurobiol 10: 535-550.

Kandel ER (1979) Cellular insights into behaviour and learning. In: Marcus D (ed) The Harvey Lectures, Ser 73. Academic, New York, pp 19-92.

Kater SB (1985) Dynamic regulators of neuronal form and connectivity in the adult snail Helisoma. In: Selverston AL (ed) Model neural networks and behavior. Plenum. New York, pp 191-209.

Kosentko MA, Geletyuk VI, Veprintsev BN (1974) Completely isolated neurons in the mollusc, Lymnaea stagnalis. A new objective for nerve cell biology investigation. Comp Biochem Physiol 49A: 89-100.

Kostenko MA, Musienko VS, Smolikhina TI (1983) Ca^{++} and pH affect the neurite formation in cultured mollusc isolated neurones. Brain Res 276: 43-50.

Oomura Y, and Maeno T (1963) Does the neurone soma actually generate action potentials? Nature 197: 358-359.

Pellegrino M, Simonneau M (1984) Distribution of receptors for acetylcholine and 5-hydroxytryptamine on identified leech neurones growing in culture. J Physiol 352: 669-684.

Rayport SG, Schacher S (1986) Synaptic Plasticity in vitro: cell culture of identified Aplysia neurons mediating short-term habituation and sensitization. J Neurosci 6: 759-763.

Ready DF, Nicholls J (1979) Identified neurones isolated from leech CNS make selective connections in culture. Nature 281: 67-69.

Reh TA, Constantine-Paton M (1985) Eye-specific segregation requires neural activity in three-eyed Rana pipiens. J Neurosci 5: 1132-1143.

Schacher S (1985) Differential synapse formation and neurite outgrowth at two branches of the metacerebral cell of Aplysia in dissociated cell culture. J Neurosci 5: 2028-2034.

Schacher S, Proshansky E (1983) Neurite regeneration by Aplysia neurons in dissociated cell culture: modulation by Aplysia hemolymph and the presence of the initial axonal segment. J Neurosci 3: 2403-2413.

Schacher S, Rayport SG, Ambron RT (1985) Giant Aplysia neuron R2 reliably forms strong chemical connections in vitro. J Neurosci 5: 2851-2856.

Wong RG, Hadley RD, Kater SB, Hauser GC (1981) Neurite outgrowth in molluscan organ and cell cultures: the role of conditioning factor(s). J Neurosci 1: 1008-1021.

Wong RG, Martel EC, Kater SB (1983) Conditioning factor(s) produced by several molluscan species promote neurite outgrowth in cell culture. J Exp Biol 105: 389-393.

Wong RG, Barker DL, Kater SB, Bodnar DA (1984) Nerve growth-promoting factor produced in culture media conditioned by specific CNS tissues of the snail <u>Helisoma</u>. Brain Res 292: 81-91.

Young SH, Poo M (1983) Spontaneous release of transmitter from growth cones of embryonic neurones. Nature 305: 634.637.

IDENTIFIED NEURONS AND CELLULAR HOMOLOGIES

ROGER P. CROLL

Department of Psychology

Dalhousie University

Halifax, Nova Scotia

Canada B3H 4J1

ABSTRACT

Studies on homologies of identified neurons offer the promise for an understanding of the evolution of gross neural structures and behaviors in terms of the evolution of single nerve cells. Strong cases now exist in the literature for cellular homologies and evidence is available that permits an initial evaluation of which specific features of nerve cells appear to be most conserved through evolution and which features appear to be plastic and therefore permit adaptive variations in the morphology of the nervous system and in its behavioral manifestations. However, due to the relatively small number of putative cellular homologies which have been studied to date, generalizations may be of questionable accuracy. Much more information is necessary in the form of more examples of identifiable cells with known functions. Such examples will possibly allow better insights into how nerve cells adapt to pressures for changes in function. New techniques must also be employed which allow for the sampling of different types of cells than have usually been identified and homologized in the past. Finally, broader phyletic surveys of such neurons are also necessary to test the generality of hypotheses on the conservation and plasticity of neuronal features through evolution.

1. INTRODUCTION

The concept of the identified neuron serves as a common cornerstone for much of the study of invertebrate nervous systems. It has now become an accepted idea that perhaps the majority of nerve cells in some invertebrates, and probably a large number of neurons in others, may be recognizable as discrete individuals or, at least, as members of discrete, small populations. These cells exhibit relative constancy of soma size, soma position, major neurite branching pattern, biochemical content, connectivity (synaptic input and output), and membrane characteristics yielding constant action potential amplitude and duration, specific current-voltage response curves, etc. The concept of identified neurons has also been extended to a degree into vertebrate systems (Bullock 1978) with the identification, for instance, of brainstem neurons which project to the spinal cords of teleosts and lampreys (Kimmel et al. 1982; Rovainen 1967, 1978; Zottoli 1978).

The historical background for the concept of identified cells in invertebrates extends back nearly a century with one of the earliest descriptions of such neurons being made by Retzius (1891) on a pair of particularly large cells in the segmental ganglia of the leech. These same cells which Retzius originally named the Kolossale Ganglienzellen have since been renamed in honor of their discoverer and are still the subject of much study (see below). Retzius's descriptions and drawings, based upon methylene blue staining, allowed for the identification of these cells not only on the bases of soma postition and size but also on the arborization of large neurites within the central ganglia. Just three years after the publication of this report, Nabias (1894) described a pair of unique, large cell bodies in the cerebral ganglia of the snail and these cells are also presently the subject of intensive study (see below).

These early anatomical descriptions of reliably identifiable elements in the nervous systems of certain invertebrates laid the foundation for subsequent studies which gradually increased the list of identified neurons throughout the early part of this century. The concept of the identified neuron later gained further strength through the work of early electrophysiologists such as Wiersma (1938, 1947, 1952), Young (1939) and Roeder (1948, 1963) who were influential in promoting the concept of identifiable physiological units which could be demonstrated to have anatomical correlates. Starting in the early 1940s Arvanitaki and her co-workers re-discovered the large nerve cell bodies in gastropod nervous systems (Arvanitaki and Cardot 1941) and in particular recognized that certain of the cell bodies in the abdominal ganlgion of Aplysia could be reliably recognized within all individuals of the species (Arvanitaki and Tchou 1942). This early work followed by research in the laboratory of Tauc (for references see Tauc 1966) foreshadowed the classic mapping study of the Aplysia nervous system by Frazier et al. (1967) in which 17 cells were identified as distinct individuals and another 13 were tentatively identified as individuals. They also identified eight discrete clusters of cells.

In 1968 intracellular dye injections were introduced (Stretton and Kravitz 1968) and this technology was subsequently modified to allow for axonal filling ('backfilling') of cell bodies with processes in nerve roots and connectives (Iles and Mulloney 1971). While these techniques are broadly applicable to many invertebrate nervous systems (Kater and Nicholson 1973), they proved to be essential for the identification of the smaller, unpigmented cells typically found in arthropods. Finally, our knowledge of identified neurons has recently been advanced greatly by histochemical and immunohistological techniques which have allowed for further desccriptions of nerve cells with distinct biochemical contents (See for example, Nässel 1987).

As the catalog of identified cells increased, several attempts were made to study some of the integrative properties of the nervous system by examining interconnections between the identified neurons. Studies on the cellular bases of behavior flourished (Kandel 1976) and yielded examples of circuits of identified cells which mediate recognizable behaviors. (For reviews see Delcomyn 1980; Roberts and Roberts 1983).

The large number of identified neurons, the increasing ease and reliability with which they could be recognized, and the demonstrated power of the cellular analyses of neural functions all logically also led to another avenue of research. Since single cells could apparently be reliably identified within all members of a species, the question soon arose as to whether such single cells could be recognized in different species. This question then lead to the question of whether neural circuits could be recognized in the brains of different animals. A cellular approach to the evolution of gross brain structures and possibly of behavior therefore appeared possible (Kandel 1979).

In more recent years, evidence has rapidly accumulated indicating that such cellular homologies do, in fact, appear to exist. However, before proceeding with a review of such evidence, it may be useful at this point to first consider the general concept of homology and to examine some criteria often employed by comparative biologists for recognizing a single structure across phylogeny.

2. DEFINITIONS OF HOMOLOGY

While usage of the term, homology, dates back at least to the early Nineteenth Century (see Boyden 1943), formalization of its definition is generally credited to the comparative anatomist Richard Owen (1843) who defined a homologue as "the same organ in different animals under every variety of form and function". Obviously at the time he was referring to gross morphological structures and not to individual cells, and also, it must be remembered that the definition was formulated in the pre-Darwinian period (Darwin 1860). However, this basic idea of comparing the "same" structure in different species has proved to be a key concept to later evolutionists, and the concept is widely applicable outside the realm of gross anatomy. Given the usefulness of the idea of homology, it might prove constructive to examine how Owen recognized "sameness" since the structures to which he referred were not necessarily the same at all; they could differ in form and/or function. Owen (1848) relied upon two key criteria for this purpose: relative position and connectivity. Using these criteria, it was possible, for example, to homologize forelimbs from different vertebrate species since a forelimb is the same structure by way of its relative postion on the animal and its characteristic connection to the axial skeleton. One can thus compare the wing of a bird with the foreleg of a horse or the arm of a man, even though all the structures are of very different forms and serve very different functions.

While these criteria certainly aid in the identification of homologies, more rigorous examination reveals that they are often ambiguous. For example, the criterion of connectivity of a forelimb to the axial skeleton is only meaningful if homologies of the scapula can be identified. For this reason, several other criteria must also be employed in conjunction with those originally proposed by Owen. One of the most frequently applied criteria is that of similarity in details and in unique features (Remane 1961; Simpson 1961). Often outward forms suggest close similarities or suggest dissimilartities between structures, however in such cases finer details of structure may often be better indicators of sameness. Again using the forelimbs of vertebrates as examples, the wings of birds, the forelimbs of terrestrial mammals and the flippers of whales outwardly appear to be very different structures, but examinations of the bony substructures of the limbs suggest a basic similarity.

Another criterion for homology that has often been advanced is that of common development. Owen (1848), in fact, suggested that homologous structures share similar developmental origins but he also noted that misleading differences can occur. Today, with a more extensive knowledge of the details of development, it appears that a stronger case for the similar development of homologous structures can be made and the correlation between homology and development might acceptably be summarized in the statement that homologous structures tend to differ less in embryonic forms that in adult forms. (For a more complete discussion of the parallels between ontogeny and homology and the concepts of von Bauer relevant to this topic, the reader is referred to Gould 1977).

Up to this point in the discussion of homology, I have not yet directly addressed the key role of the evolutionary theory, but in fact, for more that the last century the term homology has generally inferred the idea of common ancestry. That is, structures in two species are considered to be homologous if they are both derived from a common ancestral structure. Once again, however, the application of such a definition is problematic in that one must, in theory, recognize the "same" structure in the fossilized remains of the ancestor and then must be able to trace its transformation in both derived clades. Furthermore, the task of tracing the past evolutionary history of an organ becomes virtually impossible when it is applied to soft structures such as nervous tissues or even to behavior where little or no fossil evidence is available to allow for an examination of the ancestral entity. When fossil evidence is missing or is scare, it often becomes necessary to infer phylogeny by comparing the distribution of characteristics in extant species. Through such an approach one can examine intermediate forms of a structure and conclude that such forms represent a possible process by which phylogeny could take place. However, caution must be exercised when comparing extant species since it is often difficult to distinguish homology from homoplasty, that is, from structural similarities due not to common origin but rather due to convergent evolution. In the end, as Gould (1977) points out, while the theoretical implications are greatly different, the actual criteria used for the evaluation of structural sameness or homology are altered little by the addition of an evolutionary context. The primary addition of the evolutionary framework is an increased emphasis on continuity of constancy of homologous forms either through the fossil evidence or through extant species (Simpson 1961).

As should be apparent from the preceding discussion, the term homology is very difficult to define and yet virtually all modern biologists depend to some degree upon the concept. However, despite the many attempts that have been made to define the term (Boyden 1943; Haas and Simpson 1946; Zangerl 1948; Simpson 1961; Campbell and Hodos 1970; Atz 1970; Campbell 1976; Ghiselin 1976; Hodos 1976; Schaeffer 1976; Northcutt 1984), none have succeeded in presenting a single set of universally accepted criteria which can be applied to evaluate the likelihood of homology for all levels of analysis (molecular, morphological, behavioral) and for all different taxonomic distances. It appears that the best that one can suggest, then is an attempt toward a greater degree of precision in qualifying the intended use of the term. It seems prudent to first explicitly state whether entities appear to be structurally homologous or not and whether this suggests phylogenetic homology. Second, one should explicitly state which specific criteria were used in order to evaluate homology.

Given this brief overview of the confusion which has existed for greater than a century over the definition of homology, one could question the usefulness of trying to apply the concept of homology to individual cells. Furthermore, one could question any possibility of agreement on a set of criteria necessary for evaluating homologies of individual neurons, since single cells may vary in different aspects than the gross organ which they comprise. For example, would one expect Owen's criteria of homology to be applicable to single nerve cells? That is, does synaptic connectivity or relative soma position suggest homology?

The answers to these types of questions can only be realized through much more study with specific attempts being made to apply the concept of homology to single cells. Therefore, it is suggested that first, one should examine cases in which structural homology appears to exist and then try to ascertain whether phylogeny is a possible or probable mechanism for explaining the similarities. Because little is known about which specific features of single cells might be conserved through evolution, the initial identification of homologies must rely upon searches for several special characteristics which, in combination, uniquely describe a single cell in different species. If each of these characteristics is sufficiently rare, any combination of the features would be exceedingly unlikely to have occurred by chance alone. Furthermore, if one can show that alternative combinations of features might serve the same functions, one could begin to argue that the observed combination was not likely to be the result of homoplasty. A case for inheritance of features from a common ancestor might, thus, arise and once putative cellular homologies are tentatively identified, one can then determine which singular characteristics of nerve cells are most useful in such analyses and therefore which appear to be conserved through evolution. Conversely, it is of interest to note which characteristics appear by this approach to be most plastic through evolution and therefore allow for adaptations to the selective pressures of the environment.

In the remainder of this article I will use the term homology to mean fundamental similarity due to inheritance from a common ancestral form. While it is impossible to prove homology of any single neurons by this definition, I will present evidence which supports this arguement. Toward this end, I will first attempt to demonstrate that strong cases for homologous single nerve cells are already well established. There are several cases in the literature of cells in different species that share such a large number of unique characteristics that the correlations appear to be best explained in terms of common inheritance from single ancestoral origin. Within this framework I will also attempt to indicate special features of single nerve cells which appear to be especially conserved across phylogeny. I will concentrate particularly on examples of what appear to be cladistic cellular homologies, that is, cells in different species which appear to be derived from a common ancestral cell. However, other evidence is emerging from studies of so-called serial homologies within metamerically organized animals which also indicate evolutionary plasticity of certain neuronal characteristics. For instance, in relatively similar body segments within arthropods and annelids, individual cells can often be identified as equivalent serial homologs. However, in thoracic segments of insects, for instance, which have become differentiated for various roles in walking, jumping or flight certain features appear to have changed in these serially homologous cells, while other features are retained (Wilson and Hoyle 1978; Robertson et al. 1982). Metameric neuronal specializations have also been studied in the decapod abdominal motor system (Mittenthal and Wine 1978) and in the segments innervating the sexual organs in the leech (Zipser 1979, 1982). The study of cells which are thought to have been bilaterally symmetrical in an ancestral form but which have become asymmetricaly placed or which came to innervate asymmetric target tissues might also provide insights into how neurons adapt to such situations. Central neurons in gastropods subjected to torsion (Munoz et al. 1983; Hughes and Tauc 1963; Hughes 1967) and motor neurons of the asymmetric chelae of certain decapod crustaceans (Govind 1984; Mellon 1981) offer examples of such differentiated bilateral homologs. However, due to constraints on space within this volume, these examples will not be discussed further here, but rather the discussion will focus on cells which appear to be homologous between species.

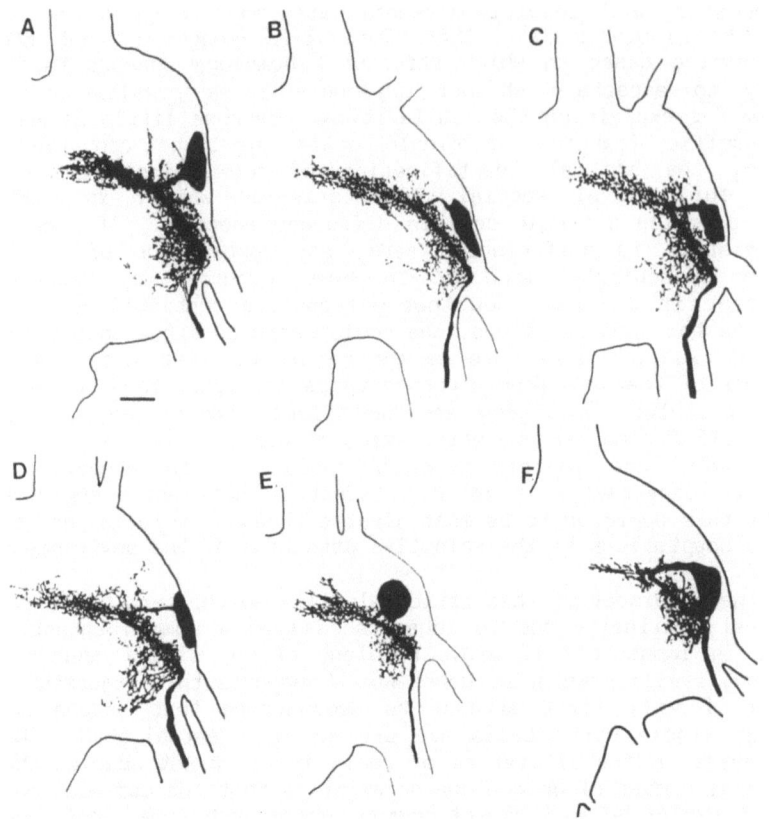

FIG. 1: Right metathoracic FETi in six species of acridids: (A) Schistocerca gregaria, (B) S. alutecea, (C) Melanopus sp., (D) M. femmurrubrum, (E) Srybula fuscouittata, (F) Gomphocerripus rufus. Calibration bar equals 100 µm. (Reproduced from Wilson et al. 1982. Copyright John Wiley and Sons).

3. CASES OF CELLULAR HOMOLOGIES

Examples of seemingly homologous neurons can be drawn from a wide spectrum of invertebrates. The arthropods, for example, can contribute several notable examples. Most orthopteran insects such as locusts and crickets are characterized by very large metathoracic legs and within these limbs large extensor tibiae muscles can easily be identified. Physiological studies have demontrated that this muscle generates the major force underlying the jumping, kicking and swimming (Heitler and Burrows 1977; Pflüger and Burrows 1978). The fast extensor tibiae motor neuron FETi) is one of only two motor neurons to innervate the muscle, (the other motor neuron, the SETi, elicits a slow rather that a fast twitch in the muscle) and because of the unique characteristics of FETi, it is a well-suited subject of the study of homology. Wilson et al. (1982) identified a unique FETi cell in six species of acridid orthopterans. They found that the cells were essentially equivalent in all these locust species on the basis of several characteristics such as cell body size and location, projection of the primary and secondary neurites, and general distribution of smaller neurites (Fig. 1), in addition to the identifying criteria of target muscle innervation and

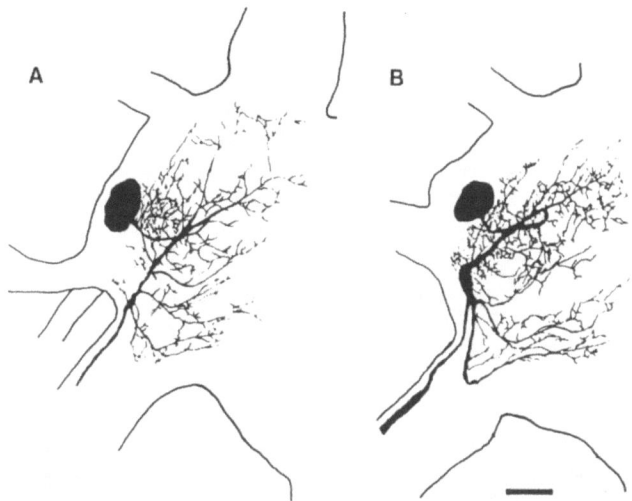

FIG. 2: Left matathoracic FETi in two species of gryllids: (A) <u>Gryllus bimaculatus</u>, (B) <u>G. assimilus</u>. Calibration bar equals 100 μm. (Reproduced from Wilson et al. 1982. Copyright John Wiley and Sons).

physiology (causing a fast muscle twitch). All differences observed in neurons between individuals of different species appeared to be within the range of intraspecific differences. A further comparison of the FETis in two gryllids (crickets) suggests the additional conservation of several features of the unique cell across orthopteran families. Specific conserved features include cell body size and location and the general projection of the primary neurite (Fig. 2). Specific differences found between the families include variations in the number of secondary neurite branches and in the density and extent of finer neurites within the ganglion. It was hypothesized by Wilson et al. (1982) that certain of these differences may reflect the fact that gryllids, while capable of jumping, primarily use their metathoracic legs for walking. While the characteristic of causing a fast twitch in the extensor tibiae muscle appeared to be conserved within this cell between the gryllids and the acridids, other evidence suggests that this characteristic may be plastic within other comparisons. Bässler and Storrer (1980) found that metathoracic FETis of the non-jumping phasmid orthopterans (walking sticks) have similar locations and morphologies as the slow extensor tibiae motor neuron (SETi) in the acridids and gryllids. Conversely, metathoracic SETis in the phasmids have similar locations and morphologies as the FETis in the acridids and gryllids.

The preceding example represents just one of several cases for homologous single cells in adult arthropods. The arthropods also provide an elegant example of extreme constancy of embryonic development of single, seemingly homologous neurons. Thomas et al. (1984) compared the early development of the segmental ganglia of certain insects and of the crayfish and found that several features appeared to be conserved over the represented evolutionary distance. While some species-specific differences in exact numbers of embryonic, neurons and neuron-precursor cells within the segmental neuroepithelia were encountered, the fundamental arragement of these cells, was found to be very similar (Fig. 3). In addition to cell body location, several neurons can also be

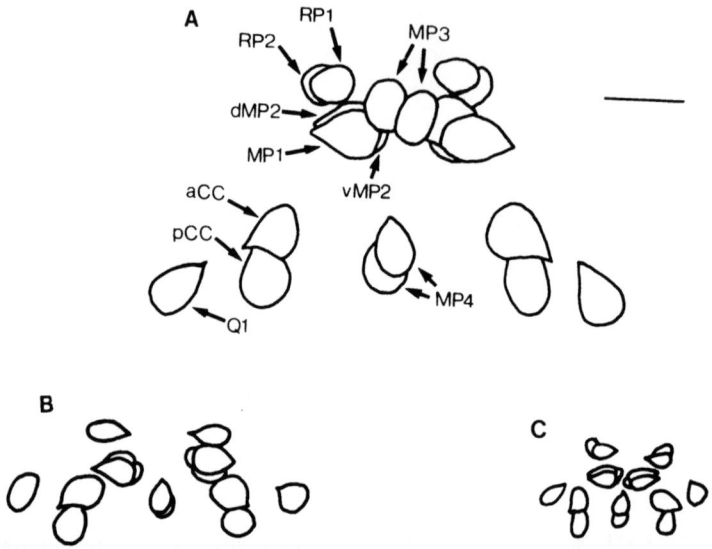

FIG. 3: Identified neurons in embryos of the locust (A), the crayfish (B) and the fruitfly (C). The distribution of cells is similar in all species with the exception that the two MP3 progeny are absent in the crayfish and in the fruitfly. In the crayfish, RP2 lies directly below RP1 and therefore is not shown. Calibration bar equals 25 nm. (Redrawn form Thomas et al. 1984).

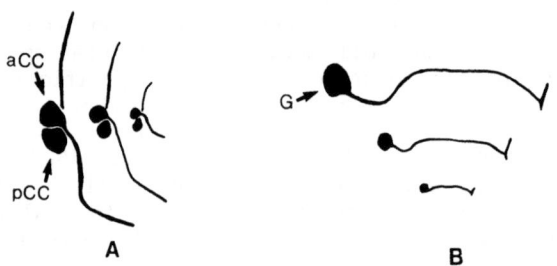

FIG. 4: Morphology of initial neurite segments of identified cells (neurons aCC and pCC in part A and neuron G in part B) in the locust (largest), hawkmoth (medium) and fruitfly (smallest) as determined by intracellular dye injections. (Redrawn from Thomas et al. 1984).

recognized in diverse arthropods on the basis of their neurite morphology (Fig. 4).

The leeches have also provided some interesting examples of probable cellular homologies. As noted earlier, the largest cells in the mid-body segmental ganglia of the leeches are the paired Retzius cells located toward the midline on the ventral surface of each ganglia. Lent (1973; 1977) found that these cells were very similar in six species from two leech families. The cells were found to be similar in morphology, amplitude of resting and action potentials, electrotonic coupling and in their characteristically high serotonin content. Keyser and Lent (1977) studied several other identified cells in three species of leeches and again reported an extreme degree of similarity. For example, they found that the T, P and N types of central mechanosensory neurons were indistinguishable in the three species with regard to the number of cells of each type, their soma sizes, soma positions, neurite arborizations, electrophysiology, and receptive fields. This degree of apparent conservation was typical also of several other cells or cell types such as the longitudinal motor neurons and the S interneuron.

Evidence for cellular homologies can also be found in a third invertebrate phylum, the Mollusca, with several attempts having been made to homologize identified cells in gastropod molluscs (Dorsett 1974; Dickinson 1980a, b; Chase and Goodman 1977). However, by far the best studied such putatively homologous cell is the metacerebral giant cell (MCG, which is also variously known as the giant cerebral neuron, the cerebral giant neuron and the serotonergic cerebral cell; see Pentreath et al. 1982). While previous work identified and characterized a single cell which was similar in several pulmonate species (Nabias 1894; Cottrell and Osborne 1970; Osborne and Cottrell 1971; Berry and Pentreath 1976), notable attempts to formalize and extend the comparsions were made by Senseman and Gelperin (1974) and Weiss and Kupfermann (1976). Later comparative reviews of the MCG were provided by Granzow and Rowell (1981) and by Pentreath et al. (1982).

The MGCs in different species are nearly ideal cells to study in this regard for several reasons. First, they are very distinctive. Their cell bodies are generally among the largest in the entire nervous system and are easily located within the cerebral ganglia. The cells are biochemically distinct (one of only 150-200 serotonergic cells in the central nervous systems of most gastropods) and they each have a prominent projection to the buccal ganglion via the cerebrobuccal connective (CBC). Combining these characteristics the MCG in each species is unique in many respects. It is the only cell projecting from the cerebral ganglion to the buccal ganglion which is sertonergic. Furthermore it is the only source of sertonergic elements in CBC, the buccal ganglia, or the intrinsic musculature of the buccal mass. Finally it is also the largest cell projecting down the CBC.

A second reason for studying the MGCs is that these cells appear to be involved in modulating the feeding behavior of gastropods (Gillette and Davis 1977; Granzow and Kater 1979; Weiss et al. 1978; Gelperin 1981; Croll et al. 1985). Feeding has diversified greatly within the gastropods and, in fact, Purchon (1977) has suggested that much of evolution of gastropods is based on adaptations of the buccal mass and radula which allow different species to feed on a wide variety of food substances. (See Fig. 5 for a summary of gastropod taxonomy and speculations on evolution). Diversification from the ancestral feeding mechanism has assisted the gastropods to colonize many marine, freshwater and terrestrial habitats as grazing and browsing hervivores, as deposit and plankton feeders, as scavengers, as parasites, as carnivorous hunters and as carnivores which graze upon sessile colonial invertebrates. See Purchon (1977) and Kohn (1983) for recent reviews of gastropod feeding behavior.

Throughout the phylogeny of the diverse gastropod feeding behaviors, the nervous system must have changed in order to innervate the modified musculature and to produce the different motor programs. With the identification of single cell like the MCGs, which can be recognized in a variety of gastropod species, one can address the question of how the fundamental units of the nervous system, the single cells, changed during

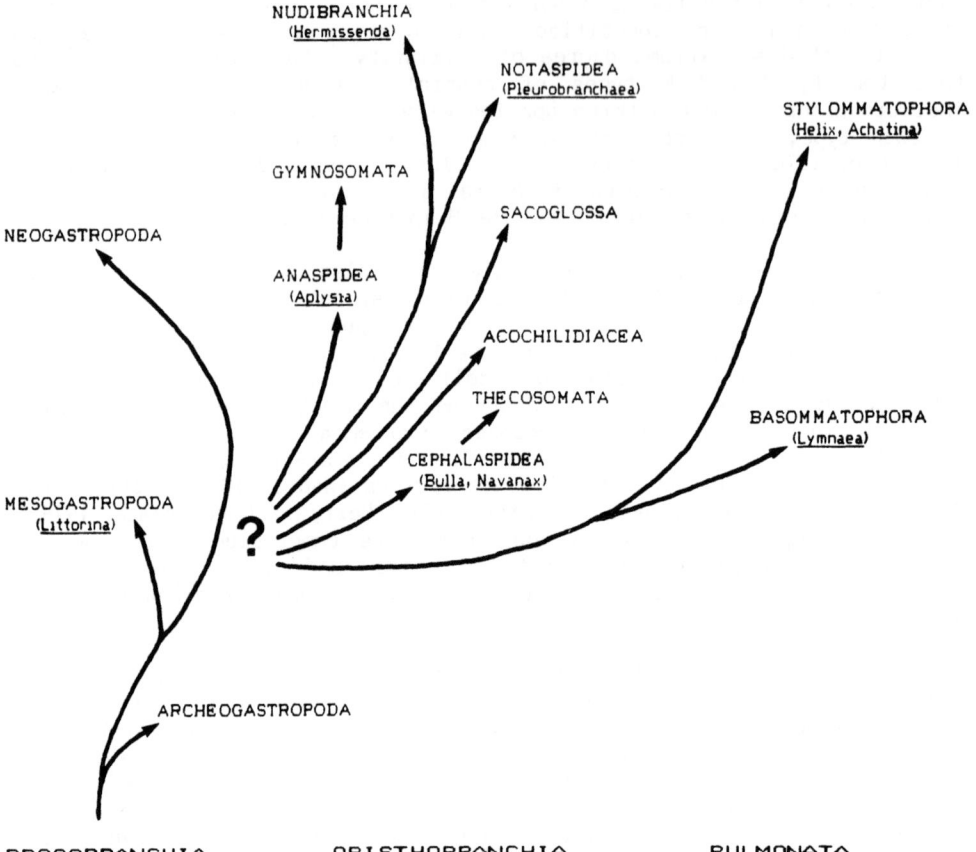

FIG. 5: Taxomony and speculations on the evolutionary relationships between the different orders of gastrpod molluscs. The names of genera mentioned within the text are included within the parentheses. (Based on Kandel 1979; Purchon 1977).

this evolutionary course. In other words, to what degree can the evolution of gastropod feeding behavior be understood in terms of the evolution of single nerve cells?

While we are far from having a complete story of the evolution of the MCG, certain facts are known. A cell which combines all of the previously listed distinct characteristics of the MCG can be found widely distributed throughout all opisthobranchs and pulmonates regardless of feeding method. It therefore appears that several fundamental features of this cell remain constant over a wide taxonomic range and over a great variety of functions. That is, only one large bilaterally symmetric pair of cells supplies all of the serotonin to the CBC, buccal ganglia and buccal mass across these two gastropod orders. The location of these cell bodies also appears to be conserved fairly although some relative positional changes can occur. In all opisthobranchs and in the basommatophore pulmonates the cell body is located on the anterior medial margin of the cerebral ganglion. In the stylommatophore pulmonates the location of the MCG is shifted toward the center of the ganglion, but this appears to be caused not by a migration of the soma within an otherwise conserved structure but

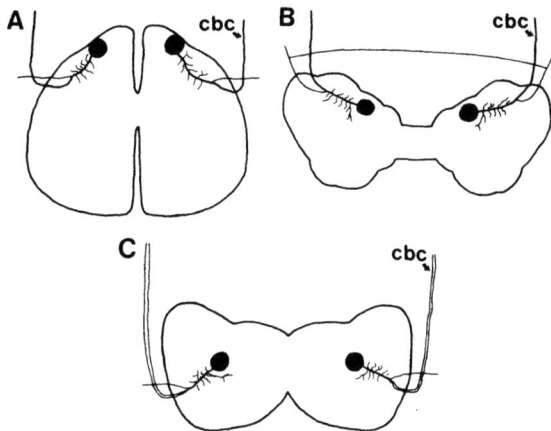

FIG. 6: Morphology of the MCG within the cerebral ganglia of (A) Hermissenda, (B) Lymnaea and (C) Achatina. In each figure the neurites directed toward the tip (anterior) lie within the CBCs. The more lateral neurites innervate the lip region via other cerebral roots. In Lymnaea an addition neurite is shown in the subcerebral commissure. (Redrawn from data presented in Croll 1987; McCrohan and Benjamin 1980a; R.P. Croll unpublished).

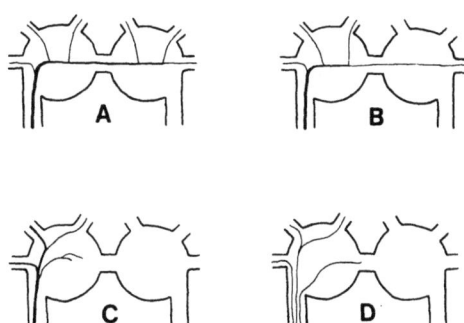

FIG. 7: Schematic diagrams of the major neurite branches of the left MCG in the buccal ganglia of Pleurobranchaea, Navanax, Lymnaea and Planorbis (in part A), of Aplysia (part B), of Hermissenda (part C), and Achatina (part D). (Based on data presented in Gillette and Davis 1977; Goldschmeding et al. 1981; Berry and Pentreath 1976; Weiss and Kupfermann 1976; Croll 1987; Granzow and Rowell 1981; R.P. Croll unpublished).

rather primarily by the addition of the procerebrum and expansion of a mesocerebrum (Bullock and Horridge 1965). In addition to the major projection of the cell into the CBC, a second neurite in the cerebral root innervating the lip also appears to be strictly conserved (see Fig. 6).

Faced with this constancy of certain features, one may question which cellular characteristics appear to be plastic over the course of evolution. For the MCG, one of the most plastic features appears to involve the neurite arborization within the buccal ganglia. Some of the different buccal arborization patterns are schematized in Fig. 7 and can be categorized by the extent of contralateral projections within the buccal ganglia. As of yet, however, no simple explanation exists for this observed variability. Differences appear to be uncorrelated with either feeding behavior or taxonomy. For example, the buccal projection of the MCG in Pleurobranchaea, Navanax, Lynmaea and Planorbis are each most accurately represented in Fig. 7A, even though these species are distantly related and their ingestive behaviors are very different (Croll and Davis 1981; Susswein et al. 1984; Benjamin 1983). Conversely, Hermissenda is thought to be more closely related to Pleurobranchaea than any of these other species (see Fig. 5) and it probably feeds in a similar manner (although it has not yet been studied in detail) and yet the buccal projections of the MCG are very different and may be best represented by Fig. 7C.

Another morphological feature of the MCG which appears to have been plastic over evolution involves the presence of a contalateral projection within the cerebral ganglion. While such a projection has been reported to be variably present in only some individuals of one nudibranch species (Bulloch 1977 as cited by Pentreath et al. 1982), it generally appears to be absent in opisthobranchs. Within the basommatophore pulmonates, the MCG in Planorbis makes a direct projection into the contralateral cerebral ganglion (Berry and Pentreath 1976). However, in Lymnaea (McGrohan and Benjamin 1980a; Goldschmeding et al. 1981)) and in Helisoma (Granzow and Kater 1977) no contralateral cerebral projection was noted. Variability in this feature also characterized the stylommatophore pulmonates. No contralateral projection is apparent in many stylommatophores (Senseman and Gelperin 1974; R.P. Croll unpublished), however, a contralateral projection is reliably encountered in the slug Ariolimax californica (Senseman and Gelperin 1974) and in the snail Helix pomatia (Pentreath 1976). Furthermore, preliminary evidence suggests that this feature may be encountered in Helix aspersa and in other Helicidae (R.P. Croll unpublished).

A final morphological characteristic of the MCG that is apparently plastic, to some degree, is the number of neurites contained within the CBC. In all of the opisthobranchs and the basommatophore pulmonates, thus far examined, there is always a single large neurite from the MCG within the ipsilateral CBC (Fig. 6). This neurite then branches in or near the buccal ganglion (Fig. 7). Within the two stylommatophore pulmonates, Achatina (R.P. Croll unpublished) and Helix (Pentreath 1976), which have so far been examined in detail, parallel branches of the MCG are found running the entire length of the CBC (Fig. 6). In Achatina these branches range in number from two to five. Preliminary evidence suggests that other stylommathophores are also characterized by parallel collaterals within the CBC (R.P. Croll unpublished). This feature thus appears to be confined to within a single gastropod order, however, no functional explanation for this phenomenon is readily apparent, since neither the feeding behavior nor the buccal musculature is greatly different between the stylommatophores and the basommatophores (Carriker 1946; Gelperin et al. 1978; Benjamin 1983; Kohn 1983).

In addition to the observations of morphological plasticity in finer neurite branching patterns of the MCG, some physiological studies suggest plasticity in synaptic connectivity. While some evidence is available as to connectivity with follower cells in the buccal ganglia and to the buccal musculature, much of the work has been confined to Aplysia (see Pentreath et al. 1982) and Lymnaea (McCrohan and Benjamin 1980b) and there is presently not enough information available to allow for general interspecific comparisons at the level of identified buccal cells with known functions. However, the connections between each MCG and its contralateral homolog has received a sufficient amount of attention in different species to form a basis for meaningful comparisons. In all species, activity in the contralateral MCGs appears to be well correlated, however, the degree of direct coupling is variable. In certain species, such as Pleurobranchaea the correlation is apparently due exclusively to similar synaptic input to the two cells (Gillette and Davis 1977). In other species, the two cells appear to be interconnected by way of a polysynaptic pathway within the cerebral ganglia (Granzow and Kater 1977) and in still other species there appears to be direct electrical coupling in the cerebral and/or buccal ganglion (McCrohan and Benjamin 1980a). Combinations of these coupling methods can be found within different species. However, once again, there is as of yet no correlation in the method of coupling with either feeding behavior or with taxonomy. For a further review of MCG interconnections refer to Granzow and Rowell (1981) and Pentreath et al. (1982).

Since the MCG appears to be generally distributed throughout the opisthobranchs and the pulmonates, an attempt was also recently made to identify it in the third gastropod sub-class, the prosobranchs (Croll 1985; Croll and Lo 1986). Littorina is a mesogastropod prosobranch and is thought to closely represent the ancestral form from which both the opisthobranchs and the pulmonates are derived (see Fig. 5). Also, this species feeds by rasping algae in a manner similar to that observed in several pulmonates. Despite apparent phylogenetic proximity and similar functional requirements, no single cell was found which could be identified as the MCG in Littorina. However, certain similarities can be found between all these gastropods. In Littorina, as in the pulmonates and the opisthobranchs, there appears to be a strong serotonergic projection from the cerebral ganglion down the CBC to the buccal ganglion. Some serotonergic fibers also course out the buccal roots directly to the intrinsic muscles of the buccal mass. However, in Littorina the neurites comprising these projections do not appear to originate from a single cell but rather they appear to come from any one of several possible sources, the most likely of which appears to be a cluster of cell bodies along the anterior medial margin of the cerebral ganglion. Since this is the location which corresponds to the sites of the MCGs in the other species it is tempting to speculate upon the possibility that the ancestral origin of the MCG is a cluster of serotonergic cells rather than a single large cell. Much more work must be accomplished, however, before the phylogeny of the MCG becomes clear.

As discussed earlier, Thomas et al. (1984) demonstrated that a study of the development of identified cells often adds persuasive evidence for homology. To date, little is currently known of the development of the MCG in any gastropod species. However, work is currently being pursued in this direciton (Longley 1985; R.P. Croll, B. Chiasson and R. Marois unpublished) and it is expected that the results of these studies will further strengthen the case for homology of the MCG in the opisthobranchs and pulmonates and may yield some insights into the phylogenetic origins of the cell.

4. CONCLUSION

Throughout the several examples of putative cellular homologies presented in this article, there have been instances of a number of features of nerve cells which appear to have been conserved over large evolutionary distances and over a variety of pressures for changes in functionality. These striking degrees of similarity between cells are stongly supportive of hypotheses of descent from common ancestral forms. These hypotheses are further strengthened in some cases by analyses of developmental cell lineages. However, it must be pointed out that degree of conservation generally occurring in the evolution of nerve cells may not be truly represented by the examples available at this time. For instance, although the three species of leeches studied by Keyser and Lent (1977) differ significantly in many regards, the cells which were studied involved functions which do not necessarily distinguish the species. More plasticity might have been noted, for instance, if feeding circuitries had been examined within the leeches. Caution must also be exercised in the interpretation of which specific neuronal features appear to be most conserved throughout evolution. Cell body size, position, primary neurite geometry and biochemical content all appear to be well conserved, however some of this apparent conservation may be a function of the sampling techniques which were employed. For example, in this article I have concentrated on cells which are distinctly identifiable as individuals since these cells most strikingly suggest cellular homologies. However, one might quite justifiably question whether these cells accurately represent all the cells within the nervous systems of invertebrates. Smaller, less distinct cells, which by virtue of these traits, have been less well studied in invertebrate nervous systems, may not show as much overall conservation or may show conservation of different features than their neighboring, large, identified cells.

5. ACKNOWLEDGEMENTS

I thank both the members of the Psychology Department at Dalhousie University and the participants at this NATO-ASI with whom I had many useful discussions on identified neurons and cellular homolgies. This work was supported by Grant U0271 from NSERC (Canada).

6. REFERENCES

Arvanitaki A, Cardot H (1941) Contribution à la morphologie du système nervuex des gastéropodes. Isolement, à l'état vivant, de corps neuroniques. CR Seances Soc Biol Fiul 135: 965-968.

Arvanitaki A, Tchou SH (1942) Les lois de la croissance relative individuelle des cellules nerveuses chez l'aplysie. Bull Histol Appl Physiol Pathol 19: 224-256.

Atz JW (1970) The application of the idea of homology to behavior. In: Aronson LR, Tobach E, Lehrman DS, Rosenblatt JS (eds) Development and evolution of behavior. W.H. Freeman and Co., San Francisco, pp 53-74.

Bässler U, Storrer J (1980) The neural basis of the femur-tibiae-control system in the stick insect (Carausius morosus. Biol Cybernet 38: 107-114.

Benjamin PR (1983) Gastropod feeding: behavioural and neural analysis of a complex multicomponent system. In: Roberts A, Roberts BL (eds) Neural origins of rhythmic movements. Cambridge Univ Press, Cambridge, pp 159-193.

Berry MS, Pentreath VM (1976) Properties of a symmetric pair of serotonin-containing neurons in the central ganglia of Planorbis. J Exp Biol 65: 361-380.

Boyden A (1943) Homology and analogy: a century after the definitions of "homologue" and "analogue" of Richard Owen. Quart Rev Biol 18: 228-241.

Bullock TH (1978) Identifiable and addressable neurons in the vertebrates. In: Faber DS, Korn H (eds) Neurobiology of the Mauthner cell. Raven Press, New York, pp 1-12.

Bullock TH, Horridge GA (1965) Structure and functions in the nervous systems of invertebrates. W.H. Freeman, San francisco.

Campbell CBG (1976) Morphological homology and the nervous system. In: Masterton RB, Hodos W, Jerison H (eds) Evolution, brain, and behavior: Persistent problems. Lawrence Erlbaum Associates, Hillsdale, NJ, pp 143-151.

Campbell CBG, Hodos W (1970) The concept of homology and the evolution of the nervous system. Brain Behav Evol 3: 353-367.

Carriker MR (1946) Morphology of the alimentary system of the snail *Lymnaea stagnalis asressa*, Say. Trans Wisc Acad 38: 1-88.

Chase R, Goodman H (1977) Homologous neurosecretory cell groups in the land snail *Achatina fulica* and the sea slug *Aplysia californica*. Cell Tissue Res 176: 109-120.

Cottrell GA, Osborne NN (1970) Subcellualr localization of serotonin in an identified serotonin-containing neurone. Nature 225: 470-472.

Croll RP (1985) Search for the metacerebral giant cell in diverse gastropods. Soc Neurosci Abstr 11: 627.

Croll RP (1987) Distribution of monoamines in the central nervous system of the nudibranch gastropod, *Hermissenda crassicornis*. Brain Res 405: 337-347.

Croll RP, Davis WJ (1981) Motor program switching in *Pleurobranchaea*: I. Behavioral and electromyographic study of ingestion and egestion in intact specimens. J Comp Physiol 145: 277-287.

Croll RP, Lo RYS (1986) Distribution of serotonon-like immunoreactivity in the central nervous system of the periwinkle, *Littorina littorea* (Gastropoda, Prosobranchia, Mesogastropoda). Biol Bull 171: 426-440.

Croll RP, Kovac MP, Davis WJ, Matera EM (1985) Neural mechanisms of motor program switching in *Pleurobranchaea*: III. Role of the paracerebral neurons and other identified brain neurons. J Neurosci 5: 64-71.

Darwin C (1860) On the origin of the species by means of natural selection. Appleton, New York.

Delcomyn F (1980) Neural basis of rhythmic behavior in animals. Science 210: 492-498.

Dickinson PS (1980a) Gill control in the notaspiden *Pleurobranchaea* and possible homologies with nudibranchs. J Comp Physiol 139: 11-16.

Dickinson PS (1980b) Neuronal control of gills in diverse *Aplysia* species: Conservative evolution. J Comp Physiol 139: 17-23.

Dorsett DA (1974) Neuronal homologies and control of branchial tuft movements in two species of *Tritonia*. J Exp Biol 61: 639-654.

Frazier WT, Kandel ER, Kupfermann I, Waziri R, Coggeshall RE (1967) Morphological and functional properties of identified neurons in the abdominal ganglion of Aplysia californica. J Neurophysiol 30: 1288-1351.

Gelperin A (1981) Synaptic modulation by identified serotonin neurons. In: Jacobs BL, Gelperin A (eds) Serotonin transmission and behavior. MIT Press, Cambridge, Mass, pp 288-304.

Gelperin A, Chang JJ, Reingold SC (1978) Feeding motor system in Limax. I. Neuromuscular correlates and control by sensory input. J Neurobiol 9: 285-300.

Ghiselin MT (1976) The nomenclature of correspondence: a new look at "homology" and "analogy". In: Masterton RB, Hodos W, Jerison H (eds) Evolution, brain, and behavior: Persistent problems. Lawrence Erlbaum Associates, Hillsdale, NJ, pp 129-142.

Gillette R, Davis WJ (1977) The role of the matecerebral giant neuron in the feeding behavior of Pleurobranchaea. J Comp Physiol 116: 129-159.

Goldschmeding JT, van Duivenboden YA, Lodder JC (1981) Axonal branching pattern and coupling mechanisms of the cerebral gaint neurones in the snail, Lymnaea stagnalis. J Neurobiol 12: 405-424.

Gould SJ (1977) Ontogeny and phylogeny. Harvard Univ Press, Cambridge, Mass.

Govind CK (1984) Development of asymmetry in the neuromuscular system of lobster claws. Biol Bull 167: 94-119.

Granzow B, Kater SB (1977) Identified higher-order neurons controlling the feeding motor program of Helisoma. Neurosci 2: 1049-1063.

Granzow B, Rowell CHF (1981) Further observations on the serotonergic cerebral neurons of Helisoma (Mollusca, Gastropoda): the case for homology with the metacerebral giant cells. J Exp Biol 90: 283-305.

Haas O, Simpson GG (1946) Analysis of some phylogenetic terms, with attempts at redefinition. Proc Amer Phil Soc 90: 319-349.

Heitler WJ, Burrows M (1977) The locust jump I. The motor programme. J Exp Biol 66: 203-219.

Hodos W (1976) The concept of homology and the evolution of behavior. In: Masterton RB, Hodos W, Jerison H (eds) Evolution, brain, and behavior: Persistent problems. Lawrence Erlbaum Associates, Hillsdale, NJ, pp 153-167.

Hughes GM (1967) Further studies on the electrophysiological anatomy of the left and right giant cell in Aplysia. J Exp Biol 46: 169-193.

Hughes GM, Tauc L (1963) An electrophysiological study of the anatomical relations of two giant nerve cells in Aplysia depilans. J Exp Biol 40: 469-486.

Iles JF, Mulloney BM (1971) Procion yellow staining of cockroach motor neurones without the use of microelectrodes. Brain Res: 397-400.

Kandel ER (1976) Cellular basis of behavior. W.H. Freeman and Co., San Francisco.

Kandel ER (1979) Behavioral biology of *Aplysia*. W.H. Freeman and Co., San Francisco.

Kater SB, Nicholson C (eds) (1973) Intracellular staining in neurobiology. Springer-Verlag, New York.

Keyser KT, Lent CM (1977) On neuronal homologies within the central nervous system of leeches. Comp Biochem Physiol 58A: 285-297.

Kimmel CB, Powell SL, Metclafe WK (1982) Brain neurons which project to the spinal cord in young larvae of the zebrafish. J Comp Neurol 205: 112-127.

Kohn AJ (1983) Feeding biology of gastropods. In: Salevddin ASM, Wilbur KM (eds) The Mollusca. Vol. 5, Physiology, Part 2. Academic Press: pp 1- 13.

Lent CM (1973) Retzius cells from segmental ganglia of four species of leeches: Comparative neuronal geometry. Comp Biochem Physiol 44A: 35-40.

Lent CM (1977) The retzius cells within the central nervous system of leeches. Prog Neurobiol 8: 81-117.

Longley AJ (1985) Development of homologous neurons in opisthobranch mollusks. Soc Neurosci Abstr 11: 918.

McCrohan CR, Benjamin PR (1980a) Patterns of activity and axonal projections of the cerebral giant cells of the snail, *Lynmaea stagnalis*. J Exp Biol 85: 149-168.

McCrohan CR, Benjamin PR (1980b) Synaptic relationships of the cerebral giant cells with motoneurons in the feeding system of *Lymneae stagnalis*. J Exp Biol 85: 169-186.

Mellon D (1981) Nerves and transformations of claw type in snapping shrimps. Trends Neurosci 4: 245-248.

Mittenthal JE, Wine JJ (1978) Segmental homology and variation in flexor motoneurons of the crayfish abdomen. J Comp Neurol 17: 311-334.

Munoz DP, Pawson PA, Chase R (1983) Symmetrical giant neurones in asymmetrical ganglia: Implications for evolution of the nervous system in pulmonate molluscs. J Exp Biol 107: 147-161.

Nabias B de (1894) Recherches histologiques et organologiques sur les centres nerveux des gasteropodes. Actes Soc Linn Bordeau 47: 11-202.

Nässel DR (1987) Neuroactive substances in the insect CNS. (This volume).

Northcutt RG (1984) Evolution of the vertebrate central nervous system: patterns and processes. Amer Zool 24: 701-716.

Osborne NN, Cottrell GA (1971) Distribution of biogenic amines in the slug, *Limax maximus*. Z Zellforsch 112: 15-30.

Owen R (1843) Lectures on the comparative anatomy and physiology of the invertebrate animals. Longman, Brown, Green, Longmans, London.

Owen R (1848) On the archetype and homolgies of the vertebrate skeleton. John van Voorst, London.

Pentreath VW (1976) Ultrastructure of the terminals of an identified 5-hydroxytryptamine-containing neurone marked by intracellular injection of radioactive 5-hydroxytryptamine. J Neurocytol 5: 43-61.

Pentreath VW, Berry MS, Osborne NN (1982) The serotonergic cerebral cells in gastropods. In: Osborne NN (ed) Biology of serotonergic transmission. John Wiley & Sons, Ltd., New York, pp 457-512.

Pflüger H-J, Burrows M (1978) Locusts use the same basic motor pattern in swimming as in jumping and kicking. J Exp Biol 75: 81-93.

Purchon R (1977) The biology of Mollusca. Pergamon Press, Oxford.

Remane A (1961) Gedanken zum Problem: Homologie und Analogie, Praeadaptation und Parallelitat. Zool Anz 166: 447-470.

Retzius G (1891) Zur Kenntnis des centralen Nervensystems der Wurner. Das Nervensystem der Annulaten. Biol Untersuch (NF) 2: 1-28.

Roberts A, Roberts BL (eds) (1983) Neural origins of rhythmic movements. Cambridge University Press, Cambridge.

Robertson RM, Pearson KG, Reichert H (1982) Flight interneurons in the locust and the origin of insect wings. Science 217: 177-179.

Roeder KD (1948) Organization of the ascending giant fiber system in the cockroach (Periplaneta americana). J Exp Zool 108: 243-261.

Roeder KD (1963) Nerve cells and insect behavior. Harvard University Press, Cambridge, Mass.

Rovainen CM (1967) Physiological and anatomical studies on large neurons of the central nervous system of the sea lamprey (Petromyzon marinus) I. Müller and Mauthner cells. J Neurophysiol 30: 1000-1023.

Rovainen CM (1978) Müller cell, "Mauthner cells," and other identified reticulospinal neurons in the lamprey. In: Faber DS, Korn H (eds) Neurobiology of the Mauthenr Cell. Raven Press, New York, pp 245-266.

Schaeffer B (1976) Practical aspects of homology recognition. In: Masterton RB, Hodos W, Jerison H (eds) Evolution, brain, and behavior: Persisten problems. Lawrence Erlbaum Associates, Hillsdale, NJ, pp 169-173.

Senseman D, Gelperin A (1974) Comparative aspects of the morphology and physiology of a single identified neuron in Helix aspersa, Limax maximus, and Ariolimax californica. Malacol Rev 7: 51-52.

Simpson GG (1961) Principles of animal taxomony. Columbia University Press, New York.

Stretton AOW, Kravitz EA (1968) Neuronal geometry: Determination with a technique of intracellular dye injection. Science 162: 132-134.

Susswein AJ, Achituv Y, Cappell MS, Spray DC, Bennett MVL (1984) Pharangeal movements during feeding sequences in Navanax inermis: a cinematographic analysis. J Comp Physiol A 155: 209-218.

Tauc L (1966) Physiology of the nervous system. In: Wilbur KM, Yonge CM (eds) Physiology of Mollusca. Vol. 2. Academic Press, New York, pp 387-454.

Thomas JB, Bastiani MJ, Bate M, Goodman CS (1984) From grasshopper to Drosophila: a common plan for neuronal development. Nature 310: 203-207.

Weiss KR, Kupfermann I (1976) Homology of the giant serotonergic neurons (metacerebral cells) in Aplysia and pulmonate molluscs. Brain Res 117: 33-49.

Weiss KR, Cohen JL Kupfermann I (1978) Modulatory control of buccal musculature by a serotonergic neuron (metacerebral cell) in Aplysia. J Neurophysiol 41: 181-203.

Wiersma CAG (1938) Function of the giant fibers of the central nervous system of the crayfish. Proc Soc Biol Med 38: 661-662.

Wiersma CAG (1947) Giant nerve fiber system of the crayfish. A contribution to the comparative physiology of synapse. J Neurophysiol 10: 23-38.

Wiersma CAG (1952) Neurons of arthropods. Cold Spring Harbor Symp Quant Biol 17: 155-163.

Wilson JA, Hoyle G (1978) Serially homologous neurones as concomitants of functional specialisation. Nature 274: 377-379.

Wilson JA, Phillips CE, Adams ME, Huber F (1982) Structural comparison of a homologous neuron in gryllid and acridid insects. J Neurobiol 13: 459-467.

Young JZ (1939) Fused neurons and synaptic contacts in the giant nerve fibres of cephalopods. Phil Trans R Soc Lond (B) 229: 465-503.

Zangerl R (1948) The methods of comparative anatomy and its contribution to the study of evolution. Evolution 2: 351-374.

Zipser B (1979) Identifiable neurons controlling penile eversion in the leech. J Neurophysiol 42: 455-464.

Zipser B (1982) Complete distribution patterns of neurons with characteristic antigens in the leech central nervous system. J Neurosci 2: 1453-1464.

Zottoli SJ (1978) Comparative morphology of the Mauthner cell in fish and amphibians. In: Faber DS, Korn H (eds) Neurobiology of the Mauthner cell. Raven Press, New York, pp 13-46.

Thomas JB, Bastiani MJ, Bate M, Goodman CS (1984) From grasshopper to
Drosophila: a common plan for neuronal development. Nature 310:
203-207.

Weiss KR, Kupfermann I (1976) Homology of the giant heads of
(metacerebral cells) in Aplysia and pulmonate molluscs. Brain Res
117: 33-49.

Weiss KR, Cohen JL, Kupfermann I (1978) Modulatory control of buccal
musculature by a serotonergic neuron (metacerebral cell) in Aplysia.
J Neurophysiol 41: 181-203.

Wiersma CAG (1938) Function of the giant fibers of the central nervous
system of the crayfish. Proc Soc Biol Med 38: 661-662.

Wiersma CAG (1947) Giant nerve fiber system of the crayfish. A contribution to the comparative physiology of synapse. J Neurophysiol 10:
23-38.

Wiersma CAG (1952) Neurons of arthropoda. Cold Spring Harbor Symp Quant
Biol 17: 155-63.

Wilson DM, Davis C (1970) Serially homologous neurones as concomitants of
functional specialization. Nature 2764: 377-379.

Wilson JA, Phillips CE, Adams ME, McKee F (1982) Structure, homotypes of
a homologous neuron in Oryllid and acridid insects. J Neurobiol 13:
429-468.

Young JZ (1939) Fused neurons and synaptic contacts in the giant nerve
fibre of cephalopods. Phil Trans R Soc Lond BB 229: 465-503.

Zangerl R (1948) The methods of comparative anatomy and its contribution
to the study of evolution. Evolution 2: 351-374.

Zawarzin A (1925) Der Mittelfuhrende Neurone nebentellulair physik Verein
4: 12. J Neurophysiol 41: 13-43.

Zucker RS (1972) Crayfish escape behavior circuits of the MLG neuron.
Synaptic and sensory lateral portions of neurons systems. J Neurophysiol 35:
599-620.

FUNCTIONS OF INVERTEBRATE GLIA

V. W. PENTREATH

Department of Biological Sciences

University of Salford

Salford M5 4WT

UK

ABSTRACT

Invertebrate glial cells exhibit a vast array of structural and functional specializations. The cells vary markedly at different sites and at different stages of development in individual species, and amongst the different groups. In a number of situations, particular glial properties have become well developed and have been studied in detail. They include mechanical support and protection of neurones, formation of occluding permeability barriers, wrapping of axons to speed up impulse conduction, removal of neurones, uptake and release of transmitters, supply and exchange of nutrients and metabolites with the neurones, and guidance of migrating neurones. The various functions are summarized and discussed.

1. INTRODUCTION

The glial cells of invertebrate nervous systems comprise an extensive variety of morphological and functional types, about which detailed knowledge is generally lacking. It is however becoming increasingly evident that the cells provide an essential environment for the normal functioning of the neurones, and that their properties are often analogous to their counterparts in the mammalian brain. Furthermore, in some situations, for example the insect compound eye and perineurium, and the leech segmental ganglia, the physiological properties of the glial cells have been fairly rigorously studied and this information may provide a basis from which similar properties can be sought in higher animals. In the following pages a summary of the data is presented, with special reference to the more recent studies that have attempted to clarify some of the functions of the different glial cell types. Much of the literature on invertebrate glia concerns their structure, and this has been comprehensively reviewed by Bullock and Horridge (1965), Roots (1978), Radojcic and Pentreath (1979), Oksche (1980), Lane (1981) and Saint Marie et al. (1984). Because of the great variety of kinds and forms of glia in the different species, many of the original descriptions are highly specialized and detailed in nature, and no attempt is made to review them further here. Instead, general or selected aspects of morphology will be described where they may be constructively related to function.

2. ORIGINS AND GENERAL CONCEPTS

Neuroglia are an obvious feature of the higher invertebrate groups such as the Arthropods, Annelids and Molluscs. They are generally recognisable, without ambiguity, by their location around and between the neurones, especially at the blood/nervous tissue interface. The cells are also generally agreed to be absent from the lower invertebrate phyla such as the Coelenterates (Horridge and MacKay 1962; Chapman 1974) and Echinoderms (Pentreath and Cobb 1972; Radojcic and Pentreath 1979; Cobb 1987. Although there are little data for many of the other lower invertebrate groups, especially the minor ones, it seems that glial cells may first appear, with some degree of certainty, in the Platyhelminthes (Bullock and Horridge 1965) and become positively identifiable in the Aschelminthes (Goldschmidt 1910; Chitwood and Chitwood 1950; Table 1). Thus glial cells are not essential for nervous activity, but appear to be necessary when the neurones become aggregated into ganglia. It is as if the grouping of the neurone perikarya, axons and their terminals (neuropile) into restricted localities imposes functional demands, met by the glial cells, that are not present in the simple nerve plexuses and pathways of Coelenterates and Echinoderms (Table 1). A factor which correlates with this is the reduced access of blood or haemolymph to the neurones when they become isolated within ganglia. Most invertebrate ganglia, apart from those of the Oligochaeta, Crustacea and Cephalopoda, are avascular, with at least one layer of glial cell processes located between the blood and the neurones.

The glial cell types of the Arthropods, Annelids and Molluscs become more diverse with the increasing complexities of the nervous systems. The different cell types also acquire distinct cytological characteristics which are presumably associated with different functions. However, several morphological characteristics occur fairly commonly. First, the glial cell nuclei in many species of all groups contain chromatin which is clumped in the periphery. This feature is rarely seen in the neuronal nuclei. Second, mitochondria, endoplasmic reticulum and Golgi structures are generally common. The cytoplasm of the glial cells also frequently contains stores of glycogen and large numbers of vesicular and granular inclusions which do not generally belong to distinct populations. Many of the membrane-bound inclusions are phagocytic, pinocytotic or lysosomal in appearance. Another common charateristic is the presence of intercellular junctions between the glial cells, and between the glial cells and neurones. These have been thoroughly discussed by Lane (1981) and are briefly summarized below and in Table 2. They occur as homocellular glial-glial junctions, or less commonly as heterocellular glial-neuron junctions. The most frequent glial-glial junction is the gap junction, which has been reported in most invertebrate ganglia. These structures are the sites of electrical and possible metabolic coupling (see below) between the glial cells. Desmosomes are also widespread. These structures are thought to protect against mechanical distortion of the nervous tissue, since they are most common in situations where stress seems most likely. Other homocellular junctions are septate junctions, which are the likely adhesive components of the perineurium that comprises the Arthropod blood-brain barrier, and tight junctions, which are thought to be responsible for the occluding permeability component of the barrier. During insect metamorphosis the septate, gap and tight junctions in the blood-brain barrier undergo cycles of disruption and reassembly. This phenomenon, with its associated permeability changes, has been fairly thoroughly studied by Lane and co-workers (see Lane 1981; Swales and Lane 1985). Another junctional type, termed scalariform, which exhibits intercellular striations, not septa, is also found between the glial cells of insects. The function(s) of scalariform junctions are not understood. The septate, tight and scalariform junctions are principal features of

TABLE 1

THE OCCURRENCE OF GLIA IN THE INVERTEBRATES

Phylum	Nervous system	Nerve ring	Aggregation of nervous tissue in simple ganglia	Organized ganglia with distinct areas of neuropile and perikarya	Connective tissue capsule	Aggregation or organized ganglia	Glia
Mesozoa	-	-	-	-	-	-	-
Porifera	-	-	-	-	-	-	-
Coelenterata	+	/	/	-	-	-	-
Echinodermata	+	+	/	-	-	-	-
Platyhelmintha	+	+	/	/	-	-	/
Aschelmintha	+	+	+	/	-	-	+
Annelida	+	+	+	+	-	-	+
Arthropoda	+	+	+	+	+	+	+
Mollusca	+	+	+	+	+	+	+

Key: - absent; + present; / occasional or in doubt (From Radojcic and Pentreath 1979).

TABLE 2

INVERTEBRATE GLIAL CELL JUNCTIONS

GLIAL-GLIAL

DESMOSOMES	(OCCUR AS PLAQUES; ANNELIDS, ARTHROPODS)
HEMI-DESMOSOMES	(WITH EXTRACELLULAR MATRIX; ANNELIDS, ARTHROPODS)
SEPTATE JUNCTIONS	(IN BOUNDARY LAYERS; COHESIVE IN ARTHROPOD PERINEURIUM)
GAP JUNCTIONS	(COMMON IN MOST INVERTEBRATE GANGLIA)
TIGHT JUNCTIONS	(OCCLUDING BARRIER IN ARTHROPOD PERINEURIUM, OTHERWISE RARE)
SCALARIFORM JUNCTIONS	(ARTHROPOD GANGLIA)

AXO-GLIAL

SEPTATE JUNCTIONS	(INSECT EYE, CRUSTACEAN NERVE)
CAPITATE JUNCTIONS	(INSECT EYE)
TROPHOSPONGIUM	(OFTEN WITH SUBSURFACE CISTERNAE; COMMON IN LARGE NEURONES IN MOST INVERTEBRATE GANGLIA)
DESMOSOMES	(ESPECIALLY ANNELIDS)

Arthropod ganglia and are infrequent or absent in the other invertebrate groups. Heterocellular junctions in the form of desmosomes or hemi-desmosomes have been described in several Annelid species (Coggeshall 1965; Coggeshall and Fawcett 1964; see Fig. 1). They are thought to have a cohesive role. Gap junctions have been described between neurone cell bodies and glial cells in the eye of the housefly (Saint Marie and Carlson 1985) and in the abdominal ganglia of the crayfish (Cuadras et al. 1985). It has been suggested that they may participate in ionic or metabolic exchanges. Other specialised regions between glial cells and neurones are the trophospongia, capitate projections, and the particulate ridges and grooves. These are thought to serve dynamic roles, probably representing sites of metabolic exchange between the two cell types.

Attempts to classify invertebrate glial cells are fraught with difficulties because of their great variation in structure and the lack of information about their functions. However, they may be divided according to certain relatively general morphological criteria (e.g. presence or absence of fibrous material) or functional roles (e.g. blood/brain barrier forming, or insulation of axons) into plasmatic, fibrous, perineurial and Schwann-like categories (Radojcic and Pentreath 1979). At the level of phyla and classes more detailed classifications are available, and these are the attempts of continual improvement by the different and subsequent groups of workers. Thus Ramón y Cajal and Sanchez's (1915) early visualisation of insect glial cells was superseded by Wigglesworth's (1959) classification into four classes, including a separate class for the perineurial cells (Wigglesworth 1960). More recently Strausfeld (1976) has described four classes of glial cells, in addition to the perineural cells, that occupy different successive, deeper strata within the ganglia. These, in turn, have been integrated into the five ultrastructurally distinct glial layers described by Saint Marie and Carlson (1983) in the housefly eye. A yet more detailed morphological description by Hoyle (1986) lists seven major types, with subdivisions, in locust metathoracic ganglia. Some other aspects of the classification of invertebrate glial cells are discussed by Radojcic and Pentreath (1979).

3. DISTRIBUTION AND QUANTITATIVE RELATIONSHIPS

A characteristic feature of many invertebrate glial cells is their presence at the blood/neurone interface, where they wrap around axons and neurone somata, projecting into the nervous tissue. Other glial cells form layers at different levels within the nervous tissue, for example, surrounding areas of neuropile. The glial cell processes frequently become attenuated into thin sheets as they extend between the neurones.

Several detailed reports are available on the quantitative neurone/glia relationships in invertebrate nervous systems. Some of the representative data is summarised in Table 3. As in vertebrates, the glial cells normally outnumber the neurones. For example, in gastropod buccal ganglia, the glia:neurone ratio is 2:1 (Radojcic and Pentreath 1981), in the stick insect the ratio is 1.5:1 (Becker 1965) and in the cricket the ratio is 8.1:1 (Gymer and Edwards 1967). The glial cells comprise approximately 50% of the volume of invertebrate ganglia. Thus in both leech (Kai-Kai and Pentreath 1981a) and snail ganglia (Radojcic and Pentreath 1981) the glial cells make up 46% of the ganglion volume. In the drone retina the glial (pigment) cells make up 57% (Coles and Tsacopoulos 1979). It has been established in several species from different invertebrate groups that the glial cells increase in number with the age of the animal (for references see Pentreath et al. 1985). However, the increases appear to be associated with growth (i.e increased volume, not numbers) of the neurones, and it seems likely that the

TABLE 3

DISTRIBUTION AND QUANTITATIVE RELATIONSHIPS OF INVERTEBRATE GLIAL CELLS

1. <u>LOCATION</u> Between blood supply and neurones, projecting into nervous tissue. Form boundary layers at the periphery, and layers isolating different levels within nervous tissue.

2. <u>NUMBERS</u> Generally outnumber neurones.
e.g. The glia:neurone ratio in locust abdominal ganglia is 4:1, in gastropod buccal ganglia it is 2:1 (cf. astrocyte:neurone ratio in human striatum is 4:1). Glia:neurone ratios correlate with neurone size and brain weight.

3. <u>VOLUMES</u> Make up to 50% of nervous tissue.
e.g. In leech ganglia glial cells make up 46%, in snail ganglia 46%.

4. <u>SURFACE:VOLUME RATIO</u> Glial cells generally have highest surface-to-volume ratios of all cells in invertebrate nervous tissue. They form a boundary up to 50% of extracellular space.

glia:neurone volume relationships may remain relatively constant with growth and ageing (Pentreath et al. 1985). A relationship also exists between neurone size and the glia-neurone ratio, with perhaps not surprisingly, large neurones being surrounded by more satellite cells than the smaller neurones. This is well illustrated by the work of Reinecke (1975), who showed that in the snail, Helix pomatia, the glial:neurone ratio ranged from 0.01 for neurones of diameter 10 μm to approximately 2.0 for neurones of 50 μm.

4. PHYSIOLOGICAL PROPERTIES OF INVERTEBRATE GLIAL CELLS

The membrane and physiological properties of certain invertebrate glial cells are fairly well understood. This has been due, in a large part, to the rigorous pioneering studies of S. W. Kuffler, J. G. Nicholls and co-workers (see Kuffler and Nicholls 1966, 1976), who developed the 'giant' glial cells of the leech as a system for analysing glial cell physiology. More recently, these studies have been extended by W. R. Schlue, W. Walz and co-workers (see Schlue and Walz 1984). Other glial cell systems studied in detail are the Schwann cells surrounding the squid giant axon (Villegas 1981) and the pigment (glial) cells in the drone retina (Coles and Tsacopoulos 1979, 1981). Some of the data from these works, especially those which appear to have general applicability, are summarised here and are also referred to in later sections.

In the leech the resting membrane potential of the outer packet glial cells may be as high as -75 mV (Kuffler and Potter 1964), which is larger than in the neurones. However, in the neuropile glial cells of the same animal the resting potential is lower, averaging -56 mV (Walz and Schlue 1982a). In the drone retina the glial resting potential is approximately -53 mV (Coles and Tsacopoulos 1979), and in the Schwann-like cells surrounding the squid giant axon and crayfish medial giant axon the resting potentials are about -40 mV (Villegas 1981; Lieberman et al. 1981), which is considerably lower than in other glial cells. The differences in resting potential are chiefly due to differences in the glial cell membrane permeabilities to potassium and other ions. The leech packet glia membranes are selectively permeable to K^+, with the membrane behaving like a K^+ electrode (i.e. the glial membrane potential accurately follows the Nernst equation), and the resting potential is determined by the ratio of $[K^+]_o/[K^+]_i$ (Kuffler 1967; Kuffler and Nicholls 1966). However, the leech neuropile glial cells' resting potential exhibits a dependence on the external concentration of both potassium and chloride; their membranes are permeable to both these ions (Walz and Schlue 1982a). Futhermore 5-hydroxytryptamine (5-HT) has the ability to modify the K^+ permeability of these cells. When the substance is applied to them they behave like potassium electrodes (Walz and Schlue 1982b). The intriguing possibility arises that 5-HT may regulate the extent to which the glial cells function as spatial buffers (see below), redistributing the potassium which accumulates around active neurones (Walz 1982; Schlue and Walz 1984). The low membrane potentials of the squid Schwann cells may be to some extent artefactual, caused by leakage current during electrode penetration into a very thin cell layer (Villegas 1981).

There is extensive evidence that under normal physiological conditions, glial cells are electrically inexcitable (i.e. do not fire action potentials). However, there is additional extensive evidence that glial cells are electrically coupled, allowing the passage of ions and small molecules (e.g. the dye Lucifer Yellow, MW 457). The phenomenon was first described by Kuffler and Potter (1964), in their early studies on the leech. If current is passed into a glial cell via a microelectrode, a graded potential which is an approximate analogue of this stimulus can be recorded from adjacent glial cells. The glial cells in the drone retina are also extensively electrically coupled (Coles and Tsacopoulos 1981). However, glial cells are not coupled to neurones. The coupling between adjacent glial cells is mediated via the specialised membrane structures (e.g. gap junctions) which join the cells together. A key question concerns the function(s) of the glial cell coupling. One likely answer, about which there is growing evidence, is that excess potassium, released from active neurones, may be redistributed through a number of glial cells in order to preserve the correct local extracellular concentrations. This could take place via a spatial buffer mechanism (Gardner-Medwin 1983a, b;) Gardner-Medwin and Nicholson 1983), or active uptake processes, or both. Another possibility is that the glial cells comprise a metabolic syncytium, providing metabolic interactions that are linked to the activities of the neurones. Both these topics are discussed further in subsequent sections.

Active neurones release potassium into the extracellular spaces which immediately surround them. Attempts have been made to measure the increases in $[K]_o$ resulting from different types of activity. In mammalian brain, extracellular K^+ is thought to reach a ceiling value of about 10 mM (from about 3.5 mM) under conditions of high but physiological activity (see Nicholson 1980; Varon and Somjen 1979). Similar maximum increments may take place in the leech (Kuffler 1967; Schlue and Walz 1984), crayfish (Smith 1983) and insect compound eye (Coles and Tsacopoulos 1979). A probable increment of about 5 mM takes place around neurones that are active under normal physiological conditions. The success in obtaining these measurement has been largely due to the use of ion-selective microelectrodes. However, it should be borne in mind that such measurements do not necessarily determine elemental composition since bound or otherwise sequestered elements are not measured, and it is assumed that intracellular ion activities are the same as in aqueous standards (Saubermann and Scheid 1985). Thus Saubermann and Scheid (1985), using x-ray microanalysis of frozen hydrated and dried sectioned leech ganglia, found that K^+ and Na^+ were present in concentrations greater than predicted by ion-selective microelectrode measurements (Deitmer and Schlue 1981; Schlue and Deitmer 1980). Nevertheless, it can be safely stated that extracellular potassiumm increases by 3-6 mM during normal nerve activity. In many situations, especially where the glial cells behave like potassium electrodes (e.g. the leech packet glial cells and the insect compound eye), the increased $[K]_o$ can be recorded as a depolarisation of the glial cells adjacent to the active neurones.

Although the membrane potentials of neurones are relatively insensitive to small alterations in external potassium, their excitability may be modified in ways that could upset the normal processes of integration. This is thought to be particularly the case around small diameter axonal processes and nerve terminals in neuropile. Small increments in $[K^+]_o$ may affect synaptic transmission by altering the calcium permeability increase in response to the depolarisation of the membrane (Cooke and Quastel 1973). In the leech the undershoots of nerve impulses are progressively reduced as more and more potassium accumulates during a train of impulses (Baylor and Nicholls 1969).

Two important consequences arise out of the above descriptions. First, the glial cells may protect against alterations in the signalling activities of the neurones, by controlling their ionic environment. Evidence for this, particularly that obtained in the insect compound eye (Coles and Tsacopoulos 1979, 1981) is discussed in a subsequent section. Second, the potassium released from active neurones may itself act as a signal to the glial cells, to produce metabolic or functional alterations that are essential for the normal functioning of the tissue (Kuffler and Nicholls 1976). Evidence for this, especially that described for leech ganglia (Pentreath 1982; Pentreath and Kai-Kai 1982), is also discussed in a later section.

An area of developing interest concerns the effects of transmitter substances and peptides on the properties of glial cells. Such substances may, like potassium, act as signals to the glial cells. A key point here is that the glial cells may not always respond by a conductance change, but by modification of one or more metabolic processes. Thus a search for electrical changes in glial cells, analogous to the transmitter/receptor mediated changes in neurones, may be meaningless. On the other hand, alterations in second messenger systems, for example, cyclic AMP or components of the phosphatidyl inositol metabolic cycle, may be of profound importance for the functioning of the glial cells, and consequently the well-being of the neurones. Some information about this potentially important field is now available from invertebrate nervous systems.

One of the best studied preparations is the glial sheath (Schwann-like cells) of the squid giant nerve fibre (Villegas 1981, 1984). The glial cells in this preparation are capable of synthesising and storing large quantities of acetylcholine (Heumann et al. 1981; Villegas and Jenden 1979). During spike activity in the giant axon a signal (possibly glutamate; see Villegas 1984) acts on the Schwann cells to cause release of the acetylcholine. The released acetylcholine has a feed-back effect on nicotinic cholinergic receptors on the Schwann cell, causing a long-lasting hyperpolarisation due to an increase in potassium conductance (Villegas 1974, 1975, 1978, 1981; Villegas and Villegas 1974, 1978). The glial cell receptors appear to mediate their actions by increasing intracellular cyclic AMP (Evans et al. 1985). More recently, it has been shown that octopamine also increases cyclic AMP levels in the Schwann cells, potentiating the actions of the cholinergic activation system (Reale et al. 1986). Unfortunately, the reasons for the existence of this complex multistep interaction between the axon and satellite glial cells are not yet clear, although they appear to be part of a complex of biochemical and physiological changes that are beneficial when the giant axon is required to fire at high frequency during stressful conditions (i.e. the escape response of the squid). Nevertheless they clearly demonstrate the presence of transmitter-activated second messenger systems in the glial cells. It appears that similar interactions may take place in the crayfish giant axon/glial cell preparation (Lieberman et al. 1981).

Transmitter substances have been shown to act on glial cells in other invertebrate nervous systems. Octopamine reduces the potassium permeability of the glial cells that form the insect blood/brain barrier (Schofield and Treherne 1985). 5-HT increases the potassium permeability of the neuropile glial cells in the leech (Walz and Schlue 1982b) and modulates glycogen metabolism in the outer (packet) glial cells of the same animal (Seal and Pentreath 1985). These topics are discussed further in sections on the functions of invertebrate glial cells.

5. FUNCTIONS OF INVERTEBRATE GLIAL CELLS

It may be predicted that the ultimate list of glial functions will be awesome, and occupy detailed volumes of work. At present, however, reliable data are available for only a few functions, although suggestions and inferences, largely resulting from the extensive anatomical descriptions, are numerous. Some of the more likely roles, listed in Table 4, are summarised below.

5.1. Support, protection against deformation

The original suggestions that neuroglia ('nerve glue') were supportive have since been given weight by many fine structural findings. The large numbers of intracellular adhesions and junctions (see Lane 1981) between adjacent glial cells, between glial cells and basement membranes and sometimes between glial cells and neurones must provide a framework for mechanical stability (Fig. 1). Many invertebrates, which lack skeletal

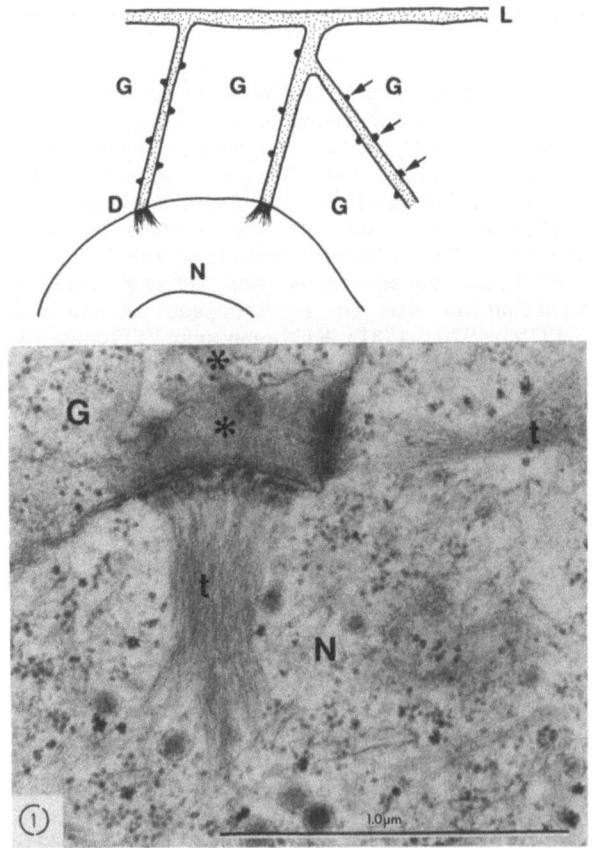

FIG. 1: The diagram (top) shows the arrangement of the matrix-filled extracellular channels that are presumed to play a supportive role in the outer regions of the segmental ganglia of the leech, Haemopis sanguisuga. The channels connect with the basal lamina (L), and extend to the neurons (N), where hemi-desmosome contacts are made (see micrograph). Smaller attachement sites are formed with the giant glial

support, undergo violent distortions during body movements, and the nervous system must be protected against damage. Not only are the cell junctions important, but the microtubules, filaments and fibres present in many glial cells also undoubtedly provide strength. Thus many glial processes surrounding the nerves and connectives in annelids and molluscs contain large amounts of fibrous material (see Radojcic and Pentreath 1979). The extensive fibrillar system in the glial tissue of <u>Nereis</u> has been described as a means for counteracting shearing and compressional forces (Baskin 1971a, b). Coggeshall (1966) has desbribed another stress-resistant specialisation in the leech. These are a form of glia occurring at the nerve/ganglional junction, which he called fasicular glia. He suggested that these cells absorb the stresses exerted at this junction by the changes in body shape. In the Arthropoda, on the other hand, filaments are frequently absent, and this may be correlated with the relatively little stress placed upon the nervous system of these animals, which possess an exoskeleton. Instead, the glial cells often contain large numbers of microtubules, and these may function as a cytoskeleton as well as in the transport of materials (Lane and Abbott 1975; Saint Marie and Carlson 1983; Saint Marie et al 1984). For further discussion on the supportive roles of invertebrate glial cells see Radojcic and Pentreath (1979).

5.2. Blood-brain and permeability barriers

One of the best documented roles of any invertebrate glial cell type is the occluding barrier formed by the arthropod perineurium. The blood-brain barrier is most effective in insects, where it maintains the appropriate ionic environment for nerve activity inside the avascular ganglia. The importance of the barrier is made clear when consideration is given to the unusual ratio of ion concentrations (i.e., low Na^+, high K^+) that exist in the haemolymph of many insects. Such composition is not found in other animal groups, and would prevent normal nerve function (Abbott and Treherne 1977; Skaer and Lane 1974). In addition, the haemolymph composition varies considerably with the different insect diets, and at different stages of development. Further significance is given to the barrier as a result of its ability to occlude toxins, pharmacological agents and wide variety of organic molecules, including many of the insecticides (see, e.g. Eldefrawi and O'Brien 1967).

Much of the present understanding of the barrier derives from the anatomical and physiological studies by Lane and Treherne and their co-workers (see Lane 1981; Treherne and Schofield 1981). The key components of the barrier, preventing the gross influx of ions, are the tight junctions joining adjacent perineurial cell processes. Thus ionic lanthanum permeates across the outer neural lamella, between the extracellular spaces, and into the gap junctions, but no further (Lane 1972: Fig. 2). This contrasts with the situation in many other invertebrate ganglia where barriers do not exist, and where tracer substances penetrate within the ganglia, between the neurones (e.g. snail,

---

cells (G) that fill the spaces between the channels and the neurones. The scale represent 5 µm. Modified from Kai-Kai and Pentreath (1981a).

The electron micrograph illustrates a hemi-desmosome joining the extracellular, matrix-filled channels (asterisks) to a neuron (N). Tonofibrils (t) radiate into the neurons' cytoplasm. The tonofibrils appear to be attached to electron dense substructures. G, part of a giant glial cell.

TABLE 4

FUNCTIONS OF INVERTEBRATE GLIAL CELLS

SUPPORT, PROTECT AGAINST DEFORMATION (VIA FILAMENTS AND INTERCELLULAR ADHESIONS).

PROVIDE PERMEABILITY BARRIERS (ESPECIALLY ARTHROPODA).

SPEED IMPULSE CONDUCTION (GLIAL WRAPPINGS LESS COMPACT THAN MYELIN).

REMOVAL OF NEURONES (DEVELOPMENT, DEGENERATION).

GUIDANCE, DIFFERENTIATION AND GROWTH OF NEURONS (DEVELOPMENT, REGENERATION).

REGULATION OF IONIC COMPOSITION OF NEURONAL ENVIRONMENT.

TRANSMITTER UPTAKE, INACTIVATION OR RELEASE.

METABOLIC INTERACTIONS WITH NEURONS (SUPPLY SUBSTRATES, REMOVE CATABOLITES).

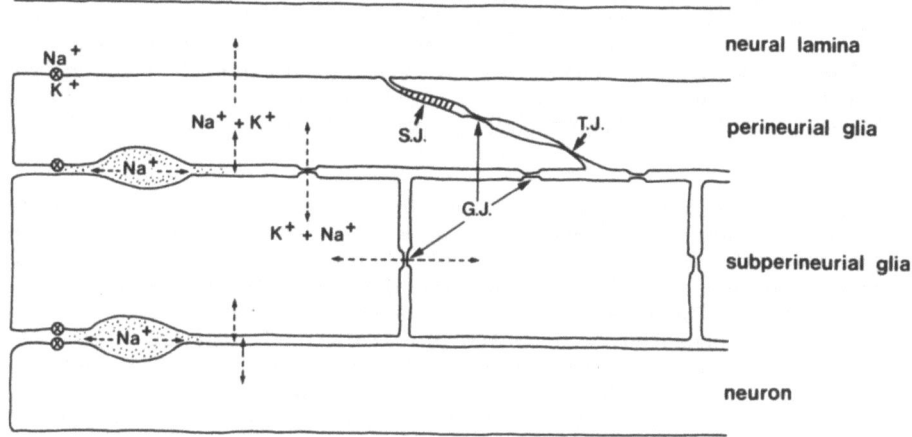

FIG. 2: The insect blood/brain barrier, and the regulation of sodium and potassium. The neurons and subperineurial glial cells are ensheathed by a layer of glial cells, the perineurium, which is in turn overlaid by the neural lamella. Tight junctions (T.J.) and septate junctions (S.J.) occur between the perineurial cells. In many regions the extracellular spaces are expanded and contain an electron opaque matrix (e.g., stipled areas at left). Sodium and potassium may transfer between blood and nervous tissue across the perineurium. Because intercellular movement is prevented by the tight junctions, transfer is via diffusion down electrochemical gradients (--->), and by means of Na^+/K^+ pumps (⊗). Diffusion will also occur between perineurial and glial cells, through gap junctions. Extracellular Na^+ is maintained high, and K^+ low, by the pumps on the inwardly facing perineurial cells and the membranes of the underlying glia and neurones, and by the properties of the extracellular matrix. Modified from Treherne and Schofield (1981), who should be studied for further details.

Pentreath and Cottrell 1970; and leech, Kai-Kai and Pentreath 1981a). Although it seems clear that the tight junction layer protects the neurones against large fluctuations in ions, and other chemicals in the haemolymph, they do not, however, account for the fine tuning of the extracellular fluid surrounding the neurones. The electrical activity of the neurones will itself also cause alterations in the ionic composition within the extracellular spaces inside the ganglia. The final composition of the extracellular fluid appears to be achieved by a combination of passive and active processes involving the neuroglia, their intercellular junctions and an extracellular anion matrix. The passive role of the perineurial junctions is augmented by glial and axonal cation pumping, which may recycle Na^+ in order to maintain the correct free $[Na^+]_o$. The extracellular anion matrix may buffer extracellular sodium, to enable increases in intraneuronal sodium to occur and thus stimulate the neuronal Na^+/K^+ pump, while keeping the free $[Na^+]_o$ constant (Fig. 2; Lane and Treherne 1980; Treherne and Schofield 1978, 1981; Treherne et al. 1982.). More recently it has been demonstrated that octopamine reduces the transperineurial potassium permeability (Schofield and Treherne 1985); thus the insect barrier system may, in part, be under humoral control.

In the crustacea the blood/brain barrier is less effective than in insects. Molecules such as horseradish peroxidase (HRP), sucrose and inulin enter the ganglia easily (Abbott 1970, 1972), although molecules larger than about 16 nm cannot (Abbott, 1970). The restriction is thought to be due to the junctions between the perineurial cells, and the mucopolysaccharide matrix (secreted by the glial cells) filling the extracellular spaces (Abbott 1972). There does not appear to be a metabolically maintained Na^+/K^+ pump component to the crustacean barrier (Abbott et al. 1977).

In the cuttlefish, Sepia, a blood/brain barrier is present which prevents the transfer of proteins, and whose tightness and properties resembles that of mammals (Abbott et al. 1982, 1985). The barrier has been attributed to the occluding junctions located between the perivascular glial cells. The cephalopod brain possesses an extensive vascular supply, which is connected to a system of glial-lined channels called the glio-vascular system (see Gray 1969). It has been suggested that the reason for this highly developed barrier is not primarily for ionic homeostasis, but that it is associated with the higher integrative functions of the cephalopod brain (Abbott et al. 1985). However, in the other molluscan groups, and indeed in most of the other non-arthropod invertebrate groups (especially the lower ones), a tight blood-brain barrier is generally lacking, with many small molecules and ions having unrestricted access to the nervous tissue (see Radojcic and Pentreath 1979).

Glial cells and their junctions may, in addition to providing barriers at the blood interface, provide for compartmentalisation within the nervous tissue. Evidence for this comes from the insect visual system, where the photoreceptors and laminar interneurones are divided by glial cells and their occluding junctions, into several different regions (see Saint Marie et al. 1984). The different regions maintain separate standing potentials and are separated by high resistance areas, which appear to correlate closely with the likely ionic barriers provided by the glial cells and their junctions. Thus a function of this glial arrangement appears to be to allow electrical activity to be modified in a relatively consistent fashion, throughout a defined area of nervous tissue or neuropile. Little is known about the occurrence of this interesting phenomenon in other invertebrates, but it may be predicted that it will be found to be of great importance and wide occurrence (see Fig. 7).

5.3. Axon wrapping, increased conduction velocity

An obvious feature in all invertebrate nervous systems which contain glial cells, is the wrapping of the axons, either singly or in groups, by a proportion of the glial cells. Furthermore, there is almost a continual gradation, in passing from the lower to the higher invertebrates (and sometimes within the individual classes or species) from axons that are wrapped simply by a single glial process, to those which are surrounded by complex, tightly packed multiple layers, resembling vertebrate myelin. This is summarised in Figs. 3-6. Differing amounts of fibrous material are present in the glial wrappings in different situations.

The problem regarding these glial wrappings is at what stage of complexity a function in speeding conduction velocity can be clearly assigned. In the earthworm (Günther 1976) and crustacean nervous systems (Heuser and Doggenweiler 1966) the situation is relatively unambiguous, for there is good anatomical and physiological evidence for an insulating sheath with nodes, and rapid saltatory conduction. Earthworm giant fibres are surrounded by dozens of lamellae of relatively uniform thickness, with distinct areas of compaction (Fig. 4). The outer layers contain greater amounts of glial cytoplasm. The sheath is believed to be spirally

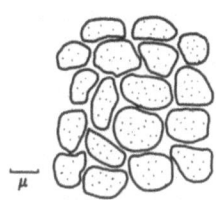
'NAKED' AXONS
(Coelenterates, Echinoderms)

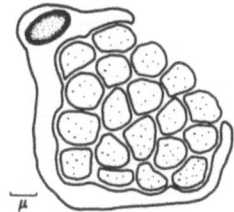
GROUP ENSHEATHMENT
(Annelids, Arthropods, Molluscs)

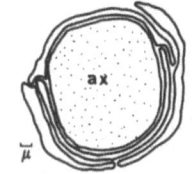
LOOSE LAYERS
(large axons in Annelids, Arthropods, Molluscs)

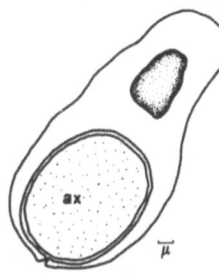
SINGLE GLIAL ENSHEATHMENT
(e.g. large axons in insect PNS)

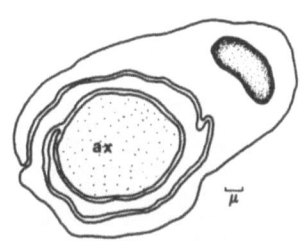
LOOSE SPIRAL
(e.g. insect PNS)

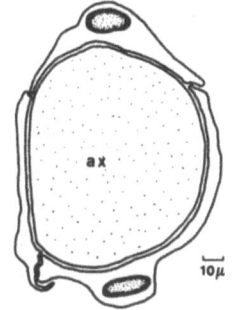
GLIAL MOSAIC
(e.g. squid giant axons)

FIG. 3: Glial sheaths around invertebrate axons. In the lower groups (coelenterates, echinoderms) the axons are naked, with no sheaths. In the other groups, glial cells ensheath axons singly or in layers. Large axons may be surrounded by a single glial cell process, as in the insect peripheral nervous system, or by multiple loose layers, as in the annelids, arthropods and molluscs. The squid giant axon has a mosaic of glial cells applied to its surface. Some of the data is modified from Bullock et al. (1977).

Large-diameter axons; annelids, arthropods, molluscs   Earthworm

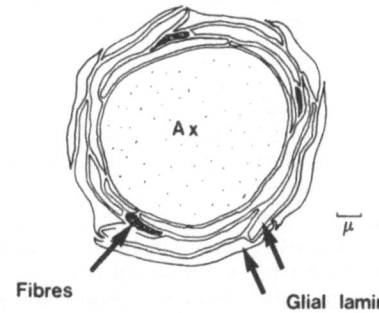
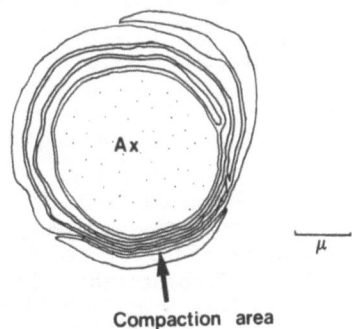

FIG. 4: The glial sheaths around some large-diameter axons in invertebrates. Multiple glial laminae (e.g., the insects) are sometimes separated by bundles of fibres (e.g., gastropods). In the earthworm (right) areas of compaction may occur between adjacent glial processes.

wrapped, and morphologically closely resembles vertebrate myelin (Hama 1959; Günther 1976; see below). Even more specialised are the sheaths of giant fibres in large shrimps and prawns (Figs. 5, 6). The conduction velocities of nerve impulses in these animals are of the highest known (up to 210 m/s at 20°C), and the glial wrappings are impressive. However, there are the following important differences from vertebrate myelin. The laminae do not connect to form a spiral, the nuclei of the sheath cells lie on the inside of the sheath (cf. vertebrate myelinated axons, where they are on the outside), desmosomes join adjacent laminae, and an extracellular gap (approximately 20 nm wide) occurs around the axon and between the innermost sheath lamina (Heuser and Doggenweiler 1966; Fig. 5). However, the sheaths are interrupted periodically to form nodes (which are analogous to the vertebrate nodes of Ranvier), where a type of cell called the nodal cell enmeshes the axon and sheath layers (Fig. 6).

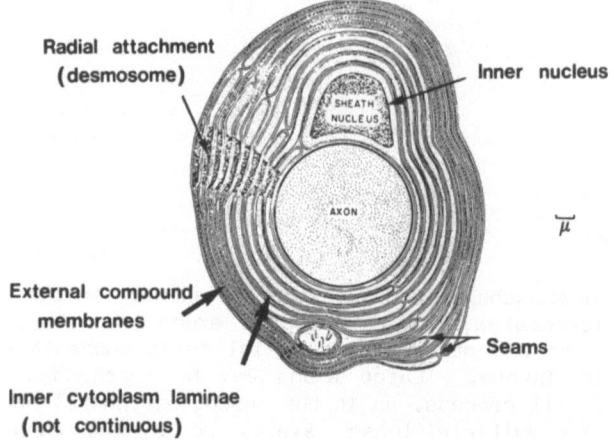

FIG. 5: Diagram of a cross-section of a prawn nerve fibre. The sheath is composed of many laminae, each containing cytoplasm and wrapped to form a seam. The seams are aligned in an organized fashion. A radial attachment zone, which is analogous to a stack of desmosomes, is formed between adjacent laminae in one region. The nucleus is located in the innermost lamina. Extracellular spaces occur around the axon and between the inner sheaths. (From Heuser and Doggenweiler 1966; reproduced from The Journal of Cell Biology, 1966, vol 30, p 390, by the copyright permission of the Rockefeller University Press).

In marked contrast, squid giant fibres have a glial sheath that can be less than 1 μm thick, which is less than 1% of the axon diameter (Fig. 3). The glial cells form a mosaic around the axon, often only a few cell processes deep. In most other invertebrates the loose glial spirals and wrappings (Fig. 3) are relatively incomplete, and there is generally no evidence of compaction. Large axons tend to be surrounded by a greater number of glial wrappings than small axons (see Radojcic and Pentreath 1979). Thin layers of cytoplasm fill the glial processes and this sometimes (e.g. the gastropods) contains an electron-opaque matrix of fibrous material (Fig. 7). In other situations an extracellular matrix may occur between adjacent glial cell processes. The matrix-filled extracellular spaces may be expanded and contain collagen-like fibrils (e.g. the crustacea) or, in the case of insects, hyaluronic acid (Ashhurst

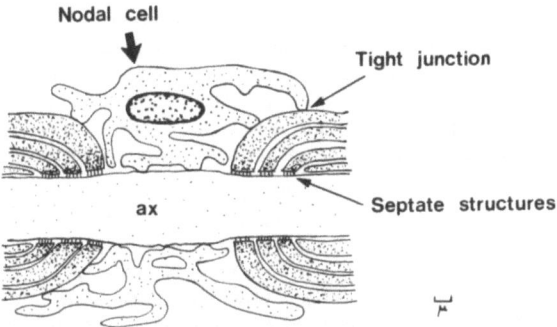

FIG. 6: Diagram of a longitudinal section of a node in the prawn nerve sheath. The glial laminae form septate structures as they terminate near the axon. The node is loosely wrapped by a characteristic type of glial cell called the nodal cell, which intermittently adheres to both axon (ax) and sheath via tight junctions. Modified from Heuser and Doggenweiler (1966).

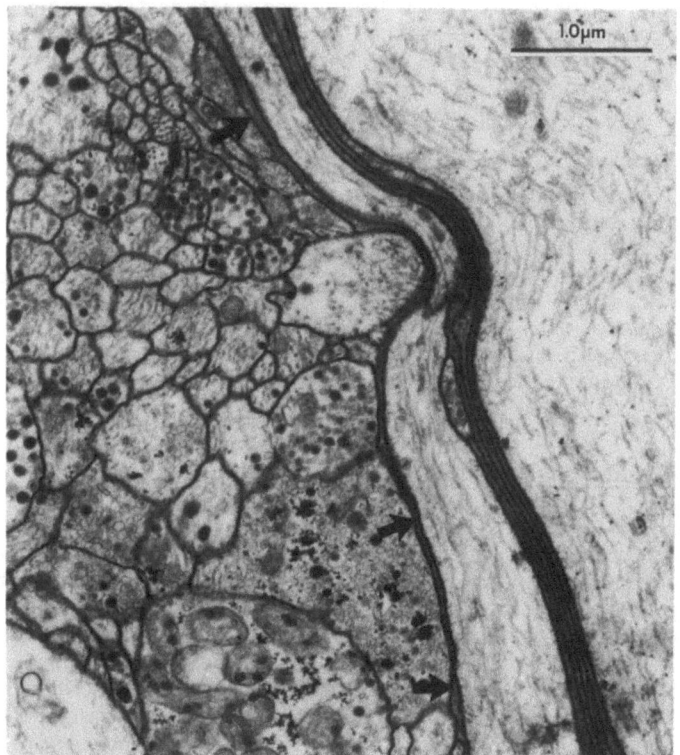

FIG. 7: Electron micrograph of part of a buccal ganglion of <u>Planorbis</u> <u>corneus</u>. Several glial lamellae containing an electron opaque matrix separate the large axon process (right) from a smaller axon process (middle). Both axons are sectioned longitudinally. An area of neuropile (left) is isolated from the middle axon by a single glial lamella (arrows). Note the general absence of glial processes from the neuropile.

and Costin 1971). Unfortunately, little is known about how these arrangements may act as insulators of the axons they surround, or whether in fact they in any way alter the cable properties of the axons. The loose glial wrappings may have little to do with increasing conduction velocities, but act as occluding barriers for ions and other substances, or for maintaining local ion homeostasis. They may also interact metabolically with the axons. These topics are discussed further in subsequent sections.

Several interesting studies have been made on the composition of invertebrate nervous tissues and glial cells, which show that there are other significant differences from the vertebrates. 2',3'-Cyclic nucleotide 3'-phosphohydrolase (CNP) has long been regarded as a myelin marker, but does not appear to be present in the nervous systems of the crab, squid, octopus, prawn or starfish (Drummond et al. 1971). The substance has however been reported as occurring in a moth, Manduca sexta, although at a much lower level than found in myelin (Taylor et al. 1976; see also Roots 1981). Galactocerebrosides and sulphatides, which are also considered to be markers for myelin are absent from the ventral nerves of earthworms (Okamura et al. 1985). The earthworms also lack sphingomyelin, which is considered to be a ubiquitous component of animal membranes, but contain cholesterol and several phosphoglycerolipids normally found in mammalian brain (Okamura et al. 1985). Shrimp nerves are rich in glucocerebrosides, albeit of an unusual composition (Shimomura et al. 1983). Squid axoplasm actively synthesises phospholipids from precursors, although it may also be supplied to the axon by a glial-axon transfer (Gould et al. 1983), similar to the one demonstrated for protein (Lasek et al. 1977; Lasek and Tytell 1981; see below).

Thus a major property of invertebrate glial cells is to isolate and wrap axons, although rarely to an extent comparable to that exhibited by the Schwann cells and oligodendrocytes in vertebrates. In only a few situations (e.g. the earthworm and shrimp) has a saltatory, impulse speeding function been demonstrated. The differences in chemical composition, anatomical arrangements and degrees of compaction show that in no instance are the glial sheaths directly comparable with the vertebrate myelin. In the majority of nervous systems the functions of the loose glial wrappings are not understood.

5.4. Removal of neurones (phagocytosis) in development, degeneration and ageing

It appears that most, if not all glial cells have the capacity to engulf and break down degenerating neurones, and other cellular materials, with which they make contact. However, it is important to realise that whilst glial cells have the capacity for phagocytosis, they are not, with a few exceptions, specialised to fulfil this as their primary role, unlike some other circulating cell types (see Varon and Somjen 1979). On the other hand, in the many invertebrate ganglia which are avascular, it would seem essential to have some cells that are locally capable of ingesting injured or senescent neuronal processes. In accord with this, there are some situations (e.g. the segmental ganglia of the leech) where a subpopulation of the glial cells (the microglial cells; Coggeshall and Fawcett 1964; Kai-Kai and Pentreath 1981a) appear to be specialised scavenger cells (see Morgese et al. 1983). It should also be remembered that the phagocytic activity may be an essential part of nervous development, removing redundant processes, and that it may take place in the normal apparently healthy tissue, as well as being an obvious feature of ageing and experimentally induced degeneration.

In the leech, the small glial cells are morphologically similar to vertebrate microglia. The cells become numerous after nerve injury and, unlike the giant 'packet' glial cells, may aid regeneration (Elliott and Muller 1981). The microglia aggregate at a site of nerve cord crush, increasing in numbers by approximately five times (Morgese et al. 1983). The cells respond to a second, later crush by re-aggregating at the new injury site. Furthermore, the changes take place when the ganglia are isolated from the blood, which shows that the microglia do not necessarily derive externally (Morgese et al. 1983). Leech microglia are illustrated in Fig. 8.

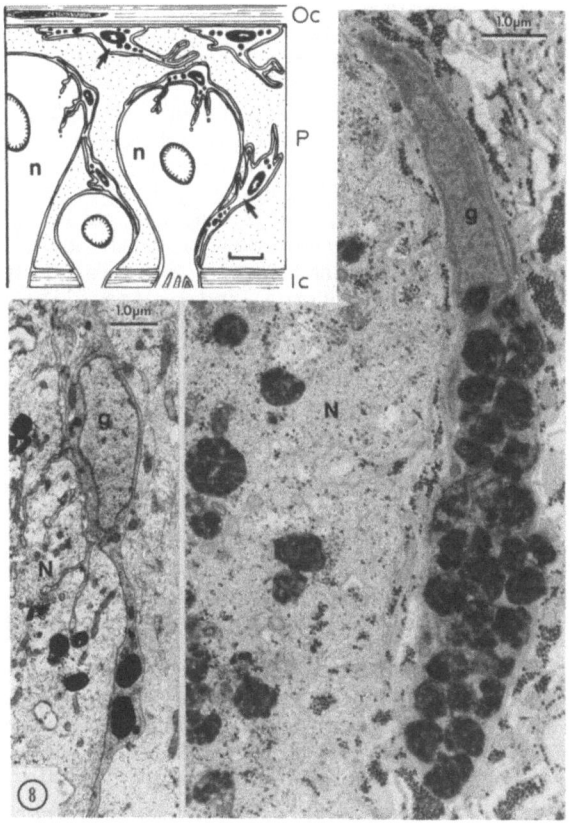

FIG. 8: Microglial cells in the segmental ganglia of the leech, <u>Haemopis sanguisuga</u>. The diagram (top left) shows the relations of the neurones (n) with the microglial cells (two arrowed) and the giant packet glial cells (p, stipled) in the rind of the ganglion. Oc, outer capsule, Ic, inner capsule connective tissue layers. The scale is 20 µm. See also Gray and Guillery (1963), and Coggeshall and Fawcett (1964), for descriptions of the arrangement of the leech ganglia. The two electron micrographs show microglial cells (g) located at the edges of neurones (n). The left micrograph shows processes from a microglial cell penetrating the neurone. In both micrographs note the similarities of the lysosomes and residual bodies in the microglial cells to those in the neurones. The microglial cells are specialised scavenger cells. Modified from Kai-Kai and Pentreath (1981a).

In the insect central nervous system, repair and regeneration of the nervous tissue appears to be chiefly mediated by the granule-containing cells, which bear a striking resemblance to the circulating hemocytes, from which they are probably derived (Smith et al. 1984, 1986). Both cell types are phagocytotically active and play a role in structural repair. Selective destruction of parts of the central nervous connectives in the cockroach produces a long-term phagocytic activity by these cells at the site of injury (Treherne et al. 1984). On the other hand, after selective glial disruption by the glial toxin ethidium bromide, the granule-containing cells produce a speedy, ordered repair of the connectives, eventually becoming indistinguishable from the normal perineurial cells (Smith et al. 1984). Thus the presence of undamaged axons and/or the extracellular matrix produces marked facilitation of the glial repair mechanisms (Treherne et al. 1984). In Musca and Calliphora, surgically induced degeneration of photoreceptor axons (by isolating them form their perikarya) is associated with the rapid production of at least two different acid phosphatases by the satellite glial cells (Griffiths 1979).

The removal of degenerating nerve processes by satellite glial cells in the other invertebrate groups has been noted fairly frequently; for example in isolated molluscan ganglia (Pentreath et al. 1985) and crayfish axons (Bittner and Mann 1976; see also Radojcic and Pentreath 1979, for other references). In particular, the phagocytic activities of the glial cells have been fairly extensively documented in the ganglia of gastropod molluscs, where neuronal residues and foreign material such as viruses and ferritin have been shown to be internalised by the glial cells (Reinecke 1976; Borovyagin et al. 1972). In the snail Planorbis, the glial cells engulf nerve endings in the apparently healthy ganglia, and this process is enhanced by experimental degeneration, and with ageing (Pentreath et al. 1985). In this animal there is a progressive increase in phagocytic activity, and lysosomal material (especially lipofuscin and other residual bodies) with age. Most of the glial cells in the ganglia show some signs of phagocytic activity, but only a small proportion of the cells appear specialised for this role. These latter cells show increased numbers and activities with degeneration. The cells sometimes contain a spectrum of axonal processes from those which appear unaltered, presumably recently engulfed, through different stages of breakdown to dense aggregates (Fig. 9a). As the cells become packed with the end-products of the breakdown process, they move to the edges of the ganglia, where they occupy subcapsular areas. Here the cells remain, with large aggregates of lipofuscin contained within them (Fig. 9b, c), conferring a yellow colour to the edges of old or degenerate ganglia (see Pentreath et al. 1985, for details and for a discussion of the phagocytic activities of glial cells). When the snails were maintained on a diet with high Vitamin E content, the lipofuscin accumulation was significantly reduced (Winstanley and Pentreath 1985). The phagocytic activities of the different invertebrate glial cells, at different stages of development and ageing, and the cues which control their activities, is an intriguing area for further study.

5.5. Roles in guidance, growth and differentiation of neurones

Barely a start has been made in the study of the potential roles of invertebrate glial cells in neurogenesis and regeneration. This is in sharp contrast to the mammalian brain, where the scaffolding and guiding functions of the radial glia during development have been fairly well established (Rakic 1974).

In the moth, Galleria, it has been suggested that glial cells may produce shortening of the connectives during metamorphosis (Pipa 1967). During this process, the glial cells become reduced, as do their associated axons (Pipa 1973). A tactic role in guiding migrating neurones has been suggested in several other arthropods (Edwards 1980; Lane 1979; Lopresti et al. 1973), and it has been suggested that glial cells may be responsible for the twisting of the pseudocartidge axon bundles in the maturing dipteran eye (Saint Marie et al. 1984).

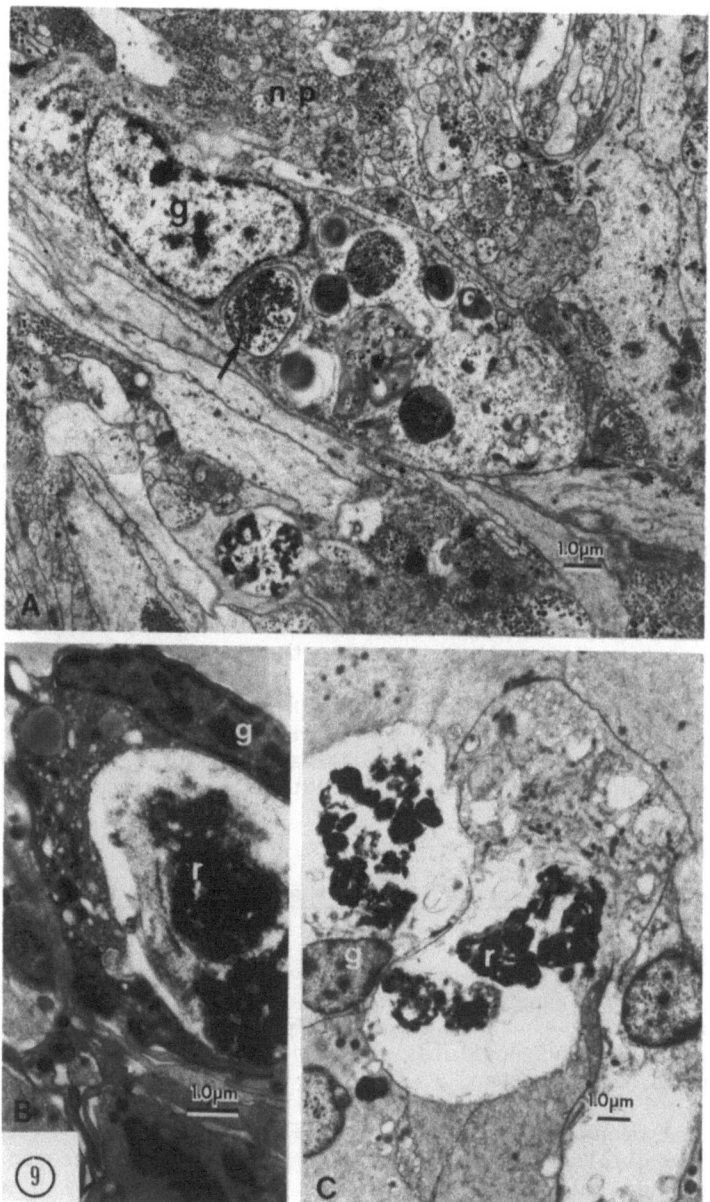

FIG. 9: Phagocytic glial cells in the buccal ganglion of the pond snail, <u>Planorbis corneus</u>. <u>A</u> The glial cell (g) contains a mixture of lysosomes, one of which (arrow) is surrounded by a double membrane, suggesting that it is in the process of being engulfed. The glial cell lies alongside an area of neuropile (np). <u>B</u> Glial cell (g), at the edge of the ganglion, containing a large residual body (r). <u>C</u> Two adjacent glial cells (one marked g) containing masses of residual lipofuscin material, in the ganglion of an aged snail. Some of the pigment has been lost during the histological processing. In this animal there is a progressive age-related increase in residual deposits in the glial cells at the edges of the ganglia (Pentreath et al. 1985; by permission of The Royal Society).

During development of the leech nervous system, the giant glial cells appear not to play an initial role in the grouping of the neurones, although their possible subsequent roles in establishing neuronal connexions (i.e. axon guidance) are not understood (see Muller et al. 1981). The giant glial cells also do not appear to be involved in the regeneration of neuronal connexions between neurones in the adult leech nervous system, although the microglial cells might (Elliott and Muller 1981, 1982).

Besides providing a guiding cellular substrate, glial cells may influence neuronal differentiation by releasing 'trophic' substances which influence the biochemical and morphological differentiation of neighbouring neurones. In mammals, several trophic substances have been shown to be released from glial cells which influence neuronal differentiation; probably the most well-known of these is nerve growth factor (NGF, see Varon and Somjen 1979). The work of Schacher and co-workers (see Schacher, 1981) suggests that similar mechanisms may take place an Aplysia. During late premetamorphic development, the abdominal ganglion of Aplysia is surrounded by support cells, which appear to later develop into glial cells. The support cells contain large numbers of secretory granules (of unknown composition), whose release appears to signal the burst of neuronal growth and maturation that occurs following metamorphosis. The granule materials seem to provide a general stimulus for neuronal differentiation including growth, spine development, and synapse formation (Schacher 1981). There is as yet no data concerning this interesting phenomenon in other invertebrates.

Thus little is known about the role of glial cells in neurogenesis in most invertebrates. This lack of information not only concerns the neurone/glial interactions, but more generally the origins, movements, and morphological identities of the different cell types at all stages of development.

5.6. Regulation of ionic composition of the neuronal environment

There seems to be now little doubt that many glial cells have an important function in the regulation of the ionic environment of neurones. This is especially the case for potassium ions which are released from active neurones, and which must subsequently be closely regulated within the extracellular spaces in order that the functioning of the neurones is not perturbed.

Two glial mechanisms are generally believed to control extracellular potassium: the glial cells may function as a 'spatial buffer' or they may actively take up potassium ions. The two mechanisms may also act in combination.

The spatial buffer theory was developed by Kuffler and co-workers (Kuffler 1967; Orkand et al. 1966), on the basis of the properties of the glial cells of the leech, and an amphibian, Necturus, where the membranes are selectively permeable to K^+. In this theory the coupled glial cells are passive channels for K^+ movement from regions of high concentration to regions of lower concentration. When K^+ is elevated locally (around an active neurone), it depolarises the adjacent glial syncytium. This depolarisation sets up ion currents within the syncytium with the depolarised parts and the more distant parts (still at resting potential) acting as poles of a current source. Only the freely permeable potassium ions can carry the charge within this circuit. K^+ thus flows into the cells in the locally depolarised region, through the syncytium where it is more negative, to emerge at a more distant region of the extracellular space. The circuit is thought to be completed by the flow of sodium ions (the major extracellular cation) in the opposite direction in the extracellular space. There is no net uptake of K^+ into the glial cells. The conditions for this simple scheme of events are that glial cell membranes be selectively permeable to K^+ (i.e., the membrane potential behave as Nernst potential for K^+), have a low resistance, and that adjacent glial cells be coupled by low resistance pathways. There is growing evidence that these conditions are fulfilled in a variety of situations.

Good evidence for the spatial buffer mechanism has been obtained in the honeybee (drone) retina by Coles, Tsacopoulos and Gardner-Medwin (Coles and Tsacopoulos 1979, 1981; Gardner-Medwin et al 1981). Some of their data is summarised in Fig. 10. The success of their work has been in part due to the orderly arrangement of the cell types in the compound eye, which are accessible to ion sensitive electrodes, and the ability to stimulate the preparation by a normal physiological stimulus (a light flash). More recently it has been demonstrated that following stimulation only small changes in sodium activity take place in the glial cells, compared to the neurones (Coles and Orkand 1985). These works have produced an impressive body of quantitative data on the ionic movements accompanying electrical activity in the drone retina.

Although a spatial buffer mechanism may operate in the outer packet glial cells of the leech (Kuffler 1967), a different situation appears to exist for the neuropile glial cells, located in the core of the segmental ganglia. These cells possess a significant contribution by chloride ions to the membrane potential under normal conditions, which rules out a function as a spatial buffer (Walz and Schlue 1982a). However, in the presence of 5-HT, the behaviour of the membrane potential of these cells is similar to that of a potassium electrode when the external potassium is altered (Walz and Schlue 1982b). Thus neuronally released 5-HT could well regulate the extent to which the neuropile glial cells function as spatial buffers in response to potassium accumulation in the extracellular spaces.

Information regarding the possible uptake of K^+ by an active process (i.e. with an exchange for Na^+ ions, via a glial Na^+-K^+-ATPase) in invertebrates is incomplete, but it generally argues against such a mechanism. Active uptake does not appear to take place to any significant extent in the packet glial cells (Hertz and Nissen 1976) or neuropile glial cells (Walz and Schlue, 1982a) of the leech, nor in the drone retina (Coles and Tsacopoulos 1981). On the other hand, a considerable weight of evidence, largely obtained from cultured glial cells, suggests that active uptake of K^+ may be important in higher vertebrates (see Walz and Hertz 1983).

Although K^+ homeostasis is now a widely accepted function of glial cells in invertebrates, there are in some situations inconsistencies, which should be pointed out. First, in some regions of the different nervous systems there may be found fairly large areas comprised of small diameter axons and nerve terminals (e.g. some neuropile regions in snail and leech ganglia; see Fig. 11), which contain hardly any glial cell processes in comparison with the regions of the nerve cell bodies. The neurone cell bodies are frequently peripherally located, in avascular ganglia, at the blood/neurone interface. This neurone/glial arrangement does not correlate with the surface-to-volume ratios of the neuronal cell membranes, where it might be predicted that the neuropile would contain the majority of glial cell processes if they were essential for K^+ homeostasis (see Pentreath, 1982). It has also been demonstrated in the leech that K^+ may clear itself by rapid diffusion through the extracellular spaces (Kuffler and Nicholls 1966). In several situations, for example the insect nervous system (Treherne and Schofield 1981), an extracellular matrix is thought to control ion (especially Na^+) movements. In

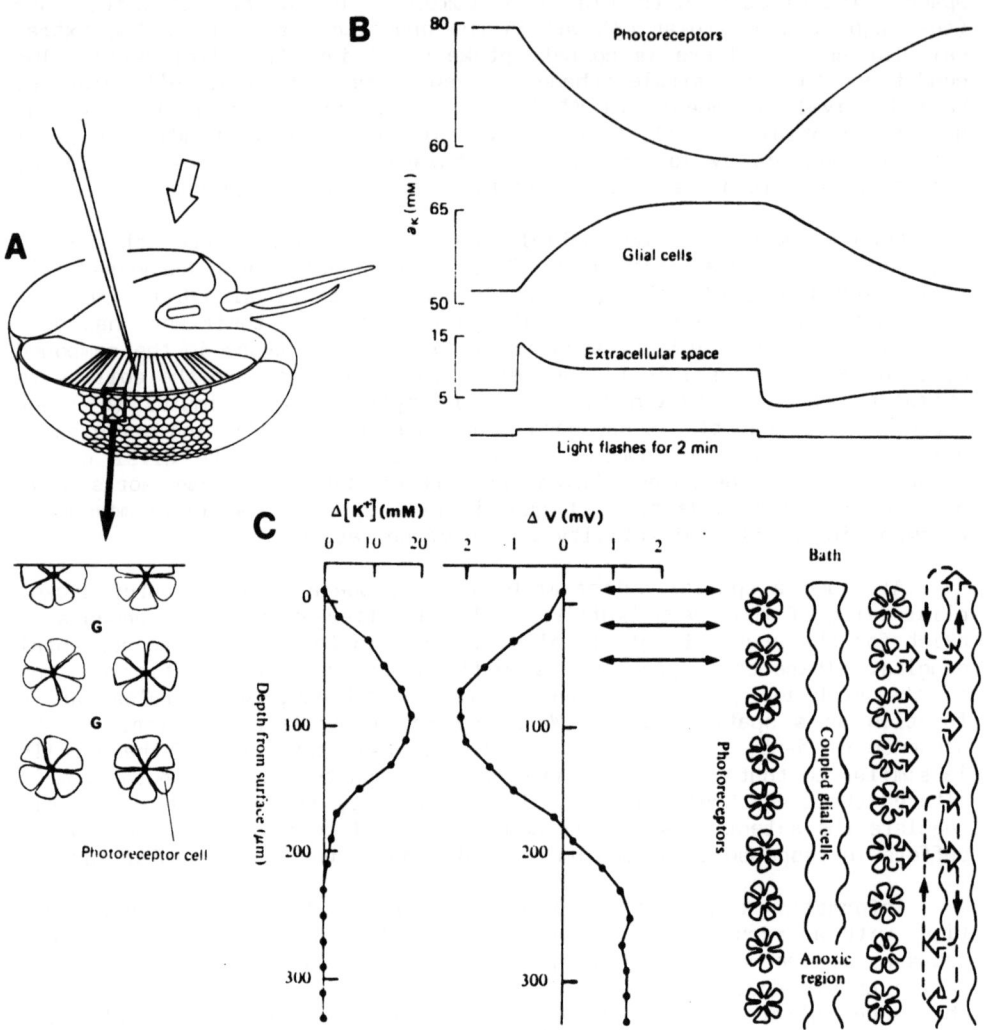

FIG. 10: Spatial bufering of K$^+$ in the compound eye (drone, honeybee) preparation. A Scheme of the cut head preparation. The head has been sliced parallel to the layers of ommatidia in each eye. Electrophysiological recordings were made by advancing a microelectrode into the retina. Stimulating light (arrow) is incident from above. The inset at bottom shows in cross-section six ommatidia; each of these is composed of six photoreceptor cells arranged in a flower-like pattern, with a small rectangular rhabdom (which absorbs light) at the centre. The space between the photoreceptors (G) is filled with glial cells (pigment cells, not drawn) which send processes between the photoreceptors to the rhabdom. B Summary of the changes in potassium activities (a_K) in the photoreceptors, the glial cells and the extracellular space to photostimulation of the photoreceptors. The increase of K$^+$ in the glial cells corresponds to most of the quantity of K$^+$ lost by the photoreceptors; if the glial cells did not take up K$^+$ the extracellular increases would be significantly greater than those shown. C Extracellular gradients and spatial buffering. A K$^+$-sensitive microelectrode was advanced into the retina and

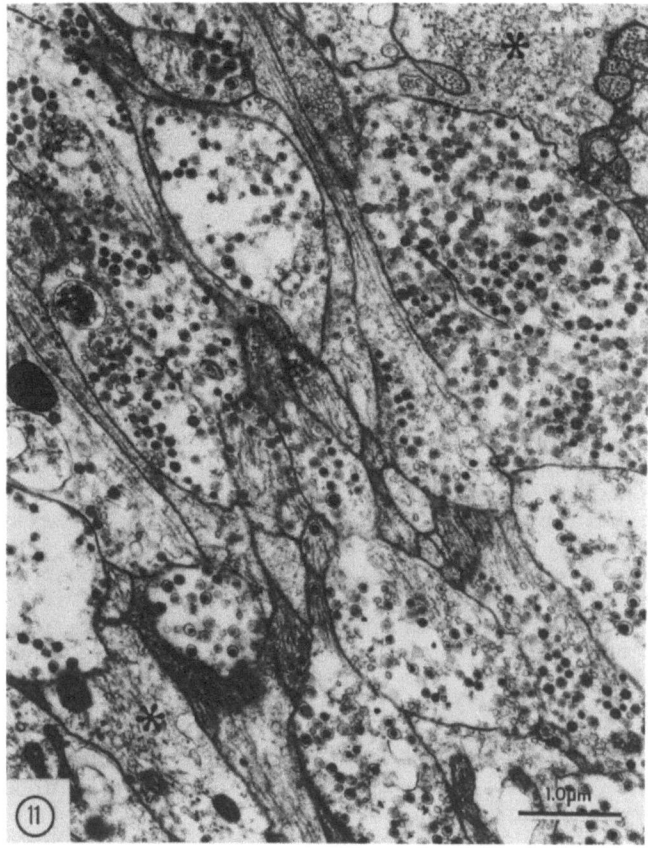

FIG. 11: Electron micrograph of part of the neuropile in the core of the visceral ganglion of the snail, Planorbis corneus. The terminal axonal processes, filled with vesicles, intermingle with preterminal or dendritic processes, which are relatively empty (e.g. asterisks). Note that no glial cell processes can be identified (see text).

withdrawn in steps (vertical scale). At each step the retina was stimulated with a train of light flashes. The left-hand plot shows the changes in $[K^+]_o$ concentration 5 s after the onset of light stimulation. The middle plot shows the extracellular potential changes due to K^+ movement. The K^+-current flows in two loops, entering the glial cells in the active regions, and leaving the glial cells either close to the bath or in the deep layers. The right-hand diagram shows the ways in which the K^+-current loops are completed. The glial cells, coupled by gap junctions, are shown as slabs running vertically between the columns of photoreceptors. The volume of the extracellular space is greatly exaggerated. Broken lines with black arrows indicate directions of current flow and the open arrows show the K^+ flux across the photoreceptor and glial cells in this tissue, although an active K^+ pump may play a minor role. (From Coles and Tsacopoulos 1979, 1981; reproduced from Pentreath, 1982).

molluscs it has been suggested that the glial cells may act together as a united system (Nicaise 1967, 1973). Many of these glial cells contain characteristic electron-dense granules, also termed gliagrana, gliosomes or gliointerstitial granules (Radojcic and Pentreath 1979), which may also bind, store and release Na^+ (Nicaise 1973). These apparently different situations suggest that it may be unwise to assume that the spatial buffering by glial cells around active neurones is always essential.

5.7. Transmitter uptake, inactivation or release

The ability of glial cells to take up and bind different transmitter substances has been recognised from a variety of autoradiographic and biochemical studies. Some fairly detailed evidence is available for glutamate and GABA in arthropods. A selective high-affinity uptake mechanism for glutamate into glia has been demonstrated at the insect neuromuscular junction. Electrical stimulation increases the uptake (Faeder and Salpeter 1970; Salpeter and Faeder 1971). Glutamate is also actively accumulated by glial cells (approximately seven times the neuronal uptake) in crustacean peripheral nerve bundles which indicates that the uptake is a general glial property in arthropods, not restricted to the neuromuscular junction (Evans 1974). It is not clear whether the glutamate is returned to the neurones as glutamine, as is thought to occur in mammals, or whether the uptake is just to inactivate neuronally released transmitter. At the lobster neuromuscular junction, exogenous GABA is taken up principally into the Schwann cells and connective tissue cells. This uptake may serve to inactivate GABA released from inhibitory neurones, or to protect against GABA circulating in the blood (Orkand and Kravitz 1971). GABA is also rapidly accumulated by glial cells in the fly's eye (Campos-Ortega 1974).

An uptake of transmitters into glial cells has been noted in several other invertebrate groups, especially the gastropoda. However, the descriptions are often brief, and are parts of studies orientated to the neurones. There are also inconsistencies regarding the cellular localisation of the uptake of some of the substances in an animal by different groups of workers. These appear to be due to the different concentrations of transmitter substances used for the uptake studies, as well as other differing experimental procedures, which are not discussed further here. A glial uptake has been shown autoradiographically for glutamate, histamine, dopamine, and 5-HT, and in some situations would seem sufficiently clear cut (e.g., Reinecke 1976; Turner and Cottrell 1978) to make further studies regarding the function(s) of the uptake worthwhile.

The glial cells surrounding the squid giant nerve fibres can synthesise and store large quantities of acetylcholine (Villegas 1981; see also above description of the physiological properties of glial cells). A signal (possibly glutamate) is released from the axon during electrical activity, which causes release of the acetylcholine. The released acetylcholine has a feed-back effect on nicotinic receptors on the glial cells, which causes an increase in K^+ conductance, and a long-lasting hyperpolarisation of the cells. The functional significance(s) of this interesting chain of events is not yet clear, but it seems likely that they form part of a complex of biochemical changes that are beneficial when the axon is required to fire at high frequency, during escape (Villegas 1978, 1981).

5.8. Metabolic interactions with neurones

Probably the most widely held belief about the functions of the non-myelin forming glial cells is that they interact metabolically with the neurones, providing nutritive (trophic) supply, exchanging metabolites and removing catabolites. This was first proposed at the beginning of the century (Holmgren 1900; Golgi 1903). The reason for the belief is based largely on intuition; the cells are located between the blood and neurones, forming an apparent uninterrupted layer, across which substances may have to pass before reaching the neurones. Although it is now clear that some substances transfer quickly via the extracellular spaces, rather than across the glial cells (see Kuffler and Nicholls 1976), the original supposition is still generally held, but with increasing evidence to back it.

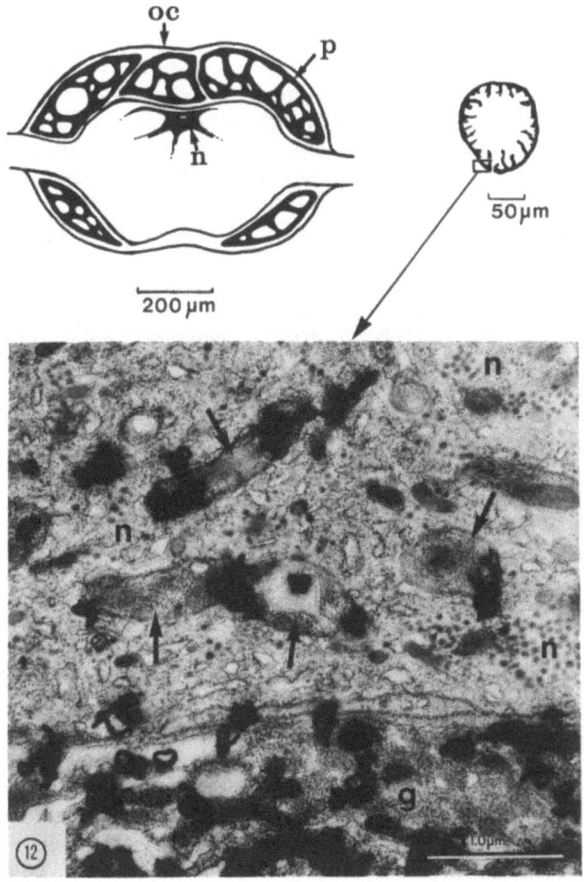

FIG. 12: The arrangement of the giant glial cells (shaded) in the leech segmental ganglia. The packet glial cells (p) encapsulate the neurone perikarya at the edges of the ganglion and neuropile glial cells (n) supply the axonal and dendritic processes in the core. Oc, outer capsule. On the right is an individual neurone soma showing the arrangement of the glial infoldings (trophospongium). These are not shown in the diagram of the whole ganglion. At the bottom is shown an electron microscope autoradiograph of tissue exposed to 10 μm [^{3}H] glucose for 60 min. The radioactive glucose is incorporated intoglycogen within the glial cell (bottom, g) and its processes which invaginate the neurone (arrows). n, neurone cytoplasm, containing vesicles.

In invertebrates the widespread presence of a trophospongium, first described by Holmgren (1901), in many neurones (see Fig. 12) adds a particularly pressing desire to accept a trophic role. Impetus was later provided by Wigglesworth (1960) who noticed that glycogen and lipid reserves in insect neuroglia were depleted during starvation, which gave rise to the hypothesis that neurones acquired food supplied via the invading trophospongium. The impressive extents of the trophospongia in insects has recently been described by Hoyle et al. (1986). In the fly optic lamina the glial trophospongia frequently terminate near the ER in the neurones (Saint-Marie and Carlsson 1983). Sugar molecules are quickly assimilated by cockroach ventral nerve cord and incorporated into amino acids (Treherne 1960), and since the outer perineurial cells are joined by occluding junctions, the sugars must first cross through these cells to reach the neurones. In most invertebrate glial cells, glycogen stores are an obvious feature, as is the case in the astrocytes of the mammalian brain.

Clear-cut evidence for a transfer of materials between glial cells and neurones has been obtained in several preparations. In the squid giant axon, radiolabelled materials are transferred from the surrounding Schwann-like cells. The axons do not contain ribosomes and cannot synthesize proteins. Radioactive amino acids are taken up first by the sheath cells, and later appear in proteins within the axon (Gainer 1978; Gainer et al. 1977; Lasek and Tytell 1981; Lasek et al. 1974, 1977). Injection or perfusion of the amino acids into the axon itself did not produce radiolabelled protein. The transfer appears to be preferential for some proteins (Lasek and Tytell 1981). Over 80 glial polypeptides appear to be involved, with a high transfer rate for actin and a fodrin-like polypeptide (Tytell and Lasek 1984). The mechanism effecting the transfer is not clearly understood, but it may involve an exocytotic/endocytotic (phagocytotic) process (see Fig. 13), or conceivably be via cytoplasmic channels joining the glial cells and axon (Tytell and Lasek 1984).

Similar phenomena have been shown to take place between the sheath glial cells and axons in crustacea. Thus radioactive amino acids, dyes, and horseradish peroxidase are transferred from glia to axons, and from neurone to neurone in the crayfish (see Bittner 1981, for summary). Substances such as Lucifer Yellow may also be passed from axoplasm to the adaxonal glia (Viancour et al. 1981). In the crayfish, pores joining axons with glial cells have been observed, and these may provide for direct communication between the axons and sheath glial cells (Peracchia 1981). The possible mechanisms for transfer of material have more recently been further studied by Cuadras and co-workers (Cuadras 1985, 1986; Cuadras and Marti-Subirana 1985). Perhaps not surprisingly, a variety of possible transfer sites have been demonstrated between glia and neurones. These include transglial channels, gap-like junctions, capitate projections, and modified neuronal/glial endocytosis (Cuadras 1986; see Table 5, Fig. 13). Unfortunately, the functional significances of the transfer processes have yet to be determined; this information is urgently required. Glial-neuronal exchange has been demonstrated in other invertebrates. In the leech segmental ganglia, radiolabelled amino acids are transferred between packet glial cells and neurones (Globus et al. 1973). In <u>Aplysia</u>, horseradish peroxidase injected into a giant 5-HT-containing neurone was transferred directly to some of the satellite glial cells (Goldstein et al. 1982). In another gastropod, <u>Lymnaea</u>, it has been suggested that labelled uridine, or a metabolite, is transferred from glia to neurones (Dyakonova 1972; Dyakonova and Veprintzev 1969). The possible significances of the transfer are again not yet clear.

The presence of large amounts of glycogen, and sometimes lipids, in

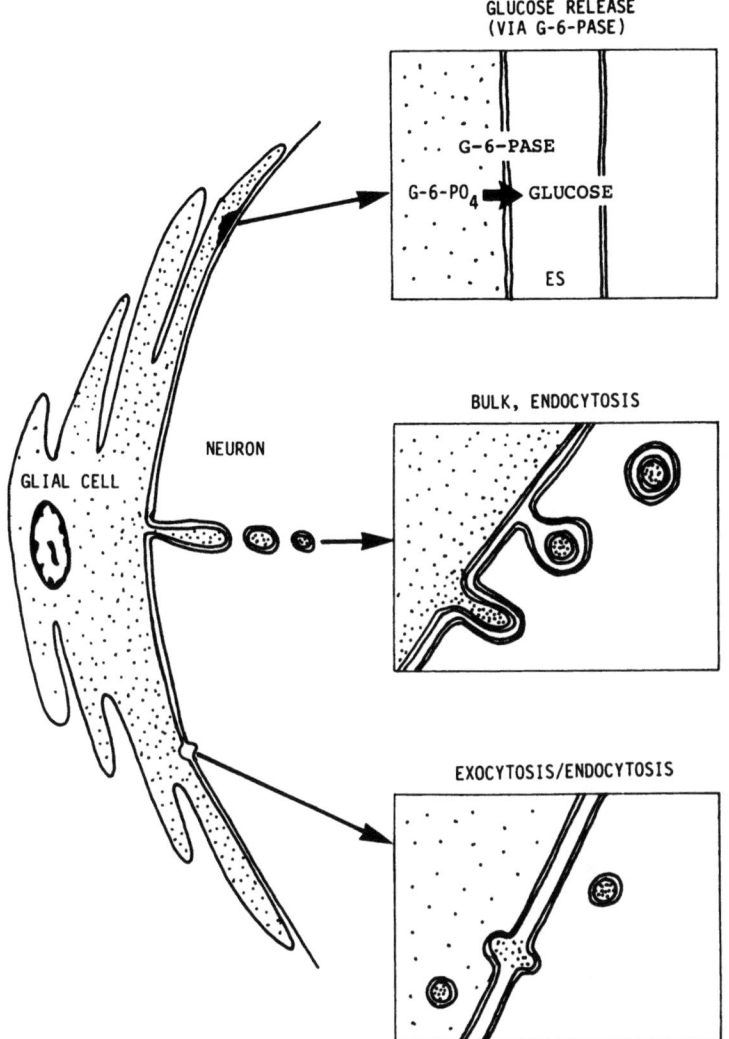

FIG. 13: Possible mechanisms involved in the transfer of materials between glial cells and neurones. Glucose-6-phosphatase (G-6-pase), localized on the glial membranes, may provide glucose to the extracellular spaces surrounding the neurones (top). G-6-pase has been demonstrated on the membranes of the glial cells in the snail, Planorbis (Pentreath et al. 1985), and in the leech, Haemopis (E.K. Winstanley and V. W. Pentreath unpublished observations). Bulk transfer of material (e.g., glycogen or lipids) may take place at the trophospongia in the large neurones of many invertebrates (e.g., annelids, molluscs), as is shown in the centre box. Exocytotic/ endocytotic coupling between neurones and glial cells (bottom box) has been suggested to effect transfer in several arthropods and molluscs (see text). Another possibility (not shown) is that substances may exchange directly via pores joining adjacent glial cells with neurones, as has been suggested in the crayfish (Peracchia 1981).

TABLE 5

STRUCTURES INVOLVED IN MACROMOLECULAR AND IONIC EXCHANGE IN CRUSTACEAN NEURONE CELL BODIES AND AXONS

(From Cuadras 1986)

	Macromolecular exchange		Ionic exchange	
	Direct	Through extracellular space	Direct	Through extracellular space
Neurone cell body	Endocytosis	Trans-glial channels	Gap junctions	
Axon	Pores	Tubular lattices Trans-glial channels Capitate projections Endocytosis	Gap junctions	Tubular lattices Capitate projections

the glial cells of many invertebrates has frequently prompted discussion that the cells provide metabolic reserves (e.g. Fahrenbach 1976; Wigglesworth 1960; Kuffler and Nicholls 1966; Wolfe and Nicholls 1967). Many invertebrate ganglia can survive isolated in the appropriate ionic saline, lacking energy substrate (e.g. glucose) for several hours, and it can be demonstrated by radiolabelling studies that the glycogen is actively synthesised or turned over (e.g. Pentreath 1982; Wolfe and Nicholls 1967). In intact animals, the glycogen stores may vary in quantity with the nutritional state of the animal (e.g. Wigglesworth 1960).

Some recent studies on the leech segmental ganglia and snail buccal ganglia have futher examined the properties of glycogen within the glial cells, as well as the effects of neuronal activity, and signals released from neurones (e.g. K^+, transmitter substances) on the glial glycogen. In the leech, glial processes penetrate the neurone perikarya and major axonal processes (Fig. 12). The glycogen may be marked with [^{3}H]2-deoxyglucose (Kai-Kai and Pentreath 1981b) or by [^{3}H]glucose (Seal and Pentreath 1985; Pentreath et al. 1986a). Antidromic activation of the neurones (which does not pass current across the glial cells) causes marked alterations in the turnover of glycogen in the glial cells (Pentreath and Kai-Kai 1982; A.J. Pennington and V.W. Pentreath unpublished observations). Small increases in external potassium (4mM), which are similar to those produced by repetitive firing of neurones (Kuffler and Nicholls 1966) also cause increased glycogen turnover (Pentreath and Kai-Kai 1982; see Fig. 14). Monoamine transmitters (e.g. 5-HT) cause glycogenolysis within the glial cells (Seal and Pentreath 1985) and these effects are modulated by a variety of neuropeptides (A. J. Pennington unpublished observations). It thus appears that a variety of neuronal signals are capable of altering carbohydrate metabolism in glial cells in intact nervous tissue. In accord with this, synaptic contacts on glial cells have been noted in several situations (e.g. gastropods; Colonnier et al. 1979; Pentreath et al. 1985). It has been suggested that some neurones may regulate energy supply in nervous

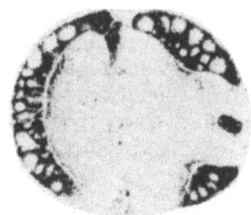
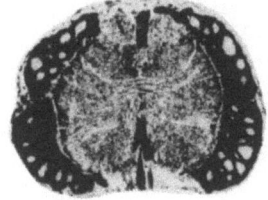
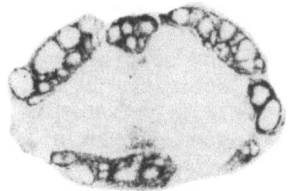

10 μM [^{3}H] GLUCOSE, 4 mM K^+ (60 min)

10 μM [^{3}H] GLUCOSE, 9 mM K^+ (60 min)

10 μM [^{3}H] GLUCOSE, 4 mM K^+ (60 min) POST. INCUB. 9 mM (20 min)

200 μm

FIG. 14: Modulation of glycogen metabolism in the glial cells of the leech, Haemopis sanguisuga, by altered K^+. On the left (control) is a light microscope autoradiograph of a section of a ganglion exposed for 60 min to 10 μM [^{3}H]glucose. This normal saline solution contains 4 mM K^+. When the K^+ is raised to 9 mM (centre) there is increased labelling (i.e., incorporation of [^{3}H]glucose into glycogen) in the glial cells. When the control ganglia were post-incubated for 60 min in 9 mM K^+, without [^{3}H]glucose (right-hand autoradiograph), the labelled glycogen was reduced. Thus elevated K^+ causes an increased turnover of the labelled glycogen.

tissue within domains defined by their projections (Pentreath et al. 1986b) and that the regulation may be both temporal and spatial (see Fig. 15), with the glial cells playing key roles. It seems likely that the transmitters may activate second messenger systems (e.g. cyclic AMP) within the glial cells (Evans et al 1985; Seal and Pentreath 1985). Activation of adenyl cyclase by monoamines has been clearly demonstrated in cultured mammalian astrocytes (see Pentreath et al. 1986). It is not however clear how the mobilised glycogen may reach the neurones, although several possibilities exist, as has been discussed above (see Fig. 13).

MODES OF REGULATION OF ENERGY METABOLISM IN NERVOUS TISSUE

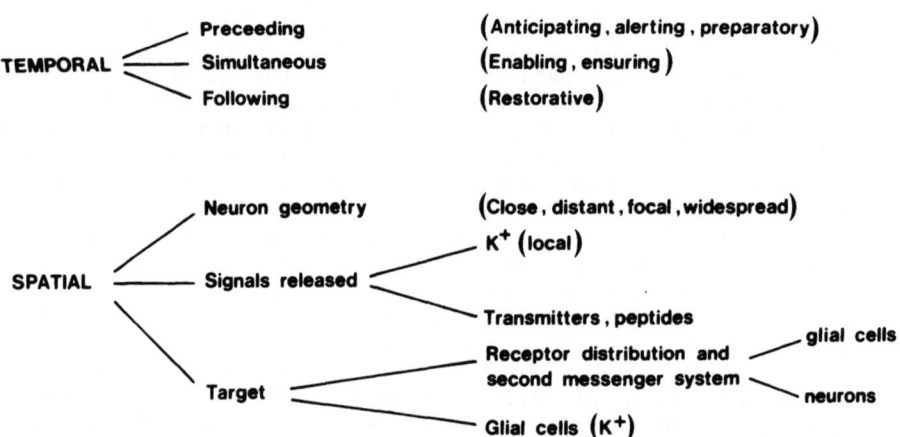

FIG. 15: Diagram summarizing ways in which energy metabolism may be controlled in nervous tissue. The different modulator substances may act synergistically (temporally or spatially), if released from overlapping projections. Each mode of temporal and spatial regulation may operate for increased or exceptional activity, or during normal activity. The modulator substances (e.g., transmit- ters, peptides) may be co-released from the same groups of nerve endings. (From Pentreath et al. 1986b, with permission from Pergamon Journals Ltd.).

In the compound eye of the drone, the mitochondria are located principally in the photoreceptors, and the glycogen within the satellite glial cells (Coles and Tsacopoulos 1981). Activation of the photo-receptors causes a breakdown of the glial glycogen, probably mediated by potassium (Evêquoz 1980; Tsacopoulos 1981).

Elevated external potassium is known to effect the metabolism of cultured glial cells in a variety of ways, as measured by enhanced O_2 consumption, glucose uptake and metabolism, and alterations in ATP concentrations, ATPases and pyruvate kinase. In <u>Necturus</u>, K^+ produces oxidation of NADH (Orkand 1973). There are however no data for comparable effects in invertebrates.

Thus there is growing evidence for a variety of metabolic and trophic interactions between neurones and glial cells. Signals released from the neurones (e.g. K^+ and transmitters) may co-ordinate these interactions in complex spatial and temporal ways. In several invertebrate preparations there is direct evidence for exchange of materials between glial cells and neurones, and for a K^+- and transmitter-mediated mobilization of glycogen reserves in the glial cells. These areas appear to hold great promise for future research into understanding the functions of invertebrate glial cells.

6. CONCLUDING REMARKS

The glial cells are a major component of the nervous systems of higher invertebrates, comprising about half the tissue volume. Much is known about their anatomy, and the varied forms the cells take in different situations. A start has been made in elucidating their physiological properties, which has been aided by the exceptional experimental advantages offered by several preparations. The cells serve a diversity of functions, about which there is increasing interest, and about which the growing information impresses one of the vital part they play in the nervous system.

7. ACKNOWLEDGEMENTS

I thank Kaye Winstanley and Sandra Pugh for expert help. Some of the work discussed in this article was supported by grants from the SERC.

8. REFERENCES

Abbott NJ (1970) Absence of blood-brain barrier in a crustacean, Carcinus maenas L. Nature 225: 291-293.

Abbott NJ (1972) Access of ferritin to the interstitial space of Carcinus brain from intracerabral blood vessels. Tissue Cell 4: 99-104.

Abbott NJ, Treherne JE (1977) Homeostasis of the brain microenvironment: A comparative account. In: Gupta BL, Moreton RB, Oschman JL, Wall BJ (eds) Transport of ions and water in animals. Academic Press, New York, pp 481-510.

Abbott NJ, Bundgaard M, Cserr HF (1982) Experimental study of the blood-brain barrier in the cuttlefish, Sepia officinalis L. J Physiol 326: 43-44.

Abbott NJ, Bundaggard M, Cserr H (1985) Tightness of the blood-brain barrier and evidence for brain interstitial fluid flow in the cuttlefish, Sepia officinalis. J Physiol 368: 213-226.

Abbott NJ, Pichon Y, Lane NJ (1977) Primitive forms of potassium homeostasis: observations on crustacean central nervous system with implications for vertebrate brain. Expt Eye Res Suppl 25: 259-271.

Ashhurst DE, Costin NM (1971) Insect mucosubstances of the central nervous system. Histochem J 3: 297-310.

Baskin DG (1971a) The supporting role of neuroglia in Nereis (Annelida, Polychaeta). Anat Rec 169: 273-274.

Baskin DG (1971b) Fine structure, functional organization and supportive role of neuroglia in Nereis. Tissue Cell 3: 579-588.

Baylor DA, Nicholls JG (1969) Changes in extracellular potassium concentration produced by neural activity in the central nervous system of the leech. J Physiol 293: 555-569.

Becker HW (1965) The number of neurons, glial and perineurium cells in an insect ganglion. Experientia 21: 719-720.

Bittner GD (1981) Trophic interactions of CNS giant axons in crayfish. Comp Biochem Physiol 68A: 299-306.

Bittner GD Mann DW (1976) Differential survival of isolated portions of crayfish axons. Cell Tissue Res 169: 301-311.

Borovyagin VL, Salanki J, Zs.-Nagy I (1972) Ultrastructural alterations in the cerebral ganglion of Anodonta cygnea L. (Mollusca, pelecypoda) induced by transection of the cerebrovisceral connective. Acta Biol Hung 23: 31-45.

Bullock TH, Horridge GA (1965) Structure and function in the nervous system of invertebrates. W.H. Freeman and co., San Francisco, London.

Bullock TH, Orkand R Grinnel A (1977) Introduction to nervous systems. W.H. Freeman and Co., San Francisco.

Campos-Ortega JA (1974) Autoradiographic localization of ^{3}H-γ-aminobutyric acid uptake in the lamina ganglionaris of Musca and Drosophila. Z Zellforsch Mikrosk anat 147: 415-431.

Chapman DM (1974) Cniderian histology. Section 1. In: Muscatin L, Lenhoff HM (eds) Coelenterate biology. Reviews and new perspectives. Academic Press, New York.

Chitwood BG, Chitwood MH (1950) Introduction to nematology. University Park Press, Baltimore, London.

Cobb JLS (1987) Neurobiology of the Echinodermata. (This volume)

Coggeshall RE (1965) A fine structural analysis of the ventral nerve cord and associated sheath of Lumbricus terrestris L. J Comp Neurol 125: 393-438.

Coggeshall RE (1966) The ganglion-connective junction in the central nervous system of the leech Hirudo medicinalis. J Morphol 119: 417-424.

Coggeshall RE, Fawcett DW (1964) The fine structure of the central nervous system of the leech Hirudo medicinalis. J Neurophysiol 27: 229-289.

Coles JA, Orkand RK (1985) Changes in sodium activity during light stimulation in photoreceptors, glia and extracellular space in drone retina. J Physiol 362: 415-435.

Coles JA, Tsacopoulos M (1979) Potassium activity in photorecptors, glial cells and extracellular space in the drone retina: Changes during photostimulation. J Physiol 290: 525-549.

Coles JA, Tsacopoulos M (1981) Ionic and possible metabolic interactions between sensory neurons and glial cells in the retina of the honeybee drone. J Exp Biol 95: 75-92.

Colonnier M, Tremblay JP, McLennan H (1979) Synaptic contacts on glial cells in the abdominal ganglion of *Aplysia californica*. J Comp Neurol 188: 391-400.

Cooke JD, Quastel DMJ (1973) The specific effect of potassium on transmitter release by motor nerve terminals and its inhibition by calcium. J Physiol 228: 435-458.

Cuadras J (1985) A mechanism for macromolecular transfer from glial to neuron cell body in crayfish. Experientia 41: 1590-1591.

Cuadras J (1986) Neuron-glia comunicatory structures in crustaceans. Comp Biochem Physiol 83A: 9-12.

Cuadras J, Marti-Subirana A (1985) Glial cells in abdominal ganglia of crayfish. Acta Zool (Stockh) 66: 217-228.

Cuadras J, Martin G, Czternasty G, Bruner J (1985) Gap-like junctions between neuron cell bodies and glial cells of the crayfish. Brain Res 326: 149-151.

Deitmer JW, Schlue WR (1981) Measurements of the intracellular potassium activity of Retzius cells in the leech central nervous system. J Exp Biol 91: 87-101.

Drummond GI, Eng DY, McIntosh CA (1971) Ribonucleoside 2',3'-cyclic phosphate diesterase activity and cerebroside levels in vertebrate and invertebrate nerve. Brain Res 28: 153-163.

Dyakonova TL (1972) Activation of RNA synthesis in glial satellite cells during electrical activity of neuron. Tsitologiya 14: 1147-1155.

Dyakonova TL, Veprintzev BN (1969) Particularities of structural and functional organization and metabolic activity of neurons of Lymnaea. Acad Sci USSR Dept Biophys, Puschino, USSR.

Edwards JS (1980) Neuronal guidance and pathfinding in the developing sensory nervous system of insects. In: Locke M, Smith DS (eds) Insect biology in the future -VBW 80. Academic Press, London, pp 667-683.

Eldefrawi ME, O'Brien RD (1967) Permeability of the adbominal nerve cord of the American cockroach, *Periplaneta americana* (L.), to aliphatic alcohols. J Insect Physiol 13: 391-398.

Elliott EJ, Muller KJ (1981) Long-term survival of glial segments during nerve regeneration in the leech. Brain Res 218: 99-113.

Elliott EJ, Muller KJ (1982) Synapses between neurons regenerate accurately after destruction of ensheathing glial cells in the leech. Science 215: 1260-1262.

Evans PD (1974) An autoradiographical study of the localization of the uptake of glutamate by the peripheral nerves of the crab, *Carcinus maenas* (L.). J Cell Sci 14: 315-367.

Evans PD, Reale V, Villegas J (1985) The role of cyclic nucleotides in modulation of the membrane potential of the Schwann cell of squid giant nerve fibre. J Physiol 363: 151-167.

Evêquoz V, Stadelmann A, Tsacopoulos M (1983) The effect of light on glycogen turnover in the retina of the intact honeybee drone (Apis mellifera). J Comp Physiol 150: 69-75.

Faeder IR, Salpeter MM (1970) Glutamate uptake by a stimulated insect nerve muscle preparation. J Cell Biol 46: 300-307.

Fahrenbach WH (1976) The brain of the horseshoe crab (Limulus polyphemus). I. Neuroglia. Tissue Cell 8: 395-410.

Gainer H (1978) Intercellular transfer of proteins from glial cells to axons. Trends Neurosci 1: 93-96.

Gainer H, Tasaki I, Lasek RJ (1977) Evidence for the glia-neuron protein transfer hypothesis from intracellular perfusion studies of squid axons. J Cell Biol 74: 524-530.

Gardner-Medwin AR (1983a) A study of the mechanisms by which potassium moves through brain tissue in the rat. J Physiol 335: 353-374.

Gardner-Medwin AR (1983b) Analysis of potassium dynamics in mammalian brain tissue. J Physiol 335: 393-426.

Gardner-Medwin AR, Nicholson C (1983) Changes of extracellular potassium activity induced by electric current ghrough brain tissue in the rat. J Physiol 335: 375-392.

Gardner-Medwin AR, Coles JA, Tsacopoulos M (1981) Clearance of extracellular potasium: evidence for spatial buffering by glial cells in the retina of the drone. Brain Res 209: 452-457.

Globus A, Lux HD, Schubert P (1973) Transfer of amino acids between neuroglia cells and neurons in the leech ganglion. Exp Neurol 40: 104-113.

Goldschmidt R (1910) Das Nervensystem von Ascaris lumbricoides and megalocephala, III. Fest Hertwigs 2: 253-354.

Goldstein RS, Weiss KR, Schwartz JH (1982) Intraneuronal injection of horseradish peroxidase lables glial cells associated with the axons of the giant metacerabral neurons of Aplysia. J Neurosci 2: 1567-1577.

Golgi C (1903) Opera Omnia, Vols I and II. U. Hoepli, Milan.

Gould RM, Pant H, Gainer H, Tytell M (1983) Phospholipid synthesis in the squid giant axon: incorporation of lipid precursors. J Neurochem 40: 1293-1299.

Gray EG (1969) Electron microscopy of the glio-vacular organization of the brain of Octopus. Phil Trans Soc B 255: 13-32.

Gray EG, Guillery RW (1963) An electron microscope study of the ventral nerve cord of the leech. Z. Zellforsch Mikrosk Anat 60: 826-849.

Griffiths G (1979) Transport of glial acid phosphatase by endoplasmic reticulum into damaged axons. J Cell Sci 36: 361-389.

Günther J (1979) Impulse conduction in the myelinated giant fibres of the earthworm. Structure and function of the dorsal nodes in the median giant fibre. J Comp Neurol 168: 505-532.

Gymer A, Edwards JS (1967) The development of the insect nervous system. I. An analysis of postembryonic growth in the terminal ganglion of Acheta domesticus. J Morphol 123: 191-198.

Hama K (1959) Some observations on the fine structure of the giant nerve fibers of the earthworm, Eisenia foetida. J Biophys Biochem Cytol 6: 61-66.

Hertz L, Nissen C (1976) Diferences between leech and mammalian nervous systems in metabolic reaction to K^+ as an indication of differences in potassium homeostasis mechanisms. Brain Res 110: 182-188.

Heumann R, Villegas J, Hertzfeld DW (1981) Acetylcholine synthesis in the Schwann cell and axon in the giant nerve fibre of the squid. J Neurochem 36: 765-768.

Heuser JE, Doggenweiler CF (1966) The fine structural organization of nerve fibers, sheaths, and glial cells in the prawn, Palaemonetes vulgaris. J Cell Biol 30: 381-403.

Holmgren E (1900) Weitere Milleilungen uber die 'Saftkanalchen' der Nervenzellen. Anat Anz 18: 290-296.

Holmgren E (1901) Beitrage zur Morphologie der Zelle. I. Nervenzellen. Anat Hefte 18: 269-325.

Horridge GA, Mackay B (1962) Naked axons and symmetrical synapses in coelenterates. Quart J Microsc Sci 103: 531-541.

Hoyle G (1986) Glial cells of an insect ganglion. J Comp Neurol 246: 85-103.

Hoyle G, Williams M, Phillips C (1986) Functional morphology of insect neuronal cell-surface/glial contacts: the trophospongium. J Comp Neurol 246: 113-128.

Kai-Kai MA, Pentreath VW (1981a) The structure, distribution and quantitative relationships of the glia in the abdominal ganglia of the horse leech, Haemopis sanguisuga. J Comp Neurol 202: 193-210.

Kai-Kai MA, Pentreath VM (1981b) High resolution analysis of [^{3}H]2-deoxyglucose incorporation into neurons and glial cells in invertebrate ganglia: histological processing of nervous tissue for selective marking of glycogen. J Neurocytol 10: 693-708.

Kuffler SW (1967) Neuroglial cells: physiological properties and a potassium mediated effect of neuronal activity on the glial membrane potential. Proc R Soc B 168: 1-21.

Kuffler SW, Nicholls JG (1966) The physiology of neuroglial cells. Ergebn Physiol 57: 1-90.

Kuffler SW, Nicholls JG (1976) From neuron to brain. Sinauer, Sunderland, Mass.

Kuffler SW, Potter DD (1964) Glia in the leech central nervous system: physiological properties and neuron-glia relationship. J Neurophysiol 27: 290-320.

Lane NJ (1972) Fine structure of a lepidopteran nervous system and its accessibility to peroxidase and lanthanum. Z Zellforsch Mikrosk Anat 131: 205-222.

Lane NJ (1979) Intramembranous particles in the form of ridges, bracelets or assemblies in arthropod tissues. Tissue Cell 11: 1-18.

Lane NJ (1981) Invertebrate neuroglia; junctional structure and development. J Exp Biol 95: 7-33.

Lane NJ, Abbott NJ (1975) The organization of the nervous system in the crayfish Procambarus clarkii, with emphasis on the blood-brain interface. Cell Tissue Res 156: 173-187.

Lane NJ, Treherne JE (1980) Junctional organization of arthropod neuroglia. In: Locke M, Smith DS (eds) Insect biology in the future VBW 80. Academic Press, London, New York, pp 765-795.

Lasek RJ, Tytell MA (1981) Macromolecular transfer from glia to the axon. J Exp Biol 95: 153-165.

Lasek RJ, Gainer H, Barker JL (1977) Cell-to-cell transfer of glial proteins to the squid giant axon. The glia-neuron protein transfer hypothesis. J Cell Biol 74: 501-523.

Lasek RJ Gainer H, Przybylski RJ (1974) Transfer of newly synthesized proteins from Schwann cells to the squid giant axon. Proc Nat Acad Sci USA 71: 1188-1192.

Lieberman EM, Villegas J, Villegas GM (1981) The nature of the membrane potential of glial cells associated with the medial giant axon of the crayfish. Neurosci 6: 261-271.

Lopresti V, Macagno ER, Levinthal C (1973) Structure and development of neuronal connections in isogenic organisms: cellular interactions in the development of the optic lamina of Daphnia. Proc Nat Acad Sci USA 70: 433-437.

Morgese VJ, Elliott EJ, Muller KJ (1983) Microglial movement to sites of nerve lesion in the leech CNS. Brain Res 272: 166-170.

Muller KJ, Nicholls JG, Stent GS (1981) Neurobiology of the leech. Cold Spring Harbor Laboratory, New York.

Nicaise G (1967) Description d'un "systeme glio-interstitiel" chez Glossodoris (Gasteropode: Opisthobranche). CR Acad Sci Paris D 246: 2793-2795.

Nicaise G (1973) The gliointerstitial system of molluscs. Int Rev Cytol 34: 251-332.

Nicholson C (1980) Dynamics of brain cell microenvironment. Neurosci Res Prog Bull 18(2): 177-322.

Okamura N, Stoskopf M, Yamaguchi H, Kishimoto Y (1985) Lipid composition of the nervous system of earthworms (Lumbricus terrestris). J Neurochem 45: 1875-1879.

Oksche A (ed) (1980) Neuroglia I. Hand Mikrosk Anat des Menschen Vol IV, part 10. Springer-Verlag, Berlin.

Orkand P, Kravitz EA (1971) Localization of the sites of γ-amino butyric acid (GABA) uptake in lobster nerve-muscle preparations. J Cell Biol 49: 75-89.

Orkand RK, Nicholls JG, Kuffler SW (1966) Effect of nerve impulses on the membrane potential of glial cells in the central nervous system of amphibia. J. Neurophysiol 19: 788-806.

Orkand PM, Bracho H, Orkand RK (1973) Glial metabolism: alteration by potassium levels comparable to those during neural activity. Brain Res 55: 467-471.

Pentreath VW (1982) Potassium signalling of metabolic interactions between neurons and glial cells. Trends Neurosci 5: 339-345.

Pentreath VW, Cobb JLS (1972) Neurobiology of echinodermata. Biol Rev 47: 363-392.

Pentreath VW, Cottrell GA (1970) The blood supply to the central nervous system of Helix pomatia. Z Zellforsch Mikrosk Anat 111: 160-178.

Pentreath VW, Kai-Kai MA (1982) Significance of the potassium signal from neurons to glial cells. Nature 295: 59-61.

Pentreath VW, Radojcic T, Seal LH, Winstanley, EK (1985) The glial cells and glia-neuron relations in the buccal ganglia of Planorbis corneus (L.): cytological, qualitaive and quantitative changes during growth and ageing. Phil Trans R Soc Lond B 307: 399-455.

Pentreath VW, Pennington AJ, Seal LH, Swift K (1986a) Modulation by neuronal signals of energy substrate in the glial cells of leech segmental ganglia. In: Althaus H, Seifert W (eds) Glial-neuronal communication in development and regeneration. Plenum, New York (in press).

Pentreath VW, Seal LH, Morrison JH, Magistretti PJ (1986b) Transmitter mediated regulation of energy metabolism in nervous tissue at the cellular level. Neurochem Int. 9: 1-10.

Perrachia C (1981) Direct communication between axons and sheath glial cells in crayfish. Nature 290: 597-598.

Pipa RL (1967) Insect neurometamorphosis. III. Nerve cord shortening in a moth, Galleria mellonella (L.), may be accomplished by humoral potentiation of neuroglial motility. J Exp Zool 164: 47-60.

Pipa RL (1973) Proliferation, movement and regression of neurons during the postembryonic development in insects. In: Young D (ed) Developmental neurobiology of arthropods. Cambridge University Press, London, pp 105-129.

Radojcic T, Pentreath VW (1979) Invertebrate glia. Prog Neurobiol 12: 115-179.

Radojcic T, Pentreath VW (1981) Quantitative analysis of neuron-glial relationships in the buccal ganglion of Planorbis: life constancy in the absence of changes in functional output. Brain Res 211: 468-475.

Rakic P (1974) Intrinsic and extrinsic factors influencing the shape of neurons and their assembly into neuronal circuits. In: Seeman P, Brown GM (eds) Frontiers in neurology and neuroscience research. Toronto Press, Toronto pp 112-132.

Ramón Y Cajal S, Sanchez D (1915) Contribution al conocimiento de los centros nerviosos de los insectos, Parte 1, Retina y centros opticos. Trab Lab Invest Biol Univ Madrid 13: 1-168.

Reale V, Evans PD, Villegas J (1986) Octopaminergic modulation of the membrane potential of the Schwann cell of the squid giant nerve fibre. J Exp Biol 121: 421-443.

Reinecke M (1975) Die Gliazellen der Cerebralganglien von Helix pomatia L. (Gastropoda: Pulmonata). I. Ultrastruktur und organisation. Zoomorphol 82: 105-136.

Reinecke M (1976) The glial cells of the cerebral ganglia of Helix pomatia L. (Gastropoda, pulmonata). II. Uptake of ferritin and ^3H glutamate. Cell Tissue Res 169: 361-382.

Roots BI (1978) A phylogenetic approach to the anatomy of glia. In: Schoffeniels E, Franck B, Hertz L, Tower LDB (eds) Dynamic properties of glial cells. Pergamon Press, New York, pp 45-54.

Roots BI (1981) Comparative studies on glial markers. J Exp Biol 95: 167-180.

Saint-Marie RL, Carlson SD (1983) The fine structure of glia in the lamina ganglionaris of the housefly, Musca domestica L. J Neurocytol 12: 243-275.

Saint-Marie RL, Carlson SD (1985) Interneuronal and glial-neuronal gap junctions in the lamina ganglionaris of the compound eye of the housefly, Musca domestica. Cell Tissue Res 241: 43-52.

Saint-Marie RL, Carlson SD, Chi C (1984) The glial cells of insects. In: King RC, Akaai H (eds) Insect ultrastructure, Vol. 2 Plenum Press, New York, pp 435-475.

Salpeter MM, Faeder IR (1971) The role of sheath cells in glutamate uptake by insect nerve muscle preparations. Prog Brain Res 34: 103-114.

Saubermann AJ, Scheid VL (1985) Elemental composition and water content of neuron and glial cells in the central neuvous system of the North American medicinal leech (Macrobdella decora). J Neurochem 44: 825-834.

Schacher SM (1981) The role of support cells in the growth and differentiation of neurones in the abdominal ganglion of Aplysia. J Exp Biol 95: 205-214.

Schlue WR, Deitmer JW (1980) Extracellular potassium in neuropile and nerve cell body region of the leech central nervous system. J Exp Biol 87: 23-43.

Schlue WR, Walz W (1984) Electrophysiology of neuropile glial cells in the central nervous system of the leech. A model system for potassium homeostasis in the brain. Cell Neurobiol 5: 143-175.

Schofield PK, Treherne JE (1985) Octopamine reduces potassium permeability of the glia that form the insect blood-brain barrier. Brain Res 360: 344-348.

Seal LH, Pentreath VW (1985) Modulation of glial glycogen metabolism by 5-hydroxytryptamine in leech segmental ganglia. Neurochem Int 7: 1037-1045.

Shimomura K, Hajura S, Ki PF, Kishimoto Y (1983) An unusual glucocerebroside in the crustacean nervous system. Science 220: 1392-1393.

Skaer HL, Lane NJ (1974) Junctional complexes, perineurial and glial-axonal relationships, and the ensheathing structures of the insect nervous system; a comparative study using conventional and freeze-cleaving techniques. Tissue Cell 6: 695-718.

Smith DO (1983) Extracellular potassium levels and axon excitability during repetitive action potentials in crayfish. J Physiol 336: 143-157.

Smith PJS, Leech CA, Treherne JE (1984) Glial repair in an insect central nervous system: effects of selective glial disruption. J Neurosci 4: 2698-2711.

Smith PJS, Howes EA, Leech CA, Treherne JE (1986) Haemocyte involvement in the repair of the insect central nervous system after selective glial disruption. Cell Tissue Res 243: 367-374.

Strausfeld NJ (1976) Atlas of an insect brain. Springer-Verlag, Berlin.

Swales LS, Lane NJ (1985) Embryonic development of glial cells and their junctions in the locust central nervous system. J Neurosci 5: 117-128.

Taylor DP, Dyer KA, Newburgh RW (1976) Cyclic nucleotides in neuronal and glial-enriched fractions from the nerve cord of _Manduca sexta_. J Insect Physiol 22: 1303-1304.

Treherne JE (1960) The nutrition of the central nervous system in the cockroach, Periplaneta americana L. The exchange and metabolism of sugars. J Exp Biol 27: 513-533.

Treherne JE, Schofield PK (1978) A model for extracellular sodium regulation in the central nervous system of an insect, _Periplaneta americana_. J Exp Biol 77: 251-254.

Treherne JE, Schofield PK (1981) Mechanisms of ionic homeostasis in the central nervous system of an insect. J Exp Biol 95: 61-73.

Treherne JE, Schofield PK, Lane NJ (1982) Physiological and ultrastructural evidence for an extracellular anion matrix in the central nervous system of an insect (_Periplaneta americana_). Brain Res 247: 255-267.

Treherne JE, Harrison JB, Treherne JM, Lane NJ (1984) Glial repair in an insect central nervous system: effects of surgical lesioning. J Neurosci 4: 2689-2697.

Tsacopoulos M, Evêquoz V (1980) L'effet de la stimulation photique sur le metabolisme du glycogene intraretinien. Klin Mbl Augenheilk 176: 519-521.

Tsacopoulos M, Poitry A, Borsellino A (1981) Diffusion and comsumption of oxygen in the superfused retina of the drone (_Apis mellifera_) in darkness. J Gen Physiol 77: 601-628.

Turner JD, Cottrell GA (1978) Cellular uptake of amines and amino acids in the central ganglion of a gastropod mollusc <u>Planorbis</u> <u>corneus</u>: an autoradiographic study. J Neurocytol 7: 759-776.

Tytell M, Lasek RJ (1984) Glial polypeptides transferred into the squid giant axon. Brain Res 324: 223-232.

Varon SS, Somjen GG (1979) Neuron-glia interactions. Neurosci Res Prog Bull 17(1):6-239.

Viancour TA, Bittner GD, Ballinger ML (1981) Selective transfer of Lucifer Yellow CH from axoplasm to adaxonal glia. Nature 293: 65-67.

Villegas GM, Villegas J (1974) Acetylcholinesterase localization in the giant nerve fibre of the squid. J Ultrastruct Res 46: 149-156.

Villegas J (1974) Effects of actylcholine and carbamylcholine on the axon and Schwann cell electric potentials in the squid nerve fibre. J Physiol 242: 647-659.

Villegas J (1975) Characterization of acetylcholine receptors in the Schwann cell membrane of the squid nerve fibre. J Physiol 249: 679-689.

Villegas J (1978) Cholinergic systems in axon-Schwann cell interactions. Trends Neurosci 1: 66-69.

Villegas J (1981) Axon/Schwann-cell relationships in the giant nerve fibre of the squid. J Exp Biol 95: 135-151.

Villegas J (1984) Axon-Schwann cell relationship. Curr Topics Membr Transport 22: 547-571.

Villegas J, Jenden DJ (1979) Acetylcholine content of the Schwann cell and axon in the giant nerve fibre of the squid. J Neurochem 32: 761-766.

Villegas J, Villegas GM (1978) Functional mechanisms in axon-Schwann cell relationships in the squid nerve fiber. Acta Cientif Venez 29: 291-294.

Walz W (1982) Do neuronal signals regulate potassium flow in glial cells? Evidence from an invertebrate central nervous system. J Neurosci Res 7: 71-79.

Walz W, Hertz L (1983) Functional interactions between neurons and astrocytes. II. Potassium homeostasis at the cellular level. Pro Neurobiol 20: 133-183.

Walz W, Schlue WR (1982a) External ions and membrane potential of leech neuropile glial cells. Brain Res 239: 119-138.

Walz W, Schlue WR (1982b) Ionic mechanism of a hyperpolarizing 5-hydroxytryptamine effect on leech neuropile glial cells. Brain Res 250: 111-121.

Wigglesworth VB (1959) The histology of the nervous system of an insect, <u>Rhodnius</u> <u>prolixus</u> (Hemiptera). II. The central ganglia. Quart J Microsc Sci 100: 299-313.

Wigglesworth VB (1960) The nutrition of the central nervous system in the cockrocah <u>Periplaneta</u> <u>americana</u> L. The role of perineurium and glial cells in the mobilization of reserves. J Exp Biol 37: 500-512.

Winstanley EK, Pentreath VW (1985) Lipofuscin accumulation and its prevention by Vitamin E in nervous tissue: quantitative analysis using snail ganglia as a simple model system. Mech Ageing Develop 29: 299-307.

Wolfe DE, Nicholls JG (1967) Uptake of radioactive glucose and its conversion to glycogen by neurons and glial cells in the leech central nervous system. J Neurophysiol 30: 1593-1609.

NEUROPEPTIDES IN INVERTEBRATES

C.J.P. GRIMMELIKHUIJZEN[1], D. GRAFF[1], A. GROEGER[1]
AND I. D. MCFARLANE[2]

[1]Zoological Institute, University of Heidelberg,

Im Neuenheimer Feld 230, 6900 Heidelberg,

Federal Republic of Germany and

[2] Department of Zoology, University of Hull,

Hull HU6 7RX, England

ABSTRACT

This chapter is divided into three parts. The first part is a general survey, describing briefly the discovery of some neuropeptides and what can be learned about their occurrence, biosynthesis, action etc. The first part will mainly deal with vertebrate neuropeptides. The second part gives an up-to-date list and a discussion of established invertebrate neuropeptides. The last part deals with the structure, localisation and action of neuropeptides in cnidarians.

1. GENERAL FEATURES OF NEUROPEPTIDE BIOLOGY

The structure of the first two sequenced neuropeptides, oxytocin and vasopressin, were published somewhat more than 30 years ago by Du Vigneau, Acher and coworkers (Du Vigneau et al. 1953 a, b; Acher and Chauvet 1953). These substances were purified from extracts of cow-pituitary gland, using milk ejection in rat as a bioassay for oxytocin, and blood-pressure rise and antidiuretic activity as bioassays for vasopressin. Because immunocytochemical techniques had not been sufficiently developed during that time, it took more than 20 years before the neuronal localisation of these peptides could be established. Oxytocin and vasopressin were found to be produced by separate populations of magnocellular neurones within the nucleus paraventricularis and nucleus supra-opticus of the hypothalamus (Vandesande and Dierickx 1975). From these locations, the peptidergic neurones project to the neurohypophysis (posterior putuitary), where they make contact with blood vessels, and, upon adequate stimuli, release their products into the blood stream (Brownstein et al. 1980). This example illustrates that neuropeptides can be genuine hormones. Later, it was found that the same hypothalamic nuclei (but not the same neurones) project oxytocin- and vasopressin-immunoreactive fibres to other brain regions, such as the limbic system and medulla oblongata, where the peptides are released at synaptic loci (Buijs 1978; Sofroniew and Weindl 1978; Mühletaler et al. 1982). This illustrates that the same peptide can be used as a neurohormone or as a neurotransmitter. Also an intermediate action of these peptides is possible: they may be released from neurones into the intercellular space, diffuse over a short distance and then act on target cells. Such an action is called "paracrine". There is no essential difference between the action of neuropeptides, whether they are released at synapses, into the intercellular space, or into the blood stream. The only difference is the distance over which they are transported, which has, of course, an influence on the onset and termination of their actions. For peptides used as neurotransmitters, the onset can be very short and comparable with that of the "classical" transmitters (Barker et al. 1980).

Peptides do not always play a direct role in signal transmission. There are also situations in which they modulate the efficiency of a synapse. Such a modulatory action often affects the presynaptic membrane by altering its response (transmitter release) upon arrival of action potentials. When a peptide acts as a modulator, its application alone, at a synaptic locus, will not elicit postsynaptic potentials.

It will be clear that the action of neuropeptides is not significantly different from that of the "classical" neurotransmitters. Adrenaline/noradrenaline are neurotransmitters in the vertebrate brain, but also hormones which are released from the adrenal medulla. Serotonin can be a genuine neurotransmitter, but also a modulator increasing the synaptic efficiency (Klein et al. 1982).

Oxytocin (sequence: CYIQNCPLGamide; see footnote of Table I for a list of one-letter abbreviations of amino acids) and vasopressin (CYFQNCPRGamide) are nonapeptides which differ only in two positions of their amino acid sequence (positions 3 and 8). This illustrates that, within one organism, a "family" of peptides can exist, and that members of this family can have totally different target cells (receptors) and therefore different actions. Not only within one organism, but also throughout the animal kingdom, families of neuropeptides occur. A variety of oxytocin- and vasopressin-like peptides, for example, which are all very similar in structure, has been isolated and sequenced from mammals, birds, reptiles, amphibians and fishes (Acher 1981). The existence of a family of neuropeptides within one organism can be explained by gene duplication of an ancestral gene, followed by mutation.

In the sixties, three more neuropeptides, the "hypothalamic-releasing-factors", were isolated and sequenced by the groups of Guillemin and Schally (Guillemin and Burgus 1972; Guillemin 1978; Schally 1978). Thyroxin-releasing-hormone (sequence: pQHPamide) stimulates the release of thyroid-stimulating-hormone and prolactin from endocrine cells of the adenohypophysis (anterior pituitary), luteinizing-hormone-releasing-hormone (pQHWSYGLRPGamide) stimulates the release of luteinizing hormone and follicle-stimulating-hormone, and somatostatin (AGCKNFFWKTFTSC) inhibits the release of growth hormone from the adenohypophysis. Like many other peptides isolated later, thyroxin-releasing- and luteinizing-hormone-releasing-hormone contain a pyroglutamate at the aminoterminus and an amidation at the carboxyterminus of their amino-acid sequence. Pyroglutamate is a cyclisation product of glutamine. It does not contain a primary amine group, and peptides containing pyroglutamate are not sensitive to digestion by "normal" aminopeptidases. Similarly, an amidation protects the peptide from breakdown by "normal" carboxy-peptidases. The presence of pyroglutamate and an amidation, therefore, extends the half-life of the peptide.

Thyroxin-releasing-hormone, luteinizing-hormone-releasing-hormone and somatostatin have been localised by immunocytochemistry in hypothalamic neurones which were contacting blood capillaries descending into the adenohypophysis (Schally 1978). These peptides, however, were also found in other brain regions and in the peripheral (autonomic) nervous system (Hökfelt et al. 1975, 1980). In these regions, the peptides have neurotransmitter or neuromodulatory functions. In addition, somatostatin was also found in endocrine cells of the stomach, intestinal tract and pancreas (Dubois 1975; Guillemin 1978). This showed for the first time that neuropeptides can be produced by endocrine cells. The reverse situation, the presence of gastro-intestinal hormones in the brain, was shown first by Vanderhaeghen and coworkers (1975). These scientists, using a radioimmunoassay for gastrin, found a small gastrin-like peptide in brain extracts of humans and other vertebrates. This material was later purified and sequenced by Dockray and collaborators (1978). The gastrin-like material turned out to be the carboxyterminal octapeptide of cholecystokinin (DYMGWMDFamide), another gastrointestinal peptide which shares a carboxyterminal sequence (GWMDFamide) with gastrin. This example, then, shows that families of peptides exist in both the nervous and endocrine tissue.

Nowadays it has become evident that nearly all peptides, which were originally discovered in endocrine tissue, have their counterparts in the central and peripheral nervous system, and vice versa, the brain peptides have their counterparts in endocrine cells. Some exceptions, however, remain, as with vasoactive-intestinal-polypeptide, which only occurs in neurones.

A third source of neuropeptide-like molecules is the amphibian skin. Starting in the early sixties, Italian pharmacologists isolated substances from the skins of frogs and toads, which had extremely potent actions when injected into mammals (see Esparmer and Melchiorri 1973, for a review). Many of these substances were peptides and were named after the species they were isolated from, thereby obtaining beautiful exotic names such as caerulein (from Hyla caerulea) and bombesin (from Bombina bombina). Caerulein, when injected into mammals, gives a strong contraction of the gall bladder (one amphibian skin contains up to 3 mg of this peptide, which is sufficient for a severe gall-bladder contraction of 30,000 humans!). This effect of caerulein (pQQDYTGWMDFamide) is not astonishing, as it shows a high degree of similarity with the carboxyterminus of cholecystokinin (see above). Bombesin causes the release of gastrin (and therefore the release of hydrochloric acid) when injected into mammals. An authentic gastrin-releasing-peptide has more recently been isolated from porcine gastro-intestinal tissue (McDonald et al. 1979). This peptide, which is 27 amino acids long, shares the carboxyterminal 8 amino-acid sequence with bombesin, which explains why the actions of these two substances are so similiar.

In the following we would like to describe shortly the biochemistry of neuropeptides. Neuropeptides are synthesised as part of a large precursor molecule on ribosomes of the rough endoplasmic reticulum (Nakanishi et al. 1979; Brownstein et al. 1980; Amara et al. 1982; Herbert 1981; Land et al. 1982). This precursor is translocated across the membrane of the endoplasmic reticulum with the aid of a "signal" region, which, during or after translocation, is cleaved off (Blobel and Dobberstein 1975; Nakanishi et al. 1980). After transport from the endoplasmic reticulum to the Golgi system the precursor is packed into vesicles together with processing enzymes (Palade 1975; Brownstein et al. 1980; Brownstein 1982). The vesicles then move along the neuronal processes (fast axonal transport) and "maturate". Among the processing enzymes within the vesicles are proteases, which cleave the precursor at the carboxyterminal side of a pair of basic amino acids (arginine and lysine). After this step, a specific carboxypeptidase (a carboxypeptidase B-like enzyme) becomes active, which removes the two carboxyterminal, basic amino acids of each fragment (Brownstein et al. 1980; Fricker and Snyder 1982). When an amidation of the carboxyterminal amino acid of the peptide is required (see above), this amino acid is always followed by glycine inside the precursor. A specialised enzyme, then, degrades the glycine, but keeps the amide for amidation of the biological active peptide (Bradbury et al. 1982; Sakata et al. 1986). As mentioned above, pyroglutamate is formed by cyclisation from glutamine. Whether this occurs spontaneously in the vesicles, or is catalysed by an enzyme, is not known. It is not so that always at a site of two basic amino acids, processing does necessarily occur. These sites are only potential cleavage sites. The brain-gut peptide neurotensin, for example, contains two adjacent arginines, where processing, obviously, has not taken place. There are also examples known (the FMRFamide precursor, see below), where processing occurs at a single basic amino acid (Schaefer et al. 1985).

The precursor molecule can contain one or more copies of the biologically active peptide. In a FMRFamide precursor of <u>Aplysia californica</u>, 28 copies of FMRFamide are present (Schaefer et al. 1985; Scheller 1986), which is the highest number of peptide copies found so far in any precursor molecule. Sometimes two or more peptides of the same family are present in the same precursor. Examples of this are substance P and neurokinin A in mammals (Nawa et al. 1983) and small cardioactive peptide -A and -B in snails (Mahon et al. 1985).

Some precursors are differentially processed in different tissues. The precursor pro-opiomelanocortin, for example, yields corticotropin and β-lipotropin in the neurohypophysis, but α-, β- and γ-melanotropin and β-endorphin in the intermediate lobe (Smyth and Zakarian 1980; Herbert 1981). This differential processing probably reflects the presence of different, tissue-specific enzymes packed in the secretory vesicles.

The formation of mRNA for the precursor molecule can involve differential splicing of the primary DNA transcript. In this way, two different neuropeptides, calcitonin and calcitonin-gene-related-product, can be formed in different tissues from the same gene (Amara et al. 1982). So far, however, differential splicing does not appear to be common during formation of peptide-precursor mRNA.

After release at synaptic loci, into the intercellular space, or into the blood stream, the peptides eventually bind to specific receptors. There, the peptides always exert their action via a "second messenger" system. This involves the activation of adenylate cyclase, the production of cAMP, and the phosphorylation of a membrane protein (the ion channel) by a specific protein kinase. This process is similar to one used by catecholamines and serotonin.

Not much is known about the enzymatic breakdown of neuropeptides. It obviously involves peptidases located near the peptide receptors but, so far, no enzyme specific for one of the peptides has been isolated. A presynaptic re-uptake of neuropeptides, as exists for the monoamines, does not occur.

Frequently, more than one peptide is produced by one neurone. The reason for this "coexistence" of neuropeptides might be that these peptides are produced by the same precursor. In other cases, the same neurone produces different precursors packed in different neurosecretory vesicles. Often, one or more neuropeptides coexist with a classical neurotransmitter (see Hökfelt et al. 1980, for a review). It is obvious that all these combinations highly increase the chemical complexity of neurones. The advantages of coexistence of signal substances in one single neurone is that these substances are released at the same time "as a cocktail". After having reached their different target cells, they can act synergetically and regulate one common process. An example of this will be shown below for snails, where peptides from one type of neurone (one precursor) will act in concert, to steer complementary parts of egg laying behaviour.

2. INVERTEBRATE NEUROPEPTIDES

Table I gives a list of invertebrate neuropeptides. This list is up-to-date (July 1986) and might even be complete. In Table I we have only listed those neuropeptides of which the amino acid sequence has been determined and of which the neuronal localisation has been established by immunocytochemistry or other means. The reason that we have used such strict conditions is, that we want to put a clear borderline somewhere: so no neuropeptides are included of which only a partial sequence or amino acid composition is known, none of the numerous substances are included which have only been demonstrated using antibodies to neuropeptides (cf. Thorpe and Duve 1987), and no peptides are listed of which there is no proof, or at least a strong indication, of a neuronal localisation (cf. Olivera et al. 1985).

The cnidarian neuropeptides of Table I will be discussed in the last part of this review. We will start, therefore, with the neuropeptides from molluscs. FMRFamide (see list of one-letter abbreviations at the bottom of Table I), was the first molluscan neuropeptide to be sequenced (Price and Greenberg 1977). This tetrapeptide has either an excitatory or an inhibitory effect on molluscan hearts, depending on the species which is investigated (Price and Greenberg 1977; Greenberg and Price 1979, 1980; Voight et al. 1981). Non-cardiac muscles, such as the radula protractor and anterior byssus retractor muscles are always excited by this peptide (Greenberg and Price 1979, 1980; Painter 1982; Nagle and Greenberg 1982). Bath application of FMRFamide to a tentacle muscle of Helix elicited the same effect (contraction) as an induced depolarisation of motorneurone C3 of the cerebral ganglion, which contained FMRFamide-like material and which projected to the muscles of the tentacles (Cottrell et al. 1983). This suggests that FMRFamide is a neurotransmitter at neuromuscular synapses. Recently, FMRFamide was also isolated from Aplysia (Lehman et al. 1984). It is likely, therefore, that this peptide is generally occurring in molluscan species. The gene encoding for the FMRFamide precursor in Aplysia has been isolated and sequenced (Schaefer et al. 1985; Scheller 1986). As mentioned earlier, 28 copies of FMRFamide are accommodated in the precursor.

FMRFamide is only one out of several cardioexcitatory peptides occurring in molluscs (Lloyd 1982). Helix aspersa and Lymnea stagnalis produce a group of similar peptides, which all have the carboxyterminal sequence RFamide in common with FMRFamide (Price et al. 1985; Ebberink and Joosse 1985; see Table I). Another cardioactive peptide, however, "small cardioactive peptide B" (SCP_B) isolated from Aplysia californica (Morris et al. 1982), is quite different in structure from the peptides belonging to the FMRFamide family. A precursor containing two small cardioactive peptides (SCP_A and SCP_B) has recently been isolated by Mahon and coworkers (1985).

The next five neuropeptides of Table I are all involved in the egg-laying behaviour of Aplysia californica. Egg-laying is a stereotype

TABLE I

PHYLOGENY OF NEUROPEPTIDE STRUCTURE, LOCALISATION, AND ACTION

Phylum + species	Peptide structure (+ trivial name)	Ref. [ ]	Neuronal localisation	Ref. [ ]	Action	Ref. [ ]
Cnidarian						
Anthopleura elegantissima	pQGRFamide (Antho-RFamide)	[1]	sensory-motorneurons in ecto- and endoderm	[2]	excites all types of endo-dermal muscles and physio-logically identified neuronal systems	[3]
Renilla köllikeri	pQGRFamide (Antho-RFamide)	[4]	sensory-motorneurons	[5]	excites endodermal muscles	[6]
Mollusca						
Macrocallista nimbosa	FMRFamide (cardioexcitatory peptide)	[7]	variety of (identified) neurons in different ganglia	[8,9]	excitation or inhibition of cardiac muscles, excita-tion of non-cardiac muscles, different effects on a variety of neurons	[10-15]
Aplysia brasiliana	FMRFamide (cardioexcitatory peptide)	[14]	"	[8,9]	"	[10-15]
Helix aspersa	pQDPFLRFamide (Helix FMRFamide)	[16]	"	[8,9]	"	[16]
Lymnea stagnalis	sDPFLRFamide	[17]	"	[8,9]	"	[17]
Lymnea stagnalis	gDPFLRFamide	[17]	"	[8,9]	"	[17]
Aplysia brasiliana	MNYLAFPRMamide (small cardioactive peptide B; SCP$_B$)	[18]	identified neurons in the buccal ganglia	[19,20]	excitation of heart and other muscles	[18,21]
Aplysia brasiliana	ARPGYLAFPRMamide (small cardioactive peptide A; SCP$_A$)	[20]	"	[19,20]	"	[18,21]

Species	Sequence	Ref.	Location	Ref.	Action	Ref.
Aplysia californica	ISINQDLKAITDM-LLTEQIRERQRYL-ADLRQRLLEKamide (egg-laying hormone; ELH)	[22,23]	bag cells and other neurons of the abdominal ganglion, some neurons in other ganglia	[23-26]	contraction of the ovo-testis, transmitter-like actions on several identi-fied neurons in abdominal, buccal and cerebral ganglia	[22,27]
Aplysia californica	APRLRFY APRLRFYS APRLRFYSL (α-bag cell peptide)	[23,28]	"	[23-26]	stimulate discharge of bag cells, exert inhibition of L_2,L_3,L_4,L_6 neurons in the abdominal ganglion (hepta-peptide is most active)	[27-29]
Aplysia californica	RLRFH (β-bag cell peptide)	[23]	"	[23-26]	excitation of L_1 and R_1 neurons in the abdominal ganglion	[27]
Aplysia californica	AVKLSSDGNY-PFDLSKEDGAQ-PYFMTPRLRFYPIamide (peptide A)	[24,25,30]	atrial gland and some neurons of the head ganglia	[24-26,30]	causes bag-cell discharge and egg laying	[30]
Aplysia californica	AVKSSSYEKY-PFDLSKEDGAQPY-FMTPRLRFYPIamide (peptide B)	[24,25,30]	atrial gland	[30]	"	[30]
Lymnea stagnalis	LSITNDLRAIADSYL-YDQHWLRERQEE-NLRRRFLELamide (caudodorsal cell hormone)	[18,31]	caudodorsal cells of the cerebral ganglion	[31]	induces ovulation	[31,32]
Aplysia californica	EAEEPSAFMTRL (peptide II)	[34,25]	identified neurons in the abdominal ganglion	[35]	?	

Crustacea

Species	Peptide		Location		Function	
Pandalus borealis	pQLNFSPGWamide (red pigment concentrating hormone; RPCH)	[36,37]	sinus glands of the eye stalk and central neurons	[38]	concentration of pigment in erythrophores	[36]
Pandalus borealis	NSGMINSILGIPRV-MTEAmide (light adapting hormone)	[39]			proximal migration of pigment in the distal retinal pigment cells	[39]
Uca pugilata	NSELINSILGLPKV-MNEAamide (pigment dispersing hormone)	[40]	neurons in the optic ganglia and brain	[41]	dispersion of pigment in melanophores	[40]
Homarus americanus	RYLPT (proctolin)	[42]	neurons of all central ganglia and interganglionic connectives	[43]	slow contractions of walking leg muscles, excitation of neurons in several ganglia	[42,44,45]
Carcinus maenas	PFCNAFTGCamide (crustacean cardioactive peptide)	[46]	nerve processes in the pericardial organ	[41]	cardioexcitation	[46]

Insecta

Species	Peptide		Location		Function	
Periplaneta americana	RYLPT (proctolin)	[47]	identified neurons and identified skeletal motorneurons in several ganglia	[48-51]	excites procteal and slow skeletal muscles	[47,50,52]
Locusta migratoria	pQLNFTPNWGTamide (adipokinetic hormone I; AKH I)	[53]	central neurons and intrinsic neurons of the corpora cardiaca	[54,55]	mobilisation of lipid or trehalose	[53]

Species	Peptide			Source	Function	
Schistocerca gregaria	pQLNFTPNWGTamide (adipokinetic hormone I; AKH I)	[53]		"		[53]
Schistocerca gregaria	pQLNFSTGWamide (adipokinetic hormone II-S; AKH II-S)	[56,57]		corpora cardiaca		[56,57]
Locusta migratoria	pQLNFSAGWamide (adipokinetic hormone II-L; AKH II-L)	[56,57]		"		[56,57]
Periplaneta americana	pQVNFSPNWamide (cockroach myoactive peptide I; periplanetin CC-1; hyperglycaemic hormone I)	[58,59]		"	contraction of skeletal muscles, cardioacceleration, hyperglycaemia	[58,59]
Periplaneta americana	pQLTFTPNWamide (cockroach myoactive peptide II; periplanetin CC-2; hyperglycaemic hormone II)	[58,59]		"	"	[58,59]
Manduca sexta	pQLTFTSSWGamide (M-AKH)	[60]		"	mobilisation of lipid or trehalose	[60]
Heliothis zea	pQLTFTSSWGamide (H-AKH)	[61]		"	"	[61]
Nauphoeta cinera	pQVNFSPGWGTamide (N-AKH)	[62]		"	"	[62]

Carausius morosus	pQLTFTPNWGTamide (C-AKH)	[63]	[63]
Leucophaea maderae	DPAFNSWGamide (leucokinin I)	head	[64]	excites proteal muscles [64]
Leucophaea maderae	DPGFSSWGamide (leucokinin II)	"	[64]	" [64]

One = letter abbreviations of amino acids: A = alanine, C = cysteine, D = aspartate (aspartic acid), E = glutamate (glutamic acid), F = phenylalanine, G = glycine, H = histidine, I = isoleucine, K = lysine, L = leucine, M = methionine, N = asparagine, P = proline, pQ = pyroglutamate (pyrrolidone carboxylic acid), Q = glutamine, R = arginine, S = serine, T = threonine, V = valine, W = tryptophan, Y = tyrosine

[1] Grimmelikhuijzen and Graff 1986, [2] Grimmelikhuijzen et al. 1986, [3] McFarlane et al. 1986, [4] Grimmelikhuijzen and Groeger 1986, [5] Grimmelikhuijzen and Anctil, in preparation, [6] Anctil and Grimmelikhuijzen, in preparation, [7] Price and Greenberg 1977, [8] Schot and Boer 1982, [9] Schaefer et al. 1985, [10] Greenberg and Price 1980, [11] Cottrell 1982, [12] Nagle and Greenberg 1982, [13] Cottrell et al. 1983, [14] Lehman et al. 1984, [15] Boyd and Walker 1985, [16] Price et al. 1985, [17] Ebberink and Joosse 1985, [18] Morris et al. 1982, [19] Mahon et al. 1985, [20] Lloyd et al. 1985, [21] Lloyd 1982, [22] Chiu et al. 1979, [23] Scheller et al. 1983 a, [24] McAllister et al. 1983, [25] Scheller et al. 1984, [26] Shyamala et al. 1986, [27] Scheller et al. 1983 b, [28] Rothman et al. 1983, [29] Sigvardt et al. 1986, [30] Heller et al. 1980, [31] Ebberink et al. 1985, [32] Dogterom et al. 1983, [33] Nambu et al. 1983, [34] Rothman et al. 1985, [35] Kreiner et al. 1984, [36] Fernlund and Josefsson 1972, [37] Fernlund 1972, [38] Mangerich et al. 1986, [39] Fernlund 1976, [40] Rao et al. 1985, [41] R. Keller, personal communication, [42] Schwartz et al. 1984, [43] Siwicki and Bishop 1986, [44] Schwartz et al. 1980, [45] Sullivan and Miller 1984, [46] Strangier et al. 1986, [47] Starratt and Brown 1975, [48] Bishop et al. 1981, [49] Eckert et al. 1981, [50] O Shea and Adams 1981, [51] Witten and O'Shea 1985, [52] Adams and O'Shea 1983, [53] Stone et al. 1976, [54] Schooneveld et al. 1983, [55] Verhaert et al. 1985, [56] Siegert et al. 1985, [57] Gäde et al. 1986, [58] Witten et al. 1984, [59] Scarborough et al. 1984, [60] Ziegler et al. 1985, [61] Jaffe et al. 1986, [62] Gäde and Rinehart 1986 a, [63] Gäde and Rinehart 1986 b, [64] Holman et al. 1986.

114

behaviour (fixed action pattern), which starts shortly after copulation of the snail as a female. It includes cessation of locomotion and feeding, increase in heart beat and respiratory pumping and expulsion of a long string of eggs (McAllister et al. 1983; Scheller et al. 1984). As the string of fertilised eggs is extruded, the animal catches the string with its mouth and makes characteristic head movements (waving) while it is depositing the string of eggs on the substrate. In earlier studies it was found that injection of an extract from two clusters of neurones situated on the top of the abdominal ganglion ("bag cells"), induced the complete egg-laying behaviour (Kupfermann and Kandel 1970). Subsequently, components responsible for this behaviour were purified from bag-cell extracts. In 1979, then the group of Strumwasser (Chiu et al. 1979) purified and sequenced a peptide, the egg-laying hormone (ELH, see Table I). ELH when injected into the haemocoel, mediates the release of eggs, suppresses feeding and also generates the characteristic head waving. ELH injected alone, however, did not elicit all the responses seen after injection of bag-cell extract, indicating, that more factors are involved. Due to the nice work of Scheller and coworkers (Scheller et al. 1982, 1983a, b, 1984), the gene for the ELH precursor was isolated and sequenced. The ELH precursor contained one copy of ELH, but also two peptide fragments which were flanked by pairs of basic amino acids (potential cleavage sites). Using radioimmunoassays against these fragments, the two peptides (α and β-bag-cell peptide) were subsequently purified from bag-cell extracts, demonstrating that the precursor was really processed to the level of these peptides (Rothman et al. 1983). The α- and β-bag-cell peptides have specific actions on several identified neurones of the abdominal ganglion (see Table I; Scheller et al. 1983b, 1984; Sigvardt et al. 1986). These neurones might be involved in the control of some visceral functions. In addition, α-bag-cell peptide stimulates the discharge of the bag-cells themselves, thereby leading to more release of the three precursor products (Scheller et al. 1983 b). A single precursor, then, produces three biologically active peptides. After release at the same sites within the abdominal ganglion, ELH, α- and β-bag-cell peptides act at their specific target neurones (transmitter and paracrine action). ELH, in addition, diffuses out of the abdominal ganglion and is transported via the haemolymph towards the ovotestis, where it mediates the release of eggs (hormonal action). The peptide is also transported to the buccal, cerebral and pedal ganglia, where it is thought to suppress feeding behaviour and to generate head waving (Stuart and Strumwasser 1980). This is a nice example of how one, single neuropeptide (ELH) governs several, complementary components of a complex behaviour (egg laying). It also illustrates that additional components of this egg laying behaviour are controlled by peptides (the α-and β-bag-cell peptides) which are produced by the same precursor protein from which ELH is made. The advantage of a common precursor for the three biologically active peptides, then, is that this mechanism garantees that all three peptides are released at the same time and in a fixed amount (i.e. in a controlled way) and that by this all components of egg laying are properly carried out.

The gene for the ELH precursor was found by hybridisation of DNA fragments of <u>Aplysia</u> with cDNA obtained after reversed transcription of mRNA isolated from the bag cells. By this method, not only the DNA fragment encoding for the ELH precursor was found, but also two other fragments which share many nucleotide sequences with the ELH-precursor gene (Scheller et al. 1984). These other fragments code for the precursors of peptide A and peptide B (see Table I). Before the discovery of their genes, peptides A and B had already been isolated from extracts of the atrial gland, an exocrine gland situated at the distal portion of the hermaphroditic duct (Heller et al. 1980). When injected into the haemocoel, peptides A and B initiated bag-cell discharge, which was followed by egg laying behaviour. It is possible that peptides A and B are released by the "strategically located" atrial gland during copulation. This would make sense, as egg laying, then, would only be initiated after fertilisation. The atrial gland, however, is exocrine and it is not quite clear how peptides A and B could cross the duct wall and reach the haemolymph. Furthermore, peptide A is also produced by neurones of the head ganglia of Aplysia (Shyamala et al. 1986). It might be, therefore, that initiation of egg laying behaviour is somewhat more complicated than a simple release of peptides A and B, caused by mechanical stimulation of the atrial gland during copulation.

From the "caudodorsal cells", two groups of endocrine cells in the cerebral ganglia of Lymnaea stagnalis, a peptide has been isolated and sequenced that induces ovulation when injected into mature, female animals (Dogterom et al. 1983; Ebberink and Joosse 1985). This peptide shows 44% homology with the sequence of Aplysia ELH (see Table I). The caudodorsal cell hormone is one of several peptide hormones controling egg laying behaviour in Lymnaea, but the other peptide hormones have not been characterised so far.

The properties of the last molluscan neuropeptide, peptide II of Aplysia californica (see Table I), are somewhat enigmatic. It was discovered on a gene which was specifically expressed in neurones R3 and R14 of the abdominal ganglion (Nambu et al. 1983). This gene encoded for a precursor protein, in which peptide II was a "theoretical peptide" (flanked by two pairs of basic amino acids). Later, this peptide was purified from extracts of Aplysia abdominal ganglia (Rothman et al. 1985). Also the neuronal localisation of peptide II has been established using immunocytochemistry (Kreiner et al. 1984), but, so far, no biological action could be found for this substance.

The first crustacean neuropeptide, red-pigment-concentrating-hormone (RPCH), was isolated and sequenced by Fernlund and Josefsson in 1972. This peptide is one of several neurohormones which are synthesised by neurones of the eye stalk (X-organ) and which are released by the sinus gland (the neurohaemal organ of the eye stalk) into the haemolymph (Mangerich et al. 1986). RPCH controls the movement of pigment in erythrophores. These cells, together with other chromatophores, enable crustaceans to match their body colour with that of the background. Another peptide, light-adapting-hormone (see Table I), regulates the pigment movement in the ommatidia of the crustaceans in order to adapt to light. This peptide was first isolated from eye-stalks of the shrimp Pandalus borealis (Fernlund 1976) and more recently a similar peptide was isolated from the fiddler crab Uca pugilator (Rao et al. 1985). The exact localisation of Pandalus light-adapting-hormone has never been established, but with antisera against the hormone from Uca, immunoreactive material could be localised in neurones of the optic ganglia, the brain and the thoracal ganglia of Carcinus maenas (R. Keller, personal communication). As the light-adapting-hormones from Pandalus and Uca are very similar in structure (Table I) and both have been isolated from neuron-rich material, it can be accepted that the Pandalus hormone is also a real neuropeptide.

The neuropeptide proctolin (Table I) was first found in insects (see below). Many years later, this same pentapeptide was also isolated from the lobster and sequenced (Schwartz et al. 1984). Proctolin is probably a neurotransmitter at crustacean neuromuscular junctions, but it also excites neurones in a variety of ganglia (Schwartz et al. 1980; Sullivan and Miller 1984). Recently, proctolin was isolated from the pericardial organs of several crabs (Strangier et al. 1986). Here, the peptide acts as a cardioexcitator, although it might also have the other two functions described above. In addition to proctolin, a second, novel cardioactive peptide could be isolated and sequenced from crab pericardial organs (Strangier et al. 1986). With antisera against this crustacean cardioactive peptide (Table I), immunoreactive material has been located in neurones (R. Keller, personal communication).

Proctolin was the first insect neuropeptide to be sequenced (Starratt and Brown 1975). As the name already suggests, this peptide has an excitatory effect on procteal (hindgut) muscles of the cockroach. In agreement with this action, is the presence of proctolin-immunoreactive material in nerve endings of the hindgut wall (Eckert et al. 1981). Proctolin, however, is also present in certain skeletal motorneurones of insects (Bishop and O'Shea 1982; Witten and O'Shea 1985). From both anatomical and electrophysiological data it is likely, now, that proctolin is an excitatory neurotransmitter used by motorneurones innervating slow skeletal muscles (Adams and O'Shea 1983; Witten and O'Shea 1985).

Adipokinetic hormone I (see Table I) was the first sequenced neuropeptide out of a family of peptides which regulate lipid and sugar (trehalose) utilisation during flight in insects. Adipokinetic hormone I (AKH I) was isolated from the corpora cardiaca of both the migratory locust, Locusta migratoria, and the desert locust, Schistocerca gregaria, (Stone et al. 1976). Later, a second, related peptide was isolated from Locusta (AKH II-L) and a third from Schistocerca (AKH II-S; Siegert et al. 1985; Gäde et al. 1986). AKH II-L and AKH II-S only differ in one amino acid from each other, and have the aminoterminal half in common with AKH I (see Table I). Further members of the AKH family were found in the cockroach Periplaneta americana (Table I). These peptides were either called myoactive peptide I and II, or hyperglycaemic hormone I and II, depending on the bioassay which was used for their isolation (Witten et al. 1984; Scarborough et al. 1984). A sixth adipokinetic hormone was isolated from the corpora cardiaca of both the tobacco hornworm Manduca sexta (Ziegler et al. 1985) and the corn borer Heliothis zea (Jaffe et al. 1986) and, more recently, a seventh and an eighth member from the cockroach Nauphoeta cinerea (Gäde and Rinehardt 1986a) and the Indian stick insect Carausius morosus (Gäde and Rinehardt 1986b).

The adipokinetic hormones show a striking similarity in structure with the crustacean red-pigment-concentrating-hormone (see Table I). AKH I, when injected into crustaceans, demonstrated RPCH activity (Mordue and Stone 1976) and RPCH, when injected into insects, markedly elevated haemolymph lipids (Mordue and Stone 1976; Dallman et al. 1981). This is an interesting example of how a peptide has been conserved during arthropod evolution, and a target cell (with receptor) has been changed.

Holman and coworkers (1986) have recently isolated and sequenced two novel, related insect peptides, which excite procteal muscles of the cockroach. These peptides, leucokinin I and II show the same action as proctolin, but are structurally very different (Table I). More similarity in structure, however, can be found between the carboxyterminal portions of the leucokinins and Manduca and Heliothis adipokinetic hormone (especially between H-AKH and Leucokinin II; cf. Table I).

From this short survey, it is clear that the family of AKH peptides is extremely important and widespread in insects or in arthropods in general. For most AKH peptides an exact localisation has not been carried out, but in those cases in which this has been done (AKH I), the peptide has always been found in neurones (Schooneveld et al. 1983; Verhaert et al. 1985). For this reason, and also because the AKH peptides were all isolated from neuronal tissue (corpora cardiaca), it can be assumed that they are true neuropeptides.

3. NEUROPEPTIDES IN CNIDARIANS

Cnidarians have the simplest nervous system of the animal kingdom, and it was probably within this group of animals that nervous systems first evolved. The general plan of the cnidarian nervous system can be described as a nerve net. In some cnidarians, however, parts of this nerve net have condensed and formed linear or circular tracts and even ganglion-like structures (cf. Grimmelikhuijzen 1985; Grimmelikhuijzen et al. 1986a, b; Satterlie and Spencer 1987). Some of the neurones of cnidarians have retained features that we must assume are primitive: Westfall has described multifunctional neurones in Hydra that show the morphological characters of both sensory- and motorneurones (Westfall 1973; Westfall and Kinnamon 1978). The first nervous systems probably used a single neuronal type like this to form an early variant of the "monosynaptic reflex arc".

Ultrastructural studies have shown that many synapses in the cnidarian nervous system are chemical. Dense-cored vesicles (70-150 nm), and pre- and postsynaptic membrane specialisations have been seen at synapses of all coelenterate classes (see Westfall 1987). Physiological studies demonstrating synaptic blockade by depletion of Ca^{2+} or by addition of excess Mg^{2+}, also suggested that classical, exocytotic release of transmitter substances occurs (Mackie 1975; McFarlane 1973; Anderson and Schwab 1982; Spencer 1982). Finally, simultaneous intracellular recordings at both pre- and postsynaptic neurones displayed latencies which are characteristic for chemical synapses (Spencer 1985).

Despite overwhelming evidence that cnidarian neurones form chemical synapses, no transmitter substance has ever been identified in these animals (see Martin and Spencer 1983 for a review). In Hydra we have been unable to show classical neurotransmitters such as the catecholamines, serotonin and acetylcholine (Grimmelikhuijzen 1986). Recently, however, using immunocytochemistry and radioimmunoassays, we could demonstrate several substances in the nervous systems of cnidarians that were related to vertebrate and invertebrate neuropeptides. In the present paper we would like to restrict ourselves to peptides related to the molluscan neuropeptide FMRFamide. The reader is referred to Grimmelikhuijzen (1984) for a review of the other neuropeptides in cnidarians.

Numerous polyclonal antisera against FMRFamide have been prepared which all gave good staining of neurones in a wide variety of cnidarians (Grimmelikhuijzen et al. 1982; Grimmelikhuijzen 1983, 1986; Grimmelikhuijzen and Spencer 1984; Mackie et al. 1985). Of antisera against different fragments of FMRFamide, those against RFamide stained much better than antisera against the complete peptide (Grimmelikhuijzen and Spencer 1984; Grimmelikhuijzen 1985; Grimmelikhuijzen et al. 1986b). This suggests that the cnidarian FMRFamide-like peptides have the sequence RFamide in common with FMRFamide, but that the aminoterminus is different (see below).

RFamide antisera have proven to be invaluable tools for determining the organisation of the cnidarian nervous system. This is especially true in hydrozoans, which are generally transparent and which can be stained as whole mounts. In the past, histological staining of cnidarian nervous tissue has always been rather difficult. With the new immunocytochemical technique, now many novel neuronal structures can be readily discovered (Grimmelikhuijzen 1985; Grimmelikhuijzen et al. 1986b; see also Fig. 1).

In addition to new and exciting information on the neuroanatomy of cnidarians, staining with RFamide antisera also gives a hint at the neurotransmitter substance which is used by the stained neurones. In

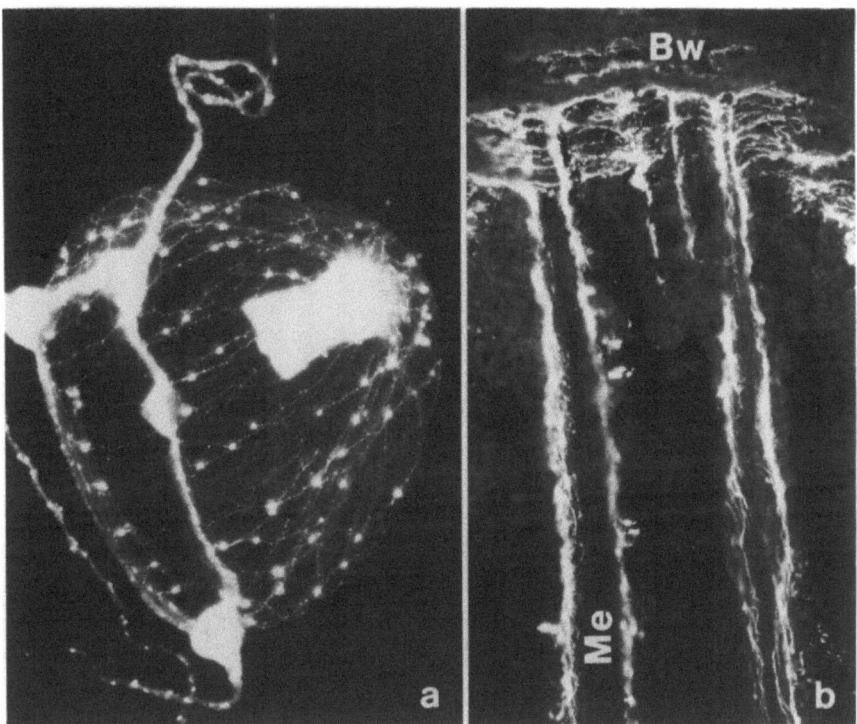

FIG. 1: Staining of the cnidarian nervous system with RFamide antiserum 146 II (1:1000). <u>a</u>. whole mount of a 2-day old medusa of <u>Eirene sp.</u>. The subumbrellar nerve net can nicely be visualised (x100). <u>b</u>. A cross section through the sea anemone <u>Calliactis parasitica</u>. The neurites innervating the radial and retractor muscles of the mesenteries (Me) and the neurites associated with the circular muscles of the body wall (Bw) are stained (x140).

order to isolate these substances, a radioimmunoassay for RFamide-like peptides has been developed (Grimmelikhuijzen and Graff 1985, 1986). This assay involves the use of ^{125}I-labelled YFMRFamide and an apppropriate RFamide antiserum (Fig. 2). The radioimmunoassay recognises free RFamide and aminoterminal elongated peptides containing the RFamide moiety, such as FMRFamide and LPLRFamide (Fig. 3). Peptides terminating with RYamide can also be measured, but with a much lower efficiency (100x less than peptides containing RFamide). Only peptides containing arginine followed by an amidated aromatic amino acid, are recognised properly in the assay. FMKFamide, a FMRFamide analogue in which the arginine is replaced by the other positively charged amino acid, lysine, reacts poorly (2000x less efficient than RFamide peptides). Other peptides in which the Phe-amide terminus is not preceded by a positively charged amino acid, such as the carboxyterminus of gastrin/cholecystokinin (WMDFamide), are not recognised in the assay (Fig. 3). The same holds for vasopressin (CYFQNCPRGamide), which contains an arginine followed by an amidated non-aromatic amino acid, and FMRF, in which the terminal amino acid is non-amidated.

With the RFamide radioimmunoassay, high amounts of immunoreactive material could be measured in acetic acid extracts of the sea anemone <u>Anthopleura elegantissima</u> (Fig. 2). Using the radioimmunoassay as a monitoring system, we have purified this material. Initial purification,

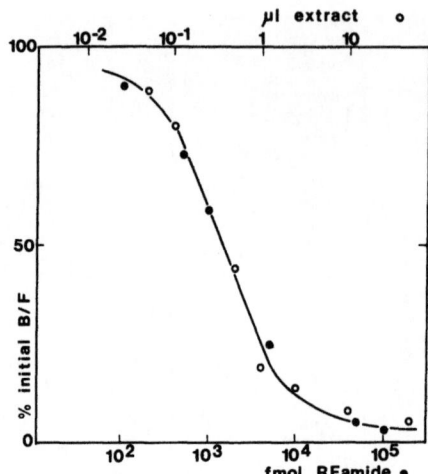

FIG. 2: RFamide radioimmunoassay of an acetic acid extract of the sea anemone <u>Anthopleura elegantissima</u>. ^{125}I-labelled YFMRFamide was used as a tracer, added to RFamide antiserum 145 IV (1:10,000) together with RFamide standard or 0.01-10 µl of the sea-anemone extract. After 2 days at 4°C, bound (B) and free (F) tracer were separated. From Grimmelikhuijzen and Graff 1985.

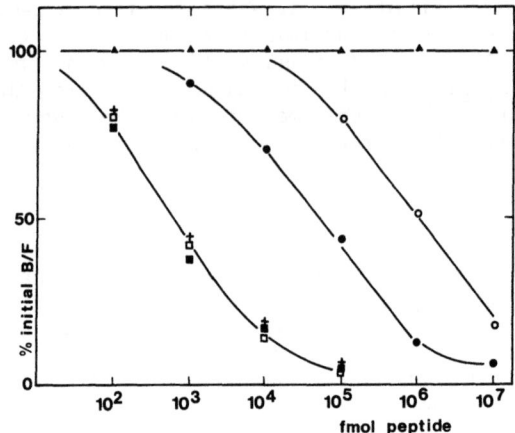

FIG. 3: Specificity of the RFamide radioimmunoassay of Fig. 2. (+) RFamide; (□) FMRFamide; (■) LPLRFamide; (●) bovine pancreatic polypeptide-(31-36)-hexapeptide (LTRPRYamide); (o) FMKFamide; (▲) FMRF, Cholecystokinin-(30-33)-tetrapeptide (WMDFamide), or [Arg^8]-vasopressin (CYFQNCPRGamide). Note, that all peptides containing the RFamide moiety are equally-well recognised. From Grimmelikhuijzen and Graff 1985.

desalting and concentration of the immunoreactive material involved passage through Sep-pak reversed-phase cartridges (Waters). A successful subsequent purification method was cation-exchange chromatography on CM-Sephadex C-25, using a salt and a pH gradient (Fig. 4). As could be

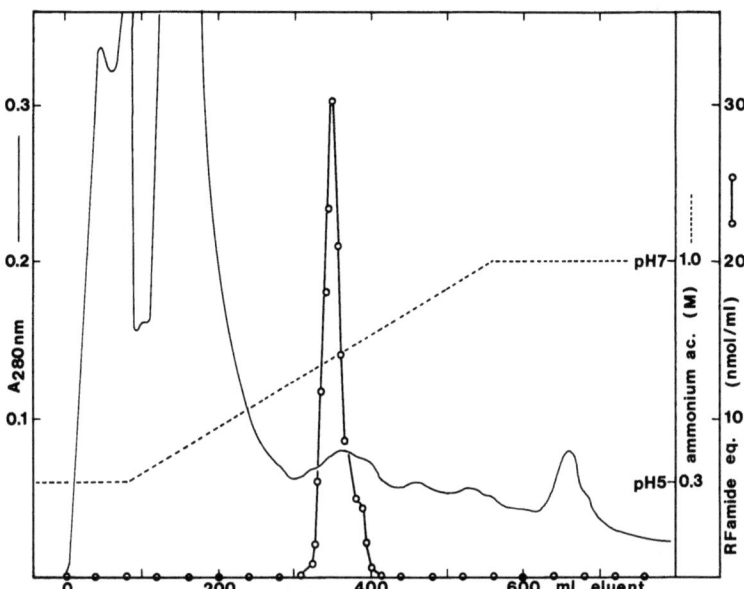

FIG. 4: Cation-exchange chromatography on CM-Sephadex C-25 of an acetic acid extract of Anthopleura (after desalting on Sep-pak and elution with 60% methanol). The column (2.6 x 41 cm) was equilibrated with 0.3 M ammonium acetate, pH 5. After elution of one void volume (80 ml), a linear gradient was started from 0.3 M (pH 5) - 1.0 M (pH 7) ammonium acetate (24 ml/h; 20h). One peak of immunoreactive material was eluted around 0.7 M ammonium acetate, pH 5.3. After Grimmelikhuijzen and Graff 1985.

expected from a positively charged peptide (see specificity of the radioimmunoassay), the immunoreactive material was strongly bound to the resin. This material could be eluted at 0.7-0.8 M ammonium acetate (pH 5.3), when most of the contaminating substances had been removed (Fig. 4). After further cation-exchange chromatography at pH 7 and pH 8.5 (Grimmelikhuijzen and Graff 1985; Grimmelikhuijzen et al. 1986a), the material was pure enough to be analysed by HPLC. On a reversed-phase column (20 min gradient of 10%-60% acetonitrile in 0.1% trifluoroacetic acid), the Anthopleura RFamide-like peptide was eluted as one single peak at 8 min (Fig. 5). This material was very pure as could be shown by subsequent HPLC (Grimmelikhuijzen and Graff 1986).

A part of the HPLC purified material was hydrolysed in 6 N HCl and subsequently dansylated (Gray 1967; Woods and Wang 1967). Separation of the dansylated mixture, using thin layer chromatography, yielded the following amino acids: Glu, Gly, Arg and Phe (Grimmelikhuijzen and Graff 1986). An end-group determination, using dansylchloride, showed that no reactive end group was present. As glutamate was one of the amino acids in the hydrolysate, the aminoterminus could by pyroglutamate, a cyclisation product of glutamine which lacks a free, primary amine. This was confirmed by treatment with the enzyme pyroglutamate aminopeptidase and subsequent purification by HPLC. The new peak of immunoreactive material contained Gly, Arg, Phe, but not Glu. Thus only one glutamate had to be present in the peptide, being in the form of an aminoterminal pGlu. At this stage we asked two collegu150s (Dr. H. Ponstingl, DKFZ, and Dr. R. Frank, EMBL, Heidelberg) to determine the stoichiometry of the

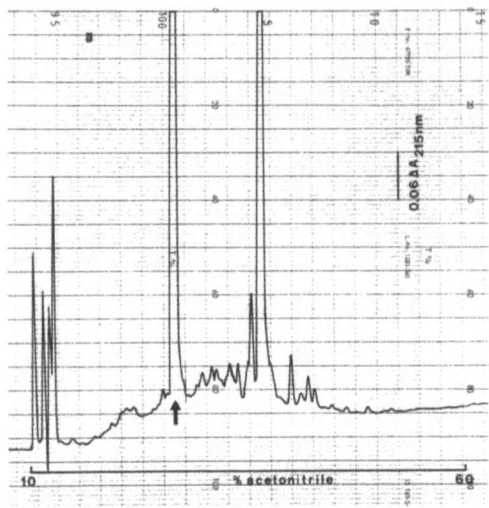

FIG. 5: HPLC of a portion (90 nmol RFamide equivalents) of immunoreactive material of <u>Anthopleura</u> purified by cation-exchange chromatography. An analytical C-18 column (Spherisorb ODS-2, 5 μm particle size, 8 nm pore size, dimensions 4 x 250 mm) was equilibrated with 10% acetonitrile in 0.1% trifluoroacetic acid. Subsequently a gradient was started from 10%-60% acetonitrile in 0.1% trifluoroacetic acid (20 min, 1 ml/min). The RFamide peptide was eluted at 8 min (arrow). After Grimmelikhuijzen and Graff 1986.

amino acids using their amino-acid analysers. Both scientists found a stoichiometry of 1:1:1:1. As there is only one Glu in the peptide, there must also be one Gly, Arg and Phe. From our knowledge of the specifity of the radioimmunoassay, by which only Gly-Arg-Phe-amide can be recognised and no other amidated combination of the three amino acids (see above), the sequence of the peptide had to be: pGlu-Gly-Arg-Phe-amide. This sequence of the <u>Anthopleura</u> RFamide peptide was confirmed by both the classical Edman degradation (Gray and Smith 1970) and the more recently developed DABITC method (Chang et al. 1978). After enzymatic removal of pGlu, we found Gly as an end group and, after degradation cycles, Arg and Phe-amide (Grimmelikhuijzen and Graff 1986). The sequence pGlu-Gly-Arg-Phe-amide was synthesised and subsequently compared with the native peptide. On six different reversed-phase HPLC columns, natural and synthetic peptide were always eluted at exactly the same positions (Grimmelikhuijzen and Graff 1986). This further proved that the sequence of the <u>Anthopleura</u>-RFamide peptide was correct.

An RFamide peptide was also isolated from the pennatulid <u>Renilla köllikeri</u>. The pennatulids (order of Pennatulacea) belong to the class of Anthozoa, as do the sea anemones (order of Actinaria), but they are phylogenetically rather remote from the latter. The <u>Renilla</u>-RFamide peptide was purified in the same way as the <u>Anthopleura</u> peptide. Analyses of HPLC purified material yielded the same sequence as that of Anthopleura-RFamide. Comparison of synthetic pGlu-Gly-Arg-Phe-amide and natural <u>Renilla</u> peptide on six different HPLC columns further demonstrated the identity of the <u>Renilla</u>- and <u>Anthopleura</u>-RFamide peptides (Grimmelikhuijzen and Groeger 1987). As two phylogenetically widely-separated anthozoans produce the same peptide, it is likely that

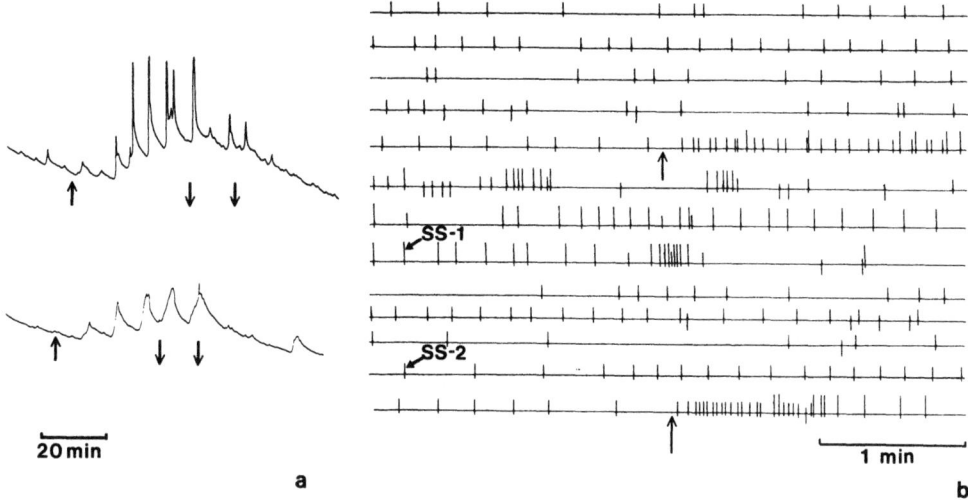

FIG. 6: Effects of synthetic Antho-RFamide on muscles and neuronal systems of the sea anemone Calliactis parasitica. a. Increase in tone, contraction amplitude, and contraction frequency of the sphincter muscle (top) and the circular muscle (bottom); ↑ indicates the addition of Antho-RFamide (final concentration: 10^{-7} M), ↓ indicates a wash. b. Increase in activity of the neuronal conduction systems SS-1 and SS-2; ↑ indicates the injection of Antho-RFamide through a canula inserted in the mouth (final concentration in the coelenteron: 10^{-6} M). After McFarlane et al. 1986.

this peptide is generally occurring in Anthozoa. For this reason we have called pGlu-Gly-Arg-Phe-amide, Antho-RFamide.

The availabity of sufficient amounts of synthetic Antho-RFamide has made it possible to test its biological action. In the sea anemone Calliactis parasitica Antho-RFamide causes a long-lasting increase in tone, contraction frequency and contraction amplitude of several endodermal muscles (McFarlane et al. 1986; Fig. 6a). In addition, the peptide increases electrical activity in an ectodermal conduction system of Calliactis, the SS1 and in an endodermal conduction system, the SS2 (Fig. 6b). Both conduction systems are supposed to have a neuronal basis (cf. also McFarlane 1982 for a review). The neuronal localisation of Antho-RFamide in sea anemones (Fig. 1b), together with its excitatory actions on muscles and neuronal systems, suggest that the peptide is a neurotransmitter or neuromodulator.

4. ACKNOWLEDGEMENTS

We would like to thank Drs. G Gäde (Düsseldorf) and R. Keller (Bonn) for access to unpublished material and the Deutsche Forschungsgemeinschaft for financial support (Gr. 762/4, 762/7). C.J.P.G. is the recipient of a Heisenberg Fellowship (Gr.762/1).

5. REFERENCES

Acher R (1981) Evolution of neuropeptides. Trends Neurosci 4: 226-230.

Acher R, Chauvet J (1953) La structure de la vasopressin de boeuf. Biochim Biophys Acta 12: 487-488.

Adams ME, O'Shea M (1983) Peptide cotransmitter at a neuromuscular junction. Science 221: 286-289.

Amara SG, Jonas V, Rosenfeld MG, Ong ES, Evans RM (1982) Alternative RNA processing in calcitonin gene expression generates mRNA's encoding different polypeptide products. Nature 298: 240-244.

Anderson PAV, Schwab WE (1982) Recent advances and model systems in coelenterate neurobiology. Prog Neurobiol 19: 213-236.

Barker JL, Vincent JD, Macdonald JF (1980) Substance P pharmacology of cultured mouse spinal neurons. In: Marsan CA, Traczk WZ (eds) Neuropeptides and neurotransmission. Raven Press, New York, pp 93-103.

Bishop CA, O'Shea M (1982) Neuropeptide proctolin (H-Arg-Try-Leu-Pro-Thr-OH): Immunocytochemical mapping of neurons in the central nervous system of the cockroach. J Comp Neurol 207: 223-238.

Bishop CA, O'Shea M, Miller RJ (1981) Neuropeptide proctolin (H-Arg-Tyr-Leu-Pro-Thr-OH): Immunological detection and neuronal localization in insect central nervous system. Proc Natl Acad Sci USA 78: 5899-5902.

Blobel G, Dobberstein B (1975) Transfer of proteins across membranes. I. Presence of proteolytically processed and unprocessed nascent immunoglobulin light chains on membrane-bound ribosomes of murine myeloma. J Cell Biol 67: 835-851.

Boyd PJ, Walker RJ (1985) Actions of the molluscan neuropeptide FMRFamide on neurones in the suboesophageal ganglia of the snail _Helix aspersa_. Comp Biochem Physiol 81C: 379-386.

Bradbury AF, Finnie MDA, Smyth DG (1982) Mechanism of C-terminal amide formation by pituitary enzymes. Nature 298: 686-688.

Brownstein MJ (1982) Posttranslational processing. Trend Neurosci 5: 318-320.

Brownstein MJ, Russel JT, Gainer H (1980) Synthesis, transport, and release of posterior pituitary hormones. Science 207: 373-378.

Buijs RM (1978) Intra- and extrahypothalamic vasopressin and oxytocin pathways in the rat. Pathways to the limbic system, medulla oblongata and spinal cord. Cell Tissue Res 192: 423-435.

Chang JY, Brauer D, Wittman-Liebold B (1978) Microsequence analysis of peptides and proteins using 4-NN-dimethylaminoazobenzene-4'-isothiocyanate/phenylisothiocyanate double coupling method. FEBS Lett 93: 205-214.

Chiu AY, Hunkapillar M, Heller E, Stuart DK, Hood LE, Strumwasser F (1979) Purification and primary structure of the neuropeptide egg-laying hormone of _Aplysia californica_. Proc Natl Acad Sci USA 76: 6656-6660.

Cottrell GA (1982) FMRFamide neuropeptides simultaneously increase and decrease K+ currents in an identified neurone. Nature 296: 87-89.

Cottrell GA, Schot LPC, Dockray GJ (1983) Identification and probable role of a single neurone containing the neuropeptide Helix FMRFamide. Nature 304: 638-640.

Croll RP (1987) Identified neurones and cellular homologies. (This volume)

Dallman SH, Herman WS, Carlsen J, Josefsson L (1981) Adipokinetic activity of shrimp and locust peptide hormones in butterflies. Gen Comp Endocrinol 43: 256-258.

Dockray GJ, Gregory RA, Hutchison JB, Harris JI, Runswick MJ (1978) Isolation, structure and biological activity of two cholecystokinin octapeptides from sheep brain. Nature 274: 711-713.

Dogterom GE, Bohlken S, Geraerts WPM (1983) A rapid in vivo bioassay of the ovulation hormone of Lymnaea stagnalis. Gen Comp Endocrinol 50: 476-482.

Dubois M (1975) Immunoreactive somatostatin is present in discrete cells of the endocrine pancreas. Proc Natl Acad Sci USA 72: 1340-1343.

Du Vigneaud V, Ressler C, Trippett S (1953a) The sequence of amino acids in oxytocin, with a proposal for the structure of oxytocin. J Biol Chem 205: 949-957.

Du Vigneaud V, Lawler HC, Popenoe EA (1953b) Enzymatic cleavage of glycinamide from basopressin and a proposed structure for this pressor-antidiuretic hormone of the posterior pituitary. J Amer Chem Soc 75: 4880-4881.

Ebberink RHM, Joosse J (1985) Molecular properties of various snail peptides from brain and gut. Peptides 6 (Supp 3): 451-457.

Ebberink RHM, Van Loenhout H, Geraerts WPM, Joosse J (1985) Purification and amino acid sequence of the ovulation neurohormone of Lymnaea stagnalis. Proc Natl Acad Sci USA 82: 7767-7771.

Eckert M, Agricola H, Penzlin H (1981) Immunocytochemical identification of proctolin-like immunoreactivity in the terminal ganglion and hindgut of the cockroach Periplaneta americana (L). Cell Tissue Res 217: 633-645.

Esparmer V, Melchiorri P (1973) Active polypeptides of the amphibian skin and their synthetic analogues. Pure Appl Chem 35: 463-494.

Fernlund P (1974) Structure of the red-pigment-concentrating hormone of the shrimp, Pandalus borealis. Biochim Biophys Acta 371: 304-311.

Fernlund P (1976) Structure of a light-adapting hormone from the shrimp, Pandalus borealis. Biochim Biophys Acta 439: 17-25.

Fernlund P, Josefsson L (1972) Crustacean color-change hormone: amino acid sequence and chemical synthesis. Science 177: 173-175.

Fricker LD, Snyder SH (1982) Enkephalin convertase: Purification and characterization of a specific enkephalin- synthezising- carboxypeptidase localized to adrenal chromaffin granules. Proc Natl Acad Sci USA 79: 3886-3890.

Gäde G, Rinehart KL (1986a) Amino-acid sequence of a hypertrehalosaemic neuropeptide from the corpus cardiacum of the cockroach, Nauphoeta cinera. Biochem Biophys Res Commun (in press).

Gäde G, Rinehart KL (1986b) Primary structure of the hypertrehalosaemic factor II from the corpus cardiacum of the Indian stick insect, Carausius morosus, determined by fast-atom bombardment mass spectrometry. Biol Chem Hoppe Seyler (in press).

Gäde G, Goldsworthy GJ, Schaffer MH, Cook JC, Rinehart KL (1986). Sequence analysis of adipokinetic hormones II from corpora cardiaca of Schistocerca nitans, Schistocerca gregaria and Locusta migratoria by fast atom bombardment mass spectrometry. Biochem Biophys Res Commun 134: 723-730.

Gray WR (1967) Dansyl chloride procedure. In: Colowick SP, Kaplan NO (eds) Methods in enzymology Vol XI. Acadimic Press, New York, pp 139-151.

Gray WR, Smith JF (1970) Rapid sequence analysis of small peptides. Anal Biochem 33: 36-42.

Greenberg MJ, Price DA (1979) FMRFamide, a cardioexcitatory neuropeptide of molluscs: an agent in search of a mission. Amer Zool 19: 163-174.

Greenberg MJ, Price DA (1980) Cardioregulatory peptides in molluscs, In: Bloom FE (ed) Peptides: Integrators of cell and tissue function. Raven Press, New York, pp 107-126.

Grimmelikhuijzen CJP (1983) FMRFamide immunoreactivity is generally occurring in the nervous systems of coelenterates. Histochem 78: 361-381.

Grimmelikhuijzen CJP (1984) Peptides in the nervous systems of coelenterates. In: Falkmer S, Hakanson R, Sundler F (eds) Evolution and tumor pathology of the neuroendocrine system. Elsevier, Amsterdam, pp 39-58.

Grimmelikhuijzen CJP (1985) Antisera to the sequence Arg-Phe-amide visualize neuronal centralization in hydroid polyps. Cell Tissue Res 241: 171-182.

Grimmelikhuijzen CJP (1986) FMRFamide-like peptides in the primitive nervous systems of coelenterates and complex nervous systems of higher animals. In: Stephano GB (ed) Handbook of comparative opioid and related neuropeptide mechanisms, Vol I. CRC Press, Boca Raton, pp 103-115.

Grimmelikhuijzen CJP, Graff D (1985) Arg-Phe-amide-like peptides in the primitive nervous systems of coelenterates. Peptides 6, (Supp 3): 477-483.

Grimmelikhuijzen CJP, Graff D (1986) Isolation of (Glu-Gly-Arg-Phe-NH_2 (Antho-RFamide), a neuropeptide from sea anemones. Proc Natl Acad Sci USA 83: 9817-9821.

Grimmelikhuijzen CJP, Groeger A (1987) Isolation of the neuropeptide pGlu-Gly-Arg-Phe-amide from the pennatulid Renilla köllikeri. FEBS Lett 211: 105-108.

Grimmelikhuijzen CJP, Spencer AN (1984) FMRFamide immunoreactivity in the nervous system of the medusa Polyorchis penicillatus. J Comp Neurol 230: 361-371.

Grimmelikhuijzen CJP, Dockray GJ, Schot LPC (1982) FMRFamide-like immunoreactivity in the nervous system of hydra. Histochem 73: 499-508.

Grimmelikhuijzen CJP, Graff D, Spencer AN (1986a) Structure, location and possible actions of Arg-Phe-amide peptides in coelenterates. In: Thorndyke MC, Goldsworthy G (eds) Invertebrate peptides and amines. Cambridge University Press (in press).

Grimmelikhuijzen CJP, Spencer AN, Carré D (1986b) Organization of the nervous system of physonectid siphonophores. Cell Tissue Res 246: 463-474.

Guillemin R (1978) Peptides in the brain: The new endocrinology of the neurone. Science 202: 390-402.

Guillemin R, Burgus R (1972) The hormones of the hypothalamus. Sci Amer 227 (no. 5): 24-33.

Heller E, Kaczmarek LK, Hunkapilar M, Hood LE, Strumwasser F (1980) Purification and primary structure of two neuroactive peptides that cause bag cell afterdischarge and egg-laying in Aplysia. Proc Natl Acad Sci USA 77: 2328-2332.

Herbert E (1981) Discovery of pro-opiomelanocortin, a cellular polyprotein. Trends Biochem Sci 6: 184-188.

Hökfelt T, Elde R, Johansson O, Luft R, Arimura A (1975) Immunohistochemical evidence for the presence of somatostatin, a powerful inhibitory peptide, in some primary sensory neurons. Neurosci Lett 1: 231-235.

Hökfelt T, Johannson O, Ljungdahl A, Lundberg JM, Schultzberg M (1980) Peptidergic neurones. Nature 284: 515-521.

Holman GM, Cook BJ, Nachman RJ (1986) Isolation, primary structure and synthesis of two neuropeptides from Leucophaea maderae: members of a new family of cephalomyotropins. Comp Biochem Physiol 84 C: 205-211.

Jaffe H, Raina AK, Riley CT, Fraser BA, Holman GM, Wagner RM, Ridgway RL, Hayes DK (1986) Isolation and primary structure of a peptide from the corpora cardiaca of Heliothis zea with adipokinetic activity. Biochem Biophys Res Commun 135: 622-628.

Klein M, Camardo J, Kandel ER (1982) Serotonin modulates a specific potassium current in the sensory neurons that show presynaptic facilitation in Aplysia. Proc Natl Acad Sci USA 79: 5713-5717.

Kreiner T, Rothbard JB, Schoolnik GK, Scheller RH (1984) Antibodies to synthetic peptides defined by cDNA cloning reveal a network of peptidergic neurons in Aplysia. J Neurosci 4: 2581-2589.

Kupfermann I, Kandel ER (1970) Stimulation of egg-laying by extracts of neuroendocrine cells (bag cells) of abdominal ganglion of Aplysia. J Neurobiol 33: 877-881.

Land H, Schütz G, Schmale H, Richter D (1982) Nucleotide sequence of cloned cDNA encoding bovine arginine vasopressin-neurophysin II precursor. Nature 295: 299-303.

Lehman HK, Price DA, Greenberg MJ (1984) The FMRFamide-like neuropeptide of Aplysia is FMRFamide. Biol Bull 167: 460-466.

Lloyd PE (1982) Cardioactive neuropeptides in gastropods. Fed Proc 41: 2948-2952.

Lloyd PE, Mahon AC, Kupfermann I, Cohen JL, Scheller RH, Weiss KR (1985) Biochemical and immunocytochemical localisation of molluscan small cardioactive peptides in the nervous system of Aplysia californica. J Neurosci 5: 1851-1861.

Mackie GO (1975) Neurobiology of Stomotoca. II. Pacemakers and conduction pathways. J Neurobiol 6: 357-378.

Mackie GO, Singla CL, Stell WK (1985) Distribution of nerve elements showing FMRFamide-like immunoreactivity in Hydromedusae. Acta Zool Stockh 66: 199-210.

Mahon AC, Lloyd PE, Weiss KR, Kupfermann I, Scheller RH (1985) The small cardioactive pepides A and B of Aplysia are derived from a common precursor molecule. Proc Natl Acad Sci USA 82: 3925-3929.

Mangerich S, Keller R, Dircksen H (1986) Immunocytochemical identification of structures containing putative red pigment-concentrating hormone in two species of decapod crustaceans. Cell Tissue Res 245: 377-386.

Martin SM, Spencer AN (1983) Neurotransmitters in coelenterates. Comp Biochem Biophys 74 C: 1-14.

McAllister LB, Scheller RH, Kandel ER, Axel R (1983) In situ hybridization to study the origin and fate of identified neurons. Science 222: 800-808.

McDonald TJ, Jörnvall H, Nilsson G, Vagne M, Ghatei M, Bloom SR, Mutt V (1979) Characterisation of a gastrin releasing peptide from porcine non-antral gastric tissue. Biochem Biophys Res Commun 90: 227-233.

McFarlane ID (1973) Spontaneous contractions and nerve-net activity in the sea anemone Calliactis parasitica. Mar Behav Physiol 2: 97-113.

McFarlane ID (1982) Calliactis parasitica. In: Shelton GAB (ed) Electrical conduction and behaviour in "simple" invertebrates. Clarendon Press, Oxford, pp 243-265.

McFarlane ID, Graff D, Grimmelikhuijzen CJP (1986) Excitatory actions of Antho-RFamide, an anthozoan neuropeptide, on muscles and conduction systems in the sea anemone Calliactis parasitica. J Exp Biol (in press).

Mordue W, Stone JV (1976) Comparison of the biological activities of an insect and a crustacean neurohormone that are structurally similar. Nature 264: 287-289.

Morris HR, Panico M, Karplus A, Lloyd PE, Riniker B (1982) Elucidation by FAB-MS of the structure of a new cardioactive peptide from Aplysia. Nature 300: 643-645.

Mühlethaler M, Dreifuss JJ, Gähwiler BH (1982) Vasopressin excites hippocampal neurones. Nature 296: 749-751.

Nagle GT, Greenberg MJ (1982) Effects of biogenic amines, FMRFamide and acetylcholine on the radula protractor muscle of a whelk. Comp Biochem Physiol 73 C: 17-21.

Nakanishi S, Inoue A, Kita T, Nakamura M, Chang ACY, Cohen SN, Numa S (1979) Nucleotide sequence of cloned cDNA for bovine corticotropin- β lipotropin precursor. Nature 278: 423-427.

Nakanishi S, Teranishi Y, Noda M, Notake M, Watanabe Y, Kakidani H, Jingami H, Numa S (1980) The protein-coding sequence of the bovine ACTH- β LHP precursor gene is split near the signal peptide region. Nature 287: 752-755.

Nambu JR, Taussig R, Mahon AC, Scheller RH (1983) Gene isolation with cDNA probes from identified Aplysia neurons: neuropeptide modulators of cardiovascular physiology. Cell 35: 47-56.

Nawa H, Hirose T, Takashima H, Inayama S, Nakanishi S (1983) Nucleotide sequence of cloned cDNAs for two types of bovine brain substance P precursor. Nature 306: 32-36.

Olivera BM, Gray WR, Zeikus R, McIntosh JM, Varga J, Rivier J, de Santos V, Cruz LJ (1985) Peptide neurotoxins from fish-hunting cone snails. Science 230: 1338-1343.

O'Shea M, Adams ME (1981) Pentapeptide (Proctolin) associated with an identified neuron. Science 213: 567-569.

Painter SD (1982) FMRFamide catch contractures of a molluscan smooth muscle: pharmacology, ionic dependence and cyclic nucleotides. J Comp Physiol 148: 491-501.

Palade G (1975) Intracellular aspects of the process of protein synthesis. Science 189: 347-358.

Price DA, Greenberg MJ (1977) Structure of a molluscan cardioexcitatory neuropeptide. Science 197: 670-671.

Price DA, Cottrell GA, Doble KE, Greenberg MJ, Jorenby W, Lehman HK, Riehm JP (1985) A novel FMRFamide-related peptide in Helix: pQDPFLRFamide. Biol Bull 169: 256-266.

Rao KR, Riehm JP, Zahnow CA, Kleinholtz LH, Tarr GE, Johnson L, Norton S, Landau M, Semmes OJ, Sattelberg RM, Jorenby WH, Hintz MF (1985) Characterization of a pigment-dispersing hormone in eye-stalks of the fiddler crab Uca pugilator. Proc Natl Acad Sci USA 82: 5319-5322.

Rothman BS, Mayeri E, Brown RO, Yuan PM, Shively JE (1983) Primary structure and neuronal effects of α -bag cell peptide, a second candidate neurotransmitter encoded by a single gene in bag cell neurons of Aplysia. Proc Natl Acad Sci USA 80: 5753-5757.

Rothman BS, Sigvardt KA, Hawke DH, Brown RO, Shively JE, Mayeri E (1985) Identification and primary structural analysis of peptide II, an end-product of precursor processing in cells R_3-R_{14} of Aplysia. Peptides 6: 1113-1118.

Sakata J, Mizuno K, Matsuo H (1986) Tissue distribution and characterization of peptide C-terminal α-amidating activity in rat. Biochem Biophys Res Commun 140: 230-236.

Satterlie RA, Spencer AN (1987) Organization of conducting systems in "simple" invertebrates: Porifera, Cnidaria and Stenophora. (This volume)

Scarborough RM, Jamieson GC, Kalish F, Kramer SJ, McEnroe GA, Miller CA, Schooley DA (1984) Isolation and primary structure of two peptides with cardioacceleratory and hyperglycaemic activity from the corpora cardiaca of Periplaneta americana. Proc Natl Acad Sci USA 81: 5575-5579.

Schaefer M, Picciotto MR, Kreiner T, Kaldany RR, Taussig R, Scheller RH (1985) Aplysia neurons express a gene encoding multiple FMRFamide neuropeptides. Cell 41: 457-467.

Schally AV (1978) Aspects of hypothalamic regulation of the pituitary gland. Science 202: 18-28.

Scheller RH (1986) Neuropeptides. Multipe regulatory mechanisms and their roles in mediating simple behaviors. In: Thorndyke MC, Goldsworthy G (eds) Invertebrate peptides and amines. Cambridge University Press (in press).

Scheller RH, Jackson JF, McAllister LB, Schwartz JH, Kandel ER, Axel R (1982) A family of genes that codes for ELH, a neuropeptide eliciting a stereotyped pattern of behavior in Aplysia. Cell 28: 707-719.

Scheller RH, Jackson JF, McAllister LB, Rothman BS, Mayeri E, Axel R (1983a) A single gene encodes multiple neuropeptides mediating a stereotyped behavior. Cell 32: 7-22.

Scheller RH, Rothman BS, Mayeri E (1983b) A single gene encodes multiple peptide-transmitter candidates involved in a stereotyped behavior. Trends Neurosci 6: 340-345.

Scheller RH, Kaldany RR, Kreiner T, Mahon AC, Nambu JR, Schaefer M, Taussig R (1984) Neuropeptides: mediators of behavior in Aplysia. Science 255: 1300-1308.

Schooneveld H, Tesser GI, Veenstra JA, Romberg-Privee HM (1983) Adipokinetic hormone and AKH-like peptide demonstrated in the corpora cardiaca and nervous system of Locusta migratoria by immunochemistry. Cell Tissue Res 230: 67-76.

Schot LPC, Boer HH (1982) Immunocytochemical demonstration of peptidergic cells in the pond snail Lymnaea stagnalis with an antiserum to the molluscan cardioactive tetrapeptide FMRFamide. Cell Tissue Res 225: 347-354.

Schwartz TL, Harris-Warrick RM, Glusman S, Kravitz EA (1980) A peptide action in a lobster neuromuscular preparation. J Neurobiol 11: 623-628.

Schwartz TL, Lee GMH, Siwicki KK, Standaert DG, Kravitz EA (1984) Proctolin in the lobster: the distribution, release and chemical characterisation of a likely neurohormone. J Neurosci 4: 1300-1311.

Shyamala M, Nambu JR, Shceller RH (1986) Expression of the egg-laying hormone gene family in the head ganglia of *Aplysia*. Brain Res 371: 49-57.

Siegert K, Morgan P, Mordue W (1985) Primary structures of locust adipokinetic hormone I and II. Biol Chem Hoppe-Seyler 366: 723-727.

Sigvardt K, Rothman BS, Brown RO, Mayeri E (1986) The bag cells of *Aplysia* as a multitransmitter system: Identification of α-bag cell peptide as a second neurotransmitter. J Neurosci 6: 803-813.

Siwicki KK, Bishop CA (1986) Mapping of proctolin-like immunoreactivity in the nervous systems of lobster and crayfish. J Comp Neurol 243: 435-453.

Smyth DG, Zakarian S (1980) Selective processing of β-endorphin in regions of porcine pituitary. Nature 288: 613-615.

Sofroniew MW, Weindl A (1978) Projections from the parvocellular vasopressin-and neurophysin-containing neurons of the suprachiasmatic nucleus. Amer J Anat 153: 391-430.

Spencer AN (1982) The physiology of a coelenterate neuromuscular synapse. J Comp Physiol 148: 353-363.

Starratt AN, Brown BE (1975) Structure of the pentapeptide proctolin, a proposed neurotransmitter in insects. Life Sci 17: 1253-1256.

Stone JV, Mordue W, Batley KE, Morris HR (1976) Structure of locust adipokinetic hormone, a neurohormone that regulates lipid utilisation during flight. Nature 263: 207-211.

Strangier J, Hilbich C, Beyreuther K, Keller R (1986) Isolation and functions of crustacean peptide hormones. In: Thorndyke MC, Goldsworthy G (eds) Invertebrate peptides and amines. Cambridge University Press (in press).

Stuart DK, Strumwasser F (1980) Neuronal sites of action of a neurosecretory peptide, egg-laying hormone, in *Aplysia californica*. J Neurophysiol 43: 499-519.

Sullivan RE, Miller MW (1984) Dual effects of proctolin on the rhytmic burst activity of the cardiac ganglion. J Neurobiol 15: 173-196.

Thorpe A, Duve H (1987) Purification, characterisation and cellular distribution of insect neuropeptides, with special emphasis on their relationship to biologically active peptides of vertebrates. (This volume)

Vanderhaeghen JJ, Signeau JC, Gepts W (1975) New peptide in the vertebrate CNS reacting with gastrin antibodies. Nature 257: 604-605.

Vandesande F, Dierickx K (1975) Identification of vasopressin producing and oxytocin producing neurons in the hypothalmic magnocellular neurosecretory system of the rat. Cell Tissue Res 164: 153-162.

Verhaert P, Grimmelikhuijzen CJP, DeLoof A (1985) Distinct localization of FMRFamide- and bovine pancreatic polypeptide-like material in the brain, retrocerebral complex and suboesophageal ganglion of the cockroach *Periplaneta americana L.*. Brain Res 348: 331-338.

Voight KH, Kiehling C, Frösch D, Schiebe M, Martin R (1981) Enkephalin-related peptides: Direct action on the octopus heart. Neurosci Lett 27: 25-30.

Westfall JA (1973) Ultrastructural evidence for a granule containing sensory-motor-interneuron in Hydra littoralis. J Ultrastruct Res 42: 268-282.

Westfall JA (1987) Ultrastructural of invertebrate synapses. (This volume)

Westfall JA, Kinnamon JC (1978) A second sensory-motor-interneuron with neurosecetory granules in Hydra. J Neurocytol 7: 365-379.

Witten JL, O'Shea M (1985) Peptidergic innervation of insect skeletal muscle: immunocytochemical observations. J Comp Neurol 242: 93-101.

Witten JL, Schaffer MH, O'Shea M, Cook JC, Hemling ME, Rinehart KL (1984) Structures of two cockroach neuropeptides assigned by fast atom bombardement mass spectroscopy. Biochem Biophys Res Commun 124: 350-358.

Woods KR, Wang KT (1967) Separation of dansyl-amino acids by polyamide thin layer chromatography. Biochim Biophys Acta 133: 369-370.

Ziegler R, Eckart K, Schwarz H, Keller R (1985) Amino acid sequence of Manduca sexta adipokinetic hormone elucidated by combined fast atom bombardement (FAB)/tandem mass spectrometry. Biochem Biophys Res Commun 133: 337-342.

PURIFICATION, CHARACTERISATION AND CELLULAR DISTRIBUTION OF INSECT NEUROPEPTIDES WITH SPECIAL EMPHASIS ON THEIR RELATIONSHIP TO BIOLOGICALLY ACTIVE PEPTIDES OF VERTEBRATES

ALAN THORPE and HANNE DUVE

School of Biological Sciences

Queen Mary College

University of London

Mile End Road

London E1 4NS

U.K.

ABSTRACT

Immunocytochemical techniques have been used on the insect nervous system to study the occurrence and distribution of neuropeptides related to the biologically active peptides of vertebrates.

Neurones immunoreactive to antibodies raised against insulin, glucagon, pancreatic polypeptide (PP), gastrin/cholecystokinin (CCK), vasopressin and the opioids including met- and leu-enkephalin and α- and β- endorphin have all been located in a range of insects.

Isolation and purification studies have provided evidence that "vertebrate-type" peptides identified by immunocytochemistry are indeed present in insect tissues. The identification of these peptides is mainly based upon radioimmunoassay but for some peptides, biological assay has also been carried out.

Insulin-like peptides have been partially purified from a number of insect tissue extracts and independently it has been shown that the insect hormone prothoracicotropic hormone (PTTH) is homologous to insulin. Other vertebrate-type peptides for which purification studies have reached a relatively advanced stage include those related to glucagon, PP, CCK, α-endorphin, met-enkephalin and vasopressin.

Studies of this type suggest a conservation of biologically active peptides over a long evolutionary period.

1. GENERAL INTRODUCTION

The nervous system is now known to produce an impressive array of neuropeptides. In mammals and other vertebrates the cellular localisation of a large number of these peptides has been determined and many have been isolated and characterised. Recent work has shown that certain biologically active peptides of the nervous system occur within the digestive tract and its associated organs and vice versa. These so-called brain-gut peptides are too numerous to mention in their entirety here, but they include such widely differing substances as cholecystokinin (CCK) and the enkephalin/endorphin peptides. Although the ubiquitous nature of peptides stems from the fact that all somatic cells contain the same DNA, it is perhaps surprising that the same genes come to be expressed in such varied locations as the duodenum and the cerebral cortex and it is intriguing to ask what role(s) the peptides play in their 'alien' site? For the most part this question remains unanswered.

Insects, as with all invertebrates, contain an equally large number of peptidergic neurones as components of their nervous system and it is widely recognised that the products of these cells control key physiological processes such as morphogenesis, reproduction, diapause, protein, lipid and carbohydrate metabolism, water and ion balance, cuticular tanning, visceral muscle functioning and heart activity. Unfortunately, despite a great deal of scientific endeavour, very few of these control 'factors' have been isolated, purified and characterised.

In this chapter we shall review, by references to certain insect neuropeptides, some of the problems associated with their cellular localisation, isolation and characterisation. At the same time, we shall address another fundamental issue of neurobiology that has been our main interest over the past decade and that is, to what extent the neuropeptides of insects (as representatives of the invertebrates) resemble those of vertebrates. In other words, are there functionally important amino acid sequences common to both groups of organisms that represent the end-products of a line of continuous evolutionary conservation of peptide structure?

Selection of the insect neuropeptides for detailed discussion in this chapter has been restricted to those that are specifically relevant to these two issues. Emphasis will be placed upon our own studies of the blowfly, Calliphora vomitoria, but reference to comparable studies of other insect species will be made where appropriate.

2. PROTHORACICOTROPIC HORMONE AND ITS RELATIONSHIP TO INSULIN

2.1. Introduction

The prothoracicotropic hormone (PTTH) of insects is known to be produced by neurosecretory cells of the brain from whence it passes by axonal flow to the corpora cardiaca/corpora allata complex. It is stored in this neurohaemal complex and released from there into the haemolymph. Its well known effect is upon the activation of the prothoracic glands, the subsequent release of ecdysone and the initiation of the moult. The titre of juvenile hormone (JH) from the corpora allata then dictates the character of the moult and the occurrence of metamorphosis.

PTTH, following its discovery by Kopec (1922), was initially known as the 'brain hormone' and it was something of a landmark in showing for the first time in any animal that the nervous system could produce a substance that has an endocrine function. Several groups of workers have since then devoted much time and energy in studies of PTTH and, as a result, a great deal of information had been obtained about its endocrine physiology and more recently its biochemistry. The most recent study of PTTH (Nagasawa et al. 1984) typifies the 'heroic' approach used to isolate, purify and characterise insect neuropeptides. It illustrates well the problems encountered and demonstrates some of the reasons why so few insect neuropeptides have been characterised to date.

One of the main problems is the small size of insects coupled with the fact that only relatively few neurones produce the active peptide(s). These two factors make it necessary to extract very large numbers of organisms, or parts of organisms, in order to obtain sufficient quantity of the pure peptide(s). In two studies of the PTTH of the Chinese silkworm Bombyx mori, Zhong et al. (1985) used a starting material comprising 2.8 million and Nagasawa et al. (1984) a total of 648,000 male moth heads. The latter number of heads weighed almost 5 kg and from this, after a 15-step purification procedure, only 50 µg of pure material was obtained. The purification method for both these studies was based on earlier studies of Nagasawa et al. (1979).

It is of significance to note that PTTH does not exist in a single molecular form but rather in a mixture of different forms with a range of molecular weights. Both the Japanese and Chinese groups seem agreed upon a Bombyx 4K-PTTH (molecular weight ~ 4400) which itself is a mixture of heterogeneous forms designated as 4K-PTTH-I, -II, -III. The other kind of Bombyx PTTH has a molecular weight ~ 22,000, the so-called 22K-PTTH. Likewise, in Manduca, the tobacco hornworm moth, purification of peptides from day-1 pupae brains has shown a 'big' PTTH (MW ~ 22,000) and a 'small' PTTH of approx. 7,000 Daltons (cf. Bollenbacher and Bowen 1983).

This pattern of heterogeneity of molecular forms within a single species of biologically active peptide is consistent with the known facts relating to the peptides of vertebrates (and to other known peptides of invertebrates). The cholecystokinin peptides isolated from vertebrates, for e.g., range from 39 amino acid residues down to the smallest known form, the tetrapeptide, CCK4. In addition to this type of macroheterogeneity of peptide structures which is due to differential post-translational processing occurring by means of specific enzyme changes, one also sees within the CCK group of peptides a form of microheterogeneity, where single amino acid residues vary as a result of post-translatory derivatisations e.g. sulphation of a tyrosine residue, or substitution of a single amino acid residue in the peptide.

The different molecular forms of PTTH found within a single insect species fit well into this general theme of peptide organisation although it is difficult, at the moment, to evaluate the reasons for the existence of such a large PTTH species as the 22K form.

2.2. The purification of PTTH

The purification procedure of Nagasawa et al. (1984) is as follows:

1. Homogenisation successively in acetone and 80% aqueous ethanol.

2. Extraction 3 times with 2% aqueous NaCl.

3. Heat treatment with boiling water; centrifugation to remove precipitate.

4. Ammonium sulphate precipitation (80% saturated solution).

5. Precipitation with 50-75% acetone.

6. Solubilisation in water, mixed with 9 vols of saturated picric acid solution and the precipitate appearing after centrifugation dissolved in 0.1 M tris-HCl pH 7.8.

7. Final extraction with acetone to give a pale brown precipitate designated as "crude" 4K-PTTH.

8. Dissolved in 0.1 M tris-HCl pH 7.8.

9. Gel filtration on Sephadex G50-Fine (6 x 67 cm) eluted with 0.2 M ammonium acetate (250 ml/h).

10. Pooled active fractions applied to DEAE-Sepharose CL-6B column (2.6 x 40 cm) equilibrated with 0.2 M ammonium acetate. The column primed with 0.4 M ammonium acetate and eluted with 1 M acetic acid (20 ml/h).

11. The elute applied to SP-Sephadex C-25 column (1.4 X 28 cm) pretreated with 0.1 M ammonium acetate buffer (pH 4.2).

12. Pooled active fractions lyophilised, dissolved in 0.1 M tris-HCl (pH 7.8) and applied to a Sephadex G50-SF column (1.8 x 170 cm) eluted with 0.2 M ammonium acetate (8 ml/h).

13. Active fractions combined (now designated as "highly purified" 4K-PTTH) and applied to a DEAE-Sepharose CL-6B column (1 x 54 cm) equilibrated with 0.05 M tris-HC1 containing 0.1 M NaCl. This column equilibrated with the same solution and then eluted with a gradient of 0.1-0.5 M sodium chloride in 0.05 M tris-HCl; pH 7.8, (8 ml/h); fractions of 3.5 ml collected.

   Three isolates 4K-PTTH-I   tubes 71-73 (0.28 M NaCl)

                    4K-PTTH-II   tubes 74-76 (0.30 M NaCl)

                    4K-PTTH-III tubes 83-89 (0.33 M NaCl)

   collected.

14. Reversed phase HPLC on Develosil Octadecylsilane (ODS)-5 Nomara (Kagaku) to give isolation of 50 µg 4K-PTTH-I, 36 µg 4K-PTTH-II and 63 µg 4K-PTTH-III.

15. The three varieties of PTTH sequenced by means of Edman degradation using an automated gas phase sequencer.

The purification of Bombyx mori 4K-PTTH reported by Zhong et al. (1985) is essentially the same as that of Nagasawa et al. (1984) and indeed, collaboration between the two groups occurred during the earlier stages of purification (cf. Suzuki et al. 1983). It appears from the literature, however, that, in this instance, purification has not proceeded beyond the "highly purified 4K-PTTH" - Step 13 above. The total reported yield of 41 mg pure material obtained is certainly impressive and should enable the Chinese group to learn more of the structure of Bombyx PTTH.

2.3. Bioassay of PTTH

The losses of material suffered at successive steps in a purification of this type are often substantial and an added difficulty is the question of how one identifies the isolated material through the different stages of purification. Three possible methods exist, viz. chemical assay, biological assay and radioimmunoassay. The first of these, chemical assay, is clearly impossible and, as yet, no radioimmunoassay has been devised for PTTH. This leaves biological assay as the only means of measuring quantities of this peptide and therein lies a complexity that has been one of the reasons for the slow progress made in identifying and characterising PTTH.

The 'in vivo' assays carried out by both Chinese and Japanese workers on Bombyx PTTH, strangely enough, do not make use of the same animal as the one from which the peptide is extracted and purified. The bioassay used routinely is the one described initially by Ishizaki and Ichikawa (1967). It involves lyophilisation of the test sample which is then taken up in 0.1 M Tris-HCl (pH 7.8) to the desired concentrations and injection of an appropriate amount (20 μl) of the final solution into debrained pupae of the Indian silkworm Philosamia cynthia ricini. The minimum amount of PTTH causing adult development in one pupa is defined as 1 Samia unit.

Such an assay is extremely time-consuming and difficult. The debrained pupae of Samia have to exist some 3-8 months after the debraining process and the normal stage of development taken to indicate PTTH activity, viz. wing apolysis, takes 4-7 days. As well as the difficulties in handling such a bioassay, the final expression of PTTH activity depends upon subjective assessment, always a serious drawback to any assay. Interestingly, Guo (1985) also describes a second action of Bombyx PTTH in Philosamia upon the mobilization of vitellogenin, and its subsequent liberation into the haemolymph and uptake by the developing oocytes - an activity most probably mediated through ecdysone secretion from the ovary). This activity has not, however, been used as the basis of a bioassay so far.

Bollenbacher and his co-workers introduced an 'in vitro' assay for the PTTH of Manduca, in which the release of ecdysone from cultured prothoracic glands under the influence of (test) samples of PTTH was measured directly by RIA (Bollenbacher et al. 1979, 1980). The contralateral gland is used as a control tissue. A similar 'in vitro' assay has been used for Bombyx PTTH (Suzuki et al. 1983).

2.4. The structure of Bombyx PTTH

The purified 4K-PTTH-I of Bombyx mori has now been subjected to amino acid composition analysis and the NH_2-terminal amino acid sequences of 4K-PTTH-I, -II and -III determined by Edman degradation in an automated gas phase sequencer (Nagasawa et al. 1984). The amino acid composition is reproduced in Table 1, where it is compared with the amino acid composition of bovine insulin and with the insulin-like peptide purified from the blowfly, Calliphora vomitoria by the authors and co-workers (Duve et al. 1979; Thorpe and Duve 1984) and with that purified from Manduca sexta by Kramer (1985).

The amino-terminal amino acid sequences of 4K-PTTH-I, -II and -III are shown in Fig. 1 (redrawn from Nagasawa et al. 1984) together with partial sequences of insulin-like growth factor (IGF-I) and the human and codfish insulin A chains.

TABLE 1. Amino acid composition of Bombyx 4K-PTTH-I and the insulin-like peptides of Calliphora and Manduca compared with that of bovine insulin. n.d.: not determined (For references see text)

	Bombyx 4K PTTH-I	Calliphora vomitoria Insulin-like peptide (Fractions 5 & 6 from analytical ODS - Step 10 of purification procedure)		Manduca sexta Insulin-like peptide	Bovine Insulin
		5	6		
Asx	4	6	6	3	3
Thr	3	3	3	Thr + Gly 8	1
Ser	1	3	3	7	3
Glx	4	7	7	6	7
Pro	1	1	1	n.d.	1
Gly	4	4	4	(See Thr)	4
Ala	3	3	3	2	3
Cys	4	n.d.	n.d.	n.d.	6
Val	4-5	3	3	3	5
Met	0	2	0	n.d.	0
Ile	0	1	1	2	1
Leu	5-6	5	5	2	6
Tyr	1-2	2	2	3	4
Phe	2	3	4	1	3
His	1	2	2	2	2
Lys	0	3	2	1	1
Arg	3	4	4	1	1
Trp	0	n.d.	n.d.	n.d.	
		Integer values normalised to 3 residues of alanine		Interger values normalised to 2 residues of alanine	

```
                              1   2   3   4   5   6   7   8   9  10  11  12  13  14  15  16  17  18  19

4K-PTTH-I              H - Gly-Val-Val-Asp-Glu-(Cys)-(Cys)-Phe-Arg-Pro-(Cys)-Thr-Leu-Asp-Val-Leu-Leu-Ser-Tyr

    -III               H - Gly-Val-Val-Asp-Glu-(Cys)-(Cys)-Leu-Gln-Pro-(Cys)-Thr- ? -Asp-Val-Val-Ala-Thr-Tyr

    -II                H - Gly-Ile-Val-Asp-Glu- Cys - Cys -Leu-Arg-Pro- Cys -Ser-Val-Asp-Val-Leu-Leu-Ser-Tyr

Codfish Insulin A chain    Gly-Ile-Val Asp Gln- Cys - Cys -His-Arg-Pro- Cys -Asp-Ile-Phe-Asp-Leu-Gln-Asn-Tyr

Human Insulin A chain
(residues 1 - 19)          Gly-Ile-Val-Glu-Gln- Cys - Cys -Thr-Ser-Ile- Cys -Ser-Leu-Tyr-Gln-Leu-Glu-Asn-Tyr

IGF - I
(residues 42 - 60)         Gly-Ile-Val-Asp-Glu- Cys - Cys -Phe-Arg-Ser- Cys -Asp-Leu-Arg-Arg-Leu-Glu-Met-Tyr
```

(Cys) - indicates that cysteine residues at positions 6, 7 and 11 in 4K PTTH-I and -III not identified fully.

FIG. 1: NH$_2$-terminal amino acid sequences of 4K-PTTH I, II and III from Bombyx mori compared with codfish and human insulin A chains (residues 1-19) and insulin-like growth factor residues 42-60. (From Nagasawa et al. 1984)

This remarkable study is interesting for several reasons. Firstly, with regard to the three PTTH molecules themselves, about one half of the amino acids in the NH$_2$-terminal sequence are common to all three peptides and any amino acid substitution that has occurred is such that the hydrophilic or hydrophobic nature of the amino acid residues is retained. Thus, Val is replaced by Ile, Phe by Leu, Thr by Ser, Leu by Val, Leu by Ala and Arg by Gln. Furthermore, except for the exchange of Leu by Ala, all these substitutions are possible by single nucleotide changes. As Nagasawa and his colleagues (1984) point out, the biological activities of the three molecular species are approximately equal and it must therefore be assumed that the substitutions can be tolerated and accepted by the PTTH receptors. It is not known whether one or more of the molecular species of PTTH occurs in indivudual organisms in a manner similar to the two species of insulin known to be possessed by the rat and mouse - differing by two amino acids in the B Chain, Pro/Ser B9 Lys/Met B29, or whether different populations of Bombyx have a single, specific type.

Secondly, in a wider context, the sequence homology of PTTH to insulin and insulin-like growth factor is an especially exciting finding and one which goes some way towards answering some of the intriguing questions concerning biochemical evolution and the conservation of DNA and protein structure. About one half of the 4K PTTH-II sequence is identical to that of the human insulin A chain and to IGF-1 and an even greater homology (58%) is seen when the sequence is compared with the codfish insulin A chain. Furthermore, when one considers the invariant residues of the known insulin A chains (cf Cutfield et al. 1979) it can be seen that Gly (1), Ile (2), Val (3), Cys (6), Cys (7), Cys (11), Leu (16) and Tyr (19) are all present in the Bombyx PTTH molecule. The only residue changes in the Bombyx molecule at an invariant insulin A chain position is Glu instead of Gln at residue 5. Also, of course we do not yet know whether the other invariant residues of the A Chain at 20 (Cys) and 21 (Asn) exist in the Bombyx PTTH molecule, nor do we know if indeed there is a second chain, equivalent to the B chain of insulin in which, of course, other important invariant residues of great significance in receptor binding exist. We await with great interest the further results from Nagasawa and his colleagues in the hope that they will clarify these important issues, but, meanwhile, it is difficult to avoid the conclusion that the Bombyx PTTH molecule is an insulin-related molecule and that it probably represents one of the earliest insulins known (see note added in proof).

It is interesting that the discovery of the insulin-like structure of PTTH came quite independently of several studies being carried out to extract and isolate an insulin-like peptide from the insect nervous system (for a review see Thorpe and Duve 1984). It is a fascinating coincidence in this regard that PTTH was discovered at the same time as insulin (Kopec 1922; Banting and Best 1922).

2.5. The purification of an insulin-like peptide from Calliphora

The studies of the insulin-like peptide of Calliphora carried out by Duve and co-workers (cf Duve and Thorpe 1979; Duve et al. 1979, 1981, 1982) arose out of earlier physiological studies which showed that animals from which the median neurosecretory cells (MNC) of the brain were extirpated became hyperglycaemic and hypertrehalocaemic (Duve 1978). Injection of extracts of the MNC or the re-implanting of these cells permitted the rising levels of both glucose and trehalose to be normalised and suggested that a hypoglycaemic 'factor' may be produced by these cells. Furthermore, immunocytochemical studies, in which antisera to bovine and porcine insulin were applied to sections of the nervous system of Calliphora, revealed a group of 4-8 insulin-immunoreactive neurones in the MNC (Duve and Thorpe 1979). Subsequently, insulin-immunoreactive material has been observed in the corpus cardiacum of colchicine-treated animals (unpublished observations). These studies initiated the biochemical studies to extract and purify the insulin-like peptide and a detailed account of the procedure will be given here, both to illustrate the problems of peptide purification in general and also to permit specific comparisons to be made between this methodology and that used in the isolation of the insulin-like PTTH molecule.

A distinct point of similarity in the two methods is the utilisation of large numbers of organisms. One study of PTTH dealt with over 3 million Bombyx heads (Zhong et al. 1985) and the definitive study of Nagasawa et al. (1984) used 648,000 heads. With Calliphora insulin-like material, after two pilot studies with ca. 100,000 organisms (Duve et al. 1979), we decided that more than 1 million animals would be necessary to generate sufficient of the insulin-like peptide with which to be able to carry out amino acid composition and possible sequence studies. Furthermore, to avoid unnecessary introduction of 'foreign' peptides and proteins into the purification procedures it was decided to use separated heads and therefore isolate, at least from the rest of the body, the specific cells known to contain the insulin-like material. Because the head of insects is separated from the thorax by such a slender 'neck' (along which the aorta is conveniently carried and from which blood sampling becomes relatively easy) it is possible, by freezing the animals with solid CO_2, to induce a state of fragility such that on shaking, the head detaches.

Calliphora vomitoria were obtained as pupae from a commercial supplier and after eclosion in the laboratory the adults were fed for up to 1 week on sugar and water. Over a 6 month period we were able to collect a sufficient number of heads with which to begin the isolation procedure. The weight of the starting material totalled 5.5 kg and the sheer bulkiness of such an amount of insect material (for whatever purification procedure might be followed) causes severe problems for the initial extraction step. In order to present a sufficiently large interface between the material to be extracted and the extraction fluid it was necessary to use more than 70 litres of acidified ethanol in the first step. The removal of this large excess volume of alcohol necessitated a large cyclical still which was operated under reduced pressure at a temperature of 26-30°C. Such large-scale extraction procedures also demand the use of large columns during the initial purification steps and, as an example, a 14 cm X 100 cm column was used during the initial Sephadex gel filtration phases. To avoid any possible contamination of the fly insulin-like peptide with mammalian insulins during extraction and purification, all apparatus, glassware and chemicals, including the chromatography columns, and their packing materials, were new. No mammalian insulins (other than the exceedingly low amounts of insulin used as standards in the radioimmunoassay) were permitted in the laboratory being used for the work and all glassware was washed separately.

The complete stages of the purification procedure are listed below.

1. Heads minced at 4°C.

2. Extraction in 74% ethanol acidified to pH 2.5 with phosphoric acid overnight at 4°C followed by re-extraction of the pellet.

3. Centrifugation (1875 x G) - supernatant filtered through a Hyflo bed.

4. Filtrate evaporated at 26-30°C under vacuum.

5. Gel filtration of a suitable residual volume (2.2 litres) from Step 4 on Sephadex G15-F. Column size 14 cm x 100 cm, eluent 3 M acetic acid, flow rate 15 ml/min, 495 ml fractions. Fractions radioimmunoassayed for insulin. [Also radioimmunoassayed for pancreatic polypeptide and glucagon - see later].

6. Fractions containing insulin immunoreactivity pooled, lyophilised and gel fitered on Sephadex G50-SF. Column 14 cm x 100 cm, flow rate 15 ml/min, 75 ml fractions.

7. Insulin-immunoreactive fractions pooled, lyophilised and applied to a DE-52-DEAE-cellulose ion exchange column 1.5 cm x 30 cm; gradient 20-400 mM ammonium acetate pH 8.5 with 20% acetonitrile, flow rate 20 ml/h, fractions 4 ml.

8. Insulin-immunoreactive fractions pooled, lyophilised and subjected to reversed phase high performance liquid chromatography (RP-HPLC). Preparative column, size 1 cm x 15 cm, column material Ultrasphere ODS, gradient 40-80% methanol/1% trifluoroacetic acid (TFA), flow rate 3 ml/min, 1 ml fractions.

9. Immunoreactive fractions reduced in volume with nitrogen and subjected to analytical RP-HPLC. Column size 0.4 cm x 15 cm, column material Ultrasphere Cyano, gradient 20-60% acetonitrile/1% TFA, flow rate 1 ml/min, 200 μl fractions.

10. A single insulin-immunoreactive fraction was finally purified by analytical RP-HPLC. Column size 0.4 cm x 25 cm, column material Ultrasphere ODS, gradient 40-80% methanol/1% TFA.

11. The two peak fractions (Nos. 5 and 6 - collected manually) were used to generate the amino acid composition analyses seen in Fig. 1.

A distinct difference in the methodology for PTTH purification compared with that for insulin is that the former requires a bioassay whereas for the insulin-like peptide a radioimmunoassay was used. Thus, throughout the isolation procedure the location and quantity of the fly insulin-like peptide was determined by means of a bovine (or porcine) insulin radioimmunoassay (RIA). Such an assay usually takes 16-24 hours to complete and compared with the bioassay for PTTH, described earlier, this can be seen to be a distinct advantage in allowing the succeeding purification stages to proceed more quickly. Indeed, at times during the purification, radioimmunoassays were performed over much shorter periods (as little as 45 min) and although equilibration was not complete, nevertheless it was possible to use such an assay to determine the specific fractions containing the insulin-like peptide and therefore to proceed to the next step almost immediately.

An important criterion applied during the purification procedure was an insistence that during radioimmunoassay, the insulin-like peptide should dilute linearly and in parallel to bovine (or porcine) insulin standards. This characteristic gives some assurance that the fly peptide bears a close relationship to insulin. Additional evidence to this effect has also come from three different bioassays of the isolated Calliphora insulin-like peptide. Firstly, in mammalian insulin bioassays we have shown that the Calliphora material is able (a) to displace ^{125}I-labelled porcine insulin from rat liver plasma membrane insulin receptors and (b) to permit the incorporation of tritiated glucose into lipid in isolated rat epididymal fat-cells. Secondly, the partially purified Calliphora insulin-like peptide is able to normalise the hyperglucaemia and hypertrehalocaemica in flies from which the MNC have been extirpated. (cf. Thorpe and Duve 1984).

The final yield of insulin-like peptide from the initial starting material of 5.5 kg was difficult to evaluate for several reasons. Firstly we have no way of knowing the precise degree of molecular homology of the fly peptide to mammalian insulin and therefore the concentrations of material that can be calculated from the radioimmunoassay may be under - or over - estimated. One certain fact is that the yield would have been much higher were it not due for the substantial rate of loss experienced with successive steps in the purification procedure. After DEAE-cellulose we calculated a yield of approximately 15 µg and yet after the final HPLC step this figure had decreased to approximately 2 µg. Just under half of this material was subjected to amino acid analysis and the other half was designated for amino acid sequence analysis. At this point however, one of the most serious causes of peptide loss that can be experienced during purification became frustratingly apparent. During the process of drying the pure sample down under nitrogen, in preparation for sequencing, the insulin-like peptide was found to have adsorbed to the surface of the glass vessel to such a degree that it could not be recovered. Subsequently, during the purification procedures followed for other peptides we have found this to be such a great problem as frequently to negate our efforts to obtain sufficient quantity of peptide for sequencing purposes.

The amino acid composition shown in Table 1 suggests a peptide with 50-52 residues (not including any cysteines or tryptophan residues that were not determined but which may be present) and if comparison is made with both the "insulin-like" Bombyx PTTH molecule and with insulin itself, then a certain degree of similarity may be observed. Whether this alone is sufficient to justify the term "insulin-like" is difficult to say since the value of amino acid composition determinations are notoriously difficult to assess and have often later been shown to be incorrect. We must wait and judge the degree of insulin-relatedness from subsequent studies. At this stage however it is tempting to speculate from the Calliphora insulin purification and bioasasay studies and from similar studies on other species that insulin-like molecules are present in insects.

As well as the biochemical evidence, there are also increasing numbers of immunocytochemical reports that certain of the median neurosecretory cells in a variety of insects are insulin immunoreactive. Thus, with an antiserum to porcine insulin it has been shown that the European cornborer, Ostrinia nubilalis has 4 immunoreactive perikarya in each brain lobe as well as having immunoreactive material in the neural part of the corpus cardiacum (Lavenseau et al. 1984). Likewise, in the hoverfly, Eristalis aeneus there are insulin-immunoreactive cells in the MNC (El-Salhy et al. 1980).

Another cytological report, of great significance in the content of the Bombyx PTTH/insulin relationship is the finding of Yui et al. (1980), that four pairs of large neurones in the MNC of the brain of the larval silkworm as well as nerve fibres descending to the corpus cardiacum and corpus allatum are insulin immunoreactive when tested with an anti-porcine insulin serum. Recently it has been shown by Ishizaki and coworkers that a monoclonal antibody raised against the 1-10 N-terminal sequence of 4K-PTTH-I detects the same 4 pairs of cells in Bombyx mori that are reactive to the anti-insulin serum (Ishizaki et al. 1985). We have confirmed this finding in a study of Bombyx mori, carried out in collaboration with Ishizaki and co-workers using the 1-10 4K-PTTH-I antibody and several different anti-porcine insulin polyclonal and monoclonal antibodies and anti-bovine insulin monoclonal antibodies.

The work of Nagasawa et al. (1984) on Bombyx PTTH shows a substantial degree of homology to the insulin A chain and we eagerly await their further findings concerning the remaining part of the PTTH molecule. Meanwhile, the debate will continue as to whether all insects have a PTTH type of molecule similar to that characterised for Bombyx and whether it is molecules of this type that are responsible for the insulin immunoreactivities that have so far been observed in other species.

3. PURIFICATION AND CHARACTERISATION OF INSECT HYPERGLYCAEMIC FACTORS – IS THERE A RELATIONSHIP TO GLUCAGON?

Interest in the hypoglycaemic factors of insects and their possible relationship to insulin (and now to PTTH) was preceded, by at least 10 years, with an interest in the hyperglycaemic hormone, so-named by Steele (1961, 1963) from studies of a "factor" present in the corpus cardiacum of the cockroach, Periplaneta americana that was capable of elevating blood sugar levels. Since then, a substantial volume of literature has accumulated on many different aspects of this "factor". However, there are still serious gaps in our knowledge and despite many attempts at isolation, very little is known about its chemical nature. One of the main problems is that we seem not to be dealing with a single peptide factor or hormone but with a variety of such compounds. Indeed, one of the few fully characterised peptide neurohormones in insects, the adipokinetic hormone, is known to have some hyperglycaemic activity in addition to its better known function in lipid mobilisation (Mordue and Morgan 1984a, b).

Allied to the variety of peptide structures that have hyperglycaemic activities in insects, there also appears to be a considerable variety in the range of physiological functions over which they have influence. The defined function of any particular purified factor in one insect, frequently is found not to exist when the same assay is performed in different species of insect and vice versa. Furthermore, many of the factors appear to be multifunctional and to have cardioacceleratory and adipokinetic activities as well as an effect on glycaemia. Considering the very small number of insect species that have actually been investigated compared with the enormous number of insect species that are known to exist this is extremely thought provoking.

A review of the chemistry of hypertrehalocaemic factors of insects hasrecently been provided by Goldsworthy and Gäde (1983) and rather than attempt to summarise this information in the present account, we propose to examine in some detail a more recent account in which, for the first time, the primary structure of a peptide with hyperglycaemic activity (and also cardioacceleratory activity) is reported (Scarborough et al. 1984).

It is the corpora cardiaca (CC) of insects that has consistently provided the richest source of hypertrehalocaemic factors, although whether these are synthesised within the brain and stored in the neurohaemal area of the CC, or whether they are produced 'in situ' within the cells of the glandular lobe is an interesting and, for some insects, an unresolved question. The study of Scarborough and his co-workers (1984) made use of some 4000 CC (with corpora allata attached) from P. americana and relied upon bioassay of the carbohydrate levels (and increased frequency of heartbeat) within animals of the same species as a means of identifying and purifying the peptide.

3.1. The purification of Periplaneta hyperglycaemic peptide

1. Homogenization in 5M acetic acid (1 ml/1000 CC).

2. Centrifugation (10 min at 10,000 x G) and re-extraction of the pellet.

3. Combined supernatants filtered on Sephadex G25 (1.3 cm x 100 cm), eluent 5M acetic acid, flow rate 6.5 ml/h.

   OR

   Supernatants applied to a Sep-Pak C18 cartridge (Waters) eluted with 30% 1-propanol/H_2O.

   * Protection of the peptide was afforded by adding 100 µg of bovine serum albumin (BSA) to the collection tubes.

4. Prepurified extracts (MW 800-1500) subjected to RP-HPLC. Column, 5 µm (Vydac 218 TP packing), column size, 25 cm x 0.46 cm, gradient 18-30% acetonitrile, 0.1% TFA, 0.2%/min increase.

   2 peaks with cardioacceleratory activity reported at 17 min = CC-1 and 34 min = CC-2.

5. Both factors (CC-1 and CC-2) rechromatographed by RP-HPLC. Column, 10 µm C_{18} (Aquapore RP-300), column size, 25 cm x 0.46 cm, gradient either: methanol in 0.1% TFA/H_2O starting at 32% methanol for CC-1 and 35% for CC-2, increasing at 0.1%/min OR 9% 1-propanol in 0.1% TFA/H_2O (for CC-1) and 12% 1-propanol (for CC-2) as eluents under isocratic conditions.

The amino acid sequences of the two peptides CC-1 and CC-2 were obtained by subjecting chymotryptic fragments to direct chemical ionisation mass spectrometry. The sequences are shown in Fig. 2 below where they are compared with adipokinetic hormones (AKH) of Locusta and Manduca and also with the N-terminal residues 1-12 of mammalian glucagon, secretin and vasoactive intestinal peptide (VIP).

The two peptides, CC-1 and CC-2, have been termed periplanetins and appear to be members of a family of peptides of which the adipokinetic hormone of Locusta, isolated and characterised by Stone et al. (1976), must be considered the archetype. This blocked decapeptide is known to be synthesised in the glandular lobes of the CC of locusts from where it is released in response to flight stimuli. In the locust it causes the mobilisation of lipid and enables long term flight to occur. In other species it has some hyperglycaemic activity (Mordue and Morgan 1984).

	1				5					10
AKH 1 (Locusta)	pGlu	Leu	Asn	Phe	Thr	Pro	Asn	Trp	Gly	Thr - NH_2
CC - 1 (Periplaneta)	pGlu	Val	Asn	Phe	Ser	Pro	Asn	Trp - NH_2		
CC - 2	pGlu	Leu	Thr	Phe	Thr	Pro	Asn	Trp - NH_2		
Glucagon (1 - 12)	His	Ser	Gln	Gly	Thr	Phe	Thr	Ser	Asp	Tyr - Ser - Lys -
AKH (Manduca)	pGlu	Leu	Thr	Phe	Thr	Ser	Ser	Trp	Gly	NH_2
Secretin (1 - 12)	His	Ser	Asp	Gly	Thr	Phe	Thr	Ser	Glu	Leu - Ser - Arg -
VIP (1 - 12)	His	Ser	Asp	Ala	Val	Phe	Thr	Asp	Asn	Tyr - Thr - Arg -

FIG. 2: Amino acid sequences of the adipokinetic hormones of Locusta and Manduca and the cardioacceleratory peptides CC-1 and CC-2 from Periplaneta compared with sequences of glucagon and secretin. (For references see text)

Exactly how many different forms of AKH-related peptides exist in insects generally or in the CC of a single insect species in particular, is difficult to judge, but it is interesting that Scarborough and co-workers (1984) suggest that the two periplanetins they have isolated are multifunctional within the cockroach. They regard as incorrect the report of Holwerda et al. (1977) in which it is stated that it is possible to separate hyperglycaemic from adipokinetic activity from the cockroach CC. They also report that CC-1 is "almost certainly" the factor originally given the name neurohormone D by Baumann and Gersch (1982).

Immunocytochemical mapping studies on the distribution of AKH-related peptides in neurones and neurohaemal organs of Locusta migratoria has been carried out by Schooneveld et al. (1983) with an antiserum raised against tyrosine-AKH [Tyr^1-AKH]. They have reported the existence of immunoreactive cells not only in the glandular lobes of the CC as would be expected, but also in neurones in the CNS. The study was carried out in all postembryonic stages and in view of the assumed function of AKH in regulating lipid mobilisation and utilisation during prolonged flight (Stone and Mordue 1980), it is interesting that they found no essential differences in the morphology and distribution of cells in nymphs (which do not fly) and adults. In a more recent study (Schooneveld et al. 1985), AKH-like immunoreactive material was reported in the nervous system of 19 species belonging to nine insect orders and except for Schistocerca (and Locusta), AKH immunoreactive material was shown to occur only in the neurones of the brain. At this stage, it is not easy to understand the function of this material, nor indeed to ascertain the precise degree to which it relates to AKH.

The question may now be considered as to whether the sequence homology shown by members of the AKH family of peptides can be extended to the family of vertebrate peptides that includes glucagon, secretin and vasoactive intestinal peptide (VIP). If the glucagon sequence beginning at the third amino acid residue from the N-terminus is lined up with AKH-related peptides, then the sequence homology shown in Fig. 2 can be seen to extend to 3 of the 8 residues of CC-1 and 4 out of the 9 residues of the AKH of Manduca. In addition, the third residue of glucagon is glutamine compared with pyroglutamate in the insect peptides and the aromatic amino acid tryptophan lines up with tyrosine in CC-1, CC-2 and AKH of Manduca. Both CC-1 and CC-2 have asparagine at position 7 compared with aspartic acid in the corresponding position on the glucagon molecule. The evidence of this structural homology may not necessarily be considered overwhelming, but it is certainly worthy of serious consideration, particularly when one considers that glucagon is known classically by virtue of the hyperglycaemia that it is albe to promote in mammals. (Secretin and VIP are also able to cause hyperglycaemia if given in sufficiently high pharmacological doses).

Before the sequence homology shown in Fig. 2 had been recognised, several workers had examined a variety of insects for evidence of glucagon-related peptides in an attempt to find some possible correlation with the hyperglycaemic 'factors'. The work of Tager and Kramer (1980) on the glucagon-like peptide of Manduca is noteworthy in this respect and the present authors, too, have located a glucagon-like immunoreactive peptide in the blowfly Calliphora vomitoria following a procedure essentially as described earlier for the insulin-like peptide. (cf. Duve and Thorpe 1984, and Thorpe and Duve 1984 for a review of insect glucagon-like peptides). Normally, we have been able to locate by immunocytochemical means, the corresponding cells thought to produce the mammalian-type peptide isolated in extracts and determined by RIA. With glucagon, however, we have not been able to do this. Certainly, we have located immunoreactive cells in the brain and thoracic ganglion with an N-terminal glucagon antiserum, but the necessary controls, in which the antiserum preabsorbed with the antigen (glucagon) should result in a lack of staining, have not been satisfactory. The only other report on the occurrence of glucagon immunoreactivity in insects is contained in a study of El Salhy et al. (1980) where a small group of immunoreactive cells were located in the brain of the hoverfly, Eristalis aeneus, but not in the corpus cardiacum.

Interestingly, cells within the corpus cardiacum of Calliphora do show immunoreactivity against secretin antisera and, furthermore, this immunoreactivity is completely absorbed out when secretin peptide is used to preabsorb the antiserum (Duve and Thorpe 1984a, b). There is certainly sufficient evidence of a link with the vertebrate peptide here to warrant further investigation of this interesting problem.

4. THE CELLULAR LOCALISATION AND PURIFICATION OF A PEPTIDE SHOWING HOMOLOGY WITH PANCREATIC PEPTIDE FROM THE BRAIN OF CALLIPHORA

4.1. Introduction

Pancreatic polypeptide (PP) is a peptide of 36 amino acids that was originally discovered as an impurity in insulin preparations of the chicken pancreas by Kimmel et al. (1975). Several mammalian variants of this peptide have now been purified and their primary structures determined (Floyd et al. 1977). The secretion of pancreatic polypeptide is regulated mainly by food intake and blood glucose concentrations through vagal cholinergic mechanisms. The physiological action of the peptide is not known for certain, but there is strong evidence that it is a hormone that inhibits pancreatic secretion and biliary tract motility (Schwartz 1983). Other members of the PP family include peptide YY (PYY), a gut hormone and neuropeptide Y (NPY), a brain peptide (Tatemoto 1982a, b).

Our earlier discovery of an insulin-like immunoreactive material within neurones of the brain of Calliphora prompted studies on the distribution of PP-immunoreactive material in this species (Duve and Thorpe 1980, 1982). By means of the peroxidase-antiperoxidase immunocytochemical method with two antisera directed against bovine PP and one against porcine PP we have shown the presence of immunoreactive neurones which have not previously been described as neurosecretory since they lack an affinity for the once commonly used neurosecretory stains. In the brain, immunoreactive cells occur in at least 6 well-defined regions as well as in certain parts of the suboesophageal ganglion. PP-immunoreactivity is also present in cells of the hypocerebral ganglion and in the thoracic and abdominal ganglia (Fig. 3). Often it is possible to trace axonal pathways within the neuropile and observations have indicated the existence of a previously unknown neurohaemal area in the dorsal part of the thoracic ganglion. Thus, considerable amounts of PP-immunoreactive material in the nerve fibres seem to be delivered to and collected within the dorsal neural sheath from where it could, presumably, be released into the surrounding haemolymph as required.

In addition to the neuronal elements, PP-immunoreactive cells have been observed within the digestive tract in several small cells with the typical features of gut endocrine cells (cf. Nishiitsutsuji-Uwo and Endo 1981). Such cells possess an extended basal, foot-like region and a slender apical filament, tapering towards the gut lumen.

PP-immunoreactivity has also been reported in 6 neurones of the brain of the larval silkworm Bombyx mori (Yui et al. 1980) and in 2 neurones of the brain, of the larval hoverfly, Eristalis aeneus (El-Salhy 1981). In a later study by the same author (El-Salhy et al. 1983), 2 groups of immunoreactive cells were demonstrated in each of the protocerebral hemispheres of adult Manduca sexta. Nerve fibres from these cells were seen to pass via the nervi corporis cardiaca to the CC where immunoreactive material was accumulated close to the cells lining the aorta.

In a study by Endo et al. (1982) on the localisation of PP-immunoreactivity in the central and visceral nervous systems of the cockroach Periplaneta americana, at least 6 pairs of cell groups were demonstrated in the brain, together with 3 pairs of immunoreactive cells in the suboesophageal ganglion. Additionally, they reported the presence of PP-immunoreactive cells in the thoracic and abdominal ganglia but found no immunoreactive cells in the retrocerebral complex.

The function of this widespread PP-immunoreactive material in the nervous system is unknown, as is the function of the PP-immunoreactive cells observed in the gut endocrine cells of Calliphora (Duve and Thorpe 1982) and in the cockroach Periplaneta americana (Iwanaga et al. 1981; Nishiitsutsuji-Uwo and Endo 1984); and mosquito (Lea and Brown 1984).

Arising out of our own studies on the cellular distribution of PP-immunoreactive material in Calliphora, the question we attempted to solve was not concerned with function, but rather the extent to which immunocytochemical data could be matched to data on composition and amino acid sequence. Is PP another example of a 'vertebrate' peptide appearing earlier in evolution than had hitherto been considered?

Fortunately, the isolation methods for insulin (described earlier in

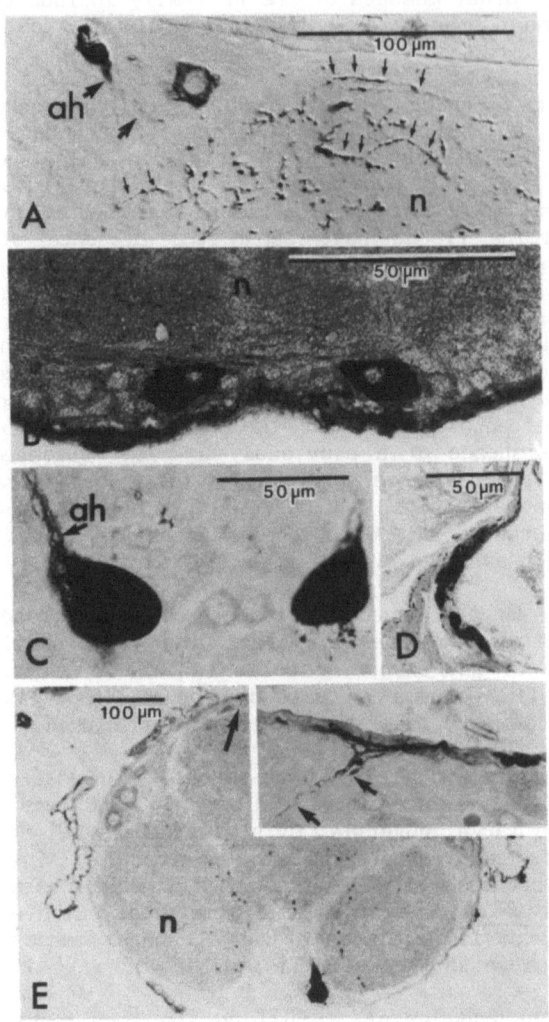

FIG. 3: A-E Distribution of cells and axons containing pancreatic polypeptide (PP)-immunoreactive material in the blowfly Calliphora vomitoria. A Transverse section through dorsolateral part of the protocerebrum showing PP-immunoreactive material in cell bodies and axons (small arrows). (Antibody dilution, 1:2000, PAP technique, DAB substrate, Nomarski optics). B Transverse section through part of the sub-oesophageal ganglion showing two PP-immunoreactive cell bodies in the rind of the ganglion. E Transverse section of thoracic ganglion treated with bovine PP antiserum, arrow indicates position of inset which shows axons containing PP-immunoreactive material joining dorsal sheath. C Detail of E showing ventrally-positioned perikarya and their axon hillocks. D Sagittal section of dorsal-posterior part of corpus cardiacum and hypocerebral ganglion. Hypocerebral ganglion showing PP-immunoreactive perikarya and axons leading into cardiac recurrent nerve.
ah axon hillock; n neuropile
C-D-E from Duve and Thorpe (1982)

detail in this chapter) were also appropriate for pancreatic polypeptide and we were able to monitor the same extracts with a radioimmunoassay for pancreatic polypeptide simultaneously to the radioimmunoassay being used to monitor the insulin-like peptide of Calliphora. In the initial stages of the purification procedure, the insulin-like and pancreatic polypeptide-like materials were seen to chromatograph in an identical manner, but complete separation was achieved at the DE-52 DEAE-cellulose stage (Step 7 - insulin purification). From then onwards the two peptides were dealt with separately although by means of almost identical chromatographic steps.

Duplicate amino acid analysis of the peak fraction of PP-like immunoreactive material derived from the final RP-HPLC step showed that the composition of the purified peptide was extremely similar to known mammalian pancreatic polypeptide molecules (Table 2).

TABLE 2. Amino acid composition of a Calliphora neuropeptide identified and monitored by means of a pancreatic polypeptide radioimmunoassay, compared with vertebrate pancreatic polypeptide molecules. (From Duve et al. 1982)

	Calliphora	Bovine	Porcine	Ovine	Human	Avian
Asx	3	3	3	3	4	6
Thr	2	2	2	2	2	2
Ser	1	0	0	1	0	1
Glx	6	6	6	6	4	4
Pro	4	5	5	4	5	4
Gly	2	1	1	1	1	3
Ala	4	5	5	5	5	1
Cys	-	0	0	0	0	0
Val	1	0	1	0	1	3
Met	1	2	2	2	2	0
Ile	2	1	1	1	1	1
Leu	3	3	3	3	3	3
Tyr	4	4	4	4	4	4
Phe	0	0	0	0	0	0
His	0	0	0	0	0	1
Lys	0	0	0	0	0	0
Arg	3	4	4	4	4	3
Trp	-	0	0	0	0	0

Two separate analyses of the Calliphora peptide were made. Integer values were normalised to four residues of tyrosine.

The homology of the Calliphora peptide with vertebrate pancreatic polypeptides is suggested by the fact that (a) the total number of amino acids is 36; (b) the individual amino acids are in proportions similar to the known vertebrate species; (c) the peptide appears to lack phenylalanine, histidine, lysine and (probably) tryptophan, and (d) the molecule elutes in the expected position of PP on HPLC. Furthermore, it would seem that the antigenic sites of the invertebrate peptide show the same conformation as those of mammalian species by the fact that immunoassay dilution curves of the fly peptide show linearity and parallelism with bovine PP standards and since the amount of material recovered from the final HPLC step as determined in amino acid analysis appears to be identical to the amount calculated from RIA.

These criteria point to the fact that the primary structure of Calliphora pancreatic polypeptide is very similar to the known pancreatic polypeptide molecules of vertebrates and suggest that evolutionary conservation of the molecule has occurred over a much longer period than hitherto believed.

5. GASTRIN/CHOLECYSTOKININ (CCK)-LIKE PEPTIDES

The knowledge that cholecystokinin (CCK) is a major vertebrate neuropeptide with an abundant and widespread distribution in both central and peripheral nervous systems (cf. Vanderhaeghen and Crawley 1985), has prompted several investigations into the possible occurrence of CCK-related peptides in the nervous system of invertebrates.

Our own studies on the gastrin/CCK-like peptides of Calliphora will serve to illustrate the approach that has been used and the progress made in insect studies.

5.1 Immunocytochemistry

Gastrin and cholecystokinin share a common C-terminal pentapeptide sequence - Gly - Trp - Met - Asp - Phe - NH_2 - that most probably stems from a common ancestral molecule (Larsson and Rehfeld 1977). For this reason, antisera that have been raised specifically against this sequence will not distinguish between the two peptides. Mainly with the use of these COOH-terminal antisera, we have mapped the distribution of gastrin/CCK immunoreactivity in both the brain and the retrocerebral complex of Calliphora and detailed reports of these findings have been published (Duve and Thorpe 1981, 1984).

Perikarya and nerve fibres immunoreactive to a range of COOH-terminal-specific antisera are widespread within the nervous system of Calliphora. In contrast, the fewer NH_2-terminal-specific antisera that have been used have given negative or non-specific results and the indication from this is that it is the amino acid sequence of the COOH-terminus rather than that of the NH_2-terminus that has been conserved during evolution. This finding, which has now been confirmed by our biochemical studies on tissue extracts of Calliphora (see later), is of interest when one considers that the COOH-terminus pentapeptide retains much of the biological activity of both gastrin and cholecystokinin and is indeed responsible for some of the overlapping gastrointestinal functions observed in studies of the two peptides.

Several regions within the brain and suboesophageal ganglion show strongly positive immunoreactive neurones (see Duve and Thorpe 1981). In addition, the three thoracic ganglia and the fused abdominal ganglia contain symmetrically arranged gastrin/CCK-immunoreactive cells both dorsally and ventrally in pairs on either side of the midline in a sagittal plane. The neuropile here has many immunoreactive nerve fibres and we have obtained evidence for the existence of a neurohaemal area within the dorsal sheath in the region of the metathoracic and abdominal ganglia where large amounts of gastrin/CCK-immunoreactive material are seen to collect (Fig. 4).

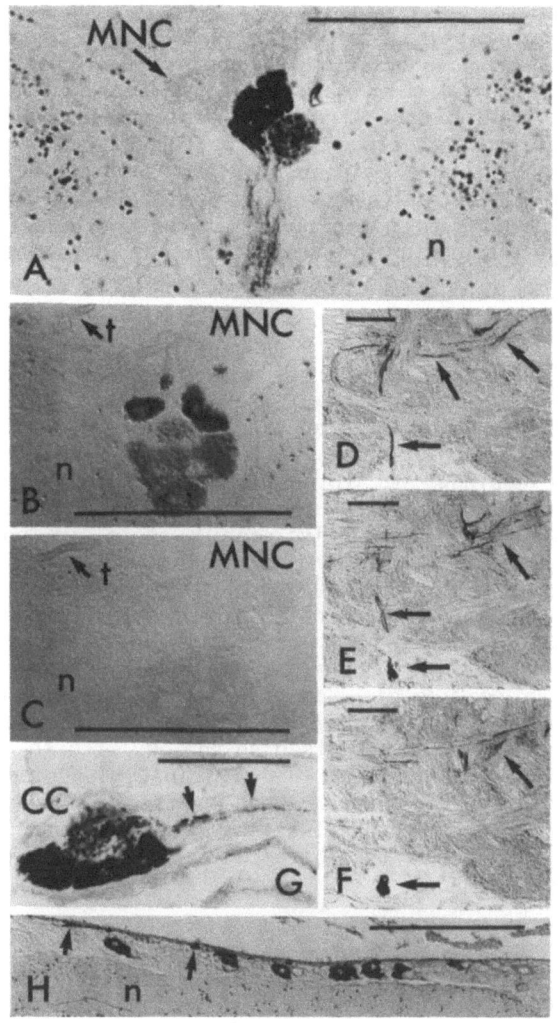

FIG. 4: Neurones of the brain, corpus cardiacum (CC) and thoracic ganglion of the blowfly, Calliphora vomitoria, showing gastric/CCK immunoreactivity (PAP technique, antiserum specific for the COOH-terminus of gastrin and CCK used at a dilution 1:1000, DAB substrate, Nomarski optics). A Transverse section through the plane of the median neurosecretory cells (MNC) of the brain showing certain of the cell bodies and axon hillocks containing immunoreactivity to gastrin/CCK. B-C Adjacent sections through the MNC region, B Section treated with gastrin/CCK antiserum and C treated with antigen-inactivated gastrin/CCK antiserum (30 μg gastrin 17 per ml diluted antiserum). D-F Adjacent sagittal sections through thoracic ganglion showing ventrally-positioned cell bodies and their axons (arrows) containing gastrin/CCK immunoreactivity. G Longitudinal section of cardiac-recurrent nerve (arrows) from MNC innervating corpus cardiacum CC. Note immunoreactivity to gastrin/CCK terminal-specific antisera in axons of nerve CRN (arrows) and in cell bodies and neuropile of CC. H Rows of gastrin/CCK immunoreactive cells and nerve terminals (arrows) just below dorsal neural sheath in abdominal ganglion.
n neuropile; t trachea; bar = 100 μm
G & H from Duve and Thorpe (1984a).

The corpus cardiacum contains immunoreactive material both within the intrinsic cells and also in the neuropile. Furthermore, we have shown that gastrin/CCK-immunoreactive material is present within the cardiac-recurrent nerve entering the CC anteriorly and also within the aortic (or cardiac) nerves leaving the gland dorso-posteriorly. The corpus allatus and the hypocerebral ganglion, on the other hand, do not contain gastrin-CCK immunoreactive material. The co-existence of gastrin-CCK/ immunoreactive material with secretin immunoreactivity in the CC and with pancreatic polypeptide immunoreactivity in the thoracic ganglion has been observed, although the significance of this is somewhat obscure at the present time.

In addition to the mapping of immunoreactive material within the nervous system of Calliphora, we have combined immunocytochemical and cobalt-backfilling techniques in an attempt to trace specific gastrin/CCK-immunoreactive peptidergic pathways (Duve et al. 1983). These types of experiment were designed to throw light on the possible function of gastrin/CCK-like peptides in an insect such as Calliphora and the results have proved highly instructive. Two peptide-containing pathways have been resolved by the combination of the two techniques, the one leading from neurosecretory cells of the mid-brain to the corpus cardiacum (and possibly beyond), the other connecting mid-brain neurosecretory cells with the central body, known to be an important integrating centre in the Diptera. If we assume that the two pathways represent two different functions for the peptide(s) - one centred in the brain itself and the other concerned with other regions of the body as the peptide is released from the CC into the haemolymph, then an analogy with the dual (brain/gut) occurrence (and dual function) of the gastrin/CCK peptides of vertebrates is apparent. In both vertebrate and invertebrate examples the neuro-transmitter/neuromodulatory role for the brain peptide appears obvious and contrasts with a possible metabolic/physiological role for the peptide found outside the brain.

Other insects shown to contain CCK/gastrin-like immunoreactivity in the nervous system include Bombyx, where at least ten CCK/gastrin-immunoreactive neurones were demonstrated in the median neurosecretory cell group, very close to the insulin-containing cells as well as in two large neurones in the thoracic ganglion (Yui et al. 1980). El-Salhy et al. (1980) reported immunoreactivity to a C-terminal gastrin antiserum in 3 groups each of 3-4 cells in areas of the brain including the MNC in Eristalis, and in Manduca they reported on the location of gastrin immunoreactive cells, despite the fact that the antiserum used cross reacts with the C-terminus of both gastrin and CCK (El-Salhy et al. 1983). No such immunoreactivity occurred in the MNC, neither did they find any immunoreactivity in the CC. Other species that have been studied include the Colorado beetle in which the existence of one pair of neurones immunoreactive to C-terminal gastrin/CCK antisera was demonstrated in each half of the dorsolateral protocerebrum (Veenstra et al. 1985). This group was unable to show immunoreactive cells with CCK mid-core specific antisera and likewise we have not detected such cells in Calliphora. This does not necessarily mean that a peptide with a CCK-related sequence does not exist in either species and, indeed, from our RIA studies of tissue extracts we have reason to believe that a peptide of this type is present. Immunocytochemistry of the CCK peptides within the brain of mammals has also provided a difficult task and there are several reasons for this, including inadequate fixation and the possible masking of the antigenic determinant during the processing and packaging of the peptides into granules within the cell.

5.2. Purification of CCK-like peptide of Calliphora

The evidence from immunocytochemistry (ICC) suggests the presence of a peptide related, at least at the COOH-terminus, to cholecystokinin and gastrin. As with all ICC studies, however, this does not constitute proof of the existence of such a molecule, but only that the same or closely similar antigenic determinants recognised by the antibody are present in the tissues. The only definite proof of a CCK-related molecule in insect tissues has, of necessity, to be derived from isolated peptides for which the amino acid composition and sequence are known.

We have now made several attempts to isolate the CCK-immunoreactive peptide of Calliphora (cf. Dockray et al. 1981; Duve et al. 1985) and whilst some progress has been made, there are still many unresolved difficulties.

In initial studies, gastrin/CCK-like material was extracted with boiling water and purified by immunoaffinity absorption to a C-terminal specific gastrin antiserum. This material was further purified by gel-filtration where it was resolved into two peaks that could be identified by antisera specific for the C-terminus of CCK and gastrin but but not by means of antisera specific for the N-terminus or for intact gastrin (Dockray et al. 1981). In a later study several different extraction media were used in addition to boiling water (Duve et al. 1985). In particular, it was found that 0.2 M HC1 resulted in a large peak of immunoreactive material with a Kav of 0.2 on Sephadex G50-SF. This was able to be further purified by means of RP-HPLC although considerable difficulties are involved due to its extremely hydrophobic nature. Thus, an unusually high percentage (between 60% and 70%) of acetonitrile or ethanol is required in order to detach it from octadecylsilane (ODS), cyanopropyl (CN) or alkyl phenyl support materials. Allied to this property, we have experienced considerable losses of peptide following its application to the HPLC columns. Some losses occur on the column itself whilst, in addition, severe losses also occur by adsorption to the walls of the collecting tubes. By means of region-specific antibodies used in radioimmunoassay we are able to say that the peptide resembles CCK more closely than gastrin although at the present time we have not obtained sufficient of the pure material from which amino acid composition or sequence analysis can be determined.

6. THE DISTRIBUTION AND PURIFICATION OF OPIOID-LIKE PEPTIDES IN INSECTS

Since the original identification of the two pentapeptides, methionine- and leucine-enkephalin (Hughes et al. 1975), a large series of opioid peptides have been demonstrated in both nervous and non-nervous tissues of vertebrates (cf. Imura et al. 1985).

An initial report suggested that invertebrates did not possess peptides of this type (Simantov et al. 1979) but since then, evidence to the contrary has been presented from studies of a range of invertebrate species including insects. The peptides investigated include most of those derived from the large precursor molecule proopiomelanocortin (POMC).

6.1 Immunocytochemical studies of opioid-like peptides in Calliphora and other insects

Within the median neurosecretory cells of the pars intercerebralis of the brain of the blowfly, we have detected 2 groups, each of approximately

6 cells, that are strongly immunoreactive to antisera raised against α-endorphin, a peptide that corresponds to the amino acid sequence present between residues 61 and 76 of the precursor molecule, β-lipotrophin (β-LPH). The immunoreactive material is also seen in the axons of these cells as they pass in the median nerve bundle ultimately to reach the corpus cardiacum. From here, some of the α-endorphin immunoreactive material can be seen to leave the CC in the cardiac-recurrent nerve, dorsal to the proventriculus in the direction of the abdomen (cf. Duve and Thorpe 1983). (Fig. 5).

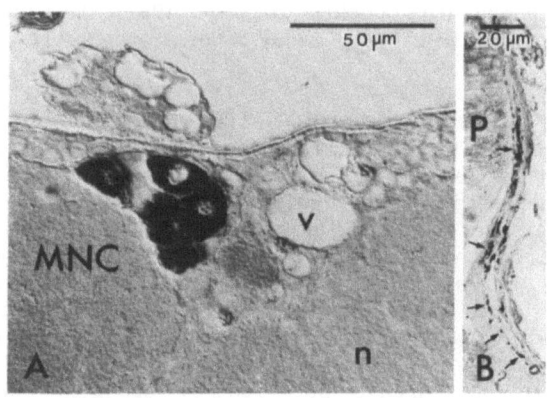

FIG. 5: A Transverse section through the MNC region of Calliphora vomitoria showing cells immunoreactive towards an α-endorphin antiserum (PAP technique - Nomarski optics). B Sagittal section showing α-endorphin-immunoreactive material in single axons in cardiac-recurrent nerve leaving CC dorsally, above proventriculus P.

Rémy and co-workers have also observed α-endorphin-immunoreactive neurones in two lepidopterans Bombyx mori and Thaumetopoea pityocampa (Rémy et al. 1978, 1979). It is of interest, however, that the reactive cells in these insects are located in the suboesophageal ganglion. Reasons for the different cellular distribution of the same or similar peptidergic substances within different insects are somewhat puzzling since one might assume that homologous peptides would occur in homologous cellular localisations. This problem has been discussed by, amongst others, Veenstra (1984).

Enkephalin-immunoreactive neurones have been detected in several insect species. In Calliphora both leu- and met-enkephalin immunoreactive cells have been observed within the CNS (Duve et al. 1986). In addition to these cytological studies, the presence of enkephalin receptors in both brain and midgut of Leucophaea maderae has been suggested in the reports of Stefano and Scharrer (1981) and Stefano et al. (1982). Immunocytochemical studies have also suggested the presence of adrenocorticotropic hormone (ACTH) and β-endorphin in insects - although it has to be pointed out that the study of Hansen et al. (1982) did not show evidence of appropriate specificity control tests.

There have been only a few studies aimed at isolating and characterising insect opioid-like substances.

6.2. The Isolation of immunoreactive opioid-like peptides in Calliphora

6.2.1. α-Endorphin

A preliminary report on the purification of an α-endorphin-like peptide has recently been published (Duve et al. 1986). In subsequent studies, the purification procedure has been slightly modified as follows:

1. 1.5 kg of whole flies homogenised in a Waring Blender (1 min) in cold 0.2 M HC1 (1:10 w/v). Extraction overnight at 4°C.

2. Filtration of extract through muslin cloth to remove crude debris, then through a Hyflo bed to remove some pigments and lipids.

3. Centrifugation (1875 x G) for 1 h.

4. Supernatant neutralised by addition of 1 M NaOH.

5. Neutralised supernatant lyophilised and gel filtered on Sephadex G50-Fine. Column 5 cm x 60 cm, eluent 1% formic acid, flow rate 74 ml/h, fraction size 12.3 ml.

6. Samples of the fraction (200 µl) were lyophilised and radioimmunoassayed for α-endorphin.

7. Immunoreactive fractions lyophilised and subjected to semi-preparative RP-HPLC. Column (Waters) µBondapak ODS 7.8 mm x 30 cm, Gradient 0-80% acetonitrile/water/0.1% TFA; flow rate 1.5 ml/min changing at 0.6% acetonitrile/min, Fraction size 1.5 ml.

8. Immunoreactive fractions applied direct (no lyophilisation!) to analytical RP-HPLC. Column, Waters µBondapak Alkyl Phenyl 3.9 mm x 30 cm, gradient 0-80% acetonitrile/water/0.1% TFA, flow rate 1.5 ml/min, fraction size 1.5 ml.

9. Immunoreactive fractions applied to analytical RP-HPLC. Column Waters µBondapak C18, size 3.9 mm x 30 cm, gradient 0-80% acetonitrile/water/0.1% TFA, flow rate 1.5 ml/min, fraction size 1.5 ml.

10. Ion-exchange chromatography of major immunoreactive fraction. Column Mono Q anion exchange (Pharmacia), gradient 0-1% acetic acid in 20 mM sodium phosphate buffer pH 6.5, flow rate 1 ml/min, fraction size (<u>either</u> various if peaks of OD material collected manually <u>or</u> 1 ml when collected automatically).

11. Immunoreactive fraction(s) re-applied to analytical RP-HPLC C18 µBondapak (Waters) [details as for Step 9 except that the counter ion heptafluorobutyric acid (HFBA) is used instead of TFA].

Several features of the purification procedure for this peptide are worthy of comment. Firstly, although after the initial gel filtration step, the α-endorphin immunoreactive material appears to be stable at room temperatures and also capable of withstanding lyophilisation, thereafter, and especially after analytical RP-HPLC steps, it becomes extremely difficult to handle and severe losses of material are experienced. One of the problems undoubtedly stems from its hydrophobic nature. Thus, we have seen that serial samples taken from a single collection vial (held at 4°C) over a period of only 3 days, show diminishing amounts of α-endorphin immunoreactivity as measured by RIA.

Losses of well over 90% have been experienced in this manner and there would seem to be only two possible explanations to account for this: (a) the material adsorbs strongly to the surfaces of collecting tubes (also the column surfaces and associated tubing), and (b) the peptide is present within the assay sample taken for the assay but the immunoreactivity is lost in some way (perhaps by oxidation?). We have shown that the material 'sticks' to polypropylene tube surfaces and this state of affairs has been remedied somewhat by the use of siliconised glass tubes. Addition of BSA to the collecting vessels in the early stages of purification also helps but, of course, in the final preparation of pure material for sequencing this is not possible. The high losses of peptide in a procedure such as this, where the active principle is present in such small quantities in the first place, continues to be a daunting and frequently occurring problem and it presents a serious hindrance to progress in the insect neuropeptide research field.

Secondly, in the isolation of this particular peptide we have had to rely completely upon radioimmunoassay as a means of identification. There is no particular physiological function that we can associate with the material in the blowfly in our present state of knowledge and it is therefore not possible to run a bioassay concurrently with the radioimmunoassay.

Thirdly, our observation that at least 5 other peptides co-elute with the α-endorphin immunoreactive peptide even after 5 chromatographic steps may be seen as a matter of some consternation to all involved in the purification of peptide from crude, whole body extracts. Even when the material has apparently been resolved into a single peak, great care must be taken to ensure that the small quantities of the immunoreactive peptide are not being masked by larger quantities of an unrelated peptide.

Lastly, purification studies of this type are made exceedingly difficult by the very small quantities of the active principle present within the correspondingly great bulk of crude proteins, peptides and pigments which forms the starting material. It is essential, therefore, that as large an amount of material, in as pure a state as possible, should be taken to begin a study of this type, even if this involves (as in the case of the Calliphora α-endorphin-immunoreactive material) frequent and time-consuming repetition of the initial steps in order to have sufficient material when one comes to the final, amino acid sequencing step.

The outcome of our studies on Calliphora is that we now have a suitable purification procedure with which to isolate sufficient material for sequencing purposes and we are currently engaged in the early phases of the procedure. We believe that Calliphora does, indeed, have a molecule similar to, and most probably a fore-runner of, the proopiomelanocortin of vertebrates. The α-endorphin-immunoreactive peptide that we have partially purified, we think, is a part of this particular precursor sequence. The immunocytochemical findings of an α-endorphin-like peptide represent the visualisation of this particular peptide within the cells that produce it and within the axons that transport it to the CC and thence to the rest of the body. If this is so, then one should also be able to locate other POMC-related sequences such as those of the enkephalins, ACTH, melanocyte-stimulating hormone (MSH) and β-endorphin. With this in mind we have been studying extracts of Calliphora (produced from 0.2 M HCl) and we now have preliminary evidence to suggest the presence of met-enkephalin, ACTH and β-endorphin as determined by radioimmunoassay (unpublished observations).

7. INSECT PEPTIDES THAT CURRENTLY ARE NOT KNOWN TO HAVE A DIRECT RELATIONSHIP WITH ANY VERTEBRATE PEPTIDE

So far, we have discussed insect neuropeptides that have an apparent relationship to known biologically-active peptides of vertebrates (either neuropeptides or peptide hormones). However, several insect peptides that bear no apparent relationship to identified vertebrate peptides have been isolated and characterised and a short summary of some of these will be given here.

7.1 Proctolin

Proctolin is a pentapeptide with the amino acid sequence H-Arg-Tyr-Leu-Pro-Thr-OH, MW 648. The first definitive isolation and purification of the peptide was carried out by Brown and Starratt (1975) and the mere thought of the 125 kg of cockroaches (Periplaneta) from which 180 µg of pure peptide was finally recovered is sufficient to restrict the field of insect neuropeptide research to a small number of scientists not prone to nightmares. The procedure adopted was as follows:
Isolation of proctolin from Periplaneta americana.

1. Whole adult cockroaches (125 kg) were chilled and blended in cold 7% perchloric acid (1 : 2 w/v) and extracted overnight.

2. The suspension was filtered through cheese-cloth and filter paper.

3. Filtrate adjusted to pH 6 with 45% KOH, precipitate removed by filtration.

4. The adjusted filtrate was brought to 50% ethanol, and precipitate removed by centrifugation.

5. Ion-exchange chromatography: Dowex 50W-x8 treated with 1N pyridine. Column 7 cm x 40 cm; eluent: a mixture of 4N NH_4OH and ethanol (1 : 1).

6. Basic fraction was taken to dryness.

7. Ion-exchange chromatography: Dowex 50W-x8, NH_4^+form. Column 3 cm x 80 cm; eluent 0.05 N NH_4OH; flow rate 75 ml/h. Recovery ~ 1.12 mg proctolin or 75%.

8. Alumina-adsorption chromatography. Column 3 cm x 40 cm; gradient 30-100% methanol; flow rate 125 ml/h. Recovery ~ 1.03 mg proctolin.

9. Ion-exchange chromatography, Rexyn 101, NH_4^+form. Column 1.3 cm x 44 cm; gradient 0-0.4 N NH_4OH; flow rate 12 ml/h. Recovery ~ 940 µg proctolin.

10. Partition procedures using a Craig counter-current unit employing butanol-acetic acid-water (4 : 1 : 5). Recovery ~ 600 µg proctolin.

11. High-voltage paper electrophoresis using Whatman 3 MM paper and solvent system pyridine-acetic acid-water in ratios of 25 : 1 : 350 (pH 6.4) and 1: 10 : 445 (pH 3.5). Recovery ~ 400 µg proctolin.

12. Gel filtration, Sephadex G 15. Column 1.6 cm x 190 cm; eluent 0.02 M ammonium formate; flow rate 11 ml/h. Recovery ~ 206 µg proctolin.

13. Ion-exchange chromatograghy, Rexyn 101, NH_4^+ form. Column 0.2 cm x 20 cm; gradient water-ammonium hydroxide, 0-0.04 N NH_4OH; flow rate 0.35 ml/h. Recovery ~ 180 µg.

Subsequently proctolin has been isolated from other insects, e.g. from the stable fly Stomoxys calcitrans (Holman and Cook 1979) and more recently from the cockroach (Leucophaea maderae) (Holman and Cook 1985). The latter account shows a great simplification in the technique.

Isolation and purification of proctolin from oviduct extracts of Leucophaea maderae.

1. Oviducts (30) adult female cockroaches homogenised in 30 ml of a mixture of methanol-water-acetic acid (90 : 9 : 1).

2. Centrifugation 30 min 13,800 x G 4°C.

3. Supernatant removed and evaporated at 40°C and finally freeze dried.

4. Sep-pak purification: a C18 Sep-pak (Waters Associates) was rinsed with 10 ml 75% acetonitrile-water-0.1% TFA. Proctolin was eluted with 4 ml 50% acetonitrile-water-0.01% TFA.

5. HPLC chromatography, µBondpak Phenyl. Column 4.6 mm x 30 cm; flow rate 1.5 ml/min; gradient 0-25% acetonitrile-water-0.1% TFA; fraction size 3 ml.

6. Quantitative bioassay with the hindgut of L. maderae demonstrated a proctolin titre of 0.93 ± 0.15 ng/oviduct

Proctolin has been shown to be present in neurones of the central and peripheral nervous systems (cf. Bishop et al. 1981) and may be assumed to have a role in neurotransmission and/or neuromodulation. It would certainly seem to have a visceral myotropic role in insects, as initially proposed by Brown (1975), although precisely for which muscles and under which particular set of physiological circumstances is not apparent for all insects. The insect bioassays for proctolin have been discussed by Miller (1983) and the subjects for these assays include the cockroach heart, the hyperneural muscle of Periplaneta (but only when connected to the ventral nerve cord), the extensor tibia muscle of Locusta, the hindgut and proctodeal muscle of either Periplaneta or Leucophaea and also the oviducts of either the horsefly, Tabanus or the cockroach Leucophaea. In addition to insect tissues, proctolin has pharmacological effects on crustacea and mammals.

The precise physiological role of proctolin in insects is further complicated by the recent discovery by Holman and Cook (1983) of a second myotropic peptide from the hindgut of the cockroach Leucophaea maderae.

7.2. The eclosion hormone of the silkworm, Bombyx mori

The elucidation of the N-terminus of the eclosion hormone from pharate adult heads of the silkworm Bombyx mori is another remarkable achievement of the Japanese group headed by Nagasawa and co-workers (Nagasawa et al. 1985). This study on the isolation and purification of a hormone long-known on account of its physiological role in pupal-adult eclosion has resulted in the identification of the amino acid sequence of the 13 N-terminal amino acid residues which are as follows: H-Ser-Pro-Ala-Ile-Ala-Ser-Ser-Tyr-Asp-Ala-Met-Glu-Ile. The procedure, which made use of 180,000 pupal heads and resulted in a yield of 1.2 nmol (10µg) eclosion hormone after a 2,000,000 fold purification is shown below:

1. Washing with acetone.
2. Washing with 80% ethanol.
3. Extraction with 2% NaCl.
4. Heat treatment.
5. Precipitation with 80% saturation of $(NH_4)_2SO_4$.
6. Precipitation with 50-75% acetone.
7. Precipitation with 90% saturation of picric acid.
8. Precipitation with 80% acetone.
9. Sephadex G-50 (fine).
10. DEAE - Sepharose CL-6B.
11. SP - Sephadex C-25.
12. QAE - Sephadex A-25.
13. Octyl - Sepharose CL-4B.
14. Sephadex G-50 (superfine).
15. DEAE - 5PW HPLC.
16. Hi-Pore RP-304 HPLC.

The bioassay method consisted of injecting aliquots of the eluates into Bombyx pharate heads whereupon one eclosion hormone (EH) unit was designated as the minimum amount necessary to elicit eclosion behaviour in more than 50% of the injected animals. It is interesting that Nagasawa and his colleagues experienced adsorption of the purified peptide to the walls of the glass tube, a common problem in insect neuropeptide isolation as already discussed. They were able to counteract the stickiness of this particular peptide by adding BSA to the aliquot before lyophilisation. Injection of 4 µg of BSA together with the purified EH appeared not to be of any detriment in the bioassay.

Unlike the PTTH hormone, there appears not to be any sequence homology between the EH and any of the known vertebrate (or invertebrate) peptide hormones, although this is not to say that homologous sequences will not ultimately be discovered, or that they do not exist in the so-far unsequenced portion of the molecule.

As the authors of this work point out, the chemical characterisation of a peptide neurohormone such as EH is the first crucial step to understanding mode of action, mechanism of biosynthesis and release and ultimately perhaps to developing new ways to control insect pests.

8. DIURETIC PEPTIDES IN INSECTS

Several research groups are currently involved in the isolation and purification of insect peptides that are believed to be responsible for water balance. These factors are variously termed diuretic and antidiuretic.

Mordue and his co-workers have published several accounts dealing with the isolation and characterisation of a locust diuretic hormone (Morgan and Mordue 1983, 1984a, b) and there is some belief from these studies that the locust diuretic hormone may exist in at least two distinct molecular forms in that methanolic extracts of the storage lobe of the corpora cardiaca subjected to RP-HPLC produces 2 peaks of diuretic activity. Continuing attempts are now being made by this group to purify sufficient of the peptide(s) in order to obtain amino acid sequence data.

Meanwhile, a most interesting study on the locust diuretic hormone [produced primarily in the suboesophageal ganglion, but also elsewhere (Proux and Rougon-Rapuzzi 1980)], is also being carried out by Proux and his collaborators and the results of these studies are possibly in conflict with the Aberdeen group. The dispute centres around whether or not the water balance hormone(s) are related to the vertebrate water balance hormones, vasopressin and/or oxytocin. Rémy et al. (1977) had earlier shown by means of immunocytochemistry that 2 neurosecretory cells present in the suboesophageal ganglion of Clitumnus showed intense immunoreactivity when treated with anti-vasopressin antisera. These observations were later extended to the brain of Bombyx and Locusta and to the entire ventral nerve cord of Locusta by Rémy et al. (1979) and by Rémy and Girardie (1980). Similar immunocytochemical results have been obtained from other insects including Periplaneta, Blabera, Gryllus, Acheta and Pollistes (Strambi et al. 1978; 1979).

Subsequently Proux's group has used both bioassay and more interestingly, a vasopressin-radioimmunoassay to measure extracted diuretic hormone in the locust. Their most recent purification study was carried out with 50,000 sub-oesophaegeal and thoracic ganglia collected over 6 years and stored in acetone at 4°C. (J. Proux personal communication). Initially, immunoabsorption was performed on gels prepared with anti-vertebrate arginine-vasopressin antibody. Subsequently RP-HPLC was carried out on C4, C8 and C18 columns with several gradient and solvent systems and throughout these procedures RIA and bioassay were used to monitor and measure the purified peptide, the recovery of which was reported to be higher than 50%. It is hoped that it will soon be possible to obtain the amino acid sequence of the arginine vasopressin (AVP)-like peptide of Locusta. It will certainly be extremely interesting to know the structural relationship (if any) to the vertebrate oxytocin/vasopressin family of peptides.

9. CONCLUDING REMARKS

The insect nervous system has much to offer in terms of basic research, since, although by no means either simple or primitive, it has, nevertheless, a complexity several orders of magnitude less than that of vertebrates. It might be assumed, therefore, that it will permit the solution of fundamental problems of neuropeptide biology such as the elucidation of chemical structure, biosynthesis, cellular processing, packaging and release, as well as neuronal circuitry and neurophysiological mechanisms. The increasing number of 'identified' insect peptidergic neurons that have been reported should speed this process in that these cells can be individually monitored by cytological techniques and manipulated in neurophysiological studies.

It may be seen from this review that there is increasing evidence that certain of the insect neuropeptides so far identified have amino acid sequence homologies with known vertebrate biologically-active peptides. This is of interest from a biochemical evolutionary viewpoint and suggests that those peptides with a functional relevance have been strongly conserved over a much greater period of time than previously supposed. As techniques of peptide purification and micro-sequencing continue to improve there seems little doubt that we shall see a great advance in the number of identified insect neuropeptides. It should soon be possible, therefore, to get a better idea of the extent to which conservation of peptide structure has occurred during evolution. The work involved in purifying and sequencing peptides will undoubtedly require a lot of effort, much of it concerned with time-consuming large-scale purification programmes such as have been detailed in this chapter. However, once we have knowledge of the complete peptide sequence, or even a part of it, then there may be some short cuts afforded by recombinant DNA techniques. If we can select a region from a partially-sequenced peptide that is rich in unambiguous codons, then a series of synthetic oligonucleotides (20 or 30 mers) could be constructed. These could then be used either in the preparation of an enriched cDNA probe or they could be $\gamma\,^{32}P$ end-labelled and used directly to screen existing cDNA or genomic banks. Expression could be tested (a) directly in Saccharomyces cerevisiae (genomic clone) or (b) under the control of suitable expression vectors containing, for example, alcohol dehydrogenase promoter sequence, again in S. cerevisiae (cDNA clone).

It is of primary importance to understand the neuropeptide chemistry of the insect brain because of the potential that information of this kind offers mankind in the continuing battles against various insect pests. Biologically-active peptide molecules are functional in minute amounts and in insects as in other organisms they are almost certainly involved in key metabolic and physiological processes. If the central neurochemical events controlling such processes can be disturbed, then the way is clearly opened for controlling and/or destroying the insects concerned.

One approach would be to introduce an inhibitor of the enzyme metabolism operating either in the formation or the degradation of a vital peptide. Such a strategy has been successfully used in medical research with the development of a drug, captopril, which is used to chelate the converting enzyme that transforms angiotensin I to angiotensin II, thus stopping the powerful vasoconstricting and thereby, hypertensive effect of this peptide hormone. Before significant progress can be made in this direction, however, it is essential to learn as much as possible about the chemical nature and the physiological roles of the peptides.

10. ACKNOWLEDGEMENTS

We are extremely grateful to the Science and Engineering Research Council of Great Britain for the award of a research grant GR/D/03383.

11. NOTE ADDED IN PROOF

Subsequent to the preparation of this chapter, Nagasawa et al. (1986) have reported the complete amino acid sequence of one of the three forms of prothoracicotropic hormone (4K-PTTH-II) of the silkworm, Bombyx mori. It is significant that the molecule has now been shown to be homologous to insulin in consisting of two non-identical peptide chains (A and B chains). The A chain consists of 20 amino acid residues and the B chain is a mixture of four microheterogeneous peptides, two of which consist of 28 residues and the other two of 26 residues. The sequences of the A and B chains are reproduced in Fig. 6 where they are compared with human insulin and porcine relaxin. The molecule shows a considerable degree of sequence homology with human insulin (eighteen sequence identities or 40%) and a lesser degree of homology to porcine relaxin (eleven sequence

```
                        1                 5                  10                 15              20
4K-PTTH-II      H-Gly-Ile-Val-Asp-Glu-Cys-Cys-Leu-Arg-Pro-Cys-Ser-Val-Asp-Val-Leu-Leu-Ser-Tyr-Cys-
Human insulin   H-Gly-Ile-Val-Glu-Gln-Cys-Cys-Thr-Ser-Ile-Cys-Ser-Leu-Tyr-Glu-Leu-Glu-Asn-Tyr-Cys-
Porcine relaxin H-Arg-Met-Thr-Leu-Ser-Glu-Lys-Cys-Cys-Glu-Val-Gly-Cys-Ile-Arg-Lys-Asp-Ile-Ala-Arg-Leu-Cys-

B chains
                        1                 5                  10                 15              20
4K-PTTH-II      pGlu-Gln-Pro-Gln-Ala-Val-His-Thr-Tyr-Cys-Gly-Arg-His-Leu-Ala-Arg-Thr-Leu-Ala-Asp-Leu-Cys-
Human insulin                H-Phe-Val-Asn-Gln-His-Leu-Cys-Gly-Ser-His-Leu-Val-Glu-Ala-Leu-Tyr-Leu-Val-Cys-
Porcine relaxin pGlu-Ser-Thr-Asn-Asp-Phe-Ile-Lys-Ala-Cys-Gly-Arg-Glu-Leu-Val-Arg-Leu-Trp-Val-Glu-Ile-Cys-

                       23         25
4K-PTTH-II      Trp-Glu-Ala-Gly-Val-Asp-OH
Human insulin   Gly-Glu-Arg-Gly-Phe-Phe-Tyr-Thr-Pro-Lys-Thr-OH
Porcine relaxin Gly-Val-Trp-Ser-OH
```

Identical residues of PTTH and insulin underlined: ___ , and with porcine relaxin ---

FIG. 6: Amino acid sequences of 4K-PTTH-II from *Bombyx mori* compared with human insulin and porcine relaxin. (From Nagasawa et al. 1986.)

identities). It can be seen that most of the hydrophobic core residues including A1 Ile, A6 Cys, A11 Cys, A16 Leu, A20 Cys, B11 Leu, B15 Leu and B19 Cys, are identical in PTTH-II and insulins from whatever source and there now seems little doubt that PTTH is a member of the insulin family of peptides. The outstanding question is the degree to which it represents the ancestral molecule of the family and it is hoped that further studies on the PTTH of Bombyx and studies of other insect insulin-like peptides will help to resolve this issue.

12. REFERENCES

Banting FG, Best CH (1922) The internal secretion of the pancreas. J Lab Clin Med 7: 251-266.

Baumann E, Gersch M (1982) Purification and identification of neurohormone D, a heart-accelerating peptide from the corpora cardiaca of the cockroach Periplaneta americana. J Insect Biochem 12: 7-14.

Bishop CA, O'Shea M, Miller RJ (1981) Neuropeptide proctolin. (H-Arg-Tyr-Leu-Pro-Thr-OH): immunological detection and neuronal localisation in insect central nervous system. Proc Natl Acad Sci USA 79: 5899-5902.

Bollenbacher WE, Bowen MF (1983) The prothoracicotrophic hormone. In: Downer GH, Laufer H (eds) Endocrinology of insects. Alan R Liss, Inc., New York, pp 89-99.

Bollenbacher WE, Agui N, Granger NA, Gilbert LI (1979) In vivo activation of insect prothoracic glands by the prothoracicotrophic hormone. Proc Natl Acad Sci USA 76: 5148-5152.

Bollenbacher WE, Agui N, Granger NA, Gilbert LI (1980) Insect prothoracic glands in vivo: A system for studying the prothoracicotrophic hormone. In: Kurstak E, Maramorosch KK, Dubendorfer A (eds) Invertebrate systems in vitro. Elsevier North-Holland, Amsterdam, pp 253-271.

Brown BE (1975) Proctolin: a peptide transmitter candidate in insects. Life Sci 17: 1241-1252.

Brown BE, Starratt AN (1975) Isolation of proctolin, a myotropic peptide, from Periplaneta americana. Insect Physiol 21; 1879-1881.

Cutfield JF, Cutfield SM, Dodson EJ, Dodson GG, Emdin SF, Reynolds CD (1979) Structure and biological activity of hagfish insulin. J Molec Biol 132: 85-100.

Dockray GJ, Duve H, Thorpe A (1981) Immunochemical characterisation of gastrin/cholecystokinin-like peptides in the brain of the blowfly Calliphora vomitoria. Gen Comp Endocrinol 45: 491-496.

Duve H (1978) The presence of a hypoglucemic and hypotrehalocemic hormone in the neurosecretory system of the blowfly, Calliphora erythrocephala. Gen Comp Endocrinol 36: 102-110.

Duve H, Thorpe A (1979) Immunofluorescent localisation of insulin-like material in the median neurosecretory cells of the blowfly, Calliphora vomitoria (Dipteria) Cell Tissue Res 200: 187-191.

Duve H, Thorpe A (1980) Localisation of pancreatic polypeptide (PP)-like immunoreactive material in neurones of the brain of the blowfly, Calliphora erythrocephala (Diptera). Cell Tissue Res 210: 101-109.

Duve H, Thorpe A (1981) Gastrin/cholecystokinin (CCK)-like immunoreactive neurones in the brain of the blowfly, Calliphora erythrocephala (Diptera) Gen Comp Endocrinol 43: 381-391.

Duve H, Thorpe A (1982) The distribution of pancreatic polypeptide in the nervous system and gut of the blowfly, Calliphora vomitoria (Diptera). Cell Tissue Res 227: 67-77.

Duve H, Thorpe A (1983) Immunocytochemical identification of α-endorphin-like material in neurons of the brain and corpus cardiacum of the blowfly, Calliphora vomitoria (Diptera). Cell Tissue Res 233: 415-426.

Duve H, Thorpe A (1984a) Immunocytochemical mapping of gastrin/CCK-like peptides in the neuroendocrine system of the blowfly Calliphora vomitoria (Dipteria). Cell Tissue Res 237: 309-320.

Duve H, Thorpe A (1984b) Comparative aspects of insect-vertebrate neurohormones. In: Bořkovec AB, Kelly TJ (eds) Insect neurochemistry and neurophysiology. Plenum Press, New York, pp 171-195.

Duve H, Thorpe A, Lazarus NR (1979) Isolation of material displaying insulin-like immunological and biological activity from the brain of the blowfly Calliphora vomitoria. Biochem J 184: 221-227.

Duve H, Thorpe A, Neville R, Lazarus NR (1981) Isolation and partial characterization of pancreatic polypeptide-like material in the brain of the blowfly Calliphora vomitoria. Biochem J 197: 767-770.

Duve H, Thorpe A, Lazarus NR, Lowry PJ (1982) A neuropeptide of the blowfly Calliphora vomitoria with an amino acid composition homologous with vertebrate pancreatic polypeptide. Biochem J 201: 429-432.

Duve H, Thorpe A, Strausfield NJ (1983) Cobalt-immunocytochemical identification of peptidergic neurons in Calliphora innervating central and peripheral targets. J Neurocytol 12: 847-861.

Duve H, Thorpe A, Rehfield JF (1985) Localisation and characterisation of cholecystokinin (CCK)-like peptides in the brain of the blowfly Calliphora vomitoria. In: Kobayashi H, Bern HA, Urano A (eds) Neurosecretion and the biology of neuropeptides. Japan Sci Soc Press, Tokyo/Springer-Verlag, Berlin, pp 401-409.

Duve H, Thorpe A, Scott A (1986) Localisation and characterisation of opioid-like peptides in the nervous system of the blowfly Calliphora vomitoria. In: Stefano GB (ed) Handbook of comparative opioid and related neuropeptide mechanisms Vol.1. CRC Press, Boca Raton, Florida, pp 197-211.

El-Salhy M (1981) Immunohistochemical localization of pancreatic polypeptide (PP) in the brain of the larval instar of the hoverfly, Eristalis aeneus (Diptera). Experientia 37: 1009.

El-Salhy M, Abou-El-Ela R, Falkmer S, Grimelius L, Wilander E (1980) Immunohistochemical evidence of gastro-entero-pancreatic neurohormonal peptides of vertebrate type in the nervous system of the larva of a dipteran insect, the hoverfly, Eristalis aeneus. Regulatory Peptides 1: 187-204.

El-Salhy M, Falkmer S, Kramer KJ, Speirs RD (1983) Immunohistochemical investigations of neuropeptides in the brain, corpora cardiaca and corpora allata of an adult lepidopteran insect, Manduca sexta (L). Cell Tissue Res 232: 295-317.

Endo Y, Iwanaga T, Fujita T, Nishiitsutsuji-Uwo J (1982) Localisation of pancreatic polypeptide (PP)-like immunoreactivity of the central and visceral nervous system of the cockroach Periplaneta. Cell Tissue Res 227: 1-9.

Floyd JC, Fajans SS, Pek S, Chance RE (1977) A newly recognised pancreatic polypeptide: plasma levels in health and disease. Recent Prog Horm Res 33: 519-570.

Goldsworthy GJ, Gäde G (1983) The chemistry of hypertrehalosemic factors. In: Downer RGH, Laufer H (eds) Endocrinology of insects. Alan R Liss Inc., New York, pp 109-119.

Guo F (1985) The effect of the Chinese silkworm brain hormone on the metamorphosis and reproduction of the Indian silkworm. In: Lofts B, Holmes WN (eds) Current trends in comparative endocrinology. Vol.I. Hong Kong, University Press, pp 322-325.

Hansen BL, Hansen GN, Scharrer B (1982) Immunoreactive material resembling vertebrate neuropeptides in the corpus cardiacum and corpus allatum of the insect Lecucophaea maderae. Cell Tissue Res 225: 319-329.

Holman GM, Cook BJ (1979) Evidence for proctolin and a second myotropic peptide in the cockroach, Leucophaea maderae determined by bioassay and HPLC analysis. Insect Biochem 9: 149-154.

Holman GM, Cook BJ (1983) Isolation and partial characterisation of a second myotropic peptide from the hindgut of the cockroach Leucophaea maderae. Comp Biochem Physiol 76C: 39-43.

Holman GM, Cook BJ (1985) Proctolin, its presence in and action on the oviduct of an insect. Comp Biochem Physiol 80C: 61-64.

Holwerda DA, Weeda E, Van Doorn JM (1977) Separation of the hyperglycaemic and adipokinetic factors from the cockroach corpus cardiacum. Insect Biochem 7: 477-481.

Hughes J, Smith TW, Kosterlitz HW, Fothergill LH, Morgan BA, Morris H (1975) Identification of two related pentapeptides from the brain with potent opiate agonist activity. Nature (Lond) 258: 577-580.

Imura H, Kato Y, Nakai Y, Nakao K, Tanaka I, Jingami H, Koh T, Yoshimasa T, Tsukada T, Tojo K, Sugawara A (1985) Endogenous opioids and related peptides: from molecular biology to clinical medicine. J Endocrinol 107: 147-157.

Ishizaki H, Ichikawa M (1967) Purification of the brain hormone of the silkworm, Bombyx mori. Biol Bull Woods Hole 133: 355-368.

Ishizaki H, Nagasawa H, Suzuki A (1985) Prothoracicotrophic hormone of the silkworm Bombyx mori: amino acid sequence homology with insulin. Diabetes Res Clin Practice Suppl 1: S262.

Iwanaga T, Fujita T, Nishiitsutsuji-Uwo J, Endo Y (1981) Immunohistochemical demonstration of PP-, somatostatin-, enteroglucagon- and VIP- like immunoreactivities in the cockroach midgut. Biomed Res 2: 202-207.

Kimmel JR, Hayden LJ, Pollock HG (1975) Isolation and characterisation of a new pancreatic polypeptide hormone. J Biol Chem 250: 9369-9376.

Kopec S (1922) Studies on the necessity of the brain for the inception of insect metamorphosis. Biol Bull Woods Hole 42: 322-342.

Kramer KJ (1985) Vertebrate hormones in insects. In: Kerkut GA, Gilbert LI (eds) Comprehensive Insect Physiology, Biochemistry and Pharmacology. Vol. 7 Endocrinology 1. Permagon Press, NY, pp 511-536.

Larsson L-I, Rehfield JF (1977) Evidence for a common evolutionary origin of gastrin and cholecystokinin. Nature (Lond) 263: 335-338.

Lavenseau L, Gadenne C, Trabelsi M (1984) Immunofluorescent localization of a substance immunologically related to insulin in the protocerebral neurosecretory cells of the European corn borer. Cell Tissue Res 238: 207-208.

Lea AO, Brown MR (1984) Endocrine-like cells in the midgut of Aedes aegypti. In: Borkovec AB, Kelly TJ (eds) Insect neurochemistry and neurophysiology. Plenum Press, New York, Lond, pp 413-415.

Miller TA (1983) The properties and pharmacology of proctolin. In: Downer RGH, Laufer H (eds) Endocrinology of insects. Alan Liss Inc., New York, pp 101-107.

Mordue W, Morgan PJ (1984a) Isolation and characterisation of neurohormones from locusts. In: Borkovec AB, Kelly TJ (eds) Insect neurochemistry and neurophysiology. Plenum Press, New York London, pp 77-91.

Mordue W, Morgan PJ (1984b) Peptides from the corpora cardiaca and CNS of insects. In: Hoffman J, Porchet M (eds) Biosynthesis, metabolism and mode of action of invertebrate hormones. Springer-Verlag, Berlin, Heidelberg, New York, Tokyo, pp 118-125.

Morgan PJ, Mordue W (1983) Separation and characteristics of diuretic hormone from the corpus cardiacom of Locusta. Comp Biochem Physiol B 75: 75-80.

Morgan PJ, Mordue W (1984a) Diuretic hormone: another peptide with widespread distribution within the insect CNS. Physiol Entomol 9: 197-206.

Morgan PJ, Mordue W (1984b) Recent progress in the isolation and characterisation of locust diuretic hormone. In: Borkovec AB, Kelly TJ (eds) Insect neurochemistry and neurophysiology. Plenum Press, New York, London, pp 439-442.

Nagasawa H, Isogai A, Suzuki A, Tamura S, Ishizaki H (1979) Purification and properties of the prothoracicotrophic hormone of the silkworm, Bombyx mori. Dev Growth Diff 21: 29-38.

Nagasawa H, Kataoka H, Isogai A, Tamura S, Suzuki A, Ishizaki H, Mizoguchi A, Fujiwara Y, Suzuki A (1984) Amino-terminal amino acid sequence of the silkworm prothoracicotrophic hormone: Homology with insulin. Science 226: 1344-1345.

Nagasawa H, Kamito T, Takahashi S, Isogai A, Fugo H, Suzuki A (1985) Eclosion hormone of the silkworm Bombyx mori. Purification and determination of the N-terminal amino acid sequence. Insect Biochem 15: 573-578.

Nagasawa H, Kataoka H, Isogai A, Tamura S, Suzuki A, Mizoguchi A, Fujiwara Y, Suzuki A, Takahashi SY, Ishizaki H (1986) Amino acid sequence of a prothoracicotropic hormone of the silkworm, Bombyx mori. Proc Natl Acad Sci 83: 5840-5843.

Nishiitsutsuji-Uwo J, Endo Y (1981) Gut endocrine cells in insects: the ultrastructure of the endocrine cells in the cockroach midgut. Biomed Res 2: 30-44.

Nishiitsutsuji-Uwo J, Endo Y (1984) Immunohistochemical demonstration of brain-midgut endocrine system in the cockroach. In: Bořkovec AB, Kelly TJ (eds) Insect neurochemistry and neurophysiology. Plenum Press, New York, London, pp 451-454.

Proux J, Rougon-Rapuzzi G (1980) Evidence for vasopressin-like molecule in migratory locust: Radioimmunological measurements in different tissues: Correlation with various states of dehydration. Gen Comp Endocrinol 42: 378-383.

Rémy C, Girardie J (1980) Anatomical organisation of two vasopressin-neurophysin-like neurosecretory cells throughout the central nervous system of the migratory locust. Gen Comp Endocrinol 40: 27-35.

Rémy C, Girardie J, Dubois MP (1977) Exploration immunocytologique des ganglions cérébroides et sous-oesophagiens du phasme Clitumnus extradentatus. Existence d'une neurosécrétion apparentée à vasopressine - neurophysine. CR Acad Sci Paris 285: 1495-1497.

Rémy C, Girardie J, Dubois MP (1978) Présence dans le ganglion sous-oesophagien de la chenille processionaire du Pin (Thaumetopoea pityocampa Schiff) de cellules révélées en immunofluorescence par un anticorps anti- α-endorphine. CR Acad Sci Paris Sér D 286: 651-653.

Rémy C, Girardie J, Dubois MP (1979) Vertebrate neuropeptide-like substances in the suboesophageal ganglion of two insects: Locusta migratoria R. & F. (Orthoptera) and Bombyx mori (Lepidoptera). Immunocytological Investigation. Gen Comp Endocrinol 37: 93-100.

Scarborough RM, Jamieson GC, Kalish F, Kramer SJ, McEnroe GA, Miller CA, Schooley DA (1984) Isolation and primary structure of two peptides with cardioacceleratory and hyperglycemic activity from the corpora cardiaca of Periplaneta americana. Proc Natl Acad Sci USA 81: 5575-5579.

Schooneveld H, Tesser GI, Veenstra JA, Romberg-Privee HM (1983) Adipokinetic hormone and AKH-like peptide demonstrated on the corpora cardiaca and nervous system of Locusta migratoria by immunocytochemistry. Cell Tissue Res 230: 67-76.

Schooneveld H, Romberg-Privee HM, Veenstra JA (1985) Adipokinetic hormone-immunoreactive peptide in the endocrine and central nervous system of several insect species: A Comparative immunocytochemical approach. Gen Comp Endocrinol 57: 184-194.

Schwartz TW (1983) Pancreatic polypeptide, a hormone under vagal control. Medicinske Doktorgrad. Kobenhavns Universitet, Denmark.

Simantov R, Goodman R, Aposhian D, Snyder SH (1979) Phylogenetic distribution of a morphine-like peptide "enkephalin" Brain Res 111: 204-211.

Steele JE (1961) Occurrence of a hyperglycaemic factor in the corpus cardiacum of an insect. Nature (Lond) 192: 680-681.

Steele JE (1963) The site of action of insect hyperglycaemic hormone. Gen Comp Endocrinol 3: 46-52.

Stefano GB, Scharrer B (1981) High affinity binding of an enkephalin analog in the cerebral ganglion of the insect Leucophaea maderae (Blattaria). Brain Res 225: 107-114.

Stefano GB, Scharrer B, Assanah P (1982) Demonstration, characterisation and localisation of opioid binding sites in the midgut of the insect Lecucophaea maderae (Blattaria). Brain Res 253: 205-212.

Stone JV, Mordue W (1980) Adipokinetic hormone. In: Miller TA (ed) Neurohormonal techniques in insects. Springer-Verlag, New York, pp 31-80.

Stone JV, Mordue W, Batley KE, Morris HR (1976) Structure of locust adipokinetic hormone, a neurohormone that regulates lipid utilisation during flight. Nature (Lond) 263: 207-211.

Strambi C, Strambi A, Cupo A, Rougon-Rapuzzi G, Martin N (1978) Etude des taux d'une substance apparentée à la vasopressine dans le système nerveux de Grillons soumis à différentes conditions hygrométiques. CR Acad Sci Paris 287: 1227-1230.

Strambi C, Rougon-Rapuzzi G, Cupo A, Martin N, Strambi A (1979) Mise en évidence immunocytologique d'un composé apparenté à la vasopressine dans le système nerveux du Grillon Acheta domesticus. CR Acad Sci Paris 288: 131-133.

Suzuki A, Nagasawa H, Matsumoto S, Kataoka H, Isogai A, Tamura S, Ishizaki H, Sakurai S, Fugo H, Sonobe H, Ogura N (1983) Neurohormones in silkworm, Bombyx mori. In: Miyamoto J et al (eds) IUPAC pesticide chemistry: Human welfare and the environment. Pergamon Press, New York, pp 97-102.

Tager HS, Kramer KJ (1980) Insect glucagon-like peptides: evidence for high molecular weight form in midgut from Manduca sexta (L). Insect Biochem 10: 617-619.

Tatemoto K (1982a) Isolation and charaterisation of peptide YY (PYY), a candidate gut hormone that inhibits pancreatic exocrine secretion. Proc Natl Acad Sci USA 79: 2514-2518.

Tatemoto K (1982b) Neuropeptide Y. The complete amino acid sequence of the brain peptide. Proc Natl Acad Sci USA 79: 5485-5489.

Thorpe A, Duve H (1984) Insulin and glucagon-like peptides in insects and molluscs. Molec Physiol 5: 235-260.

Vanderhaeghen J, Crawley R (1985) Neuronal cholecystokinin: Ann NY Acad Sci vol 448. NY Acad Sciences, New York.

Veenstra JA (1984) Immunocytochemical demonstration of a homology in peptidergic neurosecretory cells in the suboesophageal ganglion of a beetle and a locust with antisera to bovine pancreatic polypeptide, FMRFamide, vasopressin and α-MSH. Neurosci Lett 48: 185-190.

Veenstra JA, Romberg-Privee HM, Schooneveld H, Polak JM (1985) Immunocytochemical localization of peptidergic neurons and neurosecretory cells in the neuro-endocrine system of the Colorado potato beetle with antisera to vertebrate regulatory peptides. Histochem 82: 9-18.

Yui R, Fujita T, Ito S (1980) Insulin-, gastrin-, pancreatic polypeptide-like immunoreactive neurones in the brain of the silkworm, *Bombyx mori*. Biomed Res 1: 42-46.

Zhong XC, Guo F, Xia BY (1985) Purification and action of the Chinese silkworm (*Bombyx mori*) prothoracicotrophic hormone. In: Lofts B, Holmes WN (eds) Current trends in comparative endocrinology. Vol. 1. Hong Kong, Iniversity Press, pp 321-322.

NEUROACTIVE SUBSTANCES IN THE INSECT CNS

DICK R. NÄSSEL

Department of Zoology

University of Lund

Helgonavägen 3, S-223 62 Lund, Sweden

ABSTRACT

The distribution of neurons immunoreactive to antisera to various substances has been mapped in the brain and ventral ganglia of several insect species. This chapter mainly reviews the distribution of serotonin- (5-HTi), GABA-, protolin- and gastrin/CCK-like immunoreactive neurons as well as catecholamine fluorescence, octopamine occurrence and Neutral Red staining in the brain, optic lobes and ventral ganglia of the blowfly, honey bee, cockroach and locust. A few other peptides will also be discussed.

The pattern of distribution of cell bodies and processes of the chemically identified neurons vary between the substances and the insect species. There are few 5-HTi and catecholamine containing neurons in the CNS, but their processes extensively innervate most neuropils of the CNS. GABA-like immunoreactivity is seen in a very large number of neurons supplying most neuropil regions of the CNS, whereas the peptide-immunoreactivity is restricted to few neurons supplying limited portions of the CNS. Some chemically identified neurons may be identifiable also with intracellular techniques for analysis of the central action of neuroactive substances. Immunoreactive fiber systems are in some cases described in detail and possible functional aspects of chemically identified pathways can be discussed. Also, the differential segmental distribution of chemically identified neurons in the CNS is of developmental and functional interest.

1. INTRODUCTION

Immunocytochemistry has made it possible to map putative neuroactive substances in the nervous system with accuracy and relative simplicity. The accuracy, of course, depends on the antisera used and measurements taken to ensure specificity of the immunological tissue reaction. Once satisfactory tests (including chemical analysis of tissue extracts) have been made to show that the immunoreactive neurons contain the substance in question, immunocytochemistry is a powerful method for unravelling chemically specified neuronal circuits. Immunocytochemically identified cell systems can be probed electrophysiologically and pharmacologically and hence it is possible to get a handle on the functional role of different neuroactive substances.

The insect nervous system seems especially suitable for this kind of analysis because of the abundance of large identifiable neurons. A number of efferent neuron types containing e.g. acetylcholine, GABA, glutamate, proctolin or octopamine have been used for the study of peripheral actions of neuroactive substances in insects (reviewed by Klemm 1976; Evans 1980; O'Shea 1982; Nässel 1986). Little is, however, known about actions of neuroactive substances in the insect CNS (Klemm 1987; Evans 1980; Nässel 1986). One of the main aims of the present review is to summarize data on immunocytochemically identified insect neurons that are potentially identifiable also with intracellular techniques and hence useful for studies of central actions of different substances. A large portion of the neurons mapped immunocytochemically, however, do not qualify as reproducible intracellular probes. These neurons will be described in the context of chemically identified pathways or circuits. Maps of such pathways are valuable when the pharmacological-physiological role of the substances are to be probed by means of chemical or surgical lesions combined with electrophysiology and behavioral physiology. A further use of immunocytochemistry is as an anatomical marking method by which for instance the neural organization of various ganglia of the metameric nervous system can be compared or by which the differentiation and morphogenesis of neurons can be studied during normal or experimentally altered development (see Taghert and Goodman 1984; Nässel et al. 1987).

Insect CNS neurons have been mapped in some detail with antisera to the following substances: serotonin, GABA, proctolin and gastrin/cholecystokinin (Bishop and O'Shea 1982, 1983; Duve et al. 1983; Tyrer et al. 1984; Keshishian and O'Shea 1985; Nässel et al. 1985; Meyer et al. 1986; Hoskins et al. 1986; Schäfer and Bicker 1986; Nässel 1986). Several other substances have been mapped in neurosecretory cell systems or in less detail and will only be mentioned briefly below. The cellular localization of catecholamines has been investigated mainly with fluorescence histochemistry. Although the precise neuronal morphology cannot be resolved with these techniques some aspects of the distribution and projections of neurons will be presented. The insects that have been studied immunocytochemically in most detail are locusts, cockroaches, honey bees, blowflies and the Colorado potato beetle. In the following, the descriptions will be mainly based on studies of the brains of cockroach, locust, honey bee and blowfly. Other insects will only be mentioned when especially interesting. The neuroactive substances in the optic lobes will be presented separately, followed by a description of the distribution of chemically identified neurons in the ventral ganglia.

The organization of the insect brain is presented in more detail in other accounts (e.g. Bullock and Horridge 1965; Williams 1975; Strausfeld 1976; Mobbs 1985; Schürmann 1985; Nässel 1987). There will be no discussion of the stomatogastric nervous system, the neurosecretory systems and peripheral targets of efferent neurons; instead the reader is referred to papers by (Klemm 1976; Evans 1980; Veenstra 1984; O'Shea and Schaffer 1985; Nässel 1986).

2. DETERMINATION OF PUTATIVE NEUROACTIVE SUBSTANCES IN THE INSECT CNS

Neuroactive substances are used for chemical signalling (released by neurons and acting on target cells) and include three main types of intercellular communication: (1) neurotransmitters, (2) neurohormones and (3) neuromodulators. Transmitters are released at specific sites and act on receptors in the postsynaptic membranes. Neurohormones are released into the circulation or extracellularly in localized regions of the central or peripheral nervous system. Neuromodulators either change the signals transferred at conventional synapses or alter the spontaneous activity of receptive neurons or muscle cells (for further details see Dismukes 1979).

Many of the established or putative neuroactive substances of the vertebrate nervous system have been identified or tentatively identified in insects. The methods used for identification include extraction and chemical identification of substances or biosynthetic enzymes, ability of tissue to use precursors to synthesize the substances, immunoassays and histochemical or immunocytochemical localization of substances or biosynthetic enzymes in tissue. Furthermore specific uptake and release mechanisms, presence of specific receptors and physiological reponses to substances (as well as ligands or pharmacological agents interferring with transmission) have been tested (see e.g. Klemm 1976; Evans 1980; Kingan and Hildebrand 1985; O'Shea and Schaffer 1985).

The present review is mainly based on data from immunocytochemical and fluorescence histochemical mapping of different neuroactive substances. The methodology used has been described in detail in the original papers cited in the text and will not be discussed further.

3. SEROTONIN-IMMUNOREACTIVE NEURONS IN THE INSECT BRAIN

3.1. Introduction

Serotonin has been determined qualitatively and quantitatively in the CNS of many insect species (Welsh and Moorehead 1960; Gersch et al. 1961; Colhoun 1963; Hiripi and Rozsa 1973; Klemm and Axelsson 1973; Osborne and Neuhoff 1974; Clarke and Donnellan 1982; Slowley and Owen 1982; Nässel and Laxmyr 1983; Trimmer 1985). The synthesis of 5-HT in insects is from tryptophan via 5-hydroxytryptophan, catalyzed by tryptophan hydroxylase and then a dopa/5-hydroxytryptophan decarboxylase (Osborne and Neuhoff 1974; Livingstone and Tempel 1983). Metabolization of 5-HT seems not to be by means of monoamine oxidase, but by N-acetylation (see Evans 1980).

Only a few studies have focused on the central action of 5-HT in insects (see e.g. Kostowsky and Tarchalska 1972; Klemm 1976; Mercer and Menzel 1982). On peripheral targets, however, the effect is rather well known: 5-HT controls frequency of heart beat, amplitude and frequency of contractions of visceral muscle, salivation in salivary glands, excretion by Malpighian tubules (see Maddrell and Phillips 1975; Collins and Miller 1977; Trimmer 1985; Klemm et al. 1986; Nässel 1986). The 5-HTi neurons, probably acting on these peripheral targets, have been described in some insects (Nässel and Elekes 1985; Klemm et al. 1986; Nässel 1986; Bräunig 1987).

By now 5-HT has been mapped immunocytochemically in the brain of several insect species: cockroach (Biship and O'Shea 1983; Klemm et al. 1984), honey bee (Schürmann and Klemm 1984), locust (Tyrer et al. 1984) <u>Drosophila</u> and blowfly (Nässel 1986). In the following these investigations will be summarized with special emphasis on the blowfly.

3.2. Serotonin-immunoreactiove neurons in the brain of the blowfly

This summary is based on Nässel (1986) and unpublished results. Further details will be published elsewhere (see also Nässel 1987).

The total number of serotonin-immunoreactive (5-HTi) neurons in the fly brain (including optic lobes and suboesophageal ganglia) is ca. 104-108. Of these neurons there are ca. 20 in each optic lobe and 18-22

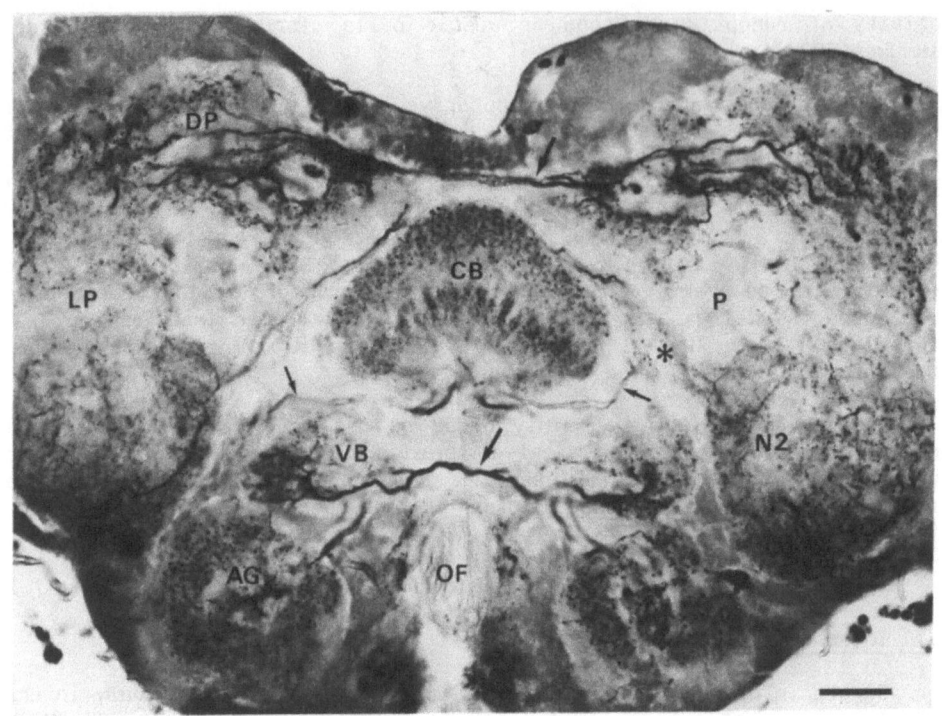

FIG. 1: Serotonin-immunoreactive neurons in the brain of <u>Calliphora</u> (frontal section, pre-incubation PAP-method, 25 μm section). Immunoreactive processes are found in both glomerular and non-glomerular neuropil. Glomerular ones seen here are central body (CB), ventral bodies (VB) and antennal glomeruli (AG). Extensively branching systems of 5-HTi fibers are found in non-glomerular neuropil of dorsal (DP) and lateral (LP) protocerebrum. In the ventro-lateral protocerebrum (N2) there are arborizations of 5-HTi processes from the neurons labeled 2 in Fig. 2. The large arrows indicate commissural 5-HTi fibers and small arrows 5-HTi fiber connections between CB and other neuropils. Scale: 50 μm. OF = oesophageal foramen; P = mushroom body peduncle. Scale: 50 μm.

in the suboesophageal ganglia. The 5-HTi cell bodies have characteristic bilateral locations in the brain (Fig. 2). They occur in small clusters, in pairs or singly in each hemisphere. In protocerebrum there is a pair of mediocaudal cell body clusters with 13 5-HTi cell bodies each and two pairs of smaller clusters. The deutocerebrum contains three lateral 5-HTi cell bodies on each side and in the tritocerebrum no cell bodies could be identified. Some of the neurons have large cell bodies in characteristic locations (Fig. 3) and would hence be possible to repeatedly impale with intracellular techniques. These potentially identifiable nerons will be described in detail below.

Almost all glomerular neuropil regions are innervated by 5-HTi processes: all parts of the central body complex, the anterior optic tubercle, antennal glomeruli, peduncles and calyces of the mushroom bodies and the optic lobes (Figs. 1-3). No immunoreactive fibers could be found in the protocerebral bridge (pons), the alfa- and beta-lobes and part of the peduncles of the mushroom bodies. Most of the non-glomerular neuropils in all the brain neuromeres are invaded by networks of varicose 5-HTi fibers.

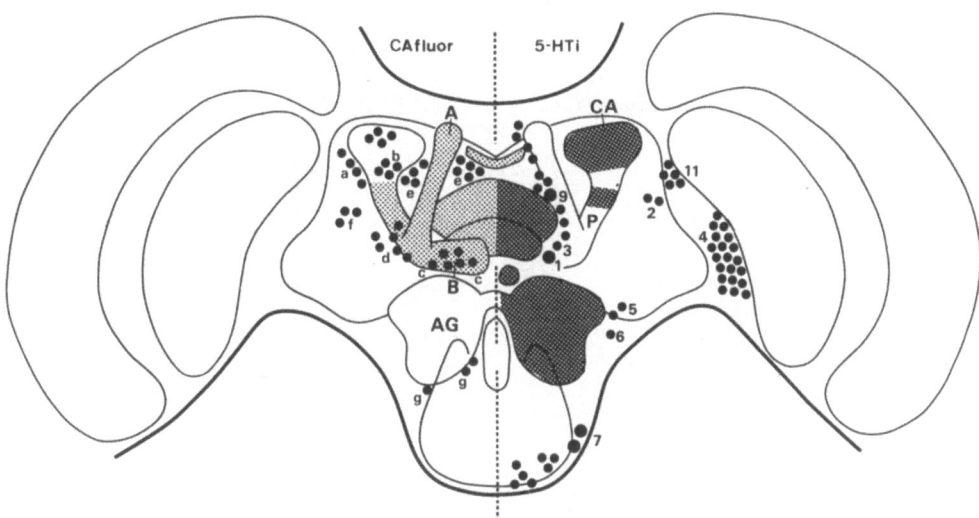

FIG. 2: Schematic presentation of distribution of catecholamine containing (CA fluor) and 5-HTi cell bodies and processes in the Calliphora brain (frontal section shown). Processes are only shown in glomerular neuropil although the surrounding non-glomerular neuropils contain large amounts of CA and 5-HTi fibers. Light stippling shows CA and darker shows 5-HTi. Small letters indicate different groups of CA cell bodies and numbers identifiable 5-HTi neurons. A = alfa lobe; B = beta lobe; CA = calyx; P = peduncle; AG = antennal glomeruli. The small CA containing neuropil above the central body is the protocerebral bridge. After Klemm (1974) and Nässel (1986).

The 5-HTi neurons are of two main types: (1) interneurons that innervate single neuropils or interconnect two or more neuropils and (2) efferent neurons that leave the CNS via cranial nerves. Most of the 5-HTi interneurons have extensive processes, some form very complex patterns others are more simple. The interneurons rarely innervate glomerular neuropils only, instead they either remain in non-glomerular neuropil or interconnect this type of neuropil with glomerular neuropil. Exceptions to this are found in the optic and antennal lobes.

In the brain and suboesophageal ganglia of Calliphora 22 5-HTi neurons (11 in each hemisphere) can be described in detail (shown schematically in Fig. 3). As will be shown below many of these are potentially identifiable. These neurons can be classified into eight types (a few exist in more than one copy in each hemisphere). (1) A pair of large neurons termed the Large Bilateral Optic lobe 5-HTi neurons (LBO5HT) innervates the lamina, medulla, lobula and lobula plate on both sides (Fig. 11) and have processes caudally in the protocerebral medio-caudal cell-body cluster. (2) Another pair of 5-HTi neurons, in the same mediocaudal cluster, bilaterally innervates several protocerebral non-glomerular neuropils connected to the central body complex. They also contribute fibers to glomerular neuropils of the central body complex and a layer of the mushroom body peduncles. (3) A third pair of 5-HTi neurons of the same cell body cluster can be resolved in detail. These neurons bilaterally innervate non-glomerular neuropil of the lateral protoce-rebrum. Their axons cross the brain in the same commissure as the LBO5HT neurons. (4) One pair of 5-HTi neurons are found dorsally in each

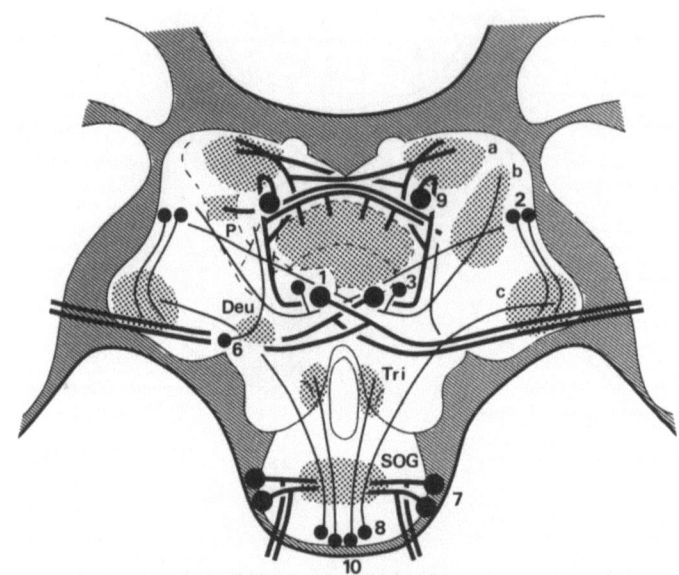

FIG. 3: Schematic representation of some identifiable 5-HTi neurons and parts of the neuropils they invade (stippled). (Calliphora brain, frontal section). 1 = the LBO5HT-neurons bilaterally innervating the optic lobes. 2 = neurons innervating ventrolateral protocerebrum (c). 3 = neurons bilaterally innervating dorsolateral protocerebrum (b). 6 = neurons connecting caudal deutocerebral (Deu) neuropil to the dorsal protocerebrum and mushroom body calyx. 7 = the giant efferent suboesophageal 5-HTi neurons innervating suboesophageal (SOG) neuropil and peripheral targets. 8 = neurons connecting SOG neuropil to ventrolateral protocerebrum (c). 9 = large 5-HTi neurons bilaterally innervating dorsal protocerbral neuropil (a), mushroom body peduncles (P) central body and other neuropils not shown here. 10 = neurons connecting SOG to tritocerebral (Tri) neuropil.

lateral protocerebrum. These neurons innervate a ventrolateral neuropil area in the ipsilateral protocerebrum. (5) On each side of the deutocerebrum there is one large 5-HTi neuron with processes in the dorsal lobe (mechanosensory antennal neuropil) and an axon running dorsally into protocerebrum. (6) Two 5-HTi neurons lateral to each antennal lobe bilaterally innervate the chemosensory antennal glomeruli. All these glomeruli appear to receive 5-HTi processes. (7) In the suboesophageal ganglia one pair of large cell bodies in each hemisphere has arborizations in suboesophageal neuropil and efferent axons supplying the neurilemma of the following structures: labro-frontal nerves, maxillary-labellar nerves, cervical connective, thoracico-abdominal ganglia and many of their nerve roots. These neurons which have been termed Giant Suboesophageal Efferent Neurons (GSEN) have been described in detail elsewhere (Nässel and Elekes 1985; Nässel 1986). They are the most extensive neurons of the whole nervous system of flies that have been reported so far and probably play a role in neurohormonal release of 5-HT into the hemolymph (Nässel and Elekes 1985; Nässel 1986; see also Trimmer 1985). (8) The last type of 5-HTi neurons traced in detail have their cell bodies ventrally in the suboesophageal ganglia and axons running to and arborizing in ventrolateral protecerebrum. There is another pair of 5-HTi neurons ventrally in the suboesophageal ganglia that has not been resolved in detail, but clearly their processes innervate tritocerebral neuropil.

Several of the 5-HTi neurons are intersegmental with cell bodies in a segment posterior to their main processes; others are intrasegmental. Both types may either supply only ipsilateral neuropil or bilaterally innervate neuropils in both hemisegments. Six 5-HTi axons connect the brain and thoracic ganglia via the cervical connective (in addition to the GSENs).

3.3. Serotonin-immunoreactive neurons in the brains of other insects

The following is a summary of the distribution and morphology of 5-HTi neurons in the brains of the locust, honey bee and cockroach, based on studies by Bishop and O'Shea (1983), Klemm and Sundler (1983), Klemm et al. (1984), Schürmann and Klemm (1984) and Tyrer et al. (1984).

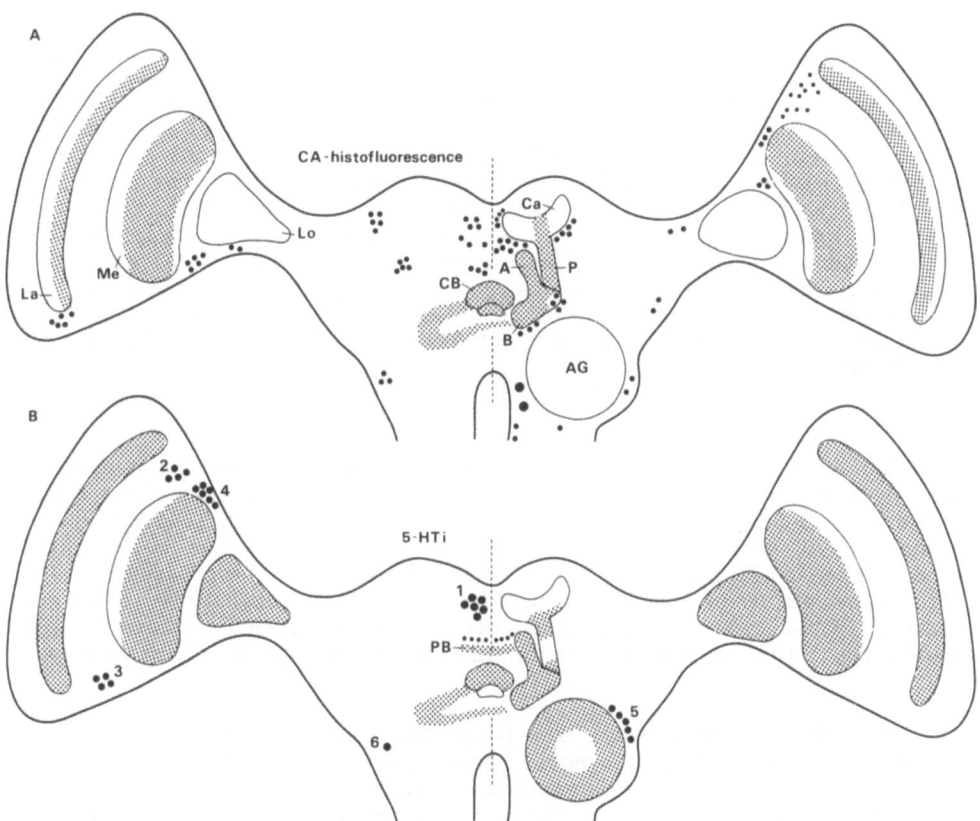

FIG. 4: CA and 5-HTi neurons in the locust brain (frontal sections; left half of each brain is a more caudal portion than the right half). The numbers refer to groups of 5-HTi neurons (Tyrer et al. 1984) and the stippling represents CA or 5-HTi processes in glomerular neuropils. A = alfa lobe; B = beta lobe; Ca = calyx; P = peduncle; AG = antennal glomeruli; La = lamina, Me - medulla; Lo = lobula; PB = protocerebral bridge; CB = central body. 4A is redrawn from Klemm and Axelsson (1973) and 4B is compiled from data by Tyrer et al. (1984).

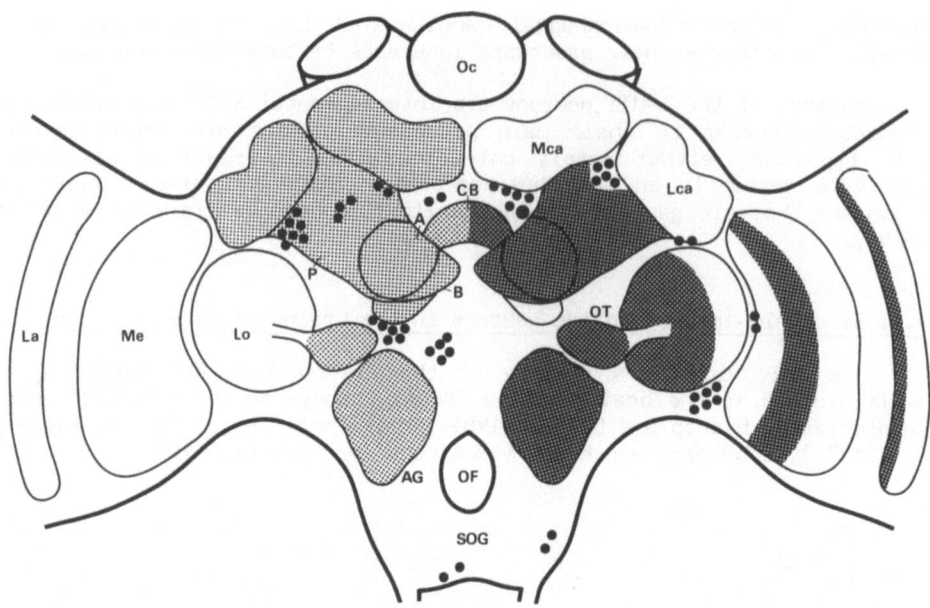

FIG. 5: CA and 5-HTi neurons in the brain of the honey bee. CA cell bodies and processes in glomerular neuropil (light stippling) to the left, 5-HTi is shown to the right (dark stippling). Note total absence of CA processes in the optic lobe, lamina (La), medulla (medulla (Me) and lobula (Lo). A = alfa-lobe; B = beta-lobe; P = peduncle; Mca = median calyx; Lca = lateral calyx; CB = central body; Oc = ocellus; OT = optic tubercle (optic focus); AG = antennal glomeruli; OF= oesophageal foramen; SOG = suboesophageal ganglia. Redrawn form accounts by Klemm (1974); Mercer et al. (1983) and Schürmann and Klemm (1984).

Also in the brains of locusts, bees and cockroaches the number of 5-HTi neurons is small (Figs. 4-6). Excluding the optic lobes the brain of the bee contains ca. 75 5-HTi cells and the cockroach (excluding the suboesophageal ganglia) ca. 110-140. The diameters of the cell bodies range between 10-30 µm in the bee and are ca. 40 µm in the cockroach. The cell bodies are arranged in characteristic bilateral clusters (6 paired clusters in the bee and 16 paired ones in the cockroach). Like in the blowfly the 5-HTi neurons form extensive arborizations in brain neuropil

lobe; CB = central body; P = pons; AN = antennal nerve; AG = antennal glomeruli; OL = optic lobe; Pro = protocerebrum; Deu = deutocerebrum; Tri= tritocerebrum. A. 5-HTi neurons (redrawn after Klemm et al. 1984). Numbers refer to groups of 5-HTi cell bodies. A few neurons are potentially identifiable (LLo, LPr, LTr) according to Bishop and O'Shea (1983).
B. CA neurons (redrawn after Klemm 1983). C Enkephalin- (Enk) and proctolin-like immunoreactive cell bodies (redrawn after Bishop and O'Shea 1982; and Verhaert and Van de Loof (1985). Proctolin immunoreactive neurons are distributed in two groups in protocerebrum (Pr1 and Pr2) and two gropus in tritocerebrum (Tr1 and Tr2). Enkephalin-immunoreactive neurons are found in pars intercerebralis (PI) above the pons (both protocerebral) and in tritocerebrum.

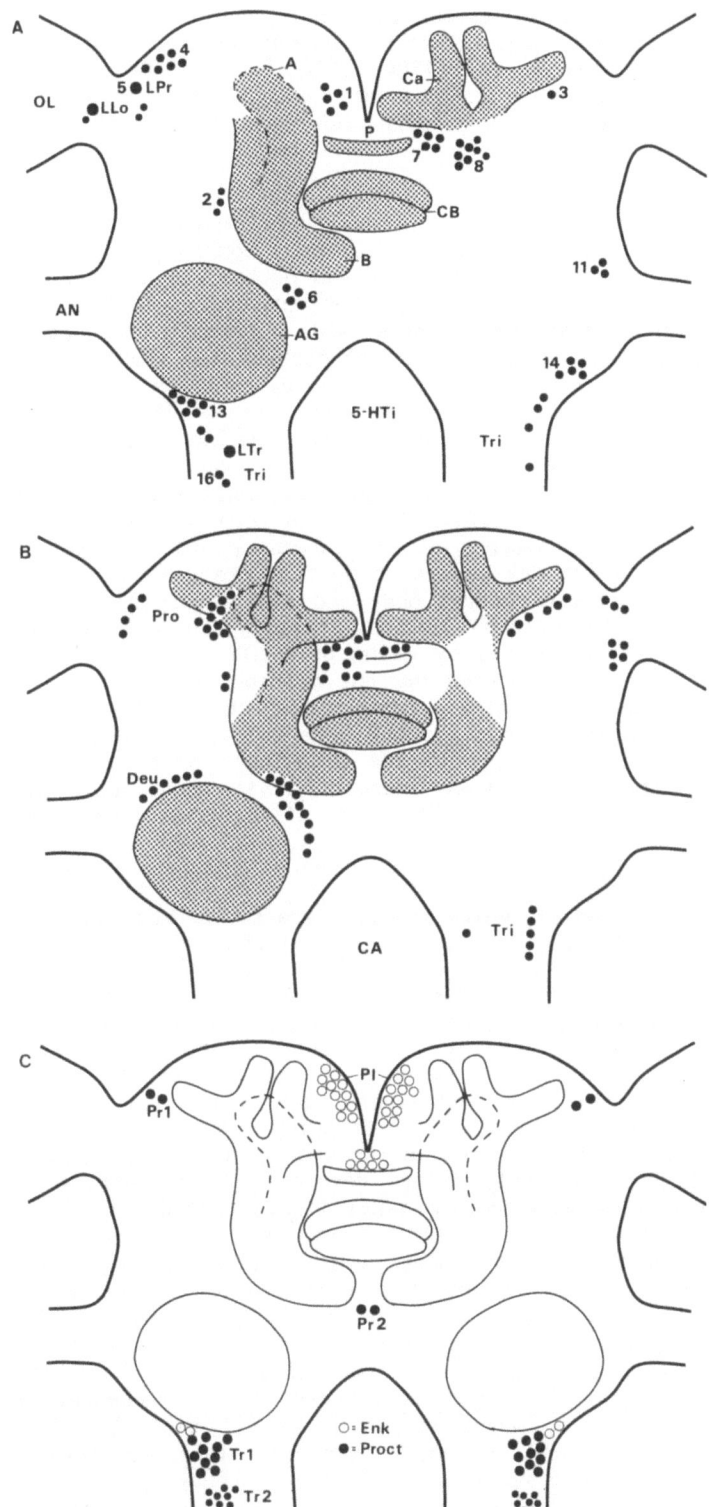

FIG. 6: Distribution of neuroactive substances in the cockroach brain (frontal views). Stippling shows distribution in glomerular neuropils. Abbreviations: A = alfa-lobe; B = beta

of both glomerular and non-glomerular type. Most of the 5-HTi neurons are interneurons, but also some efferent neurons are found in the suboesophageal ganglia. Tyrer et al. (1984) claim that in the locust the labral nerve roots of the brain contain afferent 5-HTi neurons, like several nerve roots in the ventral ganglia, but have not traced the sources of these afferents.

Serotonin-immunoreactive processes are found in most glomerular neuropil regions: optic lobes, central body complex, mushroom bodies and antennal lobes (Figs. 4-6. The detailed pattern of innervation of these neuropils, however, differ between the insects. In the central body complex all the studied species have 5-HTi fibers in the superior arch of the fan-shaped body; in bees and cockroaches also the ellipsoid body is innervated. The protocerebral bridge (pons) contains 5-HTi fibers in the locust and cockroach, but not in the bee. The 5-HTi neurons giving rise to the innervation of the central body and pons differ between the species. In locusts the fan-shaped body and superior arch are innervated by (1) small field 5-HTi neurons with their cell bodies in the anterior protocerebrum and axons running through the protocerebral bridge and (2) from wide field 5-HTi neurons of the lateral protocerebrum. In the bee the 5-HTi fibers enter the central body from its ventral aspect and there are connections between central body and non-glomerular neuropil between the alfa-lobes of the mushroom bodies. Finally, in cockroaches the protocerebral bridge is connected by means of 5-HTi fibers to the fan-shaped body via a tract also innervating the mushroom body calyces and the so called <u>stratum caudale</u> complex. Further connections from the central body are to lateral protocerebrum and the beta-lobes; and from pons to protocerebral nonglomerular neuropil, the alfa-lobe and peduncle of mushroom bodies. In none of the above insects has the anatomy of single 5-HTi neurons participating in innervation of central body complex been described.

The mushroom bodies, which in flies are rather rudimentary, form very prominent neuropil volumes in locusts, cockroaches and especially in bees. The 5-HTi innervation is correspondingly more prominent and complex in these animals than in flies (Figs. 4-6). All 5-HTi neurons supplying the mushroom bodies are of extrinsic origin. The pattern by which mushroom body neuropil is innervated by 5-HTi neurons vary substantially between species. In locusts there is only scarce innervation of calyces and peduncles, whereas alfa- and beta-lobes receive denser innervation. Klemm and Sundler (1983) found four extrinsic 5-HTi pathways that supply different parts of the mushroom bodies and connect them to other neuropil regions. In bees the calyces are devoid of 5-HTi fibers, whereas the peduncles and alfa-lobes contain 5-HTi fibers arranged in strict layers. Also the beta-lobes have regularly organized 5-HTi fibers. The 5-HTi fibers in the bee mushroom bodies have three extrinsic sources in nonglomerular protocerebral neuropil. The mushroom bodies in cockroaches have 5-HTi fibers of extrinsic origin in all the parts. The innervation is from two main systems: a caudo-dorsal system innervating the calyces and a fronto-ventral comprising the alfa- and beta-lobes and the peduncles. These systems can be divided into 8 subsystems connecting different parts of the mushroom bodies to (1) central body, (2) nonglomerular protocerebral neuropil and (3) interconnecting different regions of the mushroom bodies via extrinsic pathways.

The antennal lobes are supplied by multiglomerular 5-HTi neurons in all species (Figs. 4-6). In locusts 7-8 5-HTi cell bodies lateral to each antennal lobe invade chemosensory glomeruli. In the cockroach 6-8 medial and one lateral cell body in each hemisphere invade antennal glomeruli.

Also the dorsal mechanosensory antennal lobes contain 5-HTi fibers in these insects. In the locust, as in the blowfly, one large 5-HTi neuron in each deutocerebral hemisphere interconnects the mechanosensory antennal lobe to neuropil in the dorsal protocerebrum. Only in the blowfly commissural 5-HTi fibers interconnect the antennal lobes of the two sides.

The non-glomerular neuropils of all brain neuromeres, including tritocerebrum and suboesophageal ganglia, are packed with varicose 5-HTi fibers in the studied species.

4. GABA-LIKE IMMUNOREACTIVE NEURONS IN THE INSECT BRAIN

4.1. Introduction

The amino acid gamma-aminobutyric acid (GABA) and the enzyme glutamic acid decarboxylase (GAD) involved in GABA synthesis have been demonstrated in the CNS of insects (Frontali 1964; Ray 1965; Osborne and Neuhoff 1974; Baxter and Torralba 1975; Kingan and Hildebrand 1985). The inhibitory transmitter action of GABA in the insect CNS is well established (Gahery and Boistel 1965; Kerkut 1966; Kerkut et al. 1969; Pitman and Kerkut 1970; Gerschenfeld 1973; Callec 1974; Klemm 1976; see also Ammermüller and Weiler 1985). Only recently, however, has the cellular localization of GABA been possible to map immunocytochemically (Ammermüller and Weiler 1985; Bicker et al. 1985; Hoskins et al. 1986; Meyer et al. 1986; Schäfer and Bicker 1986). Earlier work relied on uptake of radiolabeled GABA (Frontali and Pierantoni 1973; Campos-Ortega 1974).

The mapping of GABA-like immunoreactive neurons seems especially rewarding if one can assume that GABA acts as an inhibitory transmitter in all circuits. Unfortunately it seems that only a limited number of central GABA-immunoreactive neurons can be probed electrophysiologically-pharmacologically since, as shown below, they are very numerous and relatively few qualify as identifiable neurons. Immunocytochemistry will nevertheless provide maps of chemically identified pathways and circuits.

So far GABA-like immunoreactive (GABAi) neurons have been analysed in the brain and optic lobes of the honey bee (Bicker et al. 1985; Meyer et al. 1986; Schäfer and Bicker 1986); the ocellar system of the locust (Ammermüller and Weiler 1985) and the optic lobes of flies (Meyer et al. 1986). In the following also GABAi neurons in the CNS of the flies Musca, Calliphora and Drosophila will be presented from unpublished data. Since there is a large number of GABAi neurons and their processes and terminals are densely packed, very few individual neurons can be described in detail. Instead the compound pattern of GABAi processes in glomerular and nonglomerular neuropil as well as connecting tracts will be presented.

4.2. GABAi neurons in the bee brain

The following description is based on papers by Bicker et al. (1985) and Schäfer and Bicker (1986). In the antennal lobes all the olfactory glomeruli are packed with GABAi terminals. In the lateral and dorsomedial cell body groups of the antennal lobe ca. 200 GABAi neurons were found. No immunoreactive axons were detected in the antennal verve. A few GABAi fibers were found leaving the antennal glomeruli via the mediolateral antennoglomerular tracts and projecting into the lateral deutocerebrum. Branches from these fibers also invade the pedunculus of the mushroom bodies. A commissural tract connecting the two antennal lobes consists mainly of GABAi neurons. The dorsal lobes (antennal mechanosensory neuropil of deutocerebrum) are also packed with GABAi terminals.

The mushroom bodies are innervated by extrinsic GABAi fibers. The alfa-lobe is connected to the calyx by a bundle of ca. 110 extrinsic GABAi fibers (Bicker et al. 1985). Other extrinsic neurons interconnect the alfa-lobe to the neuropil surrounding it. The beta-lobes contain fewer GABAi fibers than the alfa-lobes and are interconnected by GABAi axons in a ventral commissure.

The central body complex contains a large amount of GABAi fibers in the ventral portion (ellipsoid body) and a smaller amount in the dorsal (fan-shaped body) the noduli and protocerebral bridge are devoid of immunoreactivity. Schäfer and Bicker (1986) found about 10 thin GABAi processes invading the ocellar plexus via the ocellar tract, similar to the locust (Ammermüller and Weiler 1985). The suboesophageal ganglia contain ca. 1000 GABAi neurons. These innervate suboesophageal neuropils massively.

4.3. GABAi neurons in the fly brain

Using monoclonal (Meyer et al. 1986) and polyclonal (D.R. Nässel, L. Ohlsson and N.N. Osborne, in prep.) antibodies to GABA a mapping of GABAi neurons in the CNS of Calliphora, Musca and Drosophila has been made. A short summary will be presented here. Further details will be presented elsewhere (Nässel et al. in prep.).

A very large number of GABAi cell bodies are found in most regions of the brain. Often they are arranged in large clusters (Fig. 7b). These clusters are more easily distinguished in the larval brain (Fig. 14a). The antennal lobes (olfactory glomeruli) contain densely packed GABAi terminals (Fig. 7b). These terminals reside in the core of each glomerulus. A large proportion of the cell bodies lateral to the antennal glomeruli are GABAi. Since few fibers leave the antennal glomeruli it seems like many of the GABAi neurons are amacrines. The mushroom bodies contain extrinsic GABAi fibers in the calyces, peduncles, alfa- and beta-lobes (Figs. 7a, b). Except for the calyces the density of immunoreactive fibers is low.

In the central body complex there is a patterned dense distribution of GABAi fibers in the ellipsoid body and the bilateral tracts leaving this ventrally to end in the lateral protocerebrum. The lower part of the fan-shaped body contains a thin layer of GABAi fibers whereas the upper part (the superior arch) is devoid of immunoreactivity. Also the protocerebral bridge and noduli lack GABAi innervation.

Most other neuropils of the protocerebrum contain GABAi fibers; some of these are diffuse non-glomerular neuropils (Figs. 7a, b), others are delineated neuropils like certain optic foci. Several commissures contain GABAi axons which interconnect the two brain hemispheres. Also the nonglomerular neuropils of deuto- and tritocerebrum as well as the suboesophageal ganglia are rich in GABAi processes and terminals. The number of GABAi cell bodies in the brain of flies is high compared to other substances mapped immunocytochemically and fluorescence histochemically.

5. IMMUNOCYTOCHEMISTRY OF NEUROACTIVE PEPTIDES

5.1. Introduction

Awareness of the importance of neuroactive and myoactive peptides in insects has increased recently and seems linked to the fast development of

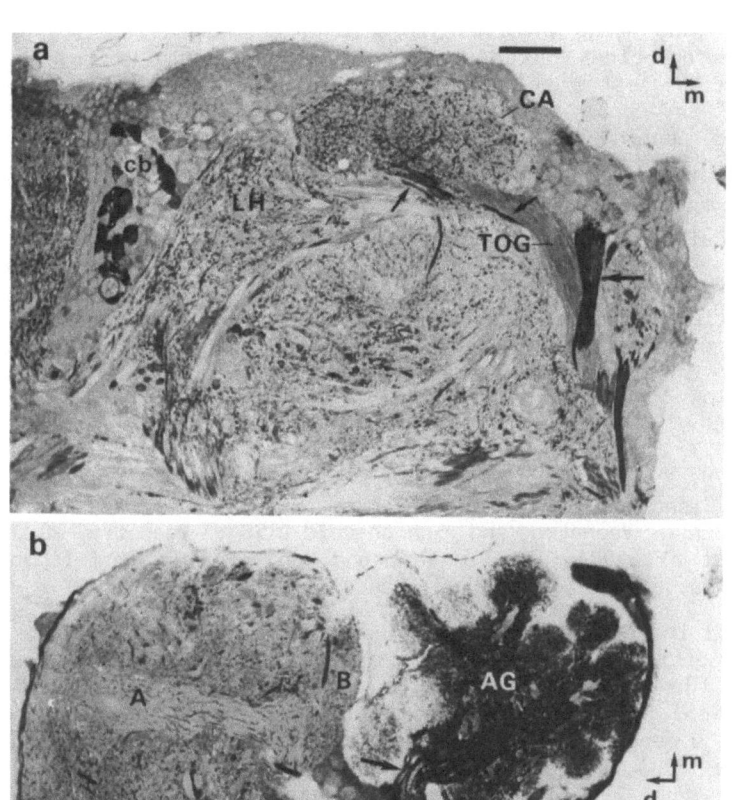

FIG. 7: Micrographs of GABA-like immunoreactive neurons in frontal sections of the brain of <u>Calliphora</u> (PAP-method, 2 μm Epon sections). d = dorsal; m = medial.
a. Left protocerebrum. Immunoreactive terminals are found in non-glomerular neuropil like the lateral horn (LH) and generally in latero-dorsal protocerebrum. The calyx (CA) of the mushroom body is invaded by terminals from fibers (small arrows) running with the antenno-glomerular tract (TOG) which also connects to the LH. Another extensive ventro-dorsal tract is indicated by a larger arrow. cb = immunoreactive cell bodies.
b. The antennal glomeruli (AG) are heavily innervated by GABA; processes from cell bodies (cb) lateral to the AG (the axonal tract is indicated by arrow). Also the mushroom body alfa-lobe (A) and beta-lobe (B) contain GABAi processes. Scales: a and b = 50 μm.

peptide immunocytochemistry and improved techniques for purification and sequencing of peptides in the last decade. Hence, not only peptides of classical neurosecretory systems, but also interneurons and motorneurons have been explored. Research has been persued along two main lines. One line has employed extensive immunocytochemical mapping, using a large number of antisera to rather well characterized mammalian bioactive peptides. The other involves isolation and characterization of specific insect neuropeptides or insect homologues to mammalian peptides. The concern of the latter approach is to determine amino acid sequences of the peptides, to employ bioassays and electrophysiology/pharmacology to understand peptide actions at targets and to isolate new bioactive peptides in the insect nervous system.

A large number of antisera to peptides have been used for immunocytochemical mapping in the insect CNS (especially in the brain neurosecretory systems). The following substances (or substances sharing antigenic determinants with these) have been tentatively localized: adipokinetic hormone, calcitonin, corticotropin, corticotropin releasing factor, enkephalins, FMRFamide, gastrin/cholecystrokinin (gastrin/CCK), glucagon, insulin, alfa-melanocyte stimulating hormone (alfa-MSH), neurophysins, oxytocin, pancreatic polypeptide, proctolin, secretin, somatostatin, substance P, vasoactive intestinal peptide, vasotocin (Remy et al. 1977; Doer-Schott et al. 1978; Remy and Dubois 1981; Duve and Thorpe 1979, 1981, 1982, 1983, 1984,; El-Sahly et al. 1980, 1981; Yui et al. 1980; Bishop and O'Shea 1982; Pages et al. 1983; Romeuf and Remy 1984; Veenstra 1984; Veenstra and Schooneveld 1984; Veenstra et al. 1984; Verhaert et al. 1984a, b; Verhaert and Van de Loof 1985; Schooneveld et al. 1986; Nässel and O'Shea 1987; Nässel et al. 1987).

Of all the peptides mapped immunocytochemically only a few have been fully or partially characterized from extracts of insect nervous tissue: proctolin (Brown and Starrat 1975; Starrat and Brown 1975; O'Shea and Shaffer 1985), adipokinetic hormone (Stone et al. 1976; Schooneveld et al. 1986), a pancreatic polypeptide-like substance (Dube et al. 1982; Duve and Thorpe 1982), a gastrin/CCK-like substance (Dockray et al. 1981; Duve and Thorpe 1981, 1984), the myoactive substance MI and MII (O'Shea et al. 1984) and bursicon (Fraenkel and Hsiao 1962; Truman and Taghert 1983). Enkephalins, which have been isolated from invertebrates (Leung and Stefano 1984; Duve et al. 1985; Davenport and Evans 1986), have also been mapped in the insect CNS (Remy et al. 1979; Duve and Thorpe 1983; Pages et al. 1983; Verhaert and Van de Loof 1985).

The invertebrate neuropeptides studied so far are involved in control of muscle contraction, control of oscillatory functions (e.g. rhythmic myogenic contractions) and they can evoke coordinated complex arrays of behavior and physiology (see O'Shea and Schaffer 1985; Evans and Myers 1986). Two of the characterized peptides have been mapped in some detail in brain interneurons (proctolin and gastrin/CCK) and will be presented below.

5.2. Gastrin/CCK-like immunoreactive neurons in the insect brain

Duve and Thorpe (1981, 1984) showed that most immunoreactive gastrin/CCK-like (G/CCKi) material in neurons of the Calliphora brain is COOH-terminal specific rather that N-terminal specific. In a biochemical study Dockray et al. (1981) further demonstrated that the immunoreactive substance is related to a common COOH-terminal sequence of gastrin and CCK. This pentapeptide sequence appears to be the biologically active part of

b. Fibers in protocerbrum as well as in deuto- and tritocerebrum.
c. In this most anterior of the three sections protocerbral fibers associated with mushroom bodies are indicated by asterisks. Large G/CCKi neurons with cell bodies ventrally in SOG (cell bodies not in micrograph) innervate SOG and deutocerbrum. The large pars intercerebralis neurons have reached tritocerebrum. Scale: a-c = 100 μm. Preparation for a-c was provided by L. Ohlsson.

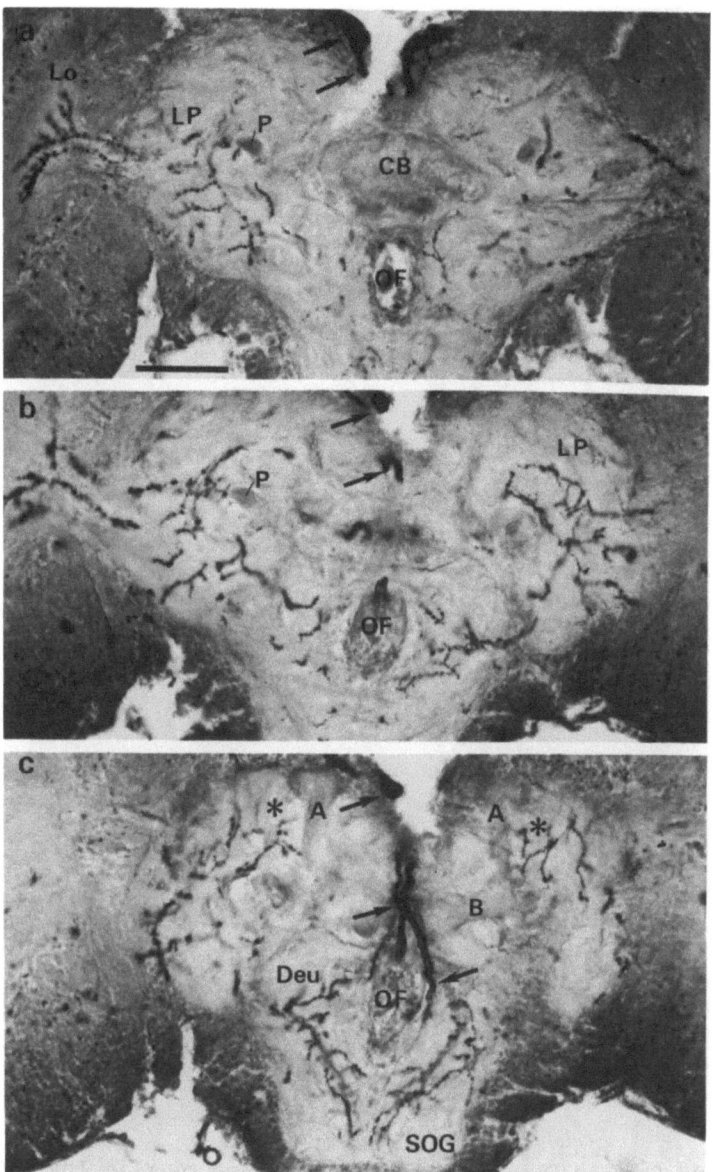

FIG. 8: Micrographs of gastrin/CCK-like immunorective neurons in the brain of <u>Calliphora</u> (preincubation PAP-method frontal sections). A 48 h pupa was used in order to show the immunoreactive processes in a more simple form. The general distribution is similar in adult insects. Abbreviations: Lo = lobula; LP = lateral protocerebrum; P = mushroom body peduncle; CB = central body; OF = oesophageal foramen; A = alfa-lobe; B = beta-lobe; Deu = deutocerebrum; SOG = suboesophageal ganglia. Arrows indicate G/CCKi neurons of pars intercerebralis projecting to tritocerebrum.

a. In this most posterior of the sections immunoreactive processes are seen in lobula, and lateral and ventral protocerebrum. The central body also contains G/CCKi fibers, although they are not visible in this micrograph.

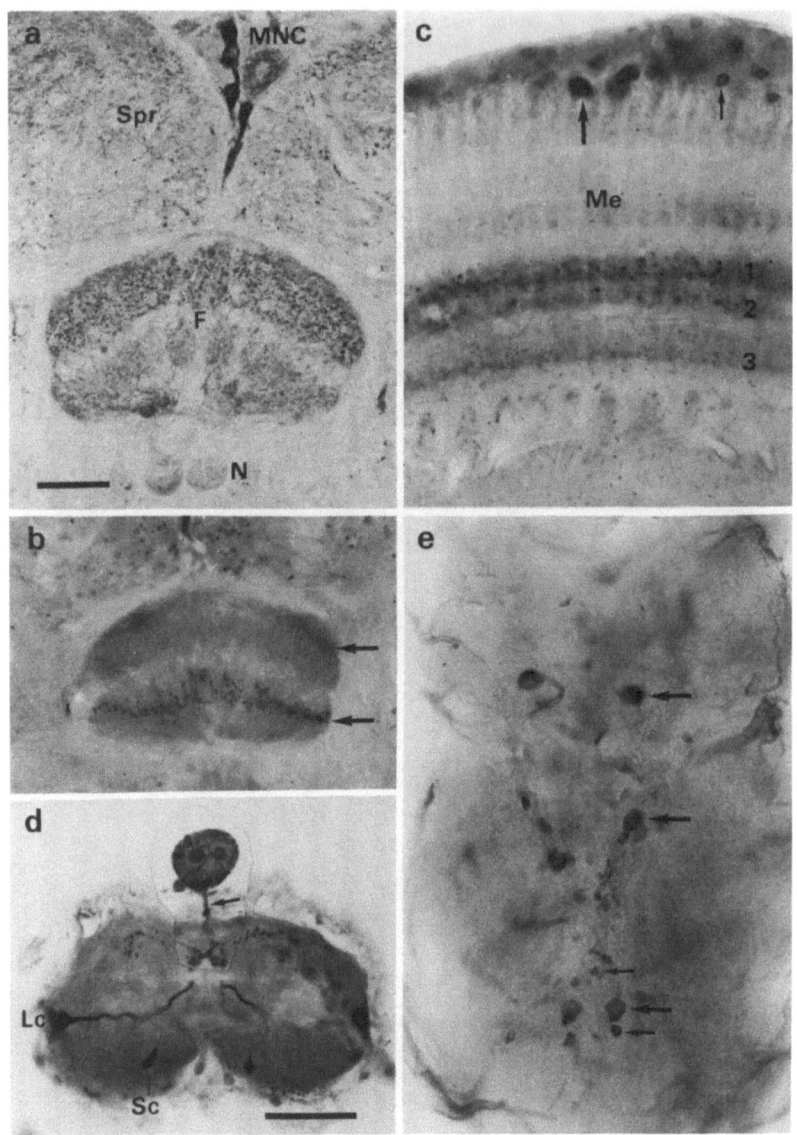

FIG. 9: Gastrin/CCK-like immunoreactive neurons in the CNS of <u>Calliphora</u> (PAP-method)
a. Cryostat section (frontal) through superior protocerebrum (Spr), the central body (fan-shaped body; F) and the cluster of median neurosecretroy cells (MNC). G/CCKi is seen in Spr, F and some of the cells in MNC. The nodula (N) do not show any G/CCKi.
b. Preembedding incubation (25 μm section) yields two main bands of G/CCKi in the central body.
c. Frontal section of the medulla showing three layers (1-3) of G/CCKi fibers. Large (large arrow) and small (small arrow) immunoreactive cell bodies appear to contribute fibers to these layers.
d. Transverse section through the mesothoracic ganglia of a third instar larva showing large (Lc) and small (Sc) immunoreactive cell bodies. The Lc neurons give rise to processes

supplying dorsal neuropil and a small dorsal neurohaemal organ sitting on a stalk (arrow).
e. Wholemount of adult thoracic ganglia viewed from the ventral side showing one pair of large G/CCKi cell bodies (large arrows) in each segment. These give rise to fibers innervating dorsal neuropil and neurohaemal regions in the dorsal neural sheath and hence they may be derived from the Lc cells in 9d. A number of smaller cell bodies (small arrows) associate the large ones. Scales: a-c = 50 μm, d-e = 100 μm. Preparation in d supplied by L. Ohlsson.

both substances. Gastrin and CCK are both present in the mammalian CNS (Dockray et al. 1978; Vanderhaeghen et al. 1975). The physiological role of gastrin/CCK-like substances in the insect CNS is not clear, but from immunocytochemical studies it may be suggested that they function as neurotransmitters/neuromodulators in the protocerebrum and thoracico-abdominal ganglia (Duve et al. 1983; Duve and Thorpe 1984) and as mediators in the retrocerebral glandular complex (pars intercerebralis-corpora cardiaca) and as a circulating neurohormone released from the neural sheath of thoracic ganglia (Duve and Thorpe 1984; Nässel and Elekes 1985).

Except for cells of the neurosecretory pathways (Fig. 8), G/CCKi neurons were found with processes in central body, lateral protocerebrum, optic lobes and in association with mushroom body neuropil (Figs. 8, 9a-c) (see also Duve et al. 1983). Using a preembedding peroxidase anti-peroxidase (PAP)-technique with postfixation in osmium and thick Araldite sections (Nässel and Elekes 1984) it has been possible to resolve G/CCKi processes in the Calliphora brain in detail (Nässel et al. in prep.). The findings of Duve et al. (1983) could be confirmed and extended as follows.

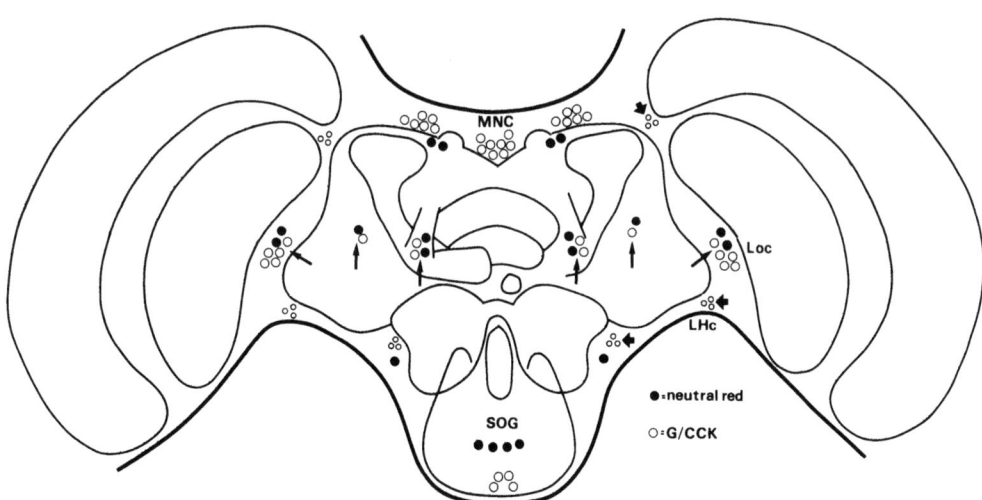

FIG. 10: Distribution of neutral red staining and Gastrin/CCK-like (G/CCK) immunoreactive cell bodies in a frontal section of the Calliphora brain. Thick arrows point at cell bodies in which G/CCK neuron clusters which do not stain consistently and in which number of cell bodies is variable (at least 6). The thin arrows indicate cell groups in which G/CCKi and neutral red staining may co-occur. MNC = median neurosecretory cell cluster; Loc = lobula cell cluster; LHc = cluster innervating lateral horn; SOG = suboesophageal ganglia.

Several groups of G/CCKi cell bodies stained consistantly (shown with large circles in Fig. 10). These represent ca. 40 neurons in the brain and suboesophageal ganglia. In addition 20-30 cell bodies were found that were more variable in their staining (in clusters indicated by thick arrows in Fig. 10). A group of ca. 8 G/CCKi cell bodies was consistantly found in the cluster of median neurosecretory cells. Axons from some of these run to the corpora cardiaca (Fig. 8) (see also Duve et al. 1983). Dorsal immunoreactive cell bodies also supply the fanshaped body of the central body with processes (Figs. 9a, b). Posteriorly in the protocerebrum six large G/CCKi cell bodies are distributed at four sites in a horizontal row (Fig. 10). About five cell bodies (Loc in Fig. 10) supply each lobula with immunoreactive processes. In the latero-ventral protocerebrum a cluster with a variable number of immunoreactive cell bodies (LHc) send processes dorsally into the lateral horn of the protocerebrum. In each half of the deutocerebrum 6-8 G/CCKi cell bodies can be found. Some of their processes invade antennal glomeruli. In the suboesophageal ganglia (SOG) four cell bodies are consistently reacting with G/CCK antisera. They supply SOG, deutocerebrum and tritocerebrum with processes (Figs. 8b, c). The large number of G/CCKi medulla neurons (Fig. 9c) give rise to three proximal layers in the medulla neuropil. As will be shown later some of the G/CCKi cell bodies are in locations very similar to Neutral Red staining cell bodies.

5.3. Proctolin-like immunoreactive neurons

The neuropeptide proctolin (H-Arg-Try-Leu-Pro-Thy-OH) was first isolated from the cockroach Periplaneta americana (Brown 1975; Brown and Starrat 1975; Starrat and Brown 1975) and is active in neuromuscular preparations of e.g. hindgut and skeletal muscle (Brown 1967; Piek and Mantel 1977; Cook and Meola 1978; Miller 1979; Adams and O'Shea 1983; O'Shea and Adams 1981) and has a biphasic effect on central monoaminergic neurons (Walker et al. 1980). Proctolin has been immunocytochemically mapped in the brain (Bishop et al. 1981; Bishop and O'Shea 1982; Veenstra et al. 1985; Nässel and O'Shea 1987) and ventral ganglia (Bishop and O'Shea 1982; Eckert et al. 1981, Agricola et al. 1985; Keshishian and O'Shea 1985; Nässel and O'Shea 1987) as well as peripherally at neuromuscular junctions (Eckert et al. 1981; Witten and O'Shea 1985).

In the brain of the cockroach Periplaneta ca. 40 proctolin-like immunoreactive (Pri) neurons were found (Bishop and O'Shea 1982). The majority of these cell bodies are in the tritocerebrum (Fig. 6c). In the protocerebrum there are three pairs of Pri neurons anteriorly: one pair in each dorso-lateral lobe and one pair ventro-medially (Fig. 6c). The tritocerebral neurons are in the lateral edges of the tritocerebral lobes. Two bilateral clusters of anterior tritocerebral Pri neurons can be distinguished: the first consists of five cells located anteriorly beneath each antennal lobe, the second is formed by eight smaller cells anteriorly in the region where the circumoesophageal connectives emerge. In the posterior part of the tritocerebrum two bilateral pairs of Pri neurons were resolved. The only immunoreactive processes that could be resolved were in bilateral neuropils in the tritocerebral lobes and fibers in the circumoesophageal connectives that run between the tritocerebrum and the suboesophageal ganglia.

In the suboesophageal ganglia of Periplaneta there are ca. 25 Pri neurons ventrally, clustered into bilateral groups near each main nerve trunk. There are also two ventro-medial groups, one anteriorly and one posteriorly. Twelve Pri neurons were found in the dorsal part of the ganglia. Most immunoreactive fibers innervate dorsal suboesophageal neuropils.

Also in the brain of the Colorado potato beetle Leptinotarsa decemlineata Pri neurons are scarce (Veenstra et al. 1985). Occasionally weakly immunoreactive neurons were found in protocerebrum and there were no immunoreactive neurons in deutocerebrum. In the tritocerebrum 8 to 10 Pri cell bodies are located in each hemisphere anteriorly below the antennal lobes. Immunoreactive processes from these cell bodies invade adjacent tritocerebral neuropil. Further immunoreactive axons are found in the antennal and frontal nerves. Suboesophageal ganglia contain more than 100 Pri cell bodies. Of these only 8 to 10 frontally located Pri neurons can be indivudually identified. In the ventral neuropils of the suboesophageal ganglia immunoreactive processes were found. In Calliphora 80-90 Pri neurons were found in the brain and ca. 200 in the lobula of the optic lobe. The brain neurons innervate pars intercerebrailis, the central body and tritocerebrum, and some groups of neurons send axons to copora cardiaca and possibly mouth-part muscles (Nässel and O'Shea 1987).

6. MAPPING OF CATECHOLAMINE-CONTAINING NEURONS

The catecholamines dopamine and noradrenaline have been determined in the CNS of a number of insect species (Frontali and Häggendahl 1969; Klemm and Björklund 1971; Klemm and Axelsson 1973; Robertson 1976; Dymond and Evans 1979; Slowley and Owen 1982; Mercer et al. 1983). The regional synthesis of these monoamines has abeen analyzed in the nervous system of Manduca sexta (Maxwell et al. 1978). Little is known about the function of catecholamines in the insect CNS (see Klemm 1976; Dymond and Evans 1979; Evans 1980; Mercer and Menzel 1982; Mercer and Erber 1983), whereas a function of dopamine as a peripheral transmitter, e.g. in salivary glands, has been proposed (Bland et al. 1973; Robertson 1975).

The cellular localization of dopamine and noradrenaline in the insect brain has mainly relied on aldehyde- or glyoxylic-acid induced histofluorescence (Frontali 1968; Klemm 1968, 1974, 1983; Klemm and Axelsson 1973; Mercer et al. 1983). A few studies have been published on immunocytochemical localization of dopamine in insect neurosecretory systems (Vieillemaringe et al. 1984) and mapping of the enzyme dopamine- β -hydroxylase-like immunoreactivity was made in two insect species in an attempt to localize putative noradrenaline containing neurons (Klemm et al. 1985). To date the most complete data available on distribution of catecholaminergic neurons is from histofluorescence studies. Hence, the following summary is based on this technique. No distinction will be made between dopamine- and noradrenaline-containing neurons, since no study has given a complete differential mapping of these neuron types.

6.2. Histofluorescence of catecholamines in the insect brain

6.2.1. Cockroach brain (Klemm 1983)

In the protocerebrum there are catecholamine containing (CA) cell bodies distributed in many regions either singly or in bilateral clusters (Fig. 6B). The following glomerular neuropils contain CA fibers (Fig. 6B): (1) the caudo-dorsal central body and the anterio-ventral ellipsoid body contain CA fibers continuing into surrounding non-glomerular neuropil, (2) the alfa-lobes of the mushroom bodies are packed with CA fibers in various layers; some of the layers invade the beta-lobes and surrounding non-glomerular neuropil, (3) beta-lobes have longitudinally oriented CA fibers some of which run into the peduncle and alfa-lobe as well as into the beta-lobe of the other hemisphere, (4) in the peduncles there are two zones of CA-fibers, the caudo-dorsal zone and the anterio-ventral. The fibers are continuous with those in alfa- and beta-lobes and non-glomerular neuropil around the mushroom bodies, (5) the calyces are invaded by CA fibers from a small number of "globuli"-cells, from the peduncle and optic lobes as well as from surrounding non-glomerular neuropil.

In the deutocerebrum the chemosensory glomeruli are surrounded and interconnected by fine meshworks of CA fibers derived from about six cell bodies fronto-dorsal to each antennal lobe. The tractus olfactorio-globularis (interconnecting antennal lobes and calyces of mushroom bodies) is surrounded by CA fibers. In each side of tritocerebrum 5 to 7 CA cell bodies are found laterally and one medially.

6.2.2. Bee brain (Klemm 1974, 1976; Mercer et al. 1983)

Catecholamine-containing cell bodies are located in clusters beneath the calyces, in lateral protocerebrum, between proto- and deutocerebrum, posteriorly in protocerebrum and in the suboesophageal ganglia (Fig. 5) CA fluorescence was found in fibers in all parts of the mushroom bodies (Fig. 5). In the peduncles and alfa-lobes CA fluorescence is layered, whereas the beta-lobes have a less pronounced layering. The CA fibers of the mushroom bodies seem to be mainly of extrinsic origin. The central body is packed with CA fibers in fan-shaped and ellipsoid bodies as well as in the noduli (Fig. 5). CA fibers of the central body invades non-glomerular protocerebral neuropil. The pons lacks CA fibers whereas the antennal chemosensory glomeurli contain such fibers peripherally.

6.2.3. Blowfly brain (Ramade and L'Hermite 1971; Klemm 1974)

CA cell bodies and processes were also mapped in Calliphora, but fiber paths were not traced in the same detail as in the cockroach (Fig. 2). Cell bodies were found mainly in clusters anteriorly and posteriorly in protocerebrum and in tritocerebrum. The mushroom bodies contain CA fibers in the alfa- and beta-lobes and lower part of the peduncles but not in the calyces. Of the central body complex the pons, fan-shaped and ellipsoid boides, noduli and ventral lobes contain CA processes (Fig. 2). Non-glomerular neuropils of proto- and tritocerebrum as well as in suboesphageal ganglia are packed with CA processes. Also in the locust CA neurons were mapped (Klemm and Axelsson 1973): The principal findings are shown in Fig. 4a.

7. OCTOPAMINERGIC NEURONS AND NEUTRAL RED STAINING IN THE BRAIN

The phenolamine octopamine has been found in relatively large amounts in the CNS of all studied insects (Robertson and Steele 1973; Robertson 1976; Evans 1978; Dymond and Evans 1979; Slowley and Owen 1982; Mercer et al. 1983; Nässel and Laxmyr 1983). The functional role of octopamine has been studied at insect neuromuscular junctions (Evans and O'Shea 1977, 1978; O'Shea and Evans 1979; Evans 1984) but its central function is poorly known (see Evans 1980; Mercer and Menzel 1982; Mercer and Erber 1983). Since octopamine is not fluorogenic in histofluorescence methods and no specific antisera have been reported so far, mapping of octo-paminergic neurons has relied on biochemical determination of single microdissected neurons from identified dorsomedial clusters in ventral ganglia and mushroom bodies (Evans and O'Shea 1977; see also Hoyle and Barker 1975; Mercer et al. 1983). Some identified neurons were found to be octopaminergic and it could be shown that these neurons selectively stained with the vital dye Neutral Red (Evans and O'Shea 1977, 1978).

Since Neutral Red stains also e.g. serotonin containing neurons (Stuart et al. 1974) this dye cannot be considered specific for octopamine-containing neurons in all cases and staining data must be interpreted cautiously. Mapping of Neutral Red staining neurons in the insect brain has been performed and correlated with microdissected brain regions (Dymond and Evans 1979; Mercer et al. 1983). In brief the following staining pattern was found in the cockroach, bee and blowfly brains.

In the cockroach and bee brain Neutral Red staining was seen in globuli cells (above calyces) of mushroom bodies and in mushroom body neuropil (Dymond and Evans 1979; Mercer et al. 1983). In the fly, incubation of whole living brains in saline with Neutral Red consistently stained cell bodies in characteristic locations (Fig. 10). These were found dorsally (two pairs) and caudally (10 cell bodies in 6 groups) in protocerebrum as well as anteriorly in deutocerebrum (one pair) and suboesophageal ganglia (four cell bodies). After prolonged incubation in the dye additional cell bodies were stained e.g. in the pars intercerebralis and caudally in the protocerebrum. The 10 posterior cell bodies in six groups in protocerebrum correspond, in size and location, to cell bodies reacting with antisera to gastrin/CCK (Fig. 10). None of the cell bodies staining with Neural Red in the first hour or two correspond to cell bodies that are 5-HTi.

8. NEUROACTIVE SUBSTANCES IN THE OPTIC LOBES

The optic lobes of insects are composed of three main columnar neuropil regions arranged in a sequence from the receptor layer: the lamina, medulla, and lobula complex (Bullock and Horridge 1965). The latter is in some insect orders divided into a posterior lobula plate and an anterior lobula. The neurons of the optic lobe are arranged in retinotopic columns (representing the projected receptor mosaic) and layers formed by dendrites or terminals of retinotopic neurons as well as tangential and amacrine neurons. This geometrically precise organization is convenient since the distribution of immunocytochemically detected substances can be related to neuropil layers or the lateral extent of the projected mosaic. The background information on morphological and physiological neuron types is extensive in insects (see Strausfeld 1976, 1984; Strausfeld and Nässel 1980; Hausen 1984; Shaw 1984) and there are good chances of matching immunocytochemically labeled neurons with previously identified neurons. In the following the general distribution of different neuroactive substances will be presented from a comparative point of view. It will be shown that the distribution of various substances vary between species; sometimes there are gross differences in other cases only the pattern of finer branches vary (see also Nässel 1986).

Biochemically, several biogenic monoamines have been determined in the optic lobes of the cockroach, locust, bee and blowfly: 5-HT, histamine, dopamine, noradrenaline and octopamine (Robertson 1976; Dymond and Evans 1979; Evans 1980; Mercer et al. 1983; Nässel and Laxmyr 1983; Elias and Evans 1983, 1984). With fluorescence histochemistry catecholamines were found in the optic lobes of locust (Fig. 4), cockroach and blowfly, but not in the honey bee; indolamines, however, were found in all the insects (Klemm 1976, 1983; Mercer et al. 1983). Serotonin-immunoreactive neurons have been mapped in more detail in these four species (Nässel and Klemm 1983; Schürmann and Klemm 1984; Tyrer et al. 1984; Nässel et al. 1985, 1987). All neuropil regions contain 5-HTi fibers but the pattern of fibers and cell bodies vary between the species (Fig. 12). Except in the cockroach, the number of cell bodies in the optic lobes is low, both for CA and 5-HTi neurons. Instead each neuron

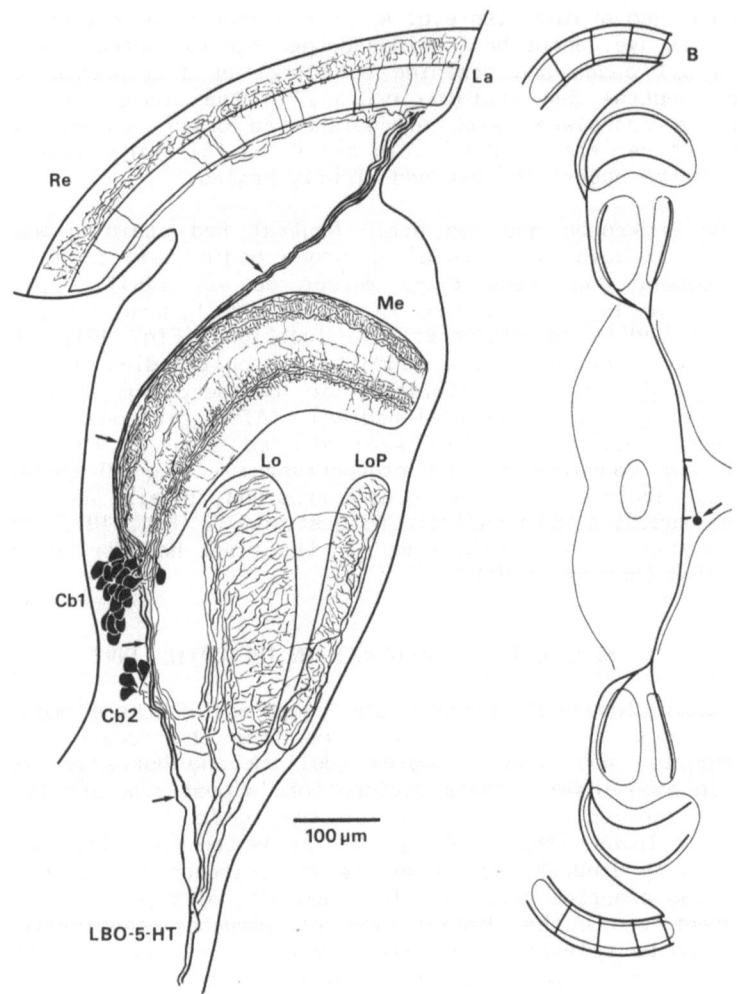

FIG. 11: Distribution of 5-HTi neurons in the optic lobe of the fleshfly Sarcophaga bullata.
A. Tracing of processes of lamina (La), medulla (Me), lobula (Lo) and lobula plate LoP). Two large neurons, LBO-5-HT (and arrows), invade all the optic lobe neuropils. The lamina processes, however, reside outside the synaptic layer, beneath the receptor retina (Re). Two clusters of amacrine neurons (Cb1 and Cb2) are 5-HTi.
B. Schematic tracing of one LBO-5-HT neuron (cell body at arrow). A and B altered after (Nässel et al. 1987).

has extensive branches, and in addition some neurons innervating optic lobe neuropil have their cell bodies in the midbrain. In Calliphora and other flies the 5-HTi neurons have been studied in detail (Fig. 11) including electron microscopy of some parts of the neurons (Nässel et al. 1985).

Localization of GABA-like immunoreactivity (GABAi) has been performed in the optic lobes of bees, a moth and three fly species (Meyer et al. 1986; Schäfer and Bicker 1986; Homberg et al. 1987). GABAi neurons are present in all neuropil regions of these insects, but some of the types of neurons displaying this reactivity are different in bees and flies.

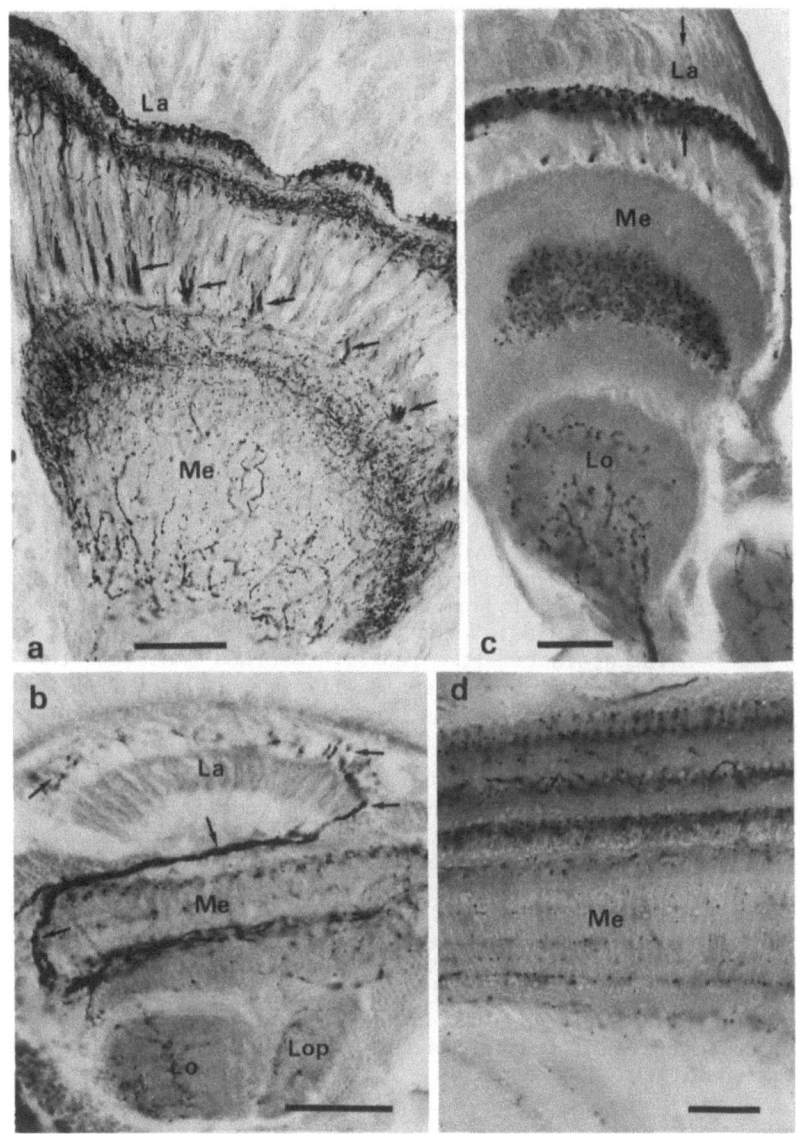

FIG. 12: 5-HTi processes in the optic lobes of four insect species. Note the different distribution of fibers in the lamina (La) and medulla (Me) between the species.
a. The cockroach Periplaneta has a bilayered innervation of the lamina and a diffuse distribution in the medulla. Connections between lamina and medulla are by means of numerous axons (arrows) running in the chiasma.
b. In Drosophila (like other higher Diptera) the lamina fibers are outside the synaptic layer and the medulla has three immunoreactive layers. Arrows point at the LBO5HT neurons.
c. The desert ant Cataglyphis bicolor has one proximal layer of 5-HTi fibers within synaptic neuropil of the lamina (borders of lamina indicated by arrows). The medulla has one broad layer.
d. In the medulla of the dragonfly Sympetrum not less than eight layers of 5-HTi processes are found.
Scales: a = 100 μm, b-d = 50 μm.

For instance lamina neurons in bees seem to be amacrines, whereas lamina processes in flies are derived from centripetal columnar medulla neurons (Meyer et al. 1986). The total number of GABAi cell bodies in the optic lobes of bees and flies is high (up to 9,000 per lobe). There are additional optic lobe neurons with cell bodies in the midbrain.

Very few neuroactive peptides have been found immunocytochemically in insect optic lobes. This is somewhat surprising since the neurons of the vertebrate retina react with a large number of antisera to peptides (see Brecha et al. 1983). In Calliphora many antisera were tested, but only gastrin/CCK-like (G/CCKi) proctolin- and FMRFamide-like immunoreactivity could conclusively be demonstrated. The G/CCKi neurons were found in the medulla and lobula (Figs. 8a, 9c). (Nässel et al. in prep.). In the cricket Acheta G/CCKi and somatostatin-like immunoreactivity was found in different types of neurons connecting the lamina and the medulla (K. Johansson and D.R. Nässel, in prep.). In the locust Veenstra et al. (1984) detected gastrin releasing peptide/bombesin-like immunoreactivity in ca. 250 small neurons innervating the lamina. These authors investigated six other insect species with the same antiserum, but found no immunoreactivity in the optic lobes. In the optic lobes of the Colorado potato beetle the following antisera revealed immunoreactive neurons: anti-FMRFamide/pancreatic polypeptide, anti-corticotropin releasing factor, anti-growth hormone, anti-rat prolactin and antimotilin (Veenstra and Schoneveld 1984; Veenstra 1984). Of these the 5-7 motilin-like and the 3-6 rat prolactin-like reactive neurons innervate the lamina, the 6-8 FMRFamide/pancreatic polypeptide-like reactive neurons innervate the medulla and the processes of the other immunoreactive neurons were not traced in detail.

With the use of antibodies to the enzyme choline acetyltransferase (ChAT; the biosynthetic enzyme for acetylcholine) a mapping of putative cholinergic neurons was made in the optic lobes of Drosophila (Salvaterra et al. 1985; Buchner et al. 1986). These authors found ChAT-like immunoreactivity in the lamina, medulla and lobula complex. The immunoreactivity was in a columnar pattern in the lamina and lobula and in a tri-layered one in the medulla. Finally Elias and Evans (1983, 1984) have demonstrated synthesis, metabolism and uptake of histamine in the optic lobes of the locust and it has been suggested from pharamacological/ physiological studies of Calliphora that photoreceptors may use histamine as their transmitter (Hardie 1987).

9. NEUROACTIVE SUBSTANCES IN THE VENTRAL GANGLIA

Several substances have been mapped immunocytochemically in the thoracical and abdominal ganglia of insects. In these ganglia the pattern of immunoreactive cell bodies and processes is relatively simple compared to the brain. The antisera employed display segmentally homologue neurons as well as neurons whose distribution vary between the segments (see Figs. 13, 15). So far mapping of chemically identifiable neurons in insect thoracic and abdominal ganglia has been more rewarding than similar studies in the brain, since the former neurons could be impaled by intracellular techniques (Evans and O'Shea 1978; Goodman and Spitzer 1979; O'Shea and Bishop 1982; O'Shea and Adams 1981; Adams and O'Shea 1983; Taghert and Goodman 1984; Witten and O'Shea 1985). The distribution of a few neuroactive substances in thoracic and abdominal ganglia will be summarized in the following: 5-HT, GABA, proctolin, gastrin/CCK, pancreatic polypeptide and FMRFamide. In addition the localization of catecholamine containing and Neutral Red staining neurons is presented.

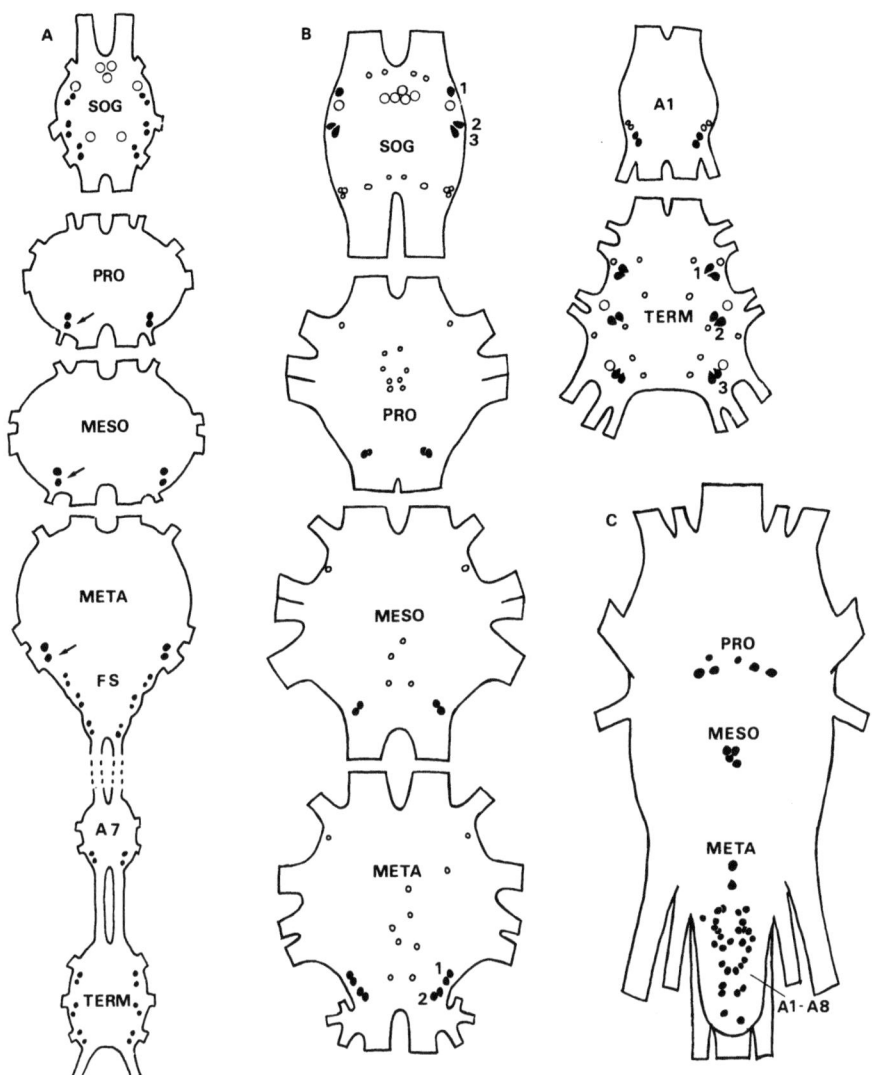

FIG. 13: Distribution of 5-HTi cell bodies in the ventral ganglia of three insects. Filled cell bodies may represent segmentally homologue neurons. Open ones are weakly staining or non-homologue neurons. A = Locust (redrawn after Tyrer et al. 1984). B = Cockroach (redrawn after Bishop and O'Shea 1982). C = _Calliphora_ (After Nässel 1987). Abbreviation: SOG = suboesophageal ganglia, FS = fused abdominal ganglia merged with metathoracic ganglion, TERM = terminal ganglion.

9.1. Serotonin-immunoreactive neurons in ventral ganglia

As shown in Fig. 13 the 5-HTi neurons have been mapped in ventral ganglia of locust, cockroach, and blowfly (Bishop and O'Shea 1983; Taghert and Goodman 1984; Tyrer et al. 1984; Nässel and Cantera 1985; Nässel 1987). Only in embryonic locusts is the complete anatomy of single 5-HTi neurons known from combined immunocytochemistry and Lucifer Yellow injection (Taghert and Goodman 1984).

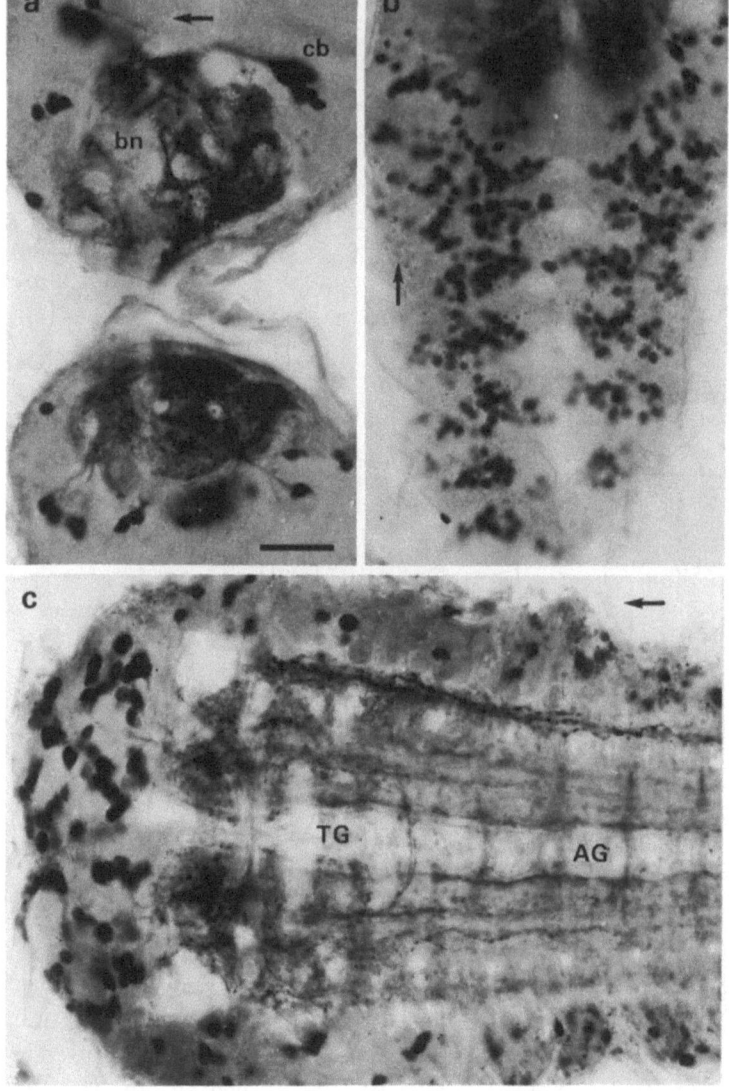

FIG. 14: GABA-like immunoreactive neurons in the CNS of larval Calliphora (horizontal 25 μm sections; pre-embedding incubation, PAP-method). Arrows point anteriorly.
a. The brain contains clusters of GABAi neurons (cb) and the brain neuropils (bn) are densely innervated by immunoreactive processes.
b. Ventral section of abdominal ganglia showing segmentally arranged bilateral clusters of GABAi cell bodies.
c. The neuropil of thoracic (TG) and abdominal (AG) ganglia contains a ladder-like distribution of GABAi processes.
Scale: a-c = 50 μm. Preparation provided by L. Ohlsson.

In the studied species each segmental ganglion contains at least one pair of 5-HTi cell bodies; most segments contain several pairs. The majority of the 5-HTi neurons appear to be interneurons. In the thoracic ganglia the following pattern of 5-HTi neurons was found. In the adult locust each of the pro- (T1), meso- (T2), and metathoracic (T3) ganglia contains two pairs of heavily staining 5-HTi neurons and three pairs of lighter staining ones (Tyrer et al. 1984). Taghert and Goodman (1984) found a different distribution of heavily staining 5-HTi neurons in embryonic locusts: three pairs in T1, two pairs in T2 and one pair in T3. The cockroach thoracic ganglia contain two pairs of 5-HTi cell bodies in each of the T1 and T2 ganglia and the fused T3 contains four pairs; further smaller 5-HTi cell bodies are located ventromedially and anterodorsally. In blowflies there are three pairs of 5-HTi cell bodies in T1, two pairs in T2 and one pair in T3. In all species the large thoracic cell bodies are located ventrally and have axons running to the next anterior ganglion and processes in neuropil of both hemispheres of the own segment.

The unfused abdominal ganglia contain two pairs of 5-HTi neurons in each segment of locusts and cockroaches. In blowfiles the first seven abdominal ganglia (A1-A7) contain two pairs and the last (A8) contains one pair of 5-HTi neurons. The fused terminal ganglia of locusts and cockroaches contain the homologues of the segmental 5-HTi neurons present in unfused ganglia. In the cockroach terminal ganglion additional neurons were found.

9.2. GABA-immunoreactive neurons in ventral ganglia

In <u>Calliphora</u> GABA-immunoreactive (GABAi) neurons were found in large numbers in all segments of the thoracic and abdominal ganglia (Nässel et al. in prep.). The segmental distribution of GABAi neurons is more apparent in ventral ganglia of larval blowflies (Figs. 14b, c). The GABAi processes invade most ventral and dorsal neuropil regions (Fig. 14c). The immunoreactive neurons are interneurons and most likely motorneurons. The interneurons are both of local nature and projection neurons whose axons interconnect segments and hemispheres. Recently a detailed description of GABAi neurons in the thoracic ganglia of a locust was provided by Watson (1986).

9.3. Proctolin-immunoreactive neurons in ventral ganglia

Proctolin-like immunoreactive (Pri) neurons were mapped in thoracic and abdominal ganglia of the cockroach (Erkert et al. 1981; Bishop and O'Shea 1982; Agricola et al. 1985), locust (Keshishian and O'Shea 1985) Colorado potato beetle (Veenstra et al. 1985 and blowfly (Nässel and O'Shea 1987). In the locust Keshishian et al. (1985) found one pair of prominent Pri neurons ventrally in each of the thoracic ganglia T1 and T2 (Fig. 15). The metathoracic ganglion T3, which is fused with three abdominal ganglia contains three pairs of ventral and one pair of dorsal heavily staining Pri neurons (Fig. 15). In addition the thoracic ganglia contain some lighter staining Pri neurons ventrally and dorsally. In the four unfused abdominal ganglia (A1-A4) no Pri cell bodies could be detected, but in the fused terminal ganglion there is six heavily staining and ca. 14-16 lighter staining neurons (Fig. 15). Some of the Pri neurons in the locust ventral ganglia are efferents (neurons labeled AVL and PDL in (Fig. 15) others are interneurons (VM in Fig. 15).

In the cockroach, blowfly and Colorado potato beetle all segments (including unfused abdominal ganglia) contain numerous Pri cell bodies in clusters ventrally and dorsally (Bishop and O'Shea 1982, Veenstra et al. 1985; Agricola et al. 1985; Nässel and O'Shea 1987). Some of the Pri neurons are motorneurons (Eckert et al. 1981; O'Shea and Bishop 1982; Adams and O'Shea 1983; Agricola et al. 1985).

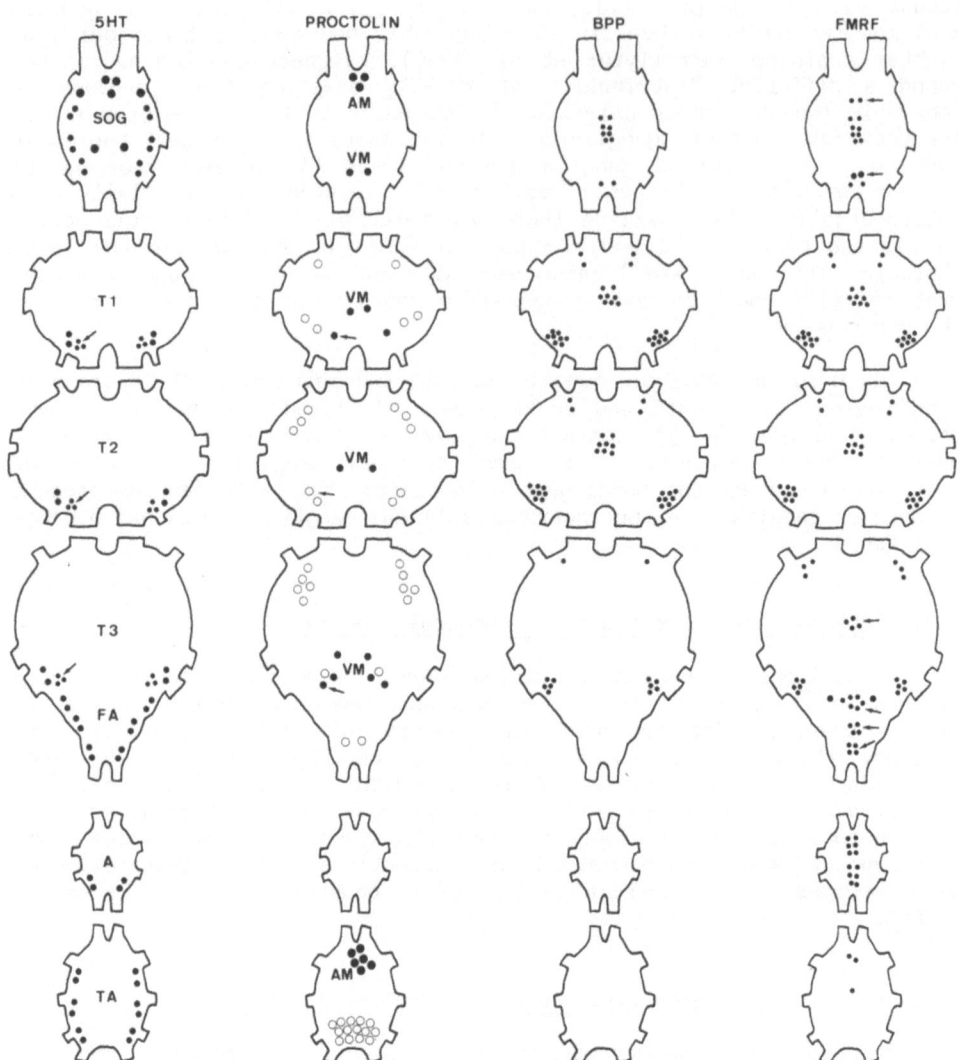

FIG. 15: The distribution of cell bodies reactive with antibodies to 5-HT, proctolin, bovine pancreatic polypeptide (BPP) and FMRFamide (FMRF) in the ventral ganglia of locust. Of the proctolin-immunoreactive cell bodies those depicted as filled profiles are consistently staining intensely, the others are variable. The VM neurons are interneurons, the posterior AM-neurons are efferents and the proctolin neurons indicated by arrows may be efferents (motorneurons). Note that all BPPi neurons also react with antibodies to FMRFamide. Some additional neurons stain with FMRFamide (indicated by arrows). Redrawn from Tyrer et al. (1984), Keshishian and O'Shea (1985) and Myers and Evans (1985).

9.4. Gastrin/CCK- and pancreatic polypeptide-like immunoreactivity in ventral ganglia

Gastrin/CCK-like (G/CCKi), FMRFamide-like and pancreatic polypeptide like (PPi) immunoreactive neurons were mapped in the thoracio-abdominal ganglia of Calliphora (Duve and Thorpe 1982, 1984; Nässel et al. in prep.). There is one pair of large ventral G/CCKi cell bodies in each thoracic neuromere and a number of smaller neurons ventrally and dorsally in each segment (Figs. 9d, e). The eight fused abdominal ganglia contain five pairs of ventral and six to seven pairs of dorsal G/CCKi cell bodies. Totally there are 40-50 G/CCKi cell bodies in the thoracic-abdominal ganglia. Many of the G/CCKi neurons appear to be interneurons with processes in ventral and dorsal neuropil. The six larger thoracic G/CCKi neurons, with cell bodies ventrally in the thoracic ganglia, send axons to dorsal neuropil where they arborize before continuing into the dorsal neural sheath (Figs. 9d, e). In the neural sheath they form an extensive plexus of immunoreactive fibers (Duve and Thorpe 1984; Nässel and Elekes 1985; Nässel et al. in prep.).

Using antisera against G/CCK and PP on adjacent sections Duve and Thorpe (1984) showed colocalization of these substances in the six large ventral thoracic neurons innervating dorsal neuropil and the neural sheath. Some of the abdominal G/CCKi neurons also contained PPi. Duve and Thorpe showed that the number and intensity of cell bodies stained depends on age and diet of the flies. The same six neurons in addition react with antisera to FMRFamide (Nässel et al. in prep.)

Pancreatic polypeptide- and FMRFamide-like immunoreactivity was also mapped in the ventral ganglia of locusts (Myers and Evans 1985a, b). Both types of immunoreactivity was in cell bodies localized in clusters medially and laterally in each segment (Fig. 15). All the bovine pancreatic polypeptide like (BPPi) neurons also react with antisera against FMRFamide. There are additional FMRFamide-immunoreactive neurons that do not react with BPP antisera (Myers and Evans 1985b).

9.5. Catecholamine-containing and Neutral Red staining neurons in ventral ganglia

Catecholamine-containing (CA) neurons have been mapped with histofluorescence techniques in the ventral ganglia of the hemipteran Rhodnius prolixus (Flanagan 1986) and in larval Drosophila (Budnik et al. 1985). In the larval CNS of Drosophila Budnik et al. (1985) found one pair of ventral CA cells per segment in thoracic and abdominal ganglia and a row of cells dorsally. This pattern was found distinct from that of 5-HTi neurons in the same ganglia (White and Valles 1984). In Rhodnius (Flanagan 1986) found seven CA cells in the prothoracic ganglion and 13 in the fused meso-metathoracic ganglion (Fig. 16). In the prothoracic ganglion three large CA cell bodies are situated along the dorsal midline and four smaller ones are located laterally. The fused thoracic ganglion contains only one large midline cell and 12 smaller lateral ones.

Neutral Red, first used in the leech CNS (Stuart et al. 1974), stains a dorsal cluster of 6-8 cells in the locust metathoracic ganglion (Evans and O'Shea 1978). These cells are the dorsal unpaired median (DUM) neurons (Hoyle and Barker 1975; Evans and O'Shea 1978). Similar Neutral Red staining DUM clusters have been shown in other thoracic and abdominal ganglia of locusts and cockroaches (Evans and O'Shea 1978; Evans 1980; Dymond and Evans 1979). The locust DUM cells are bilaterally innervating skeletal muscles and some have been shown to contain octopamine (Evans and O'Shea 1978). In Rhodnius dorsomedial clusters of Neutral Red staining cell bodies were found (Flanagan 1983). In Calliphora each thoracic ganglion has one pair of large ventral cell bodies that stain with Neutral Red. These cells are in locations similar to the large G/CCKi and BPPi cell bodies (see Fig. 9e). No dorsal Neutral Red staining cell bodies could be detected.

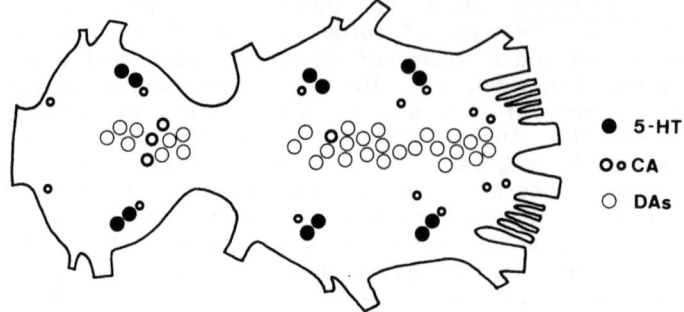

FIG. 16: The distribution of 5-HTi (5HT), catecholamine-containing (CA) and dopamine-sequestering (DAs) cell bodies in the thoracic ganglia of the hemipteran Rhodnius (redrawn from Flanagan 1986). Note the segmentally homologue 5-HTi neurons. CA and DAs are found also medially. Some of the CAs neurons appear to stain also with Neutral Red.

10. CONCLUDING REMARKS

This review has presented data on distribution of neuroactive substances in the insect CNS. The central function of the substances mapped immunocytochemically is unknown, but it can be hoped that intracellular techniques will be employed on some of the neural systems in the future. At present only GABA-, serotonin-, proctolin- and octopamine-containing neurons have been analyzed to some extent physiologically. As related in the various sections, however, the actions of these substances have been studied mainly peripherally. In the following, central functions will only be speculated upon briefly. First some comparative aspects of the distribution of different neuroactive substances in different insect species will be discussed.

10.1. Patterns of distribution of neuroactive substances

The neuroactive substances related in the previous text are distributed in quite different patterns and in different morphological classes of neurons.

10.1.1 Serotonin-immunoreactive neurons.

Serotonin-immunoreactive neurons (Figs. 2, 4-6, 13) are present in small numbers in the studied insects. The brain (excluding optic lobes) contains around 100 5-HTi neurons and ventral ganglia 2-10 neurons each. 5-HTi neurons are mainly interneurons, but also efferents exist (supplying a variety of peripheral organs (Nässel 1986)). The 5-HTi neurons form very extensive arborizations and often interconnect several neuropil regions (within and between segments). Few neuropil regions (including non-glomerular neuropil) lack 5-HTi innervation.

10.1.2. Catecholamine-containing neurons

Catecholamine-containing neurons (Figs. 2, 4-6 16) are distributed in a fashion similar to 5-HTi ones and also exist in rather small numbers.

It appears as if most CA neurons are interneurons, but some efferents may occur (innervating alimentary canal and other peripheral targets (Klemm 1876)). Like 5-HTi neurons CA-neurons invade most glomerular and non-glomerular neuropils. They interconnect neuropils and probably segments.

10.1.3. GABA-like immunoreactive neurons

GABA-like immunoreactive neurons (Figs. 7, 14) are distributed quite differently from 5-HTi and CA neurons. They exist in very large numbers. The brain of bees and flies may contain up to 20,000 GABAi neurons. These are interneurons and efferents (probably motorneurons). Many GABAi neurons are small field circuit neurons or amacrines, others are wide field projection neurons. In short, the range of morphological neuron types that are GABAi seems to be larger than for any other substance mapped so far. Also GABAi neurons supply most neuropils, but in patterns different from 5-HTi and CA neurons.

10.1.4. Gastrin/CCK-like immunoreactive neurons

Gastrin/CCK-like immunoreactive neurons (Fig. 10) are few in the Calliphora brain (ca. 60 consistently staining neurons when excluding medulla neurons). The G/CCKi neurons are interneurons and efferents (to corpora cardiaca and thoracic neurohaemal areas). Centrally the interneurons often supply only limited regions of neuropil like e.g. layers of the central body, lobula and medulla. Some more extensive G/CCKi interneurons exist that supply e.g. non-glomerular neuropil of suboesophageal and protocerebral ganglia. The overall innervation is, however, much less extensive than that of 5-HTi, CA and GABAi neurons. Some G/CCKi neurons also react with antisera to bovine pancreatic polypeptide and FMRFamide and some seem to stain with the vital dye Neutral Red.

10.1.5. Proctolin-like immunoreactive (Pri) neurons

Proctolin-like immunoreactive (Pri) neurons (Fig. 6) are also few, especially in the brain (in unfused abdominal ganglia of locusts they are missing). Pri neurons are interneurons and efferents (some are motorneurons). In the cockroach brain the main innervation appears to be in tritocerebrum. In Calliphora the central body, the pars intercerebralis, tritocerebrum and the lobula are innervated (Nässel and O'Shea 1987). In locust ventral ganglia the interneurons have wide arborizations (Keshishian and O'Shea 1985).

10.1.6 Other immunoreactive peptides

Other immunoreactive peptides (Fig. 6, 15) also occur in small numbers of neurons. In the brain many peptides have been mapped to the median neurosecretory cells, others to optic lobes or small clusters in other brain regions (see also Veenstra 1984). In ventral ganglia also small clusters are found medially or laterally.

10.1.7. Octopamine-containing and neutral staining neurons

Octopamine-containing and neutral staining neurons have been found mainly in the median unpaired dorsal neurons of locust and cockroach ventral ganglia, as well as in mushroom body globuli cells of bees and cockroaches. Since octopamine has been determined in rather large quantities in many other brain regions obviously numerous octopamine containing neurons have escaped histological detection. The known neurons are efferents (to skeletal muscle) and interneurons in mushroom bodies.

10.2 Variation in segmental distribution of neuroactive substances

It seems like there is a set of neurons in each segment characteristic for each antiserum employed (see Figs. 13 and 15) at least embryologically. Some of the chemically identifiable neurons can thus be referred to as segmental homologues. Probably due to differential development of the various segments some ganglia become more complex than others. This seems reflected in the distribution of some neuroactive substances. Looking at the ventral ganglia it is apparent that the suboesophageal ganglia often contain more immunoreactive neurons than other ganglia (Fig. 13; see also Veenstra 1984). Some peptidergic neurons (e.g. Pri, BPPi and FMRFi) are more numerous and complex in thoracic ganglia (Fig. 15). In locusts proctolin- and BPP/immunoreactive cell bodies could not be detected in unfused abdominal ganglia. The segmental brain distribution of immunoreactive neurons is more complex to analyze. In general it, however, seems like e.g. CA, 5-HTi, GABAi, G/CCKi, and enkephalin immunoreactive neurons are more numerous in protocerebral centers (including optic lobes) than other parts of the brain. In the cockroach proctolin is immunocytochemically determined in the largest number of cell bodies in tritocerebrum. This regional variation in distribution of chemically identifiable neurons is interesting from the developmental point of view. Taghert and Goodman (1984) showed the cell lineage and determination of 5-HTi neurons in the ventral ganglia of embryonic locusts. They found that the common neuroblasts in each hemisegment always produced a fixed number of progeny, but of these varying numbers acquired 5-HTi phenotype in different segments (the chemical fate of the remaining progeny is unknown). Whether this type of cell lineage exists also in brain segments is not known. It is, however, clear that the cells that are e.g. 5-HTi, CA and GABAi in the brain are extremely complex and vary substantially in number and morphology between segments.

10.3. What is next?

Obviously one direction to go from here is to use intracellular techniques, bioassays and pharmacological-behavioral approaches to study neuroactive substance functions in identifiable neurons and their targets as well as in whole circuits. Although more difficult technically, the central function of neuroactive substances need to be studied. A number of central neurons that have been immunocytochemically identified are likely to be good probes for intracellular studies of central functions. We also need to know more about the central neurons that are pre- and postsynaptic to the immununocytochemically identified neurons. Immunocytochemical double marking or combination with other tracer techniques have hardly been tried on insects, but have proven useful in studies of vertebrate neural circuits. Using immunocytochemistry as a neuroanatomical marking method, one can also study the morphogenesis of chemically identified sets of neurons during normal and experimentally (or genetically) altered development. Studies similar to that on the ventral ganglia by Taghert and Goodman (1984) may perhaps not be possible to perform on brain neurons, but it is clearly desirable to understand more about the development of the insect brain. Immunocytochemistry of differentiating chemically identified neurons may be one approach. Finally we need to characterize the endogenous substances in the peptidergic neurons that cross react with antisera to mammalian type peptides.

11. REFERENCES

Adams ME, O'Shea M (1983) Peptide cotransmitter at a neuromuscular junction. Science 221: 286-289.

Agricola H, Eckert M, Ude J, Birkenbeil H, Penzlin H (1985) The distribution of a proctolin-like immunoreactive material in the terminal ganglion of the cockroach, Periplaneta americana L. Cell Tissue Res 239: 203-209.

Ammermüller J, Weiler R (1985) S-neurons and not L-neurons are the source of GABA-ergic action in the ocellar system. J Comp Physiol A 157: 779-788.

Baxter CF, Torralba GF (1975) γ-aminobutyric acid and glutamate decarboxylase (L-glutamate 1-carboxylase E.C. 4.1.1.15) in the nervous system of the cockroach, Periplaneta americana. I. Regional distribution and properties of the enzyme. Brain Res 84: 383-397.

Bicker G, Schäfer S, Kingan TG (1985) Mushroom body feedback neurons in the honey bee show GABA-like immunoreactivity. Brain Res 360: 394-397.

Bishop CA, O'Shea M (1982) Neuropeptide proctolin (H-Arg-Try-Leu-Pro-Thr-OH): immunocytochemical mapping of neurons in the central nervous system of the cockroach. J Comp Neurol 207: 223-238.

Bishop CA, O'Shea M (1983) Serotonin immunoreactive neurons in the central nervous system of an insect (Periplaneta americana). J Neurobiol 14: 251-269.

Bishop CA, O'Shea M, Miller RJ (1981) Neuropeptide proctolin (H-Arg-Tyr-Leu-Pro-Thr-OH): Immunological detection and neuronal localization in the insect central nervous system. Proc Natl Acad Sci USA 78: 5899-6002.

Bland KP, House CR, Ginsborg BL, Laszlo I (1973) Catecholamine transmitter for salivary secretion in the cockroach. Nature New Biol 244: 26-27.

Bräunig P (1987) The satellite nervous system - an extensive neurohaemel network in the locust head. J Comp Physiol A 160: 69-77.

Brecha NC, Eldred W, Kuljis RO, Karten HJ (1983) Identification of biologically active peptides in the vertebrate retina. Prog Retina Res 3: 185-226.

Brown BE (1967) Neuromuscular transmitter substance in insect visceral muscle. Science 155: 595-597.

Brown BE (1975) Proctolin: A peptide transmitter candidate in insects. Life Sci 17: 1241-1252.

Brown BE, Starratt AN (1975) Isolation of proctolin, a myotropic peptide, from Periplaneta americana. J Insect Physiol 21: 1879-1881.

Budnik V, Valles AM, White K (1985) Histofluorescence of catecholamine-containing neurons in Drosophila. Soc Neurosci Abstr 11. 1985.

Bullock TH, Horridge GA (1965) Structure and function in the nervous system of invertebrates. San Francisco, Freeman.

Callec JJ (1974) Synaptic transmission in the central nervous system of insects. In: Treherne JE (ed) Insect neurobiology. Elsevier, Amsterdam.

Campos-Ortega JA (1974) Autoradiographic localization of ^{3}H- γ -aminobutyric acid uptake in the lamina ganglionaris of Musca and Drosophila. Z Zellforsch mikrosk Anat 147: 415-431.

Clarke BS, Donnellan JF (1982) Concentration of some putative neurotransmitters in the CNS of quick-frozen insects. Insect Biochem 12: 623-638.

Colhoun EH (1963) The synthesis of 5-hydroxytryptamine in the American cockroach. Experientia 19: 9-10.

Collins C, Miller TA (1977) Studies on the action of biogenic amines on cockroach heart. J Exp Biol 67: 1-15.

Cook BJ, Meola S (1978) The oviduct musculature of the horsefly, Tabanus sulcifrons, and its response to 5-hydroxytryptamine and proctolin. Physiol Entomol 3: 273-280.

Davenport AP, Evans PD (1986) Sex-related differences in the concentration of metenkaphalin-like immunoreactivity in the nervous system of an insect, Schistocerca gregaria, revealed by radioimmunoassay. Brain Res 383: 319-322.

Dismukes RK (1979) New concepts of molecular communication among neurons. Behav Brain Sci 2: 409-448.

Dockray GJ, Gregory RA, Hutchison JB, Harris JI, Runswick MJ (1978) Isolation, structure and biological activity of two cholecystokinin octapeptides from sheep brain. Nature 274: 711-713.

Dockray GJ, Duve H, Thorpe A (1981) Immunochemical characterization of gastrin/cholecystokinin-like peptides in the brain of the blowfly Calliphora vomitoria. Gen Comp Endocrinol 45: 491-496.

Doerr-Schott J, Joly J, Dubois MP (1978) Sur l'existence dans la Pars intercerebralis d'un insecte (Locusta migratoria R. et F.) de cellules neurosecretrices fixant un antiserum antisomatostatine. CR Acad Sci Paris 286D: 93-95.

Duve H, Thorpe A (1979) Immunoflurescent localization of insulin-like material in the median neurosecretory cells of the blowfly, Calliphora vomitoria (Diptera). Cell Tissue Res 200: 187-191.

Duve H, Thorpe A (1981) Gastrin/Cholecystokinin (CCK)-like immunoreactive neurones in the brain of the blowfly, Calliphora erythrocephala (Diptera). Gen Comp Endocrinol 43: 381-391.

Duve H, Thorpe A (1982) The distribution of pancreatic polypeptide in the nervous system and the gut of the blowfly Calliphora vomitoria (Diptera). Cell Tissue Res 227: 67-77.

Duve H, Thorpe A (1983) Immunocytochemical identification of endorphin-like material in neurones of the brain and corpus cardiacum of the blowfly, Calliphora vomitoria (Diptera). Cell Tissue Res 233: 415-426.

Duve H, Thorpe A (1984) Immunocytochemical mapping of gastrin/CCK-like peptides in the neuroendocrine system of the blowfly Calliphora vomitoria (Diptera). Cell Tissue Res 237: 309-320.

Duve H, Thorpe A, Lazarus NR, Lowry PJ (1982) A neuropeptide of the blowfly Calliphora vomotoria with an aminoacid composition homologous with vertebrate pancreatic polypeptide. Biochem J 201: 429-432.

Duve H, Thorpe A, Strausfeld NJ (1983) Cobalt-immunocytochemical identification of peptidergic neurons in Calliphora innervating central and peripheral targets. J Neurocytol 12: 847-861.

Dymond GR, Evans PD (1979) Biogenic amines in the nervous system of the cockroach Periplaneta americana: association of octopamine with mushroom bodies and dorsal unpaired neurons. Insect Biochem 9: 535-545.

Eckert M, Agricola H, Penzlin, H (1981) Immunocytochemical identification of proctolin-like immunoreactivity in the terminal ganglion and hindgut of the cockroach Periplaneta americana (L.). Cell Tissue Res 217: 633-645.

Elias MS, Evans PD (1983) Histamine in the insect nervous system, synthesis and metabolism. J Neurochem 41: 562-568.

Elias MS, Evans PD (1984) Autoradiographic localization of ^{3}H-histamine accumulation by the visual system of the locust. Cell Tissue Res 238: 105-112.

El-Sahly M (1981) Immunohistochemical localization of pancreatic polypeptide (PP) in the brain of the larval instar of the hoverfly, Eristalis aeneus (Diptera). Experientia 37: 1009-1010.

El-Sahly M, Abou-El-Ela R, Falkmer S, Grimelius L, Wilander E (1980) Immunohistochemical evidence of gastro-entero-pancreatic neurophormonal peptides of vertebrate type in the nervous system of the larva of a dipteran insect, the hoverfly, Eristalis aeneus. Regulatory Peptides 1: 187-204.

Evans PD (1978) Octopamine distribution in the insect nervous system. J Neurochem 30: 1009-1013.

Evans PD (1980) Biogenic amines in the insect nervous system. Adv Insect Physiol 15: 317-473.

Evans PD (1984) Studies on the mode of action of octopamine, 5-hydroxytryptamine and proctolin on a myogenic rhythm in the locust. J Exp Biol 110: 231-251.

Evans PD, Myers CM (1986) Peptidergic and aminergic modulation of insect skeletal muscle. J Exp Biol 124: 143-176.

Evans PD, O'Shea M (1977) The identification of an octopamine neurone which modulates neuromuscular transmission in the locust. Nature (Lond) 270: 275-279.

Evans PD, O'Shea M (1978) The identification of an octopaminergic neuron and the modulation of a myogenic rhythm in the locust. J Exp Biol 73: 235-260.

Flanagan TRJ (1983) Monoaminergic innervation in a hemipteran nervous system: a whole-mount histofluorescence survey. In: Strausfeld NJ (ed) Functional neuroanatomy. Springer-Verlag, Berlin.

Flanagan TRJ (1986) Serotonin-containing, catecholamine-containing and dopamine-sequestering neurones in the ventral nerve chord of the hemipteran Rhodnius prolixus. J Insect Physiol 32: 17-26.

Fraenkel G, Hsiao C (1962) Hormonal and nervous control of tanning in the fly. Science 138: 27-29.

Frontali N (1964) Brain glutamic acid decarboxylase and synthesis of GABA in vertebrate and invertebrate species. In: Richter D (ed) Comparative neurochemistry. Pergamon Press, Oxford.

Frontali N (1968) Histochemical localization of catecholamines in the brain of normal and drug-treated cockroaches. J Insect Physiol 14: 881-886.

Frontali N, Häeggendal J (1969) Noradrenaline and dopamine content in the brain of the cockroach Periplaneta americana. Brain Res 14: 540-542.

Frontali N, Pierantoni R (1973) Autoradiographic localization of ^{3}H-GABA in cockroach brain. Comp Biochem Physiol 44A: 1369-1372.

Gahery Y, Boistel J (1965) Study of some pharmacological substances which modify the electrical activity of the sixth abdominal ganglion of the cockroach, Periplaneta americana. In: Treherne JE, Beament JWL (eds) The physiolgoy of the insect nervous system. Academic Press, New York.

Gersch M, Fisher F, Unger H, Kabitaza W (1961) Vorkommen von Serotonin im Nervensystem von Periplaneta americana L. (Insecta). Z Naturforsch 16b: 351-352.

Gerschenfeld HM (1973) Chemical transmission in invertebrate and central nervous systems and neuromuscular junctions. Physiol Rev 53: 1-119.

Goodman CS, Spitzer NC (1979) Embryonic development of idenfified neurone: Differentiation from neuroblast to neurone. Nature (Lond) 280: 208-214.

Hardie R (1987) Is histamine a neurotransmitter in insect photoreceptors? J Comp Physiol A (in press).

Hausen K (1984) The lobula-complex of the fly: Structure, function and significance in visual behavior. In: Ali MA (ed) Photoreception and vision in invertebrates. Plenum Press, New York, London.

Hiripi L, Rozsa KS (1973) Fluorimetric determination of 5-hydroxytryptamine and catecholamines in the central nervous system and heart of Locusta migratoria migratorioides. J Insect Physiol 19: 1481-1485.

Homberg U, Kingan TG, Hildebrand JG (1987) Immunocytochemistry of GABA in the brain and suboesophageal ganglion of Manduca sexta. Cell Tissue Res 248: 1-24.

Hoskins SG, Homberg U, Kingan TG, Christensen TA, Hildebrand JG (1986) Immunocytochemistry of GABA in the antennal lobes of the sphinx moth Manduca sexta. Cell Tissue Res 244: 243-252.

Hoyle G, Barker DL (1975) Synthesis of octopamine by insect dorsal medial unpaired neurones. J Exp Zool 193: 433-439.

Kerkut GA (1966) Biochemical aspects of invertebrate nerve cells. In: Wiersma GAG (ed) Invertebrate nervous systems. Their significance for mammalian neurophysiology. Univ Chicago Press, Chicago.

Kerkut GA, Pitman RM, Walker RJ (1969) Iontophoretic application of acetylcholine and GABA onto insect central neurones. Comp Biochem Physiol 31: 611-633.

Keshishian H, O'Shea M (1985) The distribution of a peptide neurotransmitter in the postembryonic grasshopper central nervous system. J Neurosci 5: 992-1004.

Kingan TG Hildebrand JG (1985) Screening and assays for neurotransmitters in the insect nervous system. In: Breer H, Miller TA (eds) Neurochemical techniques in insect research.

Klemm N (1968) Monoaminhaltige Strukturen im Zentralnervensystem der Trichoptera (Insecta). Teil I. Z Zellforsch mikrosk Anat 92: 487-502.

Klemm N (1974) Vergleichend-histochemische Untersuchungen ber die Verteilung monoamin-haltiger Strukturen im Oberschlundganglion von Angehlrigen verschiedener Insektenordnungen. Ent germ 1: 21-49.

Klemm N (1976) Histochemistry of putative neurotransmitters in the insect brain. Prog Neurobiol 7: 99-169.

Klemm N (1983) Monoamine-containing neurons and their projections in the brain (supraoesophageal ganglion) of cockroaches. An aldehyde fluorescence study. Cell Tissue Res 229: 370-402.

Klemm N, Axelsson S (1973) Determination of dopamine, noradrenaline and 5-hydroxytryptamine in the cerebral ganglion of the desert locust, Schistocerca gregaria Forsk. (Insecta. Orthoptera). Brain Res 57: 289-298.

Klemm N, Björklund A (1971) Identification of dopamine and noradrenaline in the nervous structures of the insect brain. Brain Res 26: 459-464.

Klemm N, Sundler F (1983) the organization of catecholamine-containing and serotonin-immunoreactive neurons in the corpora pedunculata in the desert locust Schistocerca gregaria. Neurosci Lett 36: 13-17.

Klemm N, Steinbusch HWM, Sundler F (1984) Distribution of serotonin-containing neurons and their pathways in the supraoesophageal ganglion of the cockroach Periplaneta americana as revealed by immucocytochemistry. J Comp Neurol 225: 387-395.

Klemm N, Nässel DR, Osborne NN (1985) Dopamine-β-hydroxylase-like immunoreactive neurons in two insect species, Calliphora erythrocephala and Periplaneta americana. Histochem 85: 159-164.

Klemm N, Hustert R, Cantera R, Nässel DR (1986) Neurons reactive to antobodies against serotonin in the stomatogastric nervous system and in the alimentary canal of locust and crickets (Orthoptera, Insecta). Neurosci 17: 247-261.

Kostowski W, Tarchalska B (1972) the effects of some drugs affecting brain 5-HT on the aggressive behaviour and spontaneous electrical activity of the central nervous system of the ant, Formica rufa. Brain Res 38: 143-149.

Leung MK, Stefano GB (1984) Isolation and identification of enkephalins in pedal ganglia of Mytilus edulis (Mollusca). Proc Natl Acad Sci USA 81: 955-958.

Livingstone MS, Tempel BL (1983) Genetic dissection of monoamine neurotransmitter synthesis in Drosophila. Nature (Lond) 303: 67-70.

Maddrell SHP, Phillips JE (1975) Secretion of hypo-osmotic fluid by the lower Malphighian tubules of Rhodnius prolixus. J Exp Biol 62: 671-673.

Maxwell GD, Tait JF, Hildebrand JG (1978) Regional synthesis of neurotransmitter candidates in the CNS of the moth, Manduca sexta. Comp Biochem Physiol 61C: 109-119.

Mercer AR, Erber J (1983) The effects of amines on evoked potentials recorded in the mushroom bodies of the bee brain. J Comp Physiol 5: 469-476.

Mercer AR, Menzel R (1982) The effects of biogenic amines on conditioned and unconditioned responses to olfactory stimuli in the honeybee Apis mellifera. J Comp Physiol 145: 363-368.

Mercer AR, Mobbs PG, Davenport AP, Evans PD (1983) Biogenic amines in the brain of the honey bee, Apis mellifera. Cell Tissue Res 234: 655-677.

Meyer EP, Matute C, Streit P, Nässel DR (1986) Insect optic lobe neurons identifiable with monoclonal antibodies to GABA. Histochem 84: 207-216.

Miller TA (1979) Nervous versus neurohormonal control of insect heartbeat. Amer Zool 19: 77-86.

Mobbs PG (1982) The brain of the honey bee Apis mellifera. I. The connections and spatial organization of the mushroom bodies. Phil Trans R Soc Lond B 298: 309-354.

Myers CM, Evans PD (1985a) An FMRFamide antiserum differentiates between populations of antigens in the ventral nervous system of the locust, Schistocerca gregaria. Cell Tissue Res 242: 109-114.

Myers CM, Evans PD (1985b) The distribution of bovine pancreatic polypeptide FMRFamide-like-immunoreactivity in the ventral nervous system of the locust. J Comp Neurol 234: 1-16.

Nässel DR (1986) Serotonin and serotonin immunoreactive neurons in the nervous system of insects. Prog Neurobiol (in press).

Nässel DR (1987) Aspects of the functional and chemical anatomy of the insect brain. (This volume)

Nässel DR, Cantera R (1985) Mapping of serotonon-immunorective neurons in the larval nervous system of the flies Calliphora erythrocephala and Sarcophaga bullata. A comparison with ventral ganglia in adult animals. Cell Tissue Res 239: 423-434.

Nässel DR, Elekes K (1984) Ultrastructural demonstration of serotonin immunoreactivity in the nervous system of an insect, Calliphora erythrocephala. Neurosci Lett 48: 203-210.

Nässel DR, Elekes K (1985) Serotonergic terminals in the neural sheath of the blowfly nervous system: ultrastructural immunocytochemistry and 5,7-dihydryxytryptamine labelling. Neurosci 15: 293-307.

Nässel DR, Klemm N (1983) Serotonin-like immunoreactivity in the optic lobes of three insect species. Cell Tissue Res 232: 129-140.

Nässel DR, Laxmyr L (1983) Quantitative determination of biogenic amines and dopa in the CNS of adult and larval blowflies Calliphora erythrocephala. Comp Biochem Physiol 75: 259-265.

Nässel DR, O'Shea M (1987) Proctolin-like immunoreactive neurons in the blowfly central nervous system. J Comp Neurol (in press).

Nässel DR, Meyer EP, Klemm N (1985) Mapping and ultrastructure of serotonon-immunoreactive neurons in the optic lobes of three insect species. J Comp Neurol 232: 190-204.

Nässel DR, Ohlsson L, Sivasubramanian P (1987) Differentiation of serotonin-immunoreactive neurons in fly optic lobes developing in situ or cultured in vivo without eye discs. J Comp Neurol 255: 327-340.

Osborne NN, Neuhoff V (1974) Formation of serotonin in insect (Periplaneta americana) nervous tissue. Brain Res 74: 366-369.

O'Shea M (1982) Peptide neurobiology. An identified neuron approach with special reference to proctolin. Trends Neurosci 5: 69-73.

O'Shea M, Adams ME (1981) Pentapeptide (proctolin) associated with an identified neuron. Science 213: 567-569.

O'Shea M, Bishop CA (1982) Neuropeptide proctolin associated with an identified skeletal motoneuron. J Neurosci 2: 1242-1251.

O'Shea M, Evans PD (1979) Potentiation of neuromuscular transmission by an octopaminergic neurone in the locust. J Exp Biol 79: 169-190.

O'Shea M, Shaffer M (1985) Neuropeptide function: the invertebrate contribution. Ann Rev Neurosci 8: 171-198.

O'Shea M, Witten J, Schaffer M (1984) Isolation and characterization of two myoactive neuropeptides: further evidence for an invertebrate peptide family. J Neurosci 4: 521-529.

Pages M, Jimenez F, Ferrus A, Peralta E, Ramirez G, Gelpi E (1983) Enkephalin-like immunoreactivity in Drosophila melanogaster. Neuropeptides 46: 87-98.

Piek T, Mantel P (1977) Myogenic contractions in locust muscle induced by proctolin and by wasp Philanthus triangulum venom. J Insect Physiol 23: 321-325.

Pitman RM, Kerkut GA (1970) Comparison of the actions of iontophoretically applied acetylcholine and gamma aminobutyric acid with the EPSP and IPSP in cockroach central neurones. Comp Gen Pharmacol 1: 221-230.

Ramade F, L'Hermite P (1971) Mise en evidence de neurones adrenergiques par la microscopie de fuorescence dans la systeme nerveux central de Calliphora erythrocephala Meig, et Musca domestica L. CR Acad Sci D 272: 3314-3317.

Ray JW, (1965) The free aminoacid pool of cockroach (Periplaneta americana) central nervous system. In: Treherne JE, Beament JWL (eds) The physiology of the insect nervous system. Academic Press, New York.

Remy C, Dubois MP (1981) Immunohistochemical evidence of methionine enkephalin-like material in the brain of the migratory locust. Cell Tissue Res 218: 271-278.

Remy C, Girardie J, Dubois MP (1977) Exploration immunocytologique des ganglions cerebroides et sous-oesophagien du phasme Clitumnus extradentatus: Existence d'une neurosecretion apparente a la vasopressine-neurophysine. CR Acad Sci Paris Ser D 285: 1495-1497.

Remy C, Girardie J, Dubois MP (1979) Vertebrate neuropeptide-like substances in the suboesophageal ganglion of two insects: Locusta migratiria R and F (Orthoptera) and Bombyx mori (Lepidoptera). Immunocytochemical investigation. Gen Comp Endocrinol 37: 93-100.

Robertson HA (1975) The innervation of the salivary gland of the moth, Manduca sexta evidence that dopamine is the transmitter. J Exp Biol 63: 413-419.

Robertson HA (1976) Octopamine, dopamine and noradrenaline content of the brain of the locust, Schistocerca gregaria. Experientia 32: 552-553.

Robertson HA, Steele JE (1973) Octopamine in the insect central nervous system: distribution, biosynthesis and possible physiological role. J Physiol (Lond) 237: 34-35.

Romeuf M, Remy C (1984) Early immunohistochemical detection of somatostatin-like and methionine-enkephalin-like neuropeptides in the brain of the migratory locust embryo. Cell Tissue Res 236: 289-292.

Salvaterra PM, Crawford GD, Klotz JL, Ikeda K (1985) Production and use of monoclonal antibodies to biochemically defined insect neuronal antigens. In: Breer H, Miller TA (eds) Neurochemical techniques in Insect research Springer, Berlin Heidelberg, New York.

Schäfer S, Bicker G (1986) Distribution of GABA-like immunoreactivity in the brain of the honey bee. J Comp Neurol 246: 287-300.

Schooneveld H, Romberg-Privee HM, Veenstra JA (1986) Immunocytochemical differentiation between adipokinetic hormone (AKH)-like peptides in neurons and glandular cells in the corpus cardiacum of Locusta migratoria and Periplaneta americana with C-terminal and N-terminal specific antisera to AKH. Cell Tissue Res 243: 9-14.

Schürmann F-W (1985) Aspekte neuronaler Verknupfung im Zentralen Hirn der Insekten. In: Rensch B (ed) Evolution: Zelle als Organismus. Aschendorff, Munster.

Schürmann F-W, Klemm N (1984) Serotonin-immunoreactive neurons in the brain of the honey bee. J Comp Neurol 225: 570-580.

Shaw SR (1984) Early visual processing in insects. J Exp Biol 112: 225-251.

Slowley BD, Owen MD (1982) The effects of reserpine on amine concentrations in the nervous system of the cockroach (Periplaneta americana). Insect Biochem 12: 469-476.

Starrat AN, Brown BE (1975) Structure of the pentapeptide proctolin, a proposed neurotransmitter in insect. Life Sci 17: 1253-1256.

Stone JV, Mordue W, Batley KE, Morris HR (1976) Structure of locust adipokinetic hormone that regulates lipid utilisation during flight. Nature (Lond) 263: 207-211.

Strausfeld NJ (1976) Atlas of an insect brain. Springer, Heidelberg.

Strausfeld NJ (1984) Functional anatomy of the blowfly's visual system. In: Ali MA (ed) Photoreception and vision in invertebrates. Plenum, New York.

Strausfeld NJ, Nässel DR (1980) Neuroarchitecture of brain regions that subserve compound eyes in Crustacea and insects. In: Autrum H (ed) Handbook of sensory physiology. VII/6B. Springer, Berlin, Heidelberg, New York.

Stuart AE, Hudspeth AJ, Hall ZW (1974) Vital staining of specific monoamine-containing neurons in the leech nervous system. Cell Tissue Res 153: 55-61.

Taghert PH, Goodman CS (1984) Cell determination and differentiation of identified serotonin-immunoreactive neurons in the grasshopper embryo. J Neurosci 4: 989-1000.

Trimmer BA (1985) Serotonin and the control of salivation in the blowfly Calliphora. J Exp Biol 114: 307-328.

Truman JW, Taghert PH (1983) Neuropeptides in insects. In: Krieger DT, Brownstein MJ, Martin JB (eds) Brain peptides Wiley, New York.

Tyrer NM, Turner JD, Altman JS (1984) Identifiable neurons in the locust central nervous system that react with antibodies to serotonin. J Comp Neurol 227: 313-330.

Vanderhaeghen JJ, Signeau JC, Gepts W (1975) New peptide in the vertebrate CNS reacting with antigastrin antibodies. Nature (Lond) 257: 604-605.

Veenstra JA (1984) Immunocytochemical studies on peptidergic neurons in the Colorado potato beetle and some other insects. Thesis, Wageningen, Netherlands.

Veenstra JA, Schooneveld H (1984) Immunocytochemical localization of neurons in the nervous system of the Colorado potato beetle with antisera against FMRFamide and bovine pancreatic polypeptide. Cell Tissue Res 235: 303-308.

Veenstra JA, Romberg-Privee HM, Schooneveld H (1984) Immunocytochemical localization of peptidergic cells in the neuroendocrine system of the Colorado potato beetle, Leptinotarsa decemlineata, with antisera against vasopressing, vasotocin and oxytocin. Histochem 81: 29-34.

Veenstra JA, Romberg-Privee HM, Schooneveld H (1985) A proctolin-like peptide and its immunocytochemical localization in the Colorado potato beetle, Leptinotarsa decemlineata. Cell Tissue Res 240: 535-540.

Verhaert P, Van de Loof (1985) Immunocytochemical localization of a methionine-enkaphalin-resembling neuropeptide in the central nervous system of the American cockroach, Periplaneta americana L. J Comp Neurol 239: 54-61.

Verhaert P, Geysen J, Van de Loof A, Vandesande F (1984a) Immunoreactive material resembling vertebrate neuropeptides and neurophysins in the brain, suboesophageal ganglion, corpus cardiacum and corpus allatum of the dictyopteran Periplaneta americana L. Cell Tissue Res 238: 55-59.

Verhaert P, Marivoet S, Vandesande F, Van de Loof A (1984b) Localization of CRF immunoreactivity in the central nervous system of three vertebrate and one insect species. Cell Tissue Res 238: 49-53.

Vieillemaringe J, Duris P, Geffard M, Le Moal M, Delaage M, Bensch C, Girardie J (1984) Immunohistochemical localization of dopamine in the brain of the insect Locusta in comparison with the catecholamine distribution determined by histofluorescence technique. Cell Tissue Res 237: 391-394.

Walker RJ, James VA, Roberts CJ, Kerkut GA (1980) Neurotransmitter receptors in invertebrates. In: Sattelle DB, Hall LM, Hildebrand JG (eds) Receptors for neurotransmitter, hormones and pheromones in insects. Elsevier/North-Holland Biomedical Press, Amsterdam.

Watson ADH (1986) The distribution of GABA-like immunoreactivity in the thoracic nervous system of the locust Schistacerca gregaria. Cell Tissue Res 246: 331-341.

Welsh JH, Moorhead M (1960) The quantitative distribution of 5-hydroxytryptamine in the invertebrates especially in their nervous system. J Neurochem 6: 146-169.

White K, Valles AM (1984) Immunohistochemical and genetic studies of serotonin and neuropeptides in Drosophila. In: Edelman GM, Cowan WM (eds) Molecular bases of neurodevelopment. John Wiley, New York.

Williams JLD (1975) Anatomical studies of the insect central nervous system: A groundplan of the midbrain and an introduction to the central complex in the locust, Schistocerca gregaria (Orthoptera). J Zool (Lond) 176: 67-86.

Witten JL, O'Shea M (1985) Peptidergic innervation of insect skeletal muscle: Immunochemical observations. J Comp Neurol 242: 93-101.

Yui R, Fujita T, Ito S (1980) Insulin-, gastrin-, pancreatic polypeptide-like immunoreactive neurons in the brain of the silkworm, Bombyx mori. Biomed Res 1: 42-46.

ORGANIZATION OF CONDUCTING SYSTEMS IN "SIMPLE" INVERTEBRATES: PORIFERA, CNIDARIA AND CTENOPHORA

RICHARD A. SATTERLIE

Department of Zoology

Arizona State University

Tempe, Arizona 85287, USA

and

ANDREW N. SPENCER

Department of Zoology

University of Alberta

Edmonton, Alberta, Canada

T6G 2E9

ABSTRACT

The following review summarizes recent discoveries in three groups of multicellular animals in which the phrase "central nervous system" does not fit the usual conception of a centralized ganglion, or group of ganglia, located in the anterior portion of a bilaterally symmetrical animal. The radial symmetry of cnidarians and ctenophores presents unique problems for the acquisition and integration of sensory information as well as the distribution of motor output. Here we provide insights into how behavior is controlled in the "most primitive" of radially symmetrical animals, cnidarians and ctenophores. In addition, recent advances in the neurobiology of the Porifera are reviewed. This review is not presented as an encyclopedic account of all past work on these groups, but rather as a sampling of past and current studies that best illustrate general properties of these groups and highlight the most recent developments and directions of ongoing research.

1. PHYLUM PORIFERA

Although there is still considerable debate regarding the ancestry of the Metazoa (Barnes 1985), there is a consensus that the Porifera are unlikely to be the stock from which other metozoans arose (Haeckel 1874; Jagersten 1955; Marcus 1958; Hadzi 1963; Rees 1966), and that they constitute a divergence from the mainstream of eumetazoan evolution. It is not surprising therefore that within this phylum there is evidence for an unusual conducting system, although it has only been described in one class, the Hexactinellida.

From a recent ultrastructural study of the organization of the tissues of the hexactinellid Rhabdocalyptus dawsoni (Mackie and Singla 1983) it is apparent that this class of sponges is quite distinct from the other three classes (Calcarea, Demospongiae, Sclerospongiae). The differences are so striking that Reiswig and Mackie (1983) proposed a new taxonomy which recognizes two subphyla, the Symplasma and the Cellularia, while Bergquist (1985) makes a case for removing the Hexactinellida from the Porifera and thus recognizes the Symplasma as a phylum. Several of the histological features of Rhabdocalyptus that are presumed to be characteristic of the Symplasma, such as the syncytial nature of the trabecular tissue (due to the presence of cytoplasmic bridges), and the presence of specialized junctional "plugs", are probably essential structural elements for progagation of electrical events within the conducting system.

Evidence for this conducting system comes from some simple but elegant experiments by Lawn et al. (1981) and Mackie et al. (1983). In marked contrast to the claim of Bidder (1923) these studies clearly show that hexactinellids, like other sponges, are capable of actively pumping water in through the body and out through the osculum. By using a sensitive, miniature flowmeter to monitor the velocity of exhaled water at the osculum, Mackie et al (1983) were able to monitor the activity of the choanoderm flagella. Because they were unable to find any evidence that the sponge is able to rapidly control flow through the prosopyles or dermal pores they assumed that changes in the velocity of the oscular stream were entirely due to altered activity of the flagella. They were, however, unable to directly demonstrate this.

When any of the sponge's tissue is mechanically disturbed or electrically stimulated above a threshold level, a rapid reduction in the velocity of the exhalent current follows with a latency of a few seconds (Fig. 1a). Such arrests also occur spontaneously. The arrest response is all-or-none, but with paired stimuli at intervals greater than the refractory period (approximately 30 s), traces from the current velocity probe show a second response which is summed (Fig. 1b). These recordings are due to increasing delays to the resumption of pumping and not to any changes in the strength of arrests. Thus, pumping arrest is believed to be due to a sudden cessation of flagellar beating throughout the choanoderm; the time course of the flow response is a reflection of the inertia of the moving water mass and the rate at which flagella become fully active after the arrest. Flagellar arrest as a means of current control has not been described for the Calcarea or Demospongiae where contractile processes appear to be involved (Mackie 1979). Recordings of the arrest response of Rhabdocalyptus at increasing distances from the stimulating electrode in a slab of body wall (Fig. 1c) show that the response is conducted at a constant velocity (0.26 $cm.s^{-1}$) at 11°C). Experiments using pieces of body wall incised to form various circuitous pathways show that the response is conducted diffusely throughout all parts of the sponge (Mackie et al. 1983).

Only one tissue, the trabecular syncytium, is distributed uniformily throughout the animal and is therefore the primary candidate for the conducting substrate. Trabecular processes make contact with the flagellated collar bodies by cytoplasmic bridges, that may contain "plugs". A cytoplasmic pathway is thus created between the dermal membrane and all the effectors. Although there has been considerable speculation in the past as to whether sponges possess nerve cells (Pavans

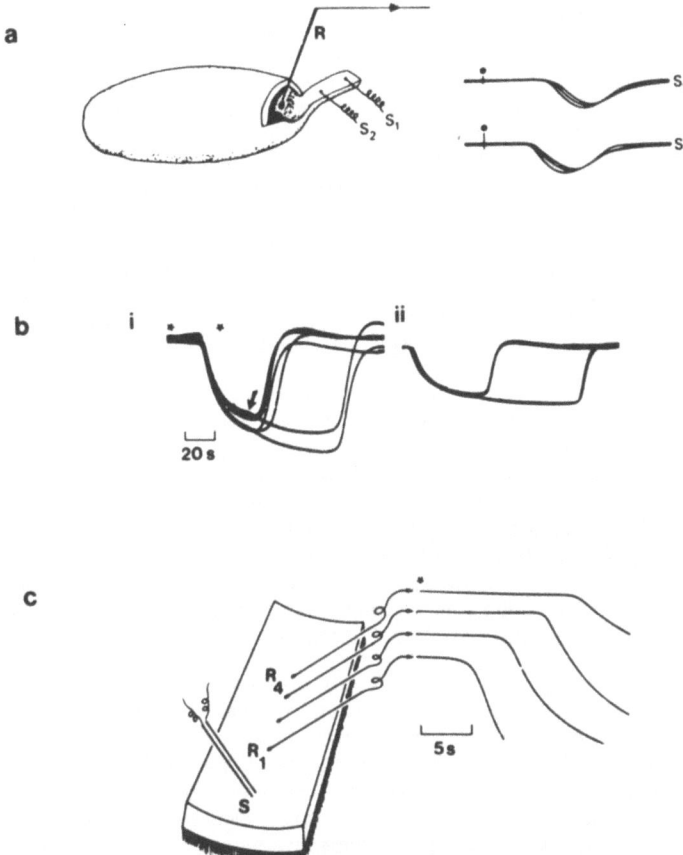

FIG. 1: Flagellar arrest in the hexactinellid sponge
<u>Rhabdocalyptus dawsoni</u>. a) Demonstration of a conducted event
which initiates flagellar arrest. A thermistor flow meter (R)
monitors the velocity of exhalant water exiting the osculum.
Arrests were evoked by electrical stimuli applied at S1 and S2
on a flap of excised body wall that is reflexed back. The delay
before a reduction in water velocity was recorded depended on
the conduction distance of the stimulating site from the body of
the sponge. No time scale given. (b) Refractoriness of the
effector mechanism. A single shock (stars) of pair of shocks up
to 28 s apart produce one family of arrest-curves. When the
inter-shock interval is longer than about 30 s a second arrest
is evoked which may be seen as an inflexion in the recording
(arrow). In another preparation (ii), where recovery had not
begun, no inflexion is apparent. (c) Arrests recorded at four
point R1 to R4, at increasing distances along a slab of body
wall following electrical stimulation at S. The stimulus
artifact is marked with a star. (Adapted from Mackie et al.
1983 and Lawn 1982).

de Ceccatty 1955, 1962, 1974; Jones 1962; Brien 1973), present opinion
favors the existence of "nerve-like" cells only in the Calcarea and
Demospongiae. Similar cells have not been found in hexactinellids (Mackie
and Singla 1983).

With regard to the conducting mechanism, Mackie et al. (1983) were forced to conclude that a conventional action potential is propagated through the trabecular syncytium by a mechanism that is analogous to that demonstrated for epithelial conducting systems in other animals (see review by Anderson 1980). They could find no evidence for conduction by chemical or mechanical means. The range of conduction velocities they measured, even assuming longer conduction distances than were actually measured, is 25% of the lowest value for cnidarian epithelia but approaches the lowest values measured in the plant Mimosa (Sibaoka 1966). The possibility of a novel conducting mechanism should therefore not be excluded. Mackie et al. (1983) believe that flagellar arrest is due to an influx of Ca++ across the choanosome membrane when it depolarized with the passage of an action potential. Recovery then proceeds as the intracellular Ca++ concentration is reduced to its former level by active pumping.

As yet there is no direct evidence to suppose that a similar conducting system exists in the other subphylum, the Cellularia. This is despite several behavioral indications that there is coordination of effectors. For example, propagated contractile waves of either the general body wall (Ephydatia, Tethya) or the oscular membranes (Euspongia, Hippospongia, Verongia) have been reported (McNair 1923; Wintermann 1951; Pavans de Ceccatty and Coraboeuf 1960; Pavans de Ceccatty 1969; Reiswig 1971), but there has been no clear-cut demonstration of coordinated control of flagellar beating in the Calcarea and Demospondiae.

Before much progess can be made in understanding the properties and potentialities of conducting systems in the hexactinellids it is essential that electrical events associated with conduction in the trabecular syncytium of Rhabdocalyptus be recorded. Most probably this will require the use of patch-recording techniques as the physical nature of this tissue precludes the use of micro-pipette impalement techniques (S.A. Arkett and G.O. Mackie personal communication). It would also be profitable to reexamine the Demospongiae for unambiguous evidence of propagated events.

2. PHYLUM CNIDARIA

Comparative neurobiologists have long been keenly interested in cnidarians due to the apparent organizational simplicity of body form, particularly that of the nervous system. Since cnidarians lack well developed organs, and thus operate at the tissue level of organization, one can argue that the morphology and physiology of present-day forms provide important insights into the form and function of ancestral stocks that quite possibly occupied a position on the main trunk of the metazoan phylogenetic tree. In terms of nervous system organization, modern cnidarians are clearly the "simplest" metazoans that possess identifiable (by contemporary morphological and/or physiological standards) nervous elements. It is impossible, then, to resist loaning our rapidly increasing body of knowledge about the cnidarian nervous system to speculation about the evolution of multicellular conducting systems in the Metazoa.

The classical view of the cnidarian nervous system has emphasized diffuse conduction via two dimensional, unpolarized nerve nets (see Bullock and Horridge 1965). We can find, in all cnidarians, subepithelial networks of bipolar or multipolar neurons with interneural chemical (polarized or symmetrical) or electrical synapses, but only in some do we find concentration of these elements into definite tracts or ganglion-like accumulations. This simplistic view of the nervous system does not match the rich behavioral repertoire of many cnidarian species. Important technological and conceptual advances over the last two and one-half decades, most notably the development of various recording techniques (Horridge 1954; Josephson 1961), the description of multiple conducting systems (i.e. McFarlane 1969a; Horridge 1956a, b) and the discovery of the role(s) of conducting epithelia in some cnidarian groups (Mackie 1970; Anderson 1980; have helped narrow the gap between our knowledge of conducting systems and descriptions of behavior.

In examining control of behavior in cnidarians one must consider the limitations imposed upon skeletal, muscular and neural systems by radial symmetry, as well as advantages gained by use of specialized or multifunctional cell types. In brief, skeletal elements can be grouped into three categories; hydrostatic skeleton, mesoskeleton and exoskeleton (Champman 1974). The hydrostatic skeleton is variable and can be regulated through muscular activity. The mesoskeleton includes the more pliable connective tissue elements (i.e. mesoglea) and may have hard spicular elements incorporated. The exoskeleton includes hardened materials such as the axial skeleton of some octocorals and the skeleton of hexacorals. Muscular systems are usually comprised of sheets of epitheliomuscular cells. As such, these cells can serve a variety of functions in addition to contractility, most notably some myoepithelia can conduct electrical events without (or with limited) intervention by neurons (i.e. Spencer and Satterlie 1981). One group of specialized cells that is intimately associated with certain behaviors, performing prey-capturing or adhesive roles, includes cnidoblasts. These will be mentioned only briefly.

Significant difference in behavioral patterns and the form and function of nervous systems exist between the cnidarian classes. Each class will be covered separately.

2.1. Class Anthozoa

Anthozoans, including sea anemones and corals, are strictly polypoid cnidarians lacking the medusoid generations found in the other classes. Colonial forms are well represented, sometimes with zooid polymorphism. Anthozoans show a rich behavioral repertoire including simple reflex reactions (i.e. single tentacle responses to direct mechanical or chemical stimulation), spontaneous activities (i.e. column shape changes) and very complex behavioral acts (i.e. swimming and shell climbing). Shelton (1982) classified anthozoan behavior at the level of effector responses as consisting of 1) local movements, 2) asymmetrical movements involving several muscle groups, and 3) symmetrical movements. If we add the possibility of fast and slow muscle contractions (Ross 1957) and different requirements for neuro-muscular facilitation (Pantin 1935b), it is quite easy to envision a wide range of possible behavioral outputs despite relatively simple neuronal assemblages (i.e. two dimensional nerve nets) and relatively small sets of distinct muscle groups. Add the possiblity of multiple conducting systems and one can begin to appreciate how these "simple" organisms can undergo such complex behavioral acts as extensive shape changes, burrowing, shell climbing, locomotion (creeping, walking and swimming), feeding and satiety (including modulation of nematocyst and spirocyst activity), and aggressive combat with neighboring individuals of the same species.

Anthozoans have been the least cooperative cnidarians for microelectrode analyses of neuronal activities underlying behavior, however a great deal of information exists linking recordable (with extracellular electrodes) conducting systems with behavioral responses. Unfortunately, the exact identity of cellular elements are known for only a few described conducting systems. Perhaps the best-known conducting systems are the Through-Conducting Nerve Net (TCNN) of sea anemones, which controls rapid withdrawal of the oral disc (Pantin 1935a; Robson and Josephson 1969), and the colonial nerve nets of various hard and soft corals, which coordinate the activities of individual polyps within the colony (see Shelton 1982). From work on these systems, it is clear that a "nerve net" must not be considered a "hard-wired" conducting unit since repetitive stimulation can yield significant changes in conduction velocity and even conduction failure (Shelton 1975a, b). Similarly, incremental conduction has been described in some corals in which distance of spread of electrical activity is dependent on the number and frequency of impulses generated (Anderson 1976).

2.1.1. Coordination in colonial anthozoans

In colonial anthozoans one can distinguish between behavior of individual polyps and that of the colony. Colonial behavior can involve effectors located in the coenasarc (colonial tissue), or can involve recruitment of polyp effectors by colonial conducting systems. An example of the latter is polyp retraction which can occur at the level of individual polyps or may involve many polyps in a coordinated manner. In a comparative study of coral species Horridge (1957) found a great deal of variability in colonial polyp withdrawal responses; ranging from colony-wide responses to a localized stimulus, to withdrawal of only a few polyps in the immediate vicinity of the stimulus. In some species, responses increase in increments with repetitive stimulation. Of these, some show responses in which the area of polyp retraction become progressivly larger with each stimulus, in some cases eventually involving the entire colony. In others, incremental spread involves increments of decreasing areas, so that a distance is reached where further stimulation does not increase the overall area of response. This last response pattern has attracted the most interest in those investigating nerve net function. One immediately wonders if the behavior mirrors underlying nerve net activity or if the colonial conducting system is "through-conducting" while the observed behavior is not.

An extreme example in colonial coordination comes from a deep water hexacoral, Lophelia pertusa (Shelton 1980). Individual polyps contain a nerve net that coordinates withdrawal of that polyp, but surprisingly no behavioral or electrophysiological evidence was found to indicate coordination between polyps. A second conducting system (slow system; see below) was found in each polyp, but again no inter-polyp connections were noted. Of those anthozoans that do possess colonial conducting systems, only one had been shown to conduct "nerve net" impulses over limited areas of the colony (Goniopora lobata; Anderson 1976; Fig. 2). Polyp withdrawal occurs in decreasing area increments from the point of repetitive stimulation, reflecting the spread of impulses in the colonial conducting system. Despite the increase in area of response with each successive stimulus, a limit was reached beyond which no further polyp withdrawal occurred. Early workers reasoned that if such behavior did mirror underlying electrical conduction, that a process of interneural facilitation (Pantin 1935a, b) could explain the increasing distance of conduction with successive stimuli (Horridge 1957; Josephson et al. 1961). It is assumed that interneural junctions (synapses) include a mixture of transmissive and non-transmissive connections. The latter require "priming" (facilitation) so that these junctions will not transmit the first or second impulse, but will be transmissive to following stimuli. Interneural facilitation, although not clearly demontrated in anthozoans, best explains the observed electrical conduction patterns in Goniopora.

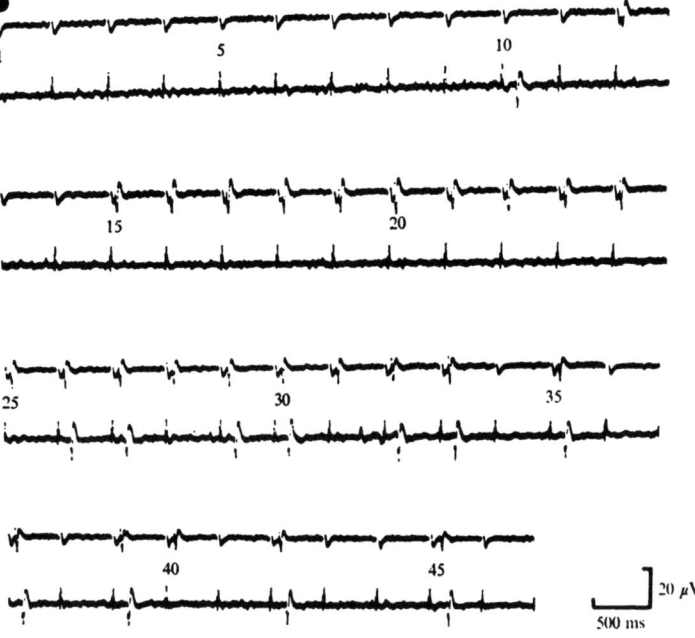

FIG. 2: Suction electrode recordings of colonial nerve net activity in the scleractinian coral Goniopora lobata. A pair of recording electrodes were placed 0.66 cm (upper traces) and 1.34 cm (bottom traces) away from the stimulating electrode. Forty-six stimuli were delivered (numbered; dot indicates stimulus artifact). Impulses were first recorded regularly at the closer electrode (triangle) and later spread to the more distant electrode. (From Anderson 1976).

In most colonial anthozoans, colonial conducting systems are through-conducting, even though polyp withdrawal patterns in some are quite limited (see Shelton 1982). The pennatulids represent one extreme where coordination of polyp withdrawal is colony wide, and can be initiated with only a few stimuli (Anderson and Case 1975; Shelton 1975b; Satterlie et al. 1976, 1980). A characteristic of pennatulids and some other anthozoans (see Shelton 1975b), is that the conduction velocity of successive colonial impulses increases with the first few stimuli in a series. Anthozoan neuro-effector systems exhibit a frequency-dependent facilitation where more than one incoming impulse is required to initiate a response, and the intensity of the response is dependent upon the frequency of successive impulses (Pantin 1935a, b, c, d; see also Shelton 1975b). In colonial anthozoans, an increase in conduction velocity of colonial nerve net impulses would contribute to a through-conducted behavioral response since polyp effectors, even at a distance from the stimulation site, would tend to receive a high frequency burst of incoming inpulses. Interneural facilitation, producing more direct conduction routes, could explain these conduction velocity increases in the colonial nerve nets.

The facilitation requirements of polyp effectors (for retraction) may be critical for those species which show limited polyp retraction despite having through-conducting colonial conducting systems (Shelton 1975 a; Satterlie and Case 1979). In these animals, conduction velocities in

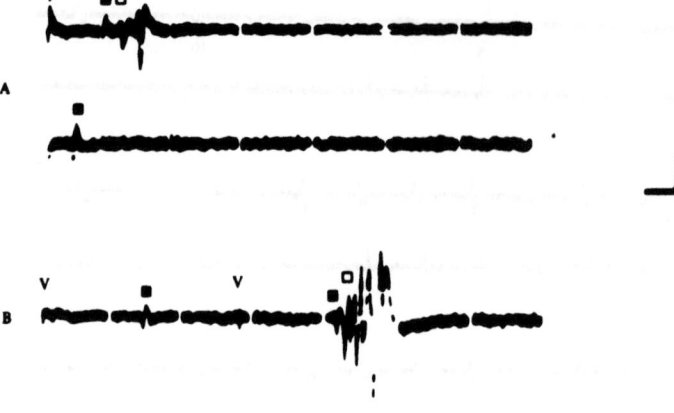

FIG. 3: Suction electrode recordings from polyps of the gorgonian coral Muricea californica. Each stimulus (v = stimulus artifact) is followed by an impulse of the colonial nerve net (solid square) and, in some cases by a burst of activity in a polyp conducting system (open squares). Each polyp burst is correlated with polyp withdrawal. Note that activity in the colonial system always precedes polyp activity. (A) dual recording with the lower electrode closer to the stimulating electrode. (B) Single electrode recording. Ordinate 5 µV, abscissa 200ms (A) and 300ms (B). (From Satterlie and Case 1979).

colonial conducting systems decrease with repetitive stimulation. Distant polyp effectors thus see a decreased frequency of incoming impulses. If the frequency exceeds the facilitation requirements of polyp effectors, polyp withdrawal will not be evoked despite through-conduction of all colonial nerve net impulses (Shelton 1975a; see also Satterlie and Case 1979). A possible explanation for increases in conduction velocity would be the build-up of fatique in interneural junctions so that successive impulses would travel progressively more circuitous routes.

In addition to conduction velocity changes, polyp responses can be limited if the junctions between the colonial conducting system and individual polyps perform a gating function so that a number of colonial impulses (at an appropriate frequency) are needed to activate polyp conducting systems. Evidence for this type of gating is found in gorgonians where several colonial impulses are required to activate a rapid burst of impulses in the polyp conducting systems, which in turn initiates polyp withdrawal (Satterlie and Case 1979; Fig. 3). It is interesting to note that the gorgonians show both the colony-polyp gating and a decreasing conduction velocity in the colonial nerve net with repetitive stimulation so that polyp withdrawal shows extremely limited spread despite through-conduction of impulses in the colonial conducting system.

In addition to the colonial nerve net which mediates protective polyp retraction as mentioned above, many colonial anthozoans possess an additional colonial conducting system (Shelton 1975b; Shelton and McFarlane 1976; Anderson and Case 1975; McFarlane 1978; Dickinson 1978; Satterlie et al. 1976, 1980). Neither the cellular identity of conducting elements nor the exact function(s) of this conducting system are known with certainty. It is believed that the "slow" conducting systems (named for their slow conduction velocity relative to the colonial nerve nets) are involved in control of colony expansion (Shelton and McFarlane 1976; Dickinson 1978; McFarlane 1978) and expansion of polyp oral discs and tentacles (McFarlane 1978).

2.1.2. Multiple conducting systems in solitary anthozoans

The seed for current work on the neuroethology of solitary anthozoans was planted by C.F.A. Pantin in his landmark papers on the morphology and physiology of anemone nerve nets (Pantin 1935a, b, c, d). He fully realized that actinian behavior involved interactions between simple and complex reflexes, widespread and local conducting systems, and spontaneous activities. His understanding of anthozoan behavior (with techniques of the day) was so complete that he wrote, "...we are perhaps nearer a complete analysis of the structural units on which behaviour is based in these than in any other animals" (Pantin 1952). While technological difficulties have hampered this quest, our knowledge of the actinian nervous system had been keenly refined and extended from Pantin's pioneering studies.

The discovery and careful description of multiple conducting systems represent major advances in our understanding of the neuroethology of solitary anthozoans. Unfortunately, we still know little about the cellular elements making up these conducting systems, and how these elements interact with one another. Thus far, five types of conducting systems have been described; the Through-Conducting Nerve Net (TCNN), Slow System 1 (SS1), Slow System 2 (SS2), Delayed Initiation System (DIS) and Local Conducting Systems. Each will be briefly considered, in turn focussing primarily on studies of the sea anemone <u>Calliactis parasitica</u> (see McFarlane 1982).

2.1.3. Through conducting nerve net (TCNN)

The TCNN includes nervous elements of the mesenteries originally described by Pantin in 1935 (a) and since investigated electrophysiologically (Josephson 1966; Robson and Josephson 1969; McFarlane 1969a; Pickens 1969). TCNN impulses can be recorded from tentacles, upper column, and oral disc as well as from mesenteries. Vigorous TCNN activity is associated with protective withdrawal of the tentacles and oral disc, although less intense activity is associated with other behaviors. The TCNN is through-conducting, with conduction velocities up to 100 cm s^{-1} (McFarlane 1982). Both fast and slow muscle contractions can be initiated by the TCNN as different muscle groups have different facilitation requirements (as noted by Pantin 1935b). The TCNN is sometimes spontaneously active but pacemaker loci are not fixed (McFarlane 1973a, b, 1974; 1983). The TCNN has primarily excitatory influences on muscles although column and sphincter muscle is sometimes inhibited by the TCNN (Lawn 1976a).

2.1.3.1. System 1 (SS1)

The SS1 is found throughout the ectoderm of anemones. SS1 impulses are through-conducted at 5-12 cm s^{-1} in <u>Calliactis</u> (McFarlane 1969a). The SS1 is labile, failing to conduct impulses at frequencies greater than about 1/3 s. This is significant since most detectable responses are only seen following repetitive firing in the SS1. SS1 effects on anemone musculature are primarily, but not exclusively inhibitory. Typical responses to SS1 activity include oral disc expansion coupled with tentacle inflation (McFarlane 1970; McFarlane and Lawn 1972), column extension (McFarlane 1982) and pedal disc detachment (McFarlane 1969b). The SS1 is thus intimately involved in prefeeding responses (activated by dissolved food chemicals) in several anemones (McFarlane 1970; Lawn 1975), shell climbing in Calliactis (McFarlane 1976), swimming in Stomphia (Lawn 1976b), and activation (excitation) of tentacle musculature in Calliactis via a pathway distinct from the TCNN - tentacle activation system (McFarlane 1984).

2.1.3.2. Slow system 2 (SS2)

The SS2 is endodermal, conducts at $3 - 6$ cm s^{-1} and fails to conduct predictably with high frequency repetitive stimulation (McFarlane 1969a). Two actions of SS2 impulses have been well documented; both are inhibitory. SS2 impulses inhibit body-wall circular and parietal muscles to produce body-wall flaccidity. In addition, SS2 activity can slow ongoing TCNN pacemaker bursts producing a decrease in overall TCNN firing frequency (McFarlane 1974). SS2 activity is most noticeable during two phases of the feeding cycle; mouth opening with pharynx protrusion and periods of digestive movements (McFarlane 1975).

2.1.3.3. Delayed initiation system (DIS)

The DIS has only been demonstrated indirectly. Stimulation to any part of a Calliactis will produce stimulus-linked SS1 impulses (direct stimulation of SS1) as well as delayed firing in the SS1 (triggered by the DIS). The two responses have different thresholds and thus can be activated separately (Jackson and McFarlane 1976). Exact behavioral roles of the DIS have not yet been noted.

2.1.3.4. Local conducting systems

Local conducting systems have long been suggested as a means of controlling many types of asymmetrical behavior in anthozoans. One such local system has recently been described in the solitary cup coral Caryophyllia smithii (Shelton and Holley 1984). A local conducting system if found in each tentacle that extends into the associated radial portion of the oral disc. Excitation spreads from tentacle to mouth at approximately 2 cm.s^{-1} and is associated with localized feeding movements of the individual tentacles (bending of the tentacle toward the mouth).

2.1.3.5. Morphological basis of conducting systems

Neurons can be morphologically identified in nearly all anthozoan tissues, and yet in many areas their profusion may be responsible for the difficulty in identifying the morphological substrates of known conducting systems. In only one system, the TCNN in the mesenteries of anemones, can we say that a reasonable correspondence between morphological and physiological evidence points to identifiability of the large bipolar nerve net as part of the TCNN (i.e. see Robson 1963, 1965). Similar evidence is accumulating suggesting that the nultipolar nerve nets of anemone tentacles may represent local conducting systems through which the TCNN and SS1 activate tentacle musculature (McFarlane 1984). In several colonial anthozoans (octocorals), colonial nerve net properties (conduction velocity and impulse amplitude) correlate with neurite diameters within colonial tissues (Satterlie et al. 1976). Despite these few encouraging examples, we still know little about the morphology of slow conducting systems. Shelton (1975c) has proposed that the SS1 and SS2 of anemones may be epithelial conducting systems although available electrophysiological evidence tends to refute this claim. Both systems are blocked by excess Mg^{++}, and both exhibit relatively small amplitude (µV range) potentials (McFarlane 1969a). In addition, gap junctions, which are found in hydrozoan epithelia known to conduct electrical impulses (see section on Hydrozoa), have not been found in the appropriate areas of anthozoan tissues that would suggest that epithelial conduction exists in this group (Mackie et al. 1984). This remains the most challenging area of anthozoan neurobiology.

Several fundamental observations and experiments in anthozoan behavioral physiology have uncovered important properties of cnidarian nervous systems. These include the pioneering work on neuro-effector facilitation; observations suggesting that a nerve net is not a simple hard-wired conducting unit but rather an integrating unit; descriptions of multiple conducting systems in individual animals; and the description of interactions between conducting systems in producing relatively complex behaviors. These properties are not unique to anthozoans, but rather can be considered fundamental properties of the cnidarian nervous system.

2.2. Classes Scyphozoa and Cubozoa

In the Scyphozoa and Cubozoa the medusoid generation is by far the most conspicuous form. Most medusae (except the stauromedusae) are free swimming bell- or disc-shaped animals that swim by periodically contracting circular subumbrellar musculature to eject water through the bell opening. Since swimming is the most noticeable behavior, and that which has been most intensely studied, this review of scyphozoan and cubozoan nervous systems will focus on neuronal control of swimming. The Scyphozoa and Cubozoa are grouped together because systems controlling swimming are essentially identical in organization. Those species most intensively studied are _Cyanea capillata_ (Scyphozoa) and _Carybdea rastonii_ (Cubozoa). In making generalizations about the two classes, both will hereafter collectively be referred to as scyphomedusae.

The predominant features of scyphomedusan nervous systems are nerve nets formed by diffuse, two-dimensional networks of bi- or multi-polar neurons (see Anderson and Schwab 1981). As in anthozoans, the nerve nets are subepidermal. Withing each network, neurons typically interact via symmetrical axo-axonal synapses (see Westfall 1987; Anderson and Schwab 1981). Scyphomedusae have many other features in common with anthozoans including multiple conducting systems (Horridge 1956a, b), frequency-dependent facilitation at neuromuscular junctions (Bullock 1943; Pantin and Vianna Dias 1952), and spontaneous activity in known conducting systems. Through recent investigations utilizing intracellular microelectrodes our knowledge of the electrical properties of individual cnidarian neurons and, more importantly, chemical synaptic junctions between these neurons has been greatly enhanced.

The first thorough descriptions of scyphomedusan behavioral physiology were completed over a century ago by Romanes (1876, 1978) and Eimer (1978). Extracellular electrical recordings were more recently used to confirm the early findings (Horridge (1954; Passano 1965, 1973). From these studies, it is clear that at least two conducting systems are present in the subumbrella of scyphomedusae. The Giant Fiber Nerve Net (GFNN), recently renamed the Motor Nerve Net (Anderson and Schwab 1981, 1982), activates the subumbrellar swimming musculature (Horridge 1954,

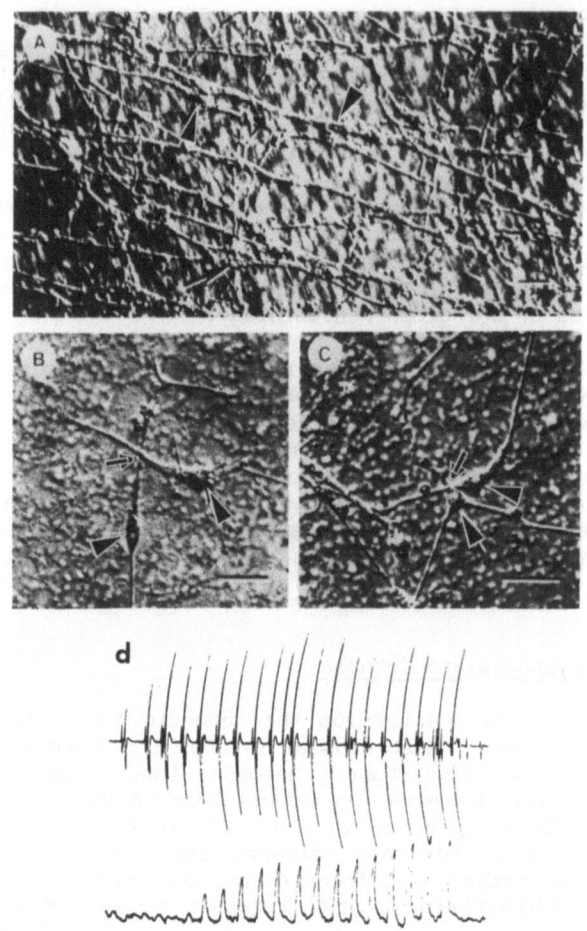

FIG. 4: (A-C) Modulation contrast photographs of living subumbrellar (perirhopalial) tissue of Cyanea capillata showing individual MNN neurons (bipolar and multipolar cells). Large arrows indicate cell bodies, and small arrows show points of suspected synaptic interaction. (D) Suction electrode (top trace) and force transducer (bottom trace) recordings from the subumbrella of the cubomedusan jellyfish Carybdea rastonii. Spontaneous electrical activity includes two separate potentials. The first invariant potential represents activity in the Motor Nerve Net and is followed by a variable potential representing electrical activity in the muscle sheet. Note that in this swimming burst the muscle potentials increase in amplitude with the first few swims, and that similar facilitation is evident in the mechanical record. Swim frequency is 1 Hz and the first MNN impulse is 15 μV. The mechanical record is uncalibrated. (A-C from Anderson 1985).

1956b; Fig. 4). Neurons of the Motor Nerve Net have large cell bodies, 10 - 15 μm in Cyanea (Anderson and Schwab 1981), which can be penetrated with microelectrodes. Axons are joined by axo-axonal synapses which are chemical as judged by their ultrastructural appearance (Horridge and Mackay 1962; Anderson and Schwab 1981; Satterlie 1979; Westfall 1987) and their magnesium sensitivity (Anderson 1985; Satterlie 1979). Network

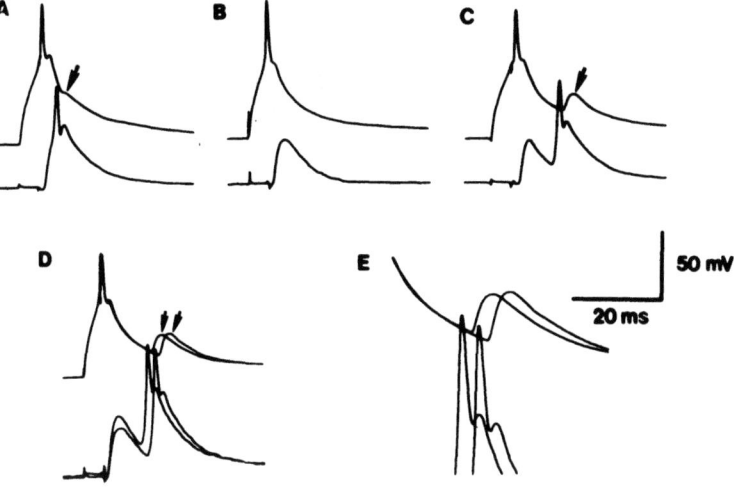

FIG. 5: Intracellular recordings from pairs of MNN neurons of the scyphozoan jellyfish Cyanea capillata. Records indicate that inter-neuronal synapses are bidirectional excitatory synapses. In all records, top trace is the "pre-synaptic" cell and lower trace represents the "post-synaptic" cell. (A) When an action potential is triggered in the post-synaptic cell, a notch (arrow) is seen in the falling phase of the pre-synaptic action potential. This notch is absent when a spike is not triggered in the post-synaptic cell (B). (C-E) Action potentials in the post-synaptic cell are delayed to show that they produce a "return" synaptic potential in the pre-synaptic cell. E is an enlargement of D. (From Anderson 1985).

conduction velocities are typically around 50 cm. s^{-1} within the Motor Nerve Net.

Despite early reports of unconventional electrical activity in neurons of the Cyanea Motor Nerve Net (Patton and Passano 1972), both Cyanea and Carybdea neurons have resting potentials in the -60 mV range and brief, overshooting action potentials (Satterlie and Spencer 1979; Satterlie 1979; Anderson and Schwab 1983; Anderson 1985; Fig. 5). Anderson has shown that the neuronal action potential in Cyanea is primarily sodium dependent with a small inward calcium component, and both voltage and calcium-dependent outward potassium currents (Anderson 1985), and that proton-activated chloride currents exist (Anderson and McKay 1985). In other words, the electrical activity of scyphomedusan neurons appears to be quite conventional. The spikes can have complex waveforms associated with the repolarizing phase which have recently been shown to be due to superimposed synaptic potentials resulting from the reflected action potentials at these bi-directional synapses (Anderson 1985). Motor Nerve Net neurons are apparently linked solely by chemical synapses as neither electrical nor dye-coupling have been demonstrated in Cyanea (Anderson and Schwab 1981, 1983; Anderson 1985). The primary output of the Motor Nerve Net is the ectodermal striated muscle (circular) that lines the subumbrellar cavity (Anderson and Schwab 1981; Satterlie 1979). Contractions of the swimming muscle, and thus pulsations of the bell, occur 1:1 with Motor Nerve Net spikes. Although the striated subumbrellar musculature exhibits well defined neuromuscular facilitation (Bullock 1943; Pantin and Vianna Dias 1952), little is known of the cellular

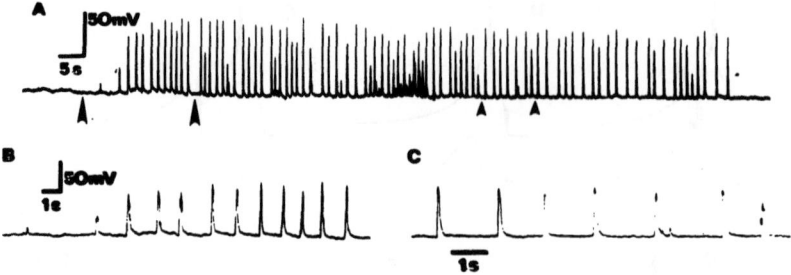

FIG. 6: Intracellular recordings from suspected striated subumbrellar muscle cells in the cubomedusa Carybdea rastonii. The area between the large arrows is enlarged in (B) and that between the small arrows in (C). Note that the first few potentials show an apparent amplitude facilitation, and that the spike-like responses have complex, variable waveforms. (From Satterlie 1979).

properties of the facilitation mechanism due to the difficulty of obtaining and maintaining microelectrode penetrations in the relatively small muscle cells during contractile activity. A few suspected muscle cells have been impaled in Carybdea revealing graded, facilitating junctional potentials phase-locked to swimming contractions (Satterlie 1979; Fig. 6).

Motor Nerve Net impulses originate in ganglion-like neuronal assemblages located in rhopalia (marginal centers) which typically contain a variety of sensory structures including statocysts, ocelli and/or lensed eyes and sensory epithelia of unknown function. Each scyphomedusan species has a characteristic number of rhopalia (four or a multiple of four) spaced around the bell margin. Removal of all rhopalia leaves the medusa unable to swim, while swimming contractions will still be generated as long as at least one rhopalium is left attached to the umbrella (Romanes 1876; Passano 1965, 1973). Pacemaker impulses originating in a rhopalium enter and are through-conducted by the Motor Nerve Net. Since each rhopalium is capable of generating impulses independently, some means of pacemaker coordination is needed. As outgoing impulses enter the Motor Nerve Net from one rhopalium, and are conducted to the other rhopalia, pacemakers of the inactive rhopalia are reset (Horridge 1959; Passano 1965, 1973). Rhopalial pacemakers thus interact via a "dominance hierarchy" in which the pacemaker with the fastest rhythm will drive swimming. When the frequency of impulse generation drops in one pacemaker, another will assume "dominance" and drive swimming (see Satterlie 1979). The apparent pacemaker redundancy provides not only an overall increase in swimming frequency when compared to preparations in which rhopalia have been removed, but also serves to decrease frequency variations (promotes swimming regularity) during long swimming bouts (Horridge 1959; Lerner et al. 1971). Even though active pacemaker sites may shift from rhopalium to rhopalium, the swimming muscles respond to each Motor Nerve Net impulse regardless of site of origin, and show normal facilitation even during these changes (Satterlie 1979).

Inputs to the rhopalial pacemakers include incoming Motor Nerve Net impulses, local inputs from rhopalial sensory structures (statocysts, ocelli, etc.), and impulses from a second subumbrellar nerve net, the Diffuse Nerve Net (see Passano 1982).

The Diffuse Nerve Net runs in parallel with the Motor Nerve Net and is subepithelial throughout the subumbrella, manubrium and oral arms. Neurons are small with neurite diameters of only 1-2 μm. Impulses are conducted through the Diffuse Nerve Net (DNN) at around 15 cm. s^{-1} (Horridge 1956b). Diffuse Nerve Net activity produces three primary effects. 1) DNN activity is associated with contraction of marginal tentacles in many scyphomedusae so that a wave of tentacle contraction can be followed as impulses are conducted around the bell (Romanes 1876; Horridge 1956b; Passano 1965, 1973). 2) Arrival of a DNN impulse at a rhopalium usually results in excitation of the swim pacemaker, thus DNN activity can have a direct modulatory effect on swimming (Romanes 1878; Passono 1965, 1973). 3) The DNN also provides modulatory inputs to the swimming musculature. Although DNN activity does not directly produce swimming contractions, it can facilitate the musculature so that subsequent Motor Nerve Net inputs will produce a large muscle contraction (Passano 1965, 1973). DNN activity can be activated by direct stimulation of any part of the subumbrella and manubrum.

Turning behavior is accomplished through asymmetrical contractions of the subumbrellar musculature. Although local conduction systems have been proposed to account for turning and righting responses, no candidate systems have been found (see Passano 1982).

A central tenet of unpolarized conduction in nerve nets is the requirement for two-way conduction junctions between network neurons. As yet, electrical junctions (gap junctions) have not been found in any scyphomedusa (Mackie et al. 1984). Morphological evidence for bidirectional chemical synapses was established some time ago (Horridge and Mackay 1962; see also Anderson and Schwab 1981; Westfall 1987, but only recently has the symmetrical nature of these junctions been demonstrated electrophysiologically (Anderson 1985). Bidirectional synapses in Cyanea function much like other excitory chemical synapses except that transmission can occur in either direction so that each cell of a communicating pair can act as either the pre-synaptic or post-synaptic cell. Transmission only occurs across bidirectional synapses when one cell of the pair produces an action potential (Fig. 5). When this occurs, the post-synaptic cell will produce a large synaptic potential (up to 70 mV) which is normally sufficient to initiate a post-synaptic action potential. The synaptic delay is around 1 ms regardless of the direction of conductions, supporting the contention that the synapses are truly non-polarized (Anderson 1985). The synapses were found to fatigue quite readily. This is possibly related to nerve net fatigue and conduction velocity changes as discussed for colonial anthozoa and may be attributed to the presence of very few synaptic vesicles (Anderson 1985).

The preceding brief treatment suggests that the scyphomedusan nervous system is constructed of the same basic building blocks as the anthozoan nervous system. Scyphomedusae have provided useful preparations for cellular-level investigations of nerve net function, and have provided important information on the conventionality of neuronal electrical activities and on the physiology of bidirectional synapses. We can only hope that these preparations can now provide answers to such questions as the nature of neuromuscular facilitation and pacemaker function.

2.3. Class Hydrozoa

Our knowledge of the organization and physiology of the cnidarian nervous system has expanded considerably in the past ten years, with many of the important findings coming from experimental studies that have used hydrozoans, particularly the hydromedusae and siphonophores. The classical model for the cnidarian nervous system, Hydra, has been less useful for obtaining physiological data since it is difficult to obtain intracellular recordings from the relatively small neurons which form a quite diffuse nervous system (Kass-Simon 1982). This appears to be true for most hydropolyps. Nevertheless, a body of information has accumulated that describes in some detail the various neuronal types found in Hydra as determined by maceration, electron microscopy, and immunohistochemistry (see for example Burnett and Diehl 1964; Davis et al. 1968; David 1973; Tardent adn Weber 1976; Epp and Tardent 1978; Kinnamon and Westfall 1981; Grimmelikhuijzen et al. 1980, 1981a, b, c, 1982a, b; Dunne et al. 1985; Grimmelikhuijzen 1985; Yu et al. 1985). In addition, there is a burgeoning interest in the role of various morphogens, some of which may be of neuronal origin, in the development and differentiation of Hydra and other hydropolyps such as Hydractinia (see for example, Burnett 1961; Wolpert 1969; Schaller and Gierer 1973; Meinhardt and Gierer 1974; Wolpert et al. 1974; Schaller 1976; Schaller et al. 1979; Heimfeld and Bode 1985; Berking 1986). No further reference will be made to Hydra and other hydropolyps since very little is known about their cellular neurophysiology. In addition, only passing reference will be made to the numerous epithelial conducting systems that have been described in this class. Readers are referred to reviews by Mackie (1970), Spencer (1974), Anderson (1980) and Spencer and Schwab (1982).

2.3.1. Basic organization of the nervous system

The dogma that the basic architecture of the hydrozoan nervous system consists of a series of discrete overlapping nerve nets (Bullock and Horridge 1965) has stood the test of time moderately well. Unfortunately the "nerve net concept" is based primarily on data obtained from anthozoans and scyphozoans and has been carried over, with little modification, to the hydrozoan condition. Some major differences in neuronal organization among the classes have been overlooked because of the apparent similarity in the general histological features of the nervous systems in the four classes; namely that they form nerve nets. Probably the most striking difference is that the member neurons of each nerve net in the Hydrozoa (e.g. Polyorchis and Aequorea; Spencer 1981; Satterlie 1985) are electrically coupled to one another through gap junctions while in the other two classes each neuron of a network is connected to its neighbors by chemical synapses (e.g. Cyanea; Anderson and Schwab 1981). Gap junctions have never been reported between any cell types in the Anthozoa and Scyphozoa (Mackie et al. 1984), yet in the Hydrozoa this junctional type seems to be the rule for intercellular communication in homogenous tissues. Like the other three classes, the Hydrozoa use chemical synapses for transmission between neurons of different nerve nets (Spencer and Arkett 1984). These differences are shown diagrammatically in Fig. 7. Many of the physiological properties of hydrozoan neurons that will be described are obviously adapted to and constrained by this organization.

2.3.2 Nature of hydrozoan action potentials

Most of the evidence presently available shows that regenerative potentials in both neurons and epithelial cells of hydrozoans are biophysically conventional (Spencer 1971, 1978, 1981, 1982; Mackie 1973, 1976b, 1978; Anderson and Mackie 1977; Anderson 1979; Roberts and Mackie 1989; Chain et al. 1981; Spencer and Arkett 1984; Satterlie 1985; Mackie and Meech 1985). They are generally all-or-none events, except for one example of a non-spiking neuron in Polyorchis, the "O" system neurons (Spencer and Arkett 1984; Arkett and Spencer 1986a, b). Action potentials

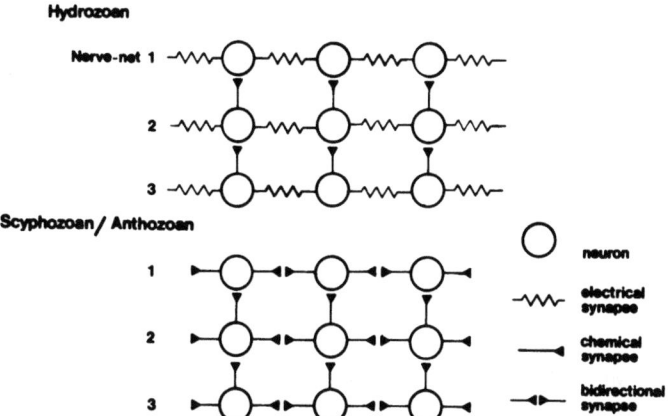

FIG. 7: A simplified, diagrammatic comparison of the organization of nervous systems in the hydrozoans with that in the scyphozoans and anthozoans. In the Hydrozoa, identifiable nerve nets are formed from electrically coupled member neurons, with unidirectional chemical synapses between different nerve nets. In the Scyphozoa and Anthozoa member neurons of each nerve net are connected by bidirectional chemical synapses. There is no evidence of electrical tranmission or gap junctions in these two classes. The synapses between diferent nerve nets are also chemical, but they tend to be unidirectional.

arise from resting potentials of from -40 to -85 mV and may have amplitudes from 30 to 145 mV. In the few cases where it has been examined, changes in resting potential approximate the predicted 58 mV per decade change in external K concentration, as if the membrane is behaving as a Nernstian K electrode (e.g. Chelophyes myoepithelium, Chain et al. 1981). Much of the inward current during the rising phase of the spike is carried by Na^+ in both neurons (Anderson 1979; Mackie and Meech 1985) and epithelia (Mackie 1976a; Spencer and Satterlie 1981; Chain et al. 1981), though there is evidence in some of these cases, that the sodium current is not blocked by tetrodotoxin (Anderson 1979; Spencer and Satterlie 1981). It is interesting that voltage clamp experiments using eggs of the anthozoan Renilla (Hagiwara et al. 1981) also show a TTX-insensitive sodium current, and in addition there are three K^+ currents: delayed rectifier; inward rectifier; and A-current. A common feature of all these spikes is the presence of a calcium current during depolarization. This current serves several important functions. In Polyorchis swimming motor neurons, a calcium current appears to play a role in producing the underlying bursting properties of these neurons by activating a K^+ current which terminates a burst (Anderson 1979). In the motor giant axon of Aglantha (Fig. 8), a low amplitude calcium spike, distinct from the usual sodium spike, activates the "slow" swimming system (Mackie and Meech 1985). In the siphonophore Hippopodius, a slow secondary calcium potential is associated with secretion in a glandular epithelium (Mackie 1976a). The action potentials recorded from swimming muscle epithelial cells have a characteristic plateau phase (Spencer 1978, 1982; Satterlie and Spencer 1983) which is believed to be due to a maintained inward calcium current that produces a maintained contraction (Spencer and Satterlie 1981).

2.3.3. Motor systems in hydromedusae

The two motor systems that have been studied most extensively are

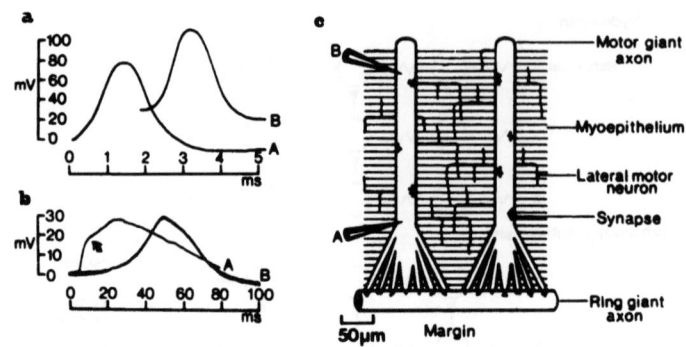

FIG. 8: Two types of action potentials recorded from the same axon in the trachyline hydromedusa Aglantha digitale. a) Intracellular recordings of a spontaneous action potential at two points, A and B, as it propagates up a motor giant axon during an escape of "fast" swimming contraction. This action potential is sodium dependent and conducts at 1 to 4 m s^{-1}. b) Intracellular recording of an action potential, associated with an endogenously generated "slow" swimming contraction, in the same axon. This type of action potential is calcium dependent and of lower amplitude and conduction velocity. It propagates at approx. 0.3 m s^{-1}. c) Diagram of an isolated quadrant of subumbrella and margin to show the morphology of the axons and the recording electrodes A and B. (Adapted from Mackie and Meech 1985).

those that control swimming and synchronous tentacle contracton. Very little is known, however, about the systems that control local and directed movements of the tentacles and manubrium during feeding or that initiate the generalized protective contractions known as "crumpling" (Mackie et al. 1967; King and Spencer 1979).

2.3.4. Motor systems controlling swimming

Located in the inner nerve-rings and radial nerves of antho- and limnomedusae are condensed motor networks of neurons that innervate the sheets of subumbrellar striated muscle, but only at their peripheries (Anderson and Mackie 1977; Satterlie and Spencer 1983). These are the muscles responsible for swimming. In leptomedusae, such as Aequorea, the arrangement is slightly different since the subumbrella is also innvervated throughout by a diffuse network of neurons (Satterlie 1985). This network may not however be associated with the swimming muscle, as Mackie et al. (1985), using immunohistochemical techniques, have shown that in another leptomedusan, Phialidium, a mostly radially oriented nerve net is associated with the ectodermal radial muscle of the subumbrella. Important features of the "swimming" motor networks are the electrical connectivity and large size of neurons. These features are associated with rapid propagation of motor spikes and synchronous excitation of the swimming muscles. The properties of this motor nerve net are best understood in the hydromesusa Polyorchis penicillatus (Spencer 1981).

It is important to appreciate that the properties that will be described are expressed by all member neurons. This homogeneity, or lack of local specialization, is presumably an adaptation to radial symmetry and a planktonic life style (Spencer and Arkett 1984). If Lucifer Yellow

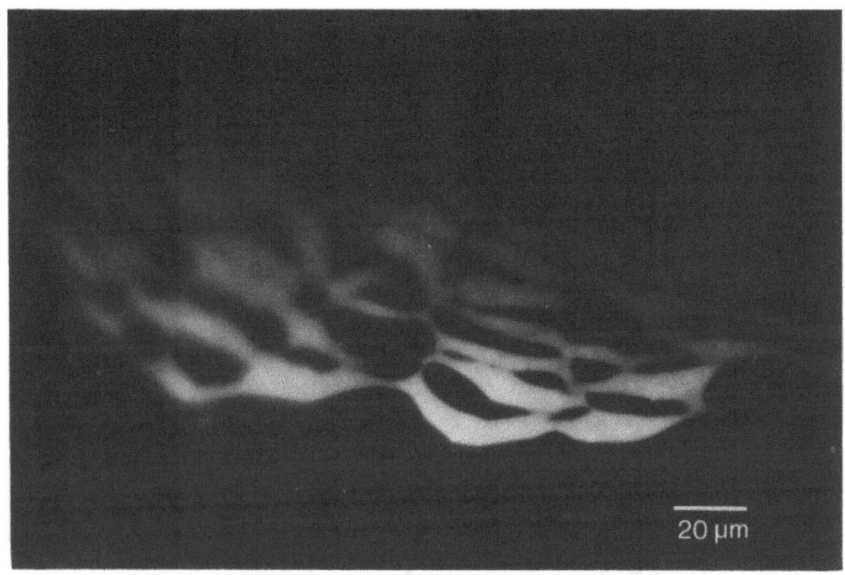

FIG. 9: Swimming motor neuron network of the inner nerve ring of _Polyorchis_ _penicillatus_. The fluorescent dye Lucifer Yellow CH was iontophoresed into one neuron and after 5 min had diffused through a large number of neurons in the network. Scale bar is 20 μm.

is iontophoresed into a swimming motor neuron (SMN) in _Polyorchis_, the dye diffuses rapidly from neuron to neuron via gap junctions to reveal a compressed nerve net formed from multipolar cells up to 25 μm in diameter (Fig. 9). Dye-coupling is associated with strong electrical coupling. Injection of constant current pulses into a neuron and recording of the membrane potential of neurons at increasing distances from the injection site (Fig. 10a) show that the space constant of this network of parallel, longitudinally and transversely connected neurons is relatively long (approx. 7 mm). Such effective coupling produces essentially synchronous membrane potential oscillations and spiking in all neurons of the network (Fig. 11) and, under conditions where all neurons are close to threshold at the same time, can result in very high conduction velocities. The network of swimming motor neurons behaves as a low-pass filter, attenuating strongly any passive potentials having frequencies above about 1 Hz. (Fig. 10a). This property is probably important for separation of local input from generalized input arriving from concentric presynaptic networks such as the "B" system (described later).

A remarkable characteristic of the swimming motor system in _Polyorchis_ is that the duration of spikes varies inversely with distance over which the spike has already travelled (Fig. 12). Anderson (1979) related changes in duration to the level of depolarization of the neurons when action potentials were triggered, while Spencer (1981) has suggested that the plateau seen in spikes recorded close to the initiation site is due to the summed electrotonic potentials arising from direct current flow from neighboring neurons that are generating asynchronous spikes. As the wave front of depolarization propagates around the network it becomes more parallel and the spike loses its plateau because of the synchronizing effect of the local current flowing between neurons. Thus at a distant site the inherent duration (i.e. very short) of the action potential is

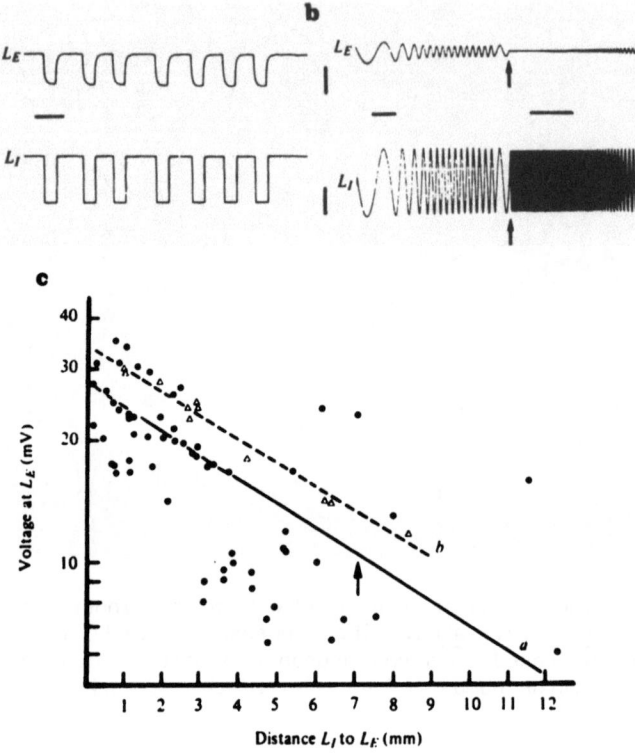

FIG. 10: Electrical coupling in the swimming motor neuron network of the anthomedusa Polyorchis penicillatus. a) Resistive coupling experiment showing hyperpolarizing current pulses injected into the network through electrode LI and recorded at a distance of 2.5 mm along the network by an intracellular electrode (LE). Vertical bar is 20 mV in trace LE and 8 nA in trace LI. b) AC coupling experiment showing a constant amplitude sinusoidal current being injected through electrode LI and recorded at a distance of 3 mm by electrode LE. At frequencies above about 1 Hz the voltage signal is rapidly attenuated. The arrow marks a change in time base. Vertical bars are 20 mV and 4.5 nA. Horizontal bar is 2 s before and after the arrow. c) Resistive coupling shown as a scatter plot of membrane voltage (log scale) against distance (linear scale) separating the current injecting electrode LI and the recording electrode LE using the pooled data from 12 preparations. Square, hyperpolarizing pulses of 12 nA and 2 s were used. (a) is the regression line using the mean of the slopes and the y intercepts. The arrow marks the average space constant value. (b) is the regression line for data (open triangles) from one jellyfish (35 mm bell diameter). (From Spencer 1981).

recorded. Nevertheless, it is also possible that these differences in duration are due to differences in the level of the resting membrane potential which influences the expression of various membrane currents. For example, if neurons ahead of the spike are strongly hyperpolarized it is possible that a stronger potassium current (such as the A current) is produced, thus repolarizing these neurons more rapidly. Variations in

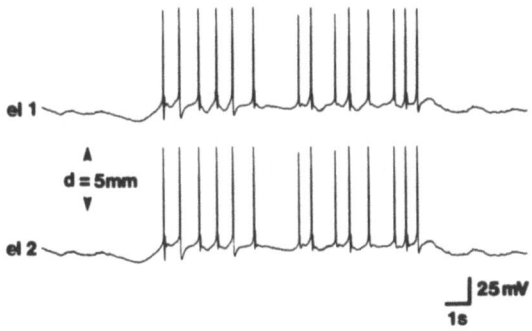

FIG. 11: Simultaneous intracellular recordings from two swimming motor neurons in the SMN nerve net of the inner nerve ring of <u>Polyorchis penicillatus</u>. The two electrodes (el1 and el2 record indentical, spontaneous electrical activity even though separated by a distance of 5 mm.

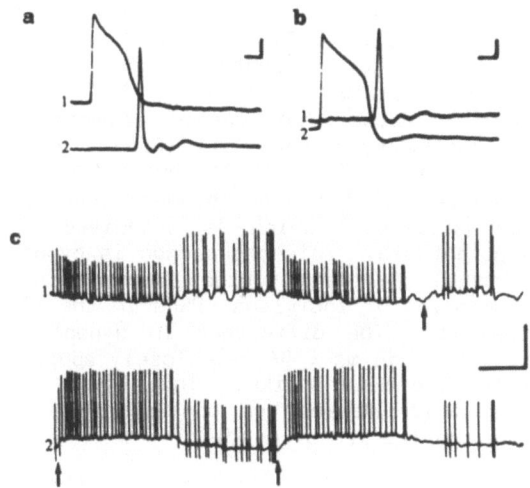

FIG. 12: Relationship between the duration of action potentials and the distance of the recording electrode from the site of A.P. initiation in the SMN network of <u>Polyorchis</u>. a) The nerve ring immediately surrounding electrode 1 is illuminated by a fiber-optic light pipe causing the neurons in this region to depolarize and generate action potentials. When action potentials arrive at electrode 2, 23 mm from electrode 1 (180° separation around the margin), they are of far shorter duration. b) The situation is reversed when the area surrounding electrode 2 is illuminated. The DC levels were not changed between a) and b). c) A similar experiment except that it is shown at a slower time scale. The arrows indicate when the light pipe illuminated the area around that electrode. In this experiment the electrodes were separated by 27 mm. Vertical bar is 20 mV in (a) and (b), 50 mV in (c). Horizontal bar is 5 ms in (a) and (b), 20 s in (c). (From Spencer 1981).

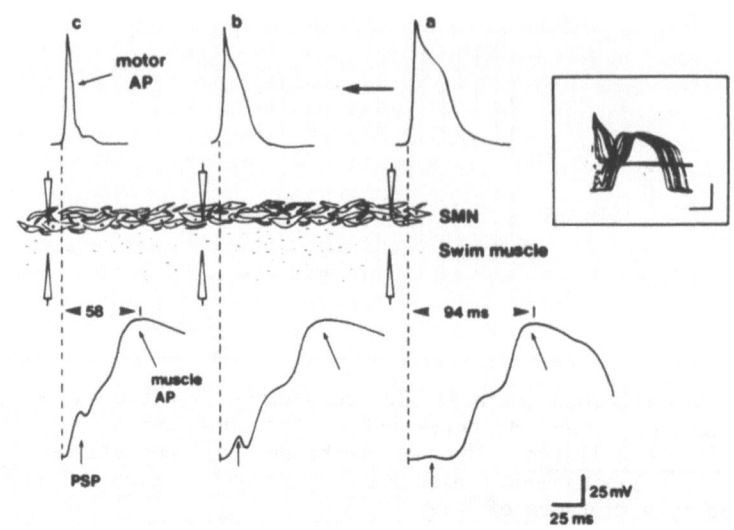

FIG. 13: Reconstruction of the causal relationship between motor action potential duration and delay at the neuromuscular synapse onto myoepithelium in the margin of Polyorchis penicillatus. A long duration action potential is initiated at point a in the swimming motor neuron network (SMN); as it propagates through the network to point c (3.6 mm form a) its duration is greatly reduced. As the motor spike propagates through the SMN network it initiates muscle action potentials in the overlying myoepithelium with a delay that varies inversely with the amplitude of the postsynaptic potential (PSP). It is assumed that the decreased synaptic delay at distant sites is due to an increase in the amplitude of the postsynaptic potential (shown in insert by the arrowhead) which is associated with short duration action potentials. This phenomenon is hypothesized to be due to an increase in the synchrony of motor spikes, and hence temporal summation of individual PSPs in the electrically coupled myoepithelium. The difference in synaptic delay at points a and c (94 ms - 58 ms = 36 ms) closely approximates the conduction time for the motor spike. Thus there is automatic compensatory delay which depends on conduction time and thus ensure synchronous contraction of muscle.

motor spike duration are closely correlated with differences in the delay measured at the neuromuscular junctions (Spencer 1982). Short duration spikes, which have necessarily been conducted a considerable distance, are associated with short delays. Thus reductions in the time taken to produce muscle action potentials exactly compensate for the conduction time of spikes, no matter where they originate (Fig. 13).

Control of swimming in the leptomedusae, exemplified by Aequorea aequorea, is somewhat modified from that seen in anthomedusae (Satterlie 1985). A full contraction of the swimming muscle requires a burst of motor spikes and the motor network can be locally inhibited by contraction of the radial muscle (Fig. 14). These differences may be linked to the presence of a subumbrellar nerve net which probably innervates the outer sheet of radial, smooth muscle.

In addition to a "slow" swimming system (Mackie 1980), which appears

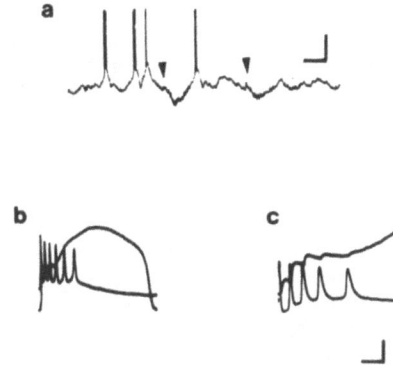

FIG. 14: Control of swimming in the leptomedusa Aequorea aequorea. a) Spontaneous bursts of action potentials recorded intracellularly from "swim" motor neurons. When the radial muscle in the region of the recording electrode is stimulated to contract by shocking an adjacent tentacle (arrowheads), long-duration hyperpolarizations are recorded from the motorneurons. Vertical bar, 30 mV and horizontal bar 3 s. b) Simultaneous intracellular recordings from a "swim" motorneuron (upper trace) and an overlying myoepithelial cell (lower trace). A spontaneous burst of motor APs produced a long duration AP in the swimming muscle. Vertical scale is 30 mV, horizontal scale is 150 ms. c) Recording as in b) at an expanded time scale to show the individual PSPs associated with each motor spike. Vertical scale is 30 mV, horizontal scale is 60 ms. (Adapted from Satterlie 1985).

to be homologous with the swimming systems of other hydromedusae, many trachyline medusae possess a system for "escape" swimming (Mills et al. 1985). In the trachyline Aglantha digitale (Roberts and Mackie 1980), escape swimming is externally stimulated, possibly via the comb pads, with the excitation first being recorded in the ring-giant axon as a burst of spikes (Fig. 15a). In contrast to the other groups of hydromedusae, this axon is located in the outer nerve ring. The ring-giant axon makes chemical synaptic contacts with each of eight syncytial motor-giant axons located outside each radial digestive canal on the subumbrellar surface (Fig. 15b). Single action potentials (Fig. 15c) propagate very rapidly (approx. 1.4 ms^{-1}) up the motor-giant axons and out onto the subumbrellar surface through circumferential lateral motor neurons which are electrically coupled to the motor-giant axons (Weber et al. 1982). Miniature synaptic potentials can be recorded in the myoepithelium close to both the motor-giants and the lateral neurons indicating that neuromuscular synapses are present throughout the sheets of swimming muscle (Kerfoot et al. 1985). This type of innvervation is presumably required in species, such as Aglantha, that lack myoid conduction. Such situations occur despite the myoepithelial cells being electrically coupled and must have evolved to allow for local control of the muscle by either restricting the spread of excitation or by locally inhibiting the muscle.

2.3.5. Outer nerve ring systems

Based on the different morphologies of the inner and outer nerve

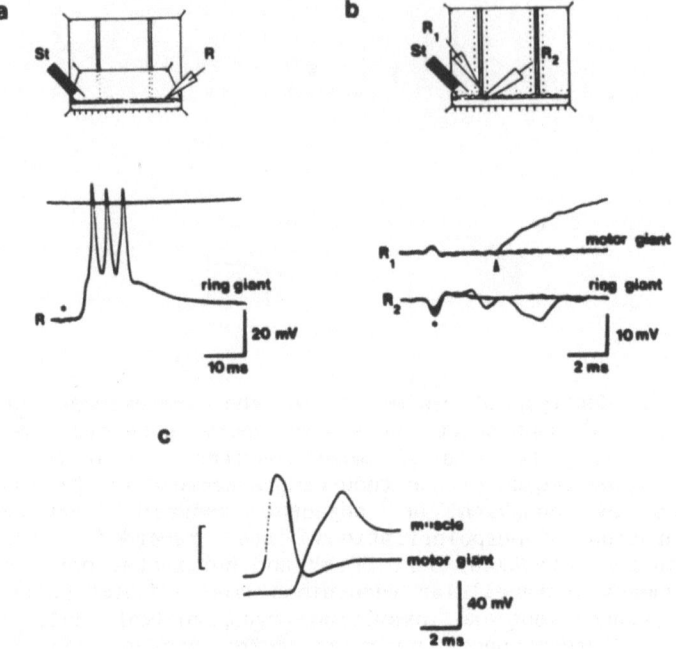

FIG. 15: Control of escape swimming in the trachymedusa Aglantha digitale. a) Intracellular recording of a burst of spikes in the ring giant axon (at position R) in response to electrical stimulation at St. on the outer nerve ring (star marks stimulus artifact). b) Postsynaptic potential (arrowhead) produced in the motor giant axon by stimulation (star marks stimulus artifact) of the presynaptic ring giant axon. The lower trace is a suction electrode recording of the ring giant axon spike. c) Intracellular recordings from the motor giant axon and a myoepithelial cell, 80 μm from the recording site in the axon, when a motor spike is stimulated. (Adapted from Roberts and Mackie 1980, and Mackie and Meech 1985).

rings, it had long been recognized that the two rings might play distinctly different roles (Bullock and Horridge 1965; Mackie 1971). the outer nerve ring receives many of the neuronal processes from sensory receptors, including the ocelli, and it is therefore not surprising that intracellular recordings from neurons in the outer nerve ring reveal that some of these neurons are pre-synaptic to motor neurons in the inner nerve ring. It has also been established that trans-mesogleal neuronal processes are present that could form the pathways for such interactions (Mackie 1971; Singla 1978a; Spencer 1979). For example, in Polyorchis, a network of electrically coupled neurons, the "bursting" or "B" system (Fig. 16a), had been described that is an interneuronal system transferring photic information synchronously to all swimming motor neurons (Spencer and Arkett 1984). Shadowing of the marginal regions causes a burst of spikes that is propagated throughout the "B" system which, in turn, leads to firing of the SMNs through summation of epsps (Fig. 16c). The "B" system is presumalby just one of the systems that have multiple excitatory inputs onto the SMNs. Collectively these presynaptic systems determine the firing frequency of the SMNs since their inputs are spatially and temporally summed throughout the nerve ring and superimposed on the endogenous "pacemaker" potentials (see above). The

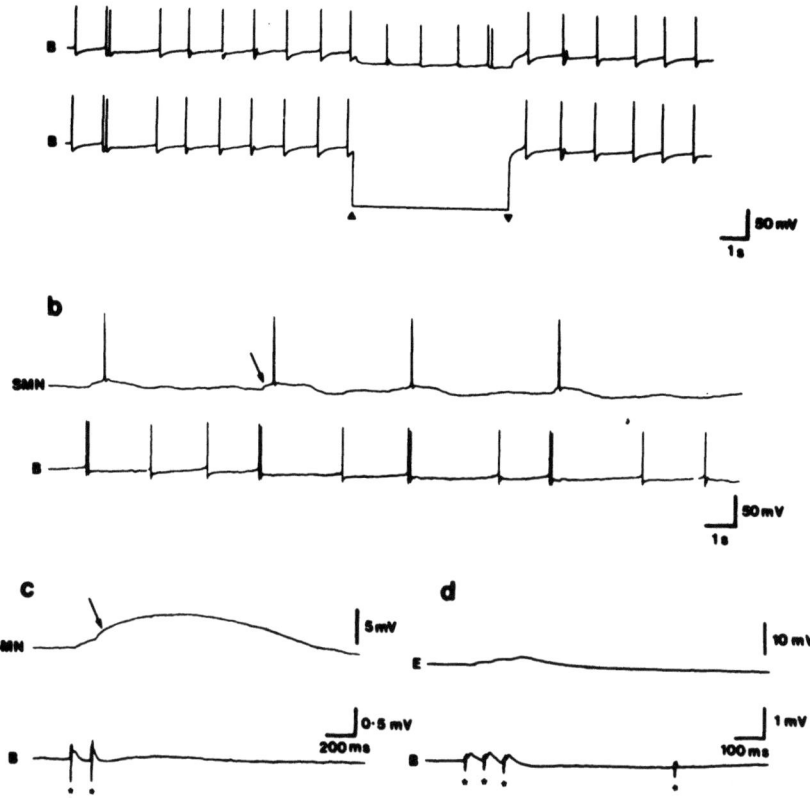

FIG. 16: The "B" or bursting system of the outer nerve ring of Polyorchis penicillatus. a) Simultaneous intracellular recordings from two "B" system neurons 380 μm apart showing synchronous, spontaneous spiking. Between the arrowheads a hyperpolarizing current pulse of 3 nA was injected into one neuron (lower trace) resulting in hyperpolarization of the distant neuron (upper trace). b) Simultaneous intracellular recordings of spontaneous activity in a swimming motor neuron (upper trace) and a "B" system neuron (lower trace) to show that "B" spikes produce epsps in SMNs that can sum and facilitate to produce APs, as shown in (c). The arrows in both (b) and (c) show the second facilitated epsp. (d) "B" system spikes produce summing epsps in an overlying epithelial cell (upper trace). The single "B" spike did not produce an epsp. In (c) and (d) the "B" spikes are recorded extracellularly with a suction electrode applied to the nerve ring. (Adapted from Spencer and Arkett 1984).

"B" system-induced epsps in SMNs also show facilitation and are almost certainly chemically mediated, since excess Mg^{++} will block these synapses. "B" system neurons are also motor neurons in their own right since they make excitatory neuromuscular synapses onto the ectodermal epithelium of the tentacles which contains the longitudinal muscle fibrils (Fig. 16d). The spontaneous firing frequency of the "B" system determines the degree of extension of the tentacles. Thus the functions of the "B" system are to produce a maintained, sychronized, tonic contraction of the tentacles and to ensure that swimming is preceded by shortening of

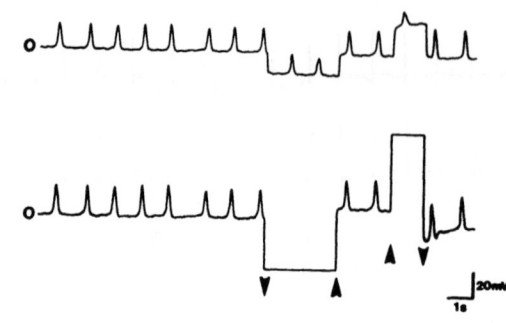

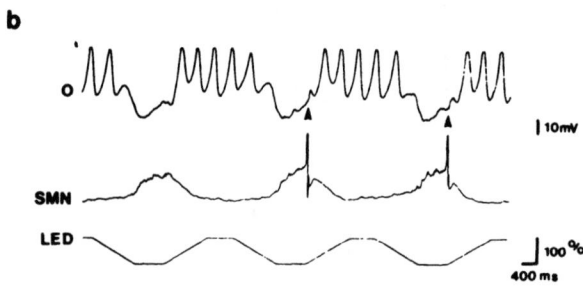

FIG. 17: The "O" or oscillator system of the outer nerve ring of Polyorchis penicillatus. a) Simultaneous intracellular recordings from two "O" system neurons 314 μm apart showing synchronous, spontaneous depolarizations. Between the arrowheads a hyperpolarizing then depolarizing current pulse of 3 nA was injected through one neuron (lower trace) and the resulting membrane potential changes recorded in the distant neuron (upper trace). b) Intracellular recording from an "O" system neuron (upper trace) to show the response of the "O" system to changes in light intensity. The bottom trace shows the current flowing through a green, light-emitting diode. At maximum current the light intensity was 2 μW cm^2 s^{-1} and at minimum current no light was emitted. As the light intensity is reduced the neuron hyperpolarized and the depolarizations cease. "O" system hyperpolarization is associated with SMN system depolarization (middle trace) which can lead to firing. There is corollary discharge of the motor neurons to produce epsps in the "O" system (arrowheads). (Adapted from Spencer and Arkett 1984, and Arkett and Spencer 1986a).

the tentacles so as to reduce drag. Extracellular recordings from the outer nerve ring in Sarsia, Spirocodon, Proboscidactyla adn Stomotoca (Passano et al. 1967; Ohtsu and Yoshida 1973; Spencer 1975; Mackie 1975) all demonstrate that synchronous tentacle contractions are associated with short duration spikes (at that time called marginal pulses), strongly suggesting that homologues of the "B" system exist in most hydromedusae.

A second identifiable network of electrically coupled neurons, the "oscillator" or "O" system, has been described in the outer nerve ring, ocelli and ocellar nerves of Polyorchis (Spencer and Arkett 1984). By inference we presume that such systems are present in other anthomedusae.

Neurons of this system possess some surprising properties which are, once again, distributed throughout all neurons of the network. Intracellular recordings from these neurons are characterized by regular (approx. 1 Hz), spontaneous membrane potential oscillations with amplitudes of about 20 mV (Fig. 17a). Spikes have never been recorded. Activity in the "O" system is strongly influenced by changes in light intensity (Fig. 17b); Spencer and Arkett 1984; Arkett and Spencer 1986a, b). In response to a shadow, "O" neurons give a rapid 10 to 20 mV hyperpolarization following a delay of some 150 ms. This hyperpolarization, which interrupts the oscillations, is maintained for as much as several minutes. Graded decreases in light intensity are accompanied by graded hyperpolarizations that are directly proportional to the rate of percentage decrease in light intensity. Increases in light intensity cause depolarizations with similar delays and are normally associated with exaggerated oscillations at initially higher frequencies. The graded nature of these responses together with the ability to record from neurons in the ocellus that have similar properties, make it likely that the "O" neurons are first-order photoreceptors in a system that is presynaptic to both the SMN and "B" systems. This conclusion is supported by the observation that the major features of the light responses are retained by only the "O" system when all synaptic input is blocked by excess Mg^{++}. Thus there are two sources for the depolarization of the swimming motor neurons following a shadow. The first is the graded depolarization associated with hyperpolarization of "O" neurons, and second is the summing "B" system-induced epsps.

Extracellular recordings, which can be treated as ERGs, from ocelli of Polyorchis (Weber 1982) and a related medusa, Spirocodon saltatrix (Ohtsu 1983), reveal similar graded potential changes in response to flashes of light of varying intensity. In the case of Spirocodon, oscillations have been recorded at "light on" and spiking at "light off". This spiking activity may be attributed to a homologue of the "B" system.

2.3.6. Immunohistochemically identified systems

Over the past six years, the development and use of antibodies directed against a variety of neuropeptides have provided a considerable amount of new information about the organization of hydrozoan nervous systems (Grimmelikhuijzen et al. 1980, 1981a, b, c, 1982a, b, 1986; Grimmelikhuijzen and Spencer 1984; Grimmelikhuijzen 1985; Mackie et al. 1985). This has not been comfortably achieved as frequently there is a lack of corroborative evidence for recognizing synonimity of systems. The only conclusive evidence for linking a particular immunoreactive system with a previously described system is to use double-labelling techniques such as Lucifer Yellow marking followed by immunohistochemical staining with a rhodamine label.

Anyone using these techniques must also be aware of the problem that several immunoreactive peptides may coexist in a neuron. For example, Grimmelikhuijzen (1983a) found that a sub-set of neurons in Hydra show both oxytocin-like and bombesin-like immunoreactivity. It is difficult to explain why it has only been possible to demonstrate immunoreactivity to the peptides, cholecystokinin, substance-P, neurotensin, bombesin, and oxytocin in neurons of Hydra while a family of peptides with the carboxy-terminus, Arg-Phe amide, are widely distributed in the nervous systems of every hydrozoan species examined so far (Grimmelikhuijzen 1983b; Grimmelikhuijzen et al. 1987).

A fairly distinct picture is emerging as to what components of the hydrozoan nervous system are immunoreactive to antibodies directed against FMRFamide or RFamide. A characteristic of many of the immunoreactive neuronal somata, for example surrounding the mouth in hydropolyps (Grimmelikhuijzen 1985), is that they resemble the classical cnidarian neurosensory cell; usually lying close to the surface of the epithelium and bearing a ciliary projection. A second feature is that the processes of these neurons often form extremely fine, beaded nerve nets that are aligned with smooth myoepithelial cells as in siphonophore gastrozooids (Grimmelikhuijzen et al. 1986). Based on this data it would be attractive to conclude that RFamide immunoreactive neurons form a discrete sensory-motor system locally controlling smooth muscle. However, some neurons in more centralized parts of the nervous system are immunoreactive (Grimmelikhuijzen and Spencer 1984; Grimmelikhuijzen 1985; Grimmelikhuijzen et al. 1986) and therefore information may be exchanged between immunoreactive and non-reactive systems. Preliminary experiments by one of us (ANS), using the double-labelling method described above, show that none of the systems so far identified physiologically and morphologically in Polyorchis, namely the SMN, "B" and "O" systems, express any RFamide immunoreactivity.

2.3.7. The siphonophora

Most of our knowledge about the functioning of the siphonophoran nervous system comes from studies of the physonectids and calycophorans. The nervous system is organized similarly in these two sub-orders despite rather different gross morphologies (Mackie and Carré 1983). In the physonects, such as Nanomia, Agalma, Halistemma, and Forskalia, a small pneumatophore acts as floatation for the remainder of the colony which is supported from a long stem. In calycophorans (e.g. Chelophyes, Hippopodius, Abylopsis, and Sulculeolaria) there is no float; the stem is ensheathed for some of its length, and there are fewer nectophores.

In 1973, Mackie was able to obtain the first intracellular neuronal recordings from any cnidarian using the giant "nerve fibres" (axons) in the stem of Nanomia (Fig. 18a). He recorded conventional, TTX-resistant action potentials with high conduction velocities (approx. 2.7 m. s^{-1}); these spikes coordinate both longitudinal muscle contraction in the stem and synchronized swimming in the nectophores. The "giant axons" are products of fused longitudinal elements of an ectodermal nerve net. Lucifer Yellow iontophoresed into a giant axon will diffuse down the length of the axon and transversely into the adjacent nerve net (Mackie 1984; Grimmelikhuijzen et al. 1986). Thus a giant axon and its associated net forms an identifiable, dye-coupled and therefore presumably electrically coupled plexus of neurons reminiscent of the organization seen in the hydromedusae. In both Namomia and Forskalia two giant axons can be seen, one above the other. Extracellular recordings from the stem show that upon stimulation, two neuronal events (n1 and n2, Fig. 18b) are conducted down the stem at different velocities (Mackie 1973, 1978). There is no evidence for direct interaction of the n1 and n2 spikes. However, at their common effector site, the ectodermal longitudinal muscle, the amplitude of the temporally summed and facilitated junctional potentials (nEJPs) depends on the interval between arrival of spikes. Thus double innervation may be a way of producing graded muscle contractions at various distances using pathways that do not conduct in a graded fashion (Mackie 1984). There are even more complexities within the stem since the endodermal epithelium constitutes a third conducting system, the slow or "S" system (Fig. 18b), which also innervates the longitudinal muscle from its endodermal surface (Spencer 1971; Mackie 1976b). The "S"-induced EJPs show different conduction velocities depending on whether or not they are conducted along the stem concurrently with neuronally induced twitch potentials (Mackie 1976b). When sEJPs are carried in a "piggyback" fashion by nEJPs the conduction velocity

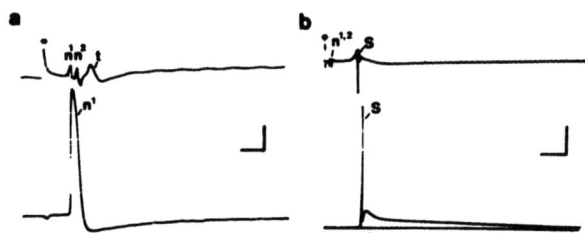

FIG. 18: Conducting systems in the stem of a physonect siphonophore, Forskalia edwardsii. a) A shock to the stem (asterisk) elicits firing in the two presumed nerve nets (n1 and n2) which causes a twitch potential (t) in the stem longitudinal muscle. The amplitude of this contraction is reduced by bathing the stem in a 1:5 solution of isotonic $MgCl_2$ and seawater. The upper trace is from an extracellular suction electrode applied to the ectoderm which records all three events while the lower trace is an intracellular recording of the n1 spike from the larger of the two stem giant axons. b) A shock to the stem also produces a slowly propagating action potential in the endoderm (s) recorded both by an extracellular (top trace) and intracellular (bottom trace) electrode in an endodermal epithelial cell. Vertical scales: a) 0.2 mV (ex), 20 mV (in); b) 0.5 mV (ex), 20 mV (in); horizontal scales a) 20 ms b) 100 ms. (Adapted from Mackie 1978).

approaches that of the nEJPs (which must be the velocity of nerve spike propagation). Mackie has explained this phenomenon by assuming that there is retroactive excitation of the "S" system through electrical coupling of the ectodermal muscle with the endoderm.

Neural innervation of the gastrozooid longitudinal muscle in Nanomia is evident from recordings of summed junctional potentials (SDs) in the ectoderm which, if supra-threshold, result in muscle action potentials. In addition there is an epithelial conducting system (BP system) in the endoderm that controls a tidal pumping activity (Mackie 1978). None of the stem systems appear to penetrate the gastrozooids, although stem activity can initiate SDs in gastrozooids and stem muscle twitches can be associated with SDs. Similar types of interactions occur between other individuals in the colony, such as the palpons, and the stem. The "relay-station" for transference of excitation between the stem and individuals is probably the collar of neuronal somata and dense neuropil seen at the bases of individuals (Grimmelikhuijzen et al. 1986). Whatever connections exist between the stem and gastrozooid nervous system, they are quite labile and are probably based on chemical trasmission as communication is lost with repetitive stimulation or excess Mg^{++}. Most of the important features of the physonectid nervous system can also be seen in the diphyids (Mackie and Carré 1983).

There are several localities in siphonophores, for example at the nectophore margin, the stem root and the bract-stem interface, where excitable epithelia are able to transfer action potentials to the nervous system (Mackie 1978; Mackie and Carré 1983). The exact mechanism and the ultrastructural basis for such transfers are unknown, while transference of excitation in the opposite direction (neuro-epithelial), at other sites, involves conventional chemical synapses.

2.3.8. Synaptic transmission and transmitters

There is overwhelming physiological and morphological evidence for both electrical and chemical synapses within the hydrozoan nervous system (see review by Martin and Spenser 1982). As yet, electrical synapses and associated coupling have only been found between member neurons of a particular network (Anderson and Mackie 1977; Spencer and Satterlie 1980; Spencer 1981; Weber et al. 1982; Satterlie and Spencer 1983; Spencer and Arkett 1984; Satterlie 1985; Grimmelikhuijzen et al. 1986); such neurons do not appear to make chemical synapses with one another. On the other hand, chemical synapses are used for communication between different nerve nets and for excitation of epithelia including myoepithelia (Singla 1978a, b; Spencer 1979, 1982; Roberts and Mackie 1980; Spencer and Arkett 1984; Satterlie 1985; Kerfoot et al. 1985). Not unexpectedly, the ultrastructure of hydrozoan electrical synapses is conventional, showing all the features characteristic of gap junctions in other taxa. In an epithelial system known to show electrical coupling, intercellular particles resembling connexin molecules are seen using lanthanum staining of extracellular space (King and Spencer 1979). When measured, the electrical coupling between neurons has proved to be exceptionally strong (Spencer 1981). Thus, electrical activity of member neurons in nerve nets tends to be synchronized.

Physiologists have been fortunate in being able to record intracellularly, both pre- and post-synaptically, at neuro-neuronal and neuromuscular junctions in hydrozoans. As described above, four major preparations have been used: the neuromuscular junctions between motor neurons and swimming muscle in Polyorchis (Spencer 1982) and Aglantha (Roberts and Mackie 1980; Kerfoot et al. 1985); the synapses from the "B" system onto the swimming motor neurons in Polyorchis (Spencer and Arkett 1984); and the synapses between the ring giant axon and motor giant axons in Aglantha (Roberts and Mackie 1980). In all cases, transmission is blocked by excess Mg^{++} and no electrical coupling can be measured. The synaptic delay at both types of synapses in Aglantha is far shorter (approx. 1.6 ms) than for Polyorchis where it varies from 1 to 7 ms at the neuromuscular junction and 5 to 8 ms at the neuro-neuronal synapse. The amplitudes of post synaptic potentials is as small as 1.5 mV for "B"-induced epsps and as large as 40 mV or more for the junctional potentials in swimming muscle of Polyorchis. The amplitudes of epsps in Aglantha are intermediate to the above values. The large variations in amplitude, and probably measured delay, are due to the spatial and temporal summation of epsps that can occur in post synaptic cells that are strongly coupled to all their neighbors, which are also postsynaptic. Recordings of spontaneous epsps in SMNs of Polyorchis show that this synapse can be strongly facilitated when the presynaptic spike interval is less than 200 ms.

Despite the experimental opportunities made available by these preparations for studying the physiological effects of putative transmitters, their agonists and antagonists, we have no evidence that would strongly support any transmitter material (Martin and Spencer 1982). This is very important information since the positive identification of a neurotransmitter in any extant cnidarian may lead to a better understanding of the phylogeny of the various transmitter groups and to an appreciation of specific properties that have been selected for during evolution. Many of the difficulties that experimenters face are usually associated with the presence of tight intercellular connections between cells in the overlying epithelia which effectively isolate the synaptic sites from the bathing medium. Even iontophoretic and pressure injection techniques have met with little sucess. Nevertheless, by partially digesting the epithelium overlying the swimming motor neuron

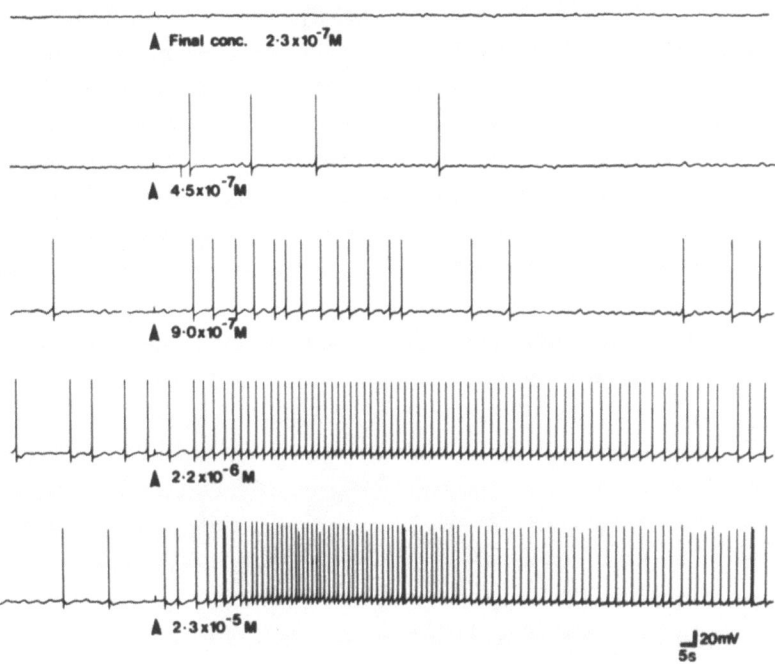

FIG. 19: Effect of Antho-RFamide peptides on the swimming motor neurons of <u>Polyorchis penicillatus</u>. Intracellular recordings were made after partial enzymatic digestion of the overlying epithelium. Peptides were dissolved in seawater and administered through a blunt micropipette immediately over the inner nerve ring. The "<u>Anthopleura</u> peptide" was sequenced by C.J.P. Grimmelikhuijzen who kindly supplied the synthetic peptide.

network of <u>Polyorchis</u> and micro-spritzing neuropeptides over the inner nerve ring it has been possible to demonstrate that peptides related to molluscan FMRFamide have excitatory effects on the swimming motor neurons (Fig. 19 and Grimmelikhuijzen et al. 1987). The long duration depolarizations produced by these peptides are similar to those seen in a number of molluscan studies (Cottrell and Davies 1987; Ruben et al. 1986). It is still too early to say whether the native RFamide peptides which have been isolated from <u>Polyorchis</u> (C.J.P. Grimmelikhuijzen, personal communication) are being used as neurotransmitters. It should be realized that the data just described do not make it any less likely that non-peptide transmitter materials will be shown to be used at other synapses. Indeed there is sufficient morphological evidence alone to suggest that there are at least two types of synaptic vesicles and hence transmitters (Spencer and Schwab 1982).

3. PHYLUM CTENOPHORA

Ctenophores are gelatinous, primarily planktonic animals that show a superficial resemblance to medusoid cnidarians. For example, both groups have biradial symmetry, canal-like gastrovascular systems, and mesogleal connective tissue. Significant specializations, unique to ctenophores, include the elaboration of cilia into comb plates, the widespread use of

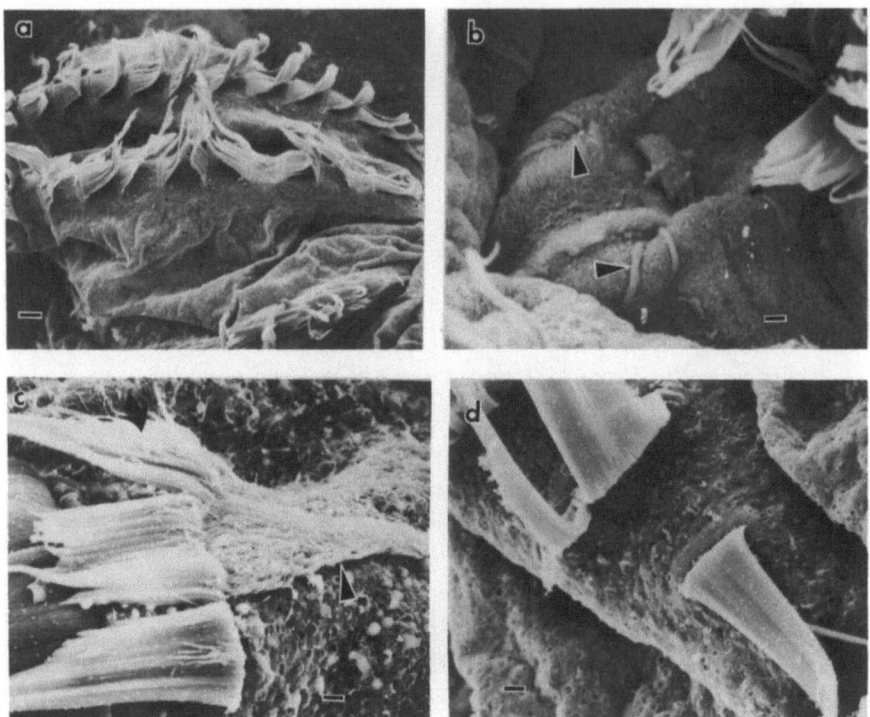

FIG. 20: Surface ultrastructure of the cydippid ctenophore *Pleurobrachia pileus*. In (a) two comb rows are shown, one of which appears to have been preserved with a metachronal wave in progress. Aboral end is to the left. b) At the aboral end, the apical rogan (large arrow) is covered by a dome of cilia. Four pairs of ciliated grooves (only two are shown clearly - small arrows) run from the apical organ to the first comb plates (not shown). c) Junction between the ciliated groove (small arrow) and the first comb plate (large arrow). d) The ciliated grooves do not extend beyond the first comb plate in cydippids and beroids. The last two (oral-most) comb plates are shown. Scale bars = 150 μm (a), 25 μm (b,c,d).

cilia for locomotory and sensory functions, and subepidermal musculature. Nervous elements of ctenophores are organized into subepidermal, diffuse nerve nets (similar to cnidarians) and show some local tract-like accumulations, for example below comb rows. Nerve nets are involved in modulating ciliary activity of the comb plates, coordination of feeding behavior and body movements, and control of bioluminescence.

Recent studies of ctenophores have centered on mechanisms of control of ciliary comb plates (comb plate coordination; see Tamm 1982). The role of the nervous system in other ctenophore behaviors has not been extensively studied.

The ctenophore body has well defined oral and aboral poles. The former contains the mouth, with extensible lips which lead to a muscular pharynx. The aboral pole contains the apical organ, a statocyst-bearing structure that is directly involved in control of locomotory movement of the comb rows. Generally, eight comb rows radiate from the aboral region

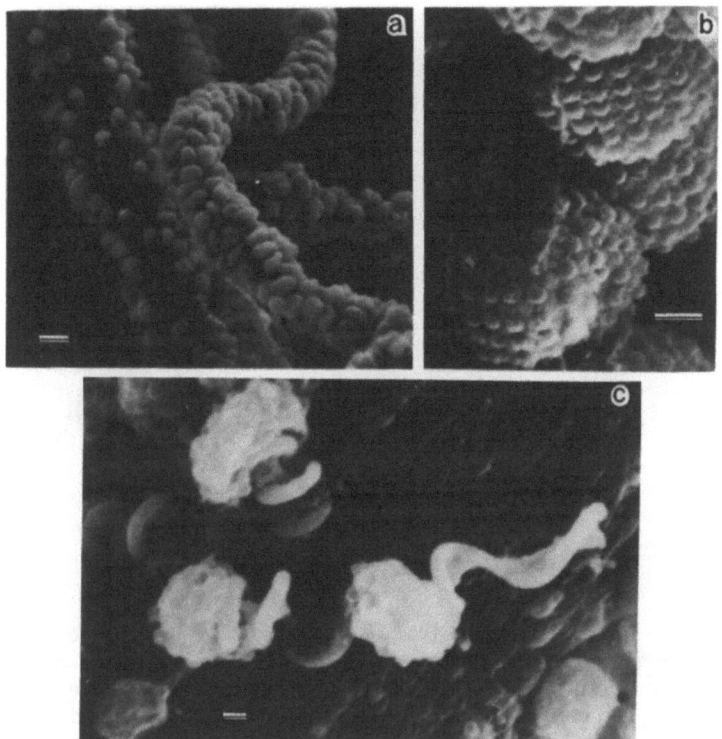

FIG. 21: Tentacle surface ultrastructure of <u>Pleurobrachia pileus</u>. a) three tentacular filaments are shown covered with colloblasts. On the largest filament, some of the colloblasts have been removed so that the coiled basal structure can be seen. b) High magnification of the "heads" of the colloblasts showing the regularly-arranged mucus packets. c) High magnification of the colloblasts in side view. Note the coiled basal structures. Scale bars = 5 μm (a) 1 μm (b,c).

in the oral-aboral plane. Each comb row is formed from several paddle-like comb plates each of which is made up to thousands of adherent cilia (Fig. 20). Each comb row is "connected" to the apical organ by a narrow band of short cilia called the ciliated groove (Fig. 20).

Most ctenophores (except beroids) have a pair of laterally placed tentacles than can retract into a tentacular sheath. Each tentacle bears numerous filaments that branch off the main trunk and are covered with specialized adhesive cells called colloblasts (Fig. 21).

Aside from cilia, effectors include muscles, photocytes and glandular cells. Muscle cells are, for the most part, isolated cells that run through the mesoglea. They lack cross-striations and can be very large (discussed later). Glandular cells are found in all regions of the external epithelium and in some areas are directly innervated by sensory cells forming local circuits (Hernandez-Nicaise 1974). Photocytes are discussed by Anctil (1987).

Ctenophores possess multiple conduction systems many of which are comprised of diffuse nerve nets (Hernandez-Nicaise 1973a, b;

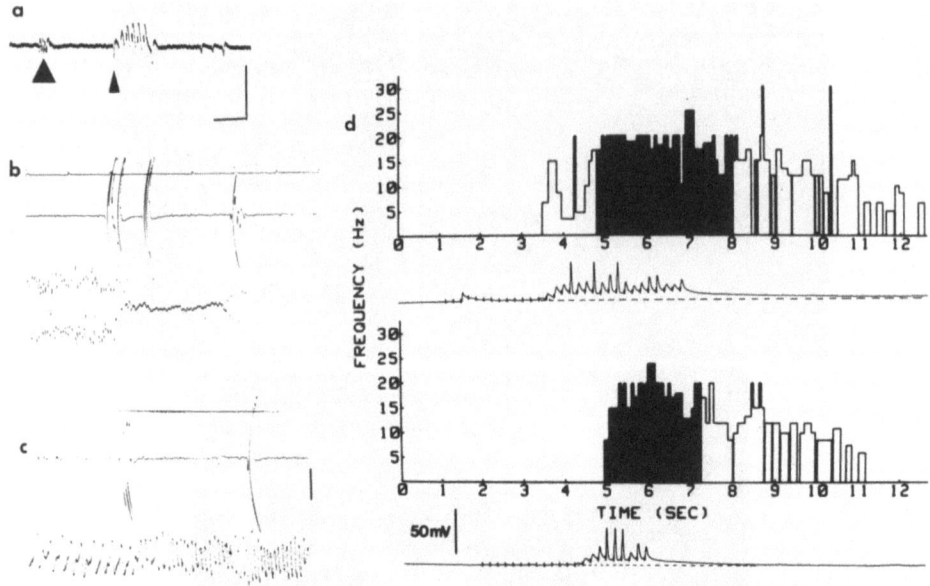

FIG. 22: Electrical activity from the cydippid ctenophore *Pleurobrachia pileus*. (a-c) Extracellular suction electrode recordings from the body surface. a) Two types of electrical potentials can be recognized; one comprised of a rapid burst of small potentials (appear to be associated with ciliary inhibition - large arrow) and the other including larger, facilitating potentials that are associated with ciliary excitation (increase comb plate beat frequency - small arrow). (b and c) Suction electrode records (middle trace) with a photoelectric motion detector placed over the comb row nearest the electrode placement (lower trace). In (b) a rapid burst of potentials (the chart recorder was not fast enough to follow the burst) caused temporary cessation of comb plate beating. In (c) a burst of surface potentials trigger an increase in comb plate beat frequency. In both (b) and (c) the top trace is a time marker (1 Hz). (d) Intracellular electrical activity from polster cells in response to extracellular stimulation of the body surface (traces below graphs). The graphs represent beating activity of a single comb plate. The open bars represent normal beating, striped bars represent intermediate beating and black bars represent reversed beating. In the top sequence a single large synaptic potential is followed by two very small synaptic potentials and a silent period. After the 13th stimulus a train of summing, facilitating synaptic potentials produce three action potentials. The lower sequence show similar activity without the early synaptic potentials. In both, rapid, reversed beating follows spiking in the polster cells. (d is from Moss and Tamm 1986). Scale bars = 30 μV, 0.5 s (a); 25 μV (b,c).

Hernandez-Nicaise et al. 1980). Little is known about the conducting properties of ctenophore nerve nets although small biphasic potentials (up to 30 μV) have been recorded from the body surface of <u>Beroe</u> (Hernandez-Nicaise et al. 1980) and two types of electrical impulses have been recorded from <u>Pleurobrachia</u> (Satterlie 1978 and Fig. 22). Mesogleal neurons have been identified in histological and ultrastructural studies (Hernandez-Nicaise 1973a, b). The orientation of neurites suggests that conduction is diffuse although some evidence for polarization (neuron tracts) is available (i.e. underneath the comb rows and ciliated grooves, and around the mouth and pharynx). Neurons are generally less than 2 µm in diameter and give clear morphological evidence for chemical synaptic transmission. This includes description of polarized, symmetrical and reciprocal synaptic contacts (Horridge and Mackay 1964; Horridge 1965a; Hernandez-Nicaise 1968, 1973c; 1974).

Ctenophores are extremely responsive to mechanical stimulation. For this reason, mechanoreception is the best studied of ctenophore senses (see Tamm 1982). Two types of suspected mechanoreceptors have been described in the epidermis, particularly in the lips and tentacles (Horridge 1965b; Hernandez-Nicaise 1974). These include cells bearing cilia with onion-like root structures and cells bearing stiff pegs. In the statocyst of the apical organ, four groups of balancer cilia support a statolith and are sensitive to changes of statolith loading (Horridge 1965a; see also Tamm 1982). Each balancer group is actually made up of 150-200 adherent cilia which are mechanically coupled to the statolith. Each balancer group gives rise to a pair of ciliated grooves each of which, in turn, runs to a comb row. Balancer cilia are motile and each beat is translated into a conducted beat of cilia in the connected ciliated grooves and comb rows. The statocyst is covered by a dome of non-motive cilia (see Fig. 20).

3.1. Apical organ and geotaxis

The apical organ balancer cilia are believed to act as motile pacemakers for the comb rows as their activity is directly translated into antiplectic metachronal beating of the comb rows (Horridge 1965a; Tamm 1982). Mechanical coupling between the statolith and the balancer cilia, and the control of only two comb rows by each balancer group, provides a means for geotactic responses by ctenophores (Tamm 1982). The beat frequency of balancer cilia varies with mechanical deformation so that in animals with the oral or aboral ends uppermost, the four groups of balancers will all beat at nearly the same rate. If an animal is tilted from this vertical plane, however, clear differences in beat frequency can be measured in the "uppermost" and "lowermost" balancers. These differences are attributed to differential loading by the statolith. Since the beat frequency is conserved in the transmission of ciliary activity from balancers to associated ciliated grooves and comb rows, the behavioral response to tilt will be a "righting" of the animal (return to a vertical plane). This righting response is predictable only in the sense that a vertical plane is achieved since individual animals have the ability to change the sign of their geotactic response. The influences of sensory inputs on these negative and positive geotactic "moods" is poorly understood. Recent work by Tamm (1982) indicates that the basic geotactic response (positive) is determined by the apical organ activity, and the changes in mood involve modulation by the nervous system.

3.2. Locomotion

Locomotory movements of ctenophores involve metachronal coordination of the comb plates within each row (Tamm 1973, 1982; Sleigh 1974). The effective stroke for comb plate movement is normally directed toward the aboral pole, thus undisturbed swimming is in a mouth-first direction. Variations in locomotory behavior result from differences in comb row beat frequencies in the four quadrants as controlled by the four balancer groups. Transmission of an individual wave of ciliary activity within a comb row is via mechanical coupling, or hydrodynamic linkage, of adjacent cilia and/or comb plates (Sleigh 1972; Tamm 1973, 1982). Despite this generalization, two patterns of comb plate-to-comb plate coordination are seen. In cydippids, beroids and cestids, the ciliated grooves run from the balancer cilia to the first comb plate, but no further (i.e. they are not found between comb plates; Fig. 20). In lobates, the ciliated grooves not only run from the balancers to the first comb plate, but also extend between successive comb plates throughout the length of the comb row. The significance of this is discussed by Tamm (1982).

Intracellular microelectrodes have been used to monitor electrical activity of polster cells (cells that provide the cilia that make up a comb plate) during comb plate beating (Horrdige 1965c; Tamm 1982; Moss and Tamm 1986). Despite early claims that membrane potential fluctuates with each beat (Horridge 1965c), recent recordings indicate that the cells are silent during normal beating (Tamm 1982; Moss and Tamm 1986).

It appears that normal swimming activity can be modified in two ways; ciliary inhibition (or decrease in beat frequency) and ciliary excitation (or increase in beat frequency; Satterlie 1978; Tamm 1979, 1982; Tamm and Tamm 1981. These responses occur following stimulation of various body parts and are independent of the fluctuations attributed to apical organ statocyst input.

Ciliary inhibition is a widespread response involving all of the cilia that are normally active during locomotion. Cutting experiments indicate that the conducting system mediating inhibition extends throughout the body and is able to conduct in any direction from the point of stimulation (non-polarized; Horridge 1966; Tamm 1982). The conducting system is magnesium sensitive, suggesting the involvement of a diffuse synaptic nerve net (Horridge 1965a, 1974). In accordance with these results, polster cells are known to receive synaptic input from underlying nerve cells (based on ultrastructural evidence; Horridge and Mackay 1964; Hernandez-Nicaise 1973a, c, 1974).

Tentacular stimulation in _Pleurobrachia_ results in retraction movements of one or both tentacles, and rapid forward swimming. Input from this excitatory system is not conducted to all comb plates, but rather is restricted to plates at the aboral end of the animal (Tamm 1982). The apical organ is not required for ciliary excitation in the half of the animal receiving the stimulus, however it is required for transmission of ciliary excitation to the non-stimulated side.

In addition to ciliary excitation and inhibition, most ctenophores exhibit ciliary reversals (see Tamm 1982). Reversals occcur in two ways). 1) Reversal in the direction of metachronal coordination of the comb plates produces symplectic waves. This is best achieved by stimulating comb plates near the oral end of the animal. 2) Reversal of ciliary beat direction has been documented in both larval and adult _Pleurobrachia_ (Tamm 1979; Tamm and Tamm 1981). Reversal of beat direction is frequently preceded by a temporary "laydown" of comb plates (Moss and Tamm 1986). Feeding responses (described below) involve both reversal of the effective stroke and reversed metachronal waves.

Intracellular recordings have revealed apparent synaptic inputs to polster cells during laydown and reversal of beat direction (Moss and Tamm 1986). Polster cells have resting potentials of around -60 mV and, as mentioned above, do not show any changes in electrical activity during ciliary beating in the normal direction (Moss and Tamm 1986). Following electrical stimulation of the body, two types of excitatory synaptic potentials are recorded (Moss and Tamm 1986; Fig. 22). Early, large amplitude potentials follow the first few stimuli, followed by summating and facilitating epsps that give rise to graded, regenerative spikes. The spikes are seldom overshooting. Repetitive firing of polster cell spikes is accompanied by ciliary laydown and beat direction reversal. Laydown alone can occur if the spike frequency is low. In addition to the epsp-induced activity, there is evidence that the rate of comb plate beating can be regulated by sub-threshold DC shifts of the membrane potential, at least in the depolarizing direction (depolarization = increase in beat frequency; Moss and Tamm 1986). The source of the DC shifts is unknown.

The coordinated movement of cilia making up a single comb plate suggests that some form of inter-ciliary coupling must exist. Ultrastructural evidence for mechanical coupling between ciliary shafts has been presented (Afzelius 1961; Tamm and Tamm 1981). In addition, the presence of gap junctions between polster cells of a single comb plate (Satterlie and Case 1978) suggests that there may be intercellular electrical coupling which would promote spread of synaptic inputs throughout a polster group during nerve net-driven modulation of swimming activity. If this is the case, the coupling may not be tight since dual intracellular recordings from polster cell pairs (underlying a single comb plate) indicate that synaptic inputs are frequently, but not always similar (Moss and Tamm 1986). It is not clear from these recordings if polster cells are indeed coupled electrically.

3.3. A complex behavioral act: Prey capture and ingestion

An interesting, complex behavior that involves many of the above responses is the feeding behavior of Pleurobrachia (Satterlie 1978; Tamm 1982; Fig. 23). In the normal fishing posture of adult Pleurobrachia, the tentacles and tentacular filaments are fully expanded. Contact of prey (i.e. a copepod) with a tentacle or filament results in the discharge of sticky mucus from colloblasts and entrapment of the prey. The basal attachment of colloblasts includes a helically-coiled process which presumably prevents detachment of the cell during struggling movements of the prey (Franc 1978; see also Fig. 21). These struggling movements eventually cause contraction of tentacular longitudinal muscle, but not full tentacle retraction. Continued stimulation of the tentacle by the prey triggers a fairly stereotypic series of responses. First, the entire body spins in the tentacular plane, and the mouth opens and bends toward the stimulated tentacle. As the ctenophore spins, the tentacles become wrapped around the body so that the mouth is in contact with, or in close proximity to a portion of the tentacles. The lips then enwrap the tentacles which contract in short jerks, pulling the tentacles through the lips. When the prey-bearing portion of the tentacle is pulled through the lips, the prey is transferred to the mouth and engulfed. Of these responses, body spinning is the most interesting since it can be constituted from a sequence of the ciliary events already described (see Tamm 1982). In closer examination, the response begins with an increase in overall beat frequency (fast forward swimming) followed by a temporary cessation of ciliary beat. Next, there is an upward curvature of the base of the quiescent comb plates in the four comb rows nearest the stimulated tentacle (comb plate laydown). Finally, high frequency ciliary activity resumes with the four comb rows involved in laydown displaying a reversal in beat direction. The interesting feature of this system is that each ciliary response can be initiated individually through activation of distinct conducting systems. During feeding behavior, all are initiated in a specific sequence following an appropriate level of stimulation that appears to have a definite threshold. The nature of this coordinating system that controls the sequential activation of conducting systems and ciliary responses awaits description.

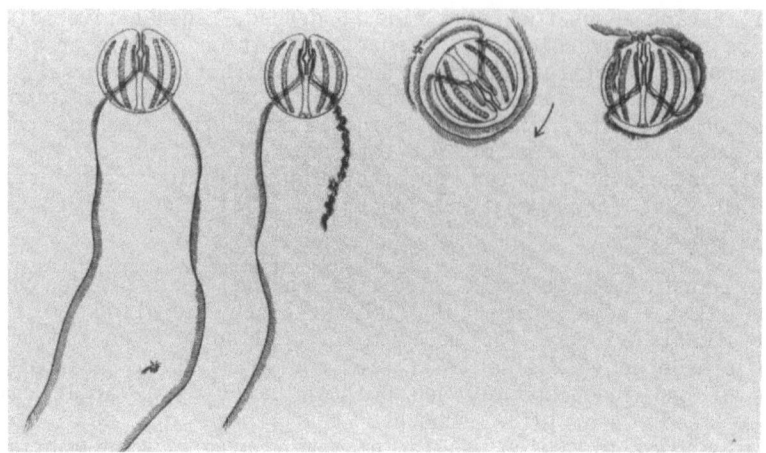

FIG. 23: Diagrammatic representation of prey capture and ingestion behavior in Pleurobrachia as explained in text. a) fishing posture, b) prey contact and tentacle contraction, c) body spinning and d) contraction of tentacles and prey ingestion.

3.4. Giant smooth muscle cells of ctenophores

One ctenophore preparation that has received considerable attention recently involves the isolated smooth muscle cells of the mesoglea and pharyngeal region (Hernandez-Nicaise and Amsellem 1980; Hernandez-Nicaise et al. 1980, 1982; Stein and Anderson 1984; Anderson 1984). The muscle cells (of Beroe) are up to 50 μm in diameter and 6 cm in length. The cells are innervated by a nerve net, and chemical neuromuscular junctions occur frequently (Hernandez-Nicaise and Amsellem 1980; Hernandez-Nicaise et al. 1980). They have resting potentials of around -60 mV and produce overshooting, progagated action potentials (Hernandez-Nicaise et al. 1980 Anderson 1984; Fig. 24). The action potentials are comprised of inward sodium and calcium currents as well as voltage- and calcium-activated potassium outward currents (Hernandez-Nicaise et al. 1980; Anderson 1984). The giant smooth muscles of ctenophores provide useful preparations for detailed studies of smooth muscle contractility, excitation-contraction coupling and ion channel function.

4. CONCLUDING COMMENTS

When describing the nervous system of cnidarians and ctenophores, the most frequently used adjective is "simple". What is "simple" about these nervous systems? Structurally, neurons of these groups resemble neurons ofhigher animals. The biophysical basis of electrogenesis in neurons, muscle and epithelial cells is certainly conventional. Intercellular junctions such as chemical and electrical synapses are similar to those found in all higher animals although the widespread use of bidirectional synapses is somewhat unusual. The "simplicity" of cnidarian and ctenophore nervous systems therefore does not appear to occur at the level of individual cells, but rather in the organization of these cells into conducting systems, in other words, the wiring. The term "simple" also carries the unfortunate connotation that these nervous systems have somehow not evolved to their full potential. For example features such as diffuseness or non-polarized conduction are used as examples of simplicity

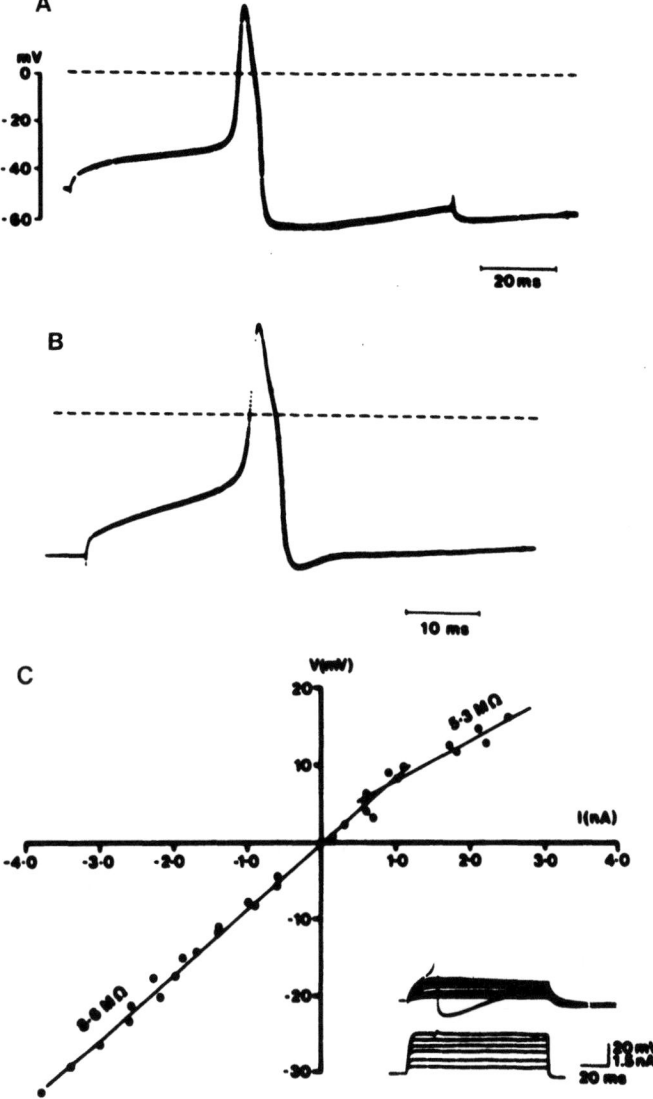

FIG. 24: Intracellular electrical activity from giant smooth muscle cells of <u>Mnemiopsis</u>. A and B are action potentials evoked by intracellular current injection. C shows the currentvoltage relationship of an in situ fiber showing outward rectification with depolarization. (From Anderson 1984).

when compared to the exterme condensation and channelled spike traffic seen in more "advanced" invertebrates. However, the types of neuronal organization that we see in these groups is more a reflection of the evolutionary constraints imposed on animals having radial symmetry than a reflection of their primitiveness. Thus unpolarized conduction should not be thought of as an archaic property, but rather an ingenious solution to problems associated with radial coordination such as the need to distinguish between local and general sensory input and to respond with either a local output or one that is synchronized throughout 360°. It is interesting to note that two types of intercellular junctions are largely responsible for unpolarized conduction in cnidarians. In anthozoans and scyphozoans, the bidirectional synapse allows for unpolarized conduction whereas in hydrozoans, the gap junction is the chief agent. We might even suggest that there is more than a casual relationship between the presence of gap junctions in hydrozoans and the dramatic condensation of nerve nets seen in many forms (i.e. marginal nerve rings of hydromedusae). A similar degree of nervous system "condensation" is not observed in the anthozoans and scyphozoans (groups that rely solely on chemical synapses) except for areas where concentration of sensory structures exist (i.e. rhopalia of scyphomedusae). We can certainly suggest that gap junctions within myoepithelia would decrease the demand for diffuse synaptic nerve nets as a necessary means of distributing excitation to all muscle cells in a two-dimentional conducting sheet. Does this mean that in terms of nervous system organization, the hydrozoans are the most advanced cnidarians? We can at least say that cnidarians have solved the problems associated with coordination of radially arranged effectors and sensory structures in more than one way.

Lentz (1968) states that, "...some characteristics probably evolved independently...their general and widespread occurrence in some cases suggests a common origin and primitive nature". If we adopt the latter approach, we can look for common features in the nervous system of the cnidarian classes (ignoring sensory systems which are quite variable) and assume that these represent "primitive" characteristics. Our list would include the presence of multifunctional neurons, the organization of neurons into nerve nets, the existence of chemical synapses, and the use of conventional electrogenesis. Are we then to assume that gap junctions represent a later, possibly hydrozoan, invention? This possibility was raised by Mackie et al. (1984) along with the possibility that gap junctions were secondarily lost in an anthozoan and scyphozoan ancestor. The latter possibility seems quite unlikely for such a major evolutionary step that is retained in virtually every other metazoan group. It therefore seems unlikely that epithelial conduction represents the most primitive form of conduction in a multicellular organism (as suggested by Mackie 1979). We would then be forced to suggest that the primary solution to information transfer in radially symmetrical animals may have been the nerve net. This is in stark contrast to the scenario drawn by Horridge (1968) where he derives the ancestral nerve net from a superficial, conducting epithelium.

Speculation aside, recent advances in the study of sponge, cnidarian and ctenophore conducting systems have greatly changed our perceptions about "primitive" nervous systems. We hope that these advances will stimulate further curiosity and investigative effort aimed at answering the many inportant questions that remain about the sensory, integrative and neuro-effector physiology of these groups.

5. AKNOWLEDGEMENTS

We would like to thank the editors and publishers of the following journals for allowing use of figures: Philosophical Transactions of the Royal Society of London, The Company of Biologists Ltd. (Journal of Experimental Biology), Journal of Neurophysiology, Nature, Journal of Neurobiology, Journal of Comparative Physiology A, and Marine Behaviour and Physiology. We would also like to thank Brock Anstine and Becky Falatka for clerical assistance, and Ray Medhus for photographic assistance.

6. REFERENCES

Afzelius BA (1961) The fine structure of the cilia from ctenophore swimming-plates. J Biophys Biochem Cytol 9: 383-394.

Anctil M (1987) Neural control mechanisms in bioluminescence. (This volume).

Anderson PAV (1976) An electrophysiologicasl study of mechanisms controlling polyp retraction in colonies of the scleractinian coral Goniopora lobata. J Exp Biol 65: 381-393.

Anderson PAV (1979) Ionic basis of action potentials and bursting activity in hydromedusan jellyfish Polyorchis penicillatus. J Exp Biol 78: 299-302.

Anderson PAV (1980) Epithelial conduction: its properties and functions. Prog Neurobiol 15: 161-203.

Anderson PAV (1984) The electrophysiology of single smooth muscle cells isolated from the ctenophore Mnemiopsis. J Comp Physiol B 154: 257-268.

Anderson PAV (1985) Physiology of a bidirectional, excitatory, chemical synapse. J Neurophysiol 53: 821-835.

Anderson PAV, Case JF (1975) Electrical activity associated with luminescence and other colonial behaviour in the pennatulid Renilla köllikeri. Biol Bull Mar Biol Lab, Woods Hole 149: 80-95.

Anderson PAV, Mackie GO (1977) Electrically coupled, photosensitive neurons control swimming in a jellyfish. Science 197: 186-188.

Anderson PAV, McKay MC (1985) Evidence for a proton-activated chloride current in coelenterate neurons. Biol Bull 169: 652-660.

Anderson PAV, Schwab WE (1981) The organization and structure of nerve and muscle in the jellyfish Cyanea capillata (Coelenterata: Scyphozoa). J Morphol 170: 383-399.

Anderson PAV, Schwab WE (1982) Recent advances and model systems in coelenterate neurobiology. Prog Neurobiol 19: 213-236.

Anderson PAV, Schwab WE (1983) Action potential in neurons of motor nerve net of Cyanea (Coelenterata). J Neurophysiol 50: 671-683.

Arkett SA, Spencer AN (1986a) Neuronal mechanisms of a hydromedusan shadow reflex I. Identified reflex components and sequence of events. J Comp Physiol 159: 201-213.

Arkett SA, Spencer AN (1986b) Neuronal mechanisms of hydromedusan shadow reflex II. Graded response of relfex components, possible mechanisms of photic integration, and function significance. J Comp Physiol 159: 215-225.

Barnes RD (1985) Current perspectives on the origins and relationships of lower invertebrates. In: Conway S, Morris SC, George JD, Gibson R, Platt HM (eds) The origins and relationships of lower invertebrates. The Systematics Association Special Volume 28. Clarendon Press, Oxford, pp 360-367.

Bergquist PR (1985) Poriferan relationships. In: Conway S, Morris SC, George JD, Gibson R, Platt HM (eds) The origins and relationships of lower invertebrates. The Systematics Association Special Volume 28. Clarendon Press, Oxford, pp 14-27.

Berking S (1986) Is homarine a morphagen in the marine hydroid Hydractinia?. Wilhelm Roux's Arch Dev Biol 195: 33-38.

Bidder GP (1923) The relationship of the form of a sponge to its currents. Q J Microsc Sci 67: 293-323.

Brien P (1973) Les demosponges: morphologie et reproduction. In: Grassé PP (ed) Traité de zoologie: Anatomie, systématique, biologie. Vol. 3. Masson, Paris, pp 133-461.

Bullock TH (1943) Neuromuscular facilitation in scyphomedusae. J Cell Comp Physiol. 22: 251-272.

Bullock TH, Horridge GA (1965) Structure and function in the nervous system of invertebrates. Vol. I. W.H. Freeman, San Francisco and London.

Burnett AL (1961) The growth process in Hydra. J Exp Zool 146: 21-84.

Burnett AL, Diehl NA (1964) The nervous system of Hydra. I. Types, distribution and origin of nerve elements. J Exp Zool 157: 217-226.

Chain BM, Bone Q, Anderson PAV (1981) Electrophysiology of a myoid epithelium in Chelophyes (Coelenterata: Siphonophora). J Comp Physiol 143: 329-338.

Chapman DM (1974) Cnidarian histology. In: Muscatine L, Lenhoff HM (eds) Coelenterate Biology. Reviews and perspectives. Academic Press, New York, pp 1-92.

Cottrell GA, Davies NW (1987) Multiple receptor sites for a molluscan peptide (FMRFamide) and related peptides of Helix. J Physiol (in press).

David CN (1973) A quantitative method for maceration of Hydra tissue. Wilhelm Roux's Arch Entwicklungsmech Org 171: 259-268.

Davis LE, Burnett AL, Haynes JF (1968) Histological and ultrastructural study of the muscular and nervous system in Hydra. II. Nervous system. J Exp Zool 167: 295-332.

Dickinson P (1978) Conduction systems controlling expansion-contraction behavior in the sea pen Ptilosarcus gurneyi. Mar Behav Physiol 5: 163-183.

Dunne JF, Javois LC, Huang LW, Bode HR (1985) A subset of cells in the nerve-net of Hydra oligactis defined by a monoclonal antibody: its arrangement and development. Dev Biol 109: 41-53.

Eimer T (1878) Die Medusen: Physiologisch und Morphologisch auf ihr Nervensystem. Tubingen.

Epp L, Tardent P (1978) The distribution of nerve cells in Hydra attenuata Pall. Wilhelm Roux's Arch Dev Biol 185: 185-193.

Franc J-M (1978) Organization and function of ctenophore colloblasts: an ultrastrunctural study. Biol Bull Mar Biol Lab, Woods Hole 155: 527-541.

Grimmelikhuijzen CJP (1983a) Coexistence of neuropeptides in Hydra. Neurosci 9: 837-845.

Grimmelikhuijzen CJP (1983b) FMRFamide immunoreactivity is generally occurring in the nervous system of coelenterates. Histochem 78: 361-381.

Grimmelikhuijzen CJP (1985) Antisera to the sequence Arg-Phe-amide visualiza neuronal centralization in hydroid polyps. Cell Tissue Res 241: 171-182.

Grimmelikhuijzen CJP, Spencer AN (1984) FMRFamide immunoreactivity in the nervous system of the medusa Polyorchis penicillatus. J Comp Neurol 230: 361-371.

Grimmelikhuijzen CJP, Sundler F, Rehfeld JF (1980) Gastrin/CCK-like immunoreactivity in the nervous system of coelenterates. Histochem 69: 61-68.

Grimmelikhuijzen CJP, Balfe A, Emson PC, Powell D, Sundler F (1981a) Substance P-like immunoreactivity in the nervous system of Hydra. Histochem 71: 325-333.

Grimmelikhuijzen CJP, Carraway RE, Rokaeus A, Sundler F (1981b) Neurotensin-like immunoreactivity in the nervous system of Hydra. Histochem 72: 199-209.

Grimmelikhuijzen CJP, Dockray GJ, Yanaihara N (1981c) Bombesin-like immunoreactivity in the nervous system of Hydra. Histochem 73: 171-180.

Grimmelikhuijzen CJP, Dierickx K, Boer GJ (1982a) Oxytocin/vasopressin-like immunoreactivity is present in the nervous system of Hydra. Neurosci 7: 3191-3199.

Grimmelikhuijzen CJP, Dockray GJ, Schot LPC (1982b) FMRFamide-like immunoreactivity in the nervous system of Hydra. Histochem 73: 499-508.

Grimmelikhuijzen CJP, Spencer AN, Carre D (1986) Organization of the nervous system of physonectid siphonophores. Cell Tissue Res 246: 463-479.

Grimmelikhuijzen CJP, Graff D, Spencer AN (1987) Structure, location and possible actions of Arg-Phe-amide peptides in coelenterates. In: Thorndyke MC, Goldsworthy G (eds) Invertebrate peptides and amines. Cambridge University Press.

Hadzi J (1963) The evolution of the Metazoa. Pergamon Press, London.

Haeckel E (1874) Die Gastraea-Theorie, die phylogenetische Klassification des Thierreichs und die Homolgie der kemiblatter. Jena Z Naturw 8: 1-55.

Hagiwara S, Yoshida S, Yoshida M (1981) Transient and delayed potassium currents in the egg cell membrane of the coelenterate Renilla köllikeri. J Physiol 318: 123-141.

Heimfeld S, Bode HR (1985) Growth regulation of the interstitial cell population in Hydra. Dev Biol 297-307.

Hernandez-Nicaise ML (1968) Distribution et ultrastructure des synapses symétriques dans le système nerveux des Cténaires. C Hebd Seanc Acad Sci, Paris 267: 1731-1734.

Hernandez-Nicaise ML (1973a) Le système nerveux des Cténaires. I. Structure et ultrastructure des réseaux épitheliaux. Z Zellforsch Mikrosk Anat 137: 223-250.

Hernandez-Nicaise ML (1973b) Le système nerveux des Cténaires. II. Les éléments nerveux intramésogléens chez les Beroides et les Cydippidés. Z Zellforsch Mikrosk Anat 143: 117-133.

Hernandez-Nicaise ML (1973c) The nervous system of Ctenophora. III. Ultrastructure of synapses. J Neurocytol 2: 249-263.

Hernandez-Nicaise ML (1974) Ultrastructural evidence for a sensory-motor neuron in ctenophora. Tissue Cell 6: 43-47.

Hernandez-Nicaise ML, Amsellem J (1980) Ultrastructure of the giant smooth muscle fiber of the Ctenophore Beroe ovata. J Ultrastruct Res 72: 151-168.

Hernandez-Nicaise ML, Mackie GO, Meech RW (1980) Giant smooth muscle cells of Beroe: Ultrastructure, innervation and electrical properties. J Gen Physiol 75: 79-105.

Hernandez-Nicaise ML, Bilbaut A, Malaval L, Nicaise G (1982) Isolation of functional giant smooth muscle cells from an invertebrate: structural features of relaxed and contracted fibers. Proc Natl Acad Sci USA 79: 1884-1888.

Horridge GA (1954) The nerves and muscles of medusae. I. Conduction in the nervous system of Aurelia aurita Lamarck. J Exp Biol 31: 594-600.

Horridge GA (1956a) The nervous system of the ephyra larva of Aurelia aurita. Q J Microsc Sci 97: 59-74.

Horridge GA (1956b) The nerves and muscles of medusae. V. Double innervation in scyphozoa. J Exp Biol 33: 366-383.

Horridge GA (1957) The co-ordination of the protective retraction of coral polyps. Phil Trans R Soc 240: 495-529.

Horridge GA (1959) The nerves and muscles of medusae. VI. The rhythm. J Exp Biol 36: 72-91.

Horridge GA (1965a) Relations between nerves and cilia in ctenophores. Amer Zool 5: 357-375.

Horridge GA (1965b) Non-motile sensory cilia and neuromuscular junctions in a ctenophore independent effector organ. Proc R Soc B162: 333-350.

Horridge GA (1965c) Intracellular action potentials associated with the beating of the cilia in ctenophore comb plate cells. Nature, 205: 602.

Horridge GA (1966) Pathways of co-ordination in ctenophores. In: Rees WJ (ed) The Cnidaria and their evolution, Symposium, Zoological Society of London. Vol. 16, pp 247-266.

Horridge GA (1968) Interneurons. Freeman, San Francisco.

Horridge GA (1974) Recent studies on the Ctenophora. In: Muscatine L, Lenhoff HM (eds) Coelenterate biology: Reviews and new perspective. Academic Press, New York, pp 439-468.

Horridge GA, Mackay B (1962) Naked axons and symmetrical synapses in coelenterates. Q J Microsc Sci. 103: 531-541.

Horridge Ga, Mackay B (1964) Neurociliary synapses in Pleurobrachia (Ctenophora). Q J Microsc Sci 105: 163-174.

Jackson AJ, McFarlane ID (1976) Delayed initiation of SS1 pulses in the sea anemone Calliactis parasitica: evidence for a fourth conducting system. J Exp Biol 65: 539-552.

Jagersten G (1955) On the early phylogeny of the Metazoa. The bilaterogastreaea theory. Zool Bidr Upps 30: 321-354.

Jones WC (1962) Is there a nervous system in sponges? Biol Rev 37: 1-50.

Josephson RK (1961) Repetitive potentials following brief electric stimuli in a hydroid. J Exp Biol 38: 579-593.

Josephson RK (1966) Neuromuscular transmission in a sea anemone. J Exp Biol 45: 305-319.

Josephson RK, Reiss RF, Worthy RM (1961) A simulation study of a diffuse conducting system based on coelenterate nerve nets. J Theor Biol 1: 460-487.

Kass-Simon G (1982) Aspects of coelenterate membrane physiology. In: Podesta RB (ed) Membrane physiology of invertebrates. Marcel Decker, NY, pp 83-120.

Kerfoot PAH, Mackie GO, Meech RW, Roberts A, Singla CL (1985) Neuromuscular transmission in the jellyfish Aglantha digitale. J Exp Biol 116: 1-25.

King MG, Spencer An (1979) Gap and septate junctions in the excitable endoderm of Polyorchis penicillatus (Hydrozoa: Anthomedusae). J Cell Sci 36: 391-400.

Kinnamon JC, Westfall J (1981) A three-dimensional serial reconstruction of neuronal distributions in the hypostome of a Hydra. J Morphol 168: 321-329.

Lawn ID (1975) An electrophysiological analysis of chemoreception in the sea anemone, Tealia felina. J Exp Biol 63: 525-536.

Lawn ID (1976a) The marginal sphincter of the sea anemone Calliactis parasitica. II. Properties of the inhibitory response. J Comp Physiol 105: 301-311.

Lawn ID (1976b) Swimming in the sea anemone Stomphia coccinea triggered by a slow conduction system. Nature 262: 708-709.

Lawn ID (1982) Porifera. In: Shelton GAB (ed) Electrical conduction and behaviour in 'simple' invertebrates. Clarendon Press, Oxford, pp 49-72.

Lawn ID, Mackie GO, Silver G (1981) A conduction system in a sponge. Science 211: 1169-1171.

Lentz TL (1968) Primitive nervous systems. Yale University Press, New Haven.

Lerner J, Mellen SA, Waldron I, Factor RM (1971) Neural redundancy and regularity of swimming beats in scyphozoan medusae. J Exp Biol 55: 177-184.

Mackie GE (1970) Neuroid conduction and the evolution of conducting tissue. Q Rev Biol 45: 319-332.

Mackie GO (1971) Neurological complexity in medusae: A report of central nervous organization in Sarsia. Actas del il Simposio Internacional de Zoofilogenia Salamance 269-280.

Mackie GO (1973) Report on giant nerve fibres in Nanomia. Publ Seto Mar Biol Lab: 745-756.

Mackie GO (1975) Neurobiology of Stomotoca. II. Pacemakers and conduction pathways. J Neurobiol 6: 357-378.

Mackie GO (1976a) Propagated spikes and secretion in a coelenterate glandular epithelium. J Gen Physiol 68: 313-325.

Mackie GO (1976b) The control of fast and slow muscle contractions in the siphonophore stem. In: Mackie GO (ed) Coelenterate ecology and behavior. Plenum Press, New York, pp 647-659.

Mackie GO (1978) Coordination in physonectid siphonophores. Mar Behav Physiol 5: 325-346.

Mackie GO (1979) Is there a conduction system in sponges? Colloques Int Cent Natn Rech Scient 291: 145-151.

Mackie GO (1980) Slow swimming and cyclical "fishing" behavior in Aglantha digitale (Hydromedusae: Trachylina). Can J Fish Aquat Sci 37: 1550-1556.

Mackie GO (1984) Fast pathways and escape behavior in Cnidaria. In: Eaton RC (ed) Neural mechanisms of startle behavior. Plenum Publ Corp, p 15-42.

Mackie GO, Carré D (1983) Coordination in a diphyid siphonophore. Mar Behav Physiol 9: 139-170.

Mackie GO, Meech RW (1985) Separate sodium and calcium spikes in the same axon. Nature 313: 791-793.

Mackie GO, Singla CL (1983) Studies on hexactinellid sponges. I. Histology of Rhabdocalyptus dawsoni (Lambe, 1883). Phil Trans R Soc Lond B 301: 365-400.

Mackie GO, Passano LM, Pavans de Ceccatty M (1967) Physiologie du comportement de l'Hydroméduse Sarsia tubulosa Sars. Les systèmes à conduction aneurale. C R Hebd Seans Acad Sci, Paris 264: 466-469.

Mackie GO, Lawn ID, Pavans de Ceccatty M (1983) Studies on hexactinellid sponges. II. Excitability, conduction and coordination of responses in Rhabdocalyptus dawsoni (Lambe, 1973). Phil Trans R Soc Lond B 301: 401-418.

Mackie GO, Anderson PAV, Singla CL (1984) Apparent absence of gap junctions in two classes of Cnidaria. Biol Bull 167: 120-123.

Mackie GO, Singla CL, Stell WK (1985) Distribution of nerve elements showing FMRFamide-like immunoreactivity in hydromedusae. Acta Zool (Stockh) 66: 199-210.

Marcus E (1958) On the evolution of animal phyla. Q Rev Biol 33: 24-58.

Martin SM, Spencer AN (1982) Neurotransmitters in Coelenterates. Comp Biochem Physiol C 74: 1-14.

McFarlane ID (1969a) Two slow conduction systems in the sea anemone Calliactis parasitica. J Exp Biol 51: 377-385.

McFarlane ID (1969b) Co-ordination of pedal-disc detachment in the sea anemone Calliactis parasitica. J Exp Biol 51: 387-396.

McFarlane ID (1970) Control of preparatory feeding behaviour in the sea anemone Tealia felina. J Exp Biol 53: 211-220.

McFarlane ID (1973a) Spontaneous electrical activity in the sea anemone Calliactis parasitica. J Exp Biol 58: 77-90.

McFarlane ID (1973) Spontaneous contractions and nerve net activity in the sea anemone Calliactis parasitica. Mar Behav Physiol 2: 97-113.

McFarlane ID (1974) Control of the pacemaker system of the nerve net in the sea anemone Calliactis parasitica. J Exp Biol 61: 129-143.

McFarlane ID (1975) Control of mouth opening and pharynx protrusion during feeding in the sea anemone Calliactis parasitica. J Exp Biol 63: 615-626.

McFarlane ID (1979) Two slow conducting systems coordinate shell-climbing behaviour in the sea anemone Calliactis parasitica. J Exp Biol 64: 431-446.

McFarlane ID (1978) Multiple conducting systems and the control of behaviour in the brain coral Meandrina meandrites (L.). Proc R Soc B 200: 193-216.

McFarlane ID (1982) Calliactis parasitica. In: Shelton GAB (ed) Electrical conduction and behaviour in 'simple' invertebrates. Clarendon Press, Oxford, pp 243-265.

McFarlane ID (1983) Nerve net pacemakers and phases of behaviour in the sea anemone Calliactis parasitica. J Exp Biol. 104: 231-246.

McFarlane ID (1984) Nerve nets and conducting systems in sea anemones: two pathways excite tentacle contractions in Calliactis parasitica. J Exp Biol 108: 137-149.

MaFarlane ID, Lawn ID (1972) Expansion and contraction of the oral disk in the sea anemone Tealia felina. J Exp Biol 57: 633-649.

McNair GT (1923) Motor reactions of the fresh-water sponge Ephydatia fluviatilis. Biol Bull 44: 153-166.

Meinhardt H, Gierer A (1974) Application of a theory of biological formation based on lateral inhibition. J Cell Sci 15: 321-346.

Mills CE, Mackie GO, Singla CL (1985) Giant nerve axons and escape swimming in Amphogona apicata with notes on other hydromedusae. Can J Zool 63: 2221-2224.

Moss AG, Tamm SL (1986) Electrophysiological control of ciliary motor responses in the ctenophore Pleurobrachia. J Comp Physiol A 158: 311-330.

Ohtsu K (1983) Antagonizing effects of ultraviolet and visible light on the ERG from the ocellus of Spirocodon saltatrix. J Exp Biol 105: 417-420.

Ohtsu K, Yoshida M (1973) Electrical activities of the anthomedusan Spirocodon saltatrix (Tilisius). Biol Bull 145: 532-547.

Pantin CFA (1935a) The nerve-net of the Actinozoa. I. Facilitation. J Exp Biol 12: 119-138.

Pantin CFA (1935b) The nerve-net of the Actinozoa. II. Plan of the nerve net. J Exp Biol 12: 139-155.

Pantin CFA (1935c) The nerve-net of the Actinozoa. III. Polarity and afterdischarge. J Exp Biol 12: 156-164.

Pantin CFA (1935d) The nerve-net of the Actinozoa. IV. Facilitation and the 'staircase'. J Exp Biol 12: 389-396.

Pantin CFA (1952) The elementary nervous system. Proc R Soc B 140: 147-168.

Pantin CFA, Vianna Dias M (1952) Rhythm and after discharge in medusae. Anais Acad Bras Cienc 24: 351-364.

Passano LM (1965) Pacemakers and activity patterns in medusae: homage to Romanes. Amer Zool 5: 465-481.

Passano LM (1973) Behavioral control systems in medusae; a comparison between hydro- and scyphomedusae. Publ Seto Mar Biol Lab 20: 615-645.

Passano LM (1982) Scyphozoa and cubozoa. In: Shelton GAB (ed) Electrical conduction and behaviour in 'simple' invertebrates. Clarendon Press, Oxford, pp 149-202.

Passano LM, Mackie GO, Pavans de Ceccatty M (1967) Physiologie du comportement de l'Hydromeduse Sarsia tubulosa Sars. Les systèmes des activités spontanées. CR Hebd Seanc Acad Sci, Paris 264: 614-617.

Patton ML, Passano LM (1972) Intracellular recording from the giant fiber nerve-net of a scyphozoan jellyfish. Amer Zool 12: 35.

Pavans de Caccatty M (1955) Le système nerveux des Eponges. Ann Sci Nat Zool 17: 203-298.

Pavans de Caccatty M (1962) Système nerveux et intégration chez les spongiaires. Ann Sci Nat Zool 4: 127-137.

Pavans de Ceccatty M (1969) Les systèmes des activités motrices, spontanées et provoquées des Eponges. C R Hebd Seanc Acad Paris 269: 596-599.

Pavans de Ceccatty M (1974) Coordination in sponges. The foundations of integration. Amer Zool 14: 895-903.

Pavans de Ceccatty M, Coraboeuf E (1960) Les réactions motrices de l'éponge siliceuse Tethya lyncurium à quelques stimulations expérimentales. Vie et Milieu 11: 594-600.

Pickens PE (1969) Rapid contractions and associated potentials in a sand-dwelling anemone. J Exp Biol 51: 513-528.

Rees WJ (1966) The evolution of the Hydrozoa. In: Rees WJ (ed) The Cnidaria and their evolution. Academic Press, pp 199-222.

Reiswig HM (1971) In situ pumping activities of tropical Demospongiae. Mar Biol 9: 38-50.

Reiswig HM, Mackie GO (1983) Studies on hexactinellid sponges. III. The taxonomic status of Hexactinellida within the Porifera. Phil Trans R Soc Lond B 301: 419-428.

Roberts A, Mackie GO (1980) The giant axon escape system of a hydrozoan medusa, Aglantha digitale. J Exp Biol 84: 303-318.

Robson EA (1963) The nerve-net of a swimming anemone, Stomphia coccinea. Q J Micrsc Sci 104: 535-549.

Robson EA (1965) Some aspects of the structure of the nervous system in the anemone Calliactis. Amer Zool 5: 403-410.

Robson EA, Josephson RK (1969) Neuromuscular properties of mesenteries from the sea-anemone Metridium. J Exp Biol 50: 151-168.

Romanes GJ (1876) The Croonian Lecture. Preliminary observations on the locomotor system of medusae. Phil Trans R Soc 166: 269-313.

Romanes GJ (1878) Further observations on the locomotor system of medusae. Phil Trans R Soc 167: 659-752.

Ross DM (1957) Quick and slow contractions in the isolated sphincter of the sea anemone, Calliactis parasitica. J Exp Biol 34: 11-28.

Ruben P, Johnson JW, Thompson S (1986) Analysis of FMRFamide effect on Aplysia bursting neurons. J Neurosci 6: 252-259.

Satterlie RA (1978) Feeding mechanisms in the ctenophore Pleurobrachia pileus. Biol Bull Mar Biol Lab, Woods Hole 155: 464.

Satterlie RA (1979) Central control of swimming in the cubomedusan jellyfish Carybdea rastonii. J Comp Physiol 133: 357-367.

Satterlie RA (1985) Central generation of swimming activity in the hydrozoan jellyfish Aequorea aequorea. J Neurobiol 16: 41-55.

Satterlie RA, Case JF (1978) Gap junctions suggest epithelial conduction within the comb plates of the ctenophore Pleurobrachia bechei. Cell Tissue Res 193: 87-91.

Satterlie RA, Case JF (1979) Neurobiology of the gorgonian coelenterates, Muricea californica and Lophogorgia chilensis. J Exp Biol 79: 191-204.

Satterlie RA, Spencer AN, (1979) Swimming control in a cubomedusan jellyfish. Nature 281: 141-142.

Satterlie RA, Spencer AN (1983) Neuronal control of locomotion in hydrozoan medusae: a comparative study. J Comp Physiol 150: 195-207.

Satterlie RA, Anderson PAV, Case J (1976) Morphology and electrophysiology of the through-conducting systems in pennatulid coelenterates. In: Mackie GO (ed) Coelenterate ecology and behavior. Plenum Publ Corp, New York, pp 619-627.

Satterlie RA, Anderson PAV, Case JP (1980) Colonial coordination in anthozoans: Pennatulacea. Mar Behav Physiol 7: 25-46.

Schaller HC (1976) Head regeneration is initiated by the release of head activator. Wilhelm Roux's Arch Dev Biol 180: 287-295.

Schaller HC, Gierer A (1973) Distribution of the head-activating substance in Hydra and its localization in membranous particles. J Embryol Exp Morphol 29: 39-52.

Schaller HC, Schmidt T, Grimmelikhuijzen CJP (1979) Separation and specificity of action of four morphogens from Hydra. Wilhelm Roux's Dev Biol 186: 139-149.

Shelton GAB (1975a) Colonial behaviour and electrical activity in the Hexacorallia. Proc R Soc B 190: 139-256.

Shelton GAB (1975b) Colonial conduction systems in the Anthozoa: Octocorallia. J Exp Biol 62: 571-578.

Shelton GAB (1975c) The transmission of impulses in the ectodermal slow conduction system of the sea anemone Calliactis parasitica (Couch). J Exp Biol 62: 421-432.

Shelton GAB (1980) Lophelia pertusa (L.): electrical conduction and behaviour in a deep water coral. J Mar Biol Assn UK 60: 517-528.

Shelton GAB (1982) Anthozoa. In: Shelton GAB (ed) Electrical conduction and behaviour in 'simple' invertebrates. Clarendon Press, Oxford, pp 203-242.

Shelton GAB, Holley MC (1984) The role of a 'local electrical conduction system' during feeding in the Devonshire cup coral Caryophyllia smithii Stokes and Broderip. Proc R Soc Lond B 220: 489-500.

Shelton GAB, McFarlane IK (1976) Electrophysiology of two parallel conducting systems in the colonial Hexacorallia. Proc R Soc 193: 77-87.

Sibaoka T (1966) Action potentials in plant organs. Symp Soc Exp Biol 20: 49-74.

Singla CL (1978a) Fine structure of the neuromuscular system of Polyorchis penicillatus (Hydromedusae: Cnidaria). Cell Tissue Res 193: 163-174.

Singla CL (1978b) Locomotion and neuromuscular system of Aglantha digitale. Cell Tissue Res 188: 317-327.

Sleigh MA (1972) Features of ciliary movement of the ctenophores Beroe, Pleurobrachia and Cestus. In: Clarkand RB, Wootton RM (eds) Essays on hydrobiology. Exeter University Press, pp 119-136.

Sleigh MA (1974) Metachronism of cilia of metazoa. In: Sleigh MA (ed) Cilia and flagella. Academic Press, New York, pp 287-304.

Spencer AN (1971) Myoid conduction in the siphonophore Nanomia bijuga. Nature 223: 490-491.

Spencer AN (1974) Non-nervous conduction in invertebrates and embryos. Amer Zool 14: 917-929.

Spencer AN (1975) Behavior and electrical activity in the hydrozoan Proboscidactyla flavicirrata (Brandt). II. The medusa. Biol Bull 148: 236-250.

Spencer AN (1978) Neurobiology of Polyorchis. I. Function of effector systems. J Neurobiol 9: 143-157.

Spencer AN (1979) Neurobiology of Polyorchis. II. Structure of effector systems. J Neurobiol 10: 95-117.

Spencer AN (1981) The parameters and properties of a group of electrically coupled neurones in the central nervous system of a hydrozoan jellyfish. J Exp Biol 93: 33-50.

Spencer AN (1982) The physiology of a coelenterate neuromuscular synapse. J Comp Physiol 148: 353-363.

Spencer AN, Arkett SA (1984) Radial symmetry and the organization of central neurones in a hydrozoan jellyfish. J Exp Biol 110: 69-90.

Spencer AN, Satterlie RA (1980) Electrical and dye coupling in an identified group of neurons in a coelenterate. J Neurobiol 11: 13-19.

Spencer AN, Satterlie RA (1981) The action potential and contraction in subumbrellar swimming muscle of Polyorchis penicillatus (Hydromedusae). J Comp Physiol 144: 401-407.

Spencer AN, Schwab WE (1982) Hydrozoa. In: Shelton GAB (ed) Electrical Conduction and behaviour in 'simple' invertebrates. Claredon Press, Oxford, pp 73-148.

Stein PG, Anderson PAV (1984) Maintenance of isolated smooth muscle cells of the ctenophore Mnemiopsis. J Exp Biol 110: 329-334.

Tamm SL (1973) Mechanisms of ciliary coordination in ctenophores. J Exp Biol 59: 231-245.

Tamm SL (1979) Ionic and Structural basis of ciliary reversal in ctenophores. J Cell Biol 83: 174a.

Tamm SL (1982) Ctenophora. In: Shelton GAB (ed) Electrical conduction and behaviour in 'simple' invernebrates. Clarendon Press, Oxford pp 266-358.

Tamm SL, Tamm S (1981) Ciliary reversal without rotation of axonemal structures in ctenophore comb plates. J Cell Biol 89: 495-509.

Tardent P, Weber C (1976) A qualitative and quantitative inventory of nervous cells in Hydra attenuata Pall. In: Mackie GO (ed) Coelenterate ecology and behavior. Plenum Press, New York, pp 501-512.

Weber C (1982) Electrical activities of a type of electroretinogram recorded from the ocellus of a jellyfish, Polyorchis penicillatus (Hydromedusae). J Exp Zool 223: 231-243.

Weber C, Singla CL, Kerfoot PAH (1982) Microanatomy of the subumbrellar motor innervation in Aglantha digitale (Hydromedusae: Trachylina). Cell Tissue Res 223: 305-312.

Westfall JA (1987) Ultrastructure of invertebrate synapses. (This volume).

Wintermann G (1951) Entwicklungsphysiologische Intersuchungen an Susswasserschwammen. Zool Jahrb Abt Anat Ont Tiere 71: 427-486.

Wolpert L (1969) Positional information and the spatial pattern of cell differentiation. J Theor Biol 25: 1-47.

Wolpert L, Hornbruch A, Clarke MRB (1974) Positional information and positional signalling in Hydra. Amer Zool 14: 647-663.

Yu SM, Westfall JM, Dunne JF (1985) Light and electron microscopic localization of a monoclonal antibody in neuron in situ in the head region of Hydra. J Morphol 184: 183-193.

ORGANISATION AND DEVELOPMENT OF THE PERIPHERAL NERVOUS SYSTEM IN ANNELIDS

SUSANNA ELIZABETH BLACKSHAW

Institute of Physiology

University of Glasgow

Glasgow G12 8QQ

Scotland

United Kingdom

ABSTRACT

In adult leeches the properties of centrally located motoneurones and skin mechanosensory neurones are known in detail as are the territories that they innervate in the periphery and the morphology of the sensory terminals. The development of receptive fields by individual mechanosensory cells has been followed in leech embryos from the time of initial axon outgrowth, as has the regeneration of fields in the adult nervous system after peripheral nerve lesions, and the way in which receptive fields spread after deletion of neighbouring neurones. The use of immunocytological techniques has enabled peptides and amines to be localised within specific neurones that act to modulate the activity of peripheral targets. In addition to the well-characterised centrally located neurones a number of peripheral neurones in Hirudo have been identified and their functions established, including large diameter stretch receptor afferents innervating body wall muscle and small diameter chemosensory, photosensory and water-movement detecting neurones located within specialised epidermal sense organs. The development of new techniques for cell lineage analysis has made it possible to study the embryonic origin of the peripheral nervous system as well as that of the CNS.

1. INTRODUCTION

The phylum Annelida is made up of 3 main classes: the Polychaeta, Oligochaeta and Hirudinea (Dales 1967). The polychaetes, the first and largest class, are essentially the marine annelids. There are 2 loose groups, the errant polychaetes such as the ragworm, Nereis virens, and the specialised tube dwellers, the sedentary fan and peacock worms. The oligochaetes include the familiar earthworm. They are a much more homogeneous group although they range in habitat from glaciers to jungles. The Hirudinea or leeches contain the smallest number of species (Mann 1962; Sawyer 1986). Their most familiar member is the medicinal leech Hirudo medicinalis. The central nervous system (CNS) of this annelid has been used extensively as a preparation to study the properties, ontogeny or regeneration of identified nerve cells (for reviews see Muller et al. 1981; Stent and Weisblat 1985; Nicholls 1987; Blackshaw 1987) because of the accessibility of neurones and glia within the ventral nerve cord, and the reliability with which individual sensory, motor and interneurones can be recognised from segment to segment, and from animal to animal. This is not the case with the CNS of polychaetes and oligochaetes where the ganglia are gradual and inconspicuous bulges in the cord and where individual nerve cells are not easily visible in the living ganglion in the light microscope.

Studies on Hirudo and related species of leeches show that each region of the adult CNS contains a characteristic number and arrangement of neurones, a unique subset of cell types and a generally unique pattern of connexions. The peripheral nervous system, though similarly stereotyped is less well characterised. Throughout the phylum there are either 3 or 4 circumferential nerve trunks which emerge bilaterally from the ganglion in each segment and form a complete ring meeting in the mid-dorsal line. These segmentally repeating nerve trunks and their branches contain the axons of CNS neurones and the centrally directed axons of neurones situated in the periphery. In leeches the properties of motoneurones and skin mechanosensory neurones whose cell bodies are located centrally are known in detail, as are the territories that they innervate in the periphery and the morphology of their terminals. Far less in known about the function and projection of neurones whose cell bodies lie outside the CNS, some of which lie along the course of peripheral nerves; others are located within the epidermis or in specialised sense organs. Earlier work on peripheral sensory neurones in polychaetes and oligochaetes is reviewed in Dorsett (1976) and Mill (1982). Recently additional small and large diameter afferent neurones have been identified in leeches and their functions established. These include stretch receptor neurones associated with body wall muscles, and ciliated chemosensory and water-movement detecting neurones located in specialised epidermal sense organs. The use of immunocytological techniques has enabled peptides and amines to be localised within specific neurones that act to modulate the activity of peripheral targets. In addition new features of the organisation of axons within peripheral nerves are emerging as a result of the production of antibodies that label specific groups of neurones; and the development of new techniques for cell lineage analysis had made it possible to study the embryonic origin of the peripheral nervous system as well as that of the CNS.

2. PERIPHERAL FIELDS AND SENSORY TERMINALS OF CENTRALLY LOCATED MECHANOSENSORY NEURONES

In the leech, a small number of centrally located neurones conveys information about mechanical stimulation of the body wall. There are three kinds of cells, totalling 14 per ganglion, responding specifically and selectively to light touch (T cells), pressure (P cells) or noxious (N) stimulation (Fig. 1). Even before the anatomical evidence was available (see below), a number of physiological experiments indicated that these cells in the leech are true sensory cells, rather than second or third order neurones driven only indirectly by sensory cells in the periphery (Nicholls and Baylor 1968). In the head ganglion of Hirudo (Yau 1976a), in the segmental ganglia of widely different species of leeches such as Haementeria (Kramer and Goldman 1981) and in the ventral nerve cord of oligochaetes (Mill and Knapp 1967), similar cells with similar properties play corresponding roles.

The striking feature of the innervation of the skin by these 14 cells is its orderliness. In each of the 21 segmental ganglia along the length of a leech individual T, P or N cells innervate the same specific areas of skin. For example, there are 6 T sensory cells in each ganglion symmetrically arranged in pairs, and each of the three cells on one side innervates a discrete area of either dorsal, lateral or ventral skin (Nicholls and Baylor 1968).

The mechanoreceptor terminals have been visualised directly in whole mounts of the body wall by injecting horse radish peroxidase (HRP) into the cell bodies of individual neurones in the ganglion and allowing it to

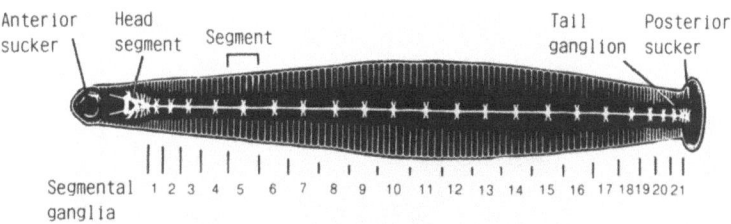

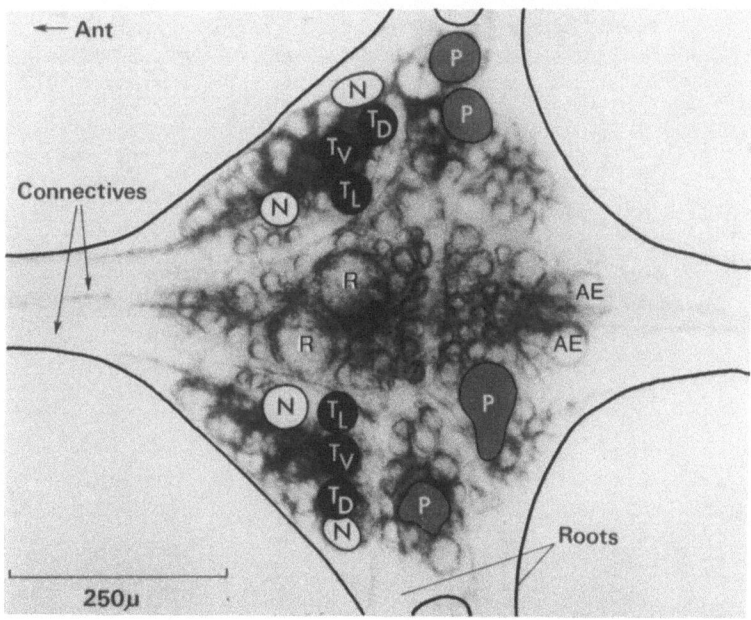

FIG. 1: The medicinal leech, like other annelids, has both the body and the nervous system made up of very similar repeating units. a. Externally each segment is divided into annuli (5 in the mid-body region). The ventral nerve cord consists of a chain of 21 segmental ganglia, with some concentration of ganglia at head and tail ends.
b. Segmental ganglion viewed from the ventral aspect. Cells labelled include the T, P and N mechanosensory cells which respond to touch, pressure of mechanosensory stimulation of the skin; the AE motoneurones which innervate the muscles responsible for raising the annuli into ridges, and the giant serotonin-containing Retzius cells (R). Individual touch cells innervate specific areas of ventral (T_V) lateral (T_L) or dorsal (T_D) skin (after Nicholls and Baylor 1968).

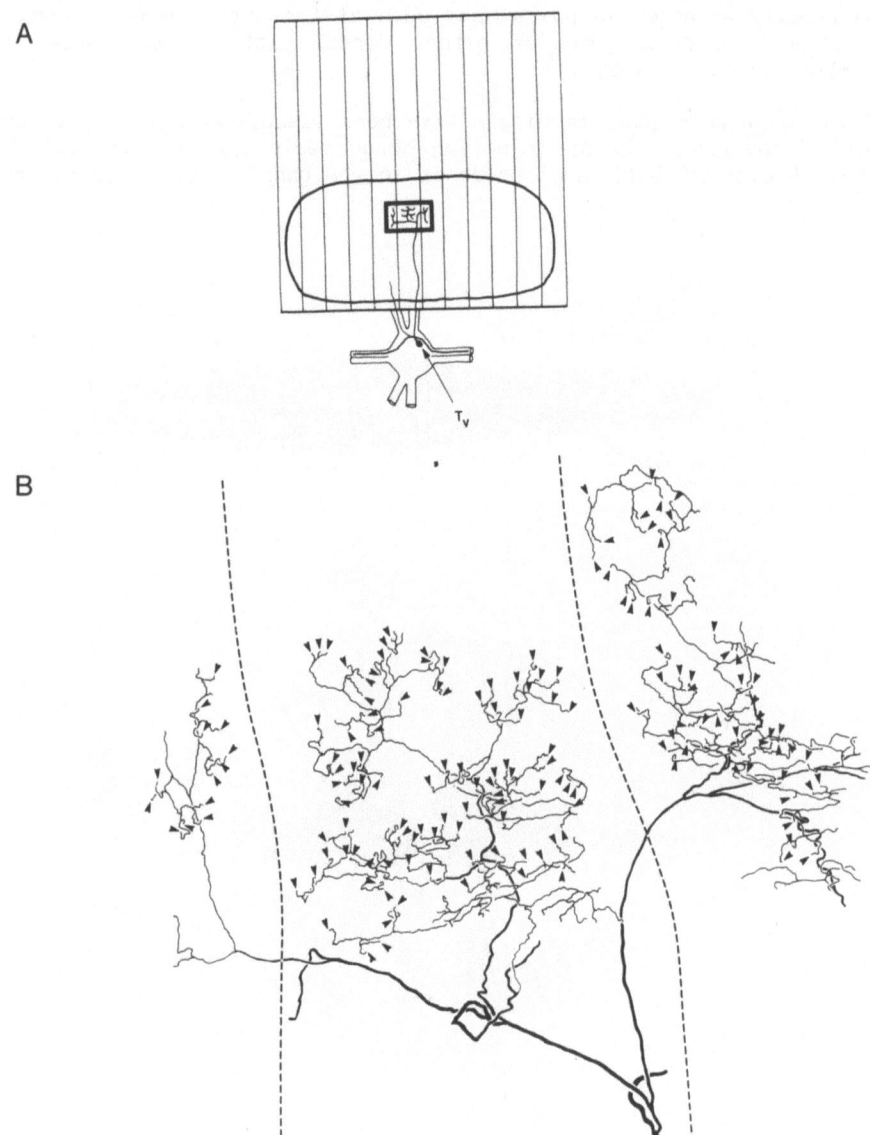

FIG. 2: Distribution of T-cell terminals in ventral skin. (A). Schematic diagram showing the cell body of the T cell, T_v in the ganglion, and its receptive field (oval) in the ventral surface of the leech. (B) Camera lucida drawing of the arborization and endings (▼) of the neurone within 1 mm^2 of skin from the centre of the cell's receptive field contained in the rectangle in A. The T cell made 178 endings within this central part of its receptive field; 108 endings were located on the central annulus, and 70 were on part of the annuli on either side. (Reprinted, with permission, from Blackshaw 1981a).

spread to the skin. In this way it has been possible to construct maps of the number of terminals and their distribution within the cells receptive fields (Blackshaw 1981a). Thus, T cell axons branch where the nerve roots divide in the body wall and become progressively finer as they dip between the layers of body wall muscle. When they reach the layer of epithelial cells in the skin, they branch extensively forming beaded chains that turn between the epithelial cells to end 1-2 μm from the skin surface in intercellular spaces immediately below the juctional complex at the outer ends of the epithelial cells. The distance at the skin surface between neighbouring terminals of the same axon (between 15 μm and 150 μm) fits well with the distribution predicted in the earlier physiological experiments. Counts of the number of terminals of one axon branch show that a T cell makes about 200 terminals within 1 mm^2 of skin in the centre of its territory (Fig. 2). Because T cell receptive fields extend longitudinally over 3 body segments, covering an area of 10-20 mm^2 of skin, a single neurone may have as many as 1000 terminals distributed throughout its receptive field (Fig. 3). Since each of these terminals can independently transduce a mechanical stimulus to the skin, this raises the problem of how the position of a stimulus is discriminated in the leech.

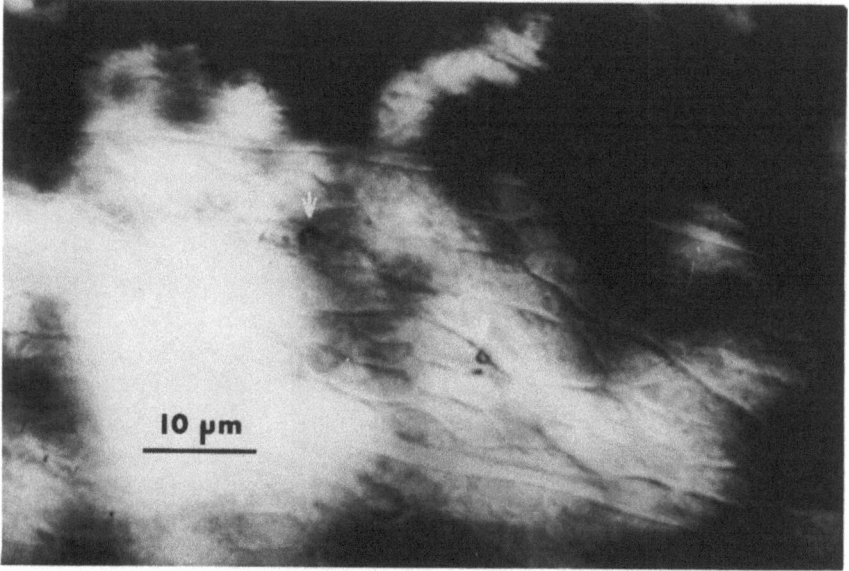

FIG. 3: Touch mechanosensory cell endings visualised in whole mounts of the body wall after injection of horseradish peroxidase into their cell bodies in the ganglion. Nomarski photograph of the surface of ventral skin showing the profiles of the skin epithelial cells and two adjacent terminals of the T cell innervating ventral skin lying between the epithelial cells at the skin surface.

The other modalities of mechanosensory neurone, the P and N cells, have receptive fields with boundaries that are different from those of T cells and sensory terminals with a different morphology (Nicholls and Baylor 1968; Blackshaw 1981b, c; Kramer and Goldman 1981; Blackshaw et al. 1982a; Johansen et al. 1984a). N cell axons in the main nerve roots give rise to fine-calibre branches about 1 μm in diameter that run within the network of peripheral nerves at deep levels of the body wall. Some of these fine branches can be traced to superficial layers of the body wall, where they run at the base of the layer of epithelial cells in the skin and end as unspecialised fine processes. Fine branches also encircle the nephridiopore in each body segment. In addition, at specific sites within the body wall, the N cell situated laterally within the ganglion makes distinctive coiled terminals over the sensory terminals of large peripheral stretch receptor neurones that lie along the course of peripheral nerves (see Section 7). Electrophysiological experiments have shown that the N cell is presynaptic to the peripheral neurones. Action potentials in the nociceptive cell are followed on a 1:1 basis and with a constant latency by excitatory synaptic potentials recorded in the cell

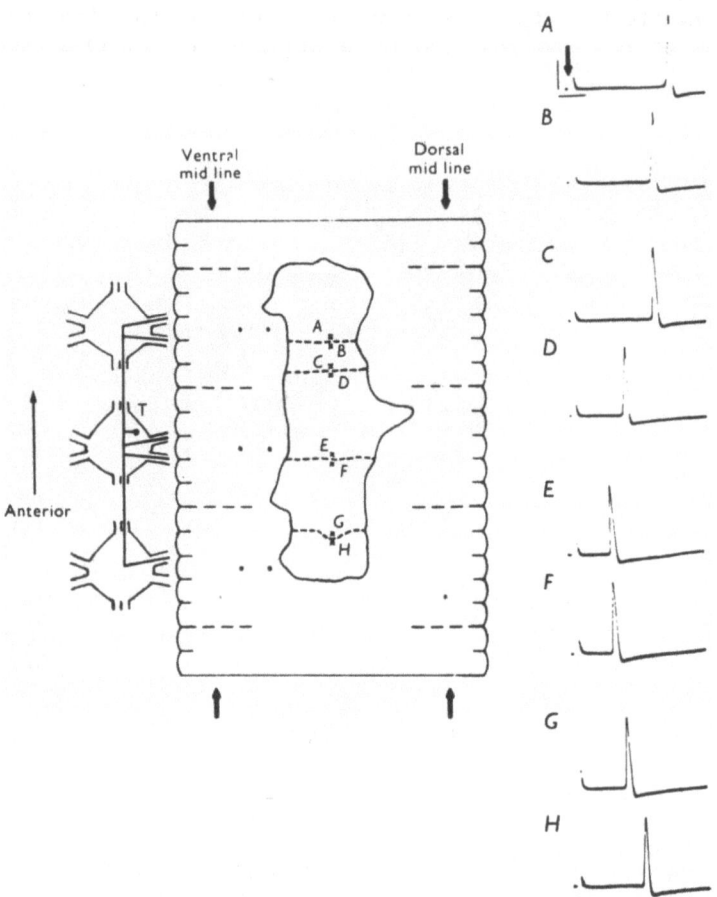

FIG. 4: Subfields of a touch cell that innervated lateral skin. Each subfield was innervated by a separate branch of the cell passing through either a root of its ganglion or that of an adjacent ganglion. Adjacent subfields had negligible overlap with each other, as indicated by discrete jumps in the time delay of intracellularly recorded action potentials (records on the right) when a mechanical stimulus was moved across the boundary between two adjoining subfields. Vertical calibration, 20 mV. Horizontal calibration, 10 ms. Arrow indicates the time when the mechanical stimulus was applied. (From Yau 1976b).

body of the peripheral neurone (Blackshaw et al. 1984; Thompson 1986. The role of this direct interaction in the periphery between the nociceptive cell and the stretch receptor is not clear but one possibility is that the N cell modulates the level of excitation of the sensory terminals of the stretch receptor cells via a peripheral axon reflex.

For touch and pressure cells, the specificity of skin innervation is carried further in that each cutaneous branch of the axon innervates a discrete area on its own (Nicholls and Baylor 1968; Yau 1976b; Kramer and Goldman 1981). The large calibre axon branches in the nerve roots of the cell's own ganglion supply major fields in that body segment. There are in addition fine calibre axon branches in the anterior and posterior connectives that innervate accessory fields in neighbouring body segments via the nerve roots of adjacent ganglia. The boundaries between the subfields are sharp and usually correspond to the edge of an annulus (Fig. 4). Thus the innervation pattern of a single neurone and its numerous branches appears like a patchwork quilt, each branch innervating a discrete territory in the skin, with no overlap by other branches of the same cell.

Although different branches of the same axon do not overlap, the fields of homologous touch cells in neighbouring ganglia overlap extensively (Fig. 5). There is also a small amount of overlap dorso-ventrally between the fields of T cells within a ganglion. Thus innervation of a patch of skin by a touch cell does not preclude the presence of other cells, or of pressure or nociceptive cells.

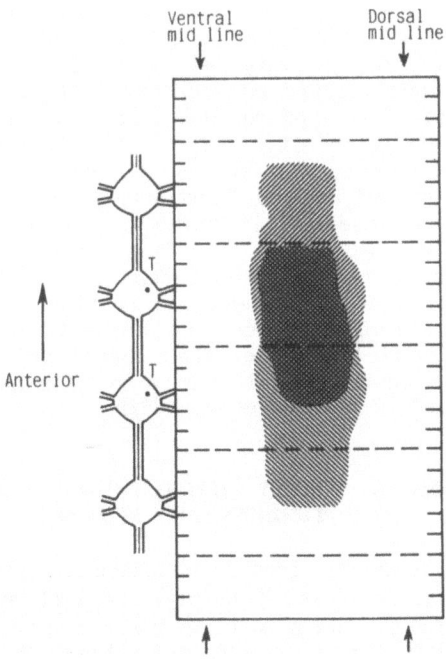

FIG. 5: Receptive fields of two T cells in adjacent ganglia that both innervate lateral skin. Each receptive field spans 12-13 annuli over three body segments, and, consequently, there is extensive overlap between the receptive fields. The size and appearance of T-cell receptive fields on ventral and dorsal skin are very comparable (From Yau 1976b).

In the suboesophageal ganglion in the head of Hirudo, neurones have been identified that are homologous to the T, P and N cells in the segmental ganglia, with similar electrical properties and responses to skin stimulation (Yau 1976a). The suboesophageal ganglion is derived by secondary fusion of ganglia during embryonic development, and in the head region the regular arrangement of body segments with 5 annuli is disrupted by the loss of annuli from particular segments and by the presence of the mouth and anterior sucker on the ventral aspects (see Sections 9 and 10). The head mechanosensory neurones innervate territories on the external surface of the head and on the non-pigmented skin of the oral folds on the interior of the mouth. Whole mounts of skin from the oral folds have a quilted appearance: the skin epithelial cells are grouped into islands and the T cell terminals emerge at the skin surface, one at the centre of each island of skin cells with a spacing between individual terminals of between 30 and 50 um. (Blackshaw 1981a).

3. REINNERVATION OF SKIN AND SPREAD OF RECEPTIVE FIELDS AFTER DELETION OF SINGLE CELLS

After cutting or crushing peripheral nerve roots, the mechanosensory axons will regenerate to reinnervate the skin (Van Essen and Jansen 1977). Moreover, the repair usually takes place with a high degree of precision. Invariably, the cells regain their appropriate modality, and although some cells show clear abnormalities in the extent or position of their receptive fields after regeneration, several T cells completely re-establish their peripheral fields so that the pattern of innervation after regeneration resembles that originally laid down during development.

These features of receptive field organisation raise a number of questions as to how the innervation is established. For example how does a neurone recognise its territory and what limits its spread? During development of the leech embryo do neurones initially innervate only the appropriate territory of skin or do they also innervate inappropriate skin from which their axons later withdraw? Do neighbouring cells of the same modality compete for territory in the skin? And if so, can the receptive field of a neurone spread if the surrounding cells are removed without damage to the neurone itself? If three out of the four N cells in a segmental ganglion are killed by protease injection and the leech is subsequently allowed to recover, the receptive field of the remaining N cell expands to take over the denervated territory on the contralateral side of the leech, a region it does not normally innervate. Similarly, the field of the T cell that innervates dorsal skin spreads across the midline to innervate contralateral skin after the three T cells on that side have been deleted (Blackshaw et al 1982b).

4. DEVELOPMENT OF SENSORY FIELDS: PRIMARY AXON OUTGROWTH BY P MECHANOSENSORY NEURONES

Innervation of skin by identified pressure sensitive neurones in embryos of the glossiphoniid leech Haementeria ghilianii has been followed from the time of initial outgrowth of their axons by injecting Lucifer Yellow intracellularly at successive developmental stages until the receptive field has achieved its adult configuration (Kuwada and Kramer 1983). Each segmental ganglion in Haementeria contains two pairs of pressure cells, each of which innervates a specific area of dorsal or ventral skin in adult leeches. During embryogenesis the ventral germinal plate near the ganglion is the target of the P_V neurone and the more distal germinal plate the target of the P_D neurone. These experiments show that P neurones send their first of 'primary' peripheral axons directly to their separate targets and begin to innervate them at approximately the same time. Thus, the P_D neurone shows an early preference in embryogenesis for dorsal skin, growing across ventral germinal plate directly to dorsal germinal plate despite an opportunity to innervate ventral germinal plate, the target of the P_V neurone. The specificity of the P neurones is therefore not accounted for by temporal differences in the outgrowth of the primary axons as has been shown for the formation of appropriate neuronal connexions in the developing arthropod visual system (Anderson 1978; Macagno 1978). The smaller secondary subfields develop later from secondary and intersegmental axons. These axons too grow directly to their appropriate skin territories and their arborisation expands until the adult receptive field pattern is established late in embryogenesis. They do not appear to initially overgrow and later trim down to the normal boundaries of the receptive field.

Not only do the neurones grow directly to target territory but they arborise in that territory in a highly stereotyped way as if the

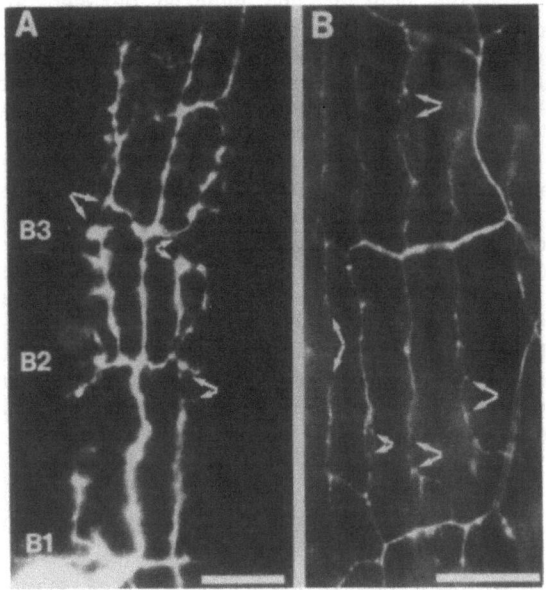

FIG. 6: Growth-limiting self-encounters of P_V neurone axons, as seen among peripheral axons of Lucifer Yellow-filled P_V neurones photographed in whole mount. A, the major field axons at stage 10(2/5). Laterally directed second-order annular branches from B1 and B2 branches encounter the B2 and B3 branches, respect- ively (at arrows). These axons are spatially separated, but focusing through the preparation revealed apparent filopodial contacts between the encountering branches; filopodia projecting from ends of axons are blurred in the photograph. B, encounter of fourth-order circular branches in a neurone in the midbody at stage 11(2/20). Circular branches growing from neighbouring third-order longitudinal branches along the same circumferential path encounter each other at arrows (that point to ends of axons) and are separated. Blurred filopodia are visible between the ends of some pairs of circular branches. The thick, faintly fluorescent circumferential stripe beneath each row of circular branches is an autofluorescent circular muscle fibre. Calibration, 50 μm. (From Kramer and Kuwada 1983).

peripheral axons were growing along predetermined pathways (Kramer and Kuwada 1983). The primary axon branch grows out circumferentially, perpendicular to the long axis of the embryo, along that area of the germinal plate from which will arise the central annulus of the segmental skin. First order longitudinal branches emerge at characteristic locations around the circumference and grow anteriorly and posteriorly along the length of the leech. As these grow longitudinally, second order annular branches emerge circumferentially along the central portions of the future skin annuli (Fig. 6). This consistently rectilinear pattern appears to match the grid-like arrangement of muscle fibres in the body wall whose development precedes the outgrowth of the primary axon and it has been suggested that the muscles might delineate the peripheral pathways of differing CNS neurones (Torrence 1984; Torrence and Stuart 1986).

A striking feature of the arborisation within the receptive fields is that throughout embryogenesis, as in the adult leech, separate axon branches of the same neurone are virtually non-overlapping, in marked contrast to the extensive contact and overlap between axon branches of homologous P neurones. It appears that the sharp boundaries between the major and minor subfields are established because the major and minor field axons of the same neurone stop growing toward each other when they meet. Thus each axon branch excludes the other from the territory it already occupies. Kramer and Kuwada (1983) suggest that the peripheral axon arborisation is constrained by a process of neuronal self-avoidance. This idea has been tested directly by delaying or surgically preventing the outgrowth of minor or major field axon branches (Kramer and Stent 1985). Interference with the outgrowth of a minor field axon results in the spread of the major field axon branch into the territory normally occupied by the absent or delayed branch and vice versa. The mutual territorial exclusion obtains only for isoneuronal branches and consequently is likely to be a different mechanism to that underlying heteroneuronal exclusion found in adult leeches between axon branches of homologous mechanosensory neurones maintaining receptive fields in adjacent skin territories (Blackshaw et al. 1982b).

5. MOTOR NEURONE PROPERTIES AND PERIPHERAL FIELDS

The numbers of motoneurones in each segmental ganglion in leeches are small and individual neurones innervate the same specific territories in each segment. Thus, on one side of a ganglion, a single motoneurone, the AE motoneurone, is responsible for raising all the annuli on one side of the segment (Fig. 7); a single motor cell innervates a well defined group of circular muscle fibres (Stuart 1970), and a single motoneurone, the HE motoneurone, innervates the ipsilateral heart tube in each segment (Thompson and Stent 1976a, b). There are similarly small numbers of motoneurones in the segmental ganglia of earthworms (Gunther 1971; Mill 1982).

The overall organisation of the body wall musculature in tubular invertebrates such as the annelids bears a superficial resemblance to the arrangement of longitudinal and circular smooth muscle layers in viscera and blood vessels of vertebrates (Fig. 8). In the annelids an outer layer of circularly arranged fibres overlies an inner layer of longitudinal fibres. These muscle layers form continuous sheets along the length of the leech with no obvious segmental boundaries. In <u>Hirudo</u> in addition to the longitudinal and circular muscle layers there <u>is</u> an oblique muscle layer whose fibres are orientated at approximately 45° to the longitudinal

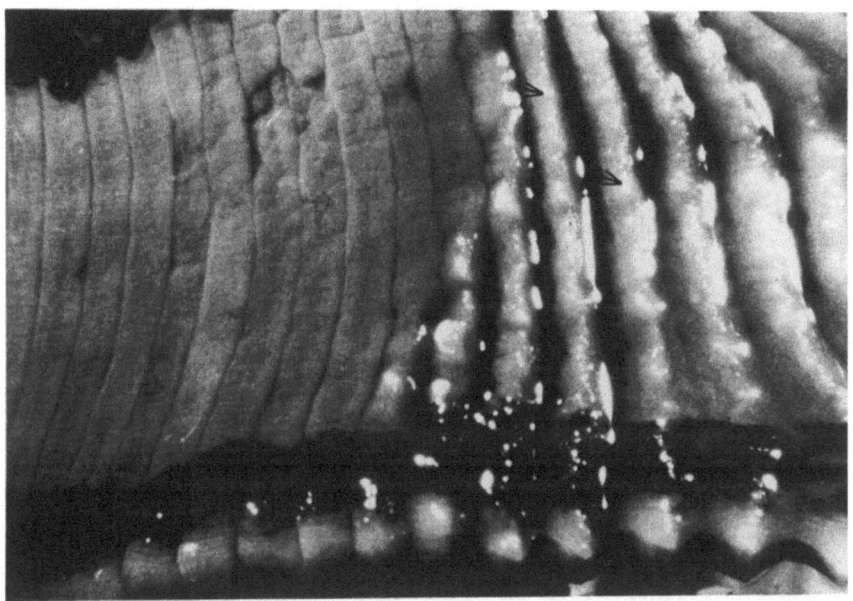

FIG. 7: Erection of annuli. The AE motor neurones on the ventral surface of the ganglion innervate muscle fibers which insert into skin on either side of an annulus. Contraction of the muscles produces erection of the annuli (↑).

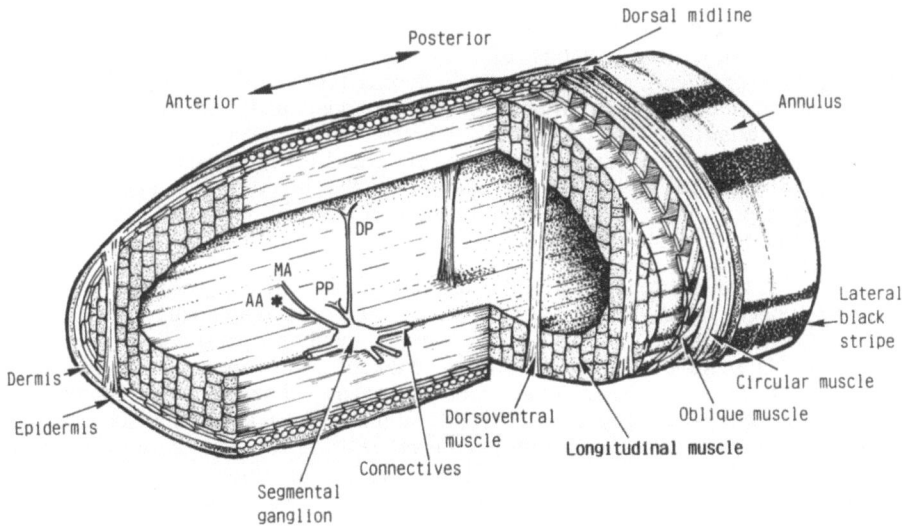

FIG. 8: Diagram of part of a leech to show the organisation of the body wall musculature and the relation of a segmental ganglion and its nerve roots (AA, MA, DP, PP: nomenclature of Ort et al. 1974) to the body wall. The viscera have been removed. Three layers of muscle make up the major part of the body wall. Immediately beneath the dermis is a layer of circularly arranged fibres. This overlies a thin oblique layer and they both overlie a layer of longitudinally arranged fibres. There are in addition dorsoventral or flattener muscles which run at right angles to the body axis.

body axis and dorso-ventral or flattener muscles that span the body from dorsal to ventral surface. Annelid muscles were long considered to be examples of smooth muscle and were classified as 'helical smooth' by Hanson and Lowy (1960). However electron microscopy shows that they comprise highly regularly ordered arrays of myofilaments and Hoyle (1983) has suggested that they should be classified as striated rather than smooth muscle cells.

The innervation of the longitudinal muscle layer in leeches is known in some detail. Each body segment of longitudinal muscle is innervated by 9 paired excitatory motoneurones; and four pairs of inhibitory motoneurones. Of the excitatory motoneurones four (cells 3, 5, 7 and 107) (Muller et al. 1981) cause contractions of dorsal longitudinal muscle fibres, one (cell 106) causes lateral contractions, three (cells 4, 8 and 108) cause ventral contractions and one, the large longitudinal motoneurone or L cell, causes simultaneous contraction of dorsal, lateral and ventral fibres on one side of a segment (Stuart 1970; Ort et al. 1974). Each of these motoneurones innervates a territory that extends longitudinally into adjacent body segments and there overlaps with the fields of homologous neurones in neighbouring ganglia; there is also a small amount of overlap between the fields of cells within a ganglion; consequently individual muscle fibres may be multiply innervated.

Excitatory transmission to leech longitudinal muscle is cholinergic (Fuhner 1917; Walker et al. 1968, 1970; Sargent 1977; Kuffler 1978; Wallace 1981a, b; Wallace and Gillon 1982). In addition to acetylcholine one of the motoneurones, the L cell that innervates fibres over the entire half segment, has been shown immunocytochemically to contain an FMRF-like peptide (Kuhlman et al. 1985a, b).

Individual motoneurones to the longitudinal muscle produce contractions with different kinetics. Each excitatory motoneurone has a characteristic functional range of firing frequencies over which changes in firing rate produce changes in peak tension or rate of rise in tension (Mason and Kristan 1982). Thus, of the motoneurones innervating dorsal longitudinal muscle, the L cell causes the strongest and fastest contractions over the narrowest range of impulse frequency; cells 5 and 107 are intermediate in frequency range, peak tension and rate of rise, and cells 3 and 7 have the broadest range of firing frequencies and evoke the weakest and slowest contractions.

The detailed distribution of axon terminals along individual muscle fibres has been studied for only one of the motoneurones, the L cell, (Kuffler 1978), and it is not known whether individual excitors innervating the same area of muscle contact the same or a different population of fibres within that general area. Enzyme and immunohistochemistry of body wall muscle in Hirudo show that within the longitudinal muscle layer fibres have different characteristics. For example the use of type-specific anti-myosin shows that fibres at the outer edge of the longitudinal muscle layer differ from the deep mass of muscle, and it is possible that these fibres may be selectively innervated by particular motoneurones (A. Rowlerson and S.E. Blackshaw, unpublished).

The effect of the inhibitory motoneurones is to reduce the force of contraction induced by the excitors rather than to reduce basal tension or alter relaxation following a contraction (Mason and Kristan 1982). Thus the excitatory motoneurones have to be active in order for the inhibitors to exert a peripheral effect. The inhibitory motoneurone also exerts effects centrally via connexions onto the excitatory motoneurones within the ganglion which are thought to be monosynaptic (Granzow et al. 1985). The different functional roles of peripheral and central inhibition by the inhibitory motoneurones have yet to be determined.

Two of the inhibitory motoneurones to the longitudinal muscle, cells 1 and 2 (Stuart 1970; Ort et al. 1974) have been shown to be GABAergic by double labelling experiments in which the inhibitors were first identified electrophysiologically and injected intracellularly with Lucifer Yellow, and then the ganglia subsequently exposed to ^{3}H-GABA - a technique which enables those neurones with high affinity uptake systems for GABA to be identified autoradiographically (Cline 1983; Cline et al. 1985). There is evidence that GABA is the inhibitory transmitter to the longitudinal musculature in the earthworm also (Ito et al. 1969; Hidako et al. 1969).

6. PERIPHERAL MODULATION OF MUSCLE

In addition to innervation by primary motoneurones, peripheral processes of the giant serotonin-containing Retzius cells end in the vicinity of the longitudinal muscle fibres. Whether Retzius axons end presynaptically on motoneurone terminals or on the muscle fibres themselves is not certain (Yaksta-Sauerland and Coggeshall 1973). A muscle modulatory role for amines has been proposed in other invertebrates (eg O'Shea and Evans 1979; Willard 1981) as well as in vertebrates (Bowman and Zaimis 1958). In Hirudo activity in the Retzius cells reduces the basal tension of leech longitudinal muscle and accelerates relaxation following contractions elicited by excitatory motoneurones (Mason and Kristan 1982) although it is not known whether they act by changing muscle fibre membrane properties or by changing the mechanical properties of the muscles. The effects of Retzius cell activity are mimicked by direct application of serotonin to the body wall (Schain 1961; Miller and Aidley 1973). Thus the effect of the Retzius cell, like that of the inhibitory motoneurones is to decrease longitudinal muscle tension. The type and time course of the effects however suggest different functions. The inhibitory motoneurones exert their peripheral effects relatively rapidly whereas the Retzius cells have effects with long latency and duration.

Other peripheral targets for 5-HT released by Retzius cells are the large unicellular glands in the body wall that secrete mucus (Mann 1962), and salivary gland cells (Marshall and Lent 1984; Lent 1985). The amount of mucus secreted by a patch of body wall increases in proportion to activity of the Retzius cells (Lent 1973). Varicosities and terminals containing large dense core vesicles are found in the body wall near mucus glands (Yaksta-Sauerland and Coggeshall 1973). Thus 5-HT released from terminals adjacent to mucus cells probably stimulates the mucus glands directly.

The tetra-peptide FMRF-amide has been localised in cholinergic motoneurones, other than the L cell notably the AE motoneurone, and the motoneurones to the heart tubes (Kuhlman et al. 1985a, b). The "heart" in Hirudo consists of two contractile lateral vessels or heart tubes that extend the entire length of the animal on either side and the heart beat is produced by the periodic contraction of the heart circular muscles (Thompson and Stent 1976a, b, c). The heart tubes are capable of myogenic activity in the absence of innervation but in the intact leech the constriction-dilation cycle of each segment of the heart tube is controlled by the activity of heart excitor (HE) motoneurones located within the ventral nerve cord. All the segmental ganglia except the first and last two contain a bilateral pair of HE cells each of which innervates the circular muscles of its ipsilateral heart tube. The heart tubes are also innervated directly by heart accessory (HA) modulatory neurones in ganglia 5 and 6 (Maranto and Calabrese 1984). Both the HE motoneurones and the HA modulatory neurones and their processes on the heart tubes show FMRF-amide-like fluorescence (Kuhlman et al. 1985a).

Bath application of FMRF-amide has a pronounced effect on the heart tubes in activating or accelerating the myogenic rhythm of an undriven heart and in increasing the beat tension of a heart entrained to HE motoneurone activity (Kuhlman et al. 1985b). These effects are very similar to those produced by activity in the HA modulatory neurones and it seems likely that the HA neurones produce their modulatory effects through release of a FMRF-amide like peptide. FMRF-amide thus may serve as a primary transmitter at these synapses. It is also clearly localised within HE and L motoneurones that use acetylcholine (ACh) as their primary transmitter and it here may serve as a co-transmitter. If co-release of ACh and FMRF-amide is a general phenomenon for leech motoneurones then there exists a role for FMRF-amide as a general modulator or regulator of muscle tension.

7. PERIPHERALLY LOCATED NEURONES

The ultrastructural organisation of the peripheral nerves in Hirudo was studied by Wilkinson and Coggeshall (1975) who showed that the majority of axons were small, only 1% having diameters greater that 1 μm. This work concentrated mainly on the discrepancy between the large numbers of small axons associated with each ganglion (20,000+; of which 9000 are in the paired anterior and posterior nerve roots and about 12000 are in the connectives) and the far smaller number of cell bodies found within a ganglion (400, Macagno 1980). A similar discrepancy between axon numbers and CNS cell body numbers was earlier noted by Ogawa (1934) for the earthworm. Since in Hirudo the majority of neurones within the ganglion are known to send only one axon into any particular nerve or connective the source of most of the axons was presumed to be extraganglionic arising from neurones located within the body wall although the modalities represented were not known (Havet 1899; Gaskell 1914; Fernandez 1978). A considerable number of these afferent axons in Hirudo have now been identified and their functions established.

7.1 Large diameter stretch receptor afferents

A prominent feature of the peripheral nerve roots in Hirudo, Macrobdella and Haemopis is the small number of large diameter axons located to one side of anterior and posterior roots (Fernandez 1978; Blackshaw et al. 1982a, b; Johansen et al. 1984b; Thompson 1986; Blackshaw and Thompson 1986, 1987). These axons are larger in diameter (10 μm) than any other in the peripheral or central nervous system, including the axon of the giant interneurone of the connective, Rohde's fibre or the S-cell, which is 3-4 μm in diameter and which produces the largest impulse that can be recorded from the connectives with extracellular electrodes (Gardner- Medwin et al. 1973; Frank et al. 1975). The large diameter axons in the anterior and posterior nerve roots in Hirudo have been shown to belong to peripheral neurones ("Ho" = Hoover or Homodromic) with characteristic fan-shaped terminals associated with the longitudinal muscle of the body wall whose morphology is highly suggestive of proprioceptive function (Blackshaw 1981b, c; 1985; Blackshaw et al. 1982a). Although there is behavioural and physiological evidence for the existence of proprioceptors signalling length or tension of body wall muscles in annelids, the appropriate structures have been elusive (Gray and Lissman 1938; Kristan and Stent 1976; Drewes and Fourtner 1976). Physiological experiments show that the large diameter afferents in Hirudo are indeed proprioceptors, signalling length or tension of body wall muscle. The cell bodies of origin occur at specific locations from segment to segment and from animal to animal along the course of particular nerve roots. They are distinct from the peripheral monoamine-containing neurones located within the body wall at the first branch point of each anterior root (Rude 1969; Lent 1981), whose function remains unknown. In Hirudo six Ho neurones innervate ventral longitudinal muscle in each segment and their axons project centrally from the body wall in a conspicuous group to one side of anterior and posterior roots, together with on each side of the leech 4 or 5 other similar sized axons which project from the dorsal body wall (see Fig. 15). In total there are between 14 and 20 of these neurones in each segment. The sensory terminals and the central projection of the axons have been visualised directly in whole mounts of body wall and ganglion by injecting Lucifer Yellow or horseradish peroxidase into the peripheral cell bodies (Fig. 9; Thompson 1986; Blackshaw and Thompson 1986, 1987). One of the neurones innervating ventral longitudinal muscle has been studied in detail. It lies along the course of nerves AA between the layers of longitudinal and oblique body wall muscle and within the sheath of the nerve. It has a distinctive morphology with two flattened, fan-shaped sensory terminals arranged in series separated by the cell body and a 300 µm long cylindrical process. Each of the sensory terminals is associated with the longitudinal muscle but with separate bands of fibres. The axon emerges from the proximal terminal and enters the segmental ganglion in the anterior nerve root where it arborises in a characteristic way within the neuropile of the ipsilateral half of the ganglion. In contrast to the cutaneous mechanosensory neurones in the leech, it does not project axon branches into the connectives to adjacent ganglia suggesting that direct synaptic connexions with motoneurones to body wall muscle are made only within the segment of origin.

The membrane properties and response characteristics of these neurones have been studied by intracellular recording from the cell body in the periphery and from the axon near the CNS. Although the peripheral cell body can generate overshooting action potentials these are not actively propagated to the CNS. Rather, imposed voltage changes spread decrementally along the axon, relaying information to CNS synapses in analogue form (Fig. 10). In this respect they resemble the non-impulsive stretch receptor neurones in the thorax of crabs which were among the earliest known examples of action at a distance as a result of electrotonic spread (Bush 1981). The large diameter leech afferents, like the crustacean non-spiking afferents, have membrane characteristics that might be expected of a decrementally conducting fibre with high values for specific membrane resistance (13,000 ohm.cm^2) and space constant (2.4 mm) (Thompson 1986; Blackshaw and Thompson 1986).

The leech afferents differ from all other stretch receptor neurones described however in that the transmembrane potential of the leech neurones increases during imposed stretch of the longitudinal muscle (Thompson 1986; Blackshaw and Thompson 1986, 1987. Stretch of the longitudinal muscle elicits a slow hyperpolarizing potential maintained for the duration of the stretch whose amplitude depends on the extent of the final displacement (Fig. 11). Release from stretch elicits a depolarizing potential. The neurones therefore communicate voltage differences of either sign, hyperpolarising to stretch and depolarising to release. The depolarising responses to ramp release have a dynamic component in which the rate of depolarisation of the neurone is determined by the rate of release from stretch, and a static component whose amplitude is dependent on the magnitude of the ramp length change (Fig. 12).

Stretch receptor neurone cell bodies in polychaetes and oligochaetes have not yet been identified although there were early reports of 'intermuscular nerve cells' in the earthworm with processes in the longitudinal and circular muscle layers (Dawson 1928), and spiking stretch-sensitive units have been recorded extracellularly from segmental nerves in the earthworm (Drewes and Fourtner 1976).

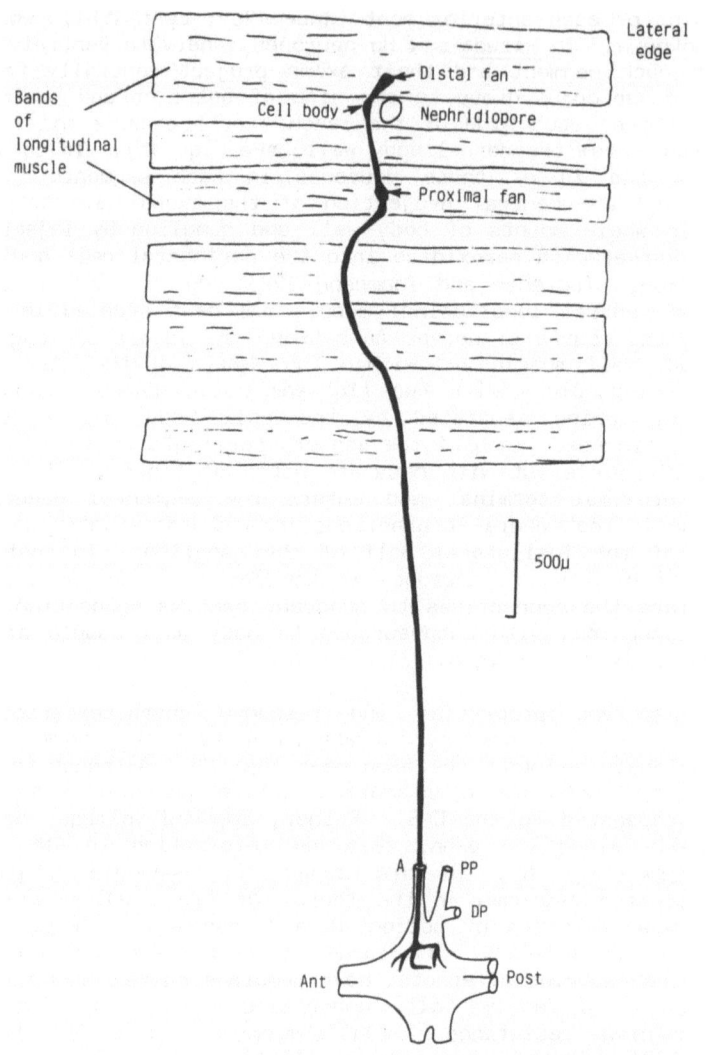

FIG. 9: Camera lucida drawing of the stretch receptor innervating ventral longitudinal muscle in the leech after intracellular injection of horseradish peroxidase. The neurone has two distinctive fan-shaped sensory terminals, the distal and proximal fans, which are approximately 70 μm across at their widest point but only a few micrometers deep. Both fans are associated with the longitudinal muscle of the ventral body wall, but with different bands of fibres. They are separated by the cell body and by a 10-15 μm diameter cylindrical process. A 10-15 μm diameter axon runs in the anterior nerve root for between 4 and 5mm before entering the segmental ganglion. It arborises in the ipsilateral half of the ganglion where it synapses with segmental motoneurones.
A, PP, DP; anterior, posterior and dorsal segmental nerve roots.

7.1.1. Small diameter afferents: Photoreceptors, chemoreceptors, water movement detectors and nephridial neurones

In addition to centrally located mechanosensory neurones which respond to mechanical stimulation of the generalised body surface via

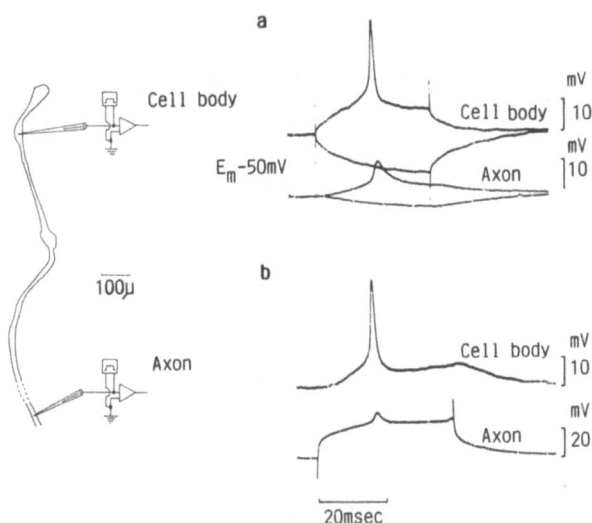

FIG. 10: Regenerative action potentials may be elicited from the cell body but not from the axon. The diagram on the left illustrates the experimental arrangement with paired intracellular recordings made from the cell body and from the axon 3mm distant. In (a) depolarising and hyperpolarising current pulses (not shown) were injected into the cell body (2 sweeps showing response of the cell body to depolarising and hyperpolarising current pulses of the same magnitude superimposed). The resulting voltage changes in the axon are shown on the bottom trace, again two sweeps superimposed. In (b) depolarising current was injected into the axon (bottom trace) while recording simultaneously in the cell body (upper trace). Current injected directly into the axon did no elicit an action potential in the axon. If sufficient current is injected, current spread from the axon elicits and action potential in the cell body which is reflected in the axon recording.

their distributed sensory terminals (Section 2) leeches have localised receptors; eyes and chemosensory structures on the head and the segmental sensilla, all of which contain the cell bodies of primary sensory neurones whose axons project centrally. The sensilla appear under the dissecting microscope as semitransparent circular areas approx 100 μm in diameter. In Hirudo there are 14 of them ranged round the middle annulus of each midbody segment. Each sensillum can be raised or retracted with respect to the body surface by papillar muscles (Whitman 1886). Both the eyes and the sensilla contain a small number of spherical refractile cells whose axons projet to the CNS and which mediate the response to light (Walther 1966; Lasanky and Fuortes 1969; Kretz et al. 1976). Friesen and colleagues have shown that the sensilla also contain mechanosensory neurones that respond specifically to water movement. There are 3 different types of ciliated cells (Derosa and Friesen 1981; Phillips and Friesen 1982) - one with a single long cilium that projects at least 12 μm from the sensillum beyond the cuticle of the body wall, and 2 kinds of multiciliated cells, one with between 2 and 4 shorted grouped cilia, the other with cilia projecting parallel to the body surface within the cuticle. Similar ciliated cells occur in earthworms (Knapp and Mill 1971; Moment and Johnson 1979) and in the cirri of nereid worms (Dorsett and Hyde 1969). Hirudo responds to stimulation produced by water waves by orientating and swimming towards the source of the waves (Young et al.

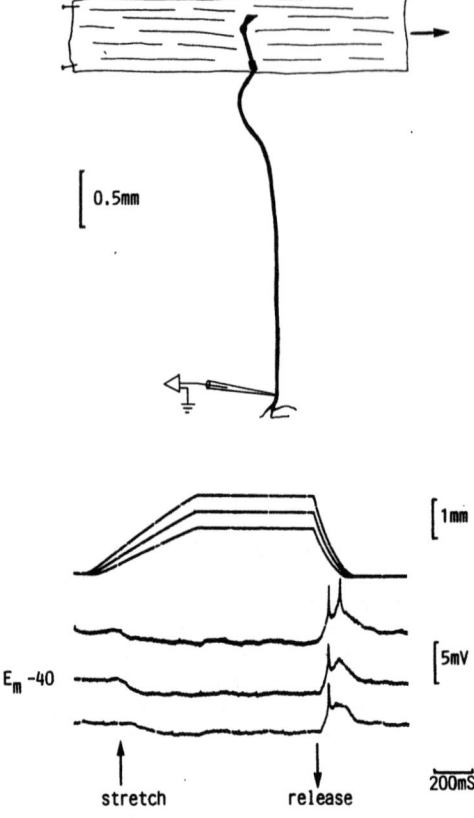

FIG. 11: Response to stretch of the longitudinal muscle. The experimental arrangement is illustrated above. Part of the ventral longitudinal muscle, between 3 and 4 body segments long with an overall length of around 2cm was used with one segmental ganglion attached by the anterior segmental nerve root. One end of the muscle was pinned to the base of a Sylgard coated dish; the other attached to the arm of a moving coil puller. Single ramp stretches were applied to the longitudinal muscle whose amplitude and velocity were controlled by a ramp function generator. The ramp stretches used gave a final displacement of between 0.5 and 3mm with velocities of stretch ranging from 1.2-6mm/s. Intracellular recordings were made from the axon near its entry to the segmental ganglion. Bottom: 3 ramp stretches with final displacements of 1.0, 1.4 and 1.8mm are shown superimposed. Below, three successive sweeps showing the response of the cell to these three ramp stretches, the bottom trace being the response to the smallest amplitude ramp stretch. The neurone responded with hyperpolarising responses maintained for the duration of the stretch with final voltage displacements of 1.2, 1.7 and 2mV from an initial E_m of -40. Marked excitatory responses were seen on release of stretch.

1981) and it has long been thought that the initiation of such directed swimming is mediated via receptors located in the sensilla (Herter 1929). Friesen (1981) has shown that a volley of small action potentials can be recorded from segmental nerves in response to low amplitude wave stimulation insufficient to excite the touch cells (Mistick 1978). Because the long single hairs project beyond the stationary boundary layer

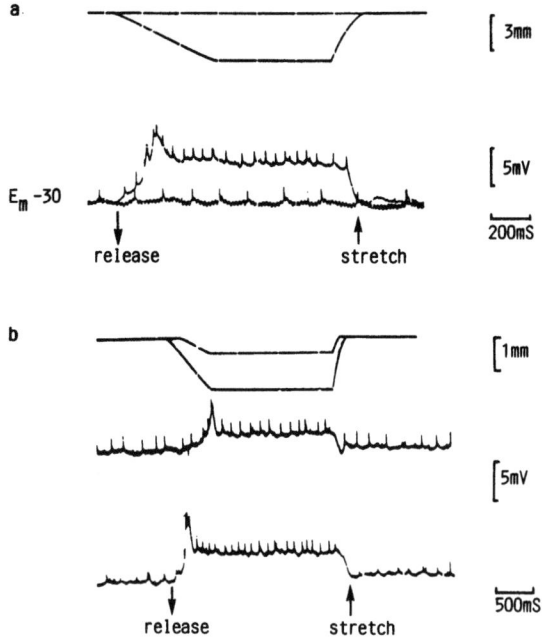

FIG. 12: Depolarising responses to release of stretch.
a. top trace, ramp function release of final amplitude 3.7mm and initial velocity of 6mm/s applied to the longitudinal muscle. Bottom trace, the response of the neurone to release is superimposed on the 'resting' E_m. A decline in the depolarisation level during the initial period of release to a lower maintained value was always seen. In recordings made from the axon near the segmental ganglion small transient events were frequently seen superimposed on the graded response to stretch. These have fast rise times, slow exponential decay and were frequently seen to summate. In addition they are blocked by 15mM Mg which is known to block central chemical synapses in the leech. Consequently they are thought to be post-synaptic potentials. The recording site in the axon in these experiments is close to the synaptic neuropile and fine sections through the axon in this region show presynaptic profiles adjacent to the outgoing synapses (S.E. Blackshaw and D.A. Mackay, unpublished). The identity of the presynaptic neurone is not known.
b. ramp releases of final displacement 0.5 and 3.7 mm (top) gave static responses of final amplitude 1.8 and 3.7mV. The dynamic phase of the response depended on the rate of release from stretch. The ramp release of velocity 1.2mm/s produced an initial component rising with a gradient of 8mV/s. The second ramp release of velocity 6mm/s gave a dynamic component rising at a rate of 13mV/s.

and because of their structural similarity to hair cells in the lateral line in vertebrates (Flock 1965), the uniciliated cells are thought to be responsible for the detection of water movements. The role of the multiciliated cells is not known.

7.1.2. Neurones associated with the nephridia

Peripherally located multipolar neurones have been described in association with the nephridia in Haemopis (Fischer 1969) and in Hirudo

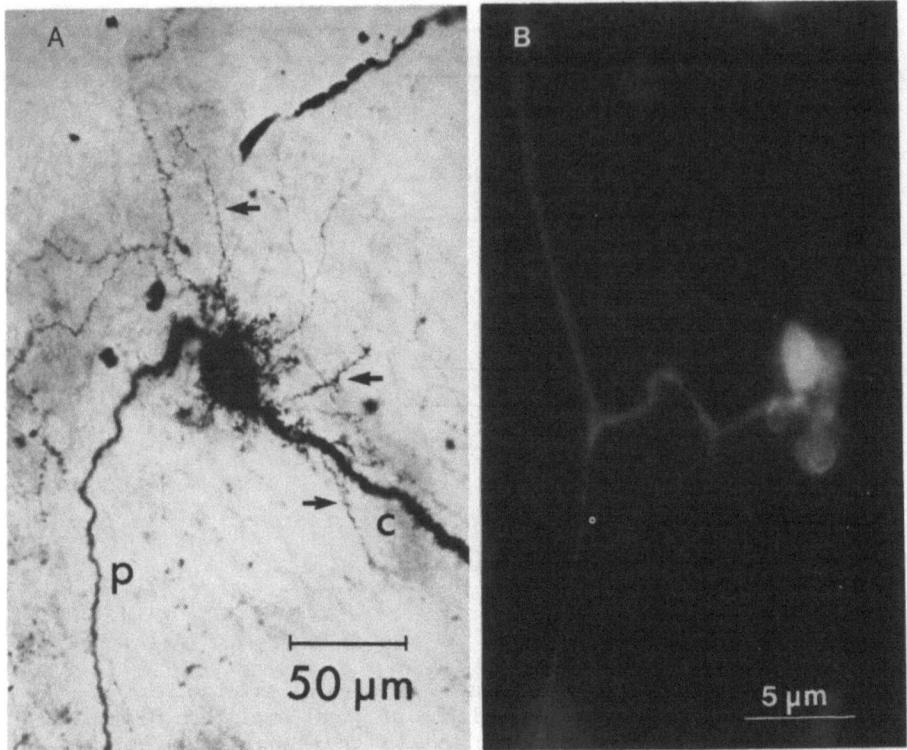

FIG. 13: A. Nephridial neurone injected with horseradish peroxidase. Note the fine branches on the urinary bladder (↓). p, peripheral branch; c, centrally projecting axon. (From Wenning 1983). B. Peripheral neurones in a 20-day Hirudo embryo labelled with Lan 3-6 (Antibody kindly donated by Dr. Birgit Zipser).

(Wenning 1983). In Hirudo one of these neurones innervates each of the 34 nephridia (Fig. 13). The cell body lies on the dorsal surface of the urinary bladder and fine processes between the urine forming cells and the blood vessels supplying the nephridium (Wenning and Cahill 1986). The axon projects in the anterior nerve root into the segmental ganglion where it divides and sends processes into anterior and posterior connectives and across the midline of the ganglion. The function of the nephridial nerve cell is not known but it has been suggested that it may be a salt receptor sensitive to changes in salt concentration in the extracellular fluid (Wenning 1985).

7.1.3. Chemosensory organs in the leech

The first physiological evidence for chemosensory function in annelids comes from experiments on the sensilla that line the upper edge of the lip in Hirudo. These contain chemoreceptors required for feeding (Elliot 1984, 1985a, b, 1986a, b). The main lip sensilla contain multiciliated cells with between 3 and 5 cilia which are different from either class of ciliated cell in the segmental sensilla (Fig. 14). A second type of smaller sensillum with similar multiciliated cells borders the band of dorsal lip sensilla. These may also be chemosensory in function. Stimulation of the dorsal lip with a mixture containing NaCl

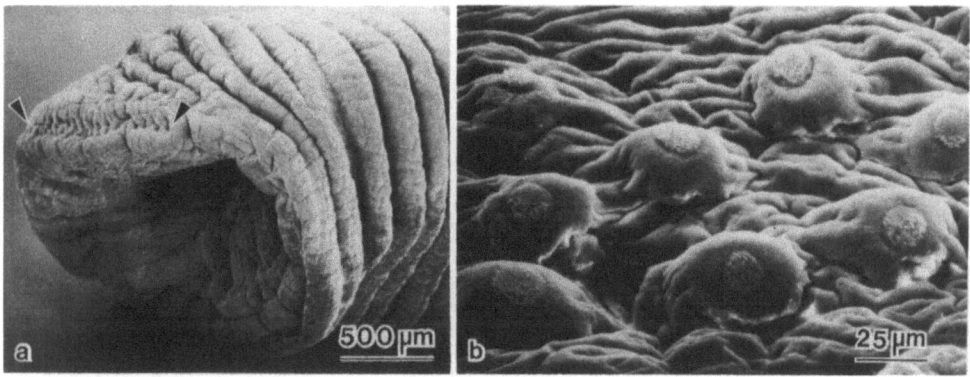

FIG. 14: a. Scanning electron micrograph of the head of <u>Hirudo medicinalis</u>, showing the location (between arrows) of the dorsal lip sensilla, which in this intact fixed preparation are retracted into pits.

b. The dorsal lip sensilla at higher magnification. In this dissected preparation, the sensilla appear as raised mounds with central ciliated 'buttons'. (Photographs courtesy of Dr. E.J. Elliot)

and arginine elicits impulses in the cephalic nerves innervating the dorsal lip sensilla.

8. ORGANISATION OF AXONS WITHIN PERIPHERAL NERVES

There is some segregation according to size in the various axon bundles in the nerve roots in leeches (Wilkinson and Coggeshall 1975; S.E. Blackshaw and D.A. Mackay, unpublished). The large axons travel singly whereasmedium sized axons tend to congregate in small fasicles defined by glial cell processes, and the large numbers of small axons are wrapped in large bundles with no intervening glial cell processes (Fig. 15). Some axons, such as the large diameter stretch receptor afferents (Section 7) that are identifiable on the basis of their size have a stereotyped location within the segmental nerves. The idea that some axons run in tracts which have specific positions is supported by experiments in which the projection of individual cutaneous mechanosensory axons was studied in cross sections of connectives and peripheral verves after labelling chosen cells with horseradish peroxidase (Johansen et al. 1984b). These experiments show that the axons of touch and pressure sensitive cells map consistently in a specific region of the connectives. What these experiments also show is that glial defined fasicles are not rigid structures, but part and fuse in a seemingly random manner. The axons of the mechanosensory cells establish a constant positional relationship to each other within the neuropile and maintain this throughout the length of the connectives without respect to the changing glial fasiculation. Such observations suggest that glial cells have only a supportive structural function and do not play any role in axon guidance or in defining specific nerve fasicles in the leech.

In the nerve roots also fasiculation is highly variable and apparently does not represent structural units such as specific pathways. In contrast to the connectives however, where axon projections of mechanosensory cells are restricted to a limited region, mechanosensory

285

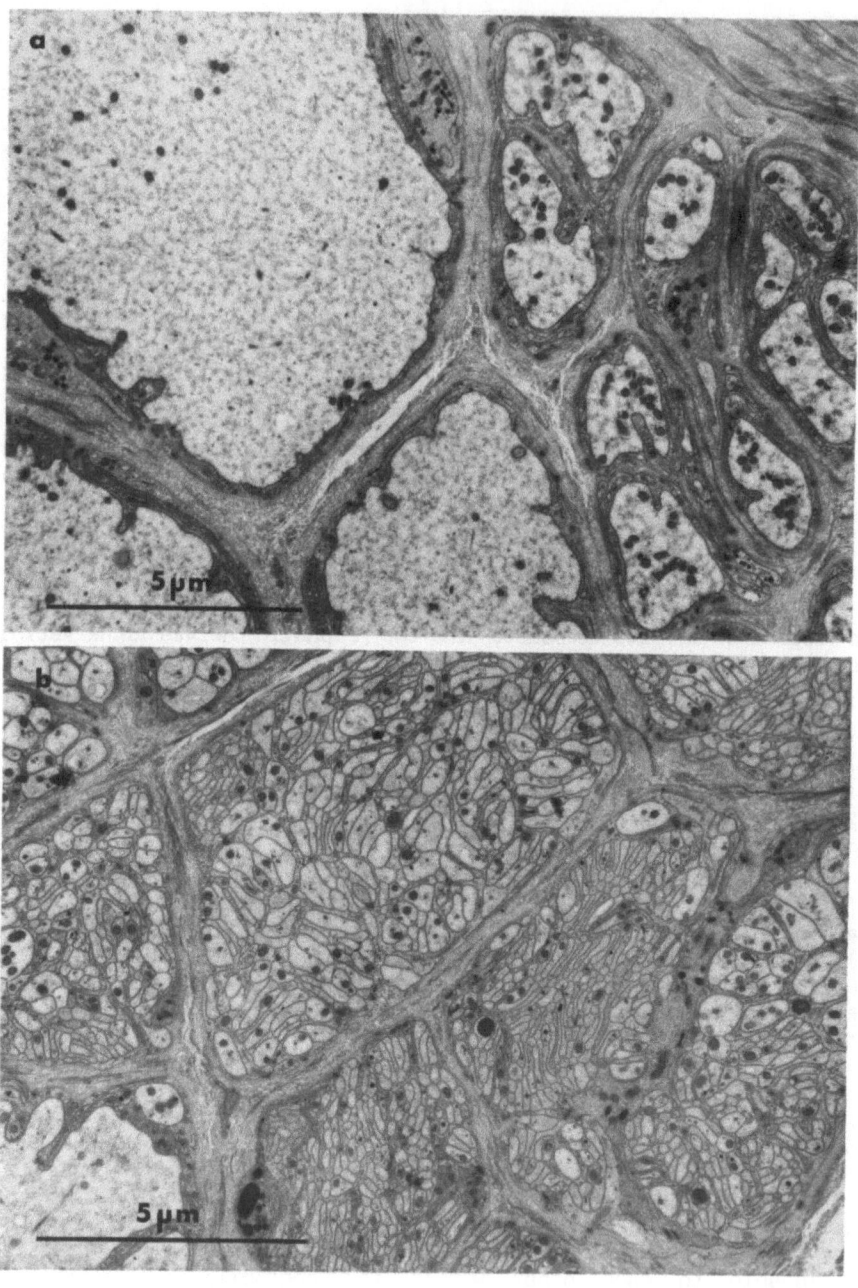

FIG. 15: a. The large stretch receptor axons have individual glial wrappings and are located in a group to one side of the anterior and posterior segmental nerve roots. b. Large numbers of small axons are wrapped in bundles with no intervening glial cell processes.

cell axons are found in all regions of the root. The P cell axons that distribute to dorsal and ventral branches of the posterior root are already located in their appropriate regions before the division of the root takes place implying that P cell axons distinguish between dorsal and ventral tracts in the root during their initial outgrowth.

8.1. Surface antigens on axon subsets

New features of the organisation of axons within the connectives and peripheral nerves are emerging as a consequence of the production of antibodies (mAbs) raised against adult leech nervous systems (Zipser and McKay 1981; Zipser 1982). In the initial studies, when the monoclonal antibodies generated were screened on ventral nerve cords, some mAbs labelled individual neurones or small groups of neurones showing that there are individual molecular signatures in the adult nervous system. For example, the antibody Lan 3-2 is specific in adult Haemopis for the two pairs of nociceptive neurone cell bodies within the ganglion which respond to noxious stimulation of the skin and gut (Nicholls and Baylor 1968; Blackshaw et al. 1982a) and for a subset of axon fasicles in the connectives. (In the related species Hirudo and Helobdella Lan 3-2 cross-reacts only with axon fasicles and not with central N cell bodies). The fact that particular mAbs bind to subsets of axons in the leech means that they can be used to study the distribution of specific immunologically identified groups of axons within the connectives and peripheral nerves (Hockfield and McKay 1983; McKay et al. 1983). For example the grouped subset of axons stained by Lan 3-2 are symmetrically located in the centrolateral part of left and right connectives and maintain their stereotyped location along the entire length of the ventral nerve cord in all animals of the same species and frequently in animals of different species (Fig. 16). Two of the antibodies generated, Lan 3-2 and Lan 4-2, bind to all of the axons within particular fasicles delineated by glial cell processes raising the possibility that there might be molecular markers for each fasicle in the connectives; another binds to single axons in fasicles that contain other unstained axons. Most investigations to date have been on the grouping of axons within the connectives, but the demonstration of antigenically related groups of axons in CNS tracts suggests that fasicles of axons carrying a particular antigen may be a feature of the organisation of peripheral nerves also.

8.2. Developmental expression of surface antigens in the embryonic peripheral nervous system

The mechanisms underlying the grouping and pathfinding of axons are largely unknown. Most proposals invoke specific molecular markers (Sperry 1963). In electron micrographs of antibody stained adult nervous system the Lan 3-2 and Lan 4-2 staining is associated with the perimeter of stained axons rather than with the axoplasm showing that these antibodies bind to surface antigens (Fig. 15). On immunoblots of proteins extracted from the leech nervous system mAb Lan 4-2 binds to a high molecular weight antigen of 130,000 daltons which comigrates with the band of antigens of high molecular size (90,000 to 130,000 daltons) recognised by the antibody Lan 3-2. The antigens recognised by both Lan 3-2 and Lan 4-2 are protease sensitive and both have been shown by lectin binding experiments to be glycosylated (McKay et al. 1983; Flanagan et al. 1986).

The fact that the axon surface can carry specific markers that distinguish axons in a given fasicle from axons in neighbouring fasicles suggests a possible role of the surface antigens in the formation of fasicles. Experiments in which the developmental appearance of the Lan

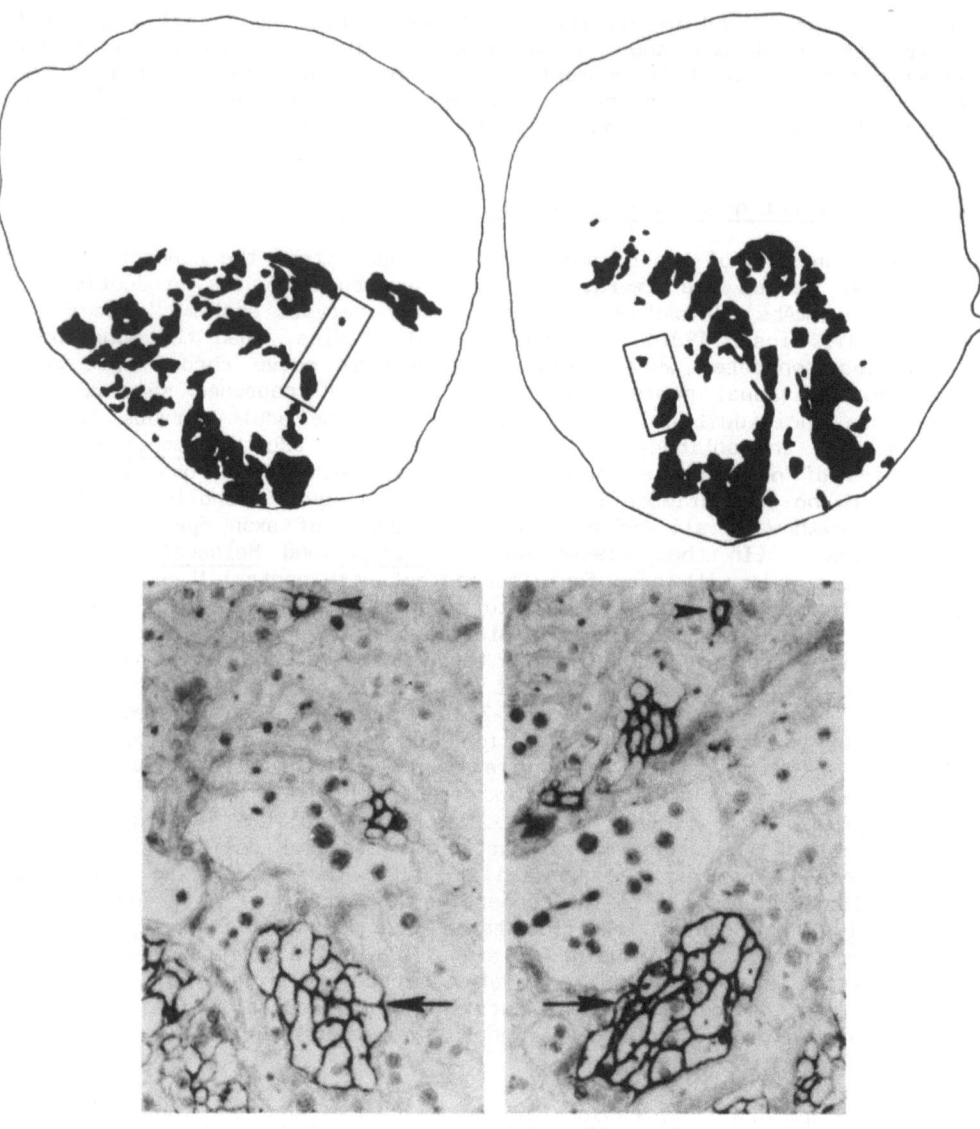

FIG. 16: Antibody binding to subsets of axons. Diagram showing the distribution of Lan 3-2 identified axon bundles in the left and right connectives. The boxes marked in the diagram are shown in the inset electron micrographs (from McKay et al. 1983).

3-2 antigen has been studied in leech embryos show that this family of protease sensitive, glycosylated antigens are differentially expressed by some neurones from the earliest stages of axon outgrowth. In the embryos of two species of leech, Haemopis marmorata and Helobdella triserialis the antigen is seen first in the peripheral nervous system in groups of cells associated with the segmental sensilla along the central annulus of each segment on the dorsal body wall (McKay et al. 1983; Johansen et al. 1985; Stewart et al. 1985). Only a few neurones in each sensillum are stained and they are associated with a 5-10 µm cilium making it likely that they correspond to uniciliate primary sensory neurones thought to be involved in water movement detection in adult leeches (see Section 7). The axons carrying these specific antigens grow into the CNS from the cell bodies in the periphery forming distinct fibre bundles with the processes or more proximally located groups of neurones. They arborise within the CNS and send processes rostrally and caudally to form antigenically positive bundles of axons in the connectives at early stages of development. The antibody Lan 3-6 (Zipser and McKay 1981) also labels cells in segmental sensilla in Haemopis embryos but a different population of ciliated cells from those labelled with Lan 3-2, indicating that the different kinds of ciliated cells in a sensillum possess different antigenic markers. Lan 3-6 also labels previously undescribed ciliated peripheral neurones of unknown modality located in the epidermis of the other annuli (Fig. 13). Peripheral neurones label first in the middle annulus of each segment beginning at the head end of the embryo and progressing rostrocaudally, and label last in the most anterior and posterior annuli of each segment. Thus there are two kinds of temporal organisations in the developing leech body wall: an overall rostrocaudal sequence which is also apparent in other tissues and a local ordering from the middle annulus towards the anterior and posterior boundaries of each segment.

9. THE EMBRYONIC ORIGIN OF THE PERIPHERAL NERVOUS SYSTEM

In spite of marked differences in their mode of life, the oligochaetes and leeches have long been recognised as related clitellate annelids whose eggs develop directly and hatch as juveniles with numerous segments. (The polychaetes are distinct among the annelids in having a trochophore larva). In both oligochaetes and leeches the ventral nerve cord arises from a teloblastic strip of cells, the germinal plate, derived from the D lineage (Anderson 1973). A glossiphoniid leech was in fact the first annelid to be described by Charles Whitman in what was to become the classical school of American lineage studies (Whitman 1878). He first stated the idea that each identified cell of the early embryo and the clone of its descendant cells plays a specific predestined role in later development. More recently there has been a revival of interest in leech embryology due in part to the detailed description of neurones of the adult nervous system and the possibility that this preparation offers for studying development of a nervous system in terms of identified neurones.

Early development of glossiphoniid leeches has been described by Weisblat et al. (1978, 1980a); Fernandez (1980); Fernandez and Stent (1980); Weisblat (1981); Fernandez and Olea (1982); and Stent et al. (1982). The initial cleavages follow the typical spiral pattern of annelids (Anderson 1973). The first three divisions divide the egg into eight blastomeres: 4 small micromeres at the animal pole marking the site of the future head, and 4 large yolky vegetal macromeres (A, B, C and D). A, B and C macromeres give rise to the endoderm. The D lineage plays a crucial role in development, as it does in other invertebrates such as molluscs, since all mesoderm and ectoderm derived tissues, including the nervous system, arise from it. Cleavages of the D macromere to give 5 bilateral pairs of teloblasts (M, N, O/P, O/P, Q) separate the embryo into 3 germinal layers. N, O/P, O/P, Q teloblast pairs are the precursors of the definitive ectoderm and the fifth pair (M) is the precursor of the mesoderm (Fig. 17).

During stage 7, each of the 5 pairs of teloblasts carries out a series of highly unequal divisions, budding off over the course of many hours, a bandlet of several dozen smaller primary blast cells. The 5 bandlets on either side of the midline merge to form a bilaterally symmetrical pair of prominent cell ridges, the right and left germinal

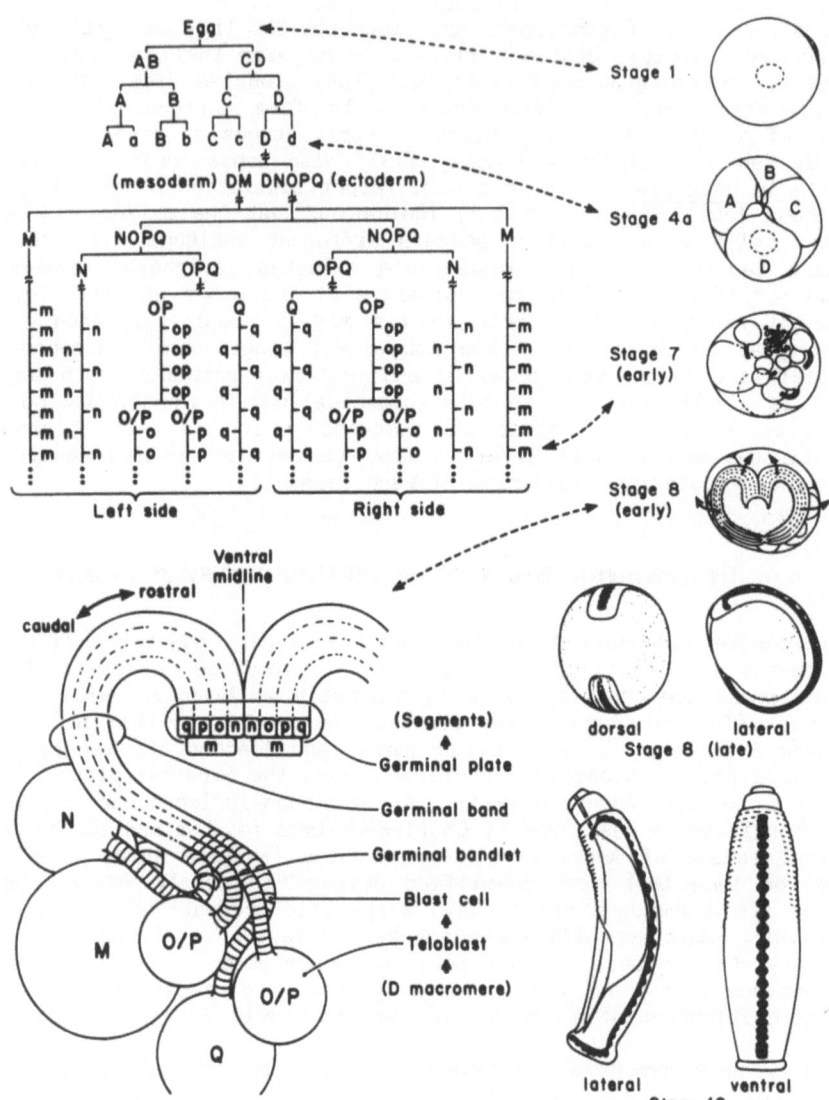

FIG. 17: Schematic summary of the development of <u>Helobdella triserialis</u>. <u>Upper left</u>: Cell pedigree leading from the uncleaved egg to the macromeres A, B and C, the micromeres a, b, c and d, the teloblast pairs M, N, O/P, O/P, and Q and the paired primary blast cell bandlets. Breaks in the lineage indicate points where additional micromeres may be produced. The number of op blast cells produced prior to cleavage of proteloblasts OP varies from four to seven (M. Shankland, unpublished observation). <u>Lower left</u>: Hemilateral disposition of the teloblasts and their primary blast cell bandlets within the germinal band and germinal plate. <u>Right margin</u>: Diagrammatic views of the embryo at various stages. The dashed circle in the uncleaved egg (stage 1) signifies the teloplasm, which is passed on mainly to the D macromere (stage 4a). In the stage 7 embryo the dashed circle signifies the right M teloblast (which is invisible from the dorsal aspect) and the many small, closed contours in the upper midportion indicate the micromere

cap. In the stage 8 (early) embryo, the heart-shaped germinal bands migrate over the surface of the embryo in the directions indicated by the arrows. The incipient larval integument is shown as a stippled area lying between the germinal bands. In the stage 8 (late) embryo the germinal plate is shown to be on the ventral midline, with the nascent ventral nerve cord and its ganglia and ganglionic primordia indicated in black. The stippled larval integument covers the entire embryo, from one edge of the germinal plate to the other. In the stage 10 embryo shown, body closure is nearly complete. Here, the stippled areas signify the yolky remnant of the macromeres and teloblasts, now enclosed in the gut of the embryo. The chain of ganglia linked via connectives, shown in black, already closely resembles the adult nerve cord. (From Weisblat et al. 1984).

bands, which overlie the macromeres. Each germinal band consists of a superficial layer of 4 ectodermal precursor bandlets designated n, o, p, q; and a deeper layer provided by the mesodermal precursor bandlet on each side, m. Within each bandlet, the first born and developmentally oldest blast cells are located at the anterior end, and their progeny will contribute to the anterior-most segments of the body. Progressively younger blast cells are located progressively more posteriorly in the bandlet, and will contribute their progeny to correspondingly more posterior segments.

Gastrulation is represented by a series of morphogenetic movements of the 5 bilateral pairs of bandlets that occur during stage 8. With ongoing production of primary blast cells the germinal bands lengthen and their mid-portions move circumferentially onto the future ventral surface. Gradually, left and right germinal bands meet and coalesce along the ventral midline, beginning at the head end and continuing rearward like a zipper. As they coalesce the germinal bands form the germinal plate which straddles the ventral midline. In the course of formation and circumferential movement the blast cells change their positions; the mediolateral order of the bandlets becomes inverted so that the left and right ectodermal bandlets come to lie together along the future ventral midline of the embryo. The midline position of the two n bandlets lead early embryologists to propose that the nervous system arises entirely from their progeny, an idea which could not be tested because cells within the germinal plate were too numerous and too small for their fates to be followed precisely with the techniques then available, and which was subsequently shown to be wrong. During stage 9 the germinal plate lengthens, broadens and thickens due to a progressive increase in the number of cells in all the bandlets, and expands circumferentially over the surface of the embryo. Eventually the leading edges of the expanding germinal plate meet and fuse along the future dorsal midline, enclosing the tubular body of the leech (stage 10).

The germinal plate becomes partitioned along its length into a series of tissue blocks that correspond to the future body segments. The first sign of segmentation of the embryo appears mid-stage 8 and continues during stage 9. Segmentation is seen first at the anterior end in the germinal band tissue derived from the m bandlet before the germinal bands begin to coalesce. When germinal band coalescence is approximately 2/3 complete the first sign of segmentation of the ectoderm appears during stage 9 as agglomeration of paired cell masses on either side of the midline. Each of these paired cell masses forms a ganglion primordium of the future ventral nerve cord. The first 4 primordia to appear are the precursors of the cephalic suboesophageal ganglion. The 21 primordia which appear next are the precursors of the midbody ganglia present by mid-stage 9, and the last primordia to appear at the caudal ends of the germinal bands eventually form the caudal ganglion. The supra-oesophageal ganglion has a different embryonic origin and arises from cells of the micromere cap (Weisblat et al. 1980a). Adjacent ganglion primordia are initially in contact but begin to move apart by the end of stage 9, linked by the establishment of longitudinal tracts of interganglionic fibres which ultimately give rise to the paired connectives and the unpaired midline Faivres nerve of the adult nervous system. The definitive anatomy is largely established at stage 10 of embryogenesis. By that point there has been a secondary fusion of the 4 most anterior segments and of the 7 most posterior segments. The 21 intervening segments show relatively little variation in structure.

To establish the line of descent of the constituent cells of the leech nervous system, Whitman's lineage studies have been extended and refined by the introduction of novel tracer techniques in which a tracer molecule such as hoseradish peroxidase (Weisblat et al. 1978) or a fluorescent compound that does not require a histochemical reaction (Weisblat et al. 1980b; Gimlich and Braun 1985) are micro-injected into an identified cell of the early embryo. The tracer is subsequently inherited by all the progeny of the injected cell. The embryo is allowed to develop to a chosen endpoint when the distribution of the labelled cells is mapped.

Use of the lineage tracer techniques has shown that the leech nervous system is derived from all 5 teloblasts (Weisblat et al. 1980b, 1984) and therefore has several embryonic sources in contrast to the single source proposed for it by early embryologists on the basis of direct observations (Whitman 1878; Schleip 1936). For the experiments, HRP was injected into each identified blastomere in turn in a stage 6 or 7 embryo (see Fig. 17), and the resulting distribution of HRP-labelled cells was then examined in the late embryo (stage 10). In this way it was shown that each of the 4 paired ectodermal teloblasts N, O/P, O/P, Q, contributes a stereotyped subset of the cells in each segmental ganglion as well as in the periphery of each segment, whereas the mesodermal M teloblast gives rise to the longitudinal, circular and oblique muscle fibres of the body wall as well as the longitudinal muscle fibres in the nerve cord and to a few presumptive neurones within the ganglion.

Of the 4 ectodermal teloblasts the N teloblast contributes most cells to the ipsilateral segmental ganglion (about 90 N-derived cells per half-ganglion) as well as a few cells outside the ganglion in the body wall. The O pattern contribution to the hemiganglion consists of about 60 cells (Fig. 18). There is also a substantial extraganglionic contribution to the ipsilateral body wall. The P pattern contributes about a dozen cells to the half ganglion and also, like the O pattern a substantial contribution to ipsilateral body wall outside the ganglion. Of the 4 ectodermal teloblasts, the Q with the most lateral blast cell bandlet in the germinal plate makes the smallest contribution - about 10 cells to the ipsilateral half-ganglion. The principal contribution of the Q teloblast is the epidermis of the future dorsal body wall (Weisblat et al. 1984; Shankland and Weisblat 1984; Kramer and Weisblat 1985).

These experiments have shown that development of the leech is highly stereotyped in the sense that a particular identifiable blastomere of the early embryo regularly gives rise to a particular part of the adult, although the particular part contributed by a given teloblast comprises a variety of cell types. In the case of the nervous system, each of the 4 paired ectodermal teloblasts contributes cells both to the CNS and to the periphery.

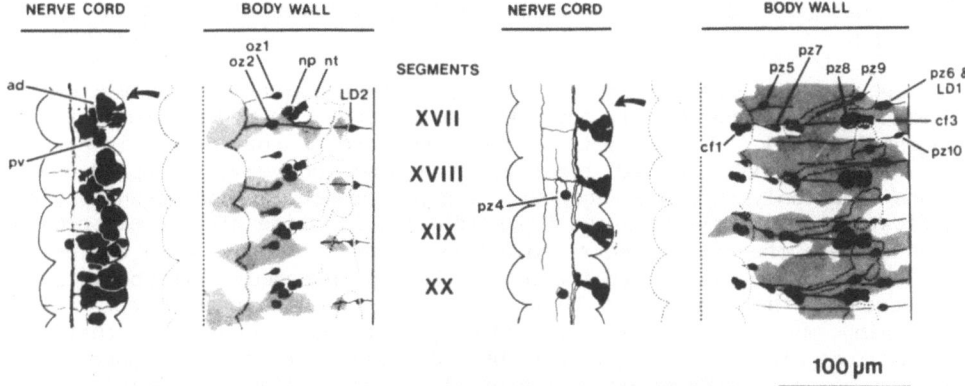

FIG. 18: Camera lucida tracings of HRP-labelled O and P descendant patterns in a series of four consecutive midbody segments. Labelled cell bodies and axons are shown in black, with the shaded regions representing labelled portions of the squamous epidermis. The ventral nerve cord is shown separately, and its location relative to the body wall is marked by a dotted outline. The nephridia are also shown by a dotted outline on the right side of each segment. Note that the labelled tissues lie on only side of the midline (dotted line), and that - with the exception of the P-derived neurone pz4 - the pattern of labelled tissues is repeated in every segment. The 32 body segments are numbered from head to tail, with the unfused abdominal segments designated V-XXV. ad, anterodorsal cell cluster; pv, posteroventral cell cluster; nt, nephridial tubule; np, nephridiopore; cf1 and cf3, cell florets 1 and 3; peripheral neurones as in the text. From Shankland and Weisblat 1984.

10. SEGMENTAL DIFFERENCES IN THE ARCHITECTURE OF THE NERVOUS SYSTEM: POSSIBLE ROLE OF TARGET INTERACTIONS

Despite the metamerically stereotyped architecture of the annelid nervous system, significant regional differentiation does exist. For example in leeches the 4 most anterior and 7 most posterior segmental ganglia which arise as separate ganglion rudiments in the embryo, are fused into compound suboesophageal and caudal ganglia in the adult (Mann, 1962; Muller et al. 1981; see Section 8) and sensory and motoneurones at head and tail ends of the cord arborise over more ganglia than their counterparts in the midbody ganglia (Yau 1976a; Gillon and Wallace 1984; Johansen et al. 1984a).

In both oligochaetes and leeches, ganglia in the genital segments contain more neurones than typical midbody ganglia (Zipser 1979; Macagno 1980). In leech embryos there is no detectable difference between sex and non-sex segments when the ganglia first condense. The sex segmental differences appear just before hatching as an increase in neurone numbers which continues postembryonically for several months in both leeches and earthworms until the adult complement of neurones is reached (Ogawa 1939; Stewart and Macagno 1984; Stewart et al. 1986).

In addition to the increased numbers of cells, regional differences also occur in the branching architecture of particular neurones. For example, the large serotonergic Retzius cells in segments 5 and 6 differ from Retzius cells in all the other segmental ganglia in their peripheral projections as well as in their central arborisation and synaptic relations.

In standard mid-body ganglia the peripheral axons of Retzius cells innervate skin mucus glands and body wall muscles (Yaksta-Sauerland and Coggeshall 1973). By contrast, in segments 5 and 6 they do not innervate the body wall but instead innervate the male and female reproductive organs found only in these segments. The differences in the morphology of Retzius cells in segments 5 and 6 apparently only arise after their growth cones contact the developing reproductive tissue. At this time the cells stop expanding neurites into the body wall and intersegmental connectives and appear rather to devote their entire peripheral innervation to the developing male and female reproductive organs, suggesting that an interaction with the target may be important in the development of segment-specific differences (Mason et al. 1984; Jellies et al. 1985; Loer et al. 1985; Glover and Mason 1986). The ability to ablate or transplant tissue in early leech embryos should make it possible to test such ideas.

11. ACKNOWLEDGEMENTS

Unpublished work by the author supported by the SERC.

12. REFERENCES

Anderson DT (1973) Embryology and phylogeny in annelids and arthropods. Pergamon Press, Oxford.

Anderson G (1978) Postembryonic development of the visual system of the locust Schistocerca gregaria. J Embryol Exp Morphol 45: 55-83.

Blackshaw SE (1981a) Morphology and distribution of touch cell terminals in the skin of the leech. J Physiol 320: 219-228.

Blackshaw SE (1981b) Morphology of nociceptive terminals in the body wall of the leech. J Physiol 317: 81-82.

Blackshaw SE (1981c) Sensory cells and motoneurons. In: Muller KJ, Nicholls JG, Sent GS (eds) Neurobiology of the leech. Cold Spring Harbor Publications, pp 51-78.

Blackshaw SE (1985) The cellular basis of coordination within and between ganglia in annelids. In: Bush BMH, Clarac F (eds) Co-ordination of motor behaviour. SEB Seminar 24, Cambridge University Press, pp 63-89.

Blackshaw SE (1987) Cell lineage and the development of identified neurones in the leech. In: Stirling RV, Stern CD, Parnavelas JG (eds) The making of the nervous system. Oxford University Press (in press).

Blackshaw SE, Thompson SWN (1986) Hyperpolarising response to stretch in neurones innervating body wall muscle in the leech. J Physiol 371: 58P.

Blackshaw SE, Thompson, SWN (1987) Hyperpolarising responses to stretch in sensory neurones innervating leech body wall muscle. (Submitted)

Blackshaw SE, Nicholls JG, Parnas I (1982a) Physiological responses, receptive fields and terminal arborisations of nociceptive cells in the leech. J Physiol 326: 251-260.

Blackshaw SE, Nicholls JG, Parnas I (1982b) Expanded receptive fields of cutaneous mechanoreceptor cells following deletion of single neurones in the CNS of the leech. J Physiol 326: 261-268.

Blackshaw SE, Mackay DA, Thompson SWN (1984) The fine structure of a leech stretch receptor neurone and its efferent input. J Physiol 360: 76P.

Bowman WC, Zaimis E (1958) The effects of adrenaline, noradrenaline and isoprenaline on skeletal muscle contractions in the cat. J Physiol 144: 92-107.

Bush BMH (1981) Non-impulsive stretch receptors in crustaceans. In: Roberts A, Bush BMH (eds) Neurones without impulses. SEB Seminar Series 6, Cambridge University Press, pp 147-176.

Cline HT (1983) ^{3}H-GABA uptake selectively lables identifiable neurons in the leech central nervous system. J Comp Neurol 215: 351-358.

Cline HT, Nusbaum MP, Kristan WB (1985) Identified GABAergic inhibitory motor neurons in the leech central nervous system take up GABA. Brain Res 348: 359-362.

Dales RP (1967) Annelids. Hutchinson. University Library, London.

Dawson AB (1928) Intermuscular cells of the earthworm. J Comp Neurol 32: 155-171.

Derosa SY, Friesen WO (1981) Morphology of leech sensilla: observations with the scanning electron microscope. Biol Bull 160: 383-393.

Dorsett DA (1976) The structure and function of proprioceptors in softbodied invertebrates. In: Mill PJ (ed) Structure and function of proprioceptors in the invertebrates. Chapman & Hall, pp 443-485.

Dorsett DA, Hyde R (1969) The fine structure of the compound sense organs on the cirri of Nereis diversicolor. Z Zellforsch 97: 512-527.

Drewes CD, Fourtner CR (1976) Stretch-sensitive neural units in the body wall of the earthworm, Lumbricus terrestris. L. J Exp Biol 65: 39-50.

Elliot EJ (1984) Chemoreception in the leech. Soc Neurosci Abstr 10: 654.

Elliot EJ (1985a) Leech lip sensilla detect NaCl and arginine in blood. Chem Senses 10: 461.

Elliot EJ (1985b) Amiloride blocks the behavioural feeding response but not the nerve response to feeding stimuli in the leech. Soc Neurosci Abstr 11: 969.

Elliot EJ (1986a) Morphology of chemosensory organs required for feeding in the leech Hirudo medicinalis. J Morphol (submitted).

Elliot EJ (1986b) Chemosensory stimuli in feeding behaviour of the leech Hirudo medicinalis. J Comp Physiol A 159: 391-401.

Fernandez J (1978) Structure of the leech nerve cord: Distribution of neurons and organisation of fiber pathways. J Comp Neurol 180: 165-192.

Fernandez J (1980) Embryonic development of the glossiphoniid leech Theromyzon rude; characterization of developmental stages. Dev Biol 76: 245-262.

Fernandez J, Olea N (1982) Embryonic development of glossiphoniid leeches. In: Harrison FW, Cowden RR (eds) Developmental biology of freshwater invertebrates. Liss, New York, pp 317-361.

Fernandez J, Stent GS (1980) Embryonic development of the glossiphoniid leech Theromyzon rude: Structure and development of the germinal bands. Dev Biol 78: 407-434.

Fischer E (1969) Morphological background of the regulation of nephridial activity in the horse leech (Haemopis sanguisuga L.). Study of nephridial innervation by means of esterase reaction. Acta Biol Acad Sci Hung 20: 381-387.

Flanagan T, Flaster MS, MacInnes J, Zipser B (1986) Probing structural homologies in cell-specific glycoproteins in the leech CNS. Brain Res 378: 152-158.

Flock A (1965) Electron microscopic and electrophysiological studies on the lateral line canal organ. Acta Oto-Laryngol Suppl 199: 1-90.

Frank E, Jansen JKS, Rinvik E (1975). A multisomatic axon in the central nervous system of the leech. J Comp Neurol 159: 1-13.

Friesen WO (1981) Physiology of water motion detection in the medicinal leech. J Exp Biol 92: 255-275.

Fuhner J (1917) Ein Vorlesungsversuch zur Demonstration der erregbarkeitssteigernden Wirkung des Physostigmins. Arch Exp Pathol Pharmakol 82: 81-85.

Gardner-Medwin AR, Jansen JKS, Taxt T (1973) The "giant" axon of the leech. Acta Physiol Scand 87: 30A-31A.

Gaskell JR (1914) The chromaffin system of annelids and the relation of this system to the contractile vascular system in the leech Hirudo medicinalis. Phil Trans R Soc Lond Series B 205: 153-207.

Gillon JW, Wallace BG (1984) Segmental variation in the arborisation of identified neurones in the leech central nervous system. J Comp Neurol 228: 142-148.

Gimlich RL, Braun J (1985) Improved fluorescent compound for tracing cell lineage. Dev Biol 109: 509-514.

Glover JC, Mason A (1986) Morphogenesis of an identified leech neuron: segmental specification of axonal outgrowth. Dev Biol 115: 256-260.

Granzow B, Friesen WO, Kristan WN (1985) Physiological and morphological analysis of synaptic transmission between leech motor neurons. J Neurosci 5: 2035-2050.

Gray J, Lissman HW (1938) Studies in animal locomotion VII. Locomotory reflexes in the earthworm.

Gunther J (1971) Mikroanatomie des Bauchmarks von Lumbricus terrestris. Z Morph Okol Tiere 70: 141-182.

Hanson J, Lowy D (1960) Structure and function of the contractile apparatus in the muscles of invertebrate animals. In: Bourne GH (ed) Structure and function of muscle. Vol I. Academic Press, New York.

Havet J (1899) Structure du système nerveux des annélides. La Cellule 17: 65-136.

Herter K (1929) Reizphysiologisches Verhalten und Parasitismus des Entenegels Protociepsis tesselata O.F. Mull. Z Vergl Physiol 10: 272-308.

Hidaka T, Ito Y, Kuriyama H, Tashiro N (1969) Neuromuscular transmission in the longitudinal layer of somatic muscle in the earthworm. J Exp Biol 50: 417-430.

Hockfield S, McKay RDG (1983) Monoclonal antibodies demonstrate the organization of axons in the leech. J Neurosci 3: 369-375.

Hoyle G (1983) Muscles and their neural control. John Wiley & Sons Inc.

Ito Y, Kuriyama H, Tashiro N (1969) Effects of γ-aminobutyric acid and picrotoxin on the permeability of the longitudinal muscle of the earthworm to various anions. J Exp Biol 51: 363-375.

Jellies J, Loer CM, Kristan WB (1985) Morphogenisis of segment specific innervation patterns in an identified leech neuron. Soc Neurosci Abstr 11: 956.

Johansen J, Hockfield S, McKay RDG (1984a) Distribution and morphology of nociceptive cells in three species of leeches. J Comp Neurol 226: 262-273.

Johansen J, Hockfield S, McKay RDG (1984b) Axonal projections of mechanosensory axons in the connectives and peripheral nerves of the leech Hemopis marmorata. J Comp Neurol 226: 255-262.

Johansen J, Thompson I, Stewart RR, McKay RDG (1985) Expression of surface antigens by the monoclonal antibody Lan 3-2 during embryonic development of the leech. Brain Res 343: 1-7.

Knapp MF, Mill PJ (1971) The fine structure of ciliated sensory cells in the epidermis of the earthworm Lumbricus terrestris. Tissue Cell 3: 623-636.

Kramer AP, Goldman JR (1981) The nervous system of the glossiphoniid leech Haementeria ghilianii I. Identification of neurones. J Comp Physiol A 144: 435-448.

Kramer AP, Kuwada JY (1983) Formation of the receptive fields of leech mechanosensory neurons during embryonic development. J Neurosci 3: 2474-2486.

Kramer AP, Stent GS (1985) Developmental arborisation of sensory neurons in the leech Haementeria ghilianii. II. Experimentally induced variations in the branching pattern. J Neurosci 5: 768-775.

Kramer AP, Weisblat DA (1985) Developmental neural kinship groups in the leech. J Neurosci 5: 388-407.

Kretz JR, Stent GS, Kristan WB (1976) Photosensory input pathways in the medicinal leech. J Comp Physiol 106: 1-37.

Kristan WB, Stent GS (1976) Peripheral feedback in the leech swimming rhythm. Cold spring Harbor Symp Quart Biol 40: 663-674.

Kuffler D (1978) Neuromuscular transmission in longitudinal muscle of the leech Hirudo medicinalis. J Comp Physiol 124: 333-338.

Kuhlman JR, Li C, Calabrese RL (1985a) FMRF-amide-like substances in the leech. I. Immunocytochemical localisation. J Neurosci 5: 2301-2309.

Kuhlman JR, Li C, Calabrese RL (1985b) FMRF-amide-like substances in the leech. II. Bioactivity on the heartbeat system. J Neurosci 5: 2310-2317.

Kuwada JY, Kramer AP (1983) Embryonic development of the leech nervous system: Primary axon outgrowth of identified neurons. J Neurosci 3: 2098-2111.

Lasanky A, Fuortes MGF (1969) The site of origin of electrical responses in visual cells of the leech, Hirudo medicinalis. J Cell Biol 42: 241-252.

Lent CM (1973) Retzius cells: neuroeffectors controlling mucus release by the leech. Science 179: 693-696.

Lent CM (1981) Morphology of neurons containing monoamines within leech segmental ganglia. J Exp Zool 216: 311-316.

Lent CM (1985) Serotonergic modulation of the feeding behaviour of the medicinal leech. Brain Res Bull 14: 643-655.

Loer CM, Jellies J, Kristan WB (1985) The possible role of target interactions in the development of segment-specific differences of an identified neuron. Soc Neurosci Abstr 11: 957.

Macagno ER (1978) A mechanism for the formation of synaptic connections in the arthopod visual system. Nature 275: 318-320.

Macagno ER (1980) Number and distribution of neurones in leech segmental ganglia. J Comp Neurol 190: 283-302.

Mann KH (1962) Leeches (Hirudinea). Pergamon Press, New York.

Maranto AR, Calabrese RL (1984) Neural control of the hearts in the leech, Hirudo medicinalis. I. Anatomy electrical coupling and innervation of the hearts. J Comp Physiol A 154: 367-380.

Marshall CG, Lent CM (1984) Calcium-dependent action potentials in leech giant salivary gland cells. J Exp Biol 113: 367-380.

Mason AJR, Glover JC, Kristan WB (1984) Embryonic development of segmentally specialised serotonergic neurones in the leech Hirudo medicinalis. Soc Neurosci Abstr 10: 1033.

Mason A, Kristan WB (1982) Neuronal excitation, inhibition and modulation of leech longitudinal muscle. J Comp Physiol 146: 527-536.

McKay RDG, Hockfield S, Johansen J, Thompson I, Frederiksen K (1983) Surface molecules identify groups of growing axons. Science 222: 788-794.

Mill PJ (1982) Recent developments in earthworm neurobiology. Comp Biochem Physiol 73A: 641-661.

Mill PJ, Knapp MF (1967) Efferent sensory impulses and the innervation of tactile receptors in Allolobophora longa Ude and Lumbricus terrestris Linn. Comp Biochem Physiol 23: 263-276.

Miller JB, Aidley DJ (1973) Two rates of relaxation in the dorsal longitudinal muscle of a leech. J Exp Biol 58: 91-103.

Mistick DC (1978) Neurones in the leech that facilitate an avoidance behaviour following nearfield water disturbances. J Exp Biol 75: 1-23.

Moment G, Johnson J (1979) The structure and distribution of external sense organs in newly hatched and mature earthworms. J Morphol 159: 1-15.

Muller KJ, Nicholls JG, Stent GS (1981) Neurobiology of the leech. Cold Spring Harbor Publications.

Nicholls JG (1987) The search for connections. Sunderland, M.A. Sinauer Press.

Nicholls JG, Baylor DA (1968) Specific modalities and receptive fields of sensory neurones in the CNS of the leech. J Neurophysiol 31: 740-756.

Ogawa F (1934) The number of ganglion cells and nerve fibres in the nervous system of the earthworm Pheretima communissima. Sci Rep Tohoku Univ 8: 345-368.

Ogawa F (1939) The nervous system of earthworm Pheretima communissima in different ages. Sci Rep Tokohu Univ 13: 395-488.

Ort CA, Kristan WB, Stent GS (1974) Neuronal control of swimming in the medicinal leech. II. Identification and connections of motor neurons. J Comp Physiol 94: 121-154.

O'Shea M, Evans PD (1979) Potentiation of neuromuscular transmission by an octopaminergic neurone in the locust. J Exp Biol 79: 169-190.

Phillips CE, Friesen WD (1982) Ultrastructure of the water-movement-sensitive sensilla in the medicinal leech. J Neurobiol 13: 473-486.

Rude S (1969) Monoamine containing neurons in the central nervous system and peripheral nerves of the leech, Hirudo medicinalis. J Comp Neurol 136: 349-371.

Sargent PB (1977) Synthesis of acetylcholine by excitatory motoneurones in central nervous system of the leech. J Neurophysiol 40: 453-460.

Sawyer R (1986) Leech biology and behaviour. Oxford University Press.

Schain RJ (1961) Effects of 5-hydroxytryptamine on the dorsal muscle of the leech Hirudo medicinalis. Br J Pharmacol 16: 257-261.

Schliep W (1936) Ontogenie der Hirudineen. In: Bronn GH (ed) Klassen und Ordnungen des Tierreichs vol 4, div. III book 4, part 2, 1-121, Akad Verlagsgesellschaft, Leipzig.

Shankland M, Wiesblat DA (1984) Stepwise commitment of blast cell fates during the positional specification of the O and P cell lines in the leech embryo. Dev Biol 106: 326-342.

Sperry RW (1963) Chemoaffinity in the orderly growth of nerve fibre patterns and connections. Proc Natl Acad Sci USA 50: 703-710.

Stent GS, Weisblat DA (1985) Cell lineage in the development of invertebrate nervous systems. Ann Rev Neurosci 8: 45-70.

Stent GS, Weisblat DA, Blair SS, Zackson SL (1982) Cell lineage in the development of the leech nervous system. In: Spitzer N (ed) Neuronal development. Plenum, New York, pp 1-44.

Stewart RR, Macagno E (1984) The development of segmental differences in cell number in the CNS of the leech. Soc Neurosci Abstr 10: 512.

Stewart RR, Macagno E, Zipser B (1985) The embryonic development of peripheral neurons in the body wall of the leech, Haemopis marmorata. Brain Res 332: 150-157.

Stewart RR, Spergel D, Macagno ER (1986) Segmental differentiation in the leech nervous system: the genesis of cell number in the segmental ganglia of Haemopis marmorata. J Comp Neurol 253: 253-259.

Stuart AE (1970) Physiological and morphological properties of motoneurones in the central nervous system of the leech. J Physiol 209: 627-646.

Thompson SWN (1986) Morphological and physiological studies of a stretch receptor neurone in the leech Hirudo medicinalis. PhD Thesis, University of Glasgow.

Thompson W, Stent, GS (1976a) Neuronal control of heartbeat in the medicinal leech. I. Generation of the vascular constriction rhythm by heart motor neurons. J Comp Physiol 111: 261-279.

Thompson W, Stent GS (1976b) Neuronal control of hearbeat in the medicinal leech. II. Intersegmental coordination of heart motoneuron activity by heart interneurons. J Comp Physiol 111: 281-307.

Thompson W, Stent GS (1976c) Neuronal control of heartbeat in the medicinal leech. III. Synaptic relations of the heart interneurons. J Comp Physiol 111: 309-333.

Torrence SA (1984) Neuroblast migration in leech embryos. Soc Neurosci Abstr 10: 512.

Torrence SA, Stuart DK (1986) Gangliogenesis in leech embryos: Migration of neural precursor cells. J Neurosci 6: 2736-2746.

Van Essen DC, Jansen JKS (1977) The specificity of reinnervation by identified sensory and motor neurons in the leech. J Comp Neurol 171: 433-454.

Walker RJ, Woodruff GN, Kerkut GA (1968) The effect of acetylcholine and 5-hydroxytryptamine on electrophysiological recordings from muscle fibres of the leech Hirudo medicinalis. Comp Biochem Physiol 24: 987-990.

Walker RJ, Woodruff GN, Kerkut GA (1970) The action of cholinergic antagonists on spontaneous excitatory potentials recorded from the body wall of the leech, Hirudo medicinalis. Comp Biochem Physiol 32: 690-701.

Wallace BG (1981a) Neurotransmitter chemistry. In: Muller KJ, Nicholls JG, Stent GS (eds) Neurobiology of the Leech. Cold Spring Harbor Publications, pp 147-172.

Wallace BG (1981b) Distribution of AChE in cholinergic and non-cholinergic neurons. Brain Res 219: 190-195.

Wallace BG, Gillon JW, (1982) Characterization of acetylcholinesterase in individual neurons in the leech central nervous system. J Neurosci 2: 1108-1118.

Walther JB (1966) Single cell responses from the primitive eyes of an annelid. In: Bernhard CG (ed) The functional organization of the compound eye. Oxford, Pergamon Press, pp 329-366.

Weisblat DA (1981) Development of the nervous system. In: Muller KJ, Nicholls JG, Stent GS (eds) Neurobiology of the leech. Cold Spring Harbor Pulications.

Weisblat DA, Sawyer R, Stent GS (1978) Cell lineage analysis by intracellular injection of a tracer enzyme. Science 202: 1295-1298.

Weisblat DA, Harper G, Stent GS, Sawyer RT (1980a) Embryonic cell lineages in the nervous system of the glossiphoniid leech Helobdella triserialis. Dev Biol 76: 58-78.

Weisblat DA, Zackson SS, Blair SS, Young JD (1980b) Cell lineage analysis by intracellular injection of fluorescent tracers. Science 209: 1538-1541.

Weisblat DA, Kim SY, Stent GS (1984) Embryonic origins of cells in the leech Helobdella triserialis. Dev Biol 104: 65-85.

Wenning A (1983) A sensory neuron associated with the nephridia of the leech Hirudo medicinalis L. J Comp Physiol 152: 455-458.

Wenning A (1985) Do the nephridial nerve cells serve as osmoreceptors in the leech? Verh Dtsch Zool Ges 78 (in press).

Wenning A, Cahill MA (1986) Nephridial innervation in the leech Hirudo medicinalis L. Cell Tissue Res (in press).

Whitman CO (1878) The embryology of Clepsine. Quart J Microsc Sci 18: 215-315.

Whitman CO (1886) The leeches of Japan. Quart J Microsc Sci 26: 317-416.

Wilkinson JM, Coggeshall RE (1975) Axonal numbers and sizes in the connectives and peripheral nerves of the leech. J Comp Neurol 162: 387-396.

Willard Al (1981) Effects of serotonin on the generation of the motor program for swimming by the medicinal leech. J Neurosci 1: 936-944.

Yaksta-Sauerland BA, Coggeshall RE (1973) Neuromuscular junctions in the leech. J Comp Neurol 151: 85-99.

Yau KW (1976a) Physiological properties and receptive fields of mechanosensory neurones in the head ganglion of the leech: comparison with homologous cells in segmental ganglia. J Physiol 263: 489-512.

Yau KW (1976b) Receptive fields, geometry and conduction blocks of sensory neurones in the CNS of the leech. J Physiol 262: 513-538.

Young SR, Dedwyler RD, Friesen WO (1981) Responses of the medicinal leech to water waves. J Comp Physiol 144: 111-116.

Zipser B (1979) Identifiable neurons controlling penile eversion in the leech. J Neurophysiol 42: 455-464.

Zipser B (1982) Complete distribution patterns of neurons with characteristic antigens in the leech central nervous system. J Neurosci 2: 1453-1464.

Zipser B, McKay RG (1981) Monoclonal antibodies distinguish identifiable neurones in the leech. Nature 289: 549-554.

ONTOGENESE DU SYSTEME NERVEUX CENTRAL DES CHELICERATES ET SA SIGNIFICATION ECO-ETHOLOGIQUE

ARTURO MUÑOZ-CUEVAS and YVES COINEAU

Laboratoire de Zoologie (Arthropodes)
Muséum National d'Histoire Naturelle
61, rue de Buffon. 75231 Paris Cédex 05. France

SUMMARY

Ontogeny of the Central Nervous System in Chelicerates and its Eco-ethological Significance.

The comparative embryogeny of the central nervous system and especially ganglia in annelids, onychophorans, chelicerates, myriapods and crustaceans emphasizes certain characteristics found in all groups and some others limited to a few related groups or even to a single one. A common characteristic of all the groups studied during early development of the nervous system is the occurrence of segmental organs of ectodermal origin which give rise to ganglia.

These organs vary morphologically from one group to another. Several morphological types are distinguishable:

- simple thickenings of the ventral embryonic ectoderm among onychophorans and Symphyla.

- spherical or subspherical organs forming follicles, detached inwards from the ectoderm, in annelids, chilopods, diplopods, xiphosurans, pycnogonids, scorpions, pseudoscorpions, solpugids, spiders and opilions.

- simple ectodermal folds unseparated from the ectoderm and opening outwards in Whip spiders (arachnids, amblypygi).

In different arthropod groups (chelicerates, xiphosurans, pycnogonids, solpugids, gonyleptid opilions), ventral organs seem to be the only ones generating nervous ganglia. In Pachylus quinamavidensis (gonyleptid opilions), six pairs of ventral organs give rise to the cerebral ganglia. This is the highest number found in arthropods; and they may hence be viewed as segmental organs related to neuromeres.

The number of ventral organs in Pachylus is correlated with the development of the neurons of the corpora pedunculata globuli which are integration centres forming 70% of the adult brain. In onychophorans, diplopods, solpugids, opilions and crustaceans, ventral organs persist during adult life.

The primitive embryonic affinities observed during the formation of ganglia in different groups are absent in Insects.

When Hanström published his work on Spiders at the beginning of the century, neuro-ethological studies were difficult to interpret since ecological, ethological and evolutionary data were lacking. Today, the classification of the spiders into four groups by protocerebral structure (Hanström 1935) corresponds to the phylogenetic pattern proposed by Platnick (1971). It would seem that it is the mechano- and chemo-receptive channels of communication of spiders which are correlated with their nocturnal habits. The visual channels and their evolutionary improvement for diurnal life, most elaborate in the Salticidae, appear to be related to the complex protocerebral structure. The change from a nocturnal to a diurnal life is doubtlessly crucial to the question. Hunting spiders (Salticidae) possess the most developed brain. It contains two pairs of optic masses and two integrating centres, the central body and the globuli of the corpora pedunculata. In sedentary spiders, the brain is much less developed; the optic masses are small and less differentiated; the corpora pedunculata are rudimentary, the glomeruli absent while the globuli and peduncle are completely lacking and the bridge rudimentary.

The Lycosid spiders tend towards a diurnal life, but are not as successfully as the Salticids. Neither do their optic structure (Munoz-Cuevas 1984) or their behavior (Tietjen and Rovner 1982) attain the degree of perfection found in the Salticids.

In the neuroanatomical classification by Hanström (1935) and that of Platnick (1971) Lycosids are placed at an intermediary level. The structure of the protocerebrum reflects this situation: the optic masses and the corpora pedunculata are less developed in lycosids than in salticids.

This evolutionary process, obvious in spiders, cannot be put forward to explain the highly developed corpora pedunculata in gonyleptid opilions. These Arachnids do not favor visual communication; their activity being, on the contrary, crepuscular or nocturnal.

1. INTRODUCTION

L'étude du développement du système nerveux central des Chélicérates a été pratiquement négligée par les biologistes pendant la première moitié de notre siècle. Il a fallu attendre la décade 1950-1960 pour qu'un regain d'intérêt pour cet important sujet se manifeste. Les travaux de Legendre et de Yoshikura sur le développement du Système Nerveux Central des Araignées datent en effet de cette époque. Ultérieurement, Aeschlimann a apporté une contribution à l'organogenèse du système nerveux des Acariens. Au cours de la décade 1960-1970, les travaux de Weygoldt montrent l'intérêt qu'il accorde aux problèmes de développement du système nerveux chez les Pseudoscorpions. Finalement durant la décade 1970-1980, sont à signaler les travaux de Babu sur le développement postembryonnaire d'Argiope aurantia (Araignées) et les travaux de Munoz-Cuevas sur le développement des corpora pedunculata chez des Opilions Gonyleptidae.

Tous ces travaux constituent l'apport de l'embryologie descriptive à la connaissance du développement du système nerveux des Chélicérates. Leur valeur réside principalement dans le fait qu'ils s'appuient sur des espèces dont la chronologie du développement est parfaitement maîtrisée en élevage. Cette condition essentielle permettra dans les années à venir de poursuivre ces recherches en utilisant des méthodes de marquage et

L'embryogénèse des Arthropodes montre trois modèles fondamentaux ce qui concerne le développement du système nerveux: le premier dérive du modèle annélidien chez qui des cordons neuraux donneront naissance aux ganglions; ce type est réalisé chez les Insectes. Le deuxième modèle se développe uniquement à partir des organes ventraux; c'est le cas des Pycnogonides. Quant au troisième, qui correspond à un type mixte dans lequel prédomine l'un ou l'autre des deux modèles précédents, il se trouve chez la plupart des Chélicérates et des Myriapodes (Juberthie et Muñoz-Cuevas 1971; Muñoz-Cuevas 1969, 1973).

2.1. Période embryonnaire chez Pachylus quinamavidensis (Arachnida, Opilion Gonyleptidae)

La quatrième phase du développement embryonnaire chez Pachylus quinamavidensis correspond à l'inversion de l'embryon; elle se déroule entre les 13e et 21e jours à 20°C. Au cours de cette phase, on observe la formation des replis oculaires, le rapprochement des chélicères vers la ligne médio-antérieure, la formation de l'orifice buccal, des griffes et des phanères, la croissance des appendices et un début de pigmentation de l'embryon (Muñoz-Cuevas 1971).

A partir du 17e jour du développement (stade IV, 4), des coupes sériées montrent la différenciation de la zone sagittale ventrale de l'ectoderme de l'embryon, et derrière le repli oculaire, la formation de plusieurs paires d'invaginations ectodermiques.

Le premier signe annonciateur de l'apparition de chaque invagination est une prolifération cellulaire; cette prolifération s'accompagne bientôt d'une invagination du feuillet ectodermique, d'abord discrète puis s'enfonçant peu à peu. La prolifération cellulaire forme autour de cette invagination une sorte de couronne cellulaire radiaire. Chaque invagination développe ensuite à sa partie distale une vésicule constituée d'amas cellulaires et qui se détache complètement de l'ectoderme. Cette vésicule est d'abord piriforme, assez allongée et, au fur et à mesure qu'elle s'éloigne de l'ectoderme, elle s'arrondit de plus en plus. L'ectoderme ventral, une fois l'organe ventral libéré, se referme aussitôt.

L'étude cytologique de cet organe montre qu'il est formé de deux à quatre assises cellulaires; les cellules ont un aspect épithélial; leur noyau, de forme arrondie, pourvu d'un ou de deux nucléoles, mesure entre 5,5 et 8,5 µm et présente une chromatine dense. Le cytoplasme, d'aspect radié est faiblement coloré par le bleu alcian. La couronne cellulaire qui entoure la cavité présente d'abondantes figures mitotiques. Depuis les premières ébauches invaginées, une substance fortement colorée par le bleu alcian après oxydation permanganique occupe une partie de la cavité centrale. Cette description correspond à la différenciation d'un organe ventral cérébral. Chez Pachylus, la différenciation de six organes ventraux cérébraux suit le même modèle.

Le processus neuroblastique débute quelques heures après la différenciation des organes ventraux. Il est reconnaissable par la grande activité mitotique des organes ventraux qui produisent dorsalement une véritable écorce ganglionnaire. En effet, il faut faire une distinction nette entre l'organe générateur de neuroblastes, qui est toujours ventral,

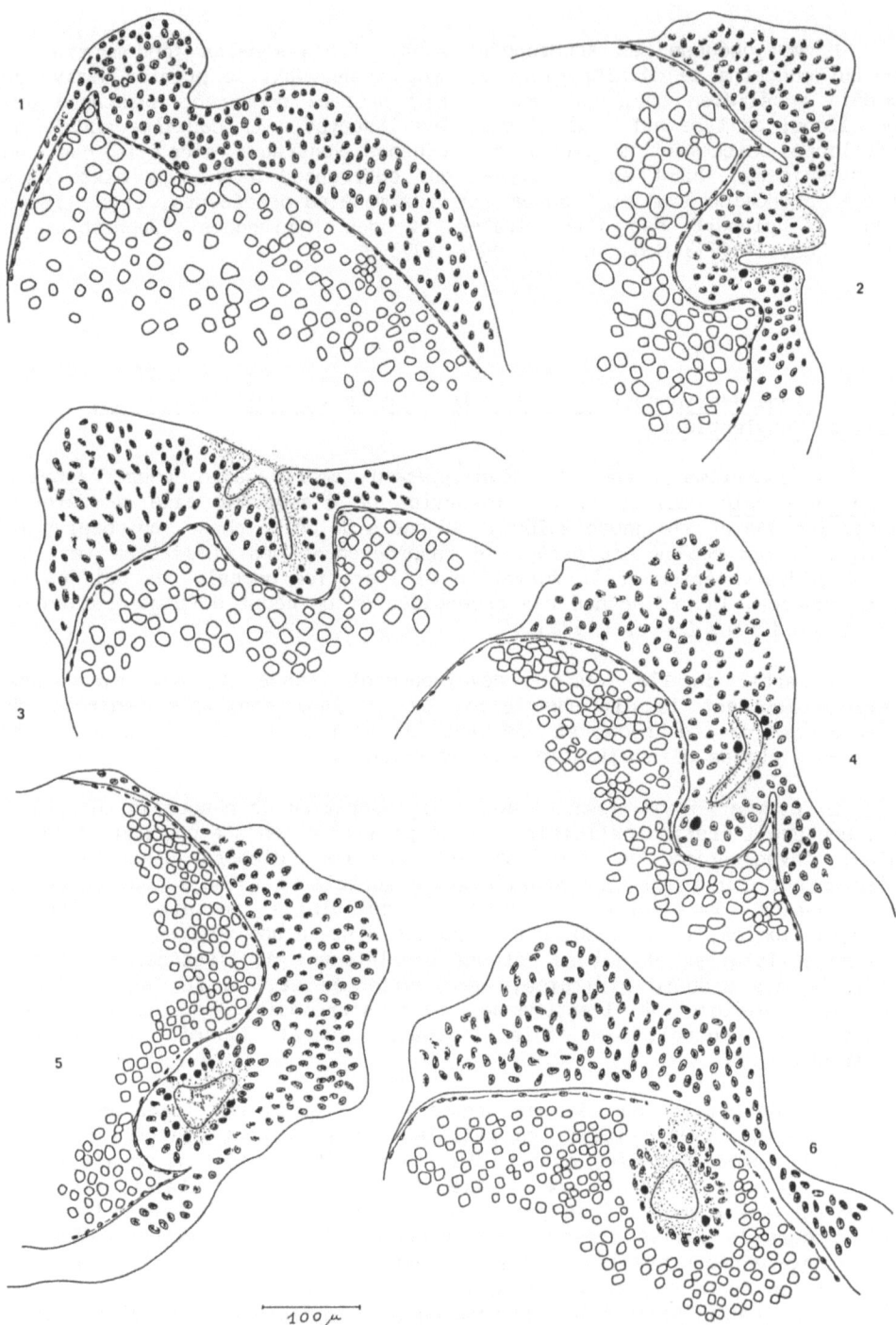

FIG. 1 - 6: Différenciation ganglionnaire et isolement à partir de l'ectoderme chez l'embryon de Pachylus quinamavidensis (stade IV, 5) (D'après Muñoz-Cuevas 1973).

FIG. 1 - 6: Ganglionic differentiation and isolation from the ectoderm of Pachylus quinamavidensis embryo (After Muñoz-Cuevas 1973).

et la zone dorsale ou ganglionnaire proprement dite. Cette dernière zone ne montre aucune activité mitotique. A mesure que les ganglions s'édifient, les organes ventraux s'entourent progressivement de cellules ganglionnaires, processus concomitant avec la condensation des ganglions. Ainsi, 24 à 48h après le début de la différenciation, les centres générateurs sont déjà entourés et profondément enfoncés dans l'écorce ganglionnaire.

2.2. Différenciation de la chaîne ventrale

La différenciation de la chaîne nerveuse ventrale se réalise aussi à partir des organes ventraux. En effet, un certain nombre de petits organes ventraux pairs d'origine ectodermique sont libérés dans le cerveau mais avec un certain décalage dans le temps. La différenciation des ganglions de la chaîne ventrale se produit à partir du 18e jour du développement à 20°C (stade IV, 5). Les organes ventraux de la chaîne sont bien plus petits que ceux du cerveau; ils sont formés par une couche cellulaire radiaire autour d'une petite cavité parfois virtuelle. Il est facile d'identifier les cinq premiers organes ventraux de la chaîne qui donnent naissance aux ganglions des pédipalpes et des pattes ambulatoires. En suivant la chaîne ventrale vers l'opisthosoma, la taille des organes ventraux diminue graduellement, dans la région postérieure, ils sont difficilement décelables.

Après la différenciation des organes ventraux des appendices, apparaissent quatre paires d'organes ventraux postérieurs, plus petits que ceux des appendices. Ici, la cavité centrale est virtuelle et n'est entourée que d'une seule assise cellulaire. Au même niveau de la chaîne ventrale, se trouve une paire d'organes ventraux de grande taille. Ces organes sont beaucoup plus développés que dans le cerveau; leur forme est sphérique. Les cellules sont disposées en trois ou quatre assises ménageant des espaces intercellulaires d'importance variable. L'aspect radié du cytoplasme est plus marqué. Quelques figures mitotiques se retrouvent dans les couches cellulaires proches de la cavité. Celle-ci mesure entre 17 et 18 µm d'épaisseur; la substance colorable au bleu alcian l'emplit à moitié, laissant au bord du cytoplasme quelques vacuoles vides.

La formation de la chaîne ventrale de Pachylus comprend la différenciation de 10 paires d'organes ventraux: 5 paires donneront naissance aux ganglions des pédipalpes et des pattes ambulatoires; 4 paires, abdominales, sont assez petites et d'observation difficile; une grande paire volumineuse correspond aux organes ventraux des derniers segments qui, restant coalescents, donnent un organe unique, énorme et multiple.

D'après ces observations il y aurait une métamérisation partielle de la chaîne nerveuse et une différenciation partielle des organes postérieurs.

2.3. Période Postembryonnaire

Les individus jeunes présentent les organes ventraux en nombre constant de 12, soit 6 dans chaque hémisphère cérébral; en outre, il en existe 2 dans la masse nerveuse sous-oesophagienne. Ceux du cerveau sont localisés symétriquement dans la région latérale et dorso-médiane des globuli, près du neurilemme. Leur morphologie rappelle exactement celle des organes ventraux de l'embryon. Aussi bien dans le cerveau que dans la masse nerveuse sous-oesophagienne, d'abondantes figures mitotiques peuvent être observées; la cavité centrale atteint 10 µm dans le cerveau et 38 µm dans la masse nerveuse; le cytoplasme cellulaire est intensément coloré et il est rare que la substance centrale emplisse complètement la cavité.

3. PRESENCE D'ORGANES EMBRYONNAIRES CHEZ L'ADULTE

Dans l'écorce cérébrale de l'adulte, 12 organes neuraux sont mis en évidence: 6 dans chaque hémisphère, localisés symétriquement à proximité du neurilemme dans les globuli des corpora pedunculata, 3 organes en position latérale et 3 en position médio-dorsale. Leur forme est sphérique ou sub-sphérique; ils sont constitués d'une à deux assises cellulaires délimitant une cavité centrale contenant un produit colorable par le bleu alcian après oxydation permanganique. Les cellules ont un aspect épithélial; leur noyau de forme arrondie mesure entre 5,5 et 8,5 µm; il présente une chromatine dense et un ou deux nucléoles. Le cytoplasme a un aspect radié orienté vers le centre de l'organe; il se colore faiblement par le bleu alcian. La cavité centrale mesurant 6,6 µm est occupée par une substance dense, fortement colorée chez l'adulte.

Les organes du cerveau de l'adulte ne montrent pas de figures mitotiques. Ces organes semblent fermés.

La masse nerveuse sous-oesophagienne comprend deux formations symétriques situées dans sa partie postérieure, à proximité du neurilemme. Ces organes sont beaucoup plus développés que dans le cerveau, leur forme est allongée et les cellules sont disposées en 3 ou 4 assises laissant des espaces variables entre les cellules. L'aspect radié du cytoplasme est assez marqué. Des figures mitotiques se trouvent dans les couches cellulaires proches de la cavité; celle-ci mesure entre 17 et 18 µm et se présente plus ou moins emplie de la substance colorable par le bleu alcian, laissant parfois au bord de la cavité quelques vacuoles incolores. Juberthie (1964) trouve chez des Opilions adultes ces formations dans le cerveau et dans certains groupes, au sein de la masse nerveuse sous-oesophagienne. Chez les Ischyropsalidae ces formations, au nombre de 8 sont localisées dans les globuli des corpora pedunculata. Chez Trogulus nepaeformis 4 formations sont localisées dans chacun des globuli des corpora pedunculata. Chez Scotolemon 3 formations sont mises en évidence dans les globuli et 2 dans la masse nerveuse. Enfin, chez Siro rubens 2 formations se trouvent dans les globuli et 2 dans la masse nerveuse.

Parmi les Diplopodes des organes neuraux ont été signalés par Sahli (1966) chez des Iulidae, sous le nom d'îlots intra-cérébraux et étudiés par Juberthie-Jupeau (1967) chez plusieurs espèces de Glomeridia; au nombre de 2, ils sont localisés dans le protocérébron.

Nguyen Duy (1971), chez un Diplopode Pénicillate Polyxenidae, a mis en évidence la présence d'une paire d'organes neuraux protocérébraux chez l'animal adulte.

Junqua (1957) admet la présence de quatre petits organes neuraux protocérébraux chez un Solifuge adulte. D'après Pflugfelder (1948), chez les Onychophores, 2 paires d'organes dérivés des 2e et 3e paires d'organes ventraux persisteraient chez l'adulte.

Bazin et Demeuzy (1968) et Bazin (1969, 1970, 1971) ont démontré la présence d'organes deutocérébraux chez les adultes de quelques espèces appartenant aux 24 genres représentant les trois sections de Crustacés Décapodes Reptantia.

4. ASPECTS COMPARATIFS AVEC LES ANNELIDES, ONYCHOPHORES ET D'AUTRES ARTHROPODES

Chez les Polychètes, Akesson (1961, 1962) a mis en évidence, au deuxième jour du développement embryonnaire, des invaginations ectodermiques dans l'éphisphère de la trocophore, qu'il a décrites sous le nom d'"invaginations cérébrales". Les invaginations se transforment rapidement en vésicules dont la cavité est pleine d'une substance d'origine cuticulaire. Akesson considère ces formations comme les centres générateurs de cellules ganglionnaires.

D'après les travaux de Kennel (1888) et Pflugfelder (1948) sur les Onychophores, la différenciation des ganglions du système nerveux des Péripates s'effectue à partir des organes ventraux. Selon Pflugfelder (1948), l'ectoderme différencie une double bandelette longitudinale qui se métamérise en des points bien précis constituant par épaississement des organes ventraux pleins dont les cellules élaborent les ganglions par prolifération active. Les ganglions, lors d'une migration profonde, se détachent des organes ventraux qui régressent puis disparaissent complètement. La deuxième paire d'organes ventraux se transforme en vésicules profondes qui, par un processus de croissance allométrique, deviennent secondairement extra-cérébrales (organes infra-cérébraux).

Morgan (1891) a été le premier à déterminer la différenciation des ganglions à partir des organes ventraux chez les Pycnogonides. Ceux-ci sont formés par une invagination de l'ectoderme ventral, invagination qui se ferme et prend la forme d'une cavité entourée d'une couronne de cellules radiaires dotées d'une grande activité mitotique. Les figures classiques des travaux de Morgan (1891) et Dogiel (1913) montrent ces organes au moment de la différenciation des ganglions. Selon Morgan, à chaque paire de ganglions correspond une paire d'organes ventraux.

Heymons (1901), dans son travail sur le développement de la Scolopendre (Myriapode), étudie la formation des ganglions. Ces ganglions proviennent d'une ébauche ectodermique paire qui donne en des points bien précis des "fossettes ganglionnaires" équivalents des organes ventraux des Péripates et des Pycnogonides. Ces fossettes ganglionnaires paires sont formées par migration en profondeur de cellules ectodermiques. Chaque migration reste limitée à une région déterminée. De semblables migrations sont observées sur tous les segments du corps. La suite du développement est caractérisée par un fort accroissement de la taille des fossettes. Libérée de l'ectoderme, chaque fossette subit une petite rotation, son ouverture, jusque-là ventrale, devient latérale. Des divisions cellulaires font augmenter la masse de la partie médiale du ganglion. Même après la fusion des ganglions en une masse impaire, on retrouve encore, de chaque côté de celle-ci, des fossettes très nettes. Au début, ces fossettes présentent une cavité réelle et se trouvent sur la partie ventrale du ganglion. Plus tard, les fossettes finissent par être tout à fait entourées par les cellules ganglionnaires et forment de petites vésicules incluses dans la masse du ganglion. Puis la lumière de la vésicule se comble de plus en plus, disparaît, et on ne voit plus à l'emplacement de ces saccules qu'un amas de petites cellules.

Chez les Symphyles, Tiegs (1940) étudie la formation d'organes ventraux et la naissance des ganglions nerveux. Dans ce groupe de Myriapodes, les organes ventraux ressemblent à ceux des Péripates. En effet, ils sont formés par un épaississement du feuillet ectodermique ventral qui prend la forme d'un petit croissant et qui ne se dégage jamais de l'ectoderme. Ces organes sont pairs et segmentaires. La multiplication active de la couche cellulaire profonde donne naissance aux neuroblastes.

Chez les Diplopodes, Pflugfelder (1932) et Dohle (1964) ont montré que les ébauches paires du protocérébron et du deutocérébron se différencient à partir de deux paires d'invaginations ectodermiques. Elles apparaissent de la façon la plus nette au début du premier stade du développement embryonnaire. Par suite de la forte prolifération cellulaire et du détachement du ganglion, les invaginations primitivement ouvertes se ferment et forment de véritables cavités entourées d'une couronne cellulaire. Pflugfeler (1932) admet que le rôle de ces formations est bien le même lors du développement embryonnaire que lors de l'anamorphose.

La formation des ganglions ventraux chez les Xiphosures a été décrite par Kishinouye (1892) et par Iwanoff (1933) comme un processus de prolifération de cellules de l'ectoderme ventral qui produit des épaississements pairs et segmentaires. Ultérieurement, ces épaississements se séparent à l'intérieur, laissant l'ectoderme à la surface.

Parmi les Arachnides, les travaux de Barrois (1896) et de Weygoldt (1964) permettent de comprendre les processus de différenciation du système nerveux des Pseudoscorpions et de suivre pas à pas la formation des ganglions. Barrois, chez <u>Geogarypus</u>, remarqua d'abord la présence d'une vésicule cérébrale qui, plus tard, se divise pour donner le ganglion chélicérien et le protocérébron. Puis apparaissent les ébauches des paires de ganglions des pédipalpes et des pattes ambulatoires et, plus tard, les ébauches de dix paires de ganglions opisthosomiens. D'après Weygoldt (1964), la paire de ganglions du prosoma, attenante au ganglion sous-oesophagien, apparaît pendant ou après la seconde phase de succion. Les ébauches ganglionnaires se forment tout d'abord à la suite de multiplications cellulaires donnant naissance à des cordons cellulaires qui traversent les segments et montrent de légers épaississements segmentaires. Ces cordons cellulaires s'invaginent sur toute leur longueur, puis les bords des invaginations se rejoignent, coupant ainsi l'invagination de l'extérieur. Ainsi se constitue une double cavité tubulaire qui montre des épaississements métamériques nets. Les invaginations des ébauches du système nerveux commencent à l'avant et progressent lentement vers l'arrière. Néanmoins, ces cavités n'atteignent jamais le quatrième segment des pattes ambulatoires. Les ébauches ganglionnaires de l'opisthosoma n'ont jamais de cavités communiquant entre elles. Elles se composent d'invaginations segmentaires isolées. Weygoldt a dénombré onze paires de ganglions opisthosomiens; le dernier, très petit, ne montre aucune invagination précise et très vite il perd son caractère de ganglion isolé. Dans le douzième segment, il n'y a pas de ganglion.

Chez les Amblypyges, Pereyaslawzewa (1901) et Weygoldt (1975) ont étudié la différenciation des ganglions. Dans ce groupe l'ectoderme qui forme les bourrelets primitifs, au lieu de se diviser en deux couches, prend part entièrement à la différenciation des systèmes céphalique ainsi que ventral.

Chez les Araignées tétrapneumones, Schimkewitsch (1911) et Yoshikura (1954) signalent l'apparition de cordons nerveux qui ne prennent part à la formation de la chaîne nerveuse que par des invaginations métamériques réparties le long de leur trajet.

Brauer (1895) signale particulièrement l'importance des invaginations métamériques dans la genèse du système nerveux des Scorpions.

Pour les Solifuges, Kästner (1952) et Junqua (1966) ont démontré que lors de la formation de la chaîne ventrale, la première esquisse n'est nullement continue, mais consiste au contraire en une succession d'ébauches parfaitement distinctes les unes des autres. Ces ébauches, manifestement métamériques, sont paires et correspondent aux futurs neuromères. Elles apparaissent sous forme d'invaginations sacciformes.

Dans l'embryogenèse du cerveau, Junqua (1966) reconnaît deux vésicules semi-lunaires qui se constituent à partir de profonds replis de l'ectoderme et qui, par fusion, donneront le ganglion occipital. Les autres ganglions cérébraux (pariétal, optique, préchélicérien et chélicérien) prendraient naissance de la même façon que les neuromères de la chaîne ventrale, à partir d'une invagination de l'ectoderme.

Cette étude sur l'embryogenèse des organes ventraux dans plusieurs groupes d'Arthropodes et chez des Annélides Polychètes met en évidence certains faits d'une valeur générale pour l'ensemble des groupes étudiés et d'autres d'une valeur plus restreinte, propres à un seul groupe ou à quelques groupes ayant des affinités.

D'abord, le caractère commun à l'ensemble des groupes étudiés réside dans la présence, aux premiers stades de l'embryogenèse du système nerveux, d'organes métamériques d'origine ectodermique qui participent à la genèse des ganglions.

La morphologie de ces organes, leur nombre et leur rôle dans la différenciation des ganglions varie d'un groupe à l'autre. Ainsi plusieurs types morphologiques d'organes ventraux apparaissent:

- un simple épaississement de l'ectoderme ventral de l'embryon chez les Onychophores et les Symphyles;

- des organes sphériques ou sub-sphériques de forme folliculaire et dégagés de l'ectoderme chez les Annélides, Chilopodes, Diplopodes, Xiphosures, Pycnogonides, Scorpions, Pseudoscorpions, Solifuges, Araignées et Opilions;

- de simple replis ectodermiques non dégagés de l'ectoderme et ouverts vers l'extérieur chez les Phrynes (Arachnides, Amblypyges).

5. CONCLUSIONS DES DONNEES EMBRYOLOGIQUES

-- La présense d'organes ventraux aux premiers stades de l'embryogenèse du système nerveux semble établie pour tous les groupes où des études embryologiques ont été réalisées (Annélides, Onychophores, Pycnogonides, Arachnides, Myriapodes, Crustacés).

-- Du point de vue de l'embryogenèse du système nerveux, il existe une différence fondamentale entre les Insectes et les groupes étudiés.

-- Dans certains groupes d'Arthropodes (Chélicérates, Xiphosures, Pycnogonides, Solifuges, Opilions Gonyleptidae), les organes ventraux semblent être les seuls organes générateurs de ganglions nerveux.

-- Chez Pachylus quinamavidensis (Opilion Gonyleptidae), le nombre des organes ventraux donnant naissance aux ganglions cérébraux est de 6 paires; c'est le nombre le plus élevé observé chez les Arthropodes.

-- Le nombre des organes ventraux chez Pachylus incite à voir en eux des formations segmentaires et donc à les rapprocher des neuromères qui participent à l'organogenèse du système nerveux.

-- L'importance du nombre d'organes ventraux chez Pachylus dans la genèse des neurones des centres d'association que sont les globuli des corpora pedunculata.

-- La persistance des organes neuraux jusqu'à la vie adulte a été mise en évidence chez les Onychophores, Diplopodes, Solifuges, Opilions et Crustacés.

-- L'aspect primitif des affinités embryologiques dans la formation de ganglions entre les Annélides, les Onychophores et les Arthropodes non Insectes apparaît sous un jour nouveau.

6. VARIATIONS DES CENTRES NERVEUX DU CERVEAU CHEZ LES CHELICERATES

Le système nerveux des Arthropodes est construit sur le type de celui des Annélides (Holmgren 1916; Hanström 1921, 1923, 1928). Il comprend trois parties: le cerveau, le système stomatogastrique et la chaîne nerveuse.

Depuis les recherches classiques de Viallanes (1882-1892), on divise le cerveau des Arthropodes en trois régions s'étageant d'avant en arrière: le protocérébron, le deutocérébron et le tritocérébron.

D'après le conception d'Hanström (1928), le cerveau des Chélicérates ne comprend qu'un protocérébron et un tritocérébron. Comme celui des autres Arthropodes, le protocérébron des Arachnides comprend des centres visuels, des centres d'association, connus sous le nom de corps central et de corps pédonculés. Dans son ensemble, le cerveau des Arachnides paraît plus rudimentaire que celui des autres groupes. Cette rudimentation a probablement commencé par la disparition du deutocérébron. Ainsi les centres optiques des Xiphosures aux Acariens présentent toute une série de stades de réduction. Chez la plupart des Arachnides, leur masse est comprise entre 0.3, et 2.8% de la masse totale du cerveau, proportion très inférieure à celle des Crustacés et des Insectes à yeux peu développés (Porcellio 7.7%, Formica 9.1%; Millot 1949).

Les corps prédonculés subissent une rudimentation considérable. Ceux-ci comprennent trois parties: des globuli, centres pairs de petites cellules; les pédoncules dont les moitiés sont reliées par un puissant pont d'union impair. Chez les Limules ils ont un développement considérable, formant de chaque côté trois groupes complexes à globuli très développés qui représentent environ 80% du cerveau entier. Hanström (1919) avait signalé ces faits et d'après les travaux d'Holmgren (1916) sur l'Opilion Gonyleptidae Acrographinotus qui ont démontré la présence de trois globuli de corpora pedunculata dans chaque hémisphère, les Opilions Gonyleeptidae sont placés immédiatement après les Limulus. En effet, chez les Gonyleptidae adultes ces centres doivent représenter environ 70% du cerveau. Cette situation, comme nous l'avons vu précédemment, n'est pas extensible à tout l'Ordre des Opilions (Juberthie 1964). Chez les Amblypyges, ces centres sont encore bien développés, représentant 50% du cerveau chez Neophrynus, mais ils ne forment qu'un seul groupe. Ils diminuent de volume chez les Uropyges, les Solifuges et les Araignées vagabondes. Les Pseudoscorpions et les Araignées sédentaires n'en possèdent que des rudiments qui vont jusqu'à disparaître complètement chez les Acariens. En ce qui concerne le corps central, d'apres Babu (1985), il apparaît comme le centre final de l'intégration visuelle pour les yeux médians et pour les yeux latéraux chez les Arachnides. A part cela, ses connexions avec le protocérébron et avec le ganglion sous-oesophagien suggèrent un rôle moteur et d'associations (Carricaburu et Muñoz-Cuevas (1986). Son développement est aussi très variable parmi les Arachnides;

il présente deux lobes chez Galeodes (Solifuge), trois chez Phrynichus (Amblypyges), quatre chez Heterometrus (Scorpion) et 4 chez Telyphonus (Uropyge). Suivant Babu (1985), son volume varie de 4.3% du volume total du cerveau chez les Araignées à 3.5% chez le Scorpion, 3.1% chez les Uropyges et 2% chez les Amblypyges. Chez les araignées primitives (Theraphosidae), le corps central est bien développé contrairement au faible développement des corpora pedunculata. Millot (1949) et Meier (1967) suggèrent une relation entre l'industrie textile des Araignées et le developpement du corps central. Le travail expérimental de Le Guelte et Witt (1971) sur l'action des rayons laser sur le corps central d'Araneus diadematus effectué afin d'évaluer le comportement constructeur après lésion du corps central n'est malheureusement pas concluant vu la complexité du corps central et l'étendue de la destruction des tissus à ce niveau.

7. SIGNIFICATION ECOLOGIQUE ET ETHOLOGIQUE DE LA STRUCTURE DU SYSTEME NERVEUX ET EVOLUTION DES ARAIGNEES

A l'époque des recherches d'Hanström, au début du siècle, les travaux anatomiques sur les Chélicérates étaient difficilement interprétables par manque de données écologiques, éthologiques et évolutives. Cette situation a beaucoup changé et il est aujourd'hui possible de tenter d'interpréter toutes ces données dans un cadre évolutif en intégrant des connaissances écologiques et éthologiques. Il s'agit uniquement d'une tentative d'approche du problème et non de sa compréhension et de sa solution définitive, encore hors de portée. Les Araignées serviront de base à cet essai synthétique car c'est dans cet Ordre d'Arachnides que les travaux d'écologie, d'éthologie, de biologie sensorielle et surtout d'évolution sont parvenus à un niveau tel que des problèmes fort complexes peuvent être abordés et associés. Le travail de Platnick (1971), par exemple, sur le comportement de cour des Araignées peut fonder en partie les propositions exposées ici. Cet auteur, spécialiste de l'évolution des Araignées, propose un tableau synthétique de la phylogénie du comportement de cour comprenant trois niveaux.

Au premier niveau, les mâles des Mygalomorphes, des Aranéomorphes haplogynes et de beaucoup de membres de la lignée Clubionides - Thomisides répondent seulement au contact direct avec la femelle.

Dans le deuxième niveau, les Araignées sont caractérisées par l'emploi de stimuli tactochimiques ou de phéromones ou des deux associés. Les familles de ce deuxième niveau comportent quelques Mygalomorphes et Aranéomorphes haplogynes, tous les Aranéomorphes qui construisent une toile et probablement les familles Lycosidae et Pisauridae.

Enfin, au troisième niveau, le canal visuel est fondamental dans le comportement de cour. Les familles Oxyopidae, Salticidae et Lyssomanidae seulement sont situées à ce niveau.

Weygoldt (1977) dans son travail de synthèse sur la communication chez les Araignées suggère également que la communication chimique est le moyen le plus primitif de communication.

Du travail de Platnick (1971) ressortent plusieurs points en relation avec la phylogénie des Araignées; le premier réside dans le fait que les deux premiers niveaux sont entièrement dépendants de la méchano- et de la chémo- communication avec toutes ses variantes. Les signaux et la communication visuels apparaissent comme le facteur primordial dans le groupe très restreint des familles du troisième niveau.

D'après Krafft (1980), les signaux sonores et vibratoires sont utilisés à des degrés différents par les araignées vagabondes et fileuses.

La transposition du modèle phylétique de Platnick sur un autre comportement tel que la prédation montre que le canal visuel du comportement correspond au groupe qui, en se libérant de la toile et de la vie nocturne et en adoptant un mode de vie diurne (Salticidae), représente une véritalbe réussite évolutive. Evidemment, la structure de l'appareil visuel des Salticidae avec ses centres visuels très sophistiqués est l'élément le plus remarquable de ce phénomène. Aujourd'hui la structure visuelle des Salticidae est de mieux en mieux connue; il s'agit d'un appareil sensoriel extrêmement complexe: rétine stratifiée à quatre niveaux de récepteurs, rétine mobile et vision probable des couleurs (Land 1969 1985; Yamashita 1985); le comportement visuel est aussi remarquable (Crane 1949; Forster 1982; Jackson 1982). Hanström (1928) avait signalé de grandes différences dans les centres visuels et les corps pédonculés selon que les Araignées sont sédentaires ou vagabondes (Figs 7,8,9). Ces schémas sont devenus classiques et se retrouvent dans tous les ouvrages qui traitent du système nerveux des Arachnides. Aujourd'hui ces études anatomiques peuvent être rapprochées tout d'abord de la phylogénie des Araignées et ensuite de l'évolution des différents comportements tels que ceux de cour ou de prédation.

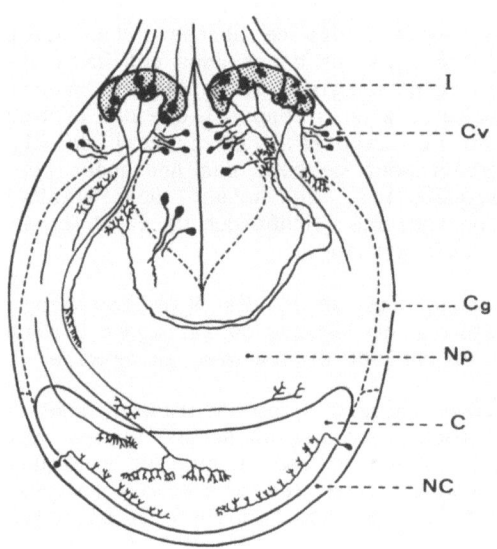

FIG. 7: Représentation schématique du cerveau d'n Aranéide (type Epeire) vu de dessus (D'après Hanström 1921, modifié). I : Centre optique des yeux latéraux; C : Corps cental; Cg : Cortex ganglionnaire; Cv : Cortex visuel; NC : Neurones du corps central; Np : Neuropile.

FIG. 7: Diagrammatic representation of an Araneide brain (Epeire = <u>Araneus</u> type) from above (After Hanström 1921, modified). I : Optic mass of the lateral eyes; C : Central body; Cg : Ganglionic cortex; Cv : Visual cortex; NC : Neurones of the central body; Np : Neuropil.

Chez les Araignées chasseresses Salticidae, les cerveau est le plus développé. Il renferme alors deux paires de centres visuels, l'une dorsale correspondant aux yeux pricipaux, l'autre, aux yeux indirects, et deux organes d'association, le corps central et les <u>globuli</u> des corps pédonculés. Chacun des centres visuels comprend un centre primaire antérieur et un centre secondaire plus ou moins postérieur. De chaque centre secondaire partent de nombreuses fibres vers le centre symétrique ou vers les organes d'association. Le corps central impair est relié aux voies optiques, particulièrement à celles qui viennent des yeux pricipaux, ce qui en fait un centre visuel d'ordre supérieur. Les <u>globuli</u> des corps pédonculés sont de petits nodules pairs, chacun relié par un puissant pédoncule à une importante commissure: le pont protocérébral. Ils sont situés entre les centres visuels secondaires des yeux indirects avec lesquels ils entrent en connexion intime. D'après Hanström (1928), ce sont des organes d'association de l'information visuelle.

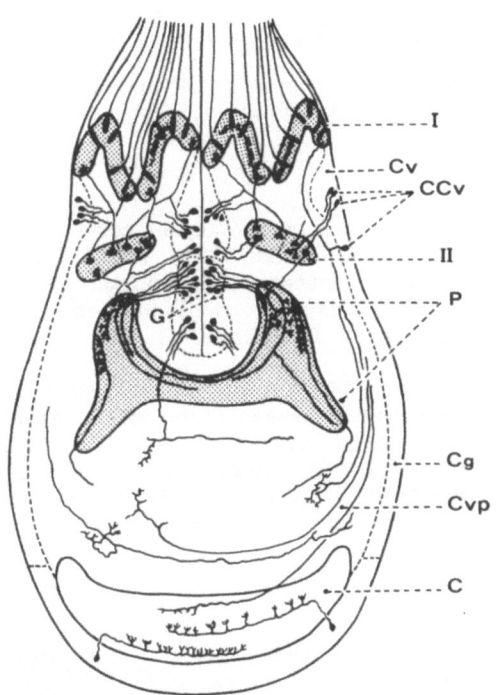

FIG. 8: Représentation schématique du cerveau d'un Lycoside vu de dessus (D'après Hanström 1921, modifié). I, II : Premiers et seconds centres optiques des yeux latéraux; C : Corps central; Cg : Cortex ganglionnaire; Cv : Cortex visuel; CCv : Cellules du cortex visuel; Cvp : Commissure visuelle postérieure; G : Globuli; P : Pédoncule du corps pédonculé.

FIG. 8: Diagrammatic representation of a Lycosid brain from above (After Hanström 1921, modified). I, II : First and second optic masses of the lateral eyes. C : Central body; Cg = Ganglionic cortex; Cv : Visual cortex; CCv : Cells of the visual cortex; Cvp : Posterior visual commissure; G : Globuli; P : Pedicel of the corpora pedunculata.

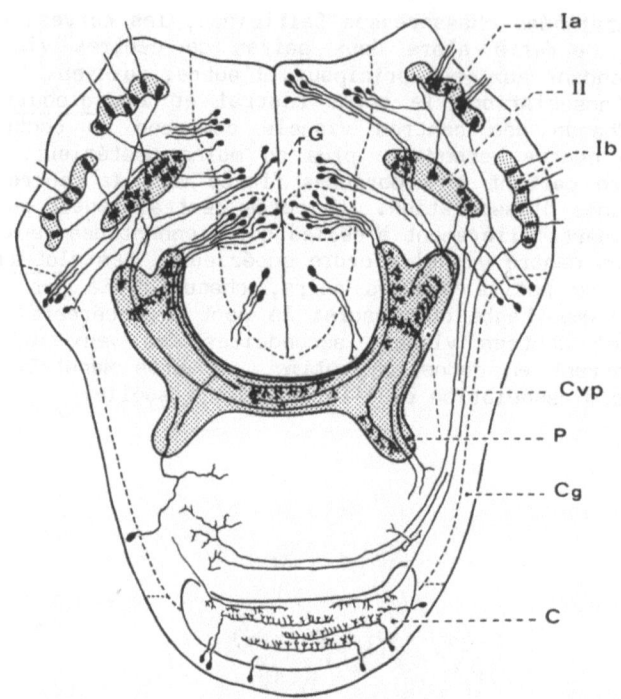

FIG. 9: Représentation schématique du cerveau d'un Salticide vu de dessus (D'après Hanström 1921, modifié). Ia, Ib, II : Premiers et seconds centres optiques des yeux latéraux; C : Corps central; Cg : Cortex ganglionnaire; Cvp : Commissure visuelle postérieure; G : Globuli; P : Pédoncule du corps pédonculé.

FIG. 9: Diagrammatic representation of a Salticid brain from above. (After Hanström 1921, modified). Ia, Ib, II : First and second optic masses of the lateral eyes; C : Central body; Cg : Ganglionic cortex; Cvp : Posterior visual commissure; G : Globuli; P : Pedicel of corpora pedunculata.

Chez les Araignées sédentaires (Agalenidentypus de Hanström comprenant: Ctenizides, Filistatides, Dysdérides, Uroctéides, Amaurobiides, Agélénides, Sicariides, Théridiides, Drassides, Clubionides), le cerveau est beaucoup moins développé. Les centres vusuels sont petits et moins différenciés. Les corps pédonculés sont très rudimentaires, les <u>glomeruli</u> ou centres visuels secondaires des yeux indirect sont disparu, les <u>globuli</u> et les pédoncules font entièrement défaut et le pont est rudimentaire.

La classificaiton des Araignées en quatre groupes d'après la structure du protocérébron (Hanström 1935) peut être rapprochée du modèle phylétique proposé par Platnick (1971). En effet, dans l'évolution des Araignées, il semble que les canaux de mécano- et chimio- communication soient liés à la vie nocturne de ces animaux. Les signaux visuels et leur perfectionnement évolutif dans la vie diurne jusqu'à la sophistication des Salticidea paraissent liés à la complexification des structures du protocérébron. Le passage de la vie nocturne à la vie diurne est sans doute le point crucial du problème.

La famille des Lycosidae est un autre groupe d'Araignées engagé vers la vie diurne, mais avec un degré de réussite bien moindre que celui des Salticidae. Chez les Lycosidae, ni les structures oculaires (Muñoz-Cuevas 1984), ni le comportement (Tietjen et Rovner 1982) n'atteignent le haut degré de sophistication des Salticidae. Dans la classification de Hanström (1935) et dans celle de Platnick (1971), les Lycosidae sont situés à un niveau intermédiaire (groupe B pour Hanström, niveau II pour Platnick). Les centres nerveux du protocérébron traduisent bien cette situation: les Lycosidae présentent, en effet, des centres visuels et des corpora pedunculata moins développés que les Salticidae. Ce phénomène évolutif, évident chez les Araignées, ne permet pas d'expliquer par ailleurs, le développement si important des corps pédonculés dans d'autres groupes d'Arachnides comme les Opilions Gonyleptidae. Ces Opilions n'ont pas privilégié le comportement visuel; il s'agit, au contraire, d'Arachnides d'activité crépusculaire et nocturne.

Fahrenbach (1979) devant l'énorme et intrigant problème de la fonction des corpora pedunculata de la Limule suggère que ces organes seraient le centre d'intégration qui gouvernerait les rythmes respiratoire, cardiaque, les rythmes de locomotion, d'alimentation et encore d'autres comportements de chémoréception. L'émumération de tous ces rôles, aussi problables que variés, ne prouve que la complexité du problème qui honnêtement reste presque entier.

8. CONCLUSIONS GENERALES

L'étude de l'ontogenèse du Système Nerveux Central des Chélicérates peut aider à comprendre certains problèmes de phylogenèse dans ce groupe si peu exploré jusqu'à présent. L'exemple du développement des corpora pedunculata parmi les Chélicérates, à l'exception des Araignées, apparaît encore comme une excitante énigme biologique. Les études comparées de neuroanatomie, de neurochimie, de physiologie du comportement, et de phylogénie dans un esprit synthétique permettront d'aborder dans le futur une interprétation valable. Dans la phylogenèse des Araignées, le passage éco- éthologique de la vie nocturne à la vie diurne en l'absence de toiles, semble être le facteur essentiel de la complexification du protocérébron.

L'unification des fonctions des centres nerveux dans divers groupes zoologiques, telle qu'elle fut suggérée dans le passé, paraît aujourd'hui difficilement acceptable.

Un aspect de première importance en neurobiologie est le fait de travailler dans la mesure du possible avec des lignées phylétiques afin de donner aux résultats et aux interprétations une dimension évolutive.

9. BIBLIOGRAPHIE

Aeschlimann A (1958) Développement embryonnaire d'Ornithodorus moubata (Murray) et transmission transovarienne de Borrelia duttoni. Acta tropica 15: 15-64.

Akesson B (1961) On the histological differentiation of the larvae of Pisione remota (Pisionidae, Polychaeta). Acta Zool intern tedskrift für zoologi, p 177-225.

Akesson B (1962) The embryology of Tomopteris helgolandica (Polychaeta). Acta Zool Stockh 43: 135-199.

Babu KS (1975) Post-embryonic development of the central nervous system of the spider Argiope aurantia (Lucas). J Morphol 146: 325-342.

Babu KS (1985) Patterns of arrangement and connectivity in the central nervous system of arachnids. In: Barth FG (ed) Neurobiology of arachnids. Berlin, Springer Verlag, p 3-19.

Barrois J (1896) Mémoire sur le développement des Chelifer. Rev Suisse Zool 3: 461-498.

Bazin F, Demeuzy N (1968) Existence d'organes intracérébraux énigmatiques chez le Crustacé Décapode Carcinus maenas. CR Acad Sci Paris 267: 356-358.

Bazin F (1969) Etude comparée d'un organe deutocérébral chez les Crustacés Décapodes Reptantia. CR Acad Sci Paris 269: 958-961.

Bazin F (1970) Etude comparée de l'organe deutocérébral des Macroures Reptantia et des Anomoures (Crustacés Décapodes). Archs Zool Exp Gén 111: 245-364.

Bazin F (1971) Le développement embryonnaire des organes deutocérébraux chez Astacus leptodactylus Esch. (Crustacés, Décapode, Reptantia). Ann Embryol Morphol 4: 137-144.

Brauer A (1895) Beiträge zur Kenntnis der Entwicklung Geschichte des Skorpions. Z wiss Zool 57: 351-435.

Carribaburu A, Muñoz-Cuevas A (1986) Spontaneous electrical activity of the suboesophageal ganglion and circadian rhythms in scorpions. Exp Biol 45: 301-310.

Crane J (1949) Comparative biology of salticid spiders of Rancho Grande, Venezuela. Part IX. An analysis of display. Zoologica 34: 159-214.

Dogiel V (1913) Embryologische Studien on Pantopoden. Z wiss Zool 107: 575-741.

Dohle W (1964) Die embryonal Entwicklung von Glomeris marginata (Villers) im Vergleich zur Entwicklung anderer Diplopoden. Zool Jahrb (Anat) 81: 241-310.

Fahrenbach WH (1979) The brain of the horseshoe crab (Limulus polyphemus) III. Cellular and synaptic organization of the corpora pedunculata. Cell Tissue Res 11: 163-199.

Forster L (1982) Visual communication in jumping spiders (Salticidae). In: Witt PN, Rovner JS (eds) Spider communication. Princeton Univ Press, Princeton New-Jersey, p 161-210.

Hanström B (1919) Zur Kenntnis der centralen Nervensystems der Arachnoiden und Pantopoden. Lund, Inuaugural Dissertation.

Hanström B (1921) Über die Histologie und vergleichende Anatomie der Sehganglien und Globuli der Araneen. Kungl Svenska Vetensk Handlingar 61 (12).

Hanström B (1923) Further notes on the central nervous system of arachnids: scorpions, phalangids and trapdoor spiders. J Comp Neurol 35: 249-274.

Hanström B (1928) Vergleichende Anatomie des Nervensystems der wirbellosen Tiere. Berlin, Springer.

Hanström B (1935) Fortgesetzte Untersuchungen über das Araneengehirn. Zool Jahrb (Anat) 59: 455-478.

Heymons R (1901) Die Entwicklungsgeschichte der Scolopender. Bibliotheca Zool 13: 1-244.

Holmgren N (1916) Zur Vergleichenden Anatomie des Gehirns von Polychaeten, Onychophoren, Xiphosuren, Arachniden, Crustaceen, Myriapoden und Insekten. Kungl Svenska Vetensk Handlingar 56: 1-303.

Iwanoff PP (1933) Die embryonale Entwicklung von Limulus moloccanus. Zool Jahrb (Anat) 56: 163-348.

Jackson RR (1982) The behavior of communicating in jumping spiders (Salticidae). In: Witt PN, Rovner JS (eds) Spider communication. Princeton Univ Press, Princeton New-Jersey, p 213-247.

Juberthie C (1964) Recherches sur la biologie des Opilions. Ann Spéléol 19: 1-238.

Juberthie C, Muñoz-Cuevas A (1971) Rôle des organes neuraux d'un Opilion Gonyleptidae, Pachylus quinamavidensis, dans la formation des globuli des corpora pedunculata. CR Acad Sci Paris 270: 1028-1031.

Juberthie-Jupeau L (1967) Existence d'organes neuraux intracérébraux chez les Glomeridia (Diplopodes) épigés et cavernicoles. CR Acad Sci Paris 264: 89-92.

Junqua C (1957) Aspects histologiques du système nerveux d'un Solufuge. Bull Soc Zool France 82: 136-138.

Junqua C (1966) Recherches biologiques et histophysiologiques sur un Solifuge saharien Othoes saharae Panouse. Thèse, Faculté des Sciences, Paris.

Kästner A (1952) Über zwei Entwicklungsstudien von Solifugen. Zool Anz 149: 8-20.

Kennel J (1888) Entwicklungsgeschichte von Peripatus edwarsi Blanch. und Peripatus torquatus, n.sp. Arb Zool Inst Würzburg 8: 1-93.

Kishinouye K (1892) On the development of Limulus longispina. J Cell Sci Tokyo 5: 53-100.

Krafft B (1980) Les Systèmes de Communication chez les Araignées. In: 8. Internat. Arachn. Kongress Wien, p 197-213.

Land M (1969) Structure of the retinae of the principal eyes of jumping spiders (Salticidae: Dendryphantinae) in relation to visual optics. J Exp Biol 51: 443-470.

Land M (1985) The morphology and optics of spider eyes. In: Barth FG (ed) Neurobiology of arachnids. Springer Verlag, Berlin, p 53-78.

Legendre R (1959) Contribution à l'étude du système nerveux des Aranéides. Thèse, Faculté des Sciences, Paris.

Le Guelte L, Witt PN (1971) Conséquences histologiques et comportementales de lésions (par laser) du ganglion susoesophagien de l'araignée: Araneus diadematus Cl. Rev Comp Animal 5: 19-26.

Meier F (1967) Beiträge zur Kenntnis der postembryonalen Entwicklung der Spinnen Araneida labidognatha. Unter besonderer Berücksichtigung der Histogenese der Zentral-Nervensystems. Rev Suisse Zool 74: 1-127.

Millot J (1949) Classe des Arachnides (Arachnida). Morphologie générale et anatomie interne. In: Grassé PP (ed) Traité de zoologie Vol 6. Masson, Paris p. 263-319.

Morgan TM (1891) A contribution to the embryology and phylogeny of the pycnogonids. Stud Biol Lab, J Hopkins Univ Baltimore 5: 1-72.

Muñoz-Cuevas A (1969) Contribution à la connaissance de la biologie des Opilions Gonyleptidae (Arachnida). Thèse Doct 3e Cycle, Faculté des Sciences Paris.

Muñoz-Cuevas A (1971) Etude du développement embryonnaire de Pachylus quinamavidansis (Arachindes, Opilions, Gonyliptidae). Bull Mus Natn Hist Nat 2e sér Zool 42: 1238-1250.

Muñoz-Cuevas A (1973) Embryogenèse, organogenèse et rôle des organes ventraux et neuraux de Pachylus quinamavidensis (Arachnides, Opilions, Gonyleptidae). Comparaison avec les Annélides et d'autres Arthropodes. Bull Mus Natn Hist Nat Paris 3e sér n° 196 Zool 128: 1517-1538.

Muñoz-Cuevas A (1984) Photoreceptor structures and vision in arachnids and myriapods. In: Ali MA (ed) Photoreception and vision in invertebrates. Plenum, New York, NATO-ASI series A 74: 335-399.

Nguyen Duy m (1971) Etude préliminaire sur la neurosécrétion céphalique chez le Diplopode Pénicillate Polyxenus lagurus (Myriapodes). CR Acad Sci Paris 42: 1238-1250.

Pereyaslawzewa S (1901) Développement embryonnaire des Phrynes. Ann Sci Nat, 8e sér 13: 123-303.

Pflugfelder O (1932) Über der Mechanismus der Segmentbildung bei Embryonalentwicklung und Anamorphose von Platyrrhacus amauros Attems. Z wiss Zool 140: 650.723.

Pflugfelder O (1948) Entwicklung von Peripatus amboinensis, n.sp. Zool Jahrb (Anat) 69: 443-492.

Platnick N (1971) The evolution of courtship behaviour in spiders. Bull Brit Arachnol Soc 2 (3): 40-47.

Sahli F (1966) Contribution à l'étude de la périodomorphose et du système neurosécréteur des Diplopodes Iulides. Thèse Doct. Sci., Dijon.

Schimkewitsch L, Schimkewitsch W (1911) Ein Beitrag zur Entwicklungsgeschichte der Tetrapneumones. Bull Acad Sci St-Pétersbourg 5: 8-10.

Tiegs O (1940) The embryology and affinities of the Symphyla based on a study of Hanseniella agilis. Q J Microsc Sci 82: 1-225.

Tietjen WJ Rovner JS (1982) Chemical communication in lycosid and other spiders. In: Witt PN, Rovner JS (eds) Spider communication. Princeton Univ Press, Princeton New Jersey p 259-279.

Viallanes H (1893) Etudes histologiques et organologiques sur les centres nerveux et les organes des sens des animaux articulés. Ann Sci Nat Zool 14: 405-456.

Weygoldt P (1964) Vergleichend-embryologische Untersuchungen an Pseudoscorpionen (Chelonethi). Z Morphol Ökol Tiere 54: 1-106.

Weygoldt P (1975) Untersuchungen zur Embryologie und Morphologie der Geisselspinne Tarantula marginemaculata C.L. Koch (Arachnida, Amblypygi, Tarantulidae). Zoomorphol 82: 137-199.

Weygoldt P (1977) Communication in crustaceans and arachnids In: Sebook ThA (ed) How animals communicate. Th.A.Sebook. Indiana Univ Press, Bloomington and London, p 303-333.

Yamashita S (1985) Photoreceptor cells in the spider eye: Spectral sensitivity and efferent control. In: Barth FG (ed) Neurobiology of arachnids. Springer-Verlag, Berlin, p 103-117.

Yoshikura M (1954) Embryological studies on the liphistiid spider: Heptathela kimurai. Part I. Kumamoto J Sci Ser B 3: 41-48.

Yoshikura M (1955) Embryological studies on the liphistiid spider: Heptathela kimurai. Part II. Kumamoto J Sci Ser B 2 (1): 1-86.

THE NERVOUS SYSTEM OF THE CRUSTACEA, WITH SPECIAL REFERENCE TO THE ORGANISATION OF THE SENSORY SYSTEM

M. S. LAVERACK

Gatty Marine Laboratory

St Andrews

Fife KY16 8LB

Scotland

ABSTRACT

An attempt is made to review the arrangement of the sensory system (excluding the eyes) in the Crustacea, with examples of types of sense organs from mechanical and chemical sensitive groups. The morphogenesis of sensory organs increases the numbers of individual units, at each moult, and ensures a distinctive pattern of sensors associated with each sensory bundle of nerves.

The mechanoreceptors of the basal insertions of limbs (body-coxal joint) form a series. Fundamentally it is likely that each segment contains a chordotonal, and a muscle receptor, organ at the base, but this primary arrangement has been modified according to the specialisation of the limb. The abdominal limbs have only large diameter, non-spiking sensors, which are mimicked in the uropods, though spiking may occur in the latter. In the thorax both types of receptor organs are found in all pereiopods and in the 2nd and 3rd maxillipeds. No chordotonal organ is located in the scaphognathite, nor the mandible, although the former has large diameter sensory neurones with central cell bodies while the latter has a muscle receptor organ (MRO) with bipolar cells. The 1st antenna possesses a chordotonal organ associated with a muscle. The number of individual receptor cells is usually small in non-spiking MROs and rather more in bipolar chordotonal organs.

Amongst chemoreceptors there is a range of organs that may take the form of sensors lying flush with the surface or alternatively as setae projecting into the environment. The number of units represented in such receptors varies from 2 - 3 in oesophageal sensors, through 12 - 14 in some amphipod setae, to 20 - 22 in funnel canals, up to 300 or so in decapod aesthetascs. Some of these receptors carry terminal apertures, whilst others do not. In some the ciliated endings of the dendrites branch but in others they do not.

In summary the sensory system of Crustacea is orderly, with varied and distributed types of end organs.

1. INTRODUCTION

The life history of a crustacean is composed of a number of stages, the precise number of which differs from group to group (see Williamson 1982) e.g. Macruran decapods such as Homarus have only three free swimming stages although the Palinuran Panulirus may have up to 11 phyllosoma, puerulus and post-puerulus stages. No matter what, or how many, stages are present the animal is at all times a crustacean.

The implications of numerous moults for the architecture and anatomy of the nervous system are immense and a great deal remains to be done in the field of growth, moulting and regeneration of the nervous system.

2. GROWTH

The succession of larval stages in the development of Crustacea leads, except perhaps in parasitic forms such as Rhizocephala cirripedes, Monstrillidae and other parasitic copepods, to a gradual increase in the complexity and size of the nervous system. The apparatus suitable for a small embryonic creature, or the first of many stages in a calanoid copepod or a palinurid decapod is not sufficient for an adult animal. Nowhere, however, has the ontogeny of the system been followed in a systematic fashion. The evidence that exists is patchy and incomplete. What may be remarked upon is that the motor system controlling muscular coordination and contraction is small in number of components in adults (fast, slow and inhibitory motor neurones are not numerous elements in the nervous system even where they are completely known), and may only change when modes of locomotion change e.g. the larval zoea of Decapoda possesses exopodites that are lost when the animal changes from a planktonic, swimming creature to a benthic, crawling animal. The exopodites beat continuously in the larval lobster (Neil et al. 1976; Macmillan et al. 1976), but disappear in stage 4 which is the stage that settles to a benthic habitat. Some loss of motor neurones may occur at this stage, or perhaps they switch to a new function. In any case the overall numbers are small.

Allen (1894a) figured various nervous pathways in the embryonic lobster and showed motor neurones, giant fibres and other intersegmental fibres in the ventral chain, and in a second paper (1894b) showed the basic anatomy of the oesophageal and stomatogastric ganglia and associated sensors. Davis (1973) and Davis and Davis (1973) showed that rhythmicity occurred in the swimmeret system before the pleopods develop and Neil et al. (1976) and Macmillan et al. (1976) showed that rhythmicity is evident from hatching onward by observation of exopodite beating. Sensory information plays little part in the establishment of these rhythms; they are generated centrally. Anatomically and physiologically, abdominal superficial flexor muscles of lobster larvae, are similar to those of adults with one inhibitory and up to three excitatory fibres (Kirk and Govind 1983).

What is interesting about these observations is that much of the additional nervous material during growth comes in the sensory system. Each moult stage increases the complexity of the sensory apparatus by adding new sensors and the attendant axons must make synaptic contact in the CNS with concomittant changes in the complexity of ganglionic neuropile. Whether additional neurones are generated centrally remains to be seen. The increase in sensors has been documented by Laverack (1976, 1978) and by Spencer (1986). Laverack (1978) showed that in Homarus gammarus at each moult additional sensillae (approx 10%) are discerned in the post-moult animal, and that numbers of cuticular articulated peg (CAP) sensillae continue to rise throughout life. We may safely presume the same occurs amongst other sensillae. The additional sensilla are undoubtedly elaborated in the intermoult period but only manifest themselves externally at the post-moult stage. Spencer (1986) quotes unpublished work with Linberg that indicates that antennular annuli are added progressively during development, but this does not affect the overall range of sensitivity of the aesthetasc setae towards amino acids. Orderliness is maintained since sensors are produced in a standard sequence and this no doubt allows correct linkage to be obtained in the central nervous system.

The orderly arrangement of the nervous system is not always immediately apparent and the tangle of the neuropile in the centre of an invertebrate ganglion may lead one to suspect that there is little rhyme or reason in the manner of construction. That this is not so is becoming evident from a number of lines of approach, in particular developmental studies and fine anatomy using dye injection techniques.

The peripheral nervous system is in its way just as ordered as the central nervous system. The eye and the arrangement of ommatidia is well known (Shaw and Stowe 1982), and other parts of the peripheral system are just as well organised. Norris and Hartman (1985) have recently shown that in crayfish the mechanosensory fields of the chela are distributed in lines and groups that are composed of particular types of sensors in circumscribed areas and whose action leads to individual and specific reflex behavioural actions. Anyone who has picked up a live lobster or crayfish and blown air gently on the surface of the chela will know that the claw closes due to stimulation of sensors carried on the surface. It is also obvious that the chela opens again. Norris and Hartman (1985) recognise four morphological types of receptors with eight different areas linearly arranged, and each innervated by a specific branch of the

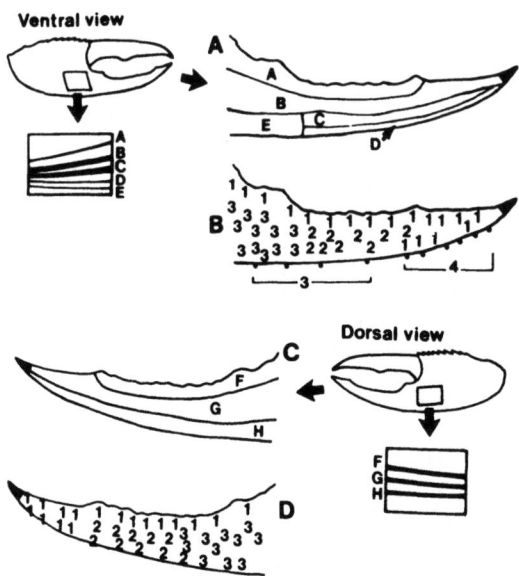

FIG. 1: Ventral and dorsal views of the crayfish claw showing the branching pattern of the afferent nerve innervating the propus, the fields of innervation and the location of the four types of cuticular hair organs (from Norris and Hartman 1985, with the permission of Pergamon Press).

afferent nerve. Of the eight branches two innervate receptors on either side of the cutting edge of the chela (propus) and stimulation leads to closing of the claw. Two other rows contain receptors that drive both opening and closing, while the remaining four are responsible for receptors that promote chela opening only (Fig. 1).

Jellies and Larimer (1986) have produced similar diagrams for the innervation fields of the body of Procambarus as represented by the activity of interneurones in the nerve cord. Information from a variety of sources on interneurone distribution and on motor cell position all indicate that the nervous system is indeed highly ordered although some individual variation may occur.

It should also be noted that sensors present in larvae may disappear when metamorphosis takes place. The complex frontal system of the nauplius larvae of barnacles for example (Walker 1974; Walker and Lee 1976) is not found in the adult since it is an adaptation to enable settlement to occur correctly. Similarly the dorsal organ (Laverack and Barrientos 1985) has not been observed in the juvenile or adult but occurs in all the larval stages of all Decapoda so far examined.

3. MOULTING

The process of moulting is, of course, a complex one and under endocrine control. Aiken (1980) gives a good review and indicates some characters that can be used to stage the moult cycle. Aiken (1973) previously described the appearance of the setae, but did not remark upon an increase in numbers, nor upon the fate of any associated sensors. This topic has been dealt with by Guse (1983), Heimann (1984) and Espeel (1986) and details of the morphogenesis of new setae should be sought there. Each author concludes that, despite the apparent inertia and insensitivity of the newly moulted animal all sensors are still in place, probably function up to the moment of moult, and in all probability still function immediately afterwards. Recent evidence (P. Borroni and M.S. Laverack, unpublished) reveals that in the lobsters, Homarus americanus and H. gammarus, in an animal in which the new cuticle is fully formed, but the old exoskeleton not yet shed, the 'fringed setae' (hedgehog hairs) are fully supplied with dendrites in the newly secreted replacement, and in the older setae about to be shed the dendrites are either absent or in the process of autolysis. This suggests that the peripheral ends are indeed lost at moult, that the final events occur close to ecdysis but precede it and hence there may be some loss of sensitivity due to destruction of receptor endings. There may, however, still be continuity of contact of the environment with the newly formed dendrites providing the central setal lumen remains open. Under these circumstances the pre-moult animal may "lose its appetite" due to loss of sensory information - but the immediately post-moult animal is already provided with a functional set of receptors. This contrasts with earlier reports in smaller Crustacea.

Only the work of Yules (1962) has investigated physiological sensory activity at the moult and he concluded that chordotonal proprioceptors do indeed function at all times. Newly added sensors must function immediately. Any lack of reaction therefore seems likely to be a central phenomenon. Internal organs such as the abdominal MROs, coxo-thoracic sensors, mouth part receptor (MPR) systems all remain constant during the moult while chordotonal organs, essentially surface structures of hypodermal origin, have the capacity to add new receptors. Indeed the constancy of numbers (usually small) in centrally located receptors may well indicate a once-for-all origin, whilst chordotonal and setal receptors add to their numbers continuously throughout life. As

Alexandrowicz (1967) and, later, Laverack (1976) pointed out there are distinct affinities between surface CAP organs (setae) and internal chordotonal organs at the joints of walking legs in Decapoda (see also Vedel 1986). The same may occur in the tension receptors found on the apodemes (Macmillan and Dando 1972). All these types may increase in cell numbers throughout growth. Species that reach a terminal size such as crabs probably have a final nervous system, but those such as lobsters that reach no final size continue to add sensors for many years.

4. AUTOTOMY AND REGENERATION

During the life of a crustacean, damage due to accident, fighting or predation may occur. Once again this is best documented amongst Decapoda. A natural reflex process of limb loss or autotomy occurs under certain circumstances and this mechanism is now well understood (McVean 1982). Any or all of the walking legs may be lost in this way though eyes, antenna, uropods are not. The pereiopods are replaced by regenerative growth which eventually leads to complete restoration of the appendage. Cuticle, muscles, blood system, superficial sensillae and proprioceptors are all replaced.

There are profound and significant implications for the nervous system in the appropriate segment. Motor fibres must regrow from the CNS centrifugally and make contact with appropriate redifferentiating muscles while sensory fibres originate from the hypodermis and migrate centripetally to the CNS. This activity necessitates re-establishment of synaptic connections within the neuropile. No work appears to have been done on natural autotomy replacement which takes place in the intermoult period. At the first moult following autotomy a limb bud appears, at the second moult a complete limb of small size, which then increases with further moults.

Experiments allied to these observations have been carried out on surgically operated nerves, especially in the laboratory of Bittner (see Hoy et al. 1967; Bittner et al. 1974; Ballinger and Bittner 1980; Bittner 1981). This group demonstrated that distal regions of sensory axons degenerate in about 20 days in a fashion similar to those of vertebrates, whilst proximal regions may re-establish contact with the CNS. Motor axons on the other hand survive morphologically and physiologically after separation for up to 200 days and outgrowing axons may re-fuse with surviving distal processes. The motor and giant axons seem constant throughout ontogeny having larger neurones and central cell bodies. There is little degeneration or regeneration even after 9 - 12 months of isolation. Redifferentiation to active neurones therefore seems limited to hypodermally originating sensors. It is noteworthy that the parts of limbs lost at autotomy do not include the coxa that is sensorily innervated from the CNS.

There is no reported difference in sensitivity between normal and regenerant limbs except in one remarkable case. If an eye is removed it is not replaced. Instead a short, stunted, antennule (called a heteromorph) regrows. This is the outer ramus portion with aesthetasc hairs etc. (Herbst 1900). So far as is known this limb functions as an antennule (Laverack 1964) and influences behaviour through the chemical sense. The alterations in the brain must be significant (Maynard and Dingle 1963).

5. MATURE FORMS

The process of cephalisation and of tagmatisation reaches a climax in the Crustacea, though it continues as a logical sequence form that of the annelids and other lowly forms. The development of an elongate body with one end usually progressing in advance of the remainder, becomes very pronounced in the Crustacea with the concentration of appendages around the mouth which are concerned with food location and presentation. Transferral of information gained anteriorly to more distant parts of the body and the coordination of other limbs for walking and swimming requires the presence of interneurones as well as motor neurones for locomotory activity.

The Crustacea are a large group with many different orders and with numerous phases to the life history. It is necessary to remember that within the Class there are all types of animals from swimming phyllopodous Branchiopoda such as Artemia and Chirocephalus which use their limbs for both swimming and feeding to the scavenging raptorial larger Decapoda with considerable limb specialisation. It should also be recalled that many representative species possess one or more larval stages in their life history.

These remarks need amplification since despite the considerable amount of progress that has been made on the understanding and knowledge of various aspects of the nervous system much of it revolves around very few species. The crayfish, in various guises (Orconectes, Pacifastacus, Austropotamobius), lobsters (Homarus), spiny lobsters (Panulirus, Palinurus), and crabs (Carcinus, Cancer) provide most of our knowledge, especially in the realms of physiology.

Very little is known about copepods, tanaids, cumaceans, or stomatopods, and even less about nauplii, zoea, megalopa, furcilia, and puerulus larvae yet they are all Crustacea. It is unfortunate that they are all rather small and hence difficult to analyse experimentally. Nonetheless in terms of behaviour and fine structure they are all likely candidates for investigation. Some work has been focussed upon these small representatives but there remain many opportunities for examination of the development of the nervous system both within the Class, the Orders and the organisms.

In order to preface these remarks it may help to outline the space of influence around the animal. This "active space" has recently received much attention and indicates how significant the immediate surroundings are for the animal and how important the local conditions are for the sensory apparatus to operate.

To take four examples;

a. Many years ago Harris and van Bergeijk (1962) qualified external mechanical signals as "near field" and "far field" in origin. This distinction relied upon the oscillation of a source that gave rise either to particle movement and actual displacement close to the source which is the "near field", or to a propagated pressure wave (sound), which is transmitted over long distances, but is reliant on pressure not on actual displacement. Obviously the distance is critical and depends upon the signal frequency and intensity. Consequently the responses may reflect sensitivity to movement in a gross manner or to pressure waves (hearing). For many years it has been uncertain whether crustaceans 'hear', or if they respond to vibration in other ways (Horch 1971). Denton and Gray (1985) have shown that the antennae of some deep sea shrimps such as Sergestes, Acetes, and Funchalia act as lateral lines in a similar way to those of fish. The antennae are trailed at a distance from the body and are covered with setae that contain mechanoreceptors, which respond to frequencies above 1 Hz and plateau at 50-100 Hz (see Laverack 1964; Tautz et al. 1981).

b. A typical crustacean standing on the sea, lake or river bottom may not be too bothered about light except as a beacon to orientate by but the immediate vicinity holds many clues as to valuable behaviour in the future. The anteriormost organs of crustaceans are the antennae, two pairs, and in the decapods these are very long and flexible. They are loaded with receptors. The animal advances with these carried before it and they are of prime importance in detecting what is at hand. The importance of the second antenna has been amply shown by Sandeman (1985) and Zeil et al. (1985) for the Australian crayfish Cherax destructor. Analysis of the antennary movements, the angles of flexion, extension, the direction of turning, the positions of tactile stimulation and the reflexive consequences of these events are well described by these authors. The properties of the mechanoreceptors with which the organ is heavily endowed are well known (Tautz et al. 1981; Wiese 1987) and the proprioceptive innervation is also known to some extent (Vedel and Monnier 1983; Vedel 1985, 1986; Wyse and Maynard 1958).

c. The significance of the first antennae, (or antennules) has been demonstrated by Atema (1985) for chemical perception for not only are these anterior limbs important for mechanoreception but they also carry a vast population of chemoreceptors (Laverack 1964; Spencer 1986). In this case the positions of the limbs are concerned with monitoring the current flow around the animal to discriminate direction and chemical content of water that may originate at a distance and be drawn towards the animal. Water to the front, and also to the sides, of the animal may be monitored and sampled. In this case it is the chemoreceptors that are stimulated rather than mechanoreceptors though, of course, in life both types of units will be in operation. It is a moot point as to which is going to be the overriding sense when an animal walks and the Zeil et al. (1985) work mentioned above has to be considered not in isolation but in conjunction with the likelihood of chemical stimulation occurring at the same time.

d. The local environment, in which sensors must work, is not always the apparent one. This is particularly so in the case of small organisms.

Atema (1987) calls attention to the various scales at which water may move, from ocean currents to micro-eddies, from patchiness to animal generated information currents. For large animals current flows and spatial variability is important but for small animals, such as copepods, nauplii, and other larvae the water is a viscous fluid and this imposes other properties on the environment. Strickler (1985) elegantly demonstrated that calanoid copepods generate flow fields with low Reynolds numbers and which are laminar. Turbulence is reduced to a minimum, and the local environment is created by the copepod and moves with it. Any disturbance of the 'capsule' can be detected against a virtually motionless background, and chemical stimuli from food particles provide clear input without degradation. The receptors, largely carried on the antennae, are placed ideally for stimulus detection. Small larval Crustacea, therefore, possess sensors that are able to utilise ambient information without degradation by environmentally induced perturbation.

As a last comment in this respect it is interesting to note that the antennal structures and activity have attracted the most attention and yet the crustacean body is heavily provided with sensors over the entire surface.

6. THE PERIPHERAL NERVOUS SYSTEM

The distinction between the central nervous system i.e. that part which is concerned with integration and with co-ordination, carried out by neurones with cell bodies located within the ganglia and with long connectives, and the peripheral nervous system i.e. that part with sensory cells situated at the edge of the body and with afferent axon tracts running centripetally towards the ganglia, is not always easily maintained.

Sensors are found both externally and internally. Much mechanical information regarding limb position and equilibrium comes from internal proprioceptors while most chemical cues are detected by external structures.

I will deal with both sets of events by reference to numbers of cells involved, types and distribution of organs.

6.1. Proprioceptors

Pringle (1961) pointed out that arthropod proprioceptors could be seemingly arranged in two groups: I) with an unbranched distal process, originating from the hypodermis and ending in cuticular material; and II) branched terminals, migrating from the central nervous system and ending in connective tissue. Type I included crustacean chordotonal organs, chelicerate slit sensillae and insect chordotonal organs. Type II comprised crustacean muscle receptor organs, innervated elastic strands, and free endings in joint membranes. Chelicerate examples were 'deeper-lying' endings, and insects possess stretch receptors (e.g. Braunig et al. 1986).

This general pattern is still acceptable (see Cobb and Heitler 1985), but one should point out that the situation has become more complicated than was realised when Pringle established his classification. Type I is still straightforward and composed of bipolar cells that insert on elastic strands with dendrites that project peripherally, with lengthy axons carrying propagated action potentials that reach the central nervous system. These originate in the hypodermis.

Type II units, however, are now known to show a range of form of which Pringle was not aware. The crustacean abdominal muscle receptor organ lies in each abdominal segment (for details of decapod examples see Bush and Laverack 1982; they are also reported in Isopoda, Stomatopoda and Amphipoda) with a large sensory cell linked to a small receptor muscle. The sensory cell has large diameter but short length dendrites that ramify amongst the muscle fibres, or in a short piece of intercalated connective tissue. Cells of similar structure exist in the peri-pharyngeal system (MPR, Laverack and Dando 1968) but they innervate sheets of connective tissue around the oesophagus (see also Laverack 1974). This type of cell also exists in the paragnath and similar cells attach to an elastic strand in the coxa-basipodite joint of the barnacle cirrus (Clark and Dorsett 1978). None of these are connected with muscle in a direct way.

On the other hand MRO systems located in the limb base sequence (see Fig. 2) in the pleopods and uropods are associated with both connective tissue, but in parallel with muscle bands of no importance in generating power. In other words they are sensors of tension and stretch relying on muscle contraction to generate a biassing when the major power muscles of the area contract. The peculiarity of these units is that the cell bodies are within the central ganglia and it is the dendrites that extend over long distances to the periphery. This is quite different to those of the abdominal MRO. In fact there are major physiological differences also. In the abdominal organ the dendrites respond to stimulation by local decrementing electrotonic potentials that give rise to an action potential

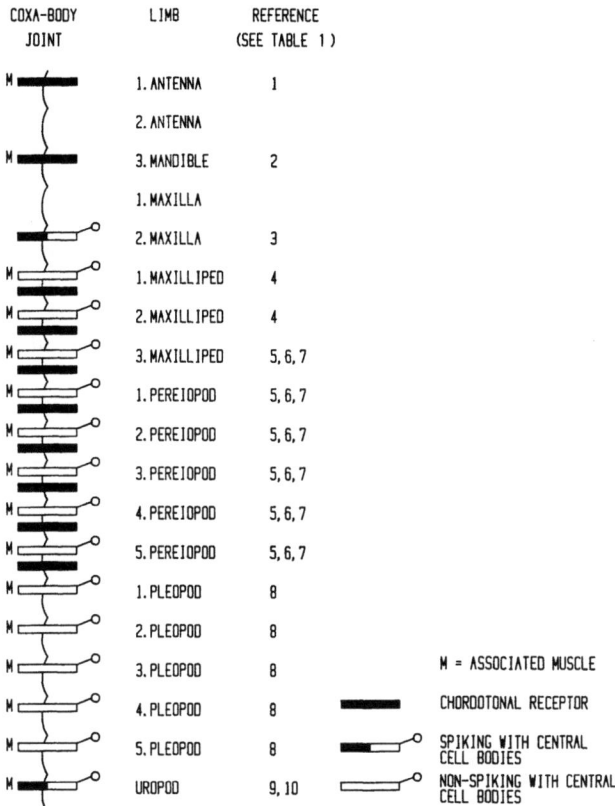

FIG. 2: The distribution of chordotonal organs and muscle receptor organs at the body-coxa joint. The MRO systems all possess central cell bodies, and are found throughout the animal with the exception of the most anterior limbs (antennae, mandible, maxillae). Chordotonal organs are almost universal but are not known in the uropods at this joint (see Table 1).

within a few hundred micrometers of the cell body within the axon. By contrast the dendrites of the uropod and pleopod receptors produce non-spiking graded potentials that are transmitted to the CNS, although in some circumstances spiking does occur in the same unit (Paul 1972; Bush 1976; Maitland et al. 1983; Pasztor and Bush 1983; Laverack this chapter).

Table 1 summarises the situation as it is known, and together with Fig. 2 shows the distribution and type of proprioceptors at the body-coxopodite joint for each limb in the segemental sequence.

Two examples of new types of proprioceptors follow;

6.1.1. Paragnath receptors

The paragnath is a small bilaterally symmetrical lobed structure that lies immediately behind the mouth. Its position in the segmental sequence has always been problematical - usually it is ignored (e.g. McLaughlin 1982). Its position is between the mandible to which it is closely

TABLE 1

Table 1 indicates information available for the proprioceptive innervation of the basal joint from body to coxa of each limb in the segmental sequence from a variety of animals. It may be hypothesized that the original sensory equipment consisted of both bipolar and an MRO arrangement, and that this basic arrangement has been modified during the process of limb specialisation. For example only an organ with bipolar cells, coupled with a muscle, and hence technically an MRO, but of a type reminiscent of the myochordotonal organ of the walking legs is present in the second antenna (Vedel and Monnier 1985). Other chordotonal organs are known in the antennules (Hartman and Austin 1972; Schöne and Schöne 1980; Wyse and Maynard 1958) but not associated with muscle or with the basal joint.

Body-Limb joint	Chordotonal			Muscle Receptor Organ		
	bipolar	spike	present	central cell body	spike	non-spike
1st antenna	yes[1]	yes[1]				
2nd antenna	yes[2]	yes[2]	yes[2] in second joint			
mandible	no	yes[3]		no	yes[3]	no
1st maxilla						
2nd maxilla	oval organ not chordotonal or MRO			yes[4]	yes[4]	yes[4]
1st max'ped	yes[5]	yes[5]		yes[5]		
2nd max'ped	yes[5]	yes[5]		yes[5]		
3rd max'ped	yes[6]	yes[6]		yes[6]		
1st pereiopod	yes[6]	yes[6]		yes[6,7]		yes[8]
2nd pereiopod	yes[6]	yes[6]		yes[6,7]		yes[8]
3rd pereiopod	yes[6]	yes[6]		yes[6,7]		yes[8]
4th pereiopod	yes[6]	yes[6]		yes[6,7]		yes[8]
5th pereiopod	yes[6]	yes[6]		yes[6,7]		yes[8]
1st pleopod		yes[9]		yes[9]		yes[9]
2nd pleopod		yes[9]		yes[9]		yes[9]
3rd pleopod		yes[9]		yes[9]		yes[9]
4th pleopod		yes[9]		yes[9]		yes[9]
5th pleopod		yes[9]		yes[9]		yes[9]
uropod		yes[10]		yes[10]	yes[11]	yes[10]

TABLE 1 (continued)

References

1. Wyse and Maynard 1958 — <u>Panulirus argus</u>
2. Vedel and Monnier 1983 — <u>Palinurus vulgaris</u>
3. Wales and Laverack 1972a,b — <u>Homarus gammarus</u>
4. Pasztor 1969;
   Pasztor and Bush 1983 — <u>Homarus gammarus</u>
5. M.S. Laverack unpublished — <u>Homarus gammarus</u>
6. Alexandrowicz and Whitear 1957 — <u>Homarus gammarus</u>
7. Wilson and Sherman 1975 — <u>Homarus</u>
8. Bush 1976 — <u>Carcinus maenas</u>
9. Heitler 1982 — <u>Pacifastacus leniusculus</u>
10. Paul 1972 — <u>Emerita</u>
11. Maitland, et al. 1983;
    Laverack this chapter — <u>Galathea strigosa</u> / <u>Homarus gammarus</u>

applied, and the first maxilla. The innervation of this structure is unclear (Bullock and Horridge 1965) and may not represent a separate neuromere. If it were a segmental appendage one could anticipate that any integral proprioceptors might fall in the chordotonal series or the typical body-coxa MRO pattern, but they do neither (Fig. 3). Rather they resemble the organs of the peripharyngeal band (MPR, Laverack and Dando 1968) which are not like limb receptors in that they are not bipolar, not attached to muscle, have peripheral cell bodies and are multi-dendritic (see also Robertson and Laverack 1979a). From such evidence one concludes that these little flaps are an extension of the posterior wall of the mouth rather than an appendage.

6.1.2. The uropod MRO in <u>Homarus</u>

A typical member of the segmental body-coxa sequence of large dendrite, central-cell body, proprioceptors is the uropod MRO (Fig. 4A). In <u>Homarus</u> this organ innervates a small tendon that joins the rather complex arrangement of uropod muscle insertion at the sclerites of the limb. The productor muscle of the uropod has two parts - the anterior portion being large and powerful while the rear part is small, attached to the larger area, but separately innervated, and with a tendinous end that extends to the telson flexor insertions (Fig. 4B). The receptor endings are of large diameter (up to 40μm) and meet the accessory productor muscle tendon. There are about 20 - 21 units in the nerve, and a small branch provides motor supply to the accessory muscle. The dendrites end in spray-like clusters on the tendon.

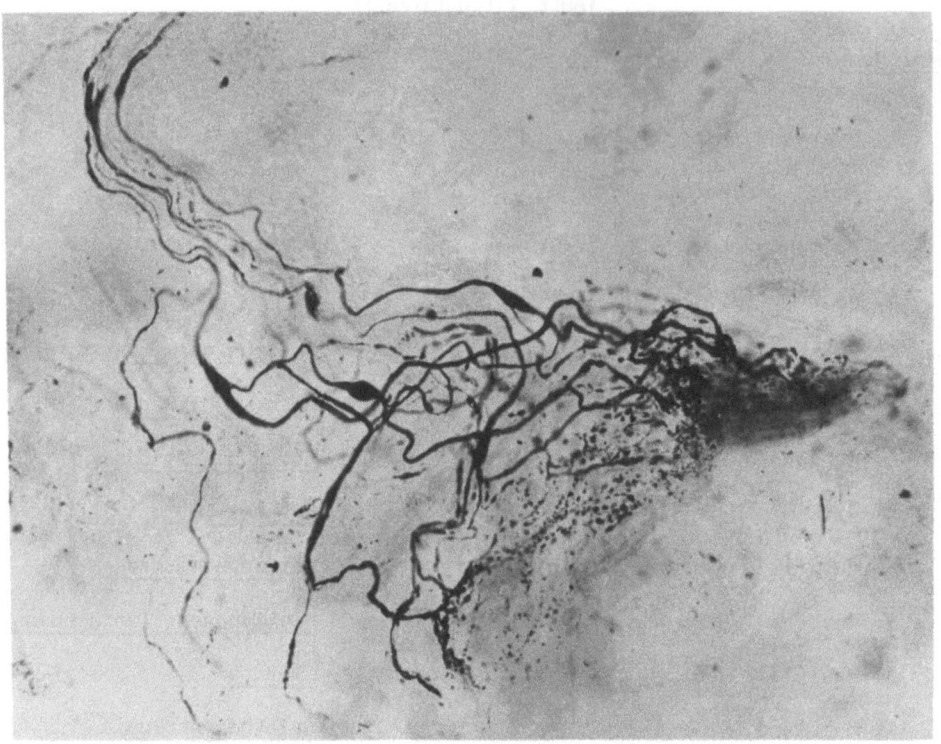

FIG. 3: A methylene blue preparation of the multidendritic elastic-strand mechanoreceptor of the paragnath of Homarus gammarus. (M.S. Laverack unpublished).

Cobalt back-fills reveal that the cell bodies are located in the last abdominal ganglion (Fig. 4C) and that 2 cells lie quite separately to the remainder. These are believed to be the motor and inhibitory neurones serving the muscle while the remainder are sensory in nature.

The physiological responses of these receptors have not been fully characterised but preliminary results suggest they have very similar properties to those of the scaphognathite oval organ (Pasztor and Bush 1983).

Three examples of uropod MROs are now known (Emerita, Paul 1972; Galathea, Maitland, et al. 1982; and Homarus, this paper). Non-spiking neurones that may be driven to spike under certain conditions seem to be standard. The details of these organs differ but the principle of construction and response is similar.

Another version of type II sensory units is an integral part of the CNS. Hughes and Wiersma (1960) demonstrated that stretch of the nerve cord led to spike discharge of mechanoreceptors apparently located within the nerve cords. Grobstein (1973) extended these observations and showed clearly that the units reponded to stretch and were to be found in each abdominal segment. Cobb and Heitler (1985) have recently elucidated the structure of these mechanoreceptors. They lie in each abdominal segment, attached to the medial side of each medial giant fibre by a connective

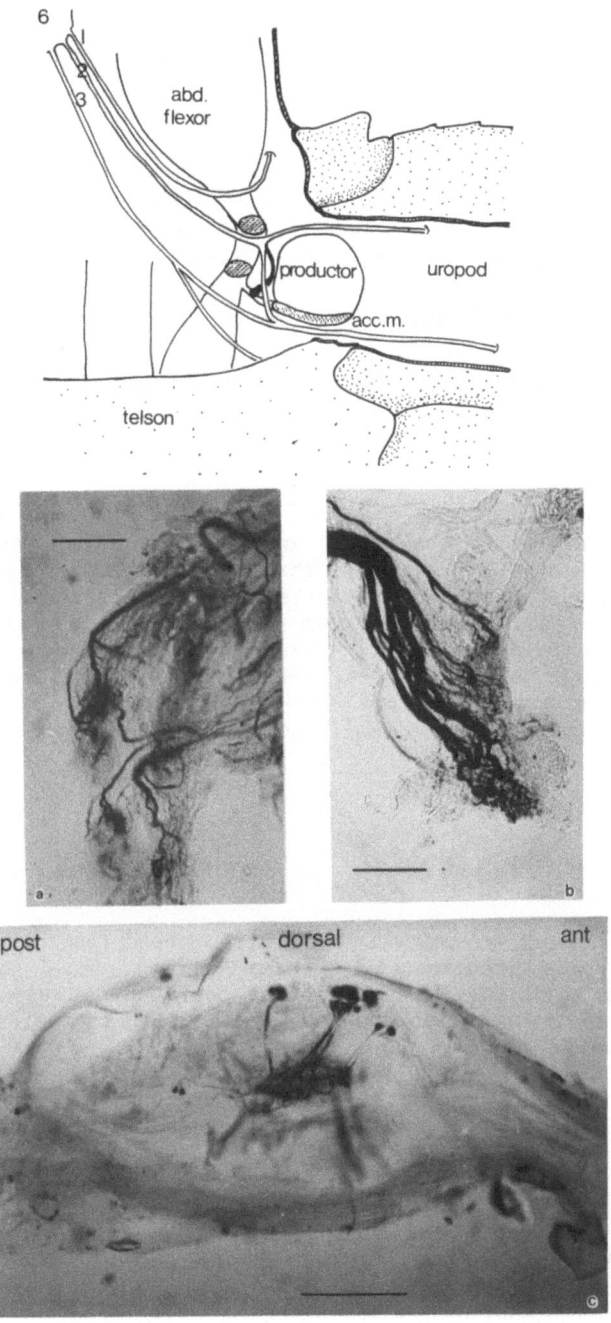

FIG. 4: The muscle receptor organ of the uropod of <u>Homarus gammarus</u>.
a. The anatomy of the organ, showing salient features of the uropod skeleton, musculature and innervation.
b. The detailed anatomy of the ending of the large diameter sensory neurones on the tendon attached to the accessory productor muscle, and the motor innervation of the accessory productor muscle. Bar = 200 μm.
c. A cobalt backfill of the sensory nerve and its central representation within the sixth abdominal ganglion.

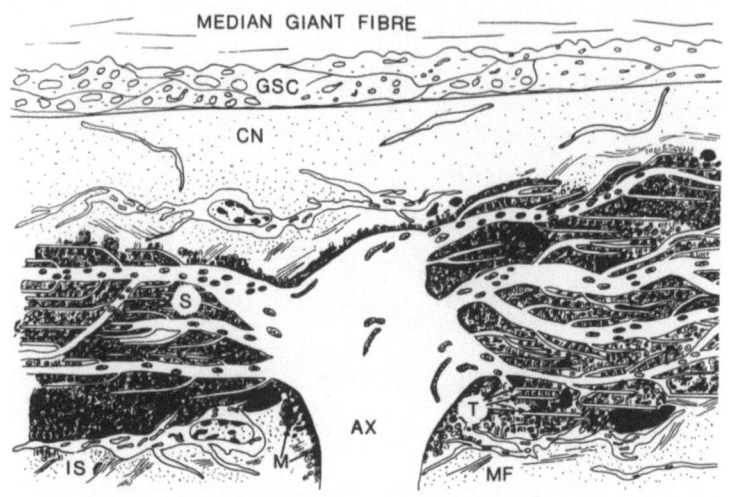

FIG. 5: Anatomy of the crayfish phasic cord stretch receptor, which lies next to the medial giant fibre. The sensory axon (AX) enters at right angles to the longitudinal orientation, and gives rise to primary dendrites (PD) that have irregular orientation but divide further into terminal dendrites (TD) lying longitudinally. CN = connective tissue, GC = glial cell, GCP = glial cell processes, DM = embedding material, FM = filamentous material (from Cobb and Heitler 1985, with the permission of Chapman and Hall Ltd.).

tissue mass in which a number of dendrites spread from a single neurone. The dendrites lie longitudinally arranged (Fig. 5). Although these authors do not indicate the site of the cell body it appears that this unit is essentially unipolar and Cobb and Heitler themselves classify it as Pringle Type II.

I would propose, therefore, that although Pringle Type II may still be tenable it should be subdivided into a number of categories.

6.1.3. Definition

6.1.3.1. <u>Type I</u>; bipolar cells, fusiform in shape, each with a single dendrite, and a spiking axon running considerable distances to the CNS. They arise in the hypodermis, and may end in cuticular structures or in scolopidia (chordotonal organs in insects and crustacea, slit sensillae in spiders).

6.1.3.2. <u>Type II</u>; multidendritic cells, ending in connective tissue, and migrating from the CNS (this last property is not always known, but assumed).

6.1.3.2.1. <u>Type IIa</u>; peripheral cell body, multidendritic with large diameter but short primary dendrites, ending in connective tissue amongst muscle fibres e.g. crustacean abdominal MRO, locust coxa-trochanteral MRO. In crustacean MRO dendrites are non-spiking but the long axon to the CNS propagates action potentials.

6.1.3.2.2. Type IIb; peripherally located cell body - multidendritic ending on sheets of connective tissue with no attached or associated muscles e.g. MPR system of crustacean foregut.

6.1.3.2.3. Type IIc; central cell body within segmental ganglion, very large diameter dendrites, that conduct usually electrotonically to the CNS, but may fire spikes as well. Dendrites end on connective tissue that is not arranged in chordotonal fashion, nor with a muscular attachment e.g. oval organ of crustacean scaphognathite.

6.1.3.2.4. Type IId; central cell body, long large diameter dendrites, non-spiking, ending on a muscle or a tendon linked to a muscle e.g. thoracico-coxal muscle receptor organ (TC MRO) in crabs, uropod MRO in Anomura and Macrura. Some examples do not propagate action potentials while others do.

6.1.3.2.5. Type IIe; presumed central cell body and central dendrites ending on connective tissue attached to a neurone within the ganglion. Spikes are recordable but the site of origin remains unknown e.g. stretch receptors of nerve cord in crayfish.

Figure 6 summarises the arrangement of the various examples of Type II endings, varying from peripheral cell bodies with short fat dendrites ending in arrays parallel to muscle fibres, exhibiting electrotonic decrementing conduction, but with a spiking axon (IIa), to peripheral

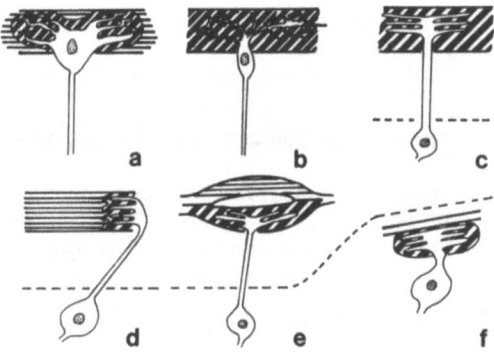

FIG. 6: Six arrangements of Type II endings (Pringle 1961).

IIa Cell body is peripheral with short, broad dendrites, ending on an intercalated tendon or connective tissue incorporated with muscle (Alexandrowicz 1967). The axon is long, and spiking, whilst the dendrites are electrotonic and non-spiking.
IIb peripheral cell body, dendrites slender and branched, axons spike.
IIc central cell body, dendrites long and of large diameter, ending with no chordotonal or muscular arrangements. Show both electrotonic and propagated potentials.
IId central cell body, long, broad dendrites ending on connective tissue in parallel or series with muscle.
IIe central cell body with long large diameter dendrites attached to a tendon lying associated with a muscle.
IIf cell body and dendrites lie within the central nervous system, attached to the sheath of the giant fibre, the organ spikes.
In all cases the boundary of the CNS is indicated by a dotted line.

cells with multiterminal dendrites ending on connective tissue and having spiking axons (IIb). The cell bodies, however, are central and peripheral dendrites long in IIc, IId, and IIe which differ in their termination on connective tissue (IIc), in parallel to muscle (IId) or within the nervous system attached to a neighbouring axon (IIe). In each case the dendrites are large, may conduct with or without spikes, and are remote from the cell body.

6.2. Limb base MROs

The limbs of decapod Crustacea are innervated with numerous proprioceptors (Bush and Laverack 1982) and the bases especially are served by a complex of both chordotonal and muscle receptor organs. Their arrangement is interesting for two reasons; the first type, chordotonal, is composed of bipolar, fusiform cells with dendrites situated on connective tissue (Pringle 1961, Type 1), while the second type, lying in parallel directly alongside, is associated with a specific muscle (usually very small and without significance in the generation of power) with sensory cells having cell bodies in the CNS and long, often large diameter, dendrites.

The basic proprioceptive information stemming from the body-coxa joint therefore is composed of both active (MRO) and passive (chordotonal) elements. Further, the bipolar cells are typically spiking, while those of the MROs appear to be often, but not invariably, non-spiking. So far these units have not been described at all basal joints in decapod limbs - but Table 1 and Fig. 6 summarise what is known. The fundamental pattern may be that each exists in each segment, and limb specialisation modifies this arrangement. It would be interesting to know the situation in smaller, less advanced, Crustacea.

7. CRUSTACEAN SURFACE SENSORY SYSTEMS

There are many types of sensillae to be found on the surface of crustaceans. Most, but not all, of these sensillae are in the form of setae or hairs that project from the surface. Various authors in the past have erected classifications and given names to these setae, but without attempting to determine functions or innervation of these differing types. It is probable that slight differences in surface structure are related to differences in sensitivity and performance. It is not my intention to attempt yet another catalogue of setal types.

The majority of setae, and other surface structures are believed to be innervated but as Bender et al. (1984) have shown some may not be directly concerned with sensors although they may have indirect links. So far as characterisation is possible those setae that have been investigated in detail fall into three main groups:

a. <u>Mechanoreceptors</u>, in which sensory neurones are attached to the base of the projecting hairs (e.g. Felgenhauer and Abele 1983). Deflection of such setae leads to depolarisation and stimulation of the afferent neurones.

b. <u>Chemoreceptors</u>, in which the dendrites of the sensory neurones ascend into the shaft of the seta and contact stimulatory substances that gain access either through the wall of the cuticular hair or via apertures, usually apically or sub-terminally placed, or through porous channels in the cuticle.

c. **Bimodal or combined receptors**; a combination of mechanoreceptor and chemoreceptor in the same hair. In this case the spatial relationships of the two types of sensors are as in types 1 and 2, but are both represented in the same structure. In some cases it seems that the mechanoreceptors may ascend the hair shaft rather than end at a basal insertion.

7.1. Mechanoreceptors

Surface placed mechanoreceptors usually project into the environment and carry one or two active sensors. Various reports, both anatomical and physiological suggest that in many cases two mechanoreceptors are found at each hair. These respond to movement (water flow, displacement) in opposite directions. In one direction one unit fires and the other is silent, in the opposite direction the second unit fires and the first is quiet (for a review see Wiese 1987).

It is interesting that in the first descriptions of these organs it was proposed that the mechanoreceptors insert on opposite sides of the hair base, but in the one case well analysed (Wiese 1976) both physiologically and anatomically both units end at the same place. More structural work on other examples would be welcome. The setae themselves take a variety of forms.

7.2. Chemoreceptors

There is no standard structure of crustacean chemoreceptors. Indeed it is difficult to generalize at the present time what a crustacean chemoreceptor is like. The usual presumption is that setae are the sites of chemoreception yet there are a number of examples that are quite different in morphology - there is also no typical innervation pattern. It is only necessary that the external chemical environment affects a neurone, sometimes despite the presence of cuticle.

Perhaps the simplest form of sensillum that can be proposed is a hole in the cuticle i.e. a place where an internally situated dendrite can make contact with the external environment. Such organs exist in the oesophagus (Robertson and Laverack 1979b; Altner et al. 1986; J. Marshall and M.S. Laverack unpublished), and have only a small number or dendrites (2 or 3 chemoreceptors plus one believed to be a mechanoreceptor). Another simple system which may be chemosensory is the dorsal organ of the larvae in which two sensors bear two cilia each that then further branch and ramify on a 'spongy' cuticle at the surface (Laverack and Barrientos 1985). There may, however, be alternative explanations for this structure. 'Funnel canals' in dactylopodites are also, in effect, holes in the cuticle although they possess a plug of material in the aperture (Gnatzy et al. 1984) and the number of dendrites may be as many as 20+ chemoreceptors and two mechanoreceptors (Schmidt and Gnatzy 1984).

Pores, or apertures, are also found carried on cuticular extensions or setae and the attendant dendrites run out into these hairs. 'Fringed setae' (Altner et al. 1983) have massive cuticle penetrated by a small lumen containing eight dendrites (interpreted as six chemo and two mechano-receptors) that end close to a sub-terminal opening. Felgenhauer and Abele (1983) have shown pores in setae in *Atya* shrimps, Laverack and Barrientos (1985) in isopod setules and cumacean aesthetascs, Heimann (1984) in isopod aesthetascs and Dahl (1973) in amphipod antennular setae.

Such pores, however, are not essential for chemoreceptors to function since some setae are very thin-walled and untanned, providing unimpeded access to molecules across the cuticle. Permeability is such that openings are not necessary, as in some decapod aesthetascs (Ghiradella et al. 1968a,b; Laverack 1976b) and the thin walled setae of some amphipods.

The amphipods are an interesting group in this respect. On the first antenna of Bathyporeia there are a small number of setae at the junction of each annulus with its neighbour (Fig. 7A). Of these sensillae one is about 50% longer (75µm) than the remainder (50µm). This unique seta displays a depressed cuticle when observed in the SEM while the others are stout and cylindrical. The TEM reveals that the longer seta is very thin walled, and distally contains many small branches of dendrites, while at the base the number of dendrites is about 100 - 150. The stouter hairs have a thick cuticle (Fig. 7B), permeated by channels close to the tip,

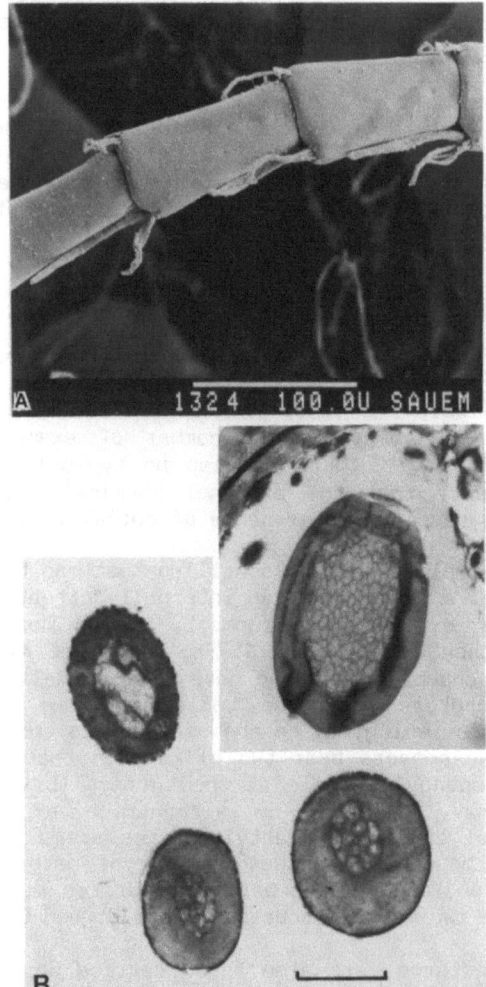

FIG. 7: Setae from the antennae of the amphipod Bathyporeia.
 A. SEM to show relative lengths of setae. One is much longer than the remainder. It is thinner-walled, and bears no pores (M.S. Laverack unpublished).
 B. The bases of the setae show 3 types of innervation. The thin-walled long seta contains over 100 dendrires, while the shorter stouter setae contain either 11 - 14 (depending upon specimen), or 2. (See text for comments on possible function) (M.S. Laverack unpublished). Scale = 1µm

with 11 - 14 dendrites internally (see also Dahl 1973), but one such hair in the group contains 2 or 3 much larger diameter dendrites (Fig. 7). Interpretation of these varied innervations is difficult without further information, but one suspects that the longer hair has a very different function (pheromone detector [?], see Dahl et al. 1970, on calceoli in Gammarus males) to the group with 11 - 14 dendrites which may be food detectors. In the meanwhile those with two or three dendrites are probably mechanoreceptors though the neurones project into the lumen of the setae rather than end at the base as is usual in such cases.

The interpretation of function is not usually possible on anatomical grounds alone. The number of microtubules and the structure of basal bodies and accessory structures may be indicative of function but not conclusive evidence without supporting physiological evidence. For example Altner et al. (1983) suggest that of eight dendrites in the lumen of the fringed seta of crayfish chelate pereiopods two may be mechanoreceptors, and the remainder chemoreceptors. Normally in crustaceans scolopale is associated with mechanical sensitive units at the insertion of the dendrites and in the crayfish this ends internal to the seta but the dendrites continue into the shaft. In Bathyporeia the seta with two large dendrites (see above) may also be a mechanoreceptor, but again scolopale has not been seen. On the other hand in Pugettia producta scolopale has been observed not only in conjuction with two or three dendritic endings, similar to various other described mechanoreceptors, but also with a larger number of endings (up to 12) of which some are thought to be chemosensors (Hamilton et al. 1985). Unfortunately the terminations of these ciliated dendrites have not been traced and the relationship of scolopale to the cuticle and to the sensors has not been described. It is unlikely that 12 units are all mechanically sensitive but the distinction between which units are, and which are not, is not clear.

Table 2 lists the range of dendrite numbers found in a variety of organs. It is possible to find any figure from 1 to over 300 in the list. Small numbers from one to four are often typical of mechanosensory organs. Chordotonal organs have many individual neurones in the collection, but the basic unit contains two (Mill and Lowe 1973). Where physiological work has been carried out two units of different directionality is the norm, and where there are more (3 or 4) the supernumerary units remain unexplained. Occasionally special examples of proprioceptors may have more basic neurone numbers e.g. coxal MRO in Emerita (4, Paul 1972) in Galathea (14, Maitland et al. 1982) and Homarus (20+, Laverack Table 1 this paper).

Chemoreceptors, however, are frequently found in much larger numbers. Two units have been described in larval decapod dorsal organ (Laverack and Barrientos 1985) in isopod aesthetascs (Risler 1977) and in oesophageal receptors of decapods (plus one mechanoreceptor Altner et al. 1986; J. Marshall and M.S. Laverack unpublished). In most cases the numbers of units are much greater ranging from 8 to over 300 (see Fig. 8). Generally aesthetascs, concerned with long range olfactory detection, contain more than other organs usually concerned with direct manipulation and assesment of food, but it is impossible to be categorical about this. Smaller Crustacea usually have small numbers of dendrites in antennular organs, but some (e.g. Bathyporeia) nonetheless have considerable aggregations.

One assumes that the catchment probability of appropriate molecules affects these numbers. Where large groups are found maybe the stimulatory molecules are less common than is the case where numbers are small. This would be the case in pheromone detectors and some of the olfactory sensors may be specialised in this direction. Food presentation requires less numbers and only a range of sensitivities necessary to determine whether the material is good. This becomes more and more refined, until in the

TABLE 2

Table 2 collects together information regarding the number of sensory neurones in crustacean sensillae of various types. They include mechanoreceptors, chemoreceptors and bimodal receptors. Where function is uncertain this is indicated by ?

Receptor Neurone Numbers

Numbers	location	species	proposed function	Reference
1	antennal scolopidia	Porcellio (Isopoda)	M	Risler 1977
	antennule MRO	Ligidium (""""""")	M	Risler 1978
2	larval dorsal organ	Decapoda zoeae and megalopa	C	Laverack and Barrientos 1985
	chordotonal organ	Cancer pagurus (Deca)	M	Mill and Lowe 1973
	"""""""""""""""	Carcinus meanas (Deca)	M	Whitear 1962
	MRO abdomen	Homarus (Decapoda)	M	Alexandrowicz 1967
	antenna	Neomysis (Mysidacea)	M	Guse 1978
	statocyst seta	Neomysis (Mysidacea)	M	Espeel 1985
	carapace setae	Homarus (Decapoda)	M	Laverack 1962
	"""""""""""""""	Procambarus """""""	M	Mellon 1963
	telson setae	"""""""""""""""	M	Wiese 1976
	antennule setae	Bathyporeia (Amphipod)	M?	M.S. Laverack unpublished
	antennular proprioceptor	Ligidium (Isopoda)	M	Risler 1978
	aesthetascs	Porcellio (Isopoda)	C	Risler 1977
	plate setae	Pugettia producta (Decapoda)	M	Hamilton et al. 1985
3	statocyst setae	Astacus (Decapoda)	M	Schone and Steinbrecht 1968
	thorax-coxa MRO	Carcinus maenas (Dec)	M	Whitear 1965

TABLE 2 (continued)

	antennule setae	Ligidium (Isopoda)	M	Risler 1978
	antenna setae	Neomysis (Mysidacea)	M	Guse 1978
	antenna setae	Acetes (Decapoda)	C	Ball and Cowan 1977
	antenna setae	Acetes """"""""""	M	""""""""""""""""""
	oesophageal pores	Austropotamobius(""")	C + M(1)	Altner et al. unpublished
	setae	Pugettia producta	M	Hamilton et al. 1985
3 - 4	"""""""""""""""""	Homarus (Decapoda)	C + M(1)	J. Marshall and M.S. Laverack unpublished
4	antennal setae	Acetes (Decapoda)	M	Ball and Cowan 1977
	aesthetascs	Daphnia (Cladocera)	C	Rieder 1978
	uropod coxa MRO	Emerita analoga(Deca)	M	Paul 1972
5	not known			
6	antennal setae	Neomysis integer (Mysidacea)	C + M	Guse 1978
7	antennule setae	Ligidium (Isopoda)	C + M	Risler 1978
	""""""""""""""""""	""""""""""""	C	""""""""""""""
8	fringed setae	Austropotamobius(Dec)	C + M(2)	Altner et al. 1983
		Homarus americanus (Decapoda)		P. Borroni and M.S. Laverack unpublished.
8 - 10	antennal setae	Acetes(Decapoda)	C	Ball and Cowan 1977
12	plate setae	Pugettia producta(Dec)	C + M	Hamilton et al. 1985
11 - 14	antennal setae	Marinogammarus(Amphi)	C?	M.S. Laverack unpublished
	"""""""""""""""""	Bathyporeia(Amphipod)	C?	""""""""""""""""
13 - 16	antennule setae	Orchestia (Amphipod)	C	Dahl 1973

TABLE 2 (continued)

14	aesthetascs	Notodromas(Ostracod)	C	Andersson 1975
	uropod MRO	Galathea (Decapoda)	M	Maitland et al. 1982
20+	uropod coxa	Homarus (Decapoda)	M	Laverack this chapter
20 - 24	funnel canals	Carcinus (Decapoda)	C + M	Schmidt and Gnatzy 1984
24 - 32	large setae	Homarus americanus (Decapoda)	C	P. Borroni and M.S. Laverack unpublished.
30 - 40	aesthetascs	Antromysis (Mysidacea)	C	Juberthie-Jupeau and Crouau 1977
	"""""""""""""	Neomysis(Mysidacea)	C	Guse 1979
	"""""""""""""	Conchoecia(Ostracod)	C	Heimann 1979
	"""""""""""""	Cylindroliberis	C	Heimann 1984
50 - 60	aesthetascs	Asellus(Isopoda)	C	"""""""""""""
60 - 80	"""""""""""""	Idotea (Isopoda)	C	Guse 1983
70 - 80	"""""""""""""	Asellus(Isopoda)	C	Heimann 1984
100	antennal seta	Bathyporeia(Amphi)	C?	M.S. Laverack unpublished
	aesthetasc	Paragrapsus(Decapod)	C	Snow 1973
300	aesthetasc	Panulirus argus(""")	C	Laverack and Ardill 1965
	"""""""""""""	Pagurus (Decapoda)	C	Ghiradella et al. 1968b
	"""""""""""""	Panulirus interruptus (Decapoda)	C	Spencer 1986

C = Chemoreceptor

M = Mechanoreceptor

C + M = Bimodal receptor (Chemo + Mechano)

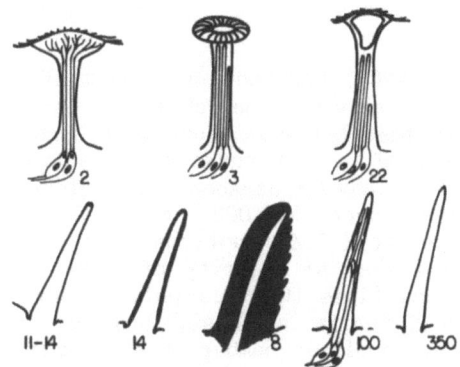

FIG. 8: Summary diagram of chemosensory sensillae and numbers of neurones.

Top row;   2 units in larval dorsal organ

3 units in oesophageal receptors

22 units in dactyl funnel canals

Bottom row; 11 - 14 units in antennal setae with pore

14 units in antennal setae (no pore)

8 units in fringed setae (hedgehog hairs)

100 units in thin-walled setae

300 units in decapod aesthetascs.

oesophagus only two units per organ are necessary to determine that the oesophagus and gut should be brought into action (Robertson and Laverack 1979b).

In bimodal receptors where both mechanical and chemoreceptors are found each modality is a significant component of the stimulus, and this provides a working definition of taste and olfaction in these aquatic organisms (Laverack 1987). Normally in such receptors the mechanically sensitive units might be expected to end at the base of any seta present, but examples are known (e.g. oesophageal pores, funnel canals) where suspected mechanoreceptors run up canals to the surface, and in at least one type of seta (the hedgehog hair) mechanoreceptors are also believed to run into the shaft of the organ.

The question of overall numbers of sensory axons which combines the information regarding individual sensillae (Table 2) with the problem of growth and development indicated in the first three sections of this paper can only be approached obliquely. Numbers of units at each kind of sensillum seem stereotyped, and vary only slightly; the numbers of sensillae increase with each moult, therefore the total population of sensors rises and is particularly notable in animals that grow throughout life (e.g. lobsters); the numbers of motor neurones meanwhile remain small and constant.

Govind and Pearce (1985) counted the number of fibres in the first and second nerve roots serving the chelae (cutter and crusher) of H.americanus (and other species) and showed that numbers of axons is very high, reaching over 120,000 in each nerve at an animal length of about 5 cm (2 years of age, C.K. Govind personal communication). Laverack and Ardill (1965) estimated over 500,000 neurones in the antennule of Panulirus argus and Spencer (1986) confirms this order of magnitude for P.interruptus. In all cases the majority of these units is composed of very small diameter axons (less than 0.5 µm). Motor neurones in decapod limbs account for 15-20 units at most, and are of large diameter. Mechanoreceptor axons are small in number at each sensillum (see Table II) though there may be many sensillae. Conduction velocities, magnitude of potentials, and TEM evidence suggests that mechanoreceptor axons are also of considerable diameter. Chemoreceptors, on the other hand, where identification is positive (antennules, pereiopod fringed setae, funnel canals, oesophagus) all have very small axons.

In view of overall numbers reported one may hypothesize that the crustacean body possesses many more chemoreceptors than previously suspected and they outnumber all other types of neurones.

This paper has attempted to review the arrangement of the sensory system of Crustacea, and shows that the numbers of elements involved is large, organised, dynamic and varies with age size and organ. Many aspects of growth, synaptic connection and variability remain to be elucidated.

8. REFERENCES

Aiken DE (1973), Proecdysis, setal development and molt prediction in the American lobster (Homarus americanus). J Fish Bd Canada 30: 1337 - 1344.

Aiken DE (1980) Molting and growth. In: Cobb JS, Phillips DF (eds) Biology and management of lobsters. Vol. I. Academic Press, New York and London, PP 91 - 163.

Alexandrowicz JS (1967) Receptor organs in thoracic and abdominal muscles of Crustacea. Biol Revs 42: 288 - 326.

Alexandrowicz JS, Whitear M (1957) Receptor elements in the coxal region of Decapoda Crustacea. J Mar Biol Assoc UK 57: 379 - 396.

Allen EJ (1894a) Studies on the nervous system of Crustacea. I. Some nerve elements of the embryonic lobster. Quart J Microsc Sci 36: 461 - 482.

Allen EJ (1894b) Studies on the nervous system of Crustacea. II. The stomatogastric system of Astacus and Homarus. III. On the beading of nerve fibres and on end swellings. Quart J Microsc Sci 36: 483 - 498.

Altner H, Hatt H, Altner I (1986) Structural and functional properties of the mechanoreceptors and chemoreceptors in the anterior oesophageal sensilla of the crayfish, Astacus astacus. Cell Tissue Res 244: 537 - 547.

Altner I, Hatt H, Altner H (1983) Structural properties of bimodal chemo and mechanosensitive setae on the pereiopod chelae of the crayfish, *Austropotamobius torrentium*. Cell Tissue Res. 228: 357 - 374.

Andersson A (1975) The ultrastructure of the presumed chemoreceptor aesthetasc 'Y' of a cypridid ostracode. Zool Scripta 4: 151 - 158.

Atema J (1985) Chemoreception in the sea; Adaptations of chemoreceptors and behaviour to aquatic stimulus conditions. In: Laverack MS (ed) Physiological adaptations of marine animals. Symp Soc Exp Biol 39: 387 - 423.

Atema J (1987) Distribution of chemical stimuli. In: Atema J, Fay RR, Popper AN (eds) Sensory physiology of aquatic animals. Springer-Verlag, Berlin and New York (in press).

Ball EE, Cowan AN (1977) Ultrastructure of the antennal sensilla of *Acetes* (Crustacea, Decapoda, Natantia, Sergestidae). Phil Trans Soc B 277: 429 - 456.

Ballinger ML, Bittner GD (1980) Ultrastructural studies of several medial and other CNS axons in crayfish. Cell Tissue Res 208: 123 - 133.

Bender M, Gnatzy W, Tautz J (1984) The antennal feathered hairs in the crayfish: a non-innervated stimulus transmitting system. J Comp Physiol A 154: 45 - 47.

Bittner GD (1981) Trophic interactions of CNS giant axons in crayfish. Comp Biochem Physiol 68 A: 298 - 306.

Bittner GD, Ballinger ML, Larimer JL (1974) Crayfish CNS: minimal degenerative changes after lesioning. J Exp Zool 189: 13 - 36.

Braunig P, Cahill MA, Hustert R (1986) The coxo-trochanteral muscle receptor organ of locusts. Dendritic tubular bodies in a non-ciliated insect mechanoreceptive neurone. Cell Tissue Res 243: 517- 524.

Bullock TH, Horridge GA (1965) Structure and function in the nervous system of invertebrates. Freeman and Co. San Francisco and London. 1719 pp.

Bush BMH (1976) Non-impulsive thoracic-coxal receptors in crustaceans. In: Mill PJ (ed) Structure and function of proprioceptors in the invertebrates. Chapman and Hall, London, pp 115 - 151.

Bush BMH, Laverack MS (1982) Mechanoreception. In: Atwood H, Sandeman DC (eds) Biology of Crustacea. Volume 3. Academic Press, New York and London, pp 399 - 468.

Clark JV, Dorsett DA (1978) Anatomy and physiology of proprioceptors in the cirri of *Balanus hameri*. J Comp Physiol 123: 229 - 237.

Cobb JLS, Heitler WJ (1985). Ultrastructure of the phasic stretch receptor in the crayfish abdominal nerve cord. J Neurocytol 14: 413- 426.

Dahl E (1973) Antennal sensory hairs in Talitrid Amphipods, Acta Zool 54: 161 - 171.

Dahl E, Emanuelsson H, von Mecklenburg C (1970) Pheromone reception in the males of the amphipod Gammarus duebeni Lilljeborg. Oikos 21: 42 - 47.

Davis WJ (1973) Development of locomotor patterns in the absence of peripheral sense organs and muscles. Proc Nat Acad Sci 70: 964 - 958.

Davis WJ, Davis KB (1973) Ontogeny of a simple locomotor system. Role of the periphery in the development of a central nervous circuitry. Amer Zool 13: 409 - 425.

Denton EJ, Gray J (1985) Lateral-line -like antennae of certain of the Penaeidae (Crustacea, Decapoda, Natantia). Proc Soc Lond B 226: 249 - 261.

Espeel M (1985) Fine structure of the statocyst sensilla of the mysid shrimp Neomysis integer. J Morphol 186: 149 - 165.

Espeel M (1986) Morphogenesis during moulting of the setae in the statocyst sensilla of the mysid shrimp Neomysis integer (Leach 1814) (Crustacea, Mysidacea). J Morphol 187: 61 - 68.

Felgenhauer BE, Abele LG (1983). Ultrastructure and functional morphology of feeding and associated appendages in the tropical freshwater shrimp Atya innocous (Herbst) with notes on its ecology. J Crust Biol 3: 336 - 363.

Ghiradella HT, Case JF, Cronshaw J (1968a). Structure of aesthetascs in selected marine and terrestrial Decapods. Chemoreceptor morphology and environment. Amer Zool 8: 603 - 621.

Ghiradella HT, Cronshaw J, Case JF (1968b) Fine structure of the aesthetasc hairs of Pagurus hirsutiusculus Dana. Protoplasma 66: 1 - 20.

Gnatzy W, Schmidt M, Rombke J (1984) Are the funnel-canal organs the 'campaniform sensilla' of the shore crab Carcinus maenas (Crustacea, Decapoda) ? Zoomorphol 104: 11 - 20.

Govind CK, Pearce J (1985) Lateralization in number and size of sensory axons to the dimorphic chelipeds of Crustaceans. J Neurobiol 16: 111 - 125.

Grobstein P (1973) Extension-sensitivity in the crayfish abdomen. 1. Neurons monitoring nerve cord length. J Comp Physiol 86: 331 - 348.

Guse G-W (1978) Antennal sensilla of Neomysis integer (Leach). Protoplasma 95: 145 - 161.

Guse G-W (1979) Feinstruktur der Aesthetasken von Neomysis integer (Leach) (Crustacea, Mysidacea). Zool Jb Jena 203: 170 - 176.

Guse G-W (1983). Ultrastructure, development and moulting of the aesthetascs of Neomysis integer and Idotea baltica (Crustacea, Malacostraca). Zoomorphol 103: 121 - 133.

Hamilton KA, Lingerg KA, Case JF (1985) Structure of dactyl sensilla in the kelp crab, Pugettia producta. J Morphol 185: 349 - 366.

Harris GG, van Bergeijk WA (1962) Evidence that the lateral line organ responds to near-field displacements of sound sources in water. J Acoust Soc Amer 34: 1831 - 1841.

Hartman HB, Austin WD (1972) Proprioceptor organs in the antennae of Decapoda Crustacea. 1. Physiology of a chordotonal organ spanning two joints in the spiny lobster Panulirus interruptus (Randall). J Comp Physiol 81: 187 - 202.

Heimann P (1979) Fine structure of sensory tubes on the antennule of Conchoecia spinirostris (Ostracoda, Crustacea). Cell Tissue Res 202: 461 - 477.

Heimann P (1984) Fine structure and moulting of aesthetasc sense organs on the antennules of the isopod Asellus aquaticus (Crustacea). Cell Tissue Res 235: 117 - 128.

Heitler WJ (1982) Non-spiking stretch-receptors in the crayfish swimmeret system. J Exp Biol 96: 355 - 366.

Herbst C (1900) Uber die Regeneration von antennenahnlichen organen an Stelle von Augen. III, IV. Arch Entwick Mech Org 9: 215 - 292.

Horch K (1971) An organ for hearing and vibration sense in the ghost crab Ocypode. Z vergl Physiol 73: 1 - 21.

Hoy RR, Bittner GD, Kennedy DM (1967) Regeneration in crustacean motoneurones; evidence for axonal fusion. Science 156: 251 - 252.

Hughes GM, Wiersma CAG (1960) Neuronal pathways and synaptic connexions in the abdominal cord of the crayfish. J Exp Biol 37: 291 - 307.

Jellies J, Larimer JL (1986) Activity of crayfish abdominal-positioning interneurons during spontaneous and sensory-evoked movements. J Exp Biol 120: 173 - 188.

Juberthie-Jupeau L, Crouau Y (1977) Ultrastructure des aesthetascs d'un Mysidace souterrain anophthalme. CR Acad Sci Paris 284: 2257 - 2259.

Kirk MD, Govind CK (1983) Innervation and motor patterns of the abdominal superficial flexor muscles in larval lobsters. J Neurobiol 14: 395 - 405.

Laverack MS (1962) Responses of cuticular sense organs of the lobster Homarus. I. Hair peg organs as water current receptors. Comp Biochem Physiol 5: 319 - 325.

Laverack MS (1964) The antennular sense organs of Panulirus argus. Comp Biochem Physiol 13: 301 - 321.

Laverack MS (1974). Comparative physiology; neurophysiology of marine invertebrates. In: Mariscal RN (ed) Experimental marine biology. Academic Press, New York and London, pp 99 - 163.

Laverack MS (1976a) External proprioceptors. In: Mill PJ (ed) Structure and function of proprioceptors in invertebrates. Chapman and Hall, London, pp 1 - 63.

Laverack MS (1976b) Properties of chemoreceptors in marine Crustacea. Olfact Taste 4: 45 - 50.

Laverack MS (1978) The organisation and distribution of CAP organs in the lobster Homarus gammarus (L.) Mar Behav Physiol 5: 201 - 208.

Laverack MS (1987) The diversity of chemoreceptors. In: Popper AN, Fay RR, Atema J (eds) Sensory biology of aquatic animals. Springer-Verlag Berlin. (in press).

Laverack MS, Ardill DJ (1965) The innervation of the aesthetasc hairs of Panulirus argus. Quart J Microsc Sci 106: 45 - 60.

Laverack MS, Barrientos Y (1985) Sensory and other superficial structures in living marine Crustacea. Trans R Soc Edin Earth Sci 76: 123 - 136.

Laverack MS, Dando MR (1968) The anatomy and physiology of mouthpart receptors in the lobster Homarus vulgaris. Z vergl Physiol 61: 176 - 195.

Macmillan DL, Dando MR (1972) Tension receptors on the apodemes of muscles in the walking legs of the crab, Cancer magister. Mar Behav Physiol 1: 185 - 208.

Macmillan DL, Neil DM, Laverack MS (1976) A quantitative analysis of exopodite beating in the larvae of the lobster Homarus gammarus (L.). Phil Trans R Soc B 274: 69 - 85.

Maitland DP, Laverack MS, Heitler WJ (1982) A spiking stretch receptor with central cell bodies in the uropod coxopodite of the squat lobster, Galathea strigosa. J Exp Biol 101: 221 - 232.

Maynard DM, Dingle H (1963) An effect of eyestalk ablation on antennular function in the spiny lobster, Panulirus argus. Z vergl Physiol 46: 515 - 540.

McLaughlin PA (1982) Comparative morphology of crustacean appendages. In: Abele LG (ed) Biology of Crustacea. Vol 2. Academic Press, New York and London. pp 197 - 256.

McVean A (1982) Autotomy. In: Sandeman DC, Atwood H (eds) Biology of Crustacea. Vol 4. Academic Press, New York and London, pp 107 - 132.

Mellon D (1963) Electrical responses from dually innervated tactile receptors in the thorax of the crayfish. J Exp Biol 40: 137 - 148.

Mill PJ, Lowe DG (1973) The fine structure of the PD proprioceptor of Cancer pagurus. 1. The receptor strand and the movement sensitive cells. Proc Soc Lond B 184: 179 - 197.

Neil DM, Macmillan DL, Robertson RM, Laverack MS (1976) The structure and function of thoracic exopodites in the larvae of the lobster Homarus gammarus. Phil Trans R Soc B 274: 53 - 68.

Norris BJ, Hartman HB (1985) Cuticular hair organs evoking reflexive closing and openong of the crayfish claw. Comp Biochem Physiol 82 A: 525 - 529.

Pasztor VM (1969) The neurophysiology of respiration in decapod Crustacea. II. The sensory system. Can J Zool 47: 435 - 441.

Pasztor VM, Bush BMH (1983) Graded potentials and spiking in single units of the oval organ, a mechanoreceptor in the lobster ventilatory organ. J Exp Biol 107: 431 - 449.

Paul DH (1972) Decremental conduction over 'giant' afferent processes in the decapod Emerita. Science 176: 680 - 682.

Pringle JWS (1961) Proprioception in Arthropods. In: Ramsay JA, Wigglesworth VB (eds) The Cell and the Organism: Essays presented to Sir James Gray. University Press, Cambridge, pp 256 - 282.

Rieder N (1978) Die Ultrastruktur der Rezeptoren auf der ersten Antennen von Daphnia magna. Verh Dtsch Zool Ges 1978: 229 G. Fischer Stuttgart.

Risler H (1977) Die Sinnessorgane der Antennula von Porcellio scaber (Crustacea, Isopoda). Zool Jb Anat 98: 29 - 52.

Risler H (1978) Die Sinnesorgane der Antennula von Ligidium hypnorum (Cuvier) (Isopoda), Crustacea). Zool Jb Anat 100: 514 - 541.

Robertson RM, Laverack MS (1979a). The structure and function of the labrum in the lobster, Homarus gammarus (L.). Proc R Soc Lond B 206: 209 - 233.

Robertson RM, Laverack MS (1979b). Oesophageal sensors and their modulatory influence on the oesophageal peristalsis in the lobster Homarus gammarus. Proc R Soc Lond B 206: 235 - 263.

Sandeman DC (1985) Crayfish antennae as tactile organs: Their mobility and the responses of their proprioceptors. J Comp Physiol A 57: 363 - 373.

Schmidt M, Gnatzy W (1984) Are the funnel-canal organs the campaniform sensilla' of the shore crab, Carcinus maenas? II. Ultrastructure. Cell Tissue Res 237: 81 - 93.

Schöne H, Schöne H (1980) Morphology and function of the antennular joint and its strand organ, instrumental to gravity reactions in the spiny lobster Panulirus argus. Zoomorphol 96: 191 - 203.

Schöne H, Steinbrecht AR (1968) Fine structure of statocyst receptor of Astacus fluviatilis. Nature 220: 184 - 186.

Shaw S, Stowe S (1982) Photoreception. In: Atwood H, Sandeman DC (eds) Biology of Crustacea Vol 3. Academic Press, New York and London.

Snow PJ (1973) Ultrastructure of the aesthetasc hairs of the littoral decapod Paragrapsus gaimardii. Z zellforsch 138: 489 - 502.

Spencer M (1986) The innervation and chemical sensitivity of single aesthetasc hairs. J Comp Physiol 158: 59 - 68.

Strickler JR (1985) Feeding currents in calanoid copepods; two new hypotheses. In: Laverack MS (ed) Physiological adaptations of marine animals. Company of Biologists, Cambridge. Symp Soc Exp Biol 39: 459 - 485.

Tautz J, Masters WM, Aicher B, Markl H (1981) A new type of water vibration receptor on the crayfish antenna. I. Sensory physiology. J Comp Physiol 144: 53 - 541.

Vedel JP (1985) Cuticular mechanoreception in the antennal flagellum of the rock lobster Palinurus vulgaris. Comp Biochem Physiol 80 A: 151 - 158.

Vedel JP (1986) Morphology and physiology of a hair plate sensory organ on the antenna of the rock lobster Palinurus vulgaris. J Neurobiol 17: 65 - 76.

Vedel JP, Monnier S (1983). A new muscle receptor organ in the antenna of the rock lobster, Palinurus vulgaris; mechanical and proprioceptive organization of the two proximal joints J0 and J1. Proc R Soc Lond B 218: 95 - 110.

Wales W, Laverack MS (1972a) The mandibular muscle receptor organ of Homarus gammarus. (L.).(Crustacea, Decapoda). Z Morph Tiere 73: 135 - 162.

Wales W, Laverack MS (1972b). Sensory activity of the mandibular muscle receptor organ of Homarus gammarus (L.). 1. Response to receptor muscle stretch. Mar Behav Physiol 1: 239 - 255.

Walker G (1974) The fine structure of the frontal filament complex of barnacle larvae (Crustacea: Cirripedia). Cell Tissue Res 152: 449 - 465.

Walker G, Lee VE (1976) Surface structures and sense organs of the cypris larva of Balanus balanoides as seen by scanning and transmission electron microscopy. J Zool 178: 161 - 172.

Whitear M (1962) The fine structure of crustacean proprioceptors. I. The chordotonal organs in the legs of the shore crab, Carcinus maenas. Phil Trans R Soc 245: 292 - 324.

Whitear M (1965) The fine structure of crustacean proprioceptors. II The thoracic-coxal organs in Carcinus, Pagurus and Astacus. Phil Trans R Soc B 248: 437 - 456.

Wiese K (1976) Mechanoreceptors for near field water displacements in crayfish. J Neurophysiol 39: 816 - 833.

Wiese K (1987) Representations of hydrodynamic movements in the nervous system of the crayfish, Procambarus. In: Popper AN, Fay RR, Atema J (eds) Sensory biology of aquatic animals. Springer-Verlag, Berlin (in press).

Williamson DI (1982) Larval morphology and diversity. In: Abele LG (ed) Biology of Crustacea. Vol 2. Academic Press, New York and London, pp 43 - 110.

Wilson AH, Sherman RG (1975) Mapping of neuron somata in the thoracic nerve cord of the lobster using cobalt chloride. Comp Biochem Physiol 50 A: 47 - 50.

Wyse GA, Maynard DM (1958) Joint receptors in the antennule of Panulirus argus. J Exp Biol 42: 521 - 535.

Yules RB (1962) Responses from a proprioceptive organ of the crab, Sesarma reticulatum, during the moult cycle. Biol Bull 123: 660 - 669.

Zeil J, Sandeman R, Sandeman DC (1985) Tactile localisation: the function of active antennal movements in the crayfish Cherax destructor. J. Comp Physiol 157: 607 - 617.

ASPECTS OF THE FUNCTIONAL AND CHEMICAL ANATOMY OF THE INSECT BRAIN

DICK R. NÄSSEL

Department of Zoology

University of Lund

Helgonavägen 3

S-223 62 Lund, Sweden

ABSTRACT

The functional organization of the insect brain is illustrated by descriptions of the optic lobes, ocellar and antennal projections and their convergence onto descending pathways. The neuropils in which the integration of signals take place are of two kinds: glomerular and non-glomerular. The former have strictly geometrically organized neural processes, the latter have been considered without discernable patterns. The dendrites of many of the descending neurons (DNs) and higher order interneurons arborize in non-glomerular neuropil and analysis of these neurons reveal patterned organization also in this type of neuropil.

Four different DN pathways in the Calliphora brain are outlined. These constitute the giant fiber pathway and three pathways subserving the large horizontal and vertical motion sensitive neurons in the lobula plate. Most DNs are arranged in clusters and it is shown that they receive multimodal inputs and are forming complex output connections.

The organization of the protocerebral mushroom bodies and central body complex and their possible integration into the sensory and descending system is discussed. These higher order neuropils form very complex connections, especially via non-glomerular neuropil and may have a role in gating of signals in descending pathways. Some of the structural analysis of the higher protocerebral centers is by means of immunocytochemistry of neuroactive substances. The possible roles of some of the neuroactive substances in modulation of neurons at different levels of the insect brain are briefly speculated upon.

1. INTRODUCTION

The brain of insects is a small, but complex aggregation of nerve centers, derived from the most anterior of the segmental ganglia. Despite its size, a relatively large number of neurons form numerous circuits. As an example, the bee brain is about 3 mm wide and less than 1 mm thick and contains almost one million neurons (Witthöft 1967) which form fine process that are densely packed in tracts and synaptic neuropil. To analyze the functional organization of such small brains, the approaches an investigator can choose from become limited compared to those available to students of the rat brain. However, as will be shown in the present review, the insect brain offers some unique opportunities to study neuronal circuits. The presence of large individually identifiable neurons and the relatively simple behaviour of most insects, as well as the advantages with small brains for serial reconstruction of central pathways are some of the features that make insect brains attractive.

Different aspects of brain organization and function in insects have been reviewed in the last three decades (e.g. Huber 1960, 1978; Bullock and Horridge 1965; Howse 1875; Klemm 1976; Strausfeld 1976a; Heisenberg 1980; Fischbach and Heisenberg 1984; Hall 1984; Strausfeld et al. 1984; Schürmann 1985, 1987; for crustacean brains see Bullock and Horridge 1965; Nässel and Elofsson 1986). Several subsystems of the brain have received considerable attention, others next to none. As a result our views of the integrated action of brain compartments in their control of motor outputs are rather generalized or in parts even based on mere speculations. This will be apparent from the present review.

There are some parts of the insect brain that have been subjected to rather extensive analysis: the visual system (including the ocellar system) (Goodman 1981; Strausfeld 1984), the higher centers termed the mushroom bodies (Heisenberg 1980; Schürmann 1985, 1987), the antennal system (Boeckh et al. 1984) and the descending neuron pathways (Rowell 1971; Strausfeld and Nässel 1980; Strausfeld et al. 1984). The present review summarizes data on the functional anatomy of these components and their convergence at the descending neuron dendrites. The detailed organization of a few descending pathways in Calliphora is described as examples of the multimodal control of neurons supplying inputs to motor circuits. Further reviews on this multimodal convergence onto descending pathways have been written recently (Strausfeld and Bacon 1983; Strausfeld et al. 1984, 1987; Schürmann 1985).

After a description of the general organization of the nervous system, and the functional organization of the sensory systems and descending pathways, I shall discuss some of the higher protocerebral centers with respect to functional and chemical anatomy. Their roles in control of other pathways including the descending ones is not clear but may be speculated upon.

The methods used for the structural analysis summarized in this review (HRP-, cobalt- and Lucifer Yellow labeling as well as histochemistry and immunocytochemistry) are described in the original papers (see also the different chapters in Strausfeld 1983). A summary of strategies for morphological analysis of arthropod brain is presented in another review (Nässel 1986b). One methodological comment is pertinent though since much of the analysis of presumed contact points between descending neurons and their presumed inputs is based on anatomical evidence. Strausfeld and Obermeyer (1976) showed that cobalt passes transneuronally in specific sets of pathways in the fly brain. This peculiarity has been extensively used by Strausfeld and coworkers for marking contiguous neurons and its reliability for proving synaptic contacts between these neurons has been discussed in their experimental papers quoted in the present review.

2. GENERAL ORGANIZATION OF THE INSECT NERVOUS SYSTEM

The insect nervous system consists of a segmentally organized central nervous system (CNS) and a peripheral nervous system which includes the stomatogastric nervous system and perisympathetic organs. The segmental ganglia of the CNS vary in number and organization between different taxonomic groups of insects. In more evolved insects the ganglia tend to fuse in different patterns. Most commonly the most anterior (or dorsal) ganglia fuse to form (1) a cephalic supraoesophageal (or preoral) ganglion mass often termed brain and (2) another cephalic mass posterior or ventral to the oesophagus, referred to as the suboesophageal ganglion (or more properly the suboesophageal ganglia). The body ganglia are either separate or fused in different constellations. Commonly the last abdominal ganglia are fused to a terminal ganglion mass and the third thoracic ganglion often fuses with a few of the most anterior abdominal ganlgia. In extreme cases, e.g. higher Diptera, all thoracic and abdominal ganglia fuse to a thoracico-abdominal ganglion mass.

The basic organization of insect ganglia is a rind of neuronal and glial cell bodies (perikarya) around a core of neuronal processes forming axonal tracts and synaptic neuropil (Bullock and Horridge 1965). Sensory neurons (with a few exceptions) have their cell bodies peripherally in the sensory structures and axons projecting to neuropil in central ganglia. Interneurons (relay neurons) and motorneurons have cell bodies located in the rind of central ganglia. In absolute terms the number of neurons is rather small in insect ganglia. The brain of the worker bee contains ca. 800,000 neurons and glial cells and the housefly ca. 300,000 neurons (Witthoft 1967; Strausfeld 1976a). A simple abdominal ganglion of a dragonfly contains ca. 600 neurons and the entire thoracico-abdominal ganglion mass in a blowfly consists of ca. 6,000-8,000 neurons (Hughes 1965; Strausfeld and Bacon 1983). Of these neurons, many are large and can be recognized individually from specimen to specimen because of their characteristic shapes, cell-body locations, electro-chemical properties, contents of neuroactive substances and synaptic connections. These neurons, termed identified neurons, can be found in all parts of the CNS and may be sensory-, motor- or interneurons. The remaining neurons do not classify as individually identifiable neurons because they (1) are present in large isomorphic clusters (e.g. columnar neurons in optic lobes or Kenyon cells of mushroom bodies described later), (2) they are small and/or have very thin processes or (3) they are part of loosely organized neuropil and have variable shapes. Some of the neurons of isomorphic clusters can be identified as assemblies. Like columnar Col A neurons of the optic lobe lobula (described below) these assemblies have their cell bodies clustered so that repeated intracellular impalements of individuals of the assembly can be performed.

Before looking into the functional organization of centers, pathways and circuits of the insect brain, a brief description of the general brain architecture may be useful. The supraoesophageal ganglia receive sensory inputs from compound eyes, ocelli, antennae, and head hairs and supply efferents to certain cephalic muscles, the anterior alimentary canal and to the stomatogastric nervous system. The suboesophageal ganglia receive sensory innervation from mouth parts, cuticle of the head and neck and innervate muscles of mouth parts, neck muscles and salivary glands.

Although there is some dispute about the matter, the brain is classically considered to consist of three segmental ganglia; (1) protocerebrum, (2) deutocerebrum, and (3) tritocerebrum, representing (1) acron (nonsegmental anterior part of the head) and a preantennal segment, (2) the antennal segment, and (3) the premandibular segment. The suboesophageal ganglia are (4) mandibular ganglion, (5) maxillary ganglion, and (6) labial ganglion, representing the segments with respective mouth parts. The tritocerebrum which actually is a circumoesophageal (or perioesophageal) ganglion, sometimes is referred to as the ganglion of the first body segment (although it is located within the adult head capsule) and hence proto- and deutocerebrum are the ganglia of the head segments (see Bullock and Horridge 1965). We can ignore the discussion about the relation of the protocerebrum and deutocerebrum to segments of the animal for now and instead look into the organization of brain neuropils.

In the brain one can distinguish between two main types of synaptic neuropil: glomerular and non-glomerular neuropil. The glomerular neuropil is well delineated and is characterized by a geometrical organization of its neuronal components (axons, terminals, dendrites and collateral branches) into columns, layers or glomeruli (spherical nests of neural processes). The non-glomerular neuropils are more or less delineated volumes of neuropil in which no obvious pattern can be resolved with regular neuroanatomical staining methods. The glomerular neuropils are often embedded in non-glomerular neuropil and furthermore there are often extensive interconnections between the two types of neuropil. In a later section it will be shown that it is possible to bring order also into non-glomerular neuropil by use of special labeling methods. Sensory-, motor- and "integrative" centers can be neuropil of either type.

The boundaries between the protocerebrum and the deutocerebrum are also a matter of discussion and hence the division of neuropil regions between the neuromeres may be confusing (see Fig. 1). Classically the deutocerebrum has been considered to consist mainly of the centers of antennal receptors (antennal chemosensory glomeruli and posterior mechanosensory antennal centers). The protocerebrum has been thought to include the optic lobes, ocellar foci, mushroom bodies, central body complex, protocerebral bridge, ventral bodies, and dorsal and lateral brain regions receiving optic lobe interneurons, some mechanosensory neurons and non-glomerular neuropil surrounding these areas (see Fig. 1). Strausfeld and Bacon (1983), however, re-defined the deutocerebrum and pushed its frontiers so that the optic lobes and all lateral centers receiving inputs from optic lobe interneurons (P-D in Fig. 1) become deutocerebral. These authors consider protocerebrum as a region without any sensory inputs and basically consisting of mushroom bodies, the central body complex, pars intercerebralis and the nonglomerular neuropil surrounding these regions. To avoid confusion these new definitions will be used only for the fly brain and only in cases where the classical definitions were unclear anyway (some regions outlined in the fly brain by Strausfeld and coworkers were previously "white spots" on the neuronal map).

The tritocerebrum is the smallest of the head ganglia. It is commonly situated lateral to the oesophageal foramen in the dorsal part of the circumoesophageal connectives. No glomeurlar neuropils can be distinguished in tritocerebrum, but some delineated non-glomerular areas are present in some species. The tritocerebrum is connected to mouth parts and foregut via the labral nerves and to the stomatogastric nervous system (frontal ganglion) via the frontal connectives (frontal nerves). Furthermore there are connections to the (post oral) suboesophageal ganglia via the circumoesophageal connectives. The three suboesophageal ganglia are connected to respective mouthparts and via the cervial connectives to thoracic ganglia. The neuropil organization in suboesophageal ganglia is basically non-glomerular.

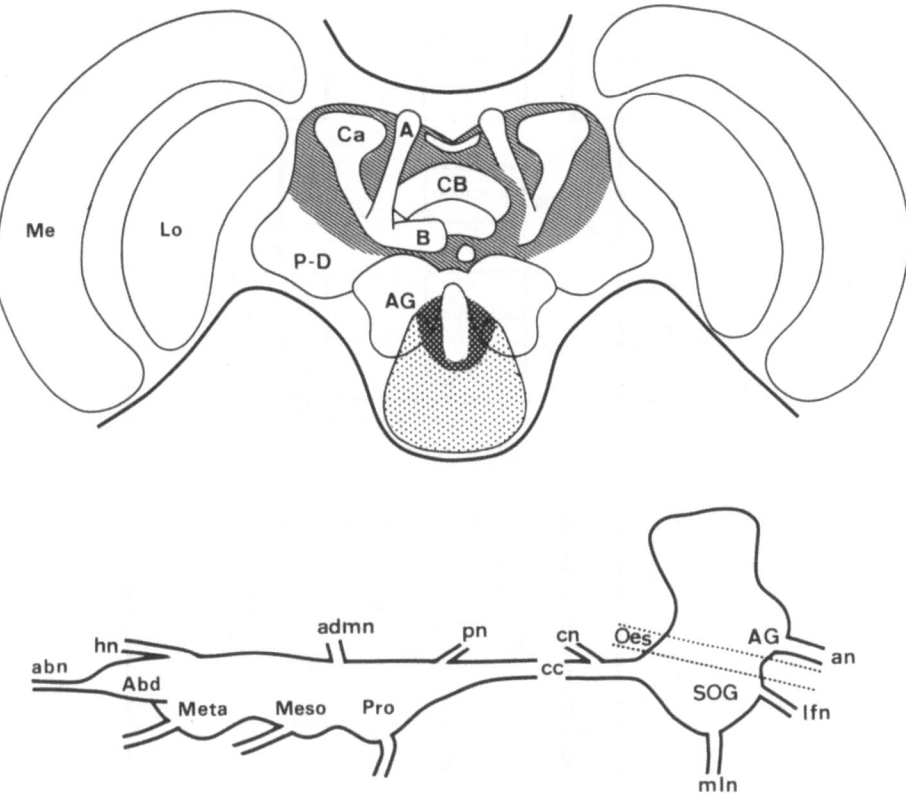

FIG. 1: Schematic drawing of the brain and ventral ganglia in Calliphora. top drawing is a frontal "section" through brain. The cross hatched region is the non-glomerular neuropil of protocerbrum. Embedded in this are the protocerebral centers: the central body (CB), the mushroom bodies with calyx (Ca), alfa-(A) and beta-(B) lobes. The white regions are deutocerebral according to Strausfeld et al. (1984) including medulla (Me), lobula (Lo), antennal glomeruli (AG) and a region laterally (P-D). The densely stippled region is tritocerebrum and the lightly stippled suboesophageal ganglia (SOG below). The Me, Lo and part of P-D are considered protocerebral by earlier investigators. The drawing below is a sagittal view of the nervous system showing some of the major peripheral nerves (an, lfn, mln, cn, admn, kn and abn) and the neuromeres of the ventral ganglia pro-, meso-, meta- and abdominal ganglia. cc = cervical connective; oes = oesophagus.

3. THE SENSORY SYSTEMS OF THE INSECT BRAIN

There are many types of sensory input to the brain. Here we are going to ignore those of the mouthparts and the head hairs and concentrate on photoreceptors of the compound eyes and ocelli and some mechanoreceptive systems of the antennae. The visual pathways from receptors to sites of interactions with descending neurons will be described. Also the antennal pathways will be followed to higher centers and to descending neurons. Some of the chemical anatomy of these systems is known and relevant data are presented. The higher order centers of the

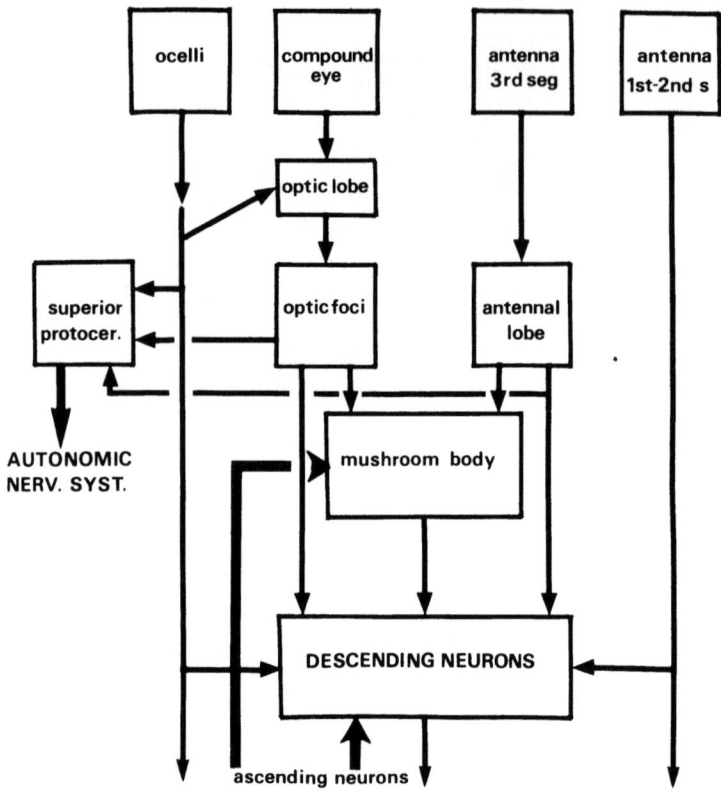

FIG. 2: Diagrammatic presentation of the sensory inputs, higher centers and their relations to descending and ascending neurons. For futher explanations see text. Based on data by Strausfeld and Bacon (1983) and Strausfeld et al. (1984).

protocerebrum that interact with sensory systems and descending neurons will be dealt with separately. A schematic representation of the relations between sensory systems, higher order centers and descending and ascending pathways is shown in Fig. 2.

3.1. The ocellar system

Most adult insects possess dorsal ocelli in addition to their compound eyes (Goodman 1981). Behavioral and physiological experiments have suggested that ocelli are involved in certain orientation behavior, including compensatory head rolling, horizon detection and navigation in polarized light patterns (Wilson 1978; Goodman 1981; Taylor 1981a, b; Rowell and Pearson 1983; Fent and Wehner 1985). Inputs from the compound eyes also contribute to these mechanisms and the ocellar system has been shown to converge with other sensory systems: mechanosensory, olfactory, and visual from the compound eyes (for references see Goodman 1981; Nässel and Hagberg 1985). As will be shown later the ocellar pathway is correspondingly complex with projections to many sites of the CNS. The relatively simple first order ocellar neuropil, with the photoreceptor synapses, is connected by means of afferent and efferent relay neurons of different types to higher order neuropil (see Goodman 1981; Milde 1984; Koontz and Edwards 1984; Nässel and Hagberg 1985; Hagberg and Nässel 1986).

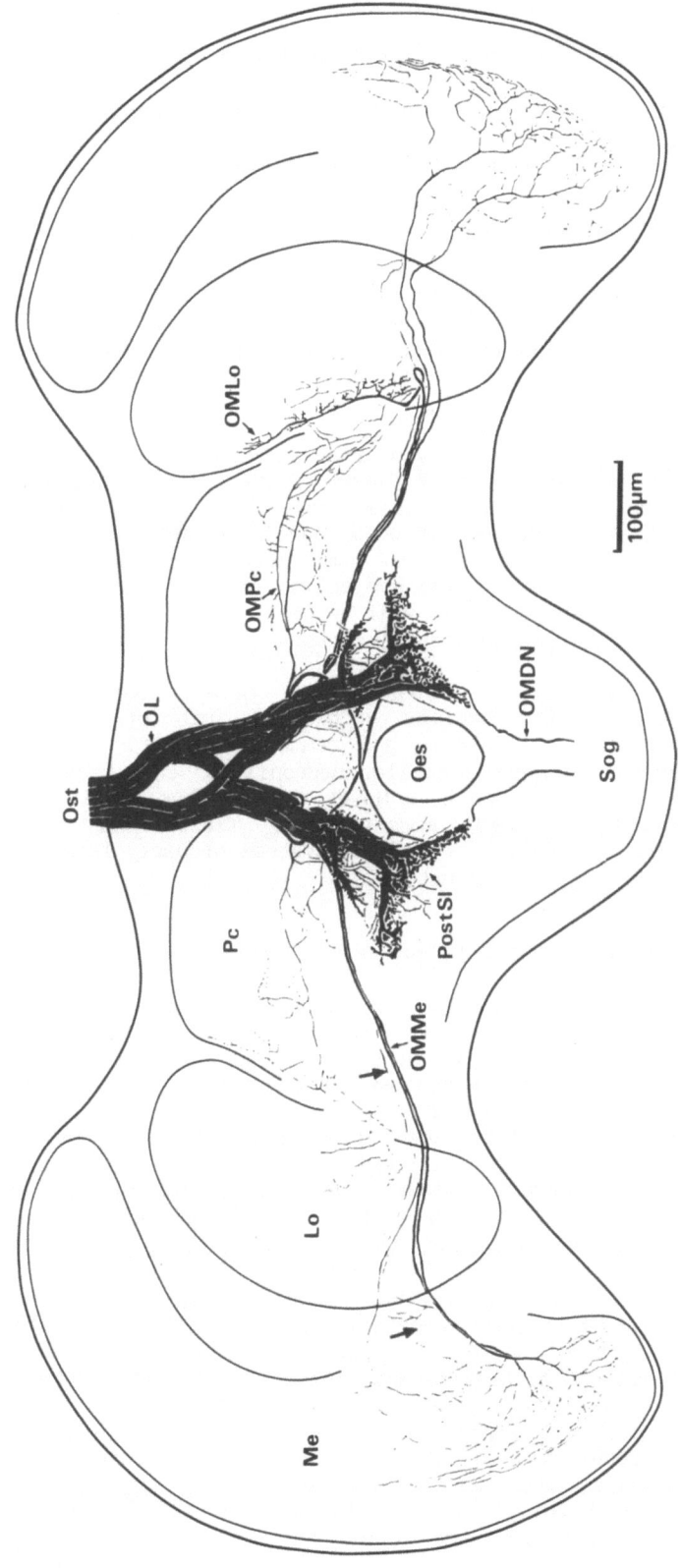

FIG. 3: The ocellar pathway of <u>Calliphora</u> revealed by cobalt backfilling from the peripheral ocellar neuropil. Large interneurons (OL) end in the posterior slope (Post Sl). Medium sized ones (OMMe, OMLo, OMPc, OMDN) project to medulla (Me), lobula (Lo), dorsal protocerebrum (Pc) and thoracic ganglia via suboesophageal ganglia (SOG). From Nässel and Hagberg (1985) by permission from Springer Verlag.

In the blowfly Calliphora three classes of ocellar relay neurons can be distinguished: large, medium and small diameter neurons (Fig. 3). There are 12 large, 10 medium and an undetermined number of small diameter neurons (Nässel and Hagberg 1985). The large neurons terminate in three subregions of the posterior slope (ocellar foci). The medium fibers arborize in several regions of the lateral and superior protocerebrum, in the posterior slope, the lobula and medulla of the optic lobe and in pro- and mesothoracic ganglia (Nässel and Hagberg 1985) (Fig. 3). Small diameter fibers arborize in the above regions as well as in tritocerebrum. Although Calliphora has three ocelli, they share a fused peripheral ocellar neuropil. Hence, it has not been possible with HRP- or cobalt-filling to reveal the neurons subserving individual ocelli (if such neurons actually exist in Calliphora). This has been possible in insects where the ocelli have separate peripheral neuropils and it could be shown that the pathways subserving the lateral ocelli differ from that of the median ocellus (see Goodman 1981; Koontz and Edwards 1984; Hagberg and Nässel 1986).

In Calliphora the 12 large relay neurons terminate among dendrites of descending neurons and terminals of large visual interneurons from the lobula plate of the optic lobe. This convergence of neurons in the posterior slope on each side of the oesophageal foramen (Strausfeld and Bassemir 1985a) will be discussed in some detail in the section on descending neurons. Another ocellar projection area of interest is that of the superior protocerebrum. This will be dealt with in connection with mushroom bodies and higher protocerebral centers.

Only two neuroactive substances have been revealed immunocytochemically in the ocellar pathway. Thin efferent fibers reacting with antisera to serotonin and gamma-aminobutyric acid (GABA) were seen entering the peripheral ocellar neuropil in honey bees (Schäfer and Bicker 1986, D.R. Nässel and E.P. Meyer, unpublished results). Also in the locust efferent GABA-immunoreactive fibers were resolved (Ammermüller and Weiler 1985). These fiber systems probably form feedback loops (GABA) or modulatory (5-HT) pathways.

3.2. The compound eyes and optic lobes

Since the visual system of the fly (i.e. Musca and Calliphora) is among the best known anatomically (see Strausfeld 1976a, b, 1984; Strausfeld and Nässel 1980) it will be used as an example of organizational principles. The visual system consists of a retina and the optic lobes. The retina is formed by optic elements, pigment cells and photoreceptors arranged in units, ommatidia; the optic lobes have three main neuropil regions, the lamina, medulla and lobula complex. The blowfly retina is composed of ca. 3,500-4,000 ommatidia with 8 photoreceptors each. In insects with fused rhabdoms (microvillar photosensitive protein of receptor cells) the receptor cells of one ommatidium innervate one column of neurons (cartridge) in the lamina (Meinertzhagen 1976). In flies the rhabdom is not fused and the photoreceptor axons of different ommatidia converge into the cartridges (Trujillo-Cenoz 1966; Braitenberg 1967; Kirschfeld 1967). The end result is that groups of six photoreceptors sampling the same point in the usual space converge onto second order neurons of each cartridge.

The columnar organization of the lamina neurons is projected to the more central neuropils, the medulla and lobula complex, by means of afferent (and efferent) relay neurons (Fig. 4). There are 10 types of second order neurons connecting the lamina and the medulla, one type of intrinsic (amacrine) neuron (Strausfeld and Nässel 1980; Strausfeld 1984)

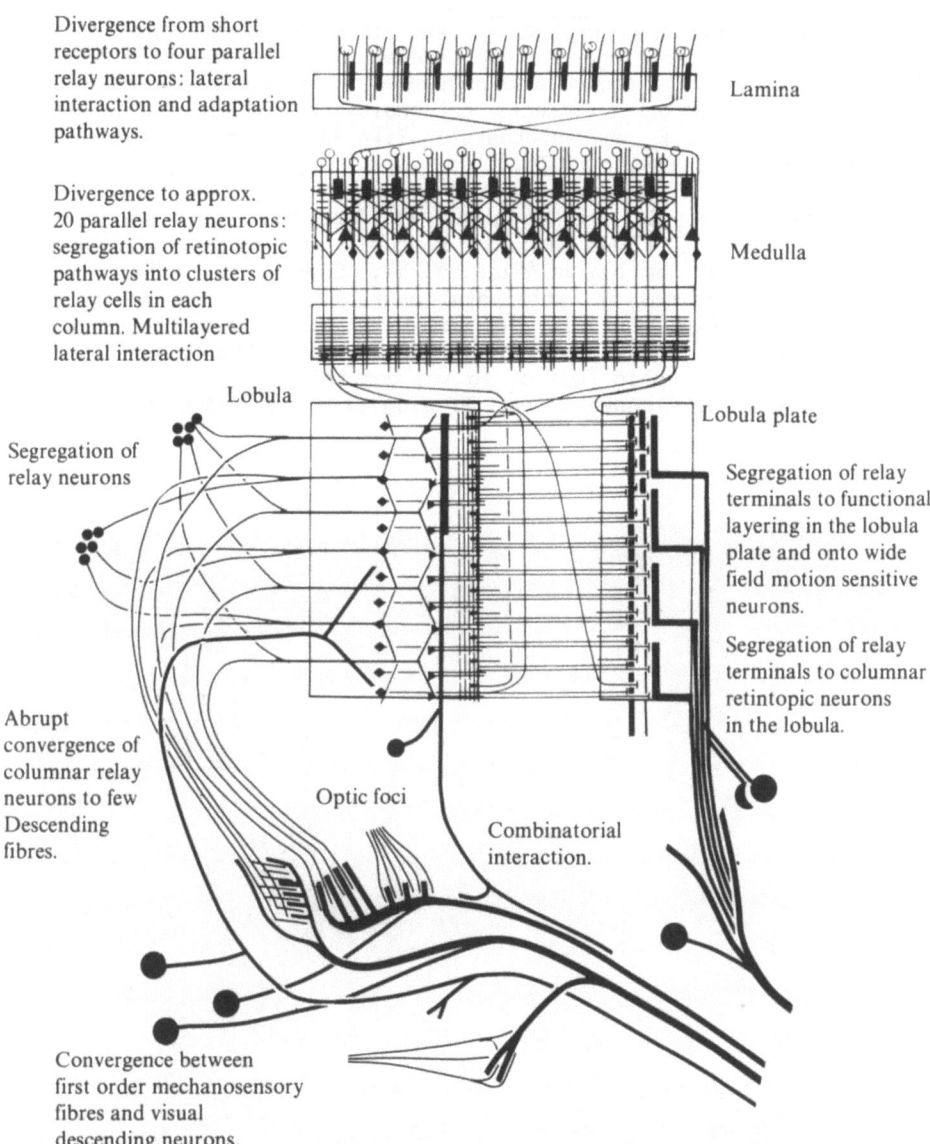

FIG. 4: Diagrammatic representation of the organization of the optic lobe and its output connections. For further details see text. From Strausfeld and Nässel (1980) by permission from Springer Verlag.

and one serotonin-immunoreactive neuron type (LBO5HT) connecting the lamina, medulla and lobula complex to the midbrain and contralateral optic lobe (Nässel et al. 1985, 1987; Nässel 1987). Of these 12 types, 8 are presnet in each column of the lamina and medulla (retinotopic columnar elements). The amacrines, the two types of tangential neurons and the LBO5HT neurons have processes spreading over large proportions of the projected retinal mosaic. The complex synaptic connections of the lamina elements have been described by Strausfeld and Campos-Ortega (1977). They proposed pathways possibly subserving neural adaptation and lateral inhibition as well as signal amplification. Only three neuroactive substances have been tentatively identified in synaptic neuropil of the lamina of <u>Calliphora</u> and <u>Drosophila</u>; choline acetyltransferase and GABA-immunoreactivity was seen in the columnar centrifugal neurons C2, which are presynaptic to cartridge elements (Meyer et al. 1986; Buchner et al. 1986) and Hardie (1987) proposed that histamine is the transmitter of the photoreceptors. In other insects catecholamine containing as well as serotonin-, gastrin/CCK- and somatostatin-like immunoreactive neurons were resolved (see Nässel 1986a, 1987).

The next neuropil, the medulla, is formed by ca. 60 types of neurons. (Strausfeld, 1976b) and the neuropil organization is far more complex than that of the lamina (Figs. 4, 5). Some of the medulla neurons are columnar centrifugal or centripetal neurons, others are amacrines or large field tangential neurons. The columnar neurons connect the medulla to the lobula and lobula plate (Figs. 4, 5) and the tangential ones project to different optic foci or to the contralateral optic lobe. Each column of the medulla contains ca. 30 axon profiles, some of which are

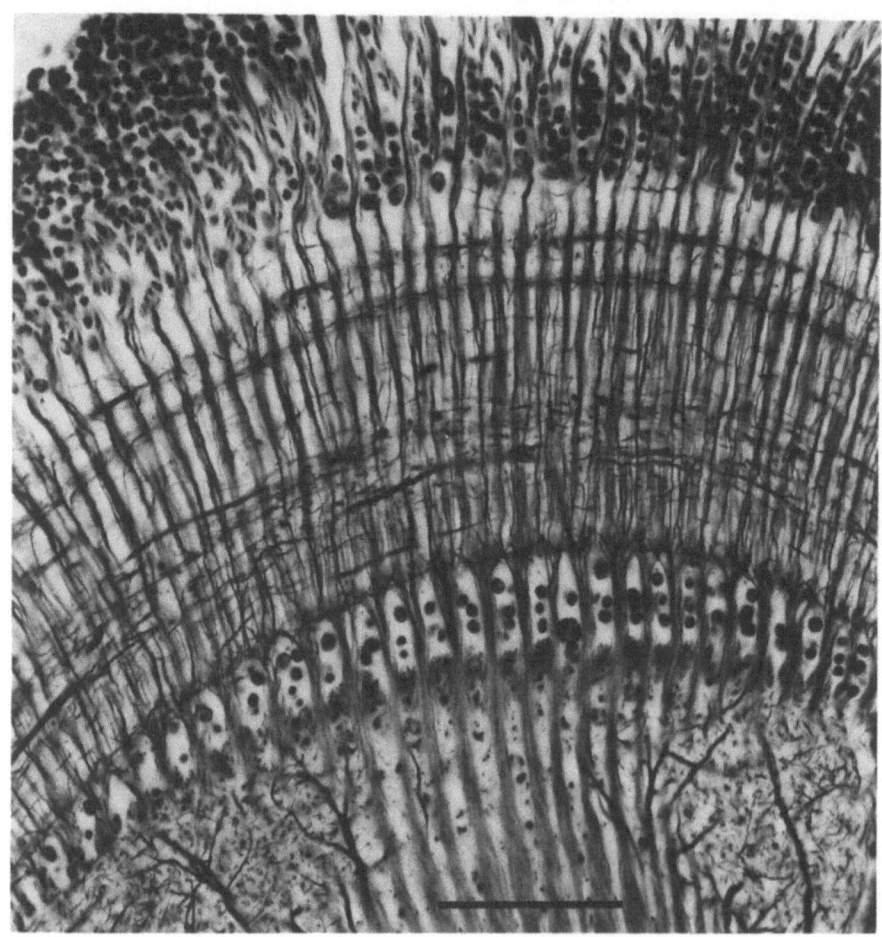

FIG. 5: Frontal section of medulla (reduced silver staining) showing the columnar and layered organization of the medulla. Note that at the lower margin of the medulla the columnar fiber bundles converge to larger bundles, thus coarsening the projected retinal mosaic, before reaching the lobula plate (at bottom of figure) and lobula (not shown). Scale: 50 μm.

terminations of lamina neurons. The medulla is further recognized by its prominent layering (Fig. 5), derived from levels of neuron terminations, dendritic arbors and tangential processes. Several of these layers contain processes of chemically identified neurons: catecholamine containing (CA), serotonin (5-HTi), GABA-like- (GABAi), FMRFamide-like and gastrin/cholecystokinin-like immunoreactivity (G/CCKi) as well as neurons reacting with antibodies to the enzyme choline acetyltransferase (see Nässel 1986a, 1987). The function of these pathways remain to be studied, but it can be speculated that GABAi neurons form inhibitory circuits and CA, 5-HTi and G/CCKi neurons form modulatory pathways (see Nässel 1986a). The morphological neuron types in the medulla of flies are listed in Strausfeld (1976a, b; 1984).

The third neuropil region, the lobula complex, is divided in flies into an anterior lobula and a posterior lobula plate (Fig. 4). In these neuropils the columnar organization is not as apparent as in the more peripheral neuropils due to more extensive arborizations of the columnar elements. The change in organization is also due to a coarsening of the projected mosaic (Fig. 5) so that each output channel of the lobula corresponds to nine converging medulla columns (Strausfeld and Nässel 1980; Strausfeld 1984). The lobula complex represents the final retinotopic regions of the optic lobe and contains columnar output neurons as well as arrays of noncolumnar large field neurons which subserve different portions of the visual field (Fig. 4). In the lobula various local neurons represent e.g. the anterio-dorsal foveal area (Hausen and Strausfeld 1980; Strausfeld 1980) or the dorsal rim area (Fig. 6) (Strausfeld and Wunderer 1985). The lobula plate is characterized by a number of large-tangential neurons that each represent a vertical or horizontal strip or portion of the projected mosaic (Pierantoni 1976; Dvorak et al. 1975; Hausen 1984). These large lobula plate as well as columnar and local lobula neurons project to specific loci in the lateral brain (area termed lateral deutocerebrum by Strausfeld and Bacon 1983) (Figs. 4, 7). A fraction of the outputs (notably from lobula and medulla) project to dorsal protocerebral optic foci or to "non-glomerular" neuropil in the superior protocerebrum also containing processes of "neurosecretory" cells (Strausfeld 1976b; Strausfeld and Bacon 1983). The terminations of the lobula complex neurons in the optic foci in the lateral brain and the functional roles of these pathways will be treated in more detail in connection with the descending pathways. Figure 7 shows the output tracts from the housefly optic lobes, some of which run to contralateral optic lobes.

In the lobula complex different layers of processes have been found that are 5-HTi, GABAi, proctolin-i, FMRFamide-i and G/CCKi and choline acetyltransferase-immunoreactive (see Nässel 1987). In the lobula plate the GABAi neurons are wide field neurons of two main types that either terminate in the ipsilateral brain or connect to the contralateral lobula plate (Meyer et al. 1986). The 5-HTi neurons are connected to the remaining optic lobe neuropils ipsi-and contralaterally and may be involved in arousal and neural activity control (see Nässel et al. 1985, Nässel 1986a).

One intriguing feature of the insect optic lobes is that single identifiable neurons and local assemblies of neurons can be found. One example is the large lobula plate neurons mentioned above which can be identified by their location, morphology and response characteristics (Hausen 1984). Another example is neurons that subserve specialized regions of the retina. In blowflies and houseflies there are local distributions of specific neurons anteriorly and anterio-dorsally (Hausen and Strausfeld 1980; Strausfeld 1980). In males the anteriodorsal retina is expanded and specialized photoreceptors can be found (Franceschini et

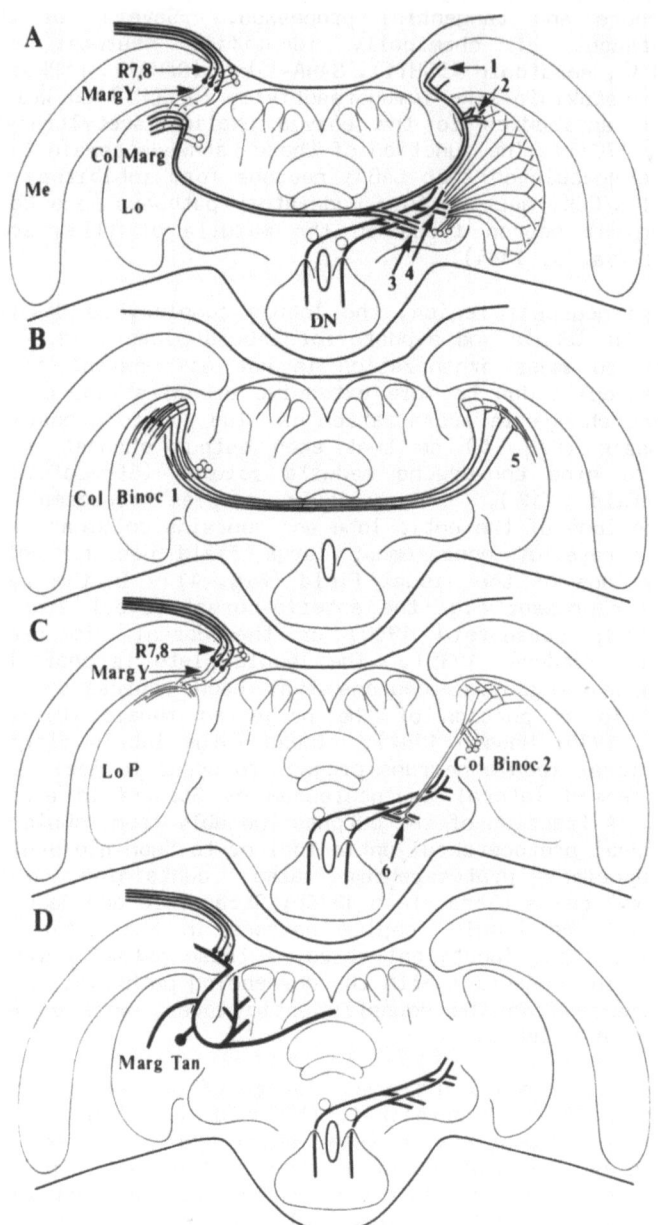

FIG. 6: Schematic representation of some fiber pathways subserving the dorsal marginal ommatidia (presumed to be specialized for polarized light detection). A and B show binocular pathways (Col Marg and Col Binoc 1). C shows local assemblies of neurons in the marginal region connecting medulla and lobula plate (Marg Y) and connecting lobula and descending neurons (Col Binoc 2). D shows a large marginal tangential neuron in the medulla-lobula (Marg Tan). From Strausfeld and Wunderer (1985) by permission of N.J. Strausfeld and Springer Verlag.

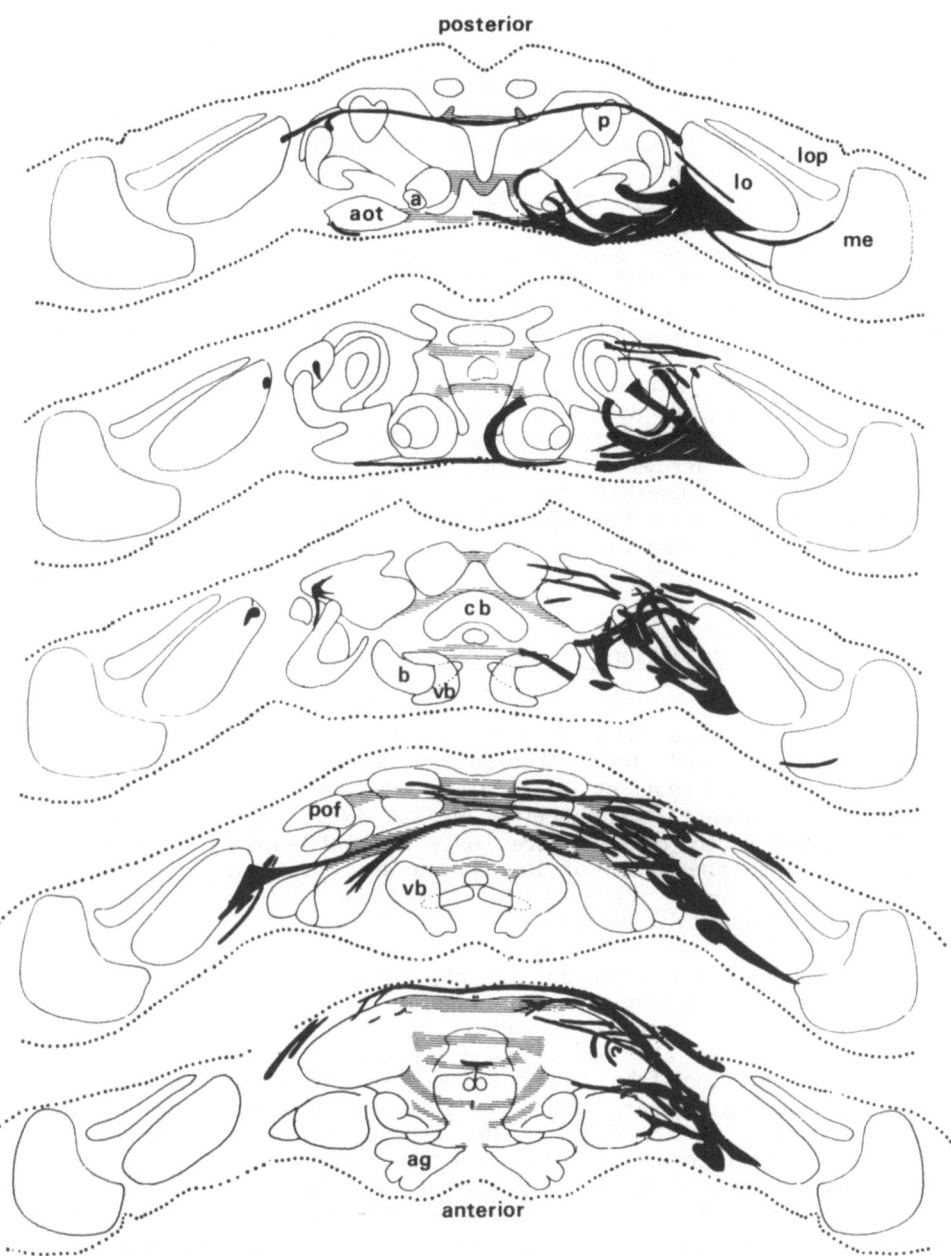

FIG. 7: The paths of optic lobe neurons to brain centers and contralateral optic lobe. Tracings of horizontal sections (some sections left out) of which the one at the top is the most dorsal. Note the extensive divergence of fiber paths from the medulla (me), lobula (lo) and lobula plate (lop). aot = anterior optic tubercle (optic focus); pof = posterior optic focus. cb = central body; a, p and b are parts of the mushroom body; vb = ventral body; ag = antennal glomeruli. From Strausfeld and Nässel (1980) by permission from Springer Verlag.

al. 1981; Hardie 1983) and specific male relay neurons subserve this region (Hausen and Strausfeld 1980; Strausfeld 1984). The males use this eye-region for tracking of females during flight (Wehrhahn 1979). In some insects the specialization of the male eyes proceed further and consequently the differences in the underlying neural organization between dorsal and ventral retinal regions may be very drastic (see e.g. Wohlburg-Buchholz 1977; Zeil 1983). In male mayflies the neural arrangements in the lamina (Wohlburg-Buchholz 1977) as well as the medulla and lobula (Nässel 1986a) are very different between the dorsal and ventral parts of the projected mosaic.

The dorsal marginal part of the retina in several insect species is composed of ommatidia that seem specialized for perception of polarized light patterns (Schinz 1975; Burghause 1979; Labhart 1980; Wunderer and Smola 1982; Meyer 1984). Also the organization of photoreceptor axons and higher order neurons is modified in this region of e.g. blowflies and desert ants (Strausfeld and Wunderer 1985; Meyer and Nässel 1986). In Calliphora the marginal ommatidia (of dorsal rim area) are represented by specialized marginal neuropil in the medulla and lobula complex, from which arise local arrays of columnar and tangential neurons which form separate pathways possibly involved in analysis of polarized light patterns (Strausfeld and Wunderer 1985). The arrangements of these pathways are illustrated in Fig. 6.

3.3. The central projections of antennal receptors

The central projections of antennal receptors and the organization of antennal centers have been studied in several insect species (e.g. Strausfeld 1976a; Ernst et al. 1977; Matsumoto and Hildebrand 1981; Stocker and Lawrence 1981; Boeckh et al. 1984; Stocker et al. 1983; Strausfeld and Bacon 1983). Here, only a brief summary of the sensory projections into the brain of Drosophila and Calliphora will be given. Special emphasis is on the possible connections with descending pathways and higher protocerebral centers.

In Drosophila the receptors (mainly chemoreceptors) of the sensilla on the third antennal segment, the flagellum, and the fourth, the arista, project into the 19 antennal glomeruli of the antennal lobe in specific patterns (Stocker et al. 1983). Some further projection areas were seen after cobalt filling from the third segment (Fig. 8): an ipsilateral posterior antennal center (iPAC) and fibers reaching the suboesophageal ganglia (Stocker et al. 1983). The receptors of the second segment, pedicellus, appear to predominatly innervate the iPAC. In Calliphora the chemoreceptor projection was not studied in the same detail, but the sensory projection from the first and second antennal segments have been described. From these segments axons enter the dorso-caudal part of deutocerebrum (correcponding to iPAC), the suboesophageal ganglia (Fig. 11a) and one large axon from the pedicellar giant campaniform sensillum continues bilaterally into the three thoracic ganglia (Strausfeld and Bacon 1983; Nässel et al. 1984). The majority of the antennal axons terminating in the mechanosensory division of the deutocerebrum could be traced back to the scolopideal sensillae of Johnston's organ in the pedicellus (Strausfeld and Bacon 1983). Johnston's organ monitors vibrations and displacements of the flagellum caused by acceleration and deceleration of air currents during flight (Gewecke 1974). The mechanosensory projections onto the ventro-lateral brain are interesting in this account since some of them terminate in the region of descending neuron dendrites.

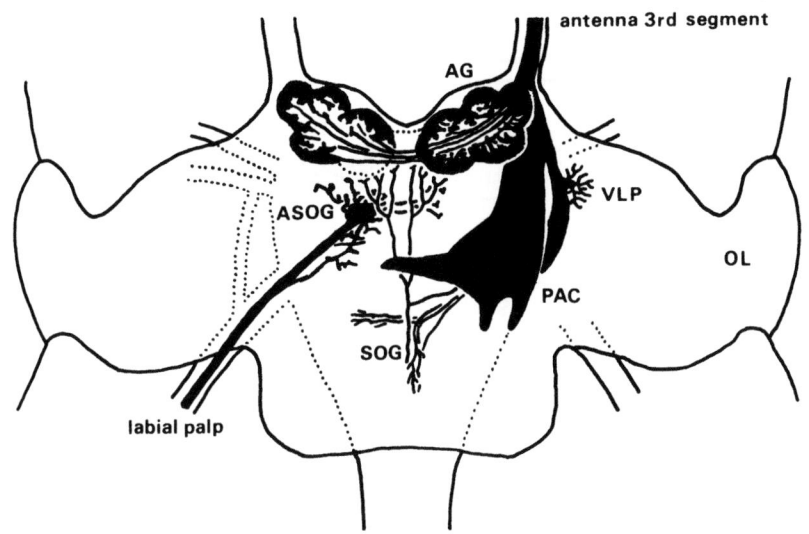

FIG. 8: Sensory pathways from 3rd antennal segment and labial palp of Drosophila seen after cobalt filling. AG = antennal glomeruli; PAC = posterior antennal center; VLP = ventrolateral protocerebrum; OL = optic lobe; SOG = suboesophageal ganglia. Some fibers from the antenna and those from the labial palp converge into a neuropil region in the anterior suboesophageal ganglia (ASOG). Redrawn after Stocker and Lawrence (1981) and Stocker (1982).

4. ORGANIZATION OF DESCENDING PATHWAYS

Descending neurons connect the brain and suboesophageal ganglia to thoracic ganglia. They, hence represent the "final" pathways from the cephalic centers (Figs. 2, 9) and are useful as probes for higher integration of sensory inputs and in the analysis of pathways involved in activating motor programs in body ganglia. We are going to consider the inputs to the descending neurons from ocelli, compound eyes, antennae, neurons ascending from ventral ganglia, and inputs from higher centers in the protocerebrum (Figs. 2, 9, 11). The scene has been partly set in previous sections which outlined the projection areas of the sensory systems in Calliphora (summarized in Fig. 11). As shown in Figs. 2 and 9, however, not all inputs to the descending neurons are directly from sensory pathways. Some are indirect via the mushroom bodies and their associated neuropils (see later section), higher protocerebral centers as well as from local interneuron circuits and modulatory pathways (some localized in non-glomerular neuropil). Before entering the organization of the descending neurons in Calliphora it may be useful to study the best characterized descending pathway in insects, the Descending Contralateral Movement Detector (DCMD) of locusts (O'Shea et al. 1974; O'Shea and Rowell 1975, 1976; Rowell et al. 1977).

As shown in Fig. 10, the DCMD receives input from the Lobula Giant Movement Detector (LGMD) as well as auditory inputs (Rowell et al. 1977). The visual input neuron LGMD, in turn, receives one retinotopic and two non-retinotopic inputs. The retinotopic input is excitatory and the non-retinotopic ones (MUB and DUB in Fig. 10) are inhibitory. The LGMD is chemically presynaptic to an ipsilateral descending neuron (DIMD in Fig.

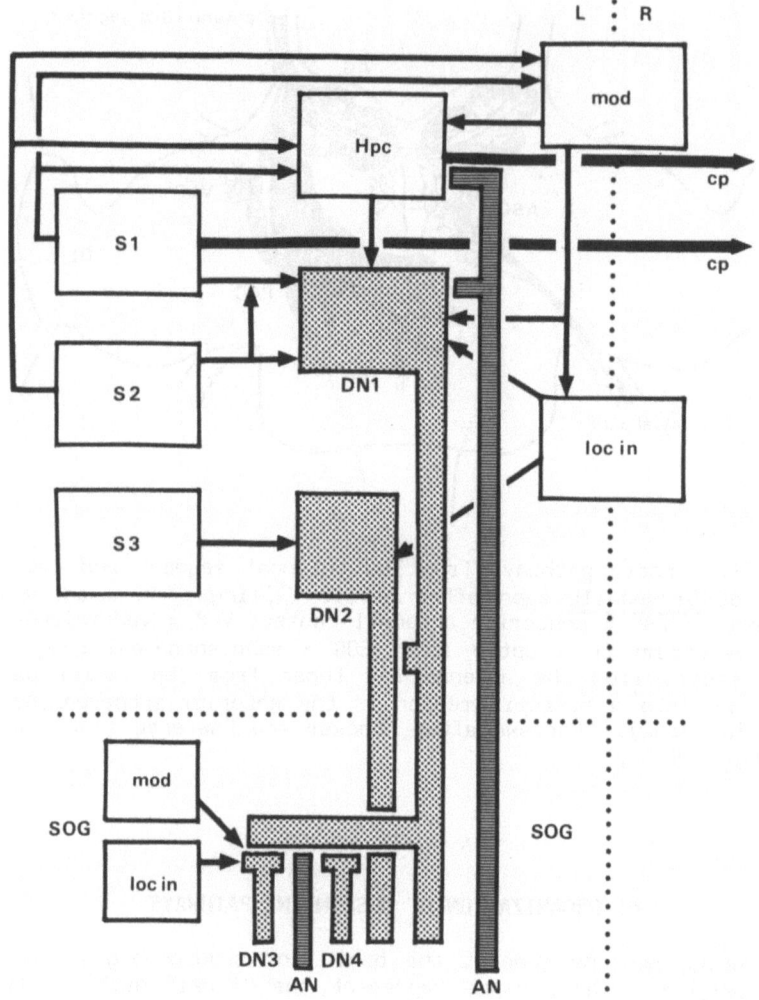

FIG. 9: Schematic diagram showing principle organization of descending neurons. Two preoral descending neurons (DN1, DN2) and two suboesophageal (SOG) descending neurons (DN3, DN4) are shown. The preoral ones receive inputs from sensory systems (S1-S3), higher protocerebral centers (Hpc), modulatory pathways (mod), local interneurons (loc in) and ascending neurons (AN). The suboesophageal ones are assumed to receive inputs from preoral descending neurons, ascending, local and modulatory neurons. L = left; R = right hemisphere; cp = contralateral connection. Futher explanations in text. Compiled from Strausfeld and Bacon (1983) and Strausfeld et al. (1984).

10) and electrically coupled to the DCMD (O'Shea and Rowell 1975). The DIMD receives additional inputs from other visual interneurons. The other sensory inputs to the DCMD are by means of chemical synapses. The DIMD and DCMD have arborizations in all neuromeres of the thoracic ganglia (Fig. 10d) The connections of the DCMD in the metathoracic ganglion are shown in Fig. 10e. The DCMD synapses onto at least six motorneurons in this ganglion and is involved in control of the escape jump reaction (Burrows and Rowell 1973; Pearson et al. 1980). This brief summary

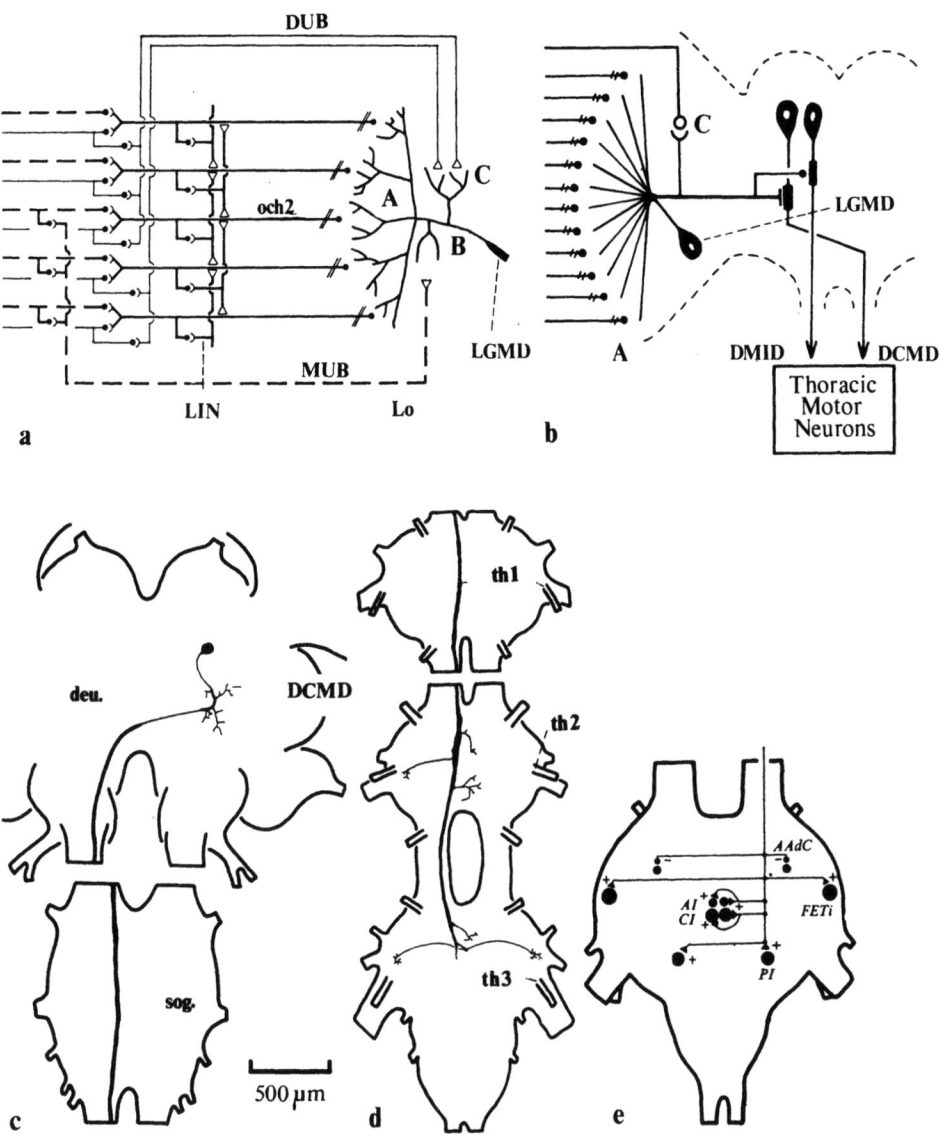

FIG. 10: The connections of the locust descending contralateral movement detector (DCMD) and the lobula giant movement detector (LGMD) and motorneurons in the thoracic ganglia (AAdC, FETi, AI, CI, PI), described by Rowell and coworkers (references in text). Further explanations in the text. From Strausfeld and Nässel (1980) by permission of Springer Verlag.

illustrates how identifiable neurons can be probed by intracellular electrophysiological-anatomical technique to reveal pathways from the lobula to the behavior output. It also shows some of the main features of descending neuron organization: (1) several types of sensory inputs synapse onto different dendritic regions of the descending neuron (e.g. A-C in Fig. 10a), (2) descending neurons often share some inputs with other decending neurons (like DCMD and DIMD in Fig. 10b), (3) the outputs in thoracic ganglia often diverge onto several postsynaptic neurons in each ganglion (Fig. 10e). Further principles will be shown below.

369

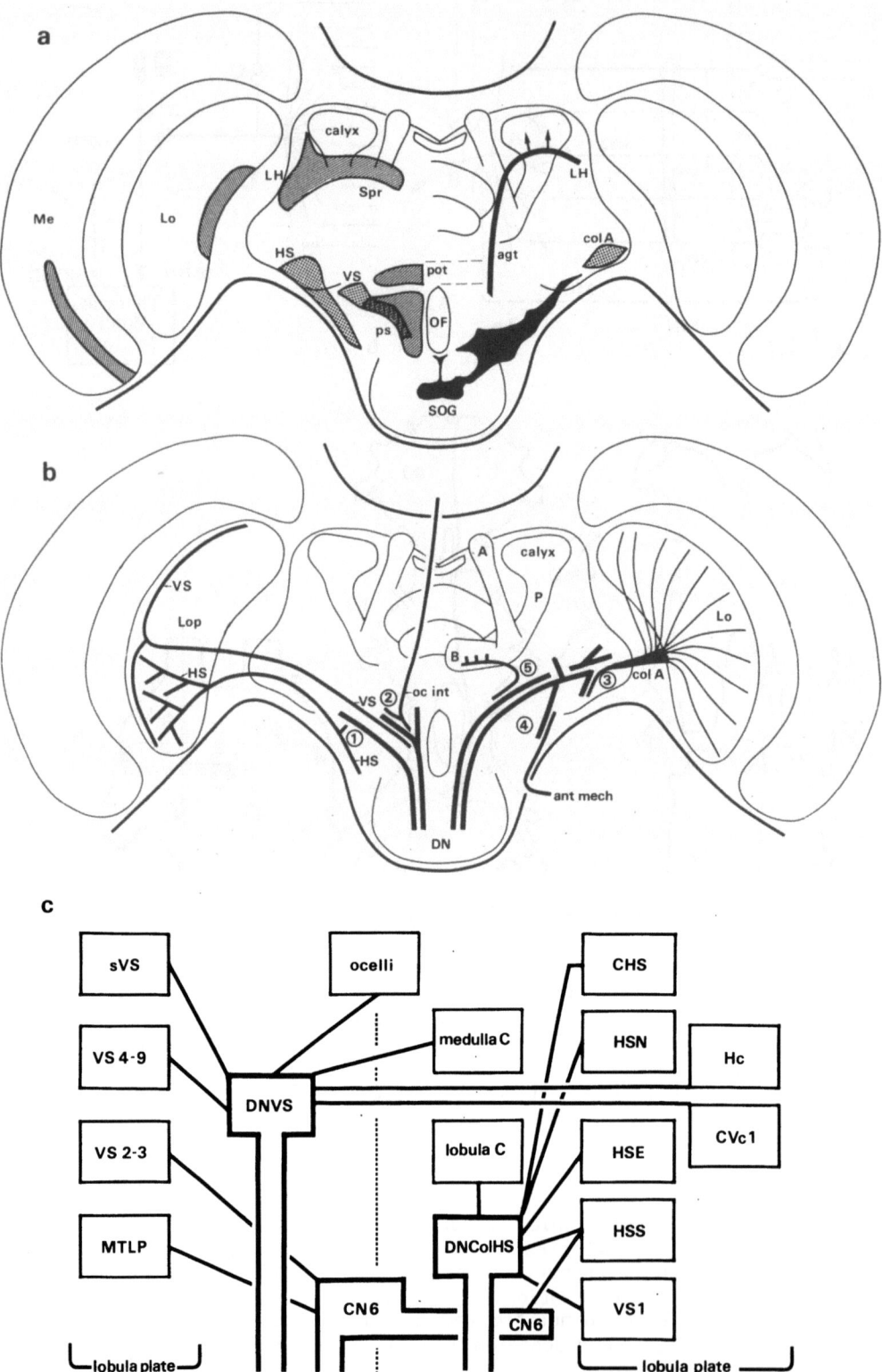

FIG. 11: The spatial relations between sensory, afferents, relay neurons and descending pathways shown schematically in the Calliphora brain.
a. Crosshatched areas are sites of terminations of ocellar interneurons in medulla (Me), lobula (Lo), lateral horn (LH) of superior protocerebrum (Spr), posterior optic tract (pot) and posterior slope (ps), dotted areas are terminations of some visual interneurons referred to in the text HS-cells, VS-cells (convergence with ocellar interneurons) and Col A cells. The black termination field is derived from the antennal mechano-sensory afferents (convergence with Col A terminals). Also the antenno-glomerular tract (agt) connecting antennal glomeruli to mushroom body calyx and lateral horn is shown (convergence between ocellar and antennal interneurons in LH).
b. Shows schematically some of the inputs to the descending neurons. 1. HS-cells of lobula plate (Lop) onto descending neuron cluster. (Further inputs at 1 shown in c). 2. VS-cells and ocellar interneurons (oc int) converge onto descending neuron cluster (further inputs in c). At 3 and 4 columnar lobula neurons (col A) and antennal mechanoreceptors converge on cluster (containing the giant descending neuron shown in more detail in Fig. 12). At 5 a connection between beta-lobe (B) of mushroom body and descending neuron is shown (discussed in text).
c. Connections between relay neurons and descending (DNVS, DNColHs) and motorneurons (CN6) in the Calliphora brain as suggested by Strausfeld and Bassemir (1985a, b) and Strausfeld and Seyan (1985). The dotted line is the midline of the brain. Medulla C, Lobula C = medulla and lobula columnar neurons. Further explanations and abbreviations in the text.

In Calliphora several types of descending neurons (DNs) have been analyzed morphologically (Strausfeld and Bacon 1983; Strausfeld et al. 1984; Strausfeld and Bassemir 1985a, b; Strausfeld and Seyan 1985; Bacon and Strausfeld 1986). These authors divided the DNs into three classes: (1) preoral DNs with cell bodies and processes mainly in deutocerebrum, (2) postoral with cell bodies in the rind of tritocerebrum or suboesophageal ganglia and processes mainly in suboesophageal ganglia and (3) protocerebral DNs, whose axons project to the autonomic nervous system (corpora cardiaca) (see Figs. 2, 9).

The deutocerebral DNs are arranged in uniquely identifiable clusters (Figs. 13, 14). The neurons of these clusters are usually individually identifiable, but share part or all of a common sensory input. Some DN clusters may, however, contain isomorphic neurons which cannot be individually identified. The dendrites of DNs of one cluster may form several subregions, which each receives a different type of input (see Fig. 12-14). Often small diameter DNs accompany and morphologically mimic the large DNs. Deutocerebral DNs receive inputs from visual pathways, mechanosensory neurons and various other interneurons of the brain. One interesting finding that has come out of the analysis of deutocerebral DNs should be mentioned in passing. Most DNs have their dendrites in non-glomerular neuropil. The organization of DN clusters, resolved e.g. after cobalt filling, however, shows that the DN dendrites are compartmentalized. These different input regions, although not showing the geometrical organization of e.g. the lamina, indicate a very precise organization of neuropil. The term non-glomerular, hence, only implies that the glomerular organization is beyond resolution with conventional staining methods.

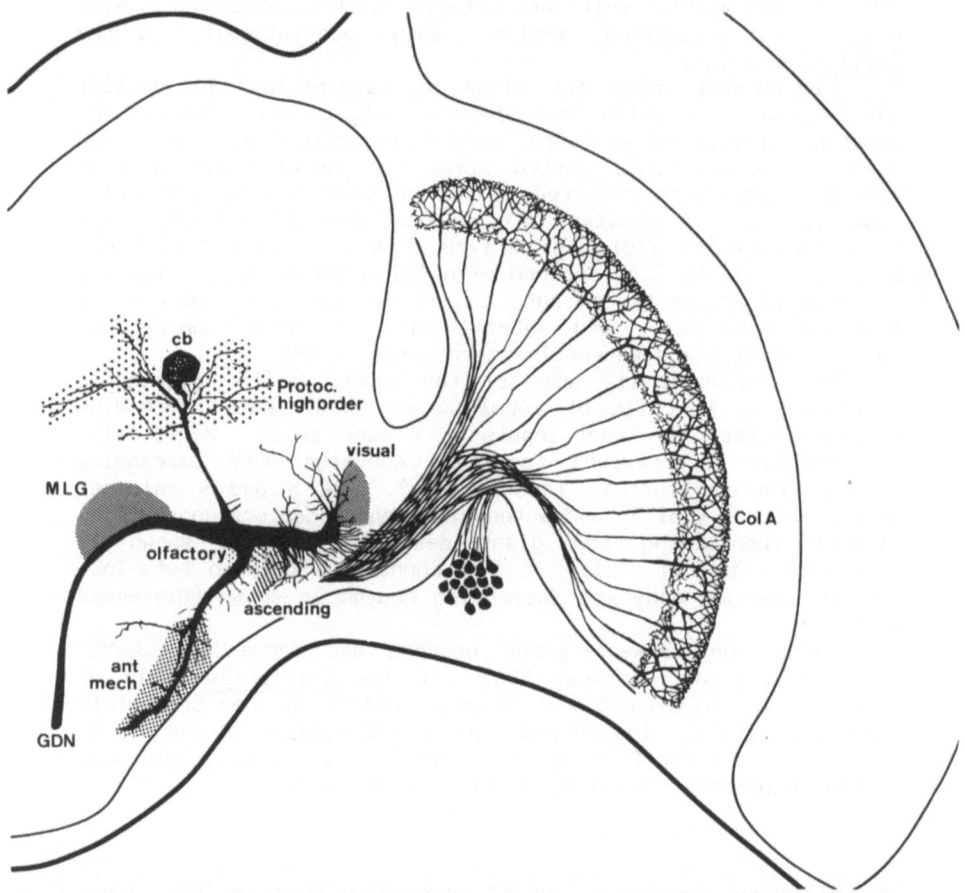

FIG. 12: Inputs onto the giant descending neuron (GDN) of <u>Musca domestica</u> as proposed by Strausfeld and Bacon (1983), Strausfeld et al. (1984) and Bacon and Strausfeld (1986). The GDN was drawn from HRP fill and the Col A from a transsynaptic cobalt fill from cervical connective (D.R. Nässel, unpublished). Further details in text.

The suboesophageal DNs have dendrites that are more diffusely arranged. It appears hard to identify individual suboesophageal DNs. The inputs to the suboesophageal DNs are presumed to be sensory afferents from mouth parts (see Fig. 8) and from axon collaterals from deutocerebral DNs (Strausfeld et al. 1984; see also Fig. 9). Strausfeld et al. (1984) also make the distiction that preoral DNs have divergent outputs in the thoracic ganglia, whereas suboesophageal DNs from the same cluster terminate in the same region of the body ganglia. Of the ca. 5,500 axons in the cervical connective, about 200 may be deutocerebral descending neurons and ca. 300 suboesophageal ones, the remainder is presumed to be motorneurons, ascending neurons and sensory afferents (descending and ascending) (Strausfeld et al. 1984; Strausfeld and Seyan 1985).

We shall now proceed to look onto a few examples of descending pathways that have been analysed in flies. These are shown schematically in Fig. 11B and in some detail in Figs. 11C-14. The first pathway constitute the so called giant fiber pathway (King and Wyman 1980;

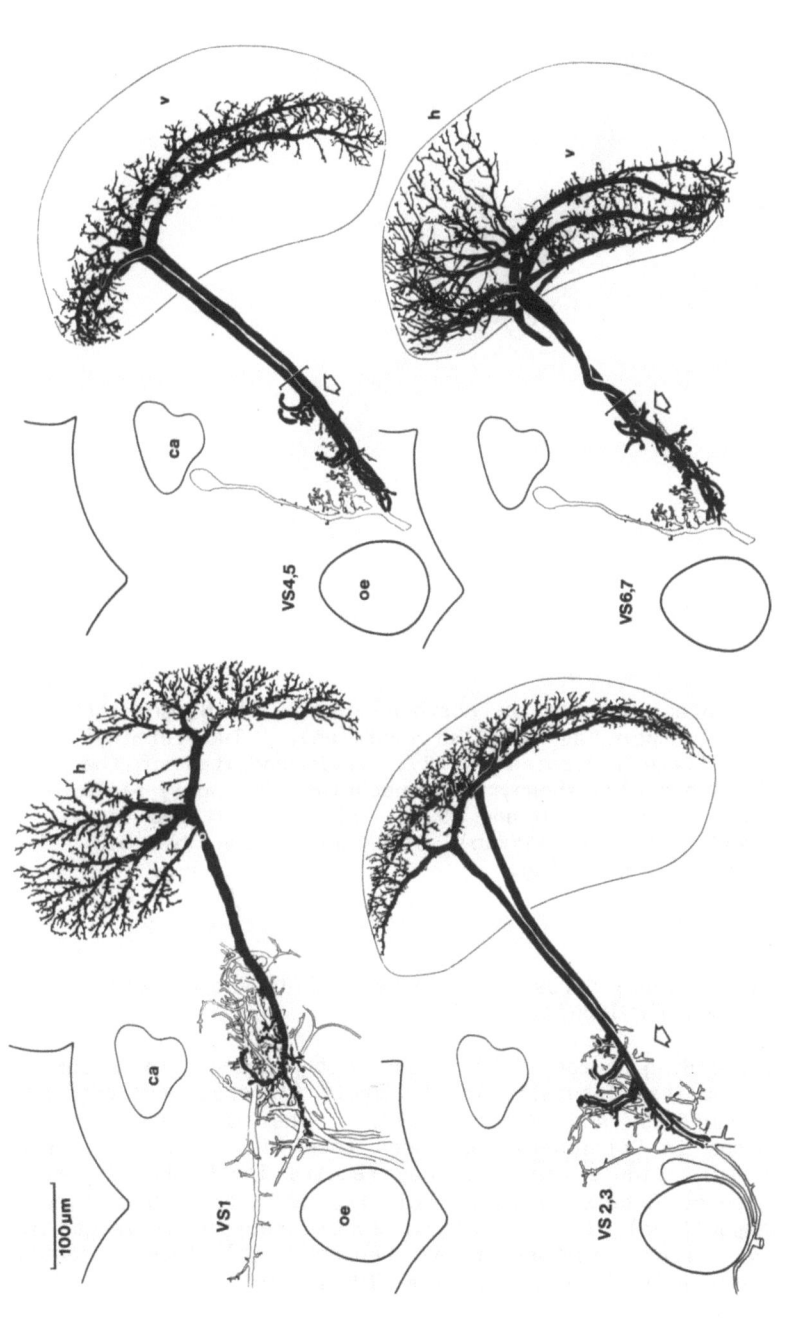

FIG. 13: Seven of the eleven VS neurons of the Calliphora lobula plate. Descending neurons and their dendrites are shown as empty profiles. The VS1 contacts the DNColHS cluster, the VS2, 3 the CN6 motorneuron and VS4-7 the DNOVS 1 (the remaining neurons of this cluster not shown). Note that VS1, 6 and 7 have processes (h) in the layer of HS dendrites, the remaining VS dendrites are in a more posterior layer (v). From Strausfeld

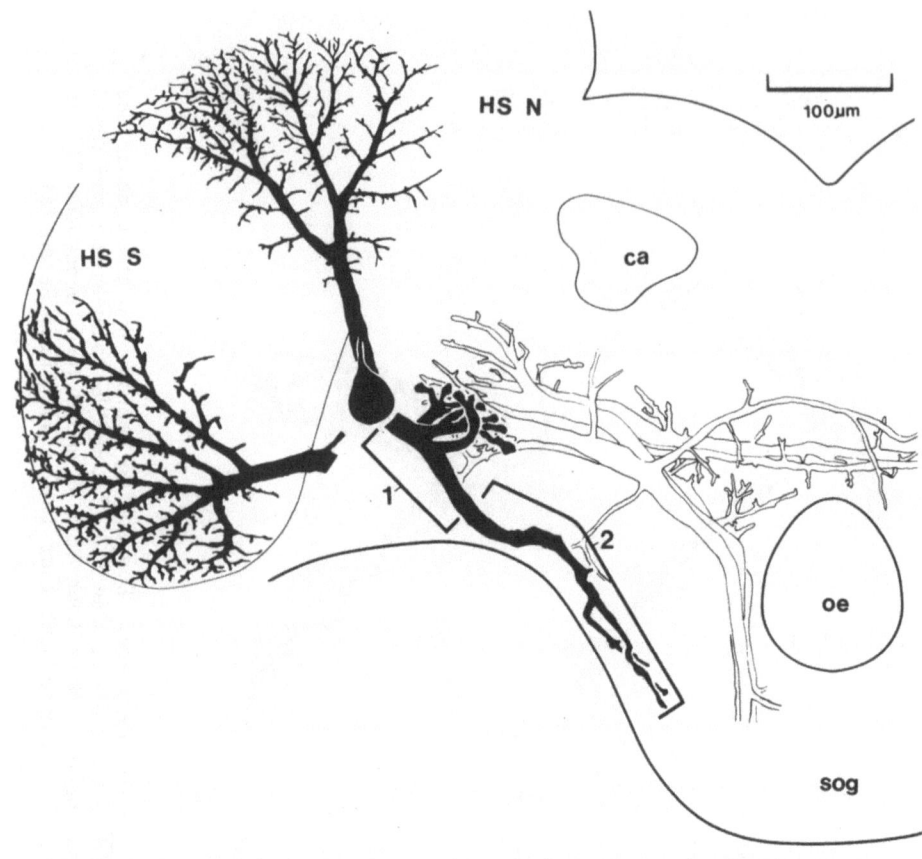

FIG. 14: The contact between the north HS neuron (HS N) and the DNCo1HS cluster (shown as unfilled profiles). Two terminal regions of HS N are indicated (1, 2). The dendrites of the south HS (HS S) are also shown, the equatorial HS roughly fills the empty field between HS N and HS S. Slightly altered from Strausfeld and Bassemir (1985b) with permission by N.J. Strausfeld and Springer Verlag.

Strausfeld and Nässel 1980; Strausfeld and Bacon 1983; Strausfeld et al. 1984; Bacon and Strausfeld 1986).

The giant descending neuron (GDN) is the most prominent in a cluster of neurons with dendrites laterally in the brain (lateral deutocerebrum according to Strausfeld and Bacon 1983). The GDN has a very characteristic shape in Calliphora and Musca (see Fig. 12). As shown schematically in this figure dendrites may receive inputs from several sources: (1) columnar lobula neurons (Col A), (2) from other visual interneurons (unspecified), (3) from local lobula neurons (enlarged in males; MLG), (4) from olfactory relay neurons, (5) from antennal mechanoreceptor afferents (presumably from Johnston's organ), (6) from ascending neurons and (7) from higher order protocerebral centers (e.g. mushroom bodies) (Strausfeld and Bacon 1983; Strausfeld et al. 1984; Bacon and Strausfeld 1986). Some of these inputs are shared by other neurons of the GDN cluster. Some of the GDN dendrites are embedded in tangled neuropil packed with processes from hitherto unidentified neurons and clearly all inputs (and outputs?) of the GDNs in the lateral deutocerebrum are not accounted for yet.

The outputs of the GDNs in thoracic ganglia are also known (King and Wyman 1980; Tanoye and Wyman 1980; Strausfeld and Bacon 1983; Strausfeld et al. 1984; Bacon and Strausfeld 1986). The Calliphora GDN contacts and is dye-coupled to the contralateral GDN and the large motorneurons to the tergotrochanter muscle (TTM) (Bacon and Strausfeld 1986). In Drosophila the GDNs also contact a large peripherally synapsing interneuron (King and Wyman 1980). In Calliphora there are two more motorneurons to the TTM which are probably not coupled to the GDN terminals (Bacon and Strausfeld 1986). These authors showed that the thoracic terminal of the GDN may receive inputs from (at least converge closely with) other descending neurons, mechanoreceptor afferents from the thorax, and the axons from the giant pedicellar campaniform sensillum. Furthermore it was found that the GDNs may contact three paris of intersegmental thoracic interneurons (Bacon and Strausfeld 1986).

The GDNs have been shown electrophysiologically to respond to light on or off stimuli (Tanoye and Wyman 1980; Bacon and Strausfeld 1986) and deflections of the antennal pedicellar and flagellar segments (Bacon and Strausfeld 1986). The GDNs furthermore appear to be electrically coupled to the TTM motorneurons (Tanoye and Wyman 1980; Bacon and Strausfeld 1986). The giant fiber pathway may be involved in, but not uniquely responsible for, triggering jump and flight initiation. The effective stimulus for triggering jump appears to be multimodal and to depend on sudden novelty features like objects invading the visual field and puffs of wind on the head and body (Bacon and Strausfeld 1986).

The next three sets of pathways involve the large horizontal (HS) and vertical (VS) motion sensitive neurons of the lobula plate. It has been suggested that the final output of the motion detection system is mediated by these HS and VS neurons (Hausen 1981, 1984). There are 11 VS neurons divided into three functional groups (Eckert and Bishop 1978, Hengstenberg 1982). One group comprising VS 2-4, is preferentially sensitive to up-down movements of horizontal gratings, the second represented by VS 8 and 9 responds to oblique movements of gratings and the third group VS 5-7, 10 and 11 seem to have intermediate response characteristics. The VS 1 neuron is unique in being up-down and front-back motion sensitive.

The VS neurons converge onto different sets of descending neurons (Strausfeld and Bassemir 1985b) (Figs. 11, 13). The VS 1 is a unique lobula plate neuron that projects to a DN cluster containing the large Descending Neuron Columnar Horizontal motion sensitive DNColHS neuron that also receives inputs from HS neurons and small field visual inputs. The VS 2 and 3 neurons are coupled to a contralateral neck muscle motorneuron (CN 6). The VS 4-9 neurons are coupled to a cluster of DNs termed to DNOVS cluster that also receives inputs from some of the large ocellar interneurons. This cluster consists of three descending neurons DNOVS 1-3. The VS 10 and 11 have not been correlated with any DNs yet.

In addition to the VS and ocellar inputs the DNOVS cluster (as shown schematically in Fig. 11) also receives inputs from (1) three types of thin-axoned ipsilateral lobula plate neurons termed small vertical cells (sVS), (2) a heterolateral neuron from the contralateral lobula plate (Hc) and a centrifugal neuron (cVSc1) connecting the DNOVS cluster to the contralateral lobula plate, (3) terminals from ascending neurons and (4) small field visual inputs from the inner layer of the contralateral medulla (via the posterior optic tract) (Strausfeld and Bassemir 1985). The DNOVS cluster like the GDN cluster hence receives a multitude of inputs, some of which (although not described in detail) are local interneurons in the lateral brain.

The output regions of the individual DNOVS neurons in the thoracic ganglia differ somewhat. One cell appears to project to motor neuropil of indirect flight muscle, the others invade all three thoracic neuromeres (Strausfeld et al. 1984). These authors originally suggested that the VS system has its main outputs onto circuits controlling wing altitude and the shape of the thoracic box. Later it was shown that part of the VS neurons (with HS neurons) have direct connections with neck motorneurons (Strausfeld and Seyan 1985; and see below). Taken together the systems (with VS inputs) may be used in visual stabilization, while in flight turning head and then body in response to displacements of visual images (Strausfeld and Bassemir 1985).

The three HS neurons can be divided into two subunits (1) the north (HSN) and equatorial (HSE) responding to horizontal rotation of the visual field and (2) the south (HSS) responding to ipsilateral progressive motion (Hausen 1981; 1984). The dendritic fields of the HS neurons correspond roughly to a third of the visual field each (dorsally, midline and ventrally). As shown in Figs. 11 and 14, the HS neurons terminate onto a cluster of descending neurons (the DNColHS cluster). This cluster also appears to be contacted by the VS 1 and columnar inputs from the lobula and the two centrifugal neurons CHS 1 and 2 that arborize in the ipsilateral lobula plate (Strausfeld and Bassemir 1985b). Also local interneurons and antennal mechanosensory afferents were resolved in the region of HS and CHS terminals and seem associated with these. Finally Strausfeld and Seyan (1985) found that the HSS neuron is coupled to a large cervical nerve motorneuron (CN 6) that also receives inputs from the VS 2 and 3 neurons (see below). It should be noted that the DNColHS neurons have more extensive arborizations (Fig. 14) that e.g. the DNOVS cluster (Fig. 13) and only a small portion of the possible neuronal connections may have been accounted for. The HS system (and the DNColHS) hence appears to be more complex than previously suggested and their exclusive role in flight steering control (visual stabilization) suggested by some experiments (Heisenberg et al. 1978; Geiger and Nässel 1981; Hausen and Wehrhahn 1983) may be questioned. If not exclusively, the HS may at least be involved in control of optokinetic and yaw-induced head movements as well as pathways subserving visually stabilized flight (Hausen 1984; Strausfeld and Bassemir 1985b; Strausfeld and Seyan 1985). This is anatomically supported by the outputs of the descending neurons coupled to the HS cells which are found in the motor centers of thoracic ganglia supplying leg, neck and flight muscle (Strausfeld and Bassemir 1985b).

The final output pathway from the brain to be discussed has already been mentioned above. It consists of motorneurons supplying neck muscles, whose dendrites in the brain receive inputs from the VS 2 and 3 neurons in one hemisphere and the HSS in the other. One of these (CN 6) has been described in more detail (Strausfeld and Seyan 1985) and is shown schematically in Figs. 11C and 15. In the brain the CN 6 in addition receives contacts from marginal tangential neurons of the lobula plate (MTL) that subserve the dorsal rim region presumed to be specialized for perception of polarized light. Also sensory afferents from the halteres terminate among the CN 6 dendrites. The axon from the CN 6 run through the cervical nerve to the neck muscles. These muscles are also supplied by axons from ten motorneurons in the prothoracic ganglion (Strausfeld and Seyan 1985) (Fig. 15). One of these motorneurons has been shown anatomically to be associated with the termination of the DNOVS 1 (ocellar- VS descending neuron), the contralateral haltere interneuron and mechanosensory afferents from the prosternal organ and the haltere (Strausfeld and Seyan 1985). In addition descending ocellar interneurons have been found weakly cobalt coupled to at least one of the large frontal nerve neck motorneurons (D.R. Nässel and M. Hagberg, unpublished; see also

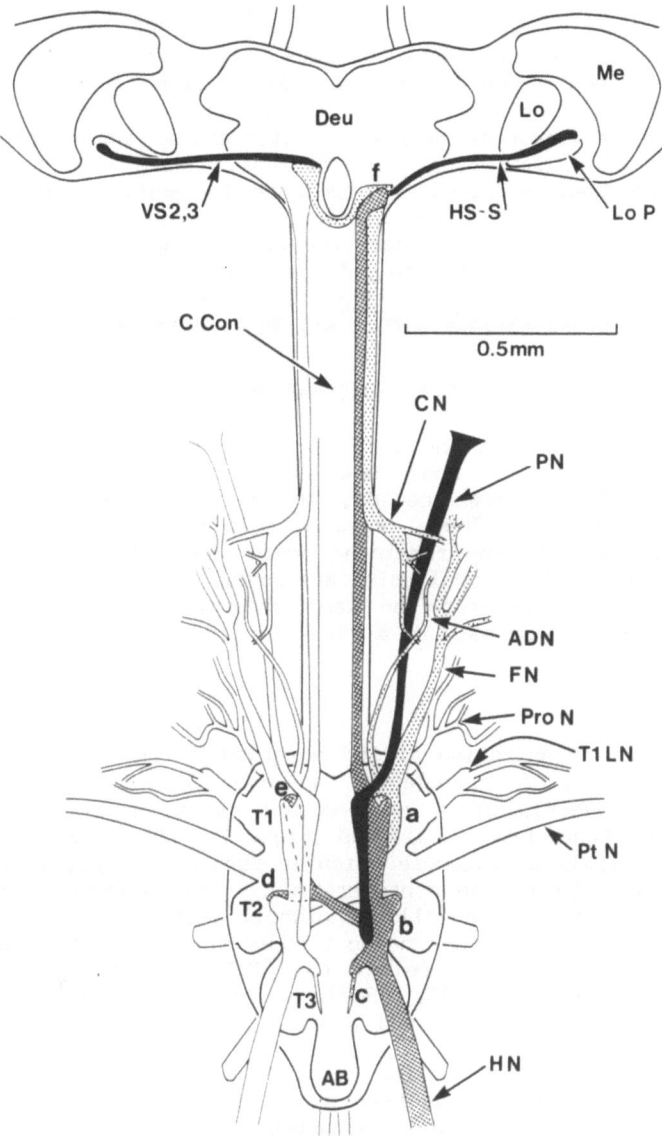

FIG. 15: The inputs of the VS2, 3 and HS S neurons onto the CN6 neck motorneuron. The CN6 (stippled) runs to neck muscle via the cervical nerve (CN). In addition prothoracic motorneurons run via anterior dorsal- (ADN) and frontal prothoracic (FN) nerves to neck muscles. In thoracic ganglia there is a convergence between prosternal organ afferents (PN, black), haltere afferents (crosshatched, HN) and the prothoracic neck motorneurons. The haltere afferents also contact CN6-HS S convergence in brain (f) and contralateral thoracic components (d, e). From Strausfeld and Seyan (1985) by permission on N.J. Strausfeld and Springer Verlag.

Nässel and Hagberg 1985). The inputs to the neck motorneurons described above have been conjectured to subserve optokinetic head movements.

In summary a few descending pathways have been described that receive different types of visual inputs. Some of these DNs are also thought to be contolled by mechanosensory inputs, ascending neurons and local circuits. Most DNs appear in clusters with complex dendritic regions and often heteromorphic output regions in the thoracic ganglia, indicating that each cluster subserves several circuits and are under complex control by afferents and higher order interneurons. With this background the concept command neuron appears to be hard to apply to single neurons. Possibly clusters of DNs execute one or several types of command functions. Nothing has so far been reported on neuroactive substances in the descending pathways. A more detailed analysis of inputs and outputs of the neck motor system in Calliphora has been published since this chapter went to press (Strausfeld et al. 1987; Milde et al. 1987).

5. HIGHER PROTOCEREBRAL CENTERS

In the previous sections reference has often been made to higher protocerebral centers. We shall now look into the organization of three such systems: the mushroom bodies, the central body complex and the superior protocerebrum. These regions are all complex and interconnect with large portions of the remaining brain neuropils and except for the mushroom bodies the functional roles are poorly understood. The following brief summary will therefore be kept general and to some extent the chemical anatomy will be used as a basis for discussion.

5.1. The mushroom bodies

The general organization of the mushroom bodies can be seen in many of the illustrations of this chapter (see e.g. Figs. 1, 11, 16, 17). Each mushroom body consists of a caudo-dorsal calyx region, which in e.g. cockroaches and honey bees is bilobed (Figs. 16, 17). The calyx region is connected to a stalk or peduncle which in turn gives rise to an alfa- and beta-lobe anteriorly in the protocerebrum. The intinsic neurons of the mushroom bodies, the Kenyon cells or globuli cells, have dentrites in the calyx and axonal projections running through the stalk into the alfa- and beta-lobes. In crickets there are ca. 50,000 and in bees ca. 340,000 Kenyon cells (Schürmann 1985, 1987). Their processes are extremely densely packed in the peduncle and lobes.

The main input region of the mushroom bodies is the calyces (Figs. 11A, 16). In the bee it has been shown that the direct inputs to the calyces are from the antennal (chemosensory) and optic lobes (Fig. 16). Other inputs to the mushroom bodies are from e.g. mechanoreceptors and acoustic interneurons (Mobbs 1982; Schürmann 1985, 1987). Also in Calliphora antennal and visual inputs to the mushroom bodies have been shown (Strausfeld et al. 1984).

The alfa- and the beta-lobes are the output regions and are connected to many neuropil regions of the brain, glomerular as well as non-glomerular. Some of the output tracts in the bee brain have been described by Mobbs (1982) (Fig. 16). The alfa-lobes connect to (1) the anterior optic tubercle (optic focus), (2) the calyces, (3) the anteriolateral protocerebrum and (4) protocerebral neuropil near the contralateral alfa-lobe. The beta-lobes connect to (1) the contralateral beta lobe, (2) to the central body and (3) to the alfa-lobe and its surrounding neuropil via the anterior dorsal protocerebral commissure (adc in Fig. 16). In addition to these clear tracts there is a large portion of the connections that connects by means of large diffuse neurons the peduncle and lobes to surrounding non-glomerular neuropil or, connects different parts of the mushroom bodies to form feed-back loops (Schürmann

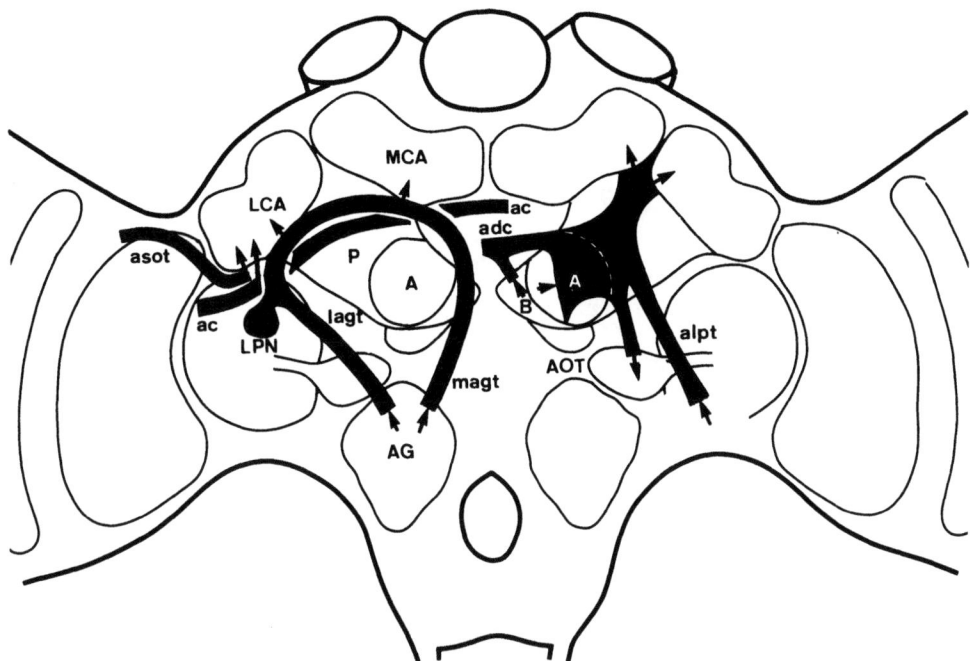

FIG. 16: The input (left) and output (right) tracts of the mushroom bodies in the honey bee based on data from Mobbs (1982). Inputs from antennal glomeruli (AG) run via the lateral (lagt) and medial (magt) antenno-glomerular tracts. These inputs invade the lateral (LCa) and medial (MCa) calyces and lateral protocerebral neuropil (LPN). Other inputs to the calyces are from the anterior superior optic tract (asot) and anterior commissure. The outputs from the alfa-(A) and beta-(B) lobes are to calyces, anterior optic tubercle (AOT), anterior dorsal protocerebral commissure (adc) and lateral protocerebrum (alpt). Note that also inputs to the mushroom body system from the alpt were inferred by Mobbs (1982). P = peduncle.

1985, 1987). The surrounding non-glomerular neuropil may hence be considered as accessory mushroom body neuropil (Schürmann 1985, 1987). A large portion of the synaptic connections between mushroom body related neurons and other circuits appears to occur in this accessory neuropil. The accessory non-glomerular neuropil contains processes from interneurons and afferent cephalic sensory systems, as well as processes from descending and ascending neurons. Studies by Schürmann and coworkers (see Schürmann 1985, 1987) show several types of ascending and descending neurons in the cricket brain whose processes lie embedded in accessory mushroom body neuropil. Some of the ascending ones resemble acoustic interneurons from ventral ganglia. Judging from the distribution of their brain processes, the connections of these ascending and descending neurons probably include circuits other than the mushroom body related ones.

The above anatomical features are derived from neuroanatomical staining methods and intracellular dye filling. Immunocytohcemistry and fluorescence histochemistry have provided further data on mushroom body connections. In Fig. 17 a number of extrinsic serotonin-immunoreactive (5-HTi) neural systems are shown (after Klemm et al. 1984). These authors found a number of 5-HTi paths interconnecting alfa- and beta-lobes with surrounding non-glomerular accessory neuropil. One tract of 5-HTi fibers

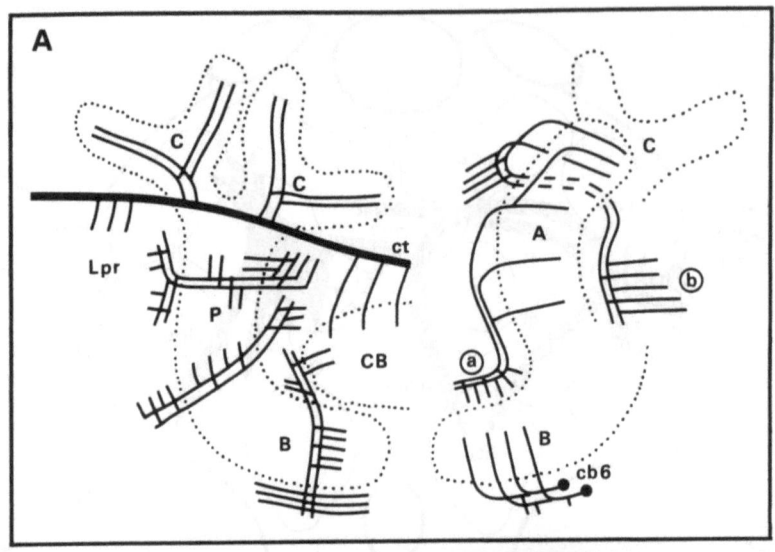

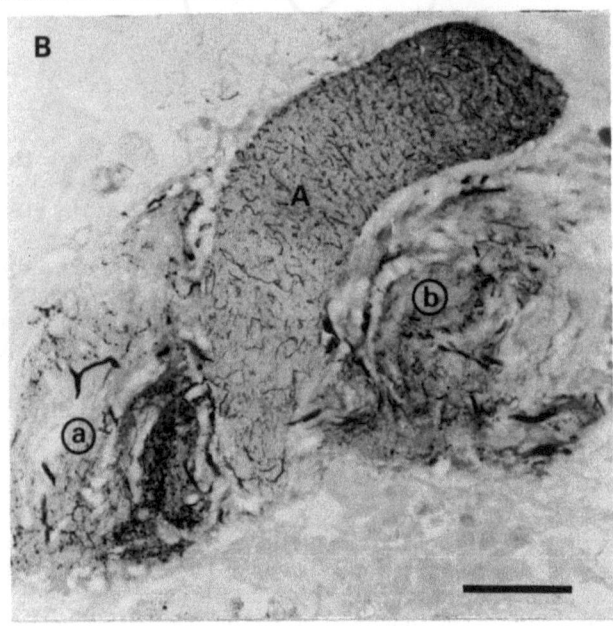

FIG. 17: Serotonin-immunorective neurons in the mushroom bodies of the cockroach (Frontal views). A. Some of the extrinsic 5-HTi pathways described by Klemm et al. (1984). A large 5-HTi tract (ct) bilaterally innervates lateral protocerebral neuropil (lpr), the calyces (C) and the central body (CB). The peduncle (P) is connected to surrounding non-glomerular neuropil. The beta-lobes (B) are connected to central body and surrounding neuropil (cb 6 are cell bodies of one set of neurons). The alfa-lobe (A) also shown in 17B is connected to surounding neuropil areas encircled a and b. B. Shows the dense 5-HTi innervation of alfa-lobe and the surrounding neuropils a and b. Scale in B = 50 μm.

was found that interconnect the calyces of both hemispheres and neuropil of the lateral protocerebrum (lateral horn, Lpr of Fig. 17) and the central body. Also connections between the beta-lobes and the central

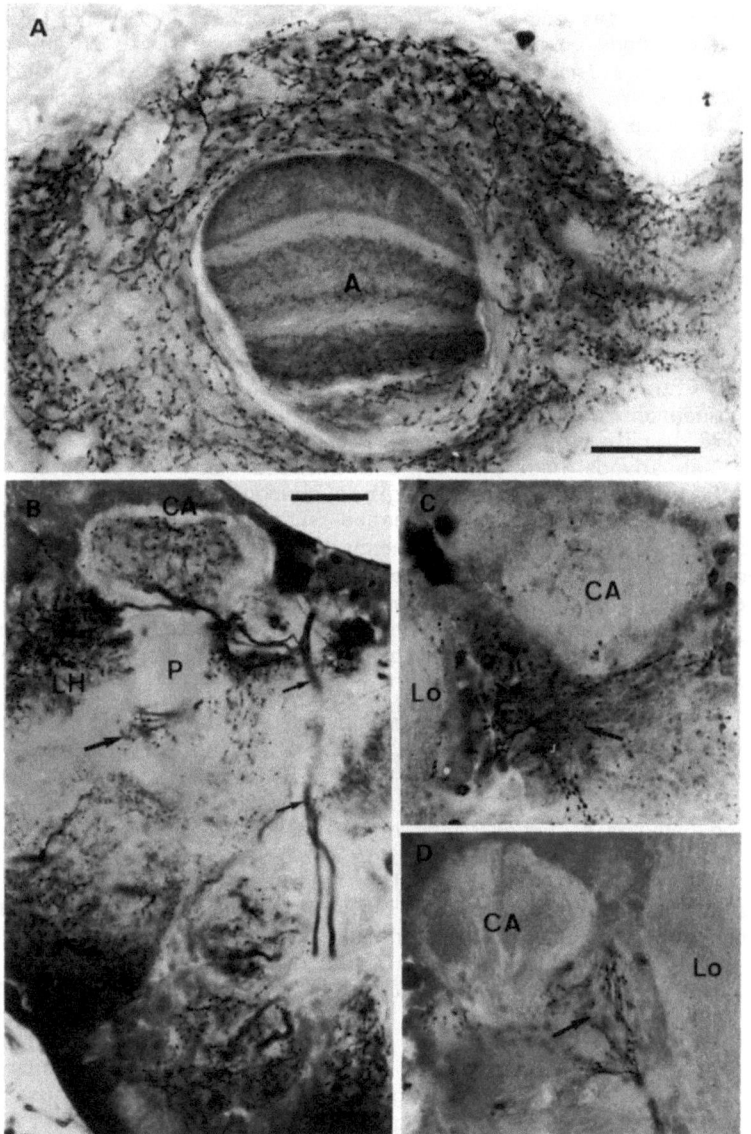

FIG. 18: Serotonin- and Gastrin/CCK-like immunoreactive fibers associated with the mushroom bodies.
A. The honey bee alfa-lobe (A) with layers of 5-HTi fibers and surrounding neuropil densely packed with immunoreactive processes.
B. In Calliphora fibers invade the calyx (CA), peduncle (P) at large arrow, lateral horn (LH) and non-glomerular neuropil. Small arrows indicate large fibers giving rise to many of the processes referred to.
C and D. Gastrin/CCK-like immunoreactive fibers invading the lateral horn (arrows) and possibly calyx (A). Lo = lobula.
Scale: A = 50 μm; B-D = 50 μm.

body was revealed. The 5-HTi innervation of alfa- and beta-lobes is very dense (Fig. 17B). Also in Calliphora (Fig. 18) the mushroom bodies are connected by means of 5-HTi neurons to other neuropil regions (Nässel

1986a, 1987): calyces and peduncle are interconnected to the lateral horn and the central body complex. No 5-HTi fibers were found in the alfa- and beta-lobes of this insect. In bees the processes of 5-HTi neurons form distinct layers in the alfa-lobes (Fig. 18A) and beta-lobes and very densely innervate the surrounding non-glomerular neuropil with varicose immunoreactive processes (Fig. 18A). Extrinsic catecholamine containing pathways similar to those of the 5-HTi ones have been resolved in the above insects by Klemm (1976). The detailed connections of these pathways, however, differ (see Nässel 1986a, 1987).

Also GABA-immonoreactive neurons have been found in all parts of the mushroom bodies and their associated neuropil of bees (Bicker et al. 1985) and flies (see Nässel 1987). In bees the GABA-immunoreative neurons form feed-back loops between at the alfa-lobe and the calyx (Bicker et al. 1985). Gastrin/CCK-like immunoreactive neurons have been found in the accessory mushroom body neuropil of Calliphora (Duve et al. 1983; Thorpe and Duve 1987). In Figs. 18C and D some gastrin/CCK-immunoreactive fibers are seen that invade neuropil of the lateral horn (associated with the calyx). The diversity of morphological neuron types associated with the mushroom bodies, hence is complemented with a chemical diversity. The classification of neuron types by means of immunocytochemistry has hardly begun and only a fraction of the neuron types have so far been accounted for. The complexity of the mushroom body circuitry is likely to increase with further application of intracellular techniques and immunocytochemistry, but we may also get closer to an understanding of functional pathways.

The functional role of the mushroom bodies and their associated neuropils have long been speculated upon. Theories have included mushroom bodies being control centers for higher behavior, sites of memory and learning, control centers similar to the mammalian cerebellum, centers controlling rhythmic behavior, and simply being a higher olfactory center. As will be shown in the following there might be a little bit of truth in all of these theories. The following summarizes some possible functions suggested by Schürmann (1974, 1985, 1987). The multimodal input convergence onto the parallel channels of the intrinsic Kenyon cells may suggest a filtering and integration of sensory information. A recombination and weighing of signals as well as sequencing of output signals may be executed. Most extrinsic mushroom body neurons are multimodal and many respond only under very specific combinations of stimuli (with respect to modality and sequence). They furthermore have long-lasting after effects (Homberg and Erber 1979). Thus mushroom body outputs seem specialized for firing under stimulus conditions relevant to initiation of specific behaviors or to mediate information about states of complex sensory situations at every given moment. This type of neurons may, as also suggested by Strausfeld and Bacon (1983) and Strausfeld et al. (1984), be appropriate for activating or gating descending pathways.

These descending neurons receive also other inputs, e.g. from sensory afferents, optic lobe interneurons, ocellar interneurons, ascending neurons (sensory afferents and interneurons) and probably from modulatory pathways. The possible additional mushroom body inputs are already multimodal and complexly integrated by the internal and external circuitry of the mushroom bodies (including local and multisegmental interneurons and possibly modulatory pathways such as serotonergic, catecholaminergic and peptidergic pathways). The firing pattern of the descending neurons will hence be dependent on many factors in the surrounding sensory world and the internal physiological state of the animal. Possible one function of the mushroom bodies is (by weighing the total input) to determine whether at a given moment certain sets of signals will go through and activate specific sets of descending neurons, thus activating specific fragments of behavior.

5.2. The central body complex and superior protocerebrum

There are some regions of the insect brain that cannot be easily fitted into the previously described convergence onto descending neuron pathways. The central body complex and the non-glomerular neuropil of the superior protocerebrum are some examples.

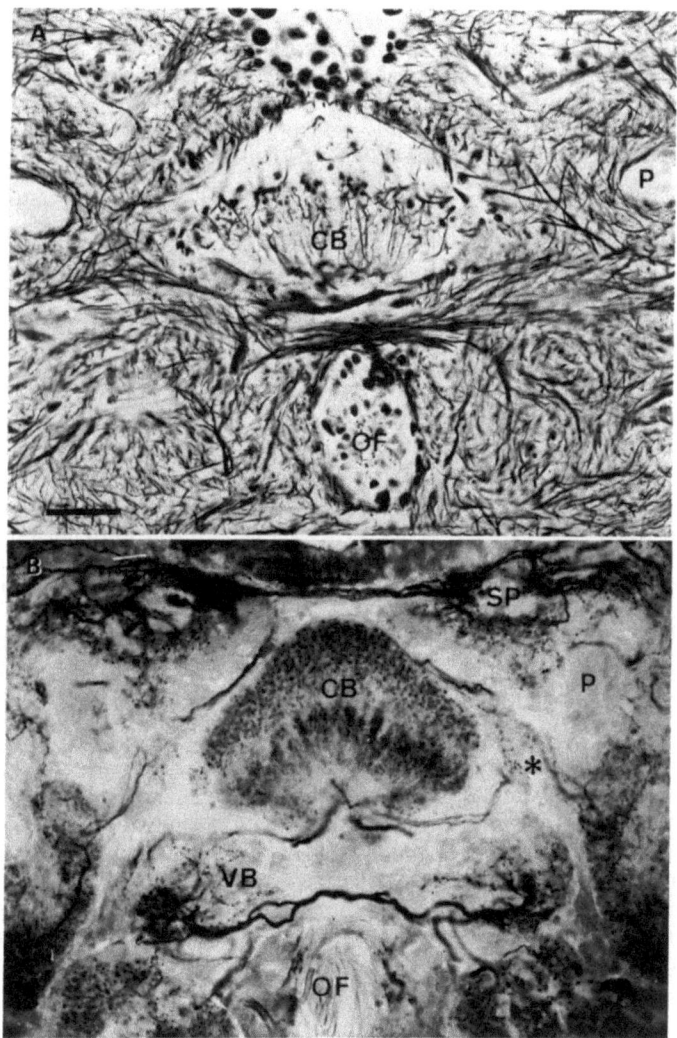

FIG. 19: The central body (CB) complex of <u>Calliphora</u> embedded in non-glomerular neuropil seen after reduced silver staining (A) and 5-HT-immunocytochemistry (B). In B the heavy, layered 5-HTi innervation of the CB is seen with the innervation of associated neuropils in superior protocerebrum (SP) and ventral bodies (VB). A lateral neuropil (asterisk) is also connected to the CB. OF = oesophageal foramen, P = mushroom body peduncle. Scale: A and B = 50 μm.

The organization of the central body complex has been described in some detail by Williams (1975) and Strausfeld (1976a). It consists of several regions, the superior arch, the fan-shaped and ellipsoid bodies and the noduli (see Figs. 19, 20). These neuropils are associated with surrounding non-glomerular neuropil as well as with the protocerebral bridge (pons) and the ventral bodies. No direct inputs from sensory afferents or optic and antennal interneurons have been resolved and we do not know where the output neurons project, except for terminations in the associated non-glomerular neuropil. The internal structure of the mushroom bodies is strictly geometrical, with layers and columns and with precise isotopic connections with the protocerebral bridge (Williams 1975; Strausfeld 1976a).

The functional role of the central body is poorly understood. Milde (1982) found interneurons in the central body of bees which respond to stimulus of the ocellar pathway; some were multimodal responding to stimulus with air puffs and ocellar or compound eye stimulus. Many neurons impaled in the central body region did not respond to any stimulus but showed spontaneous activities (Schildberger 1984; Homberg 1985; see also Schürmann 1985).

One intriguing feature with the central body of Calliphora is that subsets of its neurons react with a variety of antisera to different neural substances (some shown in Fig. 20). Antisera against the following substances display neural processes in different layer of the central body: 5-HT, Gastrin/CCK, vasoactive intestinal polypeptide, retinal S-antigen, melatonin, GABA, FMRFamide, proctolin substance P, dopamine- β -hydroxylase and histamine (Nässel 1986a, 1987; Nässel and O'Shea 1987). In addition catecholamine containing processes were revealed with histofluorescence (Klemm 1976; see Fig. 20). The layers of the central body are derived from different sets of protocerebral cell bodies for each substance. The 5-HTi and catecholamine containing neurons appear to be part of an extensive system connecting many other glomerular and non-glomerular neuropils (see the section on mushroom bodies). The peptide-, melatonin-, histamine- and S-antigen-immunoreactive neurons are derived from cell bodies in the median neurosecretroy cell body cluster or other groups in superior protocerebrum. These neurons appear to supply mainly the central body and the pars intercerebralis-superior protocerebrum with processes. The source of the GABA-immunoreactive neurons is not clear, but the axons interconnect the ellipsoid body to non-glomerular neuropil flanking the central body (see Schäfer and Bicker 1986). It should be noted that of the above substances only GABA, 5-HT, dopamine/noradrenalin, and a gastrin/CCK-like substance have been conclusively determined biochemically in Calliphora (see Nässel 1987). Except for the superior protocerebrum there is no other neuropil region in the fly brain that react with so many antisera as the central body (five antisera react exclusively in these two regions). The superior protocerebrum among other things supplies "descending neurons" to the corpora caridaca and is known to contain a variety of neurosecretory cells.

Does the possible occurrence of this large number of peptides and other substances imply that the central body (with its associated superior protocerebrum connections) is a center of modulatory action? The neurons of the superior protocerebrum are referred to as neurosecretory cells. They have processes in the superior protocerebum and axons to the corpora cardiaca where their action probably is as neurohormones. The processes in the superior protocerebrum arborize among some antennal, ocellar and optic lobe inputs (see Fig. 11A) (Strausfeld et al. 1984), but the remaining organization of this neuropil is rather poorly known, and generalized. It was considered by these authors as a higher control

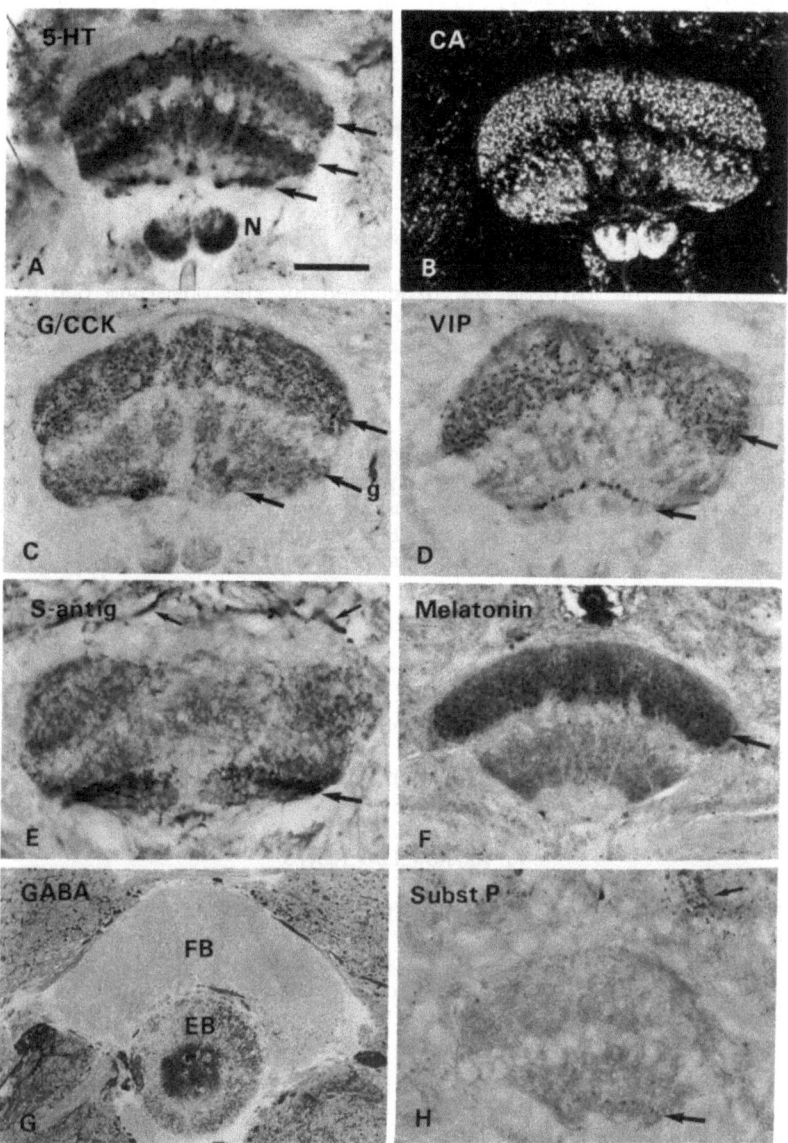

FIG. 20: The distribution of immunoreactivity and catecholamine fluorescence (CA) in the central body of Calliphora. Larger arrows indicate reactive layers. In Fig. 20G a different level of the central body is shown since GABAi is found mainly in the ellipsoid body (EB) and only a thin layer (corresponding to g in Fig. 20C can be seen in the fan-shaped body (FB). The small arrows in E and H indicate immunoreactive fibers in superior protocerebrum. N = noduli (only 5-HTi and CA reactive). Not shown is the distribution of dopamine- β-hydroxylase and histamine-like immunoreactivity. The former is similar to that of VIP (Fig. 20D) with the addition of the noduli (Klemm et al. 1984); the latter is similar to the melatonin destribution (Fig. 20F) (B by courtesy of Dr. N. Klemm, the remaining figures from unpublished studies). Scale: A-H = 50 μm.

center with outputs to the autonomic nervous system. The degree of integration of the central body and its assocated neuropils into the "neurosecretory" or modulatory systems, the autonomic system as well as sensory systems and possible interactions with descending pathways is poorly known and merits future investigation.

6. CONCLUDING REMARKS

This chapter has described the convergence of different sensory pathways onto clusters of descending neurons (DNs). Most DN clusters appear to be under multimodal sensory control. The DN pathways furthermore seem to be under influence of higher brain centers such as the mushroom bodies. The information processed by the mushroom bodies (MBs) is also multimodal and the extrinsic neurons associated with the MBs are characterized by their responses to very specific sequences of different stimuli. If the MBs are thought of as neuronal machines sampling and weighing multiple sensory inputs at every instant (and comparing to the previous instant) at a high rate, and constantly feeding their outputs to different descending pathways, the role of the MBs may be to set the levels of responsiveness of descending pathways. Hence, a given sensory stimulus specific to a descending pathway may or may not elicit a signal to thoracic ganglia, depending on the mushroom bodies' view of the signal's momentary relevance in comparison to inputs to other systems. Or, in other words, in order to accomplish sensible behavior, the descending neurons cannot all fire at once even if stimulated; some of the order may be brought about by control from higher centers (via feed-back from the effector system). Although we are still at a stage of heavy speculation, it seems that we have the means to approach the neural control of behavior with strategies like: immunocytochemical-morphological and electrophysiological analysis of identified neurons, with neurogenetics, pharmacology and further behavior studies.

7. ACKNOWLEDGEMENTS

I thank Y. Andersson and I. Norling for technical assistance, Dr. D. Byers for discussions and Dr. N.J. Strausfeld for permission to use figures. The accounts by Strausfeld and coworkers as well as the review by F.W. Schürmann (1985) listed in the reference list form the basis of the present paper and credit should be given to these authors for the original ideas.

8. REFERENCES

Ammermüller J, Weiler R (1985) S-neurons and not L-neurons are the source of GABA-ergic action in the ocellar system. J Comp Physiol A. 157: 779-788.

Bacon JP, Strausfeld NJ (1986) The dipteran 'Giant fibre' pathway: neurons and signals. J Comp Physiol A 158: 529-548.

Bicker G, Schäfer S, Kingan TG (1985) Mushroom body feedback interneurons in the honeybee show GABA-like immunoreactivity. Brain Res 360: 394-397.

Boeckh J, Ernst KD, Sass H, Waldow U (1984) Anatomical and physiological characteristics of individual neurones in the central antennal pathway of insects. J Insect Physiol 30: 15-26.

Braitenberg V (1967) Patterns of projections in the visual system of the fly. I. Retina-lamina projections. Exp Brain Res 3: 271-298.

Bullock TH, Horridge GA (1965) Structure and function in the nervous system of invertebrates. W.H. Freeman, San Francisco and London.

Buchner E, Buchner S, Crawford G, Mason WT, Salvaterra PM, Sattele DB (1986) Choline acetyltransferase-like immunoreactivity in the brain of Drosophila melanogaster. Cell Tissue Res 246: 57-62.

Burghause FM (1979) Die strukturelle Spezialisierung des dorsalen Augenteils der Grillen (Orthoptera, Grylloidea). Zool Jahrb Physiol 83: 502-525.

Burrows M, Rowell CHF (1973) Connections between visual interneurons and metathoracid motorneurons in the locust. J Comp Physiol 85: 221-234.

Duve H, Thorpe A, Strausfeld NJ (1983) Cobalt-immunocytochemical identification of peptidergic neurons in Calliphora innervating central and peripheral targets. J Neurocytol 12: 949-970.

Dvorak DR, Bishop LG, Eckert HE (1975) On the identification of movement detectors in the fly lobe. J Comp Physiol 100: 5-23.

Eckert H, Bishop LG (1978) Anatomical and physiological properties of the vertical cells in the third optic ganglion of Phaenicia sericata (Diptera, Calliphoridae). J Comp Physiol 126: 57-86.

Ernst KD, Boeckh J, Boeckh V (1977) Neuroanatomical study of the organization of the central antennal pathway in insects. Cell Tissue Res 176: 285-308.

Fent K, Wehner R (1985) Ocelli: A celestial compass in the desert ant Cataglyphis. Science 228: 192-194.

Fischbach KF, Heisenberg M (1984) Neurogenetics and behaviour in insects. J Exp Biol 112: 65-93.

Franceschini N, Hardie R, Ribi W, Kirschfeld K (1981) Sexual dimorphism in a photoreceptor. Nature 291: 241-244.

Geiger G, Nässel DR (1981) Visual orientation behaviour of flies after selective laser beam ablation of interneurons. Nature 293: 398-399.

Gewecke M (1974) The antennae of insects as air current sense organs and their relationships to the control of flight. In: Barton-Browne L (ed) Experimental analysis of insect behaviour. Springer Verlag, Heidelberg, Berlin, New York.

Goodman LJ (1981) Organization and physiology of the insect dorsal ocellar system. In: Autrum H (ed) Handbook of sensory physiology. Vol. VII/6c. Springer, Berlin.

Hagberg M, Nässel DR (1986) Interneurons subserving ocelli in two trichopterous insects: morphology and central projections. Cell Tissue Res 245: 197-205.

Hall JC (1984) Complex brain and behavioral functions disrupted by mutations in Drosophila. Develop Genetics 4: 355-378.

Hardie RC (1983) Projection and connectivity of sex-specific photoreceptors in the compound eye of the male housefly (Musca domestica. Cell Tissue Res 233: 1-21.

Hardie RC (1987) Is histamine a neurotransmitter in insect photoreceptors? J Comp Physiol (in press).

Hausen K (1981) Monocular and binocular computation of motion in the lobula plate of the fly. Verh Dtsch Zool Ges 1981: 47-70.

Hausen K (1984) The lobula-complex of the fly: Structure, function and significance in visual behaviour. In: Ali MA (ed) Photoreception and vision in invertebrates. Plenum, New York.

Hausen K, Strausfeld, NJ (1980) Sexually dimorphic interneuron arrangements in the fly visual system. Proc Soc Lond 208: 57-71.

Hausen K, Wehrhahn C (1983) Microsurgical lesion of horizontal cells change optomotor yaw responses in the blowfly Calliphora erythrocephala. Proc Soc Lond B 219: 211-216.

Heisenberg M (1980) Mutants of brain structure and function: what is the significance of the mushroom bodies for the behavior. In: Siddiqui D, Babu P, Hall LM, Hall IC (eds) Development and neurobiology of Drosophila. Plenum Press, New York.

Heisenberg M, Wonneberger R, Wolff R (1978) Optomotor-blind H31- a Drosophila mutant of the lobula plate giant neurons. J Comp Physiol 124: 287-296.

Hengstenberg R (1982) Common visual response properties of giant vertical cells in the lobula plate of the blowfly Calliphora. J Comp Physiol 149: 179-193.

Homberg U (1985) Interneurons of the central complex in the bee brain (Apis mellifera, L.). J Insect Physiol 31: 251-264.

Homberg U, Erber J (1979) Response characteristics and identification of extrinsic mushroom body neurons in the bee. Z Nautrforsch 34C: 612-615.

Howse PE (1975) Brain structure and behaviour in insects. Ann Rev Entomol 20: 612-615.

Huber F (1960) Untersuchungen uber die Funktion des Zentralnervensystems und insbesondere des Gehirnes bei der Forbewegung und der Lauterzeugung der Grillen. Z vergl Physiol 44: 60-132.

Huber F (1978) The insect nervous system and insect behaviour. Anim Behav 26: 969-981.

Hughes GM (1965) Neuronal pathways in the insect central nervous system. Treherne JE, Beament JWL (eds) The physiology of the insect nervous system. Academic Pres, New York.

King DG, Wyman RJ (1980) Anatomy of the giant fiber pathway in Drosophila. I. Three thoracic components of the pathway. J Neurocytol 9: 753-770.

Kirschfeld K (1967) Die Projektion der optischen Umwelt auf das Raster der Rhabdomere im Kimplexauge von Musca. Exp Brain Res 3: 248-270.

Klemm N (1976) Histochemistry of putative transmitter substances in the insect brain. Prog Neurobiol 7: 99-169.

Klemm N Steinbusch HWM, Sundler F (1984) Distribution of serotonin containing neurons and their pathways in the supraoesopeal ganglion of the cockroach Periplaneta americana (L) as revealed by immunocytochemistry. J Comp Neurol 225: 387-395.

Koontz MA, Edwards JS (1984) Central projections of first-order ocellar interneurons in two orthopteroid insects Acheta domestica and Periplaneta americana. Cell Tissue Res 236: 133-146.

Labhart T (1980) Specialized photoreceptors at the dorsal rim of the honeybee's compound eye: Polarizational and angular sensitivity. J Comp Physiol 141: 19-30.

Matsumoto MG, Hildebrand JG (1981) Olfactory mechanisms in the moth Manduca sexta: response charcteristics and morphology of central neurons in the antennal lobes. Proc Soc Lond B 213: 249-277.

Meinertzhagen IA (1976) The organization of perpendicular fibre pathways in the insect optic lobe. Phil Trans R Soc Lond B 274: 555-594.

Meyer EP (1984) Retrograde labelling of photoreceptors in different regions of the compound eyes of bees and ant. J Neurocytol 13: 825-836.

Meyer EP, Nässel DR (1986) Terminations of photoreceptor axons from different regions of the compound eye of the desert ant Cataglyphis bicolor. Proc R Soc Lond B (in press).

Meyer EP, Matute C, Streit P, Nässel DR (1986) Insect optic lobe neurons identifiable with monoclonal antibodies to GABA. Histochem 84: 207-216.

Milde J (1982) Elektrophysiologische und anatomische Untersuchungen and Interneuronen erster un hoher Ordnung des Biene. Doctoral dissertation. Free University of Berlin.

Milde JJ (1984) Ocellar interneurons in the honey bee. Structure and signals of L-neurons. J Comp Physiol A 154: 683-693.

Milde JJ Seyan HS, Strausfeld NJ (1987) The neck motor system of the fly Calliphora erythrocephala. II Sensory organization. J Comp Physiol A 160: 225-238.

Mobbs PG (1982) The brain of the honey bee Apis mellifica. I. The connections and spatial organization of the mushroom bodies. Phil Trans R Soc Lond B 29: 309-354.

Nässel DR (1986a) Serotonin and serotoinin-immunoreactive neurons in the nervous system of insects. Prog Neurbiol (in press).

Nässel DR (1986b) Strategies for neuronal marking in arthropod brains. In: Gupta AP (ed) The arthropod brain, its evolution, development, structure and function. Wiley, New York (in press).

Nässel DR (1987) Neuroactive substances in the insect CNS. (This volume).

Nässel DR, Elofsson R (1986) Comparative anatomy of the crustacean brain. In: Gupta AP (ed) The arthropod brain, its evolution, development, structure and function. Wiley, New York (in press).

Nässel DR, Hagberg M (1985) Ocellar interneurons in the blowfly Calliphora erythrocephala: morphology and central projections. Cell Tissue Res 242: 417-426.

Nässel DR, Klemm N (1983) Serotonin-like immunoreactivity in the optic lobes of three insect species. Cell Tissue Res 232: 129-140.

Nässel DR, O'Shea M (1987) Proctolin-like immunoreactive neurons in the blowfly central nervous system. J Comp Neurol (in press).

Nässel DR, Högmo O, Hallberg E (1984) Antennal receptors in the blowfly Calliphora erythrocephala. 1. The gigantic central projection of the pedicellar campaniform sensillum. J Morphol 180: 159-169.

Nässel DR, Meyer EP, Klemm N (1985) Mapping and ultrastructure of serotonin-immunoreactive neurons in the optic lobes of three insect species. J Comp Neurol 232: 190-204.

Nässel DR, Ohlsson L, Sivasubramanian P (1987) Differentiation of serotonin-immunoreactive neurons in fly optic lobes developing in situ or cultured in vivo without eye discs. J Comp Neurol 225: 327-340.

O'Shea M, Rowell CHF (1975) A spike-transmitting electrical synapse between visual interneurons in the locust movement detector system. J Comp Physiol 97: 143-158.

O'Shea M, Rowell CHF (1976) The neuronal basis of a sensory analyser, the acridid movement detector system. II. Response decrement, convergence, and the nature of the excitatory afferetns to the fan-like dendrites of the LGMD. J Exp Biol 65: 289-308.

O'Shea M, Rowell CHF, Williams JLD (1974) The anatomy of a locust visual interneurone; the descending contralateral movement detector. J Exp Biol 60: 1-12.

Pearson KG, Heitler WJ, Steeves JD (1980) Triggering of locust jump by multimodal inhibitory interneurons. J Neurophysiol 43: 257-278.

Pierantoni R (1976) A look into the cock-pit of the fly. The architecture of the lobula plate. Cell Tissue Res 171: 101-122.

Rowell CHF (1971) The orphopteran descendinging movement detector (DMD) neurons: a characterization and review. Z vergl Physiol 73: 167-194.

Rowell CHF, Pearson KG (1983) Ocellar input to the flight motorystem of the locust Schistocerca. Structure and function. J Exp Biol 103: 265-288.

Rowell CHF, O'Shea M, Williams JLD (1977) The neuronal basis of a sensory analyser, the acridid movement detector system. IV. The preference for small field stimuli. J Exp Biol 68: 157-185.

Schäfer S, Bicker G (1986) Distribution of GABA-like immunoreactivity in the brain of the honeybee. J Comp Neurol 246: 287-300.

Schildberger K (1983) Local interneurons associated with the mushroom bodies and the central body in the brain of Acheta domesticus. Cell Tissue Res 230: 573-586.

Schildberger K (1984) Multimodal interneurons in the cricket brain: properties of identified extrinsic mushroom body cells. J Comp Physiol A 154: 71-79.

Schinz RH (1975) Structural specialization in the dorsal retina of the bee, Apis mellifera. Cell Tissue Res 162: 23-34.

Schürmann FW (1974) Bemerkungen zur Funktion der Corpora pedunculata im Gehirn der Insekten aus morphologischer Sicht. Exp Brain Res 19: 406-432.

Schürmann FW (1985) Aspekte neuronaler Verknupfung im zentralen Hirn der Insekten. In: Rensch B (ed) Evolution: Zelle als Organismus. Aschendorff, Munster.

Schürmann FW (1987) The architecture of the mushroom bodies and related neuropils in the insect brain. In: Gupta AP (ed) The arthropod brain, its evolution, development, structure and function. Wiley, New York (in press).

Stocker RF, Lawrence PA (1981) Sensory projections from normal and homeotically transformed antennae in Drosophila. Develop Biol 82: 224-237.

Stocker RF, Singh RN, Schorderet M, Siddiqi O (1983) Projection patterns of diferent types of antennal sensilla in the antennal glomeruli of Drosophila melanogaster. Cell Tissue Res 232: 237-248.

Strausfeld NJ (1976a) Atlas of an insect brain. Springer, Heidelberg.

Strausfeld NJ (1976b) Mosaic organization, layers and visual pathways in the insect brain. In: Zettler F, Weiler R (eds) Neural priciples in vision. Springer, Heidelberg.

Strausfeld NJ (1980) Male and female visual neurones in dipterous insects. Nature 283: 381-383.

Strausfeld NJ (1983) (ed) Functional neuroanatomy. Springer, Berlin.

Strausfeld NJ (1984) Functional neuroanatomy of the blowfy's visual system. In: Ali MA (ed) Photoreception and vision in invertebrates Plenum Press, New York.

Strausfeld NJ, Bacon JP (1983) Multimodal convergence in the central nervous system of dipterous insects. In: Horn (ed) Fortschr. d. Zool. 28. gustav Fisher, New York, Stuttgart.

Strausfeld NJ, Bassemir UK (1985a) Lobula plate and ocellar interneurons converge onto a cluster of descending neurons leading to neck and leg neuropil in Calliphora erythrocephala. Cell Tissue Res 240: 617-640.

Strausfeld NJ, Bassemir UK (1985b) The organization of giant horizontal-motion-sensitive neurons and their synaptic relationships in the lateral deutocerebrum of Calliphora erythrocephala and Musca domestica. Cell Tissue Res 242: 531-550.

Strausfeld NJ, Campos-Ortega JA (1977) Vision in insects: pathways possibly underlying neural adaptation and lateral inhibition. Science 195: 894-897.

Strausfeld NJ, Nässel DR (1980) Neuroarchitecture of brain regions that subserve compound eyes in Crustacea and insects. In: Autrum H (ed) Handbook of sensory physiology. VII/6B. Spirnger, Berlin, Heidelberg, New York.

Strausfeld NJ, Obermeyer M (1976) Resolution of intraneuronal and transsynaptic migration of cobalt in the insect visual and central nervous systems. J Comp Physiol 110: 1-12.

Strausfeld NJ, Seyan HS (1985) Convergence of visual, haltere and prosternal inputs at neck motor neurons of Calliphora. Cell Tissue Res 240: 601-615.

Strausfeld NJ, Wunderer H (1985) Optic lobe projections of marginal ommatidia in Calliphora erythrocephala specialized for detecting polarized light. Cell Tissue Res 242: 163-178.

Strausfeld NJ, Seyan HS, Milde JJ (1987) The neck motor system of the fly Calliphora erythrocephala. I. Muscles and motor neurons. J Comp Physiol A 160: 205-224.

Strausfeld NJ, Bassemir U, Singh RN, Bacon JP (1984) Organizational principles of outputs from dipteran brains. J Insect Physiol 30: 73-93.

Tanoye MA, Wyman RJ (1980) Motor outputs of giant nerve fibers in Drosophila. J Neurophysiol 44: 405-421.

Taylor CP (1981) Contribution of compound eyes and ocelli to steering of locusts in flight. I. Behavioural analysis. J Exp Biol 93: 1-18.

Thorpe A, Duve H (1987) Purification, characterisation and cellular distribution of insect neuropeptides with special emphasis on their relationship to biologically active peptides of vertebrates. (This volume)

Trujillo-Cenoz O (1966) Some aspects of the structural organization of the intermediate retina of dipterans. J Ultrastruct Res 13: 1-33.

Wehrhahn C (1979) Sex specific differences in chasing behaviour of free flying houseflies (Musca). Biol Cybern 32: 239-241.

Wehrhahn C, Poggio T, Bulthoff H (1982) Tracking and chasing in houseflies (Musca). Biol Cybern 45: 123-130.

Williams JLD (1975) Anatomical studies of the insect central nervous system: a ground plan of the midbrain and an introduction to the central complex in the locust Schistocerca gregaria. J Zool Lond 176: 67-86.

Wilson M (1978) Functional organization of locust ocelli. J Comp Physiol 124: 297-316.

Witthöft W (1967) Absolute Anzahl und Verteilun der Zellen im Hirn der Honigbiene. Z Morph Tiere 61: 185-214.

Wolburg-Buchholz K (1977) The superposition eye of Cloeon dipterum: The organization of the lamina ganglionaris. Cell Tissue Res 177: 9-128.

Wunderer H, Smola U (1982) Fine structure of ommatidia at the dorsal eye margin of Calliphora erythrocephala Meigen (Diptera: Calliphoridae): An eye region specialised for the detection of polarized light. Int J Insect Morphol Embryol 11: 25-38.

Zeil J (1983) Sexual dimorphism in the visual system of flies: the compound eyes and neural superposition in Bibionidae (Diptera). J Comp Physiol 150: 379-393.

INSECT NEURONS: SYNAPTIC INTERACTIONS, CIRCUITS AND THE CONTROL OF BEHAVIOR

R. MELDRUM ROBERTSON

Department of Biology

McGill University

1205 Avenue Docteur Penfield

Montreal, Quebec

H3A 1B1, Canada

ABSTRACT

The theme of this chapter is how the properties and interactions of identified neurons underlie behavior in insects. I first present background material on the neuronal environment and on the intrinsic properties of neurons. The types of neuronal interaction are reviewed and include chemical, electrical, potassium ion accumulation and presynaptic inhibition. Neuronal circuitry is best known in the flight and jumping systems of orthopterans. This field is reviewed with special reference to gating of information flow, burst generation and mutability of the circuit. The discussion speculates on the areas of research likely to be fruitful in the near future, and on whether much of the complexity in the nervous system has adaptive function.

1. INTRODUCTION

"....it is important to sound a realistic note regarding the time-scale for the research....Regrettably, perhaps,....this is several hundred years." (Hoyle 1983).

"I think that the complete mapping of an insect's nervous system is possible....It is my guess that the map could be produced by three people in two years." (Hester 1986).

In the first quotation above, Hoyle is referring to the goal of understanding how a nervous system functions at the cellular level, with particular reference to the locust nervous system. In the second quotation, Hester refers to the task of discovering the procedures for controlling the movement of machines in a complex environment by investigating the circuitry of an insect's nervous system. The idea is laudable; insects are already the subject of research towards this end with the results published in a robotics journal (e.g. Pearson and

Franklin 1984). However, Hester's is an optimistic guess, as most people familiar with the field would immediately recognize, for many of us have been analyzing the organization and function of many insects' nervous systems for many years without the sort of complete success that he envisages. The purpose of this chapter is to review what we have found out about the physiology of insect nervous systems in the two decades since the publication of "Structure and Function in the Nervous Systems of Invertebrates" by Bullock and Horridge (1965). A wide-ranging, exhaustive review would be largely redundant thanks to the recent appearance of the monumental "Comprehensive Insect Physiology, Biochemistry and Pharmacology" edited by Kerkut and Gilbert (1985) (13 volumes, 2 of which are specifically devoted to the nervous system and most of which have chapters relevant to nervous system function). The value of a volume such as the present one is in the breadth of its scope, and each article has value primarily as a result of its context. Therefore I am going to concentrate here on features which set insects apart from other invertebrates and on other features common throughout the phyla but for which insects provide model preparations. This is an illustrative review, not an exhaustive one. Furthermore it illustrates personal judgements, certainly not shared by all, about which areas of research are most interesting and exciting.

The theme that I have chosen is one that was first championed by Wiersma (1952) and has since gained considerable support (e.g. Hoyle 1970, 1977), and that is how the properties and interactions of identified neurons underlie behavior. It reflects an intracellular experimental approach to the analysis of neural function. In spite of intracellular recordings having been made quite early in research on insects (Boistel and Coraboeuf 1954; Hagiwara and Watanabe 1956) it is relatively recently, and mainly since 1965, that this approach has borne substantial fruit. Indeed the major area of growth in our knowledge of the physiology of insect nervous systems since that time is of neural function in terms of individual neurons.

One of the advantages of insect preparations is that they will continue to "behave", or at least express complex motor patterns related to behavior, even after the sometimes radical treatment necessary to expose the nervous system for intracellular recording. The recordings made by Bentley (1969 a, b) from the neuropile of cricket thoracic ganglia during generation of flight and song motor patterns were among the first to demonstrate the potential of insects in this regard. As techniques developed, particularly the improvement of intracellular marking techniques (Pitman et al. 1972a; Bacon and Altman 1977; Stewart 1978; Strausfeld 1983), the ease of making such recordings and their significance improved dramatically. Now the number of identified neurons (and their properties and synaptic interactions) that have been described and related to insect behavior is substantial and continually increasing. After presenting some background material about the general properties of insect nervous systems and neurons, I will describe what is known about the sorts of interactions that exist between insect neurons and how, through such interactions, neurons form circuits subserving behavior.

2. THE NEURONAL ENVIRONMENT

2.1. The blood-brain barrier

With the exception of sensory neurons (see, however, Bräunig and Hustert 1980; Bräunig 1985), insect neurons are collected into discrete ganglia (see Weevers 1985), surrounded by glial cells and enclosed in a sheath of connective tissue, the neural lamella (Fig. 1; Treherne and Schofield 1981). A striking feature of insect nervous systems is that

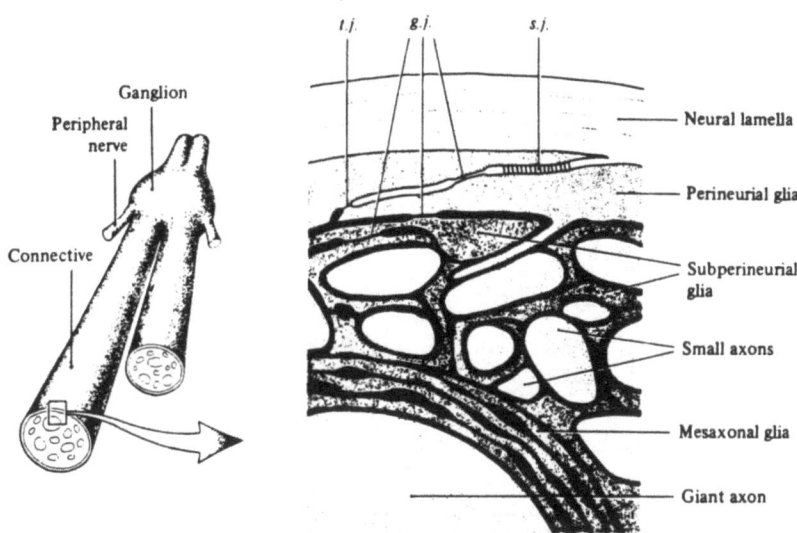

FIG. 1: Organization of neural and glial elements in the insect ventral nerve cord. The association of axons and neuroglia in the connective is ensheathed by a layer of perineurial glial cells overlaid by the neural lamella. Tight junctions (t.j.) and septate junctions (s.j.) are found between perineurial cells. Gap junctions (g.j.) interconnect perineurial and subperineurial glial cells. From Treherne and Schofield (1981).

they operate in spite of being bathed in haemolymph whose composition is very variable and often unsuitable for proper neural function. The content of the haemolymph is highly dependent on diet and contains a low concentration of sodium ions, a high concentration of potassium ions and significant amounts of various amino acids and toxic substances (Hoyle 1953; Pichon and Boistel 1963; Treherne 1985). The ability to maintain the composition of the extraneural fluid fairly constant and completely diferent from the haemolymph is vital and a property, amongst many others (see Lane 1981b), of the tissue investing the nervous system. Although the neural lamella can be an annoying barrier to the passage of glass microelectrode tips, it does not impede the rapid movement of small ions or larger macromolecules (Theherne and Pichon 1972). Recent intracellular and extracellular recordings in and around the glial cells and axons of a cockroach nerve cord put the diffusion barrier between the haemolymph and the nervous system at the layer of perineurial glial cells immediately under the lamella (Schofield et al. 1984a). The evidence for this includes the observation that the extracellular sheath potential (Pichon and Boistel 1967) is the same in very peripheral locations (i.e. immediately internal to the perineurial layer) as it is in very deep locations (i.e. immediately extracellular to a giant axon) in the nerve connective. Also, the rapid potential change which can be induced by changing the concentration of potassium in the saline bathing the nerve cord, and which indicates a barrier to the free diffusion of ions (Pichon and Treherne 1970; Pichon et al. 1971), is the same at both locations. Indeed, intracellular recordings from identified perineurial cells show that most of the potential change occurs across the basolateral membrane of these cells which is in immediate contact with the neural lamella, with only about 20% being generated across the apical membrane (Fig. 2). A modelling study has shown that an estimated trans-perineurial resistance

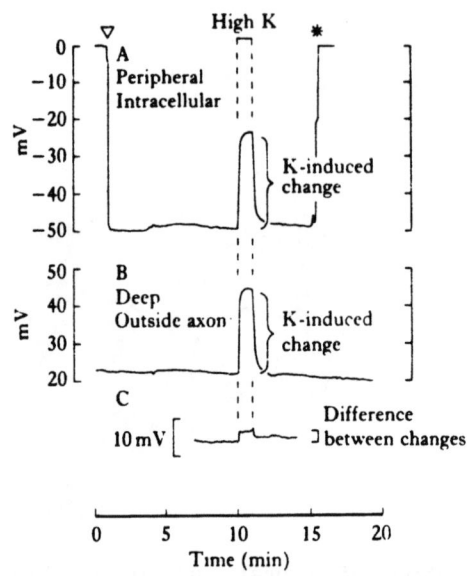

FIG. 2: Simultaneous recordings intracellularly from a perineurial cell (A) and extracellularly, deep in the connective, close to a giant axon (B) in the cockroach ventral nerve cord. Application of a high concentration of potassium induces a rapid potential change which indicates a barrier to ionic diffusion. Note that the potential change is almost the same at both locations which suggests that most of the potential change is generated across the basolateral membrane of the perineurial cell with only about 20% (C) being generated across the apical membrane. From Schofield et al. (1984).

alone is sufficient to simulate the values obtained from the physiological studies of the potential changes induced by high concentrations of potassium in the saline (Schofield and Treherne 1984). This supports the idea that perineurial glial cells and their intercellular junctions form the anatomical substrate of the blood-brain barrier. Similar intracellular and modelling studies attribute the disruption of the barrier caused by application of hypertonic urea (Treherne et al. 1973) to an action at these intercellular junctions (Schofield et al. 1984b). The anatomical evidence for the location of the barrier (obtained from electron microscope and extracellular tracer studies) seems not to be so clear and the exact nature of the barrier is still in some dispute. The accepted idea is of a seal formed by tight junctions between the perineurial cells (e.g. Lane 1981a). The evidence for this has been questioned (Shaw 1983, 1984) and an alternative idea of septate junctions connecting extensive, flat sheath cells has been proposed at least for nerve connectives in the cockroach (Shaw and Henken 1984; Shaw 1984).

2.2. The role of glia

A consequence of a restricted extracellular space surrounding neurons is that the alteration of extracellular concentrations of ions caused by neuronal activity can have significant effects (e.g. Yarom and Spira 1982; Spira et al. 1984; see below). In addition to protection from the haemolymph afforded by the perineurial glia, there are numerous homeostatic mechanisms in the sub-perineurial glia and extracellular space that act to regulate the ionic composition of the fluid surrounding nerve processes. Included in these are active ion pumps (Treherne and Schofield 1981), an extracellular anion matrix which could act as a cation (specifically sodium) reservoir (Treherne et al. 1982), and a spatial buffering mechanism mediated by glia that prevents the extracellular accumulation of potassium ions (Coles and Tsacopoulos 1981). The specific role of invertebrate glia is dealt with more fully by Pentreath 1987; see also Pentreath 1982; Hoyle 1986).

3. INTRINSIC NEURONAL PROPERTIES

The major point to be made in this section is that insect neurons do not differ greatly from neurons in any other group of animals. Neuronal cell bodies tend to be confined to a cortex primarily on the ventral surface of the ganglia. The cell body is unipolar and is invaginated with trophospongial glial processes (Hoyle et al. 1986). A single primary neurite from the cell soma gives rise to a neuropile segment from which branch numerous secondary and tertiary processes. Local (or intraganglionic) interneurons have no recognizable axons and their processes are confined to the ganglion containing the cell soma, whereas interganglionic interneurons extend one or more axons to various distances along the ventral nerve cord (Pearson 1977). The axons of interganglionic interneurons and motoneurons arise from the neuropile segment. See Weevers (1985) for an extensive review of the general form of insect neurons.

3.1. Passive properties

The rate and spread of passive decrement of an electrical signal in a neurite (or cable) are described by the time and length constants of that neurite. These constants are determined solely by the resistivity of the membrane, the capacitance per unit area of the membrane, the resistivity of the axoplasm, and the dimensions of the neurite. Pichon and Ashcroft (1985) have compiled a table of reported values of these parameters for a variety of insect nerve preparations. Membrane resistivity is between 300 and 800 $ohm.cm^2$ (excluding neurosecretory cells which have around 3000 $ohm.cm^2$); specific membrane capacitance ranges between 3 and 6 $\mu F.cm^{-2}$ for axons and is around 16 $\mu F.cm^{-2}$ for the cell soma (this is likely to be an overestimate due to inaccuracies in estimating membrane area); axoplasmic resistivity is around 46 $ohm.cm^2$. A recent analysis (Ammermüller 1986) of the passive cable properties of the L neurons (second order ocellar interneurons) in locusts gives a value for the membrane resistivity of 2,000 $ohm.cm^2$. This results in the major process of the L neuron having a length constant of 2.2 mm which accords well with the observation that electrical signals are transmitted within an L neuron with little decrement (Wilson 1978). Miller and Jacobs (1984) found that a value of 10,000 $ohm.cm^2$ for the membrane resistivity gave the best correspondence between the intracellularly recorded value of the input resistance and that computed from a compartmental model.

In addition to neuronal membrane characteristics, the impressive geometries of insect neurons (see Nässel 1987a) will have a powerful effect on the intracellular transmission of electrotonic potentials. An example of such an effect has been calculated by Rall (1981) using a model of a local, non-spiking insect interneuron described by Siegler and Burrows (1979). In the passage from one end of the neuron's neuropile segment to the other, electrical signals are likely to be attenuated by about 2 1/2 times in either direction. Attenuation will be much greater for signals travelling between the terminal branches and the neuropile segment. Further analysis using the same model shows that a 10mV signal at the terminals will be decremented to 0.5mV at the neuropile segment (Siegler 1984). These calculations highlight the possibility that different areas of an neuron may be functionally independent and operate in different local circuits. A similar modelling study has been undertaken for an intersegmental giant interneuron in the cricket in order to investigate the effect, on the activity of the neuron, of synapse location within a complex dendritic branching pattern (Miller and Jacobs 1984; Fig. 3). The conclusion reached was that all sites of possible

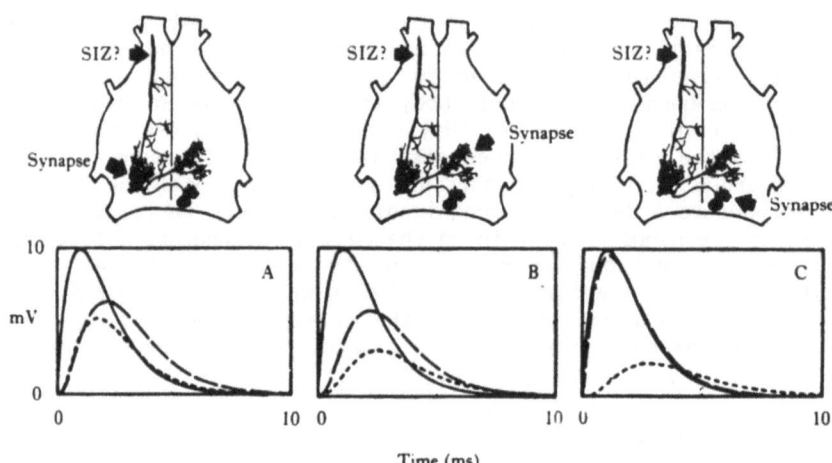

FIG. 3: Computer model of a cricket interneuron. Epsps calculated at the soma (dashed line) and at the spike initiating zone (dotted line) due to synapses of equal strength (solid line) located on different dendritic branches (arrows). SIZ? marks the putative location of the spike initiation zone. From Miller and Jacobs (1984).

synapse location were sufficiently close electrically to the spike initiating zone, to the cell soma and to each other that significant integration could occur and, for this neuron, could be monitored with an electrode in the cell body. The above results from modelling are clearly dependent upon the details of the structure of the neuron being investigated and on the values of the parameters selected for the models. Thus it is important to remember that the surface area of a neuron or neuronal segment may be much larger than is immediately apparent due to considerable infolding of glial processes (Hoyle et al. 1986). However the models do illustrate the importance of these properties in determining signal transmission within insect neurons and the way that variations in the passive properties and structures of neurons could suit different neurons for different physiological roles.

3.2. Active properties

The ability to generate and propagate action potentials is a property of active neuronal membrane which contains the appropriate voltage sensitive ionic conductances. In those instances where they have been studied in insect axons the mechanisms for action potential initiation and conduction have been found to conform to the general model provided by the squid giant axon (Narahashi 1963; Pichon 1974; Pichon and Ashcroft 1985), that is, a rapid increase in sodium conductance followed by a delayed increase in potassium conductance. The similarity extends even to the recent discovery of a population of slowly-inactivating sodium channels in cockroach axonal membrane that resemble those found in the squid axon (Yawo et al. 1985). Calcium-dependent action potentials have, however, been reported in the growing tips of regenerating axons (Meiri et al. 1981), although a subsequent study failed to confirm this (Leech and Treherne 1984), and in the axons of neurosecretory cells (Orchard and Osborne 1977; Miyazaki 1980). The distribution of active membrane on insect neurons is variable. Many neurons, including some Dorsal Unpaired Median (DUM; Plotnikova 1969; Hoyle et al. 1974; Goodman et al. 1980) neurons, are now known to be completely non-spiking (Pearson 1976, 1979; Burrows 1981; Wilson and Phillips 1983; Hisada et al. 1984; Siegler 1985). The somata of neurosecretory neurons (Zaretsky and Loher 1983; Carrow et al. 1984; Copenhaver and Truman 1986) and most DUM neurons (Jego et al. 1970; Crossman et al. 1971; Goodman and Heitler 1979; Goodman et al. 1980) do support action potentials, but, in general, the somata of motoneurons and interneurons are inexcitable (Hoyle 1970; Hoyle and Burrows 1973; Burrows 1977; Goodman and Heitler 1979; Gwilliam and Burrows 1980). A latent excitability can be unmasked in normally inexcitable somata by blocking some of the outward potassium channels with tetraethylammonium (TEA), or by treatment with colchicine (Pitman et al. 1972b; Pitman 1975a, b; Goodman and Heitler 1979). The soma spike can be generated by a rapid increase in conductance of both Na^+ and Ca^{++} ions (DUM neurons - Pitman 1975a, b; Goodman and Heitler 1979), of primarily Na^+ ions (axotomized motoneurons - Goodman and Heitler 1979), or of primarily Ca^{++} ions (neurosecretory neurons - Orchard 1976; Miyazaki 1980; Carrow et al. 1984). Axotomy and colchicine treatment are thought to increase the number of available sodium channels. The fact that some DUM somata do spike may have little functional significance and the difference between them and non-spiking somata may simply reflect the possible neurosecretory role of these neurons (Goodman and Heitler 1979). The longer duration action potentials of neurosecretory neurons (e.g. Carrow et al. 1984) that increase the amount of neurosecretory material released may be brought about by a weak potassium conductance. If the potassium conductance is similarly weak in the soma membrane of spiking DUM neurons, it may be unable to prevent the regenerative rise of the sodium/calcium conductance to produce an action potential. In normally non-spiking somata this can be mimicked with the application of TEA.

3.3. Spike initiation

The region of a neuron where passive membrane adjoins active membrane can be thought of as a spike initiation zone. The location of the spike initiation zone of insect neurons will affect the relative weights of synaptic inputs located at different sites on the dendritic branching. For insect motoneurons it was long thought that the neuropile segment was inexcitable and that spikes were initiated close to the point where the axon left the ganglion (Hoyle 1970). However the shape of extracellular recordings taken along the length of the intraganglionic portion of at least one motoneuron (Fast Extensor motoneuron of the Tibia - FETi) indicates that spikes may in fact arise from the middle of the neuropile segment (Gwilliam and Burrows 1980; Fig. 4). Other evidence is that an overshooting spike can be recorded intracellularly from this region of the neuropile segment. Intracellular recording techniques may be unreliable for such a study because local damage could render the area of membrane around the electrode tip inexcitable (Gwilliam and Burrows 1980). For interganglionic interneurons it is unclear exactly where the spike initiation zone is. The modelling study on the cricket giant interneuron mentioned above, gave a location quite distant from the dendritic branching and just before the axon exited the ganglion in the connective (see Fig. 3) but this is subject to confirmation (Miller and Jacobs 1984). My own experience with thoracic interneurons and motoneurons, other than FETi , in the locust is that overshooting spikes are never recorded unless subsequent staining shows the penetration to have

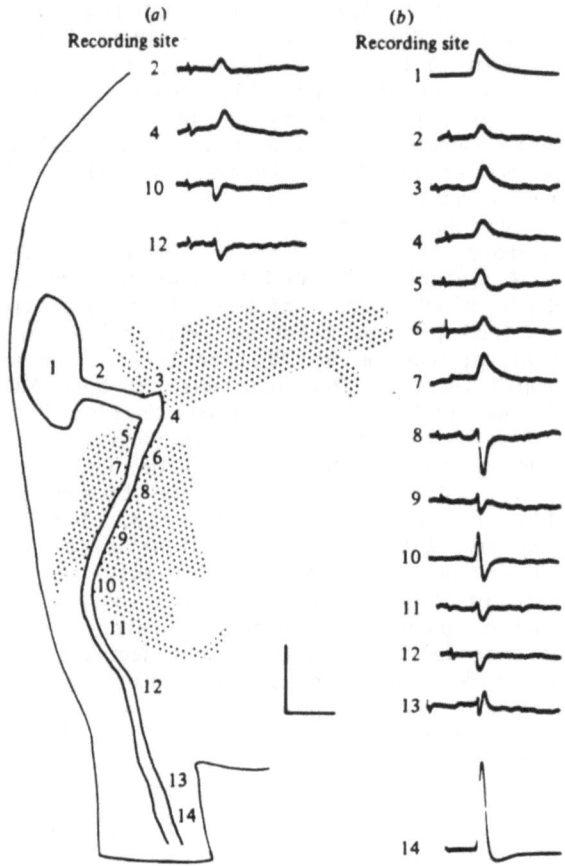

FIG. 4: Extracellular field potentials associated with an antidromic action potential in a locust motoneuron (FETi) innervating the extensor muscle of the tibia. The diagram shows the cell body and the course of the neurite in an outline of part the metathoracic ganglion. The stippled region indicates the extent of arborization from the neuropile segment and the numbers indicate the locations of the different recording sites. Extracellular recordings of the antidromic action potential taken from different sites and induced by stimulating the axon of the motoneuron in the muscle, vary in shape and amplitude. (a) Recordings aligned vertically with the stimulus artifact. (b) Recordings aligned vertically with the time of occurrence of the peak of the spike. Recordings from sites 1 and 14 are intracellular. The change in the waveform of the field potential which occurs between sites 7 and 8 is indicative of the spike initiation zone being in this region. Voltage scale: extracellular 5mV, intracellular 50mV. Horizontal scale: 4ms and 50μm. From Gwilliam and Burrows (1980).

definitely been in an axon (i.e. not in the neuropile segment or any of its branches). This is true of long term penetrations in which the amplitudes of the recorded spikes have been seen to increase, indicating a measure of recovery from the penetration damage. FETi is the only motoneuron shown physiologically to have central output connections (Hoyle and Burrows 1973) and possession of active membrane on the neuropile segment may be specifically related to this unique physiology. Pinpointing the precise spot where spikes are normally initiated is obviously a difficult task and may be impossible. There is likely to be variation in the distribution of active membrane similar to the variation in the expression of other neuronal properties (Burrows 1979; Pearson and Goodman 1979; Goodman et al. 1979).

Multiple spike initiation zones have been demonstrated in insects. The functional relevance of the 4 spike initiation zones of the DUM neuron supplying the Extensor Tibiae muscle is uncertain (Heitler and Goodman 1978). The 2 zones of the Lobular Giant Movement Detector are related to the two sensory modalities to which this neuron is responsive (O'Shea 1975).

4. NEURONAL INTERACTIONS

The scope of this section is limited to interactions between neurons and thus neglects a major part of insect neurobiology - the interactions of neurons with muscle. Insect neuromuscular systems are model systems for the study of transmitter release (e.g. Usherwood 1961, 1977), glutamatergic and GABAergic (GABA = gamma-aminobutyric acid) transmission (e.g. Cull-Candy, 1982), neuromodulation with particular reference to octopamine (Evans and O'Shea 1977; O'Shea and Evans 1979; Evans 1984), and the role of peptides as co-transmitters in motoneurons (O'Shea and Bishop 1982; O'Shea 1982; Adams and O'Shea 1983).

There are numerous ways in which the activity of one neuron can affect the activity of another neuron and, for insects, these have been thoroughly reviewed a number of times, especially with reference to chemical transmission (Pitman 1971; Gerschenfeld 1973; Hildebrand 1982; Callec 1974, 1985). Here I should like merely to catalogue these interactions and indicate features of interest and features needing resolution.

4.1. Chemical interaction

Chemical synapses in insects operate in the same fashion as they have been shown to operate in other animals. Specific transmitter substances are released from presynaptic terminals and they interact with receptors on postsynaptic membranes to increase the membrane conductance to particular ion species. There are numerous putative neuroactive substances but for the most part all that is known of many of them is their distribution in the nervous system (see Nässel 1987b). Although the stringent criteria for establishing a neurotransmitter role have not been completely fulfilled for any substance, the weight of available evidence justifies considering acetylcholine as an excitatory transmitter and GABA as an inhibitory transmitter in the central nervous system of insects (Pitman and Kerkut 1970; Sattelle 1980; Harrow et al. 1982; Sattelle et al. 1983). A central transmitter role for glutamate, octopamine and proctolin is also likely (Evans 1980; Walker et al. 1980; O'Shea 1982; Sombati and Hoyle 1984b). For practical reasons the most extensively studied type of central synapse is the excitatory monosynaptic connection from an afferent to a motoneuron or interneuron (e.g. Burrows 1975;

Pearson et al. 1976; Carr and Fourtner 1980; Shepherd and Murphey 1985). Within this general class most information has accumulated for the connection between a cercal afferent and a giant interneuron in the cockroach terminal abdominal ganglion (Callec 1974; Sattelle et al. 1983; Blagburn et al. 1984). This connection is very probably mediated by acetylcholine. Three putative acetylcholine receptors have been described in insects, and the one at this synapse is blocked by alpha bungarotoxin (Fig. 5) being similar to the vertebrate nicotinic acetylcholine receptor (Harrow et al. 1982; but see Meyer and Edwards 1980). Indeed, insect

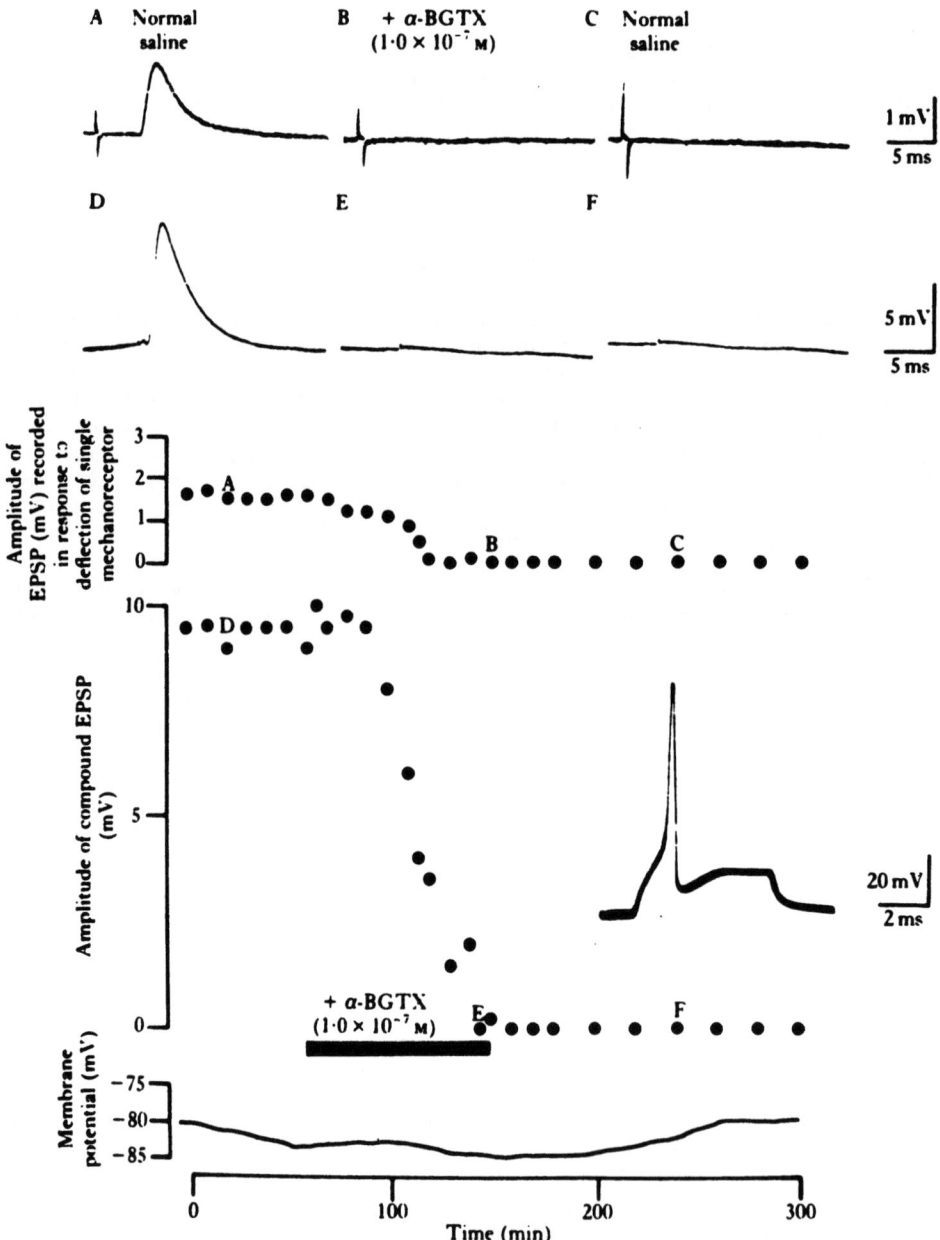

FIG. 5: α-bungarotoxin blocks epsps recorded in the cockroach giant interneuron 2. Epsps were evoked by stimulation of a

nervous tissue has provided the first identification of a functional acetylcholine receptor channel of neurons (Breer et al. 1985; Hanke and Breer 1986). Other properties of this synapse are all consistent with its chemical nature: extrapolated reversal potential, calcium dependence, depression, facilitation (Callec 1985).

Evidence suggests that inhibitory postsynaptic potentials (ipsps) are mediated by an increase in the conductance of the postsynaptic membrane to chloride ions in response to the presynaptic release of GABA (Kerkut et al. 1969; Pitman and Kerkut 1970; Nistri and Constanti 1979; Breer and Heilgenberg 1985). Consistent with this is the observation that application of picrotoxin can block ipsps (Robertson and Pearson 1985a).

One of the most significant discoveries in recent years has been that of neuronal interactions that operate via graded release of transmitter in response to tonic changes in presynaptic membrane potential (Pearson and Fourtner 1975; Burrows and Siegler 1976). The possible functional advantages of these sorts of interaction have been discussed several times and they include speed, sensitivity, accuracy and as a means of enabling local interactions of a limited portion of the neuron (Pearson 1976, 1979; Burrows 1981; Shaw 1981; Wilson and Phillips 1983; Siegler 1984, 1985). In insects, graded interactions have been extensively studied in a population of local interneurons in the metathoracic ganglion of the locust (e.g. Burrows 1978, 1980; Burrows and Siegler 1976, 1978), in the compound eye (e.g. Järvilehto and Zettler 1971; Laughlin 1973; Shaw 1979), and in the locust ocellus (e.g. Simmons 1981, 1982, 1985). Figure 6 demonstrates the chemical nature of a graded interaction between a local interneuron and a motoneuron in the locust. Essentially the discovery of graded interactions has established a range of parameters for chemical transmission. One end of this range is characterized by the classical, spike-mediated post-synaptic potential of limited duration. The other end is characterized by a connection in which the presynaptic neuron has a low membrane potential threshold for transmitter release (i.e. releases transmitter tonically at its resting potential), and the postsynaptic neuron is continually responsive to the transmitter. In graded interactions the relationship between pre- and postsynaptic voltage tends to be of high gain and linear along much of its operating range (Burrows 1980; Fig. 7). Thus maintained shifts of presynaptic potential in either direction can be transmitted to the postsynaptic neuron with little, if any, loss of information. Chemical interactions can, of course, be intermediate to these two extremes. Some presynaptic neurons have higher thresholds for transmitter release and thus cannot signal hyperpolarizations from the resting potential and in some cases a maintained presynaptic depolarization produces only a transient postsynaptic response. Inactivation of a presynaptic Ca^{++} conductance has been proposed to account for the latter phenomenon (Simmons 1985).

Local, non-spiking interneurons in the locust metathoracic ganglion have been shown to have a relatively wide sphere of influence. Depolarizing one of these interneurons can have opposite effects on different motoneurons (Burrows 1978; Wilson and Phillips 1982). Similarly, stimulation of the L neurons in the locust ocellus can cause spike-

---

single cercal mechanosensory hair or by electrically stimulating cercal nerve XI to produce a compound epsp. Blockage to the epsps with α-bungarotoxin was not reversible, was not accompanied by a change in polarization of the postsynaptic membrane, and had no effect on the amplitude and duration of action potentials recorded in the postsynaptic neuron (see inset). From Sattelle et al. (1983).

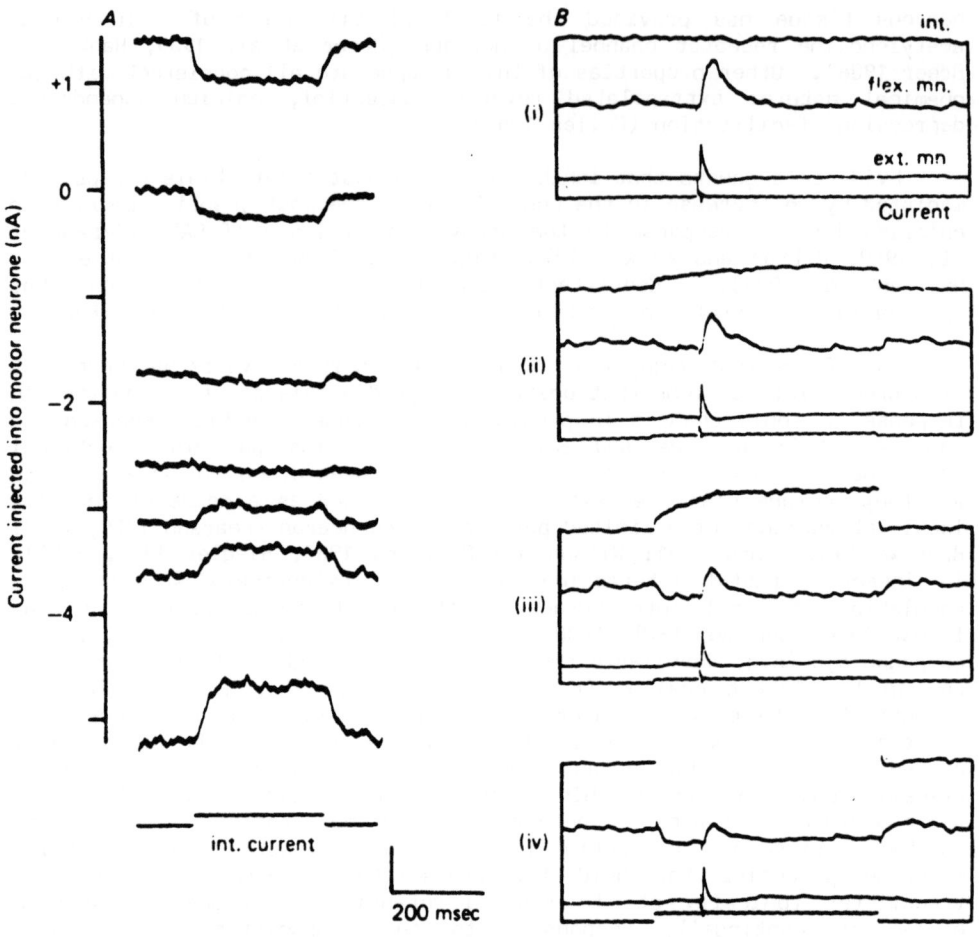

FIG. 6: Graded transmission from a local interneuron to a motoneuron in the metathoracic ganglion of the locust is chemically mediated. A. The response in the motoneuron to constant current pulses (5.6nA, 300ms) delivered to the interneuron is dependent upon the postsynaptic membrane potential which is altered by injecting current (left hand scale) into the motoneuron. Note that the postsynaptic response can be reversed by hyperpolarizing the postsynaptic membrane. B. Antidromic stimulation of an extensor motoneuron activates a central pathway to evoke a compound epsp in a flexor motoneuron. Depolarizing the local interneuron with injected current induces a hyperpolarization of the flexor motoneuron. Note that increased hyperpolarization of the flexor motoneuron, caused by increasing the current injected into the local interneuron, is associated with a decrease in the amplitude of the compound epsp. This indicates an increase in conductance of the postsynaptic membrane. Vertical scale: (A) 3.6mV; (B) interneuron 33mV, flexor motoneuron 6mV, extensor motoneuron 14mV; current (A) 28nA, (B) 66nA. From Burrows and Siegler (1978).

mediated epsps and ipsps, sustained depolarizations and transient hyper-polarizations in different follower neurons (Simmons 1982; Fig. 8). The observation of multiple effects from depolarizing a single neuron raises the question of whether a single neuron can make different types of

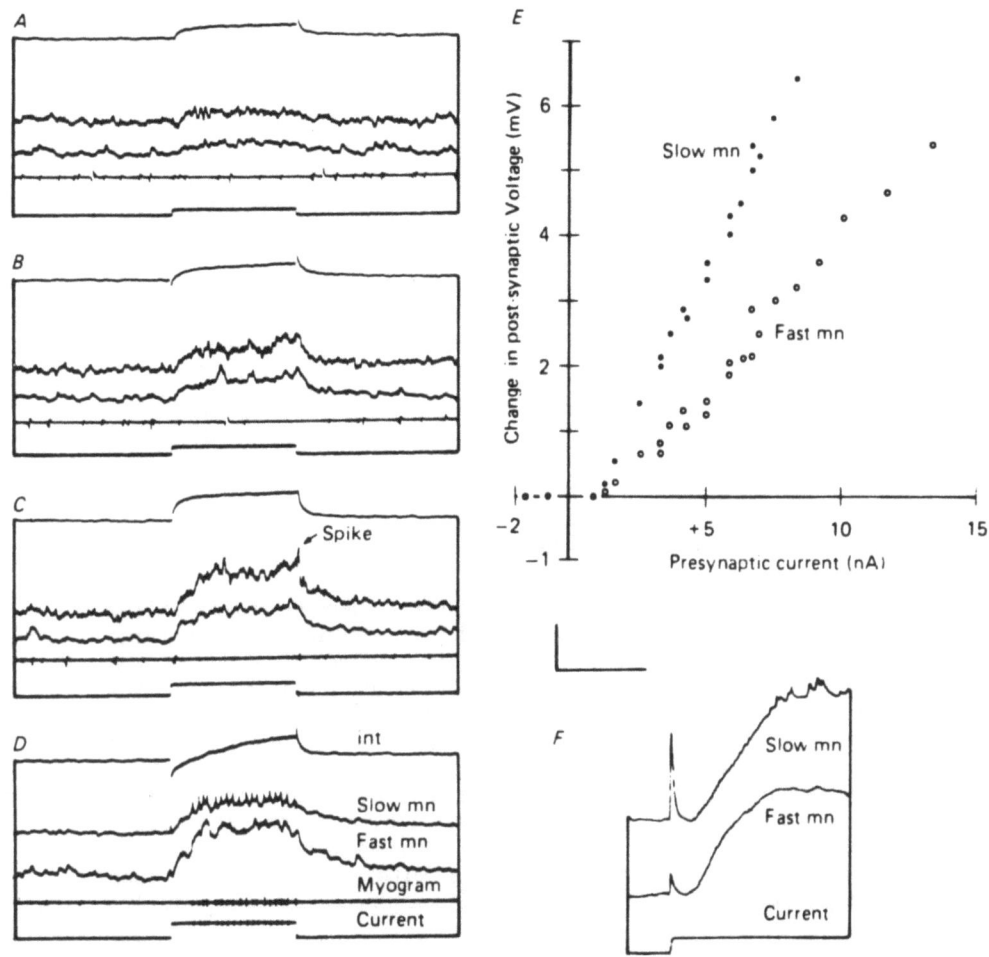

FIG. 7: Graded control of motoneurons by a local interneuron in the metathoracic ganglion of the locust. A-D. Increasing amounts of depolarizing current injected into the interneuron evokes larger depolarizations of the motoneurons. E. The relationships between the current injected into the interneuron and the postsynaptic voltage changes are linear. F. Signal average of 32 events showing the latency of the postsynaptic potentials in both motoneurons to be similar and between 8 and 10 ms. The initial upward deflection is an artifact associated with the start of current injection. Vertical scale: voltage, interneuron 40mV, (A-C) slow motoneuron 7mV, fast 4 mV, (D) slow 17mV, fast 4mV, (F) slow 2.3 mV, fast 1.5 mV; current, (A-D) 30nA, (F) 26nA. Horizontal scale: (A-D) 400ms, (F) 40ms. From Burrows (1980).

monosynaptic connection with different follower neurons. A related question concerns the possible existence of multicomponent synapses of the sort that have been described in molluscs (e.g. Wachtel and Kandel 1967; Getting 1981). In both cases validation of the phenomenon rests firmly on the demonstration that each different effect (or component of a multicomponent psp) is produced as a result of a monosynaptic connection from the same neuron. Such a demonstration is notoriously difficult to achieve. Even the criteria that have proved useful for helping to establish monosynapticity in molluscs (Berry and Pentreath 1976) are

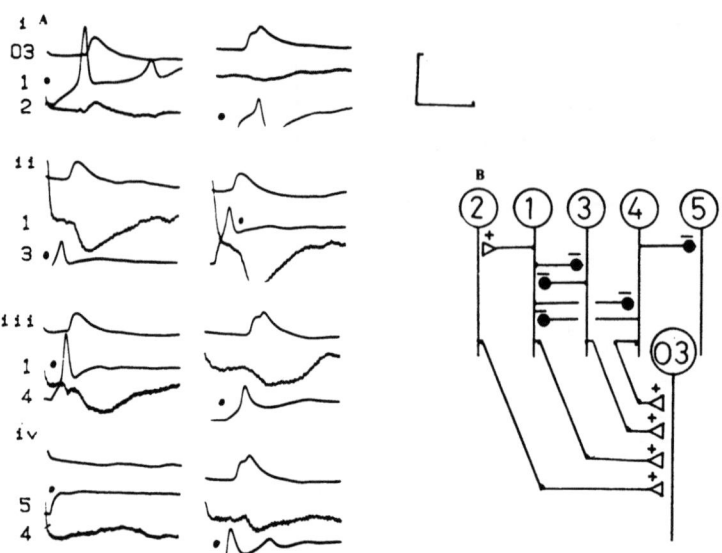

FIG. 8: Connections among five L neurons (second order ocellar neurons) in the locust and their connections to a large descending brain neuron. A. Sample records indicating the connections. In each sample one L neuron (indicated with a dot) is injected with current and the effects monitored in the descending neuron (O3) and another L neuron. B. Summary of the connectivity pattern established in this way. Open triangle and a plus sign indicate an excitatory connection, filled circle and a minus sign indicate an inhibitory connection. Vertical scale: O3, 10mV; traces with spikes, 20mV; epsp in (i), 10 mV; traces with ipsps, or no psps visible, 5mV. Horizontal scale: 10 ms. From Simmons (1982).

dependent on the communication between the neurons being spike-mediated. The situation is complicated in insect preparations by the powerful blood-brain barrier (see above) hindering the appropriate pharmacological and ion-substitution tests. To date, firmly established monosynaptic connections in insects are rare and the conclusion is substantiated with anatomical evidence (e.g. Peters et al. 1985). In physiological studies it is common to assume monosynapticity if: i) the processes of the neurons overlap; ii) the postsynaptic response follows at high frequencies; and iii) the estimated synaptic delay is short, usually less than 1 ms, and constant (Fig. 9). It is difficult to characterize any of the graded interactions even as "probably monosynaptic" because the second criterion is clearly not applicable and the observed latencies are relatively long and quite variable (Burrows and Siegler 1978; Fig. 7F). Of 32 spiking interneurons in the locust with a total of 68 known short latency connections, none has been shown to have qualitatively different short latency connections with different neurons (Pearson and Robertson 1987; Table 1). Spikes in some of these interneurons are followed by both short latency ipsps and by a slightly longer latency depolarization in different neurons (and occasionally in the same neuron raising the possibility of a multicomponent synapse) but in this case the evidence favors the interpretation that the depolarization is produced disynaptically (Robertson and Pearson 1985a: Fig. 10). It is conceivable that the synaptic repertoire of local interneurons is more complex than that of intersegmental spiking interneurons. However, at present, there is no compelling evidence for different monosynaptic actions of a single

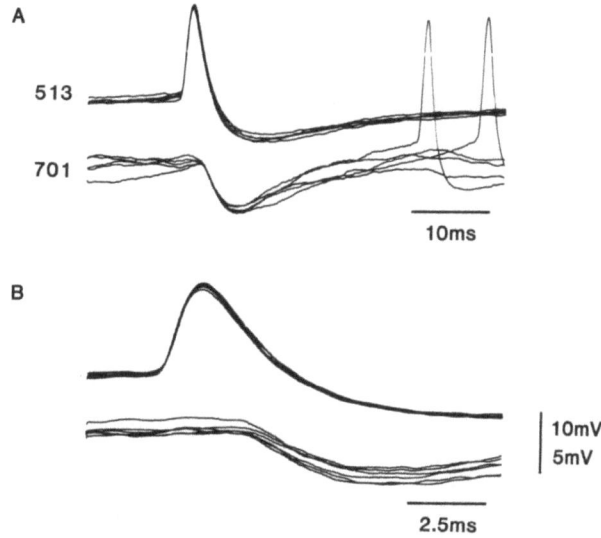

FIG. 9: A constant, short latency connection between two interneurons in the flight system of the locust. A. Five oscilloscope sweeps superimposed and triggered off the rising phase of the spike recorded in interneuron 513. Note the constancy of the latency between presynaptic spike and postsynaptic ipsp. B. As above with an expanded time scale to enable measurement of the latency. Total latency from the half-amplitude of the rising phase of the presynaptic spike to the first sign of hyperpolarization postsynaptically is 2 ms. Interneuron 513 is located in the metathoracic ganglion whereas 701 is located in the mesothoracic ganglion. Conduction delay between these ganglia is about 1 ms, giving a synaptic delay of around 1 ms. R.M. Robertson (unpublished).

interneuron either on different neurons or on a single neuron. There is evidence for complex effects from stimulation of single interneurons and much of this is likely to be mediated polysynaptically through graded interactions.

4.2. Electrical interactions

Electrical coupling between neurons is comparatively rare in insects. In fact the only direct demonstrations of electrical coupling by recording intracellularly from the pre- and postsynaptic elements are of a connection between a leg muscle motoneuron and its supernumerary partner found in only 2 out of 31 preparations (Siegler 1982; Fig. 11) and of connections between individual receptor cells in the compound eye (Shaw 1969). Indirect evidence for electrical coupling between motoneurons innervating the same muscle in the cricket has been obtained but not confirmed (Bentley 1969a). There is also indirect evidence for an electrical connection between afferents and an interneuron in the cricket (Harris and Garrison 1976) and an anecdotal report of possible dye-coupling between descending interneurons in the locust ventral nerve cord (Kien and Altman 1984). For many years the connection between the Lobula Giant Movement Detector (LGMD) and the Descending Contralateral Movement Dectector (DCMD) (O'Shea and Rowell 1975) was considered an example of a spike-transmitting electrical synapse. The evidence for this was indirect and was based on the 1:1 nature of the transmission, the following of the

TABLE I

LIST OF SPIKING INTERNEURONS IN LOCUSTS WITH KNOWN, SHORT LATENCY, POSTSYNAPTIC CONNECTIONS.

	Neuron Number*	Behavior	Number of Connections**	Previous Description
Excitatory	110	flight	1	-
	139	hearing	1	Marquart (1985)
	201	flight	5	Robertson & Pearson (1983)
	202	flight	2	-
	206	flight	1	Robertson & Pearson (1985a)
	314	jump	2	Pearson & Robertson (1981)
	503	flight	3	Robertson & Pearson 1983)
	504	flight	2	Robertson & Pearson (1983)
	505	flight	1	-
	506	flight	1	-
	514	flight	2	Robertson & Pearson (1983)
	521	flight	1	-
	529	hearing	3	Pearson et al (1985)
	530	hearing	3	Marquart (1985)
	531	hearing	1	Pearson et al (1985)
	537	-	1	-
	606	respiration	1	-
	701	flight	4	Robertson & Pearson (1983)
	714	hearing	3	-
Inhibitory	301	flight	3	Robertson & Pearson (1983)
	302	flight	2	Robertson & Pearson (1983)
	304	flight	2	-
	320	-	2	-
	401	flight	2	Robertson & Pearson (1983)
	501	flight	3	Robertson & Pearson (1983)
	511	flight	5	Robertson & Pearson (1983)
	513	flight	1	-
	515	flight	2	-
	520	flight	1	-
	535	jump	3	Pearson et al (1980)
	710	jump/flight	2	-
	719	jump	1	-

* Numbering according to the scheme of Robertson and Pearson (1983).

** Number of different postsynaptic neurons which have been found to receive input from the interneuron.

From Pearson and Robertson (1987).

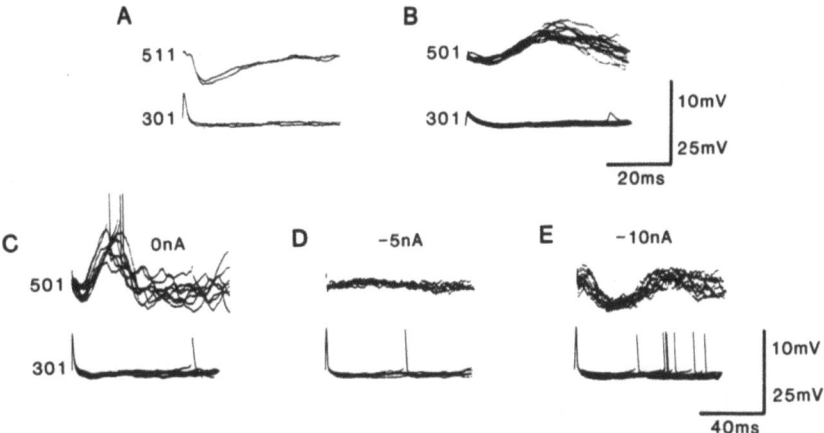

FIG. 10: Two types of postsynaptic potential following spikes in a single interneuron in the flight system of the locust. A. The short constant latency from each spike in interneuron 301 to ipsps recorded in 511 suggests that this connection may be monosynaptic. B. The delayed excitatory potentials recorded in interneuron 501 following each spike in 301 occur at a longer, though similarly constant, latency. C-E. The delayed excitatory potential is reversed by injecting hyperpolarizing current into the postsynaptic neuron indicating that it is produced by a decrease in conductance at the postsynaptic membrane. This and other evidence (not shown) suggests that the delayed excitatory potential arises as a result of a disynaptic, disinhibitory connection through an intercalated non-spiking interneuron. From Robertson and Pearson (1985b).

synapse at high firing frequencies, a very short reported latency between spikes in the two neurons, and an inability to block transmission with cold or low Ca^{++} saline. Recently it has been shown that this synapse is in fact a very high gain chemical synapse with a high voltage threshold for transmitter release at the presynaptic terminal (Rind 1984). Thus it is only in the jumping and flight systems of Diptera that the existence of electrical connections normally contributing to behavior can be claimed with any confidence (Tanouye and Wyman 1980; Koenig and Ikeda 1983; Wyman et al. 1984; Bacon and Strausfeld 1986). Even here the evidence is indirect and based on latency of transmission and following frequency. However there is additional electron microscope and dye-coupling evidence to support the conclusion (Strausfeld and Obermeyer 1976; Strausfeld and Bassemir 1983; Bacon and Strausfeld 1986).

Current flow within a neuron, produced either as an action potential or a sustained polarization, is balanced by an equal but opposite current flow extracellular to the neuron. Depending upon the extracellular resistance and the location relative to the intracellular current flow, various amplitudes and polarities of extracellular field potential will be set up (see e.g. Fig. 4). In a restricted extracellular space the field potential can be quite large and it is possible that it could indirectly affect the activity of neighboring neurons by changing their transmembrane potential. If the extracellular resistance is high enough it is possible that the route of least resistance lies across the membrane of the neighboring neuron thus directly affecting its membrane potential. Such a mechanism has been proposed to explain interactions between receptor terminals in the compound eye (Shaw, 1975, 1984; Horridge et al 1983).

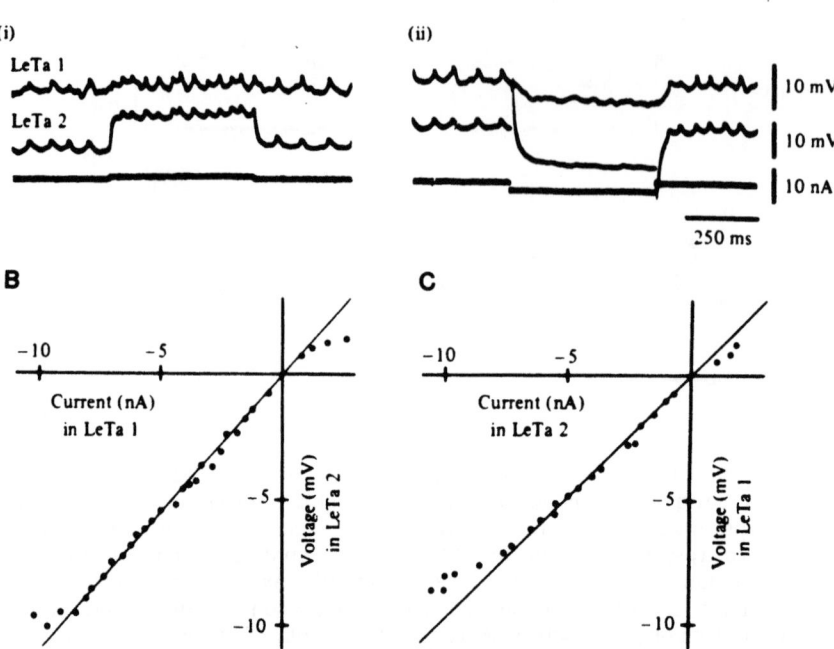

FIG. 11: Electrical junction between two motoneurons innervating the metathoracic tarsal levator muscle in the locust (LeTa 1 and 2). A. Depolarizing (i) or hyperpolarizing (ii) current is injected into LeTa 2. B. Relationship between current injected into LeTa 1 and voltage response of LeTa 2. C. As for B except current is injected into LeTa 2. From Siegler (1982).

The basis for the high extracellular resistance is the blood brain barrier at this location. In spite of an assertion to the contrary (Hoyle 1983), presynaptic inhibition between the two DCMDs (see below) is not considered to be via a field effect, and other examples of functional connections mediated by field effects in insects remain to be demonstrated.

4.3. Potassium ion accumulation

A consequence of a restricted extracellular space is that ionic concentrations can fluctuate quite rapidly. Neuronal activity can result in an extracellular accumulation of potassium ions (Spira et al. 1984). It has been shown that an interaction between two giant interneurons in the cockroach nerve cord can be attributed to an increase in extracellular potassium, presumably at an area where the distance between the two neurons is small (Yarom and Spira 1982; Fig. 12). Accumulation of extracellular potassium can also affect the safety factor for transmission at a narrow region of an axon (Spira et al. 1976). As with the extracellular field effects mentioned above, the phenomenon has not often been demonstrated and its functional significance is not clear. However, the existence of the phenomenon is not in doubt and its potential for control is recognized.

4.4. Presynaptic inhibition

Presynaptic inhibition is a mechanism whereby activity in a neuron reduces the strength of a synapse by a direct action at the presynaptic

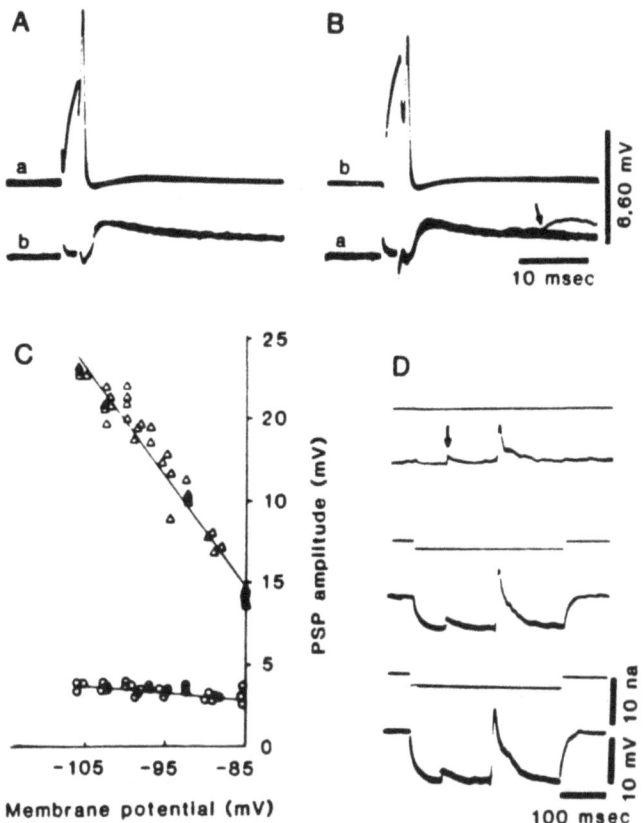

FIG. 12: Reciprocal connections between the axons of giant interneurons in the metathoracic ganglion of the cockroach. A. An action potential induced by current stimulation of interneuron a evokes a fast-rising depolarizing potential in interneuron b. B. The reciprocal connection. Note that a spontaneous chemical epsp recorded in a (arrow) has a much slower rise time. C and D. Amplitude of reciprocal connection between the giant interneurons is not substantially affected by changing the polarization level of the postsynaptic neuron (circles in C, arrow in D), whereas amplitude of a chemically mediated psp, evoked by stimulation of the contralateral connective, is increased by hyperpolarizing the postsynaptic membrane (triangles in C). Other evidence (not shown) indicates that the connection between the two giant interneurons is mediated by a local accumulation of K^+ ions. From Yarom and Spira (1982). Copyright 1982 by the AAAS.

terminals of the synapse. It could be treated in the next section (circuitry) as its full effect is realized only over two synapses. It is included here in recognition of the novel form of connection of a neuron to the axonal terminal of another. The phenomenon has been described several times in insects and its functional role seems to be as a contrast enhancer in sensory systems (Levine and Murphey 1980; Hue and Callec 1983; Shaw 1984) or to prevent summation of postsynaptic potentials to a level that might activate a motor circuit at an inappropriate time (Pearson and Goodman 1981). With the exception of the presynaptic inhibition in the compound eye which is thought to be mediated via an extracellular field effect (see above), the underlying mechanisms seem to be similar and best illustrated by the interaction between the DCMDs in the thoracic ganglion of the locust. An action potential in one DCMD is followed by a depolarization in itself and the contralateral DCMD. The amplitude of the depolarization is affected by changes in membrane potential or by injection of Cl⁻ ions, suggesting that it is mediated by a conductance increase to these ions. The latency of the depolarization is longer than would be anticipated for a monosynaptic connection although some aspects of the connection do suggest that it is direct. Pearson and Goodman (1981) suggest that an intercalated non-spiking neuron tonically releasing transmitter may be mediating the effect. Whatever the mechanism, action potentials of DCMD are reduced in amplitude during the time course of the depolarization (Fig. 13) and this corresponds in time to a reduction in the amplitude of epsps produced in response to these action potentials in other neurons (Fig. 14). In the cricket the psp reduction is associated

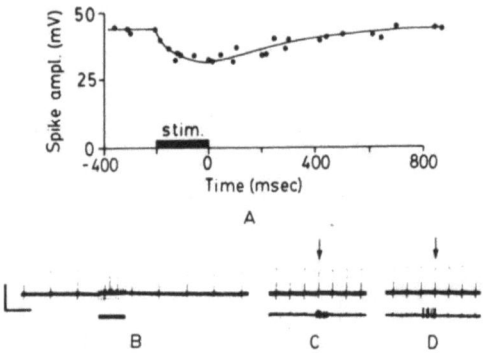

FIG. 13: Reduction in presynaptic spike amplitude in a DCMD neuron of the locust by electrical stimulation (A,B - stimulus duration marked by bar), by auditory stimulation (C, arrowed) and concurrent with activity in the contralateral DCMD (D, arrowed). Vertical scale: 50mV. Horizontal scale: B, 200ms; C and D, 400ms. From Pearson and Goodman (1981).

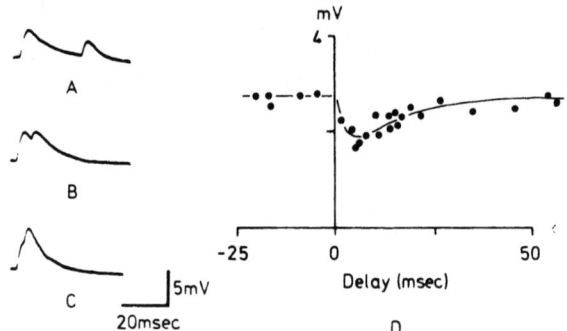

FIG. 14: Reduction in amplitude of epsps from one DCMD recorded in an M-neuron in the locust caused by activity in the other DCMD. A-C. Epsps recorded in an M-neuron. The first psp in each trace follows spikes in the left DCMD whereas the second psp was from the right DCMD. Note the reduction in the amplitude of the second psp when the delay between the two psps was 5 ms (B). D. Plot of the reduction in psp amplitude versus the delay between psps to show the time course of the reduction. From Pearson and Goodman (1981).

with a depolarization of the presynaptic terminal and a reduction in amplitude of the presynaptic action potential (Levine and Murphey 1980). In the cockroach an increase in Cl^- ion conductance (either depolarization or hyperpolarization) is associated with a reduction in the size of the epsp. Similar responses can be mimicked with GABA suggesting a role for this transmitter on the presynaptic terminal (Hue and Callec 1983). Although the presynaptic inhibitory effects described above are quite limited in their strength and effect they illustrate well the potential role that axo-axonic interactions might play in local circuits. Preliminary evidence in the crayfish indicates that an axonal terminal seems to be involved in integration and spike initiation (Wang-Bennett and Glantz 1986). There is no a priori reason why depolarization of an axon terminal should not cause transmitter release without the intermediary of spikes. If this were true of insect interganglionic interneurons, then axonal branches in distant ganglia could be functionally isolated from other parts of the neuron and subserve different behaviors.

5. CIRCUITRY UNDERLYING BEHAVIOR

The previous three sections have described aspects of the neuronal environment and how it is maintained, intrinsic properties of neurons, and the ways in which neurons have been found to interact with each other. In this section I should like to consider how particular neurons, through similar interactions, form the neural circuits that underlie behavior. Elsewhere in this volume, Altman (1987) discusses the higher order control of such networks, and Schildberger (1987) reviews the neuronal control of phonotaxis in crickets. In insects there are really only two behaviors for which a significant amount of the central circuitry is known. These are jumping and flight in orthropterans. Interestingly, some of the circuitry for jumping and flight in dipterans has also been described (e.g. Wyman et al. 1984). The advantage of dipterans is in the potential for studying the genetic and molecular bases for behavior (Thomas and Wyman 1984; Jan and Jan 1985; Wyman et al. 1985). Although it is for jumping and flight that there is most knowledge of circuitry, intracellular recordings have been made, and interneurons identified and associated with behavior in many other preparations. So before dealing specifically with features of interconnectivity I shall briefly review the expanding knowledge of identified neurons controlling behavior.

5.1. Identified neurons

Until just a few years ago it was possible to keep track of all the insect neurons that had been identified morphologically and physiologically without too much difficulty. Now their numbers are increasing so rapidly that to do so requires a constant and careful monitoring of the literature. In the majority of cases the interneurons that have been identified can be described as sensory interneurons, although their precise role is usually unknown. The reason for this is that, technically, it is often relatively straightforward to establish the spectrum of an interneuron's responses to a wide variety of sensory modalities and much more difficult to establish its integrative or motor effects. Thus an interneuron's sensory physiology is part of the set of characteristics that first identifies it. Insect interneurons have been described in all of the major sensory pathways (Table 2). For interneurons that have been described in terms of their probable motor role the list is not so impressive (Table 3). It is clear from these tables, which are almost certainly far from complete, that almost all of the identified interneurons in insects have been described in the last five years. Taking this as a trend, it is likely that very many more will be described in the

TABLE 2 REFERENCES TO IDENTIFIED INTERNEURONS (SENSORY).

System	Reference
visual	O'Shea and Williams 1974; O'Shea et al. 1974; Hertel 1980; Hengstenberg et al. 1982; DeVoe et al. 1982; Eckert 1982; Hausen 1982; Rind 1983; Olberg 1986.
ocellar	Goodman 1976; Goodman and Williams 1976; Simmons 1980; Patterson and Chappell 1980; Mizunami et al. 1982; Rowell and Pearson 1983; Milde 1984; Reichert et al. 1985.
auditory	Rehbein 1976; Casaday and Hoy 1977; Popov et al. 1978; Boyan 1980; Wohlers and Huber 1982; Popov and Markovich 1982; Romer and Marquart 1984; Schildberger 1984b; Cokl et al. 1985; Boyan and Altman 1985; Hörner and Gras 1985; Boyan and Fullard 1986.
olfactory	Matsumoto and Hildebrand 1981; Burrows et al. 1982; Ernst and Boeckh 1983; Boeckh et al. 1984.
mechanical/wind	Mendenhall and Murphey 1974; Murphey et al. 1977; Bacon and Tyrer 1978; Harrow et al. 1980; Daley et al. 1981; Sakaguchi and Murphey 1983; Pflüger 1984; Bacon and Murphey 1984; Siegler and Burrows 1984; Burrows and Siegler 1984; Murphey 1985; Hustert 1985; Collin 1985.
multimodal	Erber 1983; Schildberger 1984a; Homberg 1985.

TABLE 3 REFERENCES TO IDENTIFIED INTERNEURONS (MOTOR).

System	Reference
walking/posture	Siegler and Burrows 1979; Pearson et al. 1980; Pearson and Robertson 1981; Wilson 1981; Kien and Williams 1983; Delcomyn 1983; Kien and Altman 1984.
flight	Robertson and Pearson 1983, 1985a; Reichert and Rowell 1985; Pearson et al. 1985; Robertson 1985b.
ventilation	Pearson 1980; Burrows 1982; Komatsu 1984;
stridulation	Hedwig 1986a, b.

next five years. Several points can be made about this explosion in the catalogue of interneurons. First, it reflects the ease with which insect interneurons can be identified and this in turn reflects the idiosyncratic nature of interneuronal morphology and the highly organized structure of insect ganglia (see Nässel 1987a). A major reason for the success of the recent studies on the role of individual neurons must be because their distinctive structures allow them to be unequivocally identified in different preparations. Second, as information on identified interneurons accumulates it becomes possible to recognize common characteristics. A recent analysis of thoracic interneurons in the locust has correlated structural features with synaptic output - the possession of certain distinct features allows one to predict with some accuracy whether an interneuron will be found to make short latency inhibitory or short latency excitatory connections to other neurons (Pearson and Robertson 1987). This phenomenon may be related to the development of particular neurons. Neurons developing from the same neuroblast tend to share structural features (Goodman and Spitzer 1979; Raper et al 1983) and transmitter type (Goodman et al. 1980; Taghert and Goodman 1984). The correspondence of neuroblast of origin, structure, and transmitter type is not exact and cell lineage regulation can be overidden by other segment-specific factors to produce sibling and homologous neurons with different characteristics (Taghert and Goodman 1984; Pearson et al. 1985a). However there is a correlation, and together with the evidence relating synaptic effect with structure it suggests the hypothesis that inhibitory and excitatory neurons develop from different neuroblasts. Further information on the development of insect nervous systems is provided by Bacon (1987). It is clear that investigating the properties and behavior of single neurons can yield a substantial amount of information about neural processing in both sensory and motor pathways. Indeed, single neurons can have very powerful effects (e.g. Ritzmann et al. 1980) and can be shown to be both necessary and sufficient for particular behaviors (Nolen and Hoy 1984; Pearson et al. 1985b). This suggests a high level of functional specialization of at least some insect interneurons. However it may not be possible to specify a neuron's role(s) with such precision and care should be exercised when assigning neurons to different motor or sensory systems (see e.g. Boyan 1984; Burrows 1985).

5.2. Jumping and flight

The central circuitry for orthopteran jumping and flight has been comparatively well worked out (for reviews see Pearson 1983; Pearson and O'Shea 1984; Rowell and Reichert 1985; Robertson and Pearson 1984, 1985b; Robertson 1985a, 1986; Reichert and Rowell 1986). Briefly, for the jump two controlling interneurons have been described. One of these (the C interneuron) is activated by sensory input and causes simultaneous excitation of tibial flexor and extensor motoneurons (Pearson and Robertson 1981), the other (the M interneuron) is activated by the same modalities of sensory input and inhibits flexor motoneurons (Pearson et al. 1980). Thus C is thought to underlie the co-contraction associated with cocking prior to the jump, whereas M is thought to underlie the relaxation of the flexor muscle which allows the tibia to extend rapidly executing the jump. For flight a number of interneurons have been described and, although none of them has been ascribed a well defined role, it has been possible to explain how certain features of the flight motor pattern produced by deafferented preparations may be centrally generated (Robertson and Pearson 1983, 1985a). There is nothing startling about any of the individual synaptic connections in either of these circuits. There is no evidence for electrical connections and both excitatory and inhibitory chemical connections are involved. Inhibitory connections predominate (but are not exclusive) among interneuron to interneuron connections and this accords with the observation that the known interactions among non-spiking interneurons in the locust thoracic ganglia are all inhibitory (Siegler 1985; Pearson 1985a). The operation of the circuits does depend on their particular conformations and three aspects of this are worth further attention: gating of information flow, burst generation, and mutability of the circuit.

5.2.1. Gating of information flow

The possession of a threshold for firing in an interneuron means that communication through that neuron via spikes to another neuron can be enabled or disabled by modifying its membrane potential (i.e. taking the membrane potential closer to or further away from threshold). Such a mechanism has been proposed as regulating the activity of the jump trigger neuron (M), and as ensuring that sensory input is effective only at the appropriate phase of each flight cycle.

The jump is triggered by inhibition of flexor motoneurons (Heitler and Burrows 1977a, b). Auditory and visual stimuli can cause spikes in an M neuron and ipsps in flexor motoneurons but only if they are presented concurrently with proprioceptive information from the leg indicating that the tibia is close to full flexion (Steeves and Pearson 1982). Given that the M neuron has a particularly high threshold for spiking, this has been interpreted as a means of ensuring that the M neuron does not fire, and thus possibly disrupt normal movements, unless the leg is cocked and ready to jump - proprioceptive gating of the signal to jump. The prediction from this interpretation would be that during the co-contraction prior to a jump the membrane potential of the M neuron would gradually increase in a ramp-like manner until it crossed threshold and triggered the jump. Recently it has been possible to test this prediction by recording intracellularly from M neurons during defensive kicks (employing the same motor pattern as the jump) (Gynther and Pearson 1986). Contrary to expectations, the M neuron is inhibited throughout the co-contraction period and receives a strong, central, excitatory input causing it to fire at the time the kick is triggered (Fig. 15). An M neuron will discharge in the same fashion in the absence of proprioceptive feedback. Thus it appears that proprioceptive gating may not play a major part in regulating when a jump is triggered and central circuitry has still to be investigated to understand this process. Another recent report questioning the accepted model for the jump shows that the activity of the DCMD neuron is suppressed during motor activity associated with all phases of a defensive kick: from initial flexion, through co-contraction, to about one second after the execution of a kick (Heitler 1983). Thus an important source of input to M may not be available to it to trigger the jump. These findings also serve as another example of the gating of activity in an interneuron.

Some multimodal sensory interneurons descending from the brain in the locust respond best to the combination of inputs which signal course deviations during flight (i.e. yawing movement to the right, etc.) (Reichert et al. 1985; Reichert and Rowell 1985). It has been proposed that activation of these interneurons may mediate the changes to the flight motor pattern that underlie compensatory steering movements. The descending interneurons fire tonically in response to their preferred course deviation and they make short latency excitatory connections both with wing muscle motoneurons and with premotor interneurons. The membrane potentials of these postsynaptic neurons are also driven by the central flight circuitry to oscillate with the flight rhythm and thus the information from the deviation detectors is transmitted to the wing muscles only at particular phases of each flight cycle - central gating of

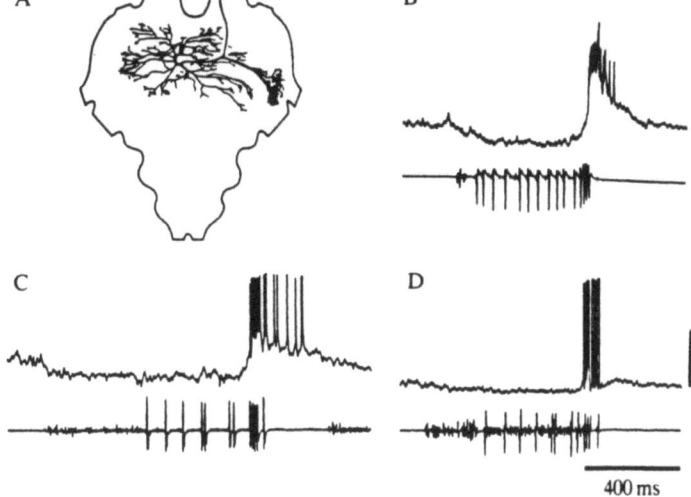

FIG. 15: Activity of the M-neuron during kicking in the locust. A. Structure of the M-neuron in the metathoracic ganglion. B-D. Three intracellular records of the activity of the M-neuron during a kick monitored with electromyograms from the flexor and extensor muscles (lower traces). In D the recording was from the large transverse process of M, near the spike initiation zone. Note in all cases that M is hyperpolarized during the cocontraction phase prior to the kick, and receives a large depolarizing input to cause a burst of spikes at the time that the kick is triggered. Vertical scale: B and D, 10mV; C, 20mV. From Gynther and Pearson (1986).

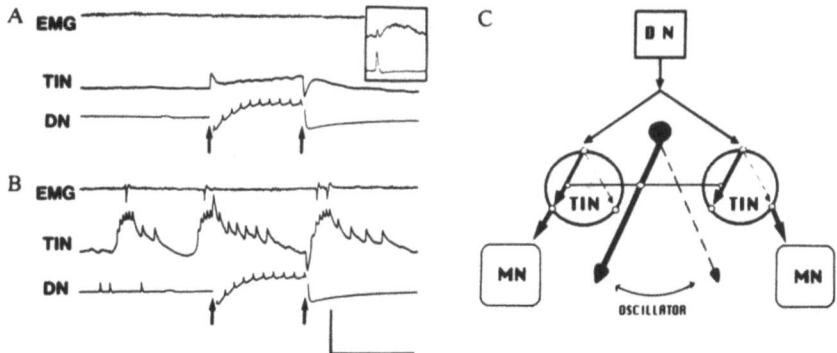

FIG. 16: Gating of a descending deviation signal in the flight system of the locust. A. Short latency connection (inset) from a descending interneuron (DN) to a thoracic premotor interneuron (TIN). A train of spikes in DN depolarizes TIN but does not cause it to spike. B. During concurrent flight activity (monitored electromyographically - EMG) the same train of spikes in DN is sufficient to change the activity of the TIN by prolonging its bust. C. Diagrammatic representation of the gating model. The flight oscillator is represented as a pendulum which switches the flow of information from the DNs to the flight motoneurons (MN) at the level of the premotor TINs. Vertical scales: TIN, 20mV; DN, 100mV. Horizontal scale: 100 ms. From Reichert and Rowell (1986).

a deviation signal. The function of this would be to ensure that the appropriate changes to the phase of activation of wing muscles, such as control steering, are efficiently produced. Because the connections from the descending neurons to the premotor interneurons are much stronger than their connections to motoneurons most of the gating is thought to occur at the interneuronal level (Reichert and Rowell 1985; Fig. 16). It is not yet clear that the observed mild effects of the descending interneurons on the spiking activity of the premotor interneurons are sufficient to change the phase of activity of wing muscles significantly during flight.

The above examples clearly illustrate the phenomenon of sensory and central gating of information flow but have not established its importance in the normal behavior of the animal. However, it would be surprising if such processes were not intimately involved in the control of normal behavior for they are the very basis of neural integration.

5.2.2. Burst generation

Many intrinsic neuronal, synaptic and network properties have been implicated in the generation of rhythmical or bursting activity (Kristan 1980; Getting 1986). Our knowledge of their existence, or absence, in insect nervous systems is extremely rudimentary. It is only for the flight system of locusts that there is any information at all on the possible neural mechanisms underlying bursting. Membrane potential oscillations of flight motoneurons and interneurons are caused by phasic excitatory and inhibitory synaptic inputs. There is no evidence to date for any intrinsic membrane property that might contribute to bursting in this system. Similarly, there is no evidence for special synaptic properties that could aid in generating bursts in a network (i.e. multicomponent synapses). However, circuits capable of burst generation have been described (Robertson and Pearson 1985a).

At the heart of the network of known connections of flight interneurons lies a circuit of delayed excitation and feedback inhibition between two neurons (interneurons 301 and 501) (Fig. 17). The delayed excitatory connection from 301 to 501 is thought to be mediated by a disynaptic disinhibition through an intercalated interneuron with tonic and graded release of transmitter at the second synapse (thus the three neurons could constitute a circuit of recurrent cyclic inhibition, e.g. Friesen and Stent 1977). The feedback inhibition is via a standard short latency inhibitory connection. Activity in 301 causes firing of 501 after a delay and this terminates the activity of 301 thus removing the excitation from 501. With a tonic excitatory input to 301 this circuit is capable of generating bursts, as are model circuits of delayed excitation and feedback inhibition (Getting 1983). Another circuit which is commonly invoked to explain rhythmicity in other systems is that of reciprocal inhibition, although this needs some other intrinsic or synaptic property to ensure the switch from one of the elements to the other (Perkel and Mulloney 1974; Satterlie 1985). Simultaneous recordings from identified flight interneurons (R.M. Robertson, unpublished) show that a circuit of reciprocal inhibition exists which could contribute to burst generation (Fig. 18).

It is becoming increasingly obvious in other systems (e.g. Selverston 1980; Robertson and Moulins 1981, 1984; Miller and Selverston 1982a, b) that attempts to locate a single, dominating, oscillator for a motor rhythm are doomed to failure. Redundancy of oscillatory mechanisms seems to be the rule. This is evident in the locust flight system also. Although the network that has been described centres around a circuit of delayed excitation and inhibitory feedback this can be disrupted (e.g.

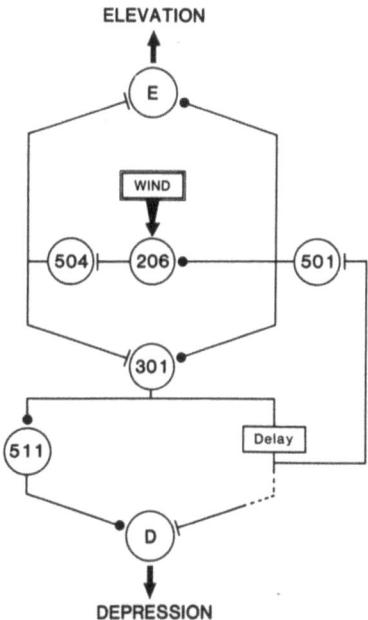

FIG. 17: Summary diagram of connections among flight neurons in the locust. D - depressor motoneurons, E - elevator motoneurons. Interneurons are represented by identifying numbers. Filled circles - inhibitory connections, 'T' -bars - excitatory connections. The delayed excitatory connection (represented by a delay box in an excitatory connecting line) is probably mediated by disynaptic disinhibition. From Robertson and Pearson (1985a).

with picrotoxin, Robertson and Pearson, 1985a) without totally abolishing rhythmicity. Furthermore there are numerous unpublished observations (e.g. G. Hoyle, personal communication; H. Wolf and K.G. Pearson, personal communication) that the isolated mesothoracic or the isolated metathoracic ganglion can produce rhythmical motor activity despite the fact that the described circuit spans the two ganglia. This redundancy can be demonstrated by stimulating several different interneurons to evoke a rhythm. Stimulating a different interneuron changes the nature of the rhythm that can be recorded as well as changing the subset of interneurons that are actively involved in generating the rhythm (Fig. 19).

Almost everything remains to be learned about burst generation in insect motor systems but it seems likely that each system will have a unique mix of overlapping, redundant mechanisms of the sort that are well known from previous studies with motor systems in other animals.

5.2.3. Mutability of the circuit

Major changes in circuitry can occur during metamorphosis and development (Levine and Truman 1982; Levine 1984, 1986), although in other cases circuitry is unaltered in spite of having the opportunity for a rearrangement at metamorphosis (Weeks and Truman 1984a, b). Even in adults, the circuits that can be described by intracellularly recording from different elements in a particular preparation should not be regarded as rigidly hard-wired. Much evidence points to a considerable flexibility

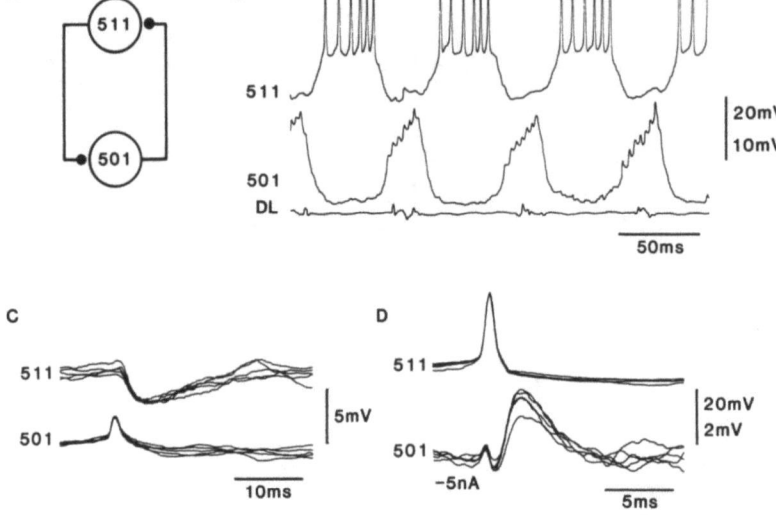

FIG. 18: Reciprocal inhibition in the flight system of the locust. A. Summary diagram of the connections demonstrated. B. Simultaneous intracellular recording of interneurons 511 and 501 during expression of the flight rhythm. The time of wing depressor activity is monitored electromyographically from the dorsal longitudinal muscles (DL). C. 5 superimposed sweeps of the oscilloscope to show the ipsp recorded in 511 a short and constant latency after a spike in 501. D. The reciprocal ipsp has been reversed, and is depolarizing, in this recording by passing about 5 nA of hyperpolarizing current into 501. This was because the membrane potential of 501 was close to the reversal potential of the ipsp which made it difficult to discern. The small membrane potential fluctuations just prior to the reversed ipsp in 501 are due to capacitative coupling with the other electrode. R.M. Robertson (unpublished).

of operation at the level of the synapse and the neuron. The nature of the reflexes elicited by stimulation of a leg proprioceptor depends upon the behavioral context (Zill and Forman 1983; Zill 1985) indicating shifting strengths of central connections. These are probably mediated by nonspiking premotor interneurons in the thoracic ganglia. Siegler (1981a, b) has shown that the membrane potential of these nonspiking neurons and the strength of their connections to motoneurons are strongly dependent on proprioceptive information and the history of prior movement. These subtle alterations in the central circuits are thought to allow different changes to the motor output depending on concurrent postures or behaviors. Changes in the strength of central connections can also be affected by the concentration of circulating neuromodulators. For example octopamine has been shown to potentiate excitatory reflexes in the locust (Sombati and Hoyle 1984a) and the moth (Claassen and Kammer 1986).

The role of phasic sensory input during flight has been a contentious issue for a long time. An early idea was that each phase of the cycle depended upon the sensory feedback caused by the previous movement (Weis-Fogh 1956). Subsequently the demonstration that a flight rhythm could be produced by a deafferented preparation (Wilson 1961) supported the central pattern generator concept whereby the motor patterns underlying behavior were centrally generated and sensory input played only a marginal,

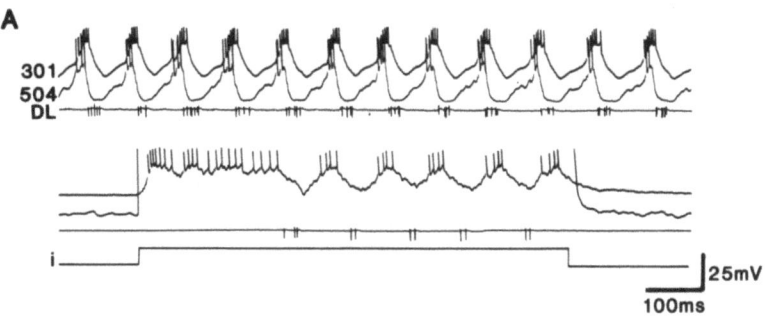

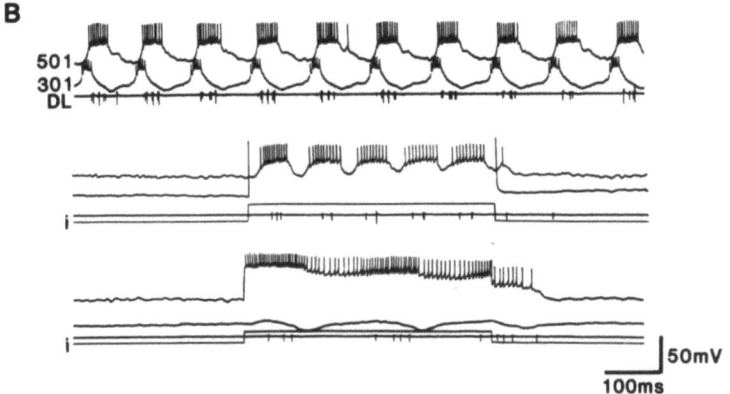

FIG. 19: Rhythm evoked by stimulating different flight interneurons in the locust. A. Top record shows normal flight activity. The bottom record shows the rhythm induced by stimulating interneuron 504. B. Top record shows normal flight activity. Middle record shows effect of stimulating 301 with a pulse of current. Bottom trace shows effect of stimulating 501 with a pulse of current. In all cases flight activity is monitored electromyographically from the dorsal longitudinal muscles (DL) and the trace labelled (i) indicates the duration of the current pulse. Note that in each case the rhythm induced is different and the subset of neurons that are active is different. For example, from this and other evidence (not shown), when a rhythm is activated by stimulating 504, 301 and 501 are also active; when a rhythm is activated by stimulating 301, 504 is not active but 501 is; when a rhythm is activated by stimulating 501, neither 504 nor 301 are active. Refer to Fig. 17. Modified from Robertson and Pearson (1985a, and R.M. Robertson unpublished).

modulatory role (Delcomyn 1980). For locust flight this idea was by no means universally accepted (Wendler 1983a, b, 1985; Möhl 1985) but it was plain that a central circuit could be identified and the belief was that it would play a major part in generating the behavior. Pearson (1985b) has recently cast doubt on this belief, and, indeed, on the central pattern generator concept. His thesis is that phasic sensory input has the ability to change the circuit of active neurons (i.e. by adding some neurons to the circuit, and by removing others) so completely as to render the central component meaningless as a functional unit. The present evidence for this is inconclusive. Elevator motoneurons receive an excitatory input in the intact animal which they do not in the deafferented preparation and this can be attributed to phasic sensory

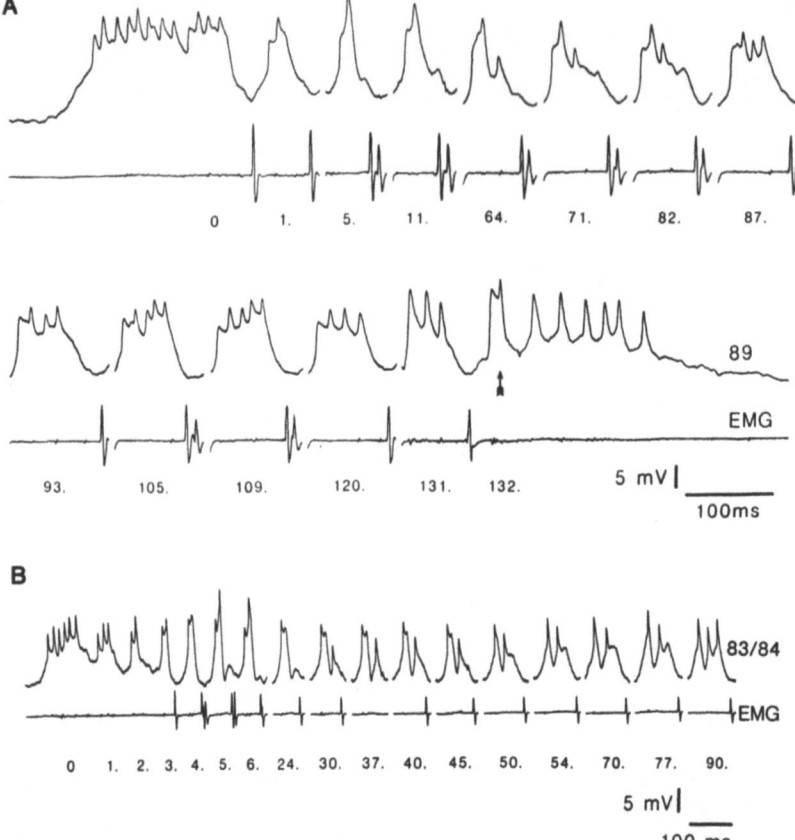

FIG. 20: Intracellular recordings from wing motoneurons during flight in the almost intact locust. A. Intracellular recording from an anterior tergocoxal motoneuron (89) during flight. Time of depression monitored with an electromyographic electrode (EMG). B. Intracellular recording from a tergosternal motoneuron (83/84) during flight. The records are not continuous and the cycle number is shown underneath the traces. In deafferented preparations the slope of the depolarization leading to spikes in elevator motoneurons tends to be shallow. In these examples taken from almost intact locusts the depolarization is very rapid. In some cases the rapid component is separate from a slower depolarization (arrow an A). Note two phases of depolarizing input (rapid and slow) in each cycle as the flight sequence continues (cycles after the 30th in B). From Wolf and Pearson (1987).

feedback (Wolf and Pearson 1987; Fig. 20). It is not yet clear if this input is a result simply of advancing a central component or a direct effect of the feedback which therefore sidesteps the central component. If the latter, then part of the motor pattern could be more the result of feedback than of any central circuit. This issue is still unresolved but three points can be made. First, it is resolvable, due to the ability to record from flight neurons in an essentially intact animal. Second, to understand the neural mechanisms whereby sensory input exerts its effects, rather than simply generating elaborate descriptions of the effects themselves, will require an intimate knowledge of the central component. Third, given the range of environmental and behavioral contexts likely to be encountered, and in view of the alterations of synaptic efficacy discussed above, any conclusion about an interneuron's importance or otherwise in flight motor pattern generation should be made with extreme caution.

6. CONCLUSIONS AND SPECULATION

Do insect nervous systems work differently from those of other animals? As one might expect the answer to this is... not in any major way. There are differences in relative abundance of particular phenomena and some phenomena have not yet been described for insect preparations. For many neural mechanisms this undoubtedly indicates more a lack of information than a fundamental difference in operation. Features that are charateristic of insects include: 1) a particularly efficient blood-brain barrier; 2) striking and very characteristic neuronal morphologies; 3) the minor role played by electrical interactions; and 4) the major role played by graded, nonspiking chemical interactions. The motor patterns of behavior are produced by nearly universal, well-described neural phenomena.

As intracellular studies on insect neurons continue it is to be expected that a lot of information will accumulate on their intrinsic and synaptic properties. For example the only example of a conductance which could remotely be considered as contributing to endogenous burstiness is a slow Na^+ conductance which can give rise to plateau-like potentials in cockroach giant axons in the presence of TEA (Yawo et al. 1984). It would be surprising if some other active properties of neurons contributing to the generation of behavior were not described. Perhaps this is more likely to be in the ventilatory system, for example, than the flight system due to the continuous nature of the production of ventilatory activity. Another field which is liable to bear fruit is the study of graded interactions. Electron microscope studies indicate a considerable wealth of synaptic interaction from and to identified motoneurons and interneurons in the neuropile (Watson and Burrows 1982, 1983). There is no theoretical opposition to functional local circuits but a physiological demonstration is still missing. Can spiking, intersegmental interneurons affect other neurons via graded release from their dendrites or at their axon terminals? It seems probable that answers to these and related questions will soon appear.

Insect ganglia are highly organized and distinct tracts and commissures can be identified (e.g. Tyrer and Gregory 1982; see also Nässel 1987a). Does this have any physiological concomitants? It may be possible to predict physiological characteristics from a knowledge of an interneuron's morphology (e.g. Bacon and Murphey 1984; Pearson and Robertson 1987), but there is another possibility. The extracellular field potentials that are recorded in the ganglia can be quite large and could affect the transmembrane potential of neurons in their vicinity as they have been shown to do in the compound eye. It is conceivable that these sorts of interactions play a significant part in the normal functioning of the nervous system.

One question that can be asked is how relevant is all of the complexity that can be described in the nervous system. For example, do the interactions mediated by high extracellular concentrations of K^+ have a significant role in generating or controlling the normal behavior of the animal? Is the possession of four spike initiating zones by DUMETi really important? The simple answer to this is that the complexity is necessary but the reasons for such fine-tuning of the control are currently unknown - sophisticated mechanisms in search of precise functions. An opposing idea is that much of the complexity is functionless at best or an impediment at worst. Specific features could be epiphenomena of vital processes selected for other reasons (e.g. a spiking soma being the result of selection for long duration action potentials). Alternatively, some features of the nervous system may be the result of nonadaptive determinants of current form and function. Thus what is described in the present may result more from a system's prior history, or from its ontogeny, or from the constraints imposed by the materials used, than from optimal design for a particular function which could be determined (Levins and Lewontin 1985; Dumont and Robertson 1986). For these reasons the organization of interneurons in the locust flight system has been attributed to its evolutionary history rather that to a possible functional adaptation specifically for flight (Robertson et al. 1982). In addition, it has been shown at the molecular level that selectively neutral mutations can occur and become passively fixed via random genetic drift (King and Jukes 1969). Mutations with no overt effect on physiology and/or behavior will not be perceived by natural selection and therefore cannot be selected against. Thus the importance of unexplained and intriguing features may lie not in their current role but in the fact that they are the substrate for subsequent evolution for they can be altered and built upon without unduly affecting the working of the nervous system.

One of the functions of sensory input to motor systems is to adapt the motor output to changes in the environment. Another one is to adapt the motor output to changes in the state of peripheral structures as an animal ages (Pearson 1981). It may be that evolutionary adaptations to the motor output, such as fit an animal to a particular life-style, are most easily accomplished by modifying the effect and role of sensory input. There are at least two possible reasons why features of the central nervous system may be evolutionarily less modifiable than features at the periphery. The first is that they may not be as exposed to selection pressure because their effect on the phenotype exposed to selection is more indirect. The second is that the development of the central nervous system is necessarily very complex, so that vital features, and features that are nonadaptive for a particular behavior, are likely to be interdependent. Mutations with the requisite limited scope may be particularly rare. Seen from this perspective it is perhaps not surprising that part of the central component generating the deafferented locust flight rhythm may be sidestepped by phasic sensory input (Wolf and Pearson 1987) and thus that the central pattern generator has no real role in producing the behavior of flight (Pearson 1985b). However this perspective does suggest the testable prediction that the central pattern generators for flight of diverse insects are similar in their organization and operation.

Accepting the fact that the physiology of insect nervous systems differs little from those of other animals, for which areas of research can they serve as model systems? There are many experimental approaches which are difficult, although not impossible, to take with insects. Studying the biophysics of interneurons is not easy. Attempts to culture insect neurons have not met with much success. The blood-brain barrier makes many ion-substitution and pharmacological experiments tedious and hard to interpret. The central role of neuromodulators and the cellular and molecular bases of learning are perhaps best studied elsewhere. It is well established that insects have unique advantages for the studies of the development of nervous tissue and the genetic control of behavior. In my opinion insects offer the best preparations for studying the structure and function of individual neurons and the cellular bases of behavior. Theirs is a complex, highly ordered nervous system and it is superbly accessible to microelectrode study even in almost intact animals. Moreover intracellular recordings can often be taken while the insect continues to behave. Insect preparations are ideally suited to at least two specific fields: the role and extent of local interactions in the central nervous system, and the central processing and role of sensory input in generating and controlling motor output patterns. Finally, I think that it is within the class Insecta that it might be possible to begin to understand how motor systems have evolved into their present conformations.

7. ACKNOWLEDGEMENTS

I thank G. Atkins, R. Chase, K.G. Pearson, G.S. Pollack and R.M. Olberg for their comments on a previous draft of this article. Supported by the Natural Sciences and Engineering Research Council of Canada.

8. REFERENCES

Adams ME, O'Shea M (1983) Peptide co-transmitter at a neuromuscular junction. Science 221: 286-289.

Altman JS (1987) How the nervous system decides what the insect will do next. (This volume).

Ammermüller J (1986) Passive cable properties of locust ocellar L-neurons. J Comp Physiol 158: 339-344.

Bacon JP (1987) The insect nervous system: in vivo and in vitro development. (This volume).

Bacon JP, Altman JS (1977) A silver intensification method for cobalt-filled neurones in whole-mount preparations. Brain Res 138: 359-363.

Bacon JP Murphey RK (1984) Receptive fields of cricket giant inter-neurones are related to their structure. J Physiol 352: 601-623.

Bacon JP, Strausfeld NJ (1986) The dipteran "Giant fibre" pathway: neurons and signals. J Comp Physiol 158: 529-548.

Bacon JP, Tyrer NM (1978) The tritocerebral commissure giant (T.C.G.): a bimodal interneurone in the locust, Schistocerca gregaria. J Comp Physiol 126: 317-325.

Bentley DR (1969a) Intracellular activity in cricket neurons during the generation of behaviour patterns. J Insect Physiol 15: 677-699.

Bentley DR (1969b) Intracellular activity in cricket neurons during generation of song patterns. Z vergl Physiol 62: 267-283.

Berry MS, Pentreath VW (1976) Criteria for distinguishing between monosynaptic and polysynaptic transmission. Brian Res 105: 1-20.

Blagburn JM, Beadle DJ, Sattelle DB (1984) Synapses between an identified giant inteneurone and a filiform hair sensory neurone in the terminal ganglion of first instar cockroaches (Periplaneta americana L.) J Exp Biol 113: 477-481.

Boeckh J, Ernst KD, Sass H, Waldow U (1984) Anatomical and physiological characteristics of individual neurones in the central antennal pathway of insects. J Insect Physiol 30: 15-26.

Boistel J, Coraboeuf E (1954) Potential de membrane et potentials d'action de nerf d'insecte recuellis à l'aide de microélectrodes intracellulaires. CR Acad Sci, Paris 238: 2116-2118.

Boyan GS (1980) Auditory neurones in the brain of the cricket Gryllus bimaculatus (De Geer). J Comp Physiol 140: 81-93.

Boyan GS (1984) What is an "auditory" neurone. Naturwiss 71: 482.

Boyan GS, Altman JS (1985) The suboesophageal ganglion: a "missing link" in the auditory pathway of the locust. J Comp Physiol 156: 413-428.

Boyan GS, Fullard JH (1986) Interneurones responding to sound in the tobacco budworm moth Heliotis virescens (Noctuidae): morphological and physiological characteristics. J Comp Physiol 158: 391-404.

Bräunig P (1985) Mechanoreceptive neurons in an insect brain. J Comp Neurol 236: 234-240.

Bräunig P, Hustert R (1980) Proprioceptors with central cell bodies in insects. Nature 283: 768-770.

Breer H, Heilgenberg H (1985) Neurochemistry of GABAergic activities in the central nervous system of Locusta migratoria. J Comp Physiol 157: 343-354.

Breer H, Kleene R, Hinz G (1985) Molecular forms and subunit structure of the acetylcholine receptor in the central nervous system of insects. J Neurosci 5: 3386-3392.

Bullock TH, Horridge GA (1965) Structure and function in the nervous systems of invertebrates. WH Freeman and Co., Sans Francisco, London, 2 Vol.

Burrows M (1975) Monosynaptic connexions between wing stretch receptors and flight motoneurones of the locust. J Exp Biol 62: 189-219.

Burrows M (1977) Flight mechanisms of the locust. In: Hoyle G (ed) Identified neurons and behavior of Arthropods. Plenum Press, New York, London. pp 339-356.

Burrows M (1978) Local interneurones and integration in locust ganglia. Verh Dtsch Zool Ges 1978: 68-79.

Burrows M (1979) Sources of variation in the output of locust spiracular motoneurones receiving common synaptic driving. J Exp Biol 74: 175-186.

Burrows M (1980) The control of sets of motoneurones by local interneurones in the locust. J Physiol 298: 213-233.

Burrows M (1981) Local interneurones in insects. In: Roberts A, Bush BMH (eds) Neurones without impulses. Cambridge University Press, Cambridge UK pp 199-221.

Burrows M (1982) Interneurones co-ordinating the ventilatory movements of the thoracic spiracles in the locust. J Exp Biol 97: 385-400.

Burrows M (1985) Nonspiking and spiking local interneurons in the locust. In: Selverston AI (ed) Model neural networks and behavior Plenum Press, New York, London. pp 109-125.

Burrows M, Siegler MVS (1976) Transmission without spikes between locust interneurones and motoneurones. Nature 262: 222-224.

Burrows M, Siegler MVS (1978) Graded synaptic transmission between local interneurones and motor neurones in the metathoracic ganglion of the locust. J Physiol 285: 231-255.

Burrows M, Siegler MVS (1984) The morphological diversity and receptive fields of spiking local interneurones in the locust metathoracic ganglion. J Comp Neurol 224: 483-508.

Burrows M, Boeckh J, Esslen J (1982) Physiological and morphological characteristics of interneurones in the deutocerebrum of male cockroaches which respond to female pheromone. J Comp Physiol 145: 447-457.

Callec J-J (1974) Synaptic transmission in the central nervous system of insects. In: Treherne JE (ed) Insect neurobiology. Elsevier, Amsterdam. pp 119-178.

Callec J-J (1985) Synaptic transmission in the central nervous system. In: Kerkut GA, Gilbert LI (eds) Comprehensive insect physiology, biochemistry and pharmacology. Pergamon Press, New York. Vol. 5, pp 139-179.

Carr CE, Fourtner CR (1980) Pharmacological analysis of a monosynaptic relex in the cokroach, Periplaneta americana. J Exp Biol 86: 259-273.

Carrow GM, Calabrese RL, Williams CM (1984) Architecture and physiology of insect cerebral neurosecretory cells. J Neurosci 4: 1034-1044.

Casaday GB, Hoy RR (1977) Auditory interneurons in the cricket Teleogryllus oceanicus - physiological and anatomical properties. J Comp Physiol 121: 1-13.

Claassen DE, Kammer AE (1986) Effects of octopamine, dopamine, and serotonin on production of flight motor output by thoracic ganglia of Manduca sexta. J Neurobiol 17: 1-14.

Cokl A, Otto C, Kalmring K (1985) The processing of directional vibratory signals in the ventral nerve cord of Locusta migratoria. J Comp Physiol 156: 45-52.

Coles JA, Tsacopoulos M (1981) Ionic and possible metabolic interactions between sensory neurones and glial cells in the retina of the honeybee drone. J Exp Biol 95: 75-92.

Collin SP (1985) The central morphology of the giant interneurons and their spatial relationships with the thoracic motorneurons in the cockroach, Periplaneta americana (Insecta). J Neurobiol 16: 249-267.

Copenhaver PF, Truman JW (1986) Identification of the cerebral neurosecretory cells that contain eclosion hormone in the moth Manduca sexta. J Neurosci 6: 1738-1747.

Crossman AR, Kerkut GA, Pitman RM, Walker RJ (1971) Electrically excitable nerve cell bodies in the central ganglia of two insect species, Periplaneta americana and Schistocerca gregaria. Investigation of cell geometry and morphology by intracellular dye injection. Comp Biochem Physiol 40: 579-596.

Cull-Candy SG (1982) Properties of postsynaptic channels activated by glutamate and GABA in locust muscle fibres. In: Neuropharmacology of Insects. Ciba Foundation Symposium 88. Pitman Books, London. pp 70-87.

Daley DL, Vardi N, Appignami B, Camhi JM (1981) Morphology of giant interneurons and cercal nerve projections of the american cockroach. J Comp Neurol 196: 41-52.

Delcomyn F (1980) Neural basis of rhythmic behavior in animals. Science 210: 492-498.

Delcomyn F (1983) Activity and structure of movement-signalling (corollary discharge) interneurons in a cockroach. J Comp Physiol 150: 185-193.

DeVoe RD, Kaiser W, Ohm J, Stone LS (1982) Horizontal movement detectors of honeybees: directionally-selective visual neurons in the lobula and brain. J Comp Physiol 147: 155-170.

Dumont JPC, Robertson RM (1986) Neuronal circuits: an evolutionary perspective. Science 233 : 849-853.

Eckert H (1982) The vertical-horizontal neurone (VH) in the lobula plate of the blowfly, Phaenicia. J Comp Physiol 149: 195-205.

Erber J (1983) The search for neural correlates of learning in the honeybee. In: Huber F, Markl H (eds) Neuroethology and behavioral physiology. Springer-Verlag, Berlin, Heidelberg. pp 216-227.

Ernst KD, Boeckh J (1983) A neuroanatomical study on the organization of the central antennal pathways in insects. III. Neuroanatomical characterization of physiologically defined response types of deutocerebral neurons in Periplaneta americana. Cell Tissue Res 229: 1-22.

Evans PD (1980) Biogenic amines in the insect nervous system. Adv Insect Physiol 15: 317-473.

Evans PD (1984) The role of cyclic nucleotides and calcium in the mediation of the modulatory effects of octopamine on locust skeletal muscle. J Physiol 348: 325-340.

Evans PD, O'Shea M (1977) An octopaminergic neurone modulates neuromuscular transmission on the locust. Nature 270: 257-259.

Friesen WO, Stent GS (1977) Generation of a locomotory rhythm by a neural network with recurrent cyclic inhibition. Biol Cybernet 28: 27-40.

Gerschenfeld HM (1973) Chemical transmission in invertebrate central nervous systems and neuromuscular junctions. Physiol Rev 53: 1-19.

Getting PA (1981) Mechanisms of pattern generation underlying swimming in Tritonia. I. Neuronal network formed by monosynaptic connections. J Neurophysiol 46: 65-79.

Getting PA (1983) Mechanisms of pattern generation underlying swimming in Tritonia. II. Network reconstruction. J Neurophysiol 49: 1017-1035.

Getting PA (1986) Comparative analysis of invertebrate central pattern generators. In: Cohen AH, Rossignol S, Grillner S (eds) Neural control of rhythmic movements. John Wiley and Sons, Inc., New York. in press.

Goodman CS (1976) Anatomy of the ocellar interneurones of acridid grasshoppers. I. The large interneurones. Cell Tissue Res 175: 166-183.

Goodman CS, Heitler WJ (1979) Electrical properties of insect neurones with spiking and non-spiking somata: normal, axotomized, and colchicine-treated neurones. J Exp Biol 83: 95-121.

Goodman CS, Spitzer NC (1979) Embryonic development of identified neurones: differentiation from neruoblast to neurone. Nature 280: 208-213.

Goodman CS, Williams JLD (1976) Anatomy of the ocellar interneurones of acridid grasshoppers. II. The small interneurones. Cell Tissue Res 175: 203-225.

Goodman CS, Pearson KG, Heitler WJ (1979) Variability of identified neurons in grasshoppers. Comp Biochem Physiol 64: 455-462.

Goodman CS, Pearson KG, Spitzer NC (1980) Electrical excitability: a spectrum of properties in the progeny of a single embryonic neuroblast. Proc Natl Acad Sci USA 77: 1676-1680.

Gwilliam GF, Burrows M (1980) Electrical characteristics of the membrane of an identified insect motor neurone. J Exp Biol 86: 49-61.

Gynther IC, Pearson KG (1986) Intracellular recording from interneurones and motoneurones during bilateral kicks in the locust: implications for mechanisms controlling the jump. J Exp Biol 122: 323-343.

Hagiwara S, Watanabe A (1956) Discharges in motoneurons of cicadas. J Cell Comp Physiol 47: 415-428.

Hanke W, Breer H (1986) Channel properties of an insect neuronal acetylcholine receptor protein reconstituted in planar lipid bilayers. Nature 321: 171-174.

Harris CL, Garrison W (1976) Electronic coupling between cercal afferents and giant interneurons in the American cockroach. J Insect Physiol 22: 31-40.

Harrow ID, Hue B, Pelhate M, Sattelle DB (1980) Cockroach giant interneurones stained by cobalt-backfilling of dissected axons. J Exp Biol 84: 341-343.

Harrow ID, David JA, Sattelle DB (1982) Acetylcholine receptors of identified insect neurons. In: Neuropharmacology of Insects. Ciba Foundation Symposium #88. Pitman, London. pp 12-31.

Hausen K (1982) Motion sensitive interneurons in the optomotor system of the fly. I. Horizontal cells: structure and signals. Biol Cybernet 45: 143-156.

Hedwig B (1986a) On the role in stridulation of plurisegmental interneurons of the acridid grasshopper Omocestus viridulus L. I. Anatomy and physiology of descending cephalothoracic interneurons. J Comp Physiol 158: 413-427.

Hedwig B (1986b) On the role in stridulation of plurisegmental interneurons of the acridid grasshopper Omocestus viridulus L. II. Anatomy and physiology of ascending and T-shaped interneurons. J Comp Physiol 158: 429-444.

Heitler WJ (1983) Suppression of a locust visual interneurone (DCMD) during defensive kicking. J Exp Biol 104: 203-215.

Heitler WJ, Burrows M (1977a) The locust jump. I. The motor programme. J Exp Biol 66: 203-219.

Heitler WJ, Burrows M (1977b) The locust jump. II. Neural circuits of the motor programme. J Exp Biol 66: 221-241.

Heitler WJ, Goodman CS (1978) Multiple sites of spike initiation in a bifurcating locust neurone. J Exp Biol 76: 63-84.

Hengstenberg R, Hausen K, Hengstenberg B (1982) The number and structure of giant vertical cells (VS) in the Lobula plate of the blowfly Calliphora erythrocephala. J Comp Physiol 149: 163-177.

Hertel H (1980) Chromatic properties of identified interneurons in the optic lobes of the bee. J Comp Physiol 137: 215-231.

Hester R (1986) From insects to robotics. Byte (letters column) 11: 26.

Hildebrand JG (1982) Chemical signalling in the insect nervous system. In: Neuropharmacology of Insects. Ciba Foundation Symposium #88. Pitman, London. pp 5-11.

Hisada M, Takahata M, Nagayama T (1984) Local non-spiking interneurons in the arthropod motor control systems: characterization and their functional significance. Zool Sci 1: 681-700.

Homberg U (1985) Interneurones of the central complex in the bee (Apis mellifera, L.). J Insect Physiol 31: 251-264.

Hörner M, Gras H (1985) Physiological properties of some descending neurons in the cricket brain. Naturwiss 72: 603.

Horridge GA, Marcelja L, Jahnke R, Matic T (1983) Single electrode studies on the retina of the butterfly Papilio. J Comp Physiol 150: 271-294.

Hoyle G (1953) Potassium ions and insect nerve muscle. J Exp Biol 30: 121-135.

Hoyle G (1970) Cellular mechanisms underlying behavior - neuroethology. Adv Insect Physiol 7: 349-444.

Hoyle G (ed) (1977) Identified neurons and behavior of Arthropods. Plenum Press, New York, London.

Hoyle G (1983) On the way to neuroethology: the identified neuron approach. In: Huber F, Markl H (eds) Neuroethology and behavioral physiology. Springer-Verlag, Berlin, Heidelberg. pp 9-25.

Hoyle G (1986) Glial cells of an insect ganglion. J Comp Neurol 246: 85-103.

Hoyle G, Burrows M (1973) Neural mechanisms underlying behavior in the locust Schistocerca gregaria. I. Physiology of identified motorneurons in the metathoracic ganglion. J Neurobiol 4: 3-41.

Hoyle G, Dagan D, Moberly B, Colquhoun W (1974) Dorsal unpaired median insect neurones make neurosecretory endings on skeletal muscle. J Exp Zool 187: 159-165.

Hoyle G, Williams M, Phillips C (1986) Functional morphology of insect neuronal cell-surface/glial contacts: the trophospongium. J Comp Neurol 246: 113-128.

Hue B, Callec JJ (1983) Presynaptic inhibition in the cercal-afferent giant-interneurone synapses of the cockroach Periplaneta americana. J Insect Physiol 29: 741-748.

Hustert R (1985) Multisegmental integration and divergence of afferent information from single tactile hairs in a cricket. J Exp Biol 118: 209-227.

Jan YN, Jan LY (1985) Genetic and molecular studies of a potassium channel gene in Drosophila. In: Selverston AI (ed) Model neural networks and behavior. Plenum Press, New York, London. pp 537-546.

Järvilehto M, Zettler F (1971) Localised intracellular potentials from pre- and postsynaptic components in the external plexiform layer of an insect retina. Z Vergl Physiol 75: 422-440.

Jego P, Callec JJ, Pichon Y, Boistel J (1970) Etude électrophysiologique de corps cellulaires excitables de VIème ganglion abdominal de Periplaneta americana: aspects électriques et ioniques. CR Hedb Seanc Acad Sci, Paris 164: 893-904.

Kerkut GA, Gilbert LI (eds) (1985) Comprehensive insect physiology, biochemistry and pharmacology. Pergamon Press, Oxford, New York. 13 Volumes.

Kerkut GA, Pitman RM, Walker RJ (1969) Sensitivity of the insect central nervous system to iontophoretically applied acetylcholine and GABA. Nature 222: 1075-1076.

Kien J, Altman JS (1984) Descending interneurones from the brain and sub-oesophagel ganglia and their role in the control of locust behavior. J Insect Physiol 30: 59-72.

Kien J, Williams M (1983) Morphology of neurons in locust brain and suboesophageal ganglion involved in initiation and maintenance of walking. Proc R Soc Lond B 219: 175-192.

King JL, Jukes TH (1969) Non-Darwinian evolution. Science 164: 788-798.

Koenig JH, Ikeda K (1983) Reciprocal excitation between identified flight motor neurons in Drosophila and its effect on pattern generation. J Comp Physiol 150: 305-317.

Komatsu A (1984) Ascending interneurons that convey a respiratory signal in the central nervous system of the dragonfly larva. J Comp Physiol 154: 331-340.

Kristan WB (1980) Generation of rhythmic motor patterns. In: Pinsker HM, Willis WD (eds) Information processing in the nervous system. Raven Press, New York. pp 241-261.

Lane NJ (1981a) Tight junctions in arthropods. Int Rev Cytol 73: 243-318.

Lane NJ (1981b) Invertebrate neuroglia - junctional structure and development. J Exp Biol 95: 7-33.

Laughlin SB (1973) Neural integration in the first optic neuropile of dragonflies. I. Signal amplification in dark-adapted second order neurons. J Comp Physiol 84: 335-355.

Leech CA, Treherne JE (1984) Growth and ion-specificity of excitability in regenerating cockroach giant interneurones. J Exp Biol 110: 311-318.

Levine RB (1984) Changes in neuronal circuits during insect metamorphosis. J Exp Biol 112: 27-44.

Levine RB (1986) Reorganization of the insect nervous system during metamorphosis. Trends Neurosci 9: 315-319.

Levine RB, Murphey RK (1980) Pre- and postsynaptic inhibition of identified giant interneurons in the cricket (Acheta domesticus). J Comp Physiol 135: 269-282.

Levine RB, Truman JW (1982) Metamorphosis of the insect nervous system: changes in morphology and synaptic interactions of identified neurons. Nature 299: 250-252.

Levins R, Lewontin R (1985) The dialectical biologist. Chapter 2. Adaptation. Harvard University Press, Cambridge Mass. pp 65-84.

Marquart V (1985) Local interneurons mediating excitation and inhibition onto ascending neurons in the auditory pathway of grasshoppers. Naturwiss 72: 42.

Matsumoto SG, Hildebrand JG (1981) Olfactory mechanisms in the moth Manduca sexta - response characteristics and morphology of central neurons in the antennal lobes. Proc R Soc Lond B 213: 249-277.

Meiri H, Spira ME, Parnas I (1981) Membrane conductance and action potential of a regenerating axonal tip. Science 211: 709-712.

Mendenhall B, Murphey RK (1974) The morphology of cricket giant interneurons. J Neurobiol 5: 565-580.

Meyer MR, Edwards JS (1980). Muscarinic cholinergic binding sites in an orthopteran nervous system. J Neurobiol 11: 215-219.

Milde JJ (1984) Ocellar interneurons in the honeybee. Structure and signals of L-neurons. J Comp Physiol 154: 683-693.

Miller JP, Jacobs GA (1984) Relationships between neuronal structure and function. J Exp Biol 112: 129-145.

Miller JP, Selverston AI (1982a) Mechanisms underlying pattern generation in lobster stomatogastric ganglion as determined by selective inactivation of identifed neurons II. Oscillatory properties of pyloric neurons. J Neurophysiol 48: 1378-1391.

Miller JP, Selverston AI (1982b) Mechanisms underlying pattern generation in lobster stomatogastric ganglion as determined by selective inactivation of identifed neurons. IV. Network properties of pyloric system. J Neurophysiol 48: 1416-1432.

Miyazaki S (1980) The ionic mechanism of action potentials in neurosecretory cells and non-neurosecretory cells of the silkworm. J Comp Physiol 140: 43-52.

Mizunami M, Yamashita S, Tateda H (1982) Intracellular stainings of the large ocellar second order neurons in the cockroach. J Comp Physiol 149: 215-219.

Möhl B (1985) Sensory aspects of flight pattern generation in the locust. In: Gewecke M, Wendler G (eds) locomotion. Verlag Paul Parey, Berlin, Heidelberg. pp 139-148.

Murphey RK (1985) A second cricket cercal sensory system: bristle hairs and the interneurons they activate. J Comp Physiol 156: 357-367.

Murphey RK Palka J, Hustert R (1977) The cercus-to-giant interneuron system of crickets. II. Response characteristics of the giant interneurons. J Comp Physiol 119: 285-300.

Narahashi T (1963) The properties of insect axons. Adv Insect Physiol 1: 175-256.

Nässel DR (1987a) Aspects of the functional and chemical anatomy of the insect brain. (This volume)

Nässel DR (1987b) Neuroactive substances in the insect CNS. (This volume)

Nistri A, Constanti A (1979) Pharmacological characterization of different types of GABA and glutamate receptors in vertebrate and invertebrates. Prog Neurobiol 13: 117-235.

Nolen TG, Hoy RR (1984) Initiation of behavior by single neurons: the role of behavioral context. Science 226: 992-994.

Olberg RM (1986) Identified target-selective visual interneurons descending from the dragonfly brain. J Comp Physiol 159: 827-840.

Orchard I (1976) Calcium dependent action potentials in a peripheral neurosecretory cell of a stick insect. J Comp Physiol 112: 95-102.

Orchard I, Osborne MP (1977) The effects of cations upon the action potentials recorded from neurohaemal tissue of the stick insect. J Comp Physiol 118: 1-12.

O'Shea M (1975) Two sites of axonal spike initiation in a bimodal interneuron. Brain Res 96: 93-98.

O'Shea M (1982) Peptide neurobiology: an identified neurone approach with special reference to proctolin. Trends Neurosci 5: 69-73.

O'Shea M, Bishop CA (1982) Neuropeptide proctolin associated with an identified skeletal motoneuron. J Neurosci 2: 1242-1251.

O'Shea M Evans PD (1979) Potentiation of neuromuscular transmission by an octopaminergic neurone in the locust. J Exp Biol 79: 169-190.

O'Shea M, Rowell CHF (1975) A spike transmitting electrical synapse between visual interneurones in the locust movement detector system. J Comp Physiol 97: 143-153.

O'Shea M, Williams JLD (1974) Anatomy and output connection of the lobular giant movement detector (LGMD) of the locust. J Comp Physiol 41: 257-266.

O'Shea M, Rowell CHF, Williams JLD (1974) The anatomy of a locust visual interneuron: the descending contralateral movement detector. J Exp Bio 60: 1-12.

Patterson JA, Chappell RL (1980) Intracellular responses of procion filled cells and whole nerve cobalt impregnations in the dragonfly ocellus. J Comp Physiol 139: 25-39.

Pearson KG (1976) Nerve cells without action potentials. In: Fentress JC (ed) Simpler networks and behavior. Sinauer Associates, Inc., Massachussetts. pp 99-110.

Pearson KG (1977) Interneurons in the ventral nerve cord of insects. In: Hoyle G (ed) Identified neurons and behavior of Arthropods. Plenum Press, New York, London. pp 329-337.

Pearson KG (1979) Local neurons and local interactions in the nervous systems of invertebrates. In: Schmitt FO, Worden FG (eds) The neurosciences fourth study program. MIT Press, Massachusetts. pp 145-157.

Pearson KG (1980) Burst generation in co-ordinating interneurons of the ventilatory system of the locust. J Comp Physiol 137: 305-313.

Pearson KG (1981) Function of sensory input in insect motor systems. Can J Physiol Pharmacol 59: 660-666.

Pearson KG (1983) Neural circuits for jumping in the locust. J Physiol (Paris) 78: 765-771.

Pearson KG (1985a) Neuronal circuits for patterning motor activity in invertebrates. In: Cohen MJ, Strumwasser F (eds) Comparative neurobiology: Modes of communication in the nervous system. John Wiley and Sons Inc., New York. pp 225-244.

Pearson KG (1985b) Are there central pattern generators for walking and flight in insects? In: Barnes WJP, Gladden MH (eds) Feedback and motor control in invertebrates and vertebrates. Croom Helm, Ltd., London. pp 307-315.

Pearson KG, Fourtner CR (1975) Non-spiking interneurons in the walking system of the cockroach. J Neurophysiol 38: 33-52.

Pearson KG, Franklin R (1984) Characteristics of leg movements and patterns of coordination in insects walking on rough terrain. Int J Robotics Res 3: 101-112.

Pearson KG, Goodman CS (1979) Correlation of variability in structure with variability in synaptic connections of an identified interneuron in locusts. J Comp Neurol 184: 141-166.

Pearson KG, Goodman CS (1981) Presynaptic inhibition of transmission from identified interneurons in locust central nervous system. J Neurphysiol 45: 501-515.

Pearson KG, O'Shea M (1984) Escape behavior of the locust. The jump and its initiation by visual stimuli. In: Eaton RC (ed) Neural mechanisms of startle behavior. Plenum Press, New York, London. pp 163-178.

Pearson KG, Robertson RM (1981) Interneurons coactivating hindleg flexor and extensor motoneurons in the locust. J Neurophysiol 144: 391-400.

Pearson KG, Robertson RM (1987) Structure predicts synaptic action of two classes of interneurons in locust thoracic ganglia. Cell Tissue Res (in press).

Pearson KG, Wong RKS, Fourtner CR (1976) Connexions between hair plate afferents and motoneurones in the cockroach leg. J Exp Biol 64: 251-266.

Pearson KG, Heitler WJ, Steeves JD (1980) Triggering of locust jump by multimodal inhibitory interneurons. J Neurophysiol 43: 257-278.

Pearson KG, Boyan GS, Bastiani M, Goodman CS (1985a) Heterogeneous properties of segmentally homologous interneurons in the ventral nerve cord of locusts. J Comp Neurol 233: 133-145.

Pearson KG, Reye DN, Parsons DW, Bicker G (1985b) Flight-initiating interneurons in the locust. J Neurophysiol 53: 910-925.

Pentreath VW (1982) Potassium signalling of metabolic interactions between neurons and glial cells. Trends Neurosci 5: 339-345.

Pentreath VW (1987) Functions of invertebrate glia. (This volume)

Perkel DH, Mulloney B (1974) Motor pattern production in reciprocally inhibitory neurons exhibiting postinhibitory rebound. Science 185: 181-183.

Peters BH, Altman JS, Tyrer NM (1985) Synaptic connections between the hindwing stretch receptor and flight motor neurones in the locust revealed by double cobalt labelling for electron microscopy. J Comp Neurol 233: 269-284.

Pflüger H-J (1984) The large fourth abdominal intersegmental interneuron: a new type of wind-sensitive ventral cord interneuron in locusts. J Comp Neurol 222: 343-357.

Pichon Y (1974) Axonal conduction in insects. In: Treherne JE (ed) Insect neurobiology. Elsevier, Amsterdam. pp 73-117.

Pichon Y, Ashcroft FM (1985) Nerve and muscle: electrical activity. In: Kerkut GA, Gilbert LI (eds) Comprehensive insect physiology, biochemistry and pharmacology. Pergamon Press, New York. Vol 5, pp 85-113.

Pichon Y, Boistel J (1963) Modifications of the ionic content of the haemolymph and of the activity of Periplaneta americana in relation to diet. J Insect Physiol 9: 887-891.

Pichon Y, Boistel J (1967) Microelectrode study of the resting and action potentials of the cockroach giant axon with special reference to the role played by the nerve sheath. J Exp Biol 47: 357-373.

Pichon Y, Treherne JE (1970) Extraneuronal potentials and potassium depolarization in cockroach giant axons. J Exp Biol 53: 485-493.

Pichon Y, Moreton RB, Treherne JE (1971) A quantitative study of the ionic basis of extraneuronal potential changes in the central nervous system of the cockroach (Periplaneta americana, L.). J Exp Biol 54: 757-777.

Pitman RM (1971) Transmitter substances in insects: a review. Comp Gen Pharmacol 2: 347-371.

Pitman RM (1975a) The ionic dependence of action potentials induced by colchicine in an insect motorneurone cell body. J Physiol 247: 511-520.

Pitman RM (1975b) Calcium-dependent action potentials in the cell body of an insect motoneurone. J Physiol 251: 62-63P.

Pitman RM, Kerkut GA (1970) Comparison of the actions of iontophoretically applied acetylcholine and gamma-aminobutyric acid in cockroach central neurones. Comp Gen Pharmacol 1: 221-230.

Pitman RM Tweedle CD, Cohen MJ (1972a) Branching of central neurons: intracellular cobalt injection for light and electron microscopy. Science 176: 412-414.

Pitman RM, Tweedle CD, Cohen MJ (1972b) Electrical properties of insect central neurons: augmentation by nerve section or colchicine. Science 178: 507-509.

Plotnikova SI (1969) Effector neurones with several axons in the ventral nerve cord of the Asian grasshopper, Locusta migratoria. J Evol Biochem Physiol 5: 276-278.

Popov AV, Markovich AM (1982) Auditory interneurons in the prothoracic ganglion of the cricket, Gryllus bimaculatus. II. A high frequency ascending neurone (HF1AN). J Comp Physiol 146: 351-359.

Popov AV, Markovich AM, Andjan AS (1978) Auditory interneurons in the prothoracic ganglion of the cricket Gryllus bimaculatus. I. The large segmental auditory neuron (LSAN). J Comp Physiol 126: 183-192.

Rall W (1981) Functional aspects of neuronal geometry. In: Roberts A, Bush BMH (eds) Neurones without impulses. Cambridge University Press, Cambridge, UK pp 223-254.

Raper JA, Bastiani M, Goodman CS (1983). Pathfinding by neuronal growth cones in grasshopper embryos. I. Divergent choices made by the growth cones of sibling neurons. J Neurosci 3: 20-30.

Rehbein H (1976) Auditory neurons in the ventral cord of the locust: morphological and functional properties. J Comp Physiol 110: 233-250.

Reichert H, Rowell CHF (1985) Integration of non-phaselocked exteroceptive information in the control of rhythmic flight in the locust. J Neurophysiol 53: 1201-1218.

Reichert H, Rowell CHF (1986) Neuronal circuits controlling flight in the locust: how sensory information is processed for motor control. Trends Neurosci 9: 281-283.

Reichert H, Rowell CHF Griss C (1985) Course correction translates feature detection into behavioural action in locusts. Nature 315: 142-144.

Rind FC (1983) A directionally sensitive motion detecting neurone in the brain of a moth. J Exp Biol 102: 253-271.

Rind FC (1984) A chemical synapse between two motion detecting neurones in the locust brain. J Exp Biol 110: 143-167.

Ritzmann RE, Tobias ML, Fourtner CR (1980) Flight activity initiated via giant interneurons of the cockroach: evidence for bifunctional trigger interneurons. Science 210: 443-445.

Robertson RM (1985a) Central neuronal interactions in the flight system of the locust. Gewecke M, Wendler G (eds) Insect locomotion. Paul Parey, Berlin, Hamburg. pp 183-194.

Robertson RM (1985b) Interneurons in the flight system of the cricket, Teleogryllus oceanicus. Soc Neurosci Abstr 11: 512.

Robertson RM (1986) Neuronal circuits controlling flight in the locust: central generation of the rhythm. Trends Neurosci 9: 278-280.

Robertson RM, Moulins M (1981) Oscillatory command input to the motor pattern generators of the crustacean stomatogastric ganglion. I. The pyloric rhythm. J Comp Physiol 143: 453-463.

Robertson RM, Moulins M (1984) Oscillatory command input to the motor pattern generators of the crustacean stomatogastric ganglion. II. The gastric rhythm. J Comp Physiol 154: 473-491.

Robertson RM, Pearson KG (1983) Interneurons in the flight system of the locust: distribution, connections and resetting properties. J Comp Neurol 215: 33-50.

Robertson RM, Pearson KG (1984) Interneuronal organization in the flight system of the locust. J Insect Physiol 30: 95-101.

Robertson RM, Pearson KG (1985a) Neural circuits in the flight system of the locust. J Neurphysiol 53: 110-128.

Robertson RM, Pearson KG (1985b) Neural networks controlling locomotion in locusts. In: Selverston AI (ed) Model neural networks and behavior. Plenum Press, New York, London. pp 21-35.

Robertson RM, Pearson KG, Reichert H (1982) Flight interneurons in the locust and the origin of insect wings. Science 217: 177-179.

Romer H, Marquart V (1984) Morphology and physiology of auditory interneurons in the metathoracic ganglion of the locust. J Comp Physiol 155: 249-262.

Rowell CHF, Pearson KG (1983) Ocellar input to the flight motor system of the locust: structure and function. J Exp Biol 103: 265-288.

Rowell CHF, Reichert H (1985) Compensatory steering in locusts: integration of non-phase locked input with a rhythmic motor output. In: Gewecke M, Wendler G (eds) Insect locomotion. Paul Parey, Berlin, Hamburg. pp 175-182.

Sakaguchi DS, Murphey RK (1983) The equilibrium detecting system of the cricket: physiology and morphology of an identified interneuron. J Comp Physiol 150: 141-152.

Sattelle DB (1980) Acetylcholine receptors of insects. Adv Insect Physiol 15: 215-315.

Sattelle BD, Harrow ID, Hue B, Pelhate M, Gepner JI, Hall LM (1983) α-Bungarotoxin blocks excitatory synaptic transmission between cercal sensory neurones and giant interneurone 2 of the cockroach, Periplaneta americana. J Exp Biol 107: 473-489.

Satterlie RA (1985) Reciprocal inhibition and postinhibitory rebound produce reverberation in a locomotor pattern generator. Science 229: 402-404.

Schildberger K (1884a) Multimodal interneurones in the cricket brain: properties of identified extrinsic mushroom body cells. J Comp Physiol 154: 71-79.

Schildberger K (1984b) Temporal selectivity of identified auditory neurons in the cricket brain. J Comp Physiol 155: 171-185.

Schildberger K (1987) Acoustic communication in crickets: Behavioral and neuronal mechanisms of song recognition and localization. (This volume)

Schofield PK, Treherne JE (1984) Localization of the blood-brain barrier of an insect: electrical model and analysis. J Exp Biol 109: 319-331.

Schofield PK, Swales LS, Treherne JE (1984a) Potentials associated with the blood-brain barrier of an insect: recordings from identified neuroglia. J Exp Biol 109: 307-318.

Schofield PK, Swales LS, Treherne JE (1984b) Quantitative analysis of cellular and paracellular effects involved in disruption of the blood-brain barrier of an insect by hypertonic urea. J Exp Biol 109: 333-340.

Selverston AI (1980) Are central pattern generators understandable? Behav Brain Sci 3: 535-571.

Shaw SR (1969) Interreceptor coupling in ommatidia of drone honeybee and locust compound eyes. Vision Res 9: 999-1029.

Shaw SR (1975) Retinal resistance barriers and electrical lateral inhibition. Nature 255: 480-483.

Shaw SR (1979) Signal transmission by graded slow potentials in the arthropod visual system. In: Schmitt FO, Worden FG (eds) The neurosciences: fourth study program. MIT Press, Cambridge, Massachusetts. pp 275-295.

Shaw SR (1981) Anatomy and physiology of identified non-spiking cells in the photreceptor-lamina complex of the compound eye of insects, especially Diptera. In: Roberts A, Bush BMH (eds) Neurones without impulses. Cambridge University Press, Cambridge, UK. pp 61-116.

Shaw SR (1983) Is the blood-brain barrier of insects just a single seal of tight junctions, as in vertebrates? Soc Neurosci Abstr 9: 885.

Shaw SR (1984) Early visual processing in insects. J Exp Biol 112: 225-251.

Shaw SR, Henken DB (1984) The formation of the insect blood-brain barrier: evidence from the cockroach nerve cord against the tight junction hypothesis. In: Borkovec AB, Kelly TJ (eds) Insect neurochemistry and neurophysiology. Plenum Press, New York. pp 471-473.

Shepherd D, Murphey RK (1985) Competition controls quantal release at an identified insect synapse. Soc Neurosci Abstr 11: 958.

Seigler MVS (1981a) Posture and history of movement determine membrane potential and synaptic events in nonspiking interneurons and motor neurons of the locust. J Neurophysiol 46: 296-309.

Siegler MVS (1981b) Postural changes alter synaptic interactions between nonspiking interneurons and motor neurons of the locust. J Neurophysiol 46: 310-323.

Siegler MVS (1982) Electrical coupling between supernumerary motor neurones in the locust. J Exp Biol 101: 105-119.

Siegler MVS (1984) Local interneurones and local interactions in arthropods. J Exp Biol 112: 253-281.

Siegler MVS (1985) Non-spiking interneurons and motor control in insects. Adv Insect Physiol 18: 249-304.

Siegler MVS, Burrows M (1979) The morphology of local non-spiking interneurones in the metathoracic ganglion of the locust. J Comp Neurol 183: 121-147.

Siegler MVS, Burrows M (1984) The morphology of two groups of spiking local interneurons in the metathoracic ganglion of the locust. J Comp Neurol 224: 463-482.

Simmons PJ (1980) A locust wind and ocellar brain neurone. J Exp Biol 85: 281-294.

Simmons PJ (1981) Synaptic transmission between second- and third-order neurones of a locust ocellus. J Comp Physiol 145: 265-276.

Simmons PJ (1982) Transmission mediated with and without spikes at connexions between large second-order neurones of locust ocelli. J Comp Physiol 147: 401-414.

Simmons PJ (1985) Postsynaptic potentials of limited duration in visual neurones of a locust. J Exp Biol 117: 193-213.

Sombati S, Hoyle G (1984a) Central nervous sensitization and dishabituation of reflex action in an insect by the neuromodulator octopamine. J Neurobiol 15: 455-480.

Sombati S, Hoyle G (1984b) Glutamatergic central nervous transmission in locusts. J Neurobiol 15: 507-516.

Spira ME, Yarom Y, Parnas I (1976) Modulation of spike frequency by regions of special axonal geometry and by synaptic inputs. J Neurophysiol 39: 882-899.

Spira ME, Yarom Y, Zeldes D (1984) Neuronal interactions mediated by neurally evoked changes in the extracellular potassium concentration. J Exp Biol 112: 179-197.

Steeves JD, Pearson KG (1982) Proprioceptive gating of inhibitory pathways to hindleg flexor motoneurons in the locust. J Comp Physiol 146: 507-515.

Stewart WW (1978) Fuuntional connections between cells, as revealed by dye-coupling with a highly fluorescent naphthalimide tracer. Cell 14: 741-759.

Strausfeld NJ (ed) (1983) Functional neuroanatomy. Springer series in experimental entomology. Springer-Verlag, New-York, Heidelberg, Berlin.

Strausfeld NJ, Bassemir UK (1983) Cobalt-coupled neurons of a giant fibre system in Diptera. J Neurocytol 12: 971-991.

Strausfeld NJ, Obermeyer M (1976) Resolution of intraneuronal and transynaptic migration of cobalt in the insect visual and central nervous system. J Comp Physiol 110: 1-12.

Taghert PH, Goodman CS (1984) Cell determination and differentiation of identified serotonin-immunoreactive neurons in the grasshopper embryo. J Neurosci 4: 989-1000.

Tanouye MA, Wyman RJ (1980) Motor outputs of the giant nerve fibre in Drosophila. J Neurophysiol 44: 405-421.

Thomas JB, Wyman RJ (1984) Mutations altering synaptic connectivity between identified neurons in Drosophila. J Neurosci 4: 530-538.

Treherne JE (1985) Blood-brain barrier. In: Kerkut GA, Gilbert LI (eds) Comprehensive insect physiology, biochemistry and physiology. Pergamon Press, New York, Vol 5, pp 115-137.

Treherne JE, Pichon Y (1972) The insect blood-brain barrier. Adv Insect Physiol 9: 257-313.

Treherne JE, Schofield PK (1981) Mechanisms of ionic homeostasis in the central nervous system of an insect. J Exp Biol 95: 61-73.

Treherne JE, Schofield PK, Lane NJ (1973) Experimental disruption of the blood-brain barrier system of an insect (Periplaneta americana L.) J Exp Biol 59: 711-723.

Treherne JE, Schofield PK, Lane NJ (1982) Physiological and ultrastructural evidence for an extracellular anion matrix in the central nervous system of an insect (Periplaneta americana). Brain Res 247: 255-267.

Tyrer NM, Gregory GE (1982) A guide to the neuroanatomy of locust suboesophageal and thoracic ganglia. Phil Trans R Soc Lond B 297: 91-123.

Usherwood PNR (1961) Spontaneous miniature potentials from insect muscle fibres. Nature 191: 814-815.

Usherwood PNR (1977) Neuromuscular transmission in insects. In: Hoyle G (ed) Identified neurons and behavior of Arthropods. Plenum Press, New York. pp 31-48.

Wachtel H, Kandel ER (1967) A direct synaptic connection mediating both excitation and inhibition. Science 158: 1206-1208.

Walker RJ, James VA, Roberts CJ, Kerkut GA (1980) Neurotransmitter receptors in invertebrates. In: Sattelle DB, Hall LM, Hildebrand JG (eds) Receptors for neurotransmitters, hormones and pheromones in insects. Elsevier, Amsterdam. pp 41-57.

Wang-Bennett LT, Glantz RM (1986) Integration and spike initiation in neuronal terminals. J Neurosci 6: 1726-1732.

Watson AHD, Burrows M (1982) The ultrastructure of identified locust motor neurones and their synaptic relationships. J Comp Neurol 205: 383-397.

Watson AHD, Burrows M (1983) The morphology, ultrastructure and distribution of synapses on an intersegmental interneurone of the locust. J Comp Neurol 214: 154-169.

Weeks JC, Truman JW (1984a) Neural organization of peptide-activated ecdysis behaviors during the metamorphosis of Manduca sexta. I. Conservation of the peristalsis motor pattern at the larval-pupal transformation. J Comp Physiol 155: 407-422.

Weeks JC, Truman JW (1984b) Neural organization of peptide-activated ecdysis behaviors during the metamorphosis of Manduca sexta. II. Retention of the proleg motor pattern despite loss of the prolegs at pupation. J Comp Physiol 155: 423-433.

Weevers R de G (1985) The insect ganglia. In: Kerkut GA, Gilbert LI (eds) Comprehensive insect physiology, biochemistry and pharmacology. Pergamon Press, New York. Vol 5, pp 213-297.

Weis-Fogh T (1956) Biology and physics of locust flight. IV. Notes on sensory mechanisms in locust flight. Phil Trans R Soc LOnd B 239: 553-584.

Wendler G (1983a) The interaction of peripheral and central components in insect locomotion. In: Huber F, Markl H (eds) Neuroethology and behavioral physiology. Springer Verlag, Berlin, Heidelberg. pp 42-53.

Wendler G (1983b) The locust flight system: functional aspects of sensory input and methods of investigation. In: Nachtigall W (ed) BIONA - report 2. Gustav Fisher, Stuttgart. pp 113-125.

Wendler G (1985) Insect locomotory system: control by proprioceptive and exteroceptive inputs. In: Gewecke M, Wendler G (eds) Insect locomotion. Verlag Pual Parey, Berlin, Heidelberg. pp 245-254.

Wiersma CAG (1952) Neurons of arthropods. Symp Quant Biol 17: 155-163.

Wilson DM (1961) The central nervous control of flight in a locust. J Exp Biol 38: 471-490.

Wilson JA (1981) Unique, identifiable local non-spiking interneurons in the locust mesothoracic ganglion. J Neurobiol 12: 353-366.

Wilson JA, Phillips CE (1982) Locust local non-spiking interneurons drive antagonistic motor neurons: physiology, morphology and ultrastructure. J Comp Physiol 204: 21-31.

Wilson JA, Phillips CE (1983) Pre-motor non-spiking interneurons. Prog Neurobiol 20: 89-107.

Wilson M (1978) Generation of graded potential signals in the second order cells of locust ocellus. J Comp Physiol 124: 317-331.

Wohlers DW, Huber F (1982) Processing of sound signals by six types of neurons in the prothoracic ganglion of the cricket Gryllus campestris L. J Comp Physiol 146: 161-173.

Wolf H, Pearson KG (1987) Comparison of motor patterns in the intact and deafferented flight system of the locust. II. Intracellular recordings from flight motoneurons. J Comp Physiol A (in press).

Wyman RJ, Thomas JB, Salkoff L, King DG (1984) The _Drosophila_ giant fiber system. In: Eaton RC (ed) Neural mechanisms of startle behavior. Plenum Press, New York, London. pp 133-161.

Wyman RJ, Thomas JB, Salkoff L, Costello W (1985) The _Drosophila_ thorax as a model system for neurogenetics. In: Selverston AI (ed) Model neural networks and behavior. Plenum Press, New York, London. pp 513-535.

Yarom Y, Spira ME (1982) Extracellular potassium ions mediate specific neuronal interaction. Science 216: 80-82.

Yawo H, Kojima H, Kuno M (1985) Low-threshold, slow-inactivating Na^+ potentials in the cockroach giant axon. J Neurophysiol 54: 1087-1100.

Zaretsky M, Loher W (1983) Anatomy and electrophysiology of individual neruosecretory cells of an insect brain. J Comp Neurol 216: 253-263.

Zill SN (1985) Plasticity and proprioception in insects. II. Modes of reflex action of the locust metathoracic femoral chordotonal organ. J Exp Biol 116: 463-480.

Zill SN, Forman RR (1983) Proprioceptive reflexes change when an insect assumes an active, learned posture. J Exp Biol 107: 385-390.

ONTOGENESIS OF THE NERVOUS SYSTEM IN CEPHALOPODS

H.-J. MARTHY

Laboratoire Arago - U.A. 117 C.N.R.S.

Université Pierre et Marie Curie (Paris VI)

66650 Banyuls-sur-mer, France

"Sieh da einen Menschen, der in die Erde gräbt,
und dort einen, der auf der Erde steht und geht
und dort einen, der zum Himmel schaut -
allüberall, wo Menschen leben, sind sie gerufen,
mit ihrem Geist die Dinge zu erforschen,
zu hinterfragen und zu deuten."
Richard Thalmann.
(From "Geist, der Erde umfasst", Rat Verlag)

ABSTRACT

Our knowledge on the ontogenisis of the nervous system in cephalopods is rather fragmentary. Many major questions such as the precise origin of the head ganglia from the (neur-)ectoderm in early embryogenesis, the chronology of ganglia "assemblage", the developmental events in nerve cell differentiation and the "overlap" of simultaneous growth and functioning of the brain are essentially still open. However, thanks to histological studies made on developing embryos of a decapodan (Loligo vulgaris) and and octopodan species (Octopus vulgaris) the general course of the nervous system formation in cephalopods can be recognised. Occasional observations on ganglion formation are found dispersed in the literature. The differentiation of the giant fibre system has been studied also and the first experimental approach for studying neurogenesis has been made. So far, hardly any physiological work has been done on cephalopod embryos. Some insight into the postembryonic development of the cephalopod brain is also available. (In contrast to the few studies available on embryos, an important literature exists dealing with descriptive and experimental observations made on all parts of the nervous system of adults).

In a squid hatchling all the central ganglia (cerebral, pedal visceral, peduncular, olfactory, basal and optic ganglia) are present and form the supra- and suboesophageal ganglia mass (=head). Except for the optic ganglia, they are divided into varous subunits, the lobes. The giant fibre system is distinct by large nuclei and its differentiation state clearly precedes that of the other parts of the nervous system. Also, most of the peripheral ganglia (brachial, buccal, subradular, gastric and stellate) are present.

The morphological situation found in hatchlings (resembling essentially that of the adults) is reached during embryogenesis from about stage VII onwards. The period until stage XII is the phase of actual ganglion formation. From stage XIII until hatching stage XX, the final topology of the ganglia and their relations are established and the lobular and cellular differentiation progresses. Based on the different results given by various authors, on personal communications from colleagues and on personal observations, the crucial developmental steps in the formation of the nervous system in decapodan and octopodan cephalopods are reviewed.

1. INTRODUCTION

Despite the fact that the embryogenesis of cephalopods is well known from a general point of view, our knowledge on the ontogenesis of the central and peripheral nervous systems is still rather fragmentary. Even major questions such as the precise (ectodermal) origin of the various "brain" forming ganglia, the mode of ganglia "assemblage" in relation to yolk reduction and the growing neighbouring organs, the specific chronology of differentiation of neural tissue structures or of individual neurones and/or the considerable "overlap" between brain growth and simultaneous functioning could still be answered in much more detail. Experimental studies performed on the embryonic or postembryonic nervous system are lacking (only very recently has a successful approach been made in culturing in vitro embryonic ganglion fragments (Marthy and Aroles 1983; 1987)). In this review I therefore present essentially descriptive anatomical data concerning developmental aspects of: A) the central and the peripheral nervous system and B) the "giant nerve fibre system" of decapods. In reviewing the embryonic development of the nervous system in both decapods and octopods, an attempt is made to provide a somewhat easier access to the highly interesting but poorly explored field of investigation.

2. THE DEVELOPMENT OF THE CENTRAL AND PERIPHERAL NERVOUS SYSTEMS

2.1. The central nervous system

2.1.1. Gross morphology of the brain of hatchlings

In hatchlings, the central nervous system of decapodan and octopodan Cephalopods is already "adultlike". That is, four paired neurone centres are arranged into a morphological and physiological unit for which the term "brain" appears fully adequate. The morphological situation of such a brain is shown in Fig. 1 (A and B).

Three central pairs of ganglia surround the oesophagus behind the buccal mass. A supra-oesophageal mass (cerebral ganglia) is distinct from the sub-oesophageal complex (pedal and visceral ganglia). Laterally to these ganglia and connected with the dorsal and ventral brain mass by the optical tract, one finds two large optic ganglia. The central head ganglia (excepting the optic ganglia) are divided into several subunits, the lobes. The various lobes are interconnected, often in a complex manner, by transverse and longitudinal fibre tracts. Within the central brain one is reminded of the so-called "vertical lobe system" (comprising the upper frontal, sub-vertical and vertical lobes) which is the highest integration centre of the brain (Boycott and Young 1955). The other functional parts of the brain are (Frösch 1971): the intermediate motor centres (sub-oesophageal lobes); the higher motor centres (anterior and posterior basal lobes); the primary sensory centres (brachial, buccal and optic lobes); - in the peripheral nervous system some lower motor centres are present: arm nerves, stellate ganglia. The supra-oesophageal brain mass is directly connected to the sub-oesophageal part by complex fibre tracts and connectives.

The sub-oesophageal ganglia mass is divided into the brachial and th pedal lobes. Whereas the former are oriented towards the arms, the latte innervate the funnel, the eye muscles and the statocysts. Within the sub-oesophageal brain, one also finds the "magnocellular lobe" which is, after Marquis (1981) an enlarged connective and at least partially derived from the visceral ganglion. Finally, one also may mention the "chromatophore lobes". (For numerous details on the structures of the adult brain as well as on the experimentation performed on the nervous system of adults one should consult for instance: Wells 1962, 1978; Young 1971; Messenger 1979, 1982; Boyle 1983, 1986; Dubas et al. 1986). The visceral ganglia also belonging to the sub-oesophageal brain mass control: via the mantle connectives running to the peripheral stellate ganglia, the mantle movements; via the collar nerves, the funnel valves and via the visceral nerves, the intestinal mass.

The optic lobes are the largest primary sensory centres. They are connected by large commissures and they store essentially visual information (memory function; Young 1974) but also selectively transmit information to other parts of the brain (e.g. to the vertical lobe system; they are not however directly connected to the vertical lobes) and, from there, via the magnocellular lobe to the sub-oesophageal brain section. Within the central nervous system of decapods at the moment of hatching, the "giant cells" of the "giant nerve fibre system" (section 3) are the most differentiated neural structures.

2.1.2. Formation of the brain

Korschelt (1892) was the first to recognise that the nervous tissue in Caphalopods is derived from the ectodermal germ layer. Others have confirmed this observation and shown that the anlagen of the various ganglia are cells which delaminate locally from the ectoderm in the course of early organogenesis. The differentiation of the nervous system and of the primary sense organs occur simultaneously. The most recent investigations on the ontogenesis of the nervous system have been done on embryos of Loligo vulgaris (Meister 1972) and of Octopus vulgaris (Marquis 1981). This review is effectively based on the results presented by these authors (who also review the older literature). Whereas the descriptions for Loligo are incorporated into a broader context and are thus less complete in certain details, those for Octopus deal exclusively with the subject sensu stricto and thus provide considerable details. Both are important.

2.1.2.1. Loligo vulgaris

Plotted against the developmental stages of Naef (1928), the main observations of Meister (1972) can be summarised as follows:

Stage VIII : Formation of a common anlage (!) of the cerebral and optic ganglia. Formation of the visceral ganglia.
" IX : Formation of the pedal and the brachial ganglia. The two parts of the cerebral ganglia become connected.
" X : Ganglia increase in size. Formation of the peripheral stellate ganglia.
" XI : Cells "preforming" the pallial nerves are recognized.

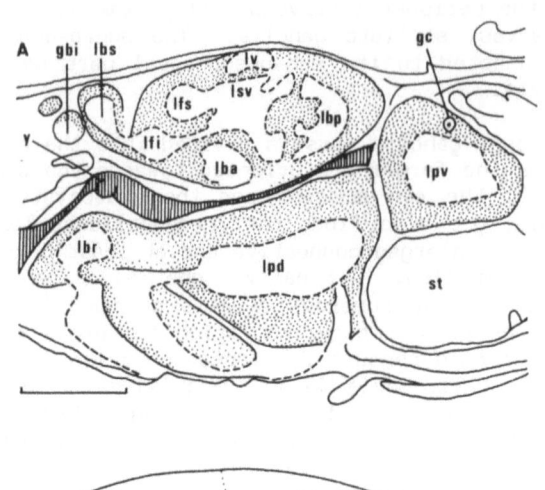

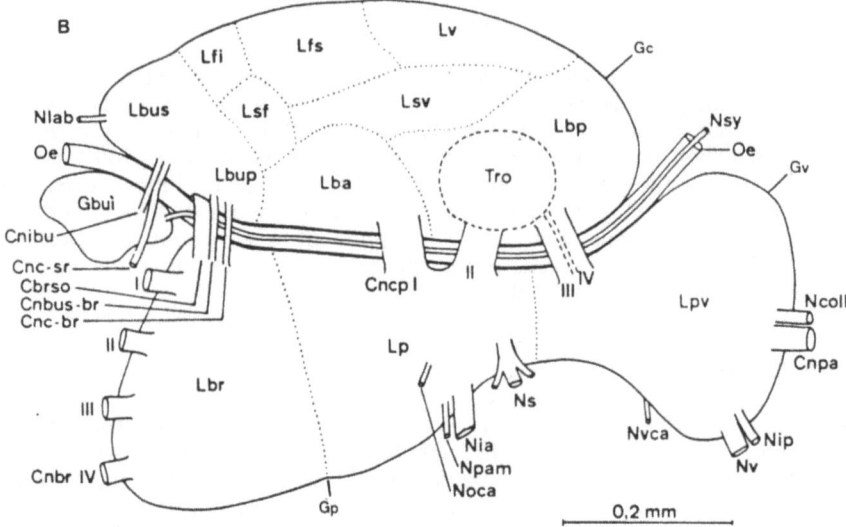

FIG. 1: A) Schematic view of a sagittal section through the brain of a hatchling of the squid <u>Loligo vulgaris</u>.
Scale bar: 0.5 mm. (After a photograph of Frösch 1971).

B) Reconstruction of the brain of a hatchling of the Octopod <u>Octopus valgaris</u>. (From Marquis 1981).

Abbrevations for A and B: Cbrso: Commissura brachialis supraoesophagealis; Cnbr I-IV: Connectivum (Cn.) brachiale I-IV; Cnbus-br: Cn. buccale superius-brachiale; Cnc-br: Cn. cerebrobrachiale; Cncp I-IV: Cn. cerebrale posterius I-IV; Cnc-sr: Cn. cerebro-subradulare; Cnibu: Cn. interbuccale; Cnpa: Cn. palliale; Gbui, gbi: Inferior buccal ganglion; Gc: Cerebral ganglion; gc: giant cell; Gp: Pedal ganglion; Gv: Visceral ganglion; Lba: Lobus (L.) basalis anterior; Lbp: L. basalis posterior; Lbr: L. brachialis; Lbus, Lbs: L. buccalis superior; Lbup: L. buccalis posterior; Lfi: L. frontalis inferior; Lfs: L. frontalis superior; Lpd, Lp: L. pedalis; Lpv: L. pallioviceralis; Lsf: L. subfrontalis; Lsv: L. subverticalis; Lv: L. verticalis; Ncoll: Nervus (N) collaris; Nia: N. infundibuli anterior (funnel nerve); Nip: N. infundibuli

"   XII : Fibres become visible in the optic, visceral and stellate ganglia as well as in the medulla of the arms. Head ganglia enlarge rapidly.
"   XIII : Formation of the peripheral gastric ganglion. Connexions between the head ganglia become fibrous. Pallial nerves also become fibrous.
"   XIV : Formation of the subradular ganglion.
"   XV   : Formation of the peripheral branchial ganglia.
"   XVI : All the main ganglia are formed and the morphological situation corresponds approximately to that found in the brain of hatchlings (Fig. 1A). Further development of the brain and the peripheral ganglia is essentially characterised by an increase of the volume (important fibre mass) and the progression of specific cellular differentiation. (The reader is also referred to Figs. 15-17 of Meister (1972) which show drawings of histological sections through the head region of embryos at stages IX, XII and XIV).
"   XVII : Formation of the subacetabular ganglia (sucker ganglia).

2.1.2.2. Octopus vulgaris

Marquis (1981) in a careful thesis work distinguished two phases in the ontogenesis of the nervous system: a) the phase of gangliogenesis, lasting from stage VI to stage XII and b) the phase of lobular and cellular differentiation, becoming particularly evident from stage XIII onwards. The observations of Meister (1972) also fit fairly well into this subdivision which is, of course, done for didactical reasons. In reality, the processes of gangliogenesis, growth and differentiation of the ganglia overlap.

2.1.2.2.1. The phase of gangliogenesis

The essential steps of gangliogenesis and the gross arrangment of the ganglia into a brain are summarised in Fig. 2 (A-F).

Firstly, the paired anlagen of the optic ganglia appear in the dorso-lateral blastoderm region of an embryo at stage VI (A). Histology reveals that only a few cells delaminating from the ectoderm give rise to such rudiments. This is generally true also for the other ganglia. In stage VII (when the anlagen of the optic vesicles become distinct; Marthy 1973), the anlagen extends below the future retina and towards the stomodeum region (B). At stage VIII, a similar situation is observed. At stage VIII too, the anlagen of the cerebral ganglia become first visible (C). They form in the laterocaudal portion of the mouth lip, on both sides. Shortly after their appearance, they already extend towards the growing optic ganglia but do not yet contact them directly. Thus, in Octopus embryos, it appears evident that optic and cerebral ganglia form independently. At stage IX, the optic ganglia come in direct contact with the cerebral ganglia (D); furthermore, above the stomodaeum some cells "connect" with the two cerebral ganglia, "preforming" the future cerebral commissures. As early as stage VII (B) the anlagen of the peripheral brachial ganglia become visible (see section: peripheral ganglia). Around

---

posterior; Nlab: N. labialis; Noca: N. oculomotorius anterior; Npam: N. musculi retractoris pallii mediani; Ns: N. staticus; Nsy: N. sympathicus; Nv: N. visceralis; Nvca: N. venae cavae anterior; Oe: oesophagus; St: statocyst; Tro: Tractus opticus; Y: Yolk vessel.

stages VII-VIII, two new pairs of ganglia appear in the ventral region near the funnel rudiment and behind the statocysts: the pedal and the visceral ganglia (E), that is, the future sub-oesophageal brain mass. The pedal ganglia come in contact laterally with the optic and the brachial ganglia. At stage XI, they make contact with the cerebral ganglia. Also around stage XI, the optic ganglia differentiate directly below the optic vesicles a fibrous layer which is the future " plexiform layer" (Young 1974). The more the statocysts invaginate, the better the exact topography of the visceral ganglia can be seen. Around stage XII the paired visceral ganglia "fuse" and the rostral part comes into close contact with the pedal ganglia. The zone of contact between the pedal and the visceral ganglia then becomes fibrous; this is the future brachio-palliovisceral connective. In contrast to Loligo (Meister 1972), no autonomous brachial ganglia are found in Octopus.

From stage X onwards all ganglia grow considerably. Once stage XII is reached (F), the head ganglia join together to form a ring around the oesophagus (and the yolk "tube"). The brain proper is essentially present at this stage. The following embryonic stages are characterised by a progressive increase of the brain mass and in particular by its differentiation.

2.1.2.2.2. The phase of lobular and cellular differentiation

It is not possible to repeat here all histological details on differentiating nervous tissue made by Meister (1972), Marquis (1981) or others such as Faussek (1901), Sacarrao (1956), Boletzky (1968), Frösch (1971) etc. For this purpose a short commentary is sufficient.

In octopodan and decapodan embryos, from stage XIII onwards, the ganglionic character of the central nervous system progressively disappears. As related earlier, the four brain centres become divided into numerous subunits (about 30), the lobes. According to Marquis (1981), the subdivision of the ganglia into the lobes is histologically first recognisable as a weak concentration of neuropiles. As development continues, the cell bodies of the cortical regions become arranged in a specific manner and the fibre mass becomes relatively large. The differentiated cells of the various lobes will finally greatly vary in their size: large neurones are typically motor in function (e.g. found in the suboesophageal lobes) and the smaller ones (5 µm and less) are found in regions which are concerned in the analysis of sensory retention (e.g. in the vertical lobe system) - (Wells 1962; Young 1963; Packard and Albergoni 1970). At stage XIV, the supraoesophageal brain mass is mainly marked by five paired neuropile centres: the "lobus buccalis superior, l. basalis anterior, l. basalis posterior, and l. verticalis". In the suboesophageal brain portion, the "lobus brachialis" becomes more prominent. The various commissures and connectives differentiate from cells which are derived from the ganglia themselves and not from the ectoderm as suggested by earlier authors such as Korschelt (1892). From the various brain regions the different nerves grow out. At stage XVI, in Octopus as in Loligo, the lobular differentiation of the brain is far advanced and the morphological situation is similar to that found in hatchlings (Fig. 1, A and B). However, cell proliferation and, in particular, individual neuroblast differentiation continue throughout embryogenesis and through post-embryogenesis and juvenile life (Wirz-Mangold 1959; Frösch 1971).

As to the problem of neuroblast differentiation, a first attempt has recently been made for a better understanding of this process (Marthy and Aroles 1983, 1987).

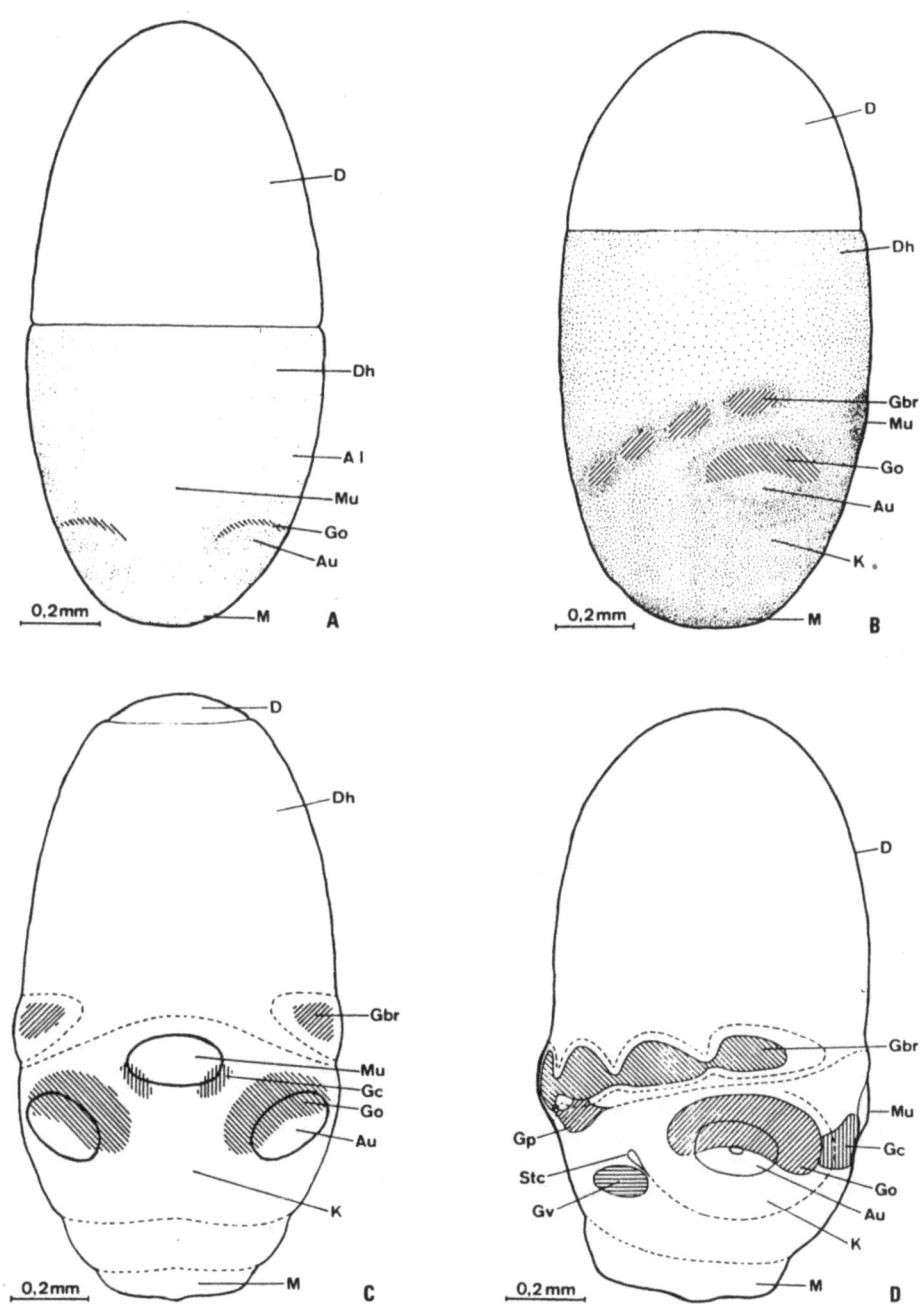

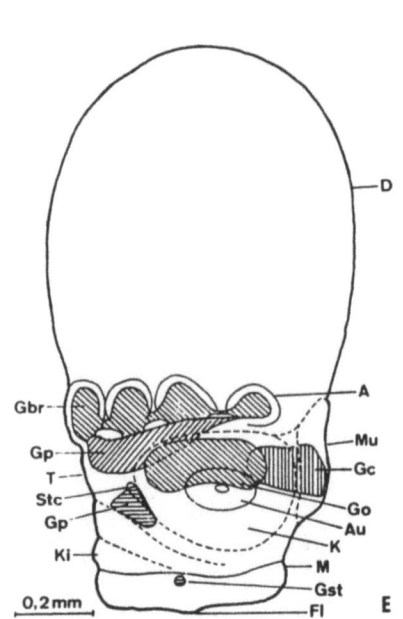 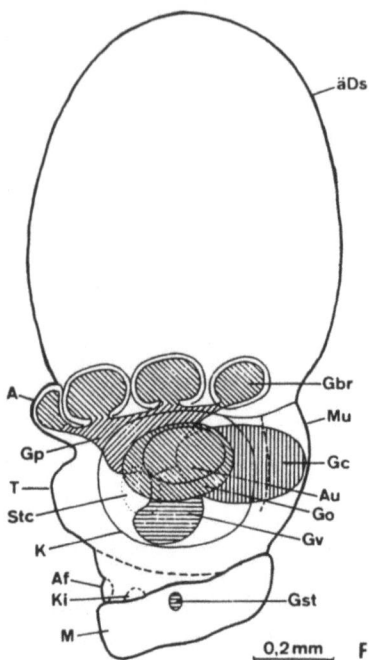

Fig. 2: Embryos of Octopus vulgaris at different developmental stages, showing the formation of the main ganglia and their assemblage into a "brain". (From Marquis 1981).
A) Stage VI. Dorsal view. Anlagen of the paired optic ganglia (Go).
B) Stage VII. Lateral view. Anlagen of the optic ganglion (Go) and the brachial ganglia (Gbr).
C) Stage VIII. Dorsal view. Anlagen of the paired cerebral ganglia (Gc), the optic ganglia (Go) and the arm ganglia (Gbr).
D) Stage IX. Lateral view. Anlagen of the central head ganglia (cerebral, optic, pedal, visceral ganglia) and the brachial ganglia.
E) Stage X. Lateral view. Central head ganglia increased in size. Peripheral nervous system: brachial ganglia (medulla) and stellate ganglia (Gst).
F) Stage XII. Lateral view. The four pairs of head ganglia of the central nervous system forming a ring around the oesophagus. Thus, the ganglia are assembled into a "brain". Peripheral nervous system: Brachial and stellate ganglia.
Abbreviations: A: arms; Af: anus; Au: eye, optic vesicle; D: yolk; Dh: yolk sac envelope; äDs: outer yolk sac; Fl: fins; Gbr: arm ganglion; brachial ganglion, medulla; Gc: cerebral ganglion; Go: optic ganglion, optic lobe; Gp: pedal ganglion; Gst: stellate ganglion; Gv; visceral ganglion; K: head region; Ki: gills; M: mantle; Mu: mouth; Stc: statocyst; T: funnel.

When culturing in vitro the "oculo-ganglionar complex" of Loligo vulgaris, embryos at about stage XIII-XIV, cells migrate off from the explant and differentiate into bi- and multipolar cell types (Fig. 3) which strongly resemble in their external morphology the "visual cells" shown by Young (1974) for the optic lobes of adult squids. This experiment provides evidence that a) ganglion-derived cells of stage XIII-XIV embryos are able to migrate, to proliferate and finally to

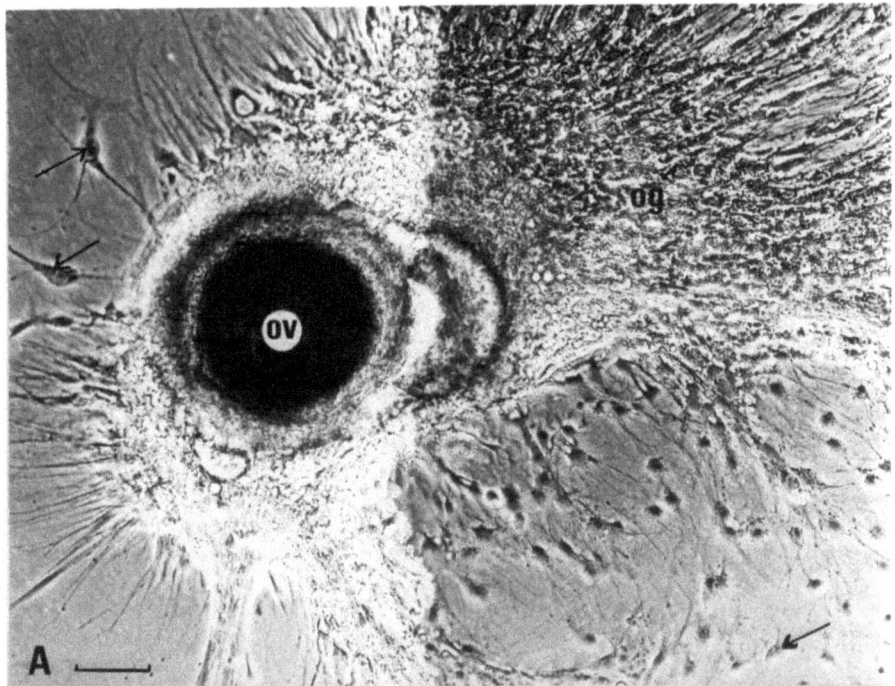

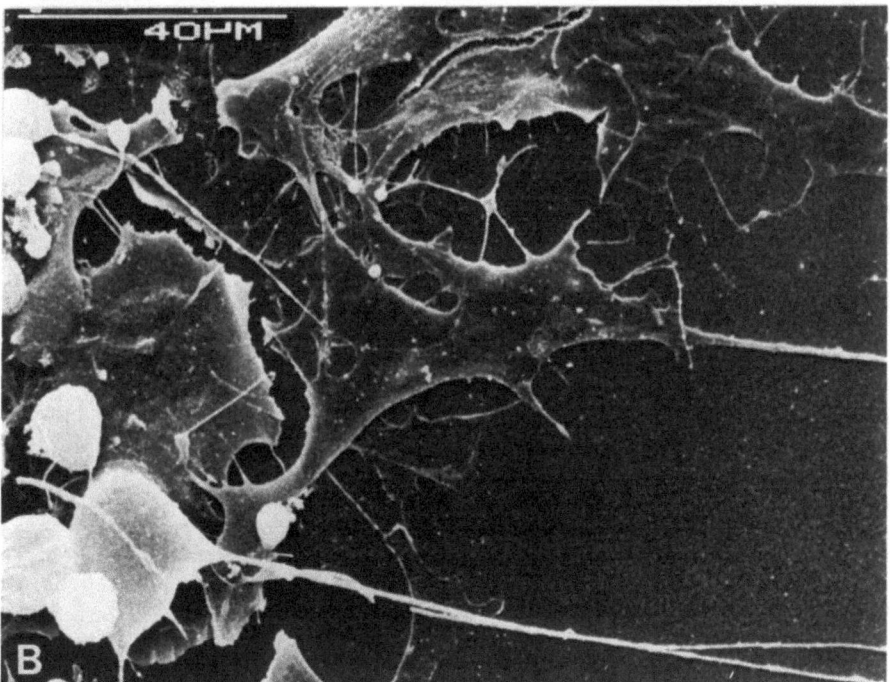

FIG. 3: A) Oculo-ganglionar complex from an embryo of <u>Loligo vulgaris</u> of stage XIII-XIV after 15 days in culture. Individual cells migrate off from the optic ganglion and differentiate into bi- and multipolar (nerve) cells (arrows). og: optic ganglion; ov: optic vesicle. Scale bar: 0.1 mm.
B) Scanning electron micrograph of ganglion-derived cells differentiating in vitro.

differentiate in vitro and b) ganglion-derived cells differentiate in vitro apparently according to their determination. The origin of the cells as well as the modification of their shape suggest that they are in a "neuroblast-state" at the onset of the in vitro culture and that they differentiate into several types of neurones (neglecting other cell types in this mixed cell culture). Further studies are under way. Also experiments on photoreceptor cells would be particularly interesting following the recent and precise descriptions of Yamamoto (1985) and Yamamoto et al. (1985).

No ultrastructural studies are available on the differentiating nervous system except those performed by Martin (from 1965 onwards) on th "giant cells" of the "giant nerve fibre system" (see section 3).

2.2. The peripheral nervous system

Parts of the peripheral nervous system appear early in organogenesis. The brachial ganglia can be seen in histological sections already at stage VII in Octopus (Fig. 2B) and at stage IX in Loligo. Other peripheral ganglia (buccal, branchial, subradular ganglia etc.) become distinct only if organogenesis is more advanced.

2.2.1. The brachial ganglia

As early as stage VII some cells below the arm buds represent the anlagen of the brachial or arm ganglia. At stage VIII, the ganglia establish cellular connexions amongst themselves and, at stage IX/X the connexion with the brain is made in establishing a contact with the pedal ganglia. A thin band of cells situated below the arm bases is the anlage of the brachial lobe. Through this band, the dorso-lateral and dorsal arms also become linked to the pedal ganglia. At stage XII, the neurones begin to extend fibres. At stage XV, the typical "medullar cord" has formed although it still contains some neurones within the fibre mass. At stage XVI, all the arm ganglia are connected above the oesophagus by the "interbrachial ring commissure".

2.2.2. The stellate ganglia

The anlagen of the stellate ganglia become visible in the mantle tissue around stage X (Fig. 2E). Their development has been studied in several cephalopod species (Faussek 1901; Sacarrao 1956; Meister 1972; Marquis 1981). It is derived from the ectoderm but, in Octopus, no ectodermal cells are added to the anlagen after stage X. At stage XI, a cellular strand extends from the stellate ganglia in the rostral direction, preforming the mantle connective. Fibre formation within the globular stellate ganglia starts around stage XII; the first fibres run through the mantle connective to the suboesophageal brain mass. Around stage XIV, nerves going into the mantle tissue become visible. These "stellar nerves" are particularly prominent in hatchlings. The "epistellar body" on the periphery of the stellate ganglia forms around stage XIV. This stucture containing photoreceptors (Mauro 1977) exists in octopodan species only.

2.2.3. The lower buccal ganglia

At stage XIII, together with the genesis of the buccal mass (stomodeum, radula, foregut), the buccal ganglia (probably originating from the oesophagus wall) differentiate on the ventro-lateral wall of the oesophagus. At stage XIV, the buccal ganglia have fused and at stage XV a central fibre complex is formed. The control of the ganglia by the brain is possible by the "connectivum interbuccale". The buccal ganglion is connected, via the "nervus sympathicus" to the gastric ganglion. The latter <u>unpaired</u> ganglion becomes visible in histological sections from stage XVII-XVIII onwards.

2.2.4. Other ganglia

No precise information is available on the formation of the subradular ganglion and of the gastric ganglion. As to the latter, Marquis (1981) assumes that it originates from migrating cells. The small branchial, gill and fusi-form ganglia around the visceral nerve cannot be distinguished in embryos. The anlagen of the gill ganglia apparently can be seen in older <u>Loligo</u> embryos (Meister 1972). - The subacetabular ganglia are small <u>reflex</u> centres at the bases of the suckers. Even in hatchlings no anlagen of these ganglia have been observed.

3. THE GIANT NERVE FIBRE SYSTEM

The term "giant-fibre system" or "giant nerve fibre system" is "confined to the especially large (nerve) fibres which operate the contractions of the (longitudinal) retractor muscles of the head and funnel and of the (circular) muscles of the mantle, movements by means of

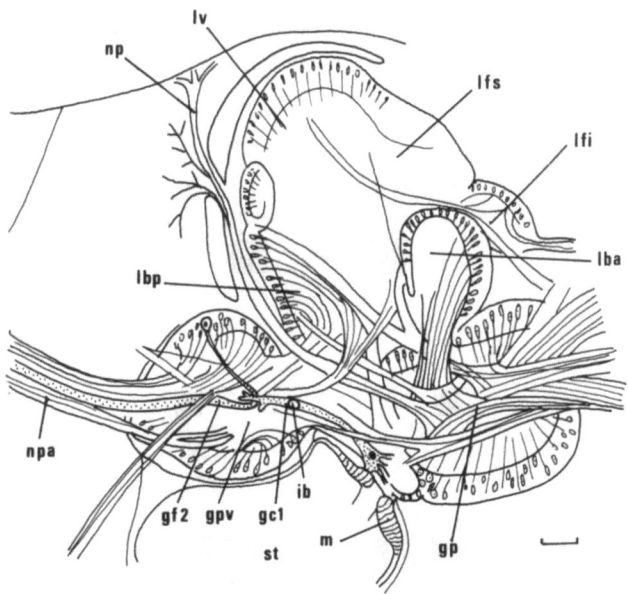

FIG. 4: Diagrammatic view of a sagittal section through the lateral portion of the nervous system of a young <u>Loligo</u>. Note the position of the first- and second-order giant axons. (After a drawing of Young 1939).
Abbreviations: gc 1: first-order giant cell; gf: 2 second-order giant fibre; gp: Pedal ganglion; gpv: Palliovisceral ganglion; ib: Interaxonic bridge; lba: Lobus (L.) basalis anterior; lbp: L. basalis posterior; lfi: L. frontalis inferior; lfs: L. frontalis superior; lv: Lobus verticalis; m: macula; np: Nervus postorbitalis; npa: Nervus pallialis; st: Statocyst. Scale bar: 200 μm.

which the animal shoots rapidly through the water" (Young 1939). Giant fibres only are present in teuthoid and sepioid decapodan cephalopods (e.g. Loligo, Sepia) but not in octopods, in which the mantle is no longer the main locomotory organ. Young (1939) described the general arrangment of the giant fibre system in adults as well as numerous details of various parts of it. His description deals with the species Loligo pealei but it appears to correspond to the situation in other squid species.

Within the central nervous system of an adult squid, close to the meeting point of the optic, cerebral, pedal and palliovisceral ganglia and just above the statocysts and amongst some other large cells, lie two very distinct giant cells (Fig. 4). The cells are found in a clearly demarked lobe and due to their presence the lobe is called "magnocellular lobe" or "lobus magnocellularis". The two giant cells, one on each side, are large motor neurones, termed "first-order giant cells". One should remember that the whole giant fibre system, including the effecting muscles which it innervates, can be set in action by impulses generated in either one or both of these first-order giant cells. The axons of the two cells extend into the palliovisceral lobe where they form a chiasma. They then form numerous branches and make synapses with "second-order giant cells" (about 7 in Loligo), the cell bodies of which also lie in this lobe. The axons of the second-order giant neurones run to motor endings in the funnel muscle and the retractor muslces of the funnel and head as well as to the stellate ganglia via the mantle connective. In the stellate ganglia they form synapses (the giant synapse amongst others) with "third-order giant axons" which form by fusion and which pass out as giant fibres through the stellate nerves to innervate the circular muscle fibres of the mantel. The third-order giant fibres are the well-known "giant axons of the squid". Martin (1977) provides a schematic representation of the various connexions in the giant fibre system of Loligo (Fig. 5).

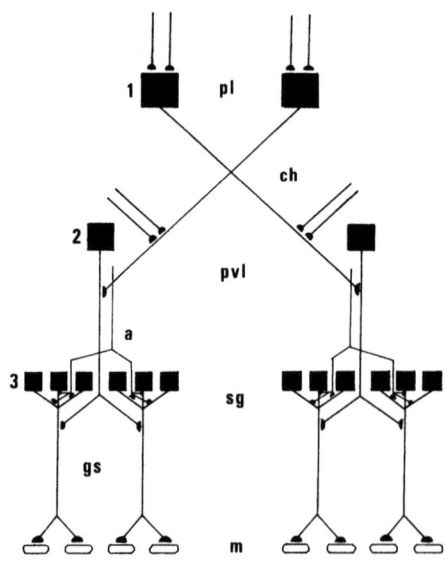

FIG. 5: Schematic representation of the connexions in the giant fibre system of Loligo vulgaris. (After Martin 1977).
Abbreviations: 1) first-order giant cell; 2) second-order giant cell; 3) third-order giant fibres; - a: access; ch: chiasma; gs: giant synapse; m: mantle; pl: pedal lobe; pvl: palliovisceral lobe;—•: afferent boutons.

The first elements in the motor-neurone-chain of the giant fibre system are the two first-order giant cells of the magnocellular lobe. They are in an ideally central position within the brain and can receive (and transmit) impulses from a great variety of sources (Fig. 5: afferent boutons). Impulses can then pass directly to all the muscles affecting the rapid movements of the animal such as when swimming, attacking, escaping, combined with directed water jets by rapidly orienting the funnel and often accompanied by particular chromatophore displays. An impulse generated in the first-order giant cell(s) must pass one synapse to reach the retractor muscles of the funnel and head but pass over two synapses to reach the muscles of the mantle (Fig. 5). The synapse in the stellate ganglion is the well-known "giant synapse of the squid" (for detailed review see Martin and Miledi 1986).

3.1. The formation of the giant nerve fibre system

As is the case for the entire central and peripheral nervous systems, extensive studies are still necessary in order to fully understand (in anatomical, cytological, chronological and physiological terms) the process of differentiation of the giant nerve fibre system. However, highly valuable information is already available for some elements of this process and in particular, on the development of the first-order system and its chiasma (Martin 1965 onwards; Martin and Rungger 1966).

Referring to the studies of Martin (1965, 1969, 1977 and personal communication) and Martin and Miledi (1986) nerve fibres in the brain of Loligo vulgaris embryos belonging to the giant fibre system can be first recognised at "day 12" of embryogenesis. This corresponds approximately to stage XII/XIII at a water temperature around 20°C. This event roughly coincides with the appearance of the first weak and irregular contractions of the differentiating mantle musculature. The giant fibre system is not yet functional at this stage however and it is likely that small nerve fibres are responsible for these contractions. Within the pedal ganglion, two bi-polar cells can be distinguished by their large size, the "first-order giant cells". Their axons run backwards forming a commissure in the palliovisceral lobe but appear not to be connected to second-order giant axons. Third-order giant axons can be identified in the peripheral stellate ganglion by their size. The cells in the giant fibre lobe of Loligo are not really "giant" but of usual size; the giant axons form by fusion of a large number of smaller axons.

In the course of subsequent embryonic stages, growth and cytological differentiation of the giant fibre system continue. One observes a clearcut differentiation gradient, that is, the differentiation of first-order giant cells precedes that of second and third-order giant fibres. The nuclei of the first-order giant cells increase progressively in size, showing diameters of about 8 µm at stage XIV and of 12-13 µm at stage XX. The nuclear diameters of the surrounding cells are around 6 µm throughout embryogenesis. In adults, the average diameter of the nucleus of a first-order giant cell is approximately 35 µm and that of the nuclei from surrounding cells of the magnocellular lobe about 26 µm). The diameters of the giant axons also increase at the same time; at stage XIV-XV they are about 1 µm in diameter and at stage XX and in early post-embryogenesis about 6-8 µm. The thickening of the axons appears to be accelerated during the late phase of embryogenesis, that is, from stage XVIII onwards, but increases more slowly after hatching. In hatchlings, the first-order giant cells show a large soma, thick dendrites and thick axons with several ramifications. The connexions with the sense organs also appear to be essentially established: afferent boutons can be localised on the first-order giant axons.

The giant fibre system therefore may be considered as functional - and thus "adultlike" - from stage XIX-XX onwards. In fact, from stage XVIII onwards, embryos, when removed from their chorion and placed in sea water, respond strongly to external stimuli by rapid swimming, flight reactions and pronounced chromatophore displays. While, still within the chorion however, embryos close to hatching are in a "tranquil state" and are surprisingly insensitive to external stimuli, due to the presence of a natural tranquilliser (Marthy et al. 1976; Weischer and Marthy 1983). This tranquillising factor prevents effectively premature hatching; it may interfere either with parts of the giant nerve fibre system and/or afferent fibres. Compared to the differentiation state of the central and peripheral nervous systems in general, the giant fibre system is clearly more developed at the moment of hatching. A crucial question raised by Martin (1965) is still timely: how the brain simultaneously can grow and operate in growing "larvae" and juveniles? No electrophysiological studies have been performed so far on the giant nerve fibre system of embryos hatchlings or small juveniles.

The process of differentiation of the second- and third-order giant fibres has not yet been studied. However the formation of the "interaxonic bridge" of first-order axons (Figs. 4, 5), the structure which is considered to be responsible for the functional bilaterality of the giant fibre system has been studied by Young (1939) and Martin (1965 and later). The description of Martin from histological and ultrastructural studies can be summarised as follows: the chiasma of the first-order giant cells becomes syncytial only in the course of post-embryogenesis. In embryos, the axons form a commissure in the palliovisceral lobe. This situation lasts until stage XVIII, each axon after the intersection then splits into a descending branch which proceeds alongside the branch of the contralateral axon. At hatching the crossing fibres are still not fused but appear to form synaptic contacts (Martin 1969: Figs. 6, 7). Thus in hatchlings and in young juveniles it appears that the functional bilaterality of the giant-fibre system is guaranteed even though the structure is not yet an adultlike "bridge" (Fig. 6).

In hatchlings of Loqigo vulgaris, a gradient of differentiation in the developing giant fibre system is evident also in the giant synapse in the stellate ganglion; the presynaptic second-order giant fibre, in striking contrast to the situation in adults, is about as thick as the post-synaptic (third-order) giant motor axon (Martin and Miledi 1986; Fig. 37).

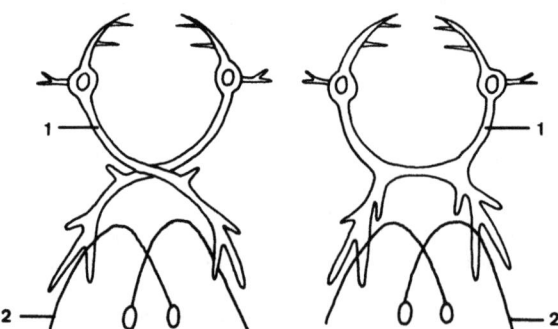

FIG. 6: Schematic drawing of the course of fibres in the chiasm of the first-order axons in the "larval" and adult squid Loligo vulgaris. In the hatchlings the "interaxonic bridge" of fused first-order giant axons is not yet complete. 1: first-order giant axons; 2: second-order giant axons. (After Martin 1977).

4. CONCLUSION

As stated earlier, the intention of this review is to give a somewhat easier access to the recent literature on the ontogenesis of the nervous sytem in cephalopods, a highly interesting field of investigation but one which is poorly explored. From the essentially anatomical and histological data given here at least two main points emerge: First, the often cited authors such as Marquis, Martin and Meister have created (based on earlier studies) a solid basis for further descriptive and experimental investigations on the central and peripheral nervous systems. Second, the numerous open questions, particularly those which may be solved only by experimental approaches, are a real "challenge" for undertaking or continuing the analysis of the development of the nervous system in cephalopods. There is no doubt that the experimentation in vivo on older embryos and small juveniles will be extremely difficult; last but not least, these technical difficulties may be an additional challenge for younger scientists!

5. ACKNOWLEDGMENTS

The kind permission of Fred Marquis (Basel, CH) to use some illustrations from his thesis work is gratefully acknowledged. Rainer Martin (Ulm, D) critically read Section 3. Ursula Marthy (Banyuls, F) prepared the illustrations and Richard Tait (Banyuls, F) made the text corrections. Many thanks to all of them.

6. REFERENCES

Boletzky SV (1968) Untersuchungen über die Organogenese des Kreislaufsystems von Octopus vulgaris Lam. Rev Suisse zool 75: 765-812.

Boycott BB Young JZ (1955) A memory system in Octopus vulgaris Lam. Proc R Soc Lond B 143: 449-480.

Boyle PR (ed) (1983) Cephalopod life cycles. Vol I. Academic Press, London.

Boyle PR (1986) Neural control of cephalopod behavior. In: Willows AOD (ed) The Mollusca. Vol 9, Part 2, Neurobiology and behavior. Academic Press, New York, pp 1-99.

Dubas F, Hanlon RT, Ferguson GP, Pinsker HM (1986) Localization and stimulation of chromatophore motorneurones in the brain of the squid, Lolliguncula brevis. J Exp Biol 121: 1-25.

Faussek V (1901) Untersuchungen über die Entwicklung der Cephalopoden Mitt Zool Sta Neapel 14: 83-237.

Frösch D (1971) Quantitative Untersuchungen am Zentralnervensystem der Schlüpfstadien von zehn mediterranen Cephalopodenarten. Rev Suisse Zool. 78: 1069-1122.

Korschelt E (1892) Beiträge zur Entwicklungsgeschichte der Cephalopoden. I. Die Entstehung des Darmkanals und des Nervensystems in Beziehung zur Keimblätterfrage. Verh d Zool - Bot Ges, Leipzig, Festschr Leukart: 345-373.

Marquis F (1981) Untersuchungen über die Entwicklung des Nervensystems im Embryo von Octopus vulgaris Lam. Inauguraldissertation, Universität Basel (CH).

Marthy H-J (1973) An experimental study of eye development in the cephalopod Loligo vulgaris: determination and regulation during formation of the primary optic vesicle. J Embryol Exp Morphol 29: 347-361.

Marthy H-J, Aroles L (1983) Culture in vitro du complexe oculoganglionnaire de l'embryon du cephalopode Loligo vulgaris. Biol Cell 49: 16a.

Marthy H-J, Aroles L (1987) In vitro culture of embryonic organ and tissue fragments of the squid Loligo vulgaris with special reference to the establishment of a long term culture of ganglion-derived nerve cells. Zool Jb Physiol 91: 189-202.

Marthy H-J, Hauser R, Scholl A (1976) Natural tranquilliser in cephalopod eggs. Nature 261: 496-497.

Martin R (1965) On the structure and embryonic development of the giant fibre system of the squid Loligo vulgaris. Z Zellforsch 67: 77-85.

Martin R (1969) The structural organization of the intracerebral giant fibre system of cephalopods. I. The chiasma of the first order giant axons. Z Zellforsch 97: 50-68.

Martin R (1977) The giant nerve fibre system of cephalopods, recent structural findings. Symp Zool Soc Lond 38: 261-275.

Martin R, Miledi R (1986) The form and dimensions of the giant synapse of squids. Phil Trans R Soc Lond B 312: 355-377.

Martin R, Rungger D (1966) Zur struktur und Entwicklung des Riesenfasersystems erster Ordnung von Sepia officinalis L. Z Zellforsch 74: 454-463.

Mauro A (1977) Extra-ocular photoreceptors in cephalopods. Symp Zool Soc Lond 38: 287-308.

Meister G (1972) Organogenese von Loligo vulgaris Lam. Zool Jb Anat 89: 247-300.

Messenger JB (1979) The neuron system of Loligo. IV. The peduncle and olfactory lobes. Phil Trans R Soc Lond B 285: 275-309.

Messenger JB (1982) Multimodal convergence and the regulation of motor programs in cephalopods. Fortschr Zool 28: 77-98.

Naef A (1928) Die Cephalopoden. Fauna Flora del Golfo di Napoli. V-IX, 35 (2): 1-357. Berlin.

Packard A, Albergoni V (1970) Relative growth, nucleic acid content and cell numbers of the brain in Octopus vulgaris (Lamarck). J Exp Biol 52: 539-552.

Sacarrao GF (1956) Contribution à l'étude du développement embryonnaire du ganglion stellaire et de la glande endocrine des Céphalopodes. Arqu Mus Bocage, Lisboa 27: 137-152.

Weischer M-L, Marthy H-J (1983) Chemical and physiological properties of the natural tranquilliser in the cephalopod eggs. Mar Behav Physiol 9: 131-138.

Wells MJ (1962) Brain and behaviour in cephalopods. Heinemann Studies in Biology, London, Melbourne, Toronto.

Wells MJ (1978) Octopus. Physiology and behaviour of an advanced invertebrate. Chapman and Hall, London.

Wirz - Mangold K (1959) Etude biométrique du système nerveux des Céphalopodes. Bull Biol 93: 78-117.

Yamamoto M (1985) Ontogeny of the visual system in the cuttlefish, Sepiella japonica. I. Morphological differentiation of the visual cell. J Comp Neurol 232: 347-361.

Yamamoto M, Takasu N, Uragami I (1985) Ontogeny of the visual system in the cuttlefish, Sepiella japonica. II. Intramembrane particles, histofluorescence, and electrical responses in the developing retina. J Comp Neurol 232: 362-371.

Young JZ (1939) Fused neurones and synaptic contacts in the giant nerve fibres of cephalopods. Phil Trans R Soc Lond B 229: 465-503.

Young JZ (1963) The number and sizes of nerve cells in octopus. Proc Zool Soc Lond 140: 229-254.

Young JZ (1971) The anatomy of the nervous system of Octopus vulgaris. Clarendon Press, Oxford.

Young JZ (1974) The central nervous system of Loligo. I. The optic lobe. Phil Trans R Soc Lond B: 267: 263-302.

NERVOUS SYSTEM IN CHAETOGNATHA

T. GOTO and M. YOSHIDA

Ushimado Marine Laboratory

130-17 Kashino, Ushimado

Okayama 701-43, Japan

ABSTRACT

The nervous system of chaetognatha consists of 6 ganglia in the head region, one ventral ganglion in the body, nerves connecting these ganglia and peripheral nerves passing out of these ganglia. The ganglia in the head region are the cerebral ganglion, a pair of vestibular ganglia, a pair of oesophageal ganglia, and a suboesophageal ganglion. Putative sensory organs are a pair of eyes, the corona (a ciliary loop) behind the eyes, ciliary tufts on the body surface and papillae around the mouth. Decapitated worms lose photoresponsiveness but can swim normally. When the ventral ganglion is ablated, the worms can no longer swim, suggesting that a motor center is located in this ganglion. The nerves from the eyes and the corona run directly to the cerebral ganglion where the sensory information must be integrated. Except for the swimming movements, we can see movements of hooks which participate in grasping prey. A pair of vestibular ganglia have been considered to control these movements. Neither large neurons nor thick fibers have been detected. However, axonal pathways of relatively thick fibers are detected by vital staining using methylene blue. Immunohistochemical studies also gave information on axonal pathways, including neurotransmitter candidates in the nervous system. Serotonin-, methionin-enkephalin-, and neurotensin-like immunoreactivities exist in the nervous system, suggesting that these substances are neurotransmitter candidates in chaetognatha.

1. INTRODUCTION

Chaetognaths, commonly called arrowworms, move very quickly and show complex behaviors such as phototactic movements (Goto and Yoshida 1981, 1983), feeding (Horridge and Boulton 1967; Feigenbaum and Reeve 1977), egg-laying (unpublished) and mating (Goto and Yoshida 1985). However, the neural mechanisms controlling their behavior are uncertain.

Chaetognaths have a well-developed nervous system. The general structure of the nervous system studied by earlier workers (Hertwig 1880;

Grassi 1883; Burfield 1927; John 1933) have been reviewed by Hyman (1959) and Bullock and Horridge (1965). Recently, Bone and Pulsford (1984) examined the nervous system, with special reference to the ventral ganglion, by using techniques of vital staining with methylene blue, silver impregnation and electron microscopy. Rehkamper and Welsch (1985) have showed the fine structure of the cerebral ganglion. As regards the synaptic organization at the neuromuscular junctions, Duvert and Barets (1983) have proposed that the motor terminals of the body muscles lie external to an acellular thick basement membrane.

In this article, we describe the neural architecture (mainly in Sagitta crassa), studied at the light and electron microscopic levels. In addition, neurotransmitter candidates have been sought immunohistochemically. This work gave information on axonal pathways for neurons which contain specific neurotransmitters.

2. ARRANGEMENT OF THE NERVOUS SYSTEM

2.1. General Plan

The central nervous system consists of 6 ganglia in the head region, one ventral ganglion in the body, nerves connecting these ganglia and peripheral nerves passing out of these ganglia (Fig. 1). The ganglia in the head region are the cerebral ganglion; a pair of vestibular ganglia; a pair of oesophageal ganglia; and a suboesophageal ganglion (Fig. 1C). Earlier works describe a pair of frontal ganglia located near each vestibular ganglion but we have not detected them in Sagitta crassa. We have found, however, an undescribed ganglion beneath the oesophagus and propose to call it the suboesophageal ganglion. Connective nerves are found between the cerebral and the vestibular, the cerebral and the ventral, and the vestibular and the oesophageal ganglia. A commissural nerve is found only between the vestibular ganglia.

According to the terminology of earlier investigators, all the nerves connecting each ganglion have been called "commissures" without distinguishing the connective and the commissure nerves. In this article we propose to modify the Burfield's terminology in accordance with the neuroanatomical terminology as defined by Bullock and Horridge (1965). Thus, the antero-posteriorly running nerves which connect the cerebral and the ventral ganglia will be called the main connectives, and those which connect the cerebral and the vestibular ganglia, the frontal connectives.

2.2. Functional Implications

2.2.1. Sensory

Several kinds of sensory organs have been proposed, but, lacking functional verification, most of them must be taken with some reserve. The most obvious organs are the paired eyes, deprivation of both of which abolishes phototactic movements (Goto and Yoshida 1983). Using the method of Ozaki et al. (1983), a rhodopsin-like substance is histochemically detected in the eye (unpublished). Other organs supposed to be sensory are the ciliary tuft (Figs. 2A and C, ct) on the body surface, a ciliated loop called the corona (Figs. 2A and C, c) behind the eyes, and the papillae (Figs. 2B and D, pa) around the mouth and on the vestibular ridge lying behind the posterior teeth. All these organs possess ciliated cells and send nerves to ganglia specific for each organ: the optic nerve and the coronal nerve to the cerebral ganglion, the nerves from the papillae to the vestibular ganglion, and those from the ciliary tuft probably to

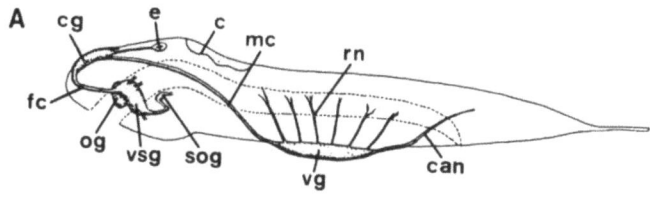
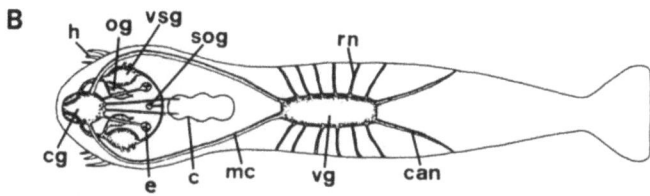
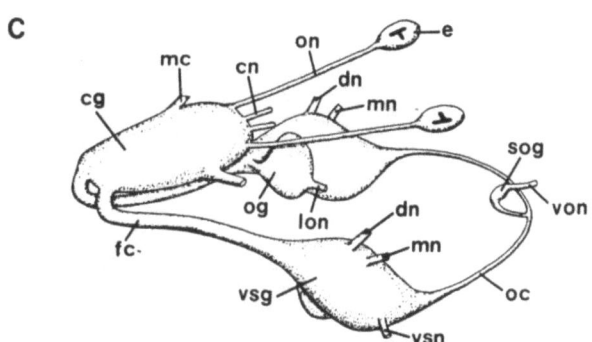

FIG. 1: General scheme of the nervous system of an arrowworm. A: Lateral view. B: Dorsal view. C: Three-dimensional drawing of the nervous system in the head region. c, corona; can, caudal nerve; cg, cerebral ganglion; cn, coronal nerve; dn, dorsal nerve; e, eye; fc, frontal connective; h, hooks; lon, lateral oesophageal nerve; mc, main connective; mn, mandibular nerve; oc, oesophageal commissure; og, oesophageal ganglion; on, optic nerve; rn, radial nerve; sog, suboesophageal ganglion; vg, ventral ganglion; von, ventral oesophageal nerve; vsg, vestibular ganglion; vsn, vestibular nerve.

the ventral ganglion. Figures 2E and F show axons running out of the corona and the papillae, respectively.

2.2.2. Motor

Swimming movements are produced by bending the body. Spreading and closing the hooks for grasping prey are accomplished by some of the head muscles. Probably, radially extending nerves from the ventral ganglion innervate the body muscles and nerves from the vestibular ganglia, the muscles related to the hook movements.

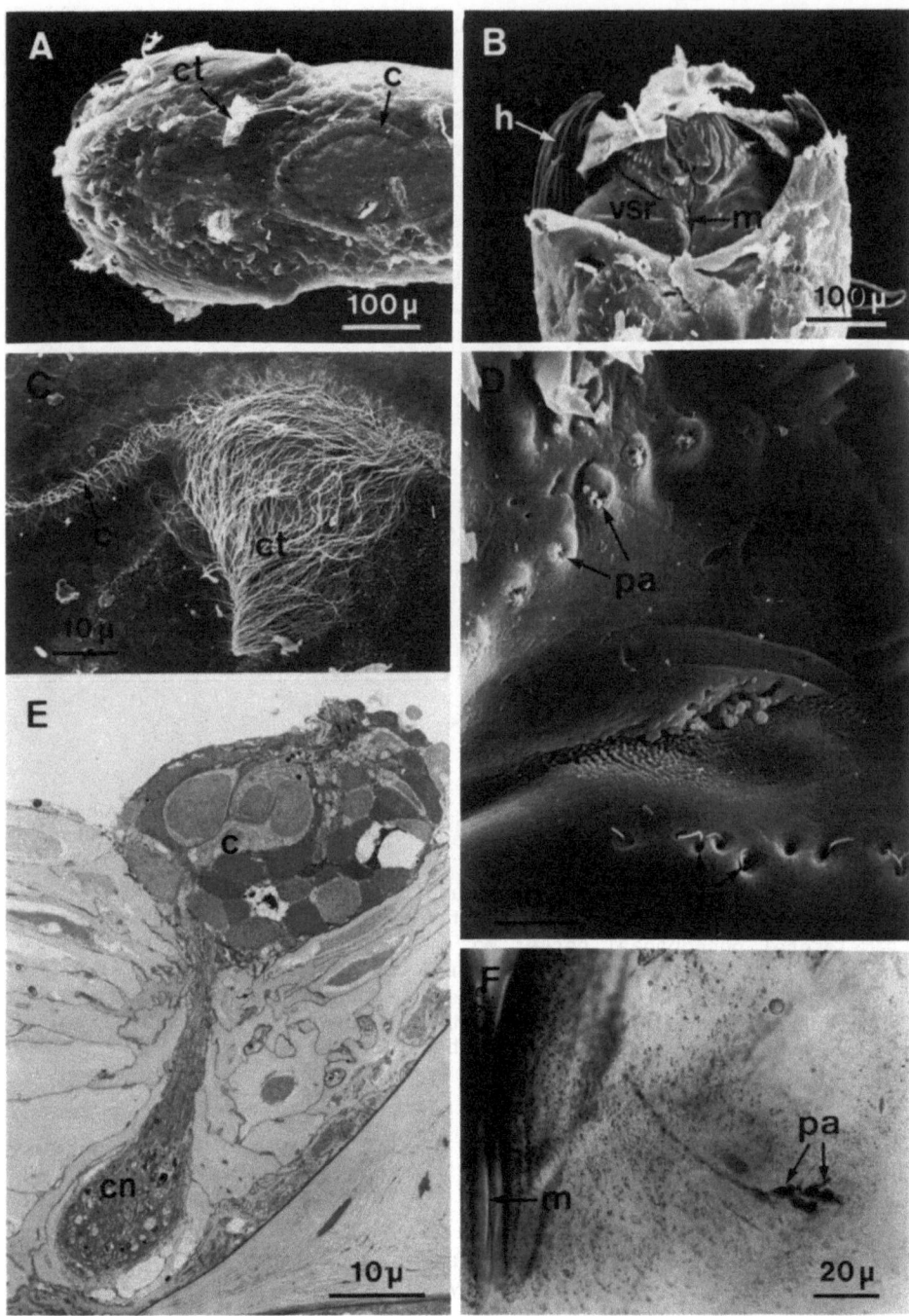

FIG. 2: Putative sensory organs. A and B: SEM views of the dorsal and the ventral surface of the head region, respectively. C: Higher magnification of corona (c) and ciliary tuft (ct). D: Papillae (pa) found near the mouth which is located on the left of this figure. E: Axons from the corona (c) to the coronal nerve (cn). F: Sensory cells of papillae and their axons stained with methylene blue. h, hooks; m, mouth; vsr, vestibular ridge.

3. DETAILED STRUCTURE OF EACH GANGLION

3.1 The Cerebral Ganglion

This ganglion is situated near the anterior end beneath the epidermis of the dorsal surface and slightly projects over the body surface, forming a flattened hump (Fig. 3). The ganglion is composed of an extensive area of neuropil and a distinct aggregate of somata. From the cerebral ganglion, connective nerves pass to the vestibular and the ventral ganglia (Fig. 1). Peripheral nerves arising from this ganglion go to the pair of eyes and the corona.

At the anterior end the frontal connectives pass out of this ganglion downward and outward to the vestibular ganglia (Fig. 1C) so that, in serial transverse sections cut from anterior to posterior (Fig. 4), neuropils appear first lateral to a mass of medially situated somata (Fig. 4A). The neuropils then fuse with one another (Fig. 4B) and another component of neuropil appears dorsally (Fig. 4C). After all these are combined together, lateral components begin to depart from the central one (Fig. 4D). The lateral components become the main connectives which run ventro-laterally and descend to the ventral ganglion. Then somata appear at the dorsal region (Fig. 4E). In this profile, we should pay attention to the somata extending their axons to the neuropil (arrowheads). On both sides at a more posterior region, there appears a different structure called the retrocerebral organ (Fig. 4F). This structure is composed of closely packed microvilli, each containing a central thread (Fig. 5A, confirmation of Scharrer 1965). The cell bodies bearing microvilli are rich in mitochondria (Fig. 5C). An opening named the retrocerebral pore appears at a medial region (Fig. 4H) where several cells extend microvilli (Fig. 5B, arrowheads), similar to those in the retrocerebral organ. Scharrer (1965) pointed out a structural similarity between the retrocerebral organ and masses of microvilli found in the cerebral

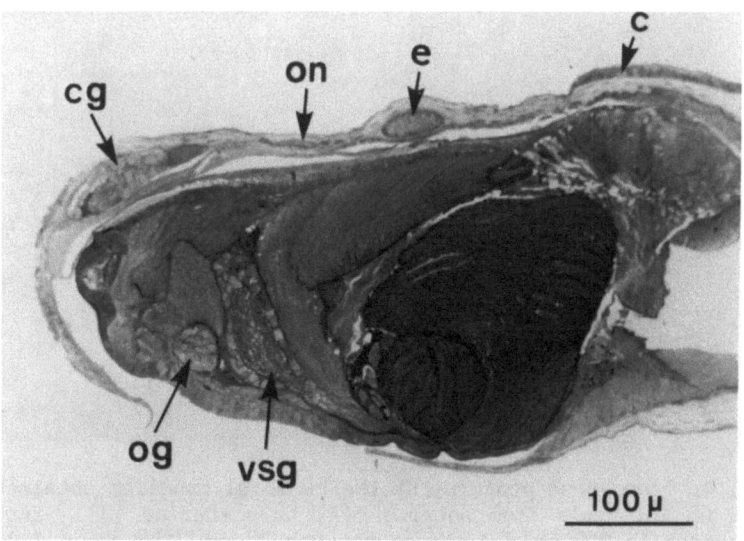

FIG. 3: An obliquely sagittal profile showing nervous elements in the head region of <u>Sagitta crassa</u>, 8-9 mm in body length. c, corona; cg, cerebral ganglion; e, eye; og, oesophageal ganglion; on, optic nerve; vsg, vestibular ganglion.

ganglion of the crustacean Leptodora. However, the function of the retrocerebral organ is unknown as yet. In Fig. 4G, the optic nerves become separated from the neuropil. The coronal nerves are not clear at this point but become evident a little more posteriorly (Fig. 4H). At the posterior end of the ganglion, the two pairs of peripheral nerves come out (Figs. 4I and J) and extend to the eyes and the corona.

Though many axons in the cerebral ganglion are derived from sensory cells, many must also be interneuronal (Fig. 6A). As mentioned above, somata sending their axons to the neuropil are mostly seen in the region

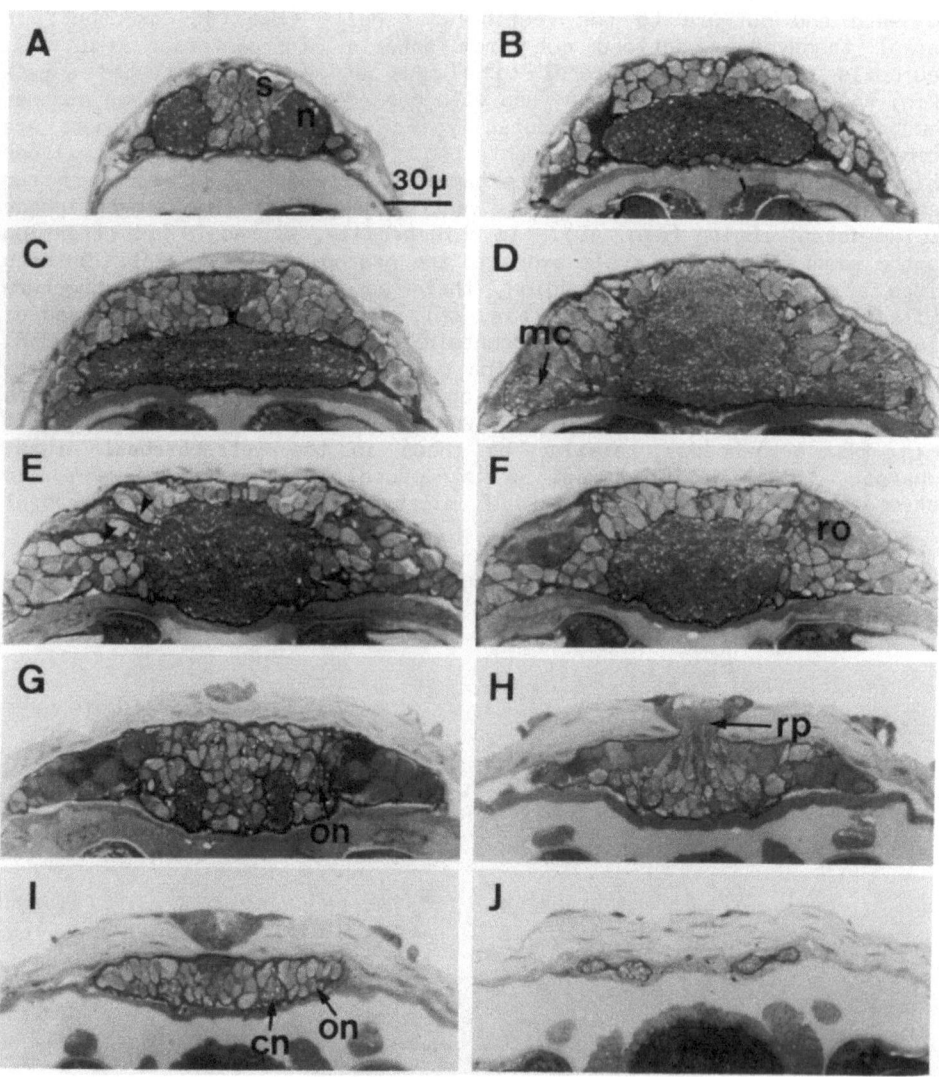

FIG. 4: Transverse profiles of the cerebral ganglion obtained at 5-20 μm intervals from anterior (A) to posterior (J). Section thickness is 0.5 μm. Distance between A and J is about 120 μm. Arrowheads in E indicate the somata extending their axons to neuropil. cn, coronal nerve; mc, main connective; n, neuropil; on, optic nerve; ro, retrocerebral organ; rp, retrocerebral pore; s, somata.

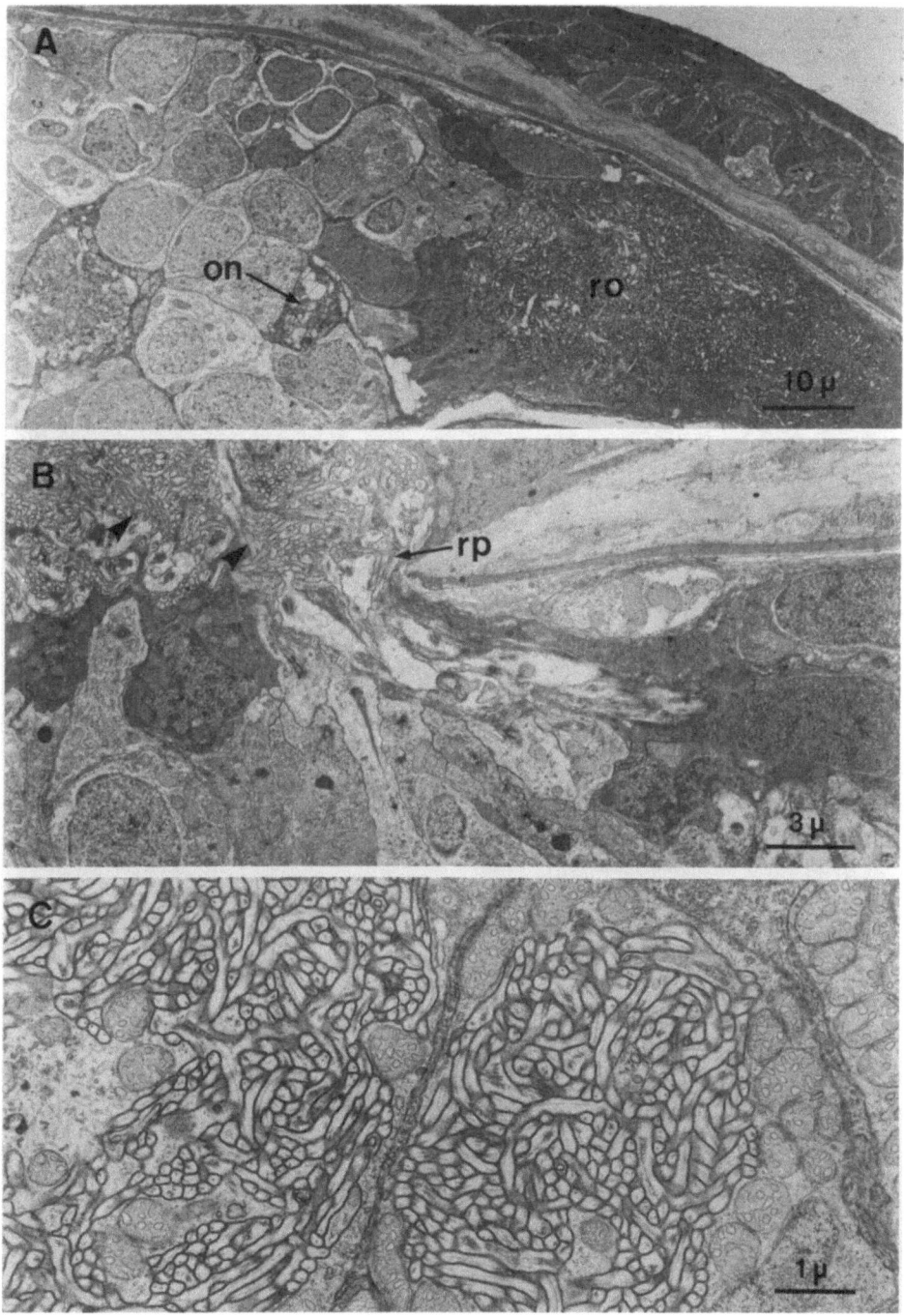

FIG. 5: EM profiles of the retrocerebral organ. A: Retrocerebral organ (ro) appears in the lateral region. B: Retrocerebral pore (rp) region. Arrowheads indicate microvilli. A and B correspond to Fig. 4G and H, respectively. C: Cells composing retrocerebral organ. on, optic nerve.

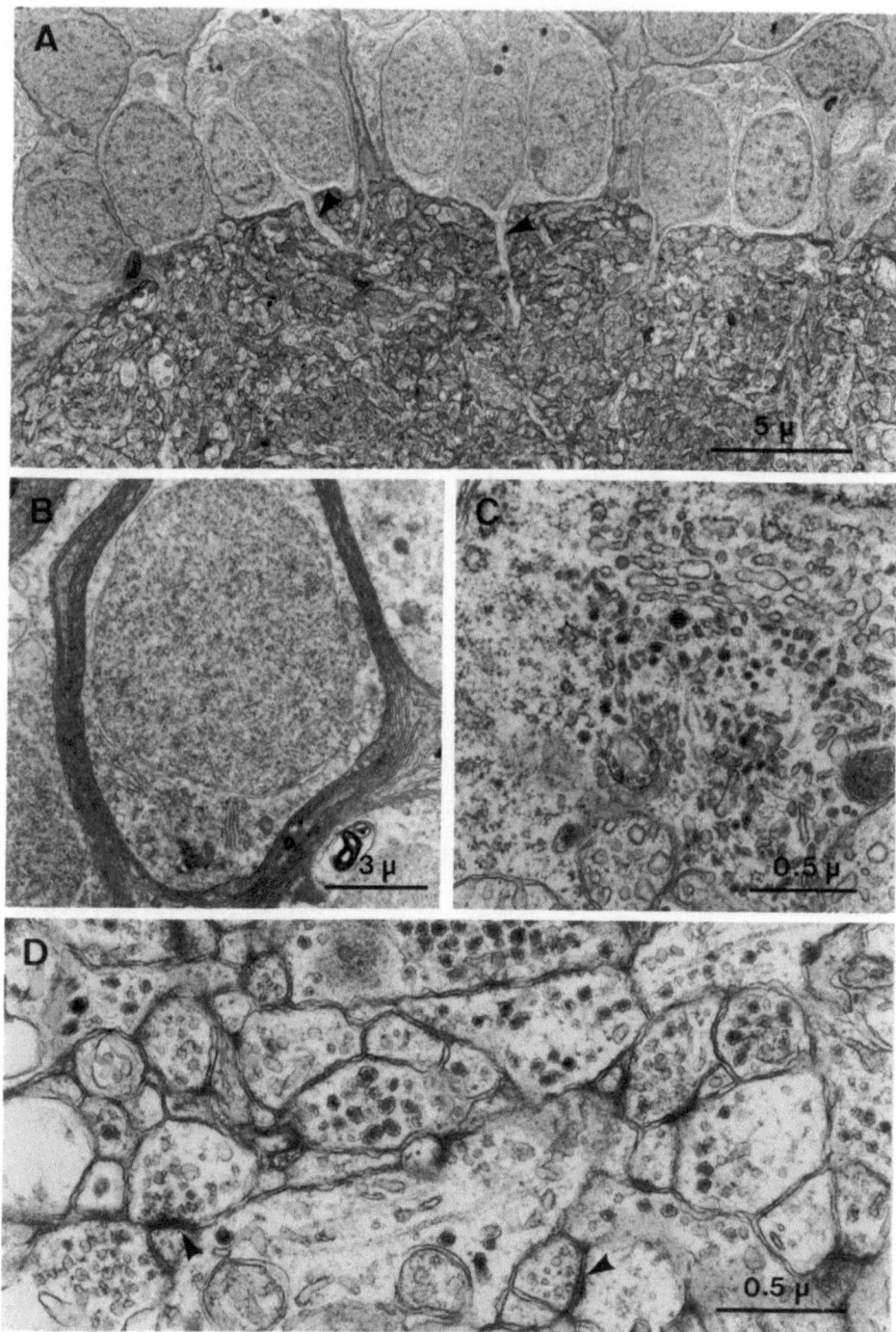

FIG. 6: EM profiles of somata and neuropil in the cerebral ganglion. A (correspond to Fig. 4E): A few somata extend their axons to neuropil (arrowheads). B: Somata covered with lamellar sheath remarkable in <u>Sagitta hexaptera</u>. C: Soma containing many vesicles. D: Higher magnification of neuropil showing fibers containing vesicles and synaptic contacts (arrowheads).

anterior to the retrocerebral organ (Fig. 4E). There are no large neurons or fibers. The somata are up to 5 μm in cross section. Rehkamper and Welsch (1985) pointed out in Sagitta setosa that each soma is covered with a lamellar sheath. We confirmed the presence of this sheath in several local species and Fig. 6B shows a markedly developed example found in Sagitta hexaptera. Some of the somata are provided with many vesicles, probably containing a neuroactive substance (Fig. 6C). The axons are small (average diameter, about 0.5 μm) and contain microtubules, mitochondria and vesicles (dense cored and clear) (Fig. 6D). Morphologically specialized chemical synapses similar to those in higher animals are also present (Fig. 6C, arrowheads).

Immunohistochemical studies have demonstrated that the cerebral ganglion contains a serotonin (5-HT)-like substance (Fig. 7). The positively reacting somata (Fig. 7A) and fibers (Fig. 7B) are present at the posterior region of the ganglion. Varicose fibers run randomly and spread widely in this area (Fig. 7B). It appears that these cells are a class of interneuron, because the positive fibers are confined within the ganglion.

Spike activities in this ganglion have only been observed spontaneously (Fig. 8). Though sensory inputs to this ganglion may be expected from the anatomical features, no responses to light or mechanical (water flow by a pipette) stimuli have been recorded as yet.

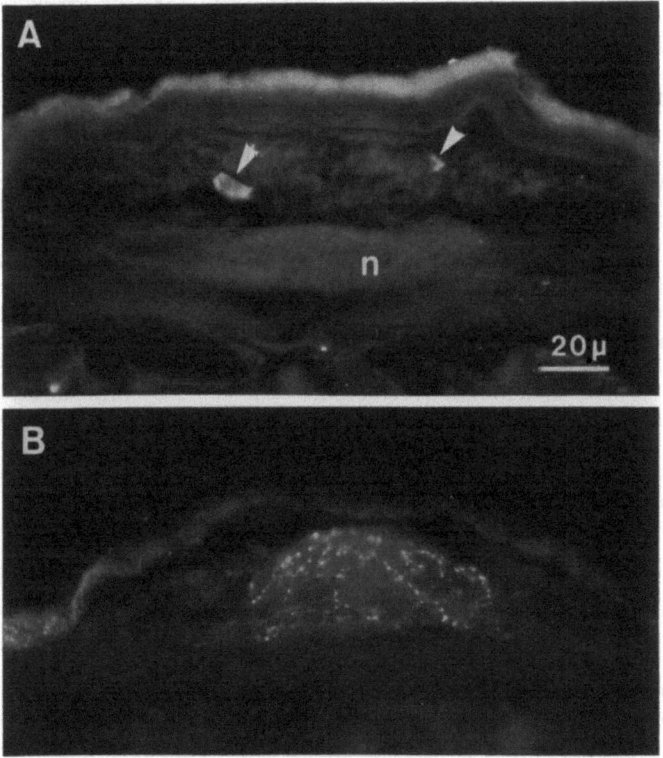

FIG. 7: 5-HT-like immunofluorescence produced in the cerebral ganglion by an indirect immunofluorescent method using FITC. Sections were cut transversely. A (correspond to Fig. 4E): Positively reacting somata (arrowheads). B (correspond to Fig. 4D): Positively reacting fibers. n, neuropil; s, somata.

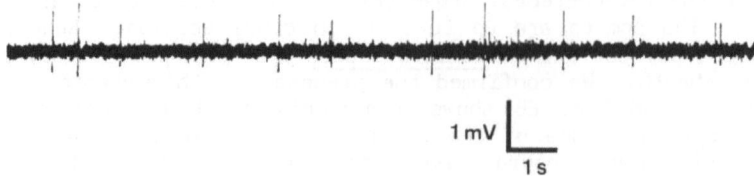

FIG. 8: Spontaneous electrical activity recorded from the cerebral ganglion. A specimen of Spadella schizoptera placed on Sylgard was immobilized with a piece of plankton net (60 μm mesh). A glass microelectrode filled with 2M KCl was placed in the ganglion and extracellular activity was recorded through an AC amplifier.

3.2 The Vestibular Ganglia

A pair of vestibular ganglia lie on both sides of the oesophagus, posterior to the cerebral ganglion (Fig. 1). These ganglia are connected to each other by the oesophageal commissure and each ganglion is connected anteriorly with the cerebral ganglion by the frontal connective and medially with the oesophageal ganglion by a short nerve.

The complicated shape of the vestibular ganglion was followed by semiserial transverse sections (Fig. 9). The frontal connectives which run out of the ganglion anteriorly (Fig. 1) appear first (Fig. 9A). Earlier works describe a frontal ganglion which sends axons to the posterior teeth near the antero-lateral region of the vestibular ganglion. Burfield (1927) claimed, however, in Sagitta bipunctata that the frontal ganglion is hardly distinguishable from the vestibular ganglion. We also failed to find any structure other than the vestibular ganglion in the corresponding region of Sagitta crassa. A nerve connecting with the oesophageal ganglion arises medially from the base of the frontal connective (Figs. 9B and 11A). Two other nerves pass out dorso-laterally; one is the dorsal nerve (Fig. 9C) which extends to head muscles and the other is the mandibular nerve (Fig. 9D) which extends to the muscles for hook movements. Ventro-laterally, the vestibular nerve goes to a region near the mouth (Fig. 9E) where the papillae exist (Figs. 2B and D). Bone and Pulsford (1984) found that the sensory cells of the papillae extend their axons to the brain ganglia (this may be the vestibular ganglion). From the posterior end of the ganglion, the oesophageal commissure arises (Fig. 9F). The nerves run near inner surface of the ventral epithelium (Fig. 10A) and meet together at the center (Fig. 10D) where a pair of eyes appear on the dorsal side.

Somata exist mainly in the dorso-medial area and the neuropil occupies a large area where large fibers are scattered. Vesicles of different kinds are found in the neuropil but the nature of neuroactive substance(s) is unknown as yet.

Animals deprived of the cerebral ganglion are able to spread and close the hooks on both sides together, which may probably be achieved by the commissure nerve between the vestibular ganglia.

3.3 The Oesophageal Ganglia

A pair of small oesophageal ganglia lie on both sides of the oesophagus and connect with the vestibular ganglia existing more laterally

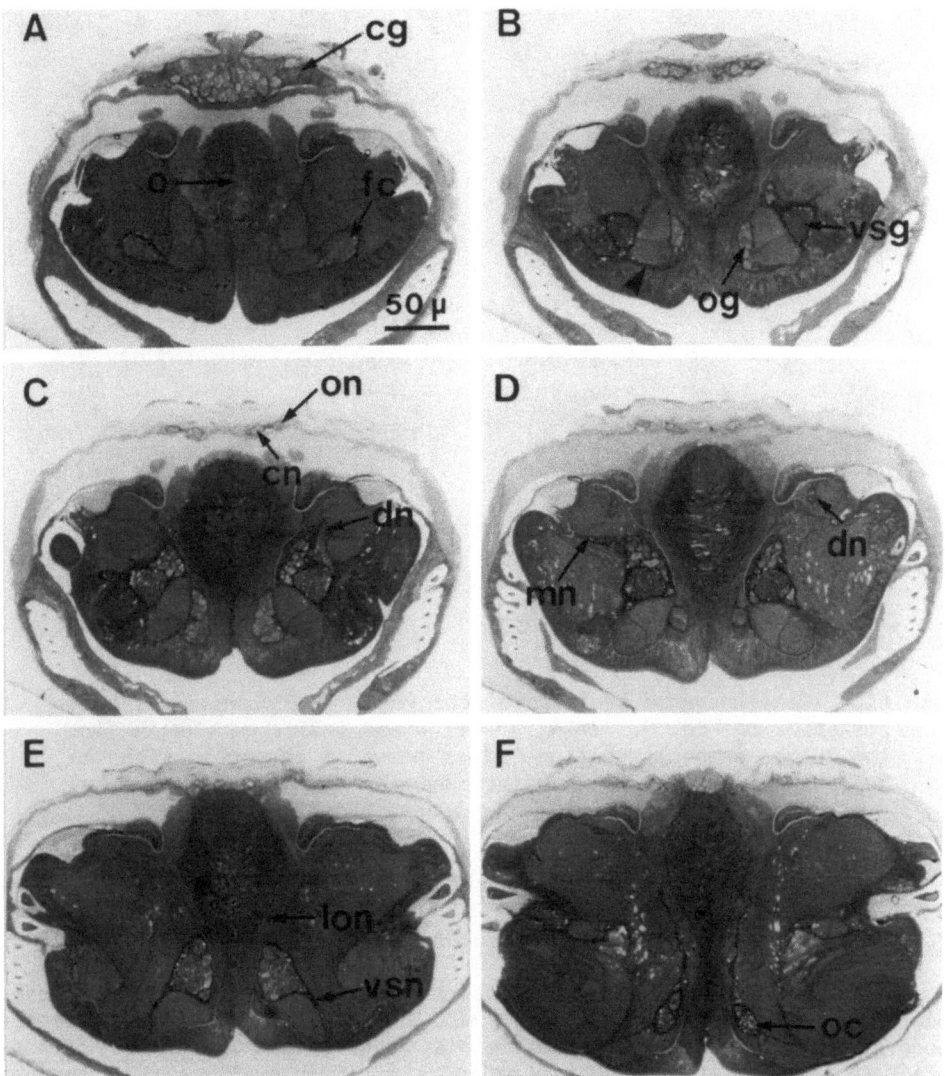

FIG. 9: Transverse profiles of the vestibular and the oesophageal ganglia. From anterior (A) to posterior (F). A, B and C correspond to H, I, and J in Fig. 4, respectively. An arrowhead in B indicates a nerve connecting between the vestibular and the oesophageal ganglia. cg, cerebral ganglion; cn, coronal nerve; dn, dorsal nerve; fc, frontal connective; lon, lateral oesophageal nerve; mn, mandibular nerve; o, oesophagus; oc, oesophageal commissure; og, oesophageal ganglion; on, optic nerve; vsg, vestibular ganglion; vsn, vestibular nerve.

(Figs. 9B and 11A). In transverse sections, the oesophageal ganglia appear together with the posterior end of the cerebral ganglion (Fig. 10B). Unlike the other ganglia, the oesophageal ganglion is only composed of somata (Fig. 11A) and gives off lateral oesophageal nerves (Fig. 9D) from the posterior region.

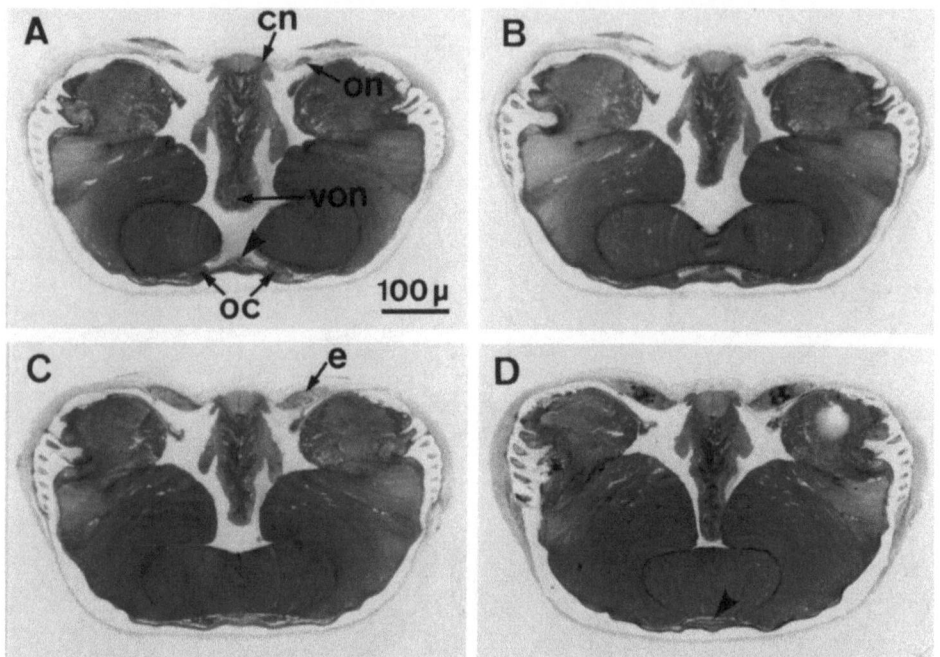

FIG. 10: Transverse profiles of the vestibular commissure, more posterior to Fig. 9. In A, an arrowhead indicates a nerve from the suboesophageal ganglion. This nerve is connected with the oesophageal commissure in D (arrowhead). cn, coronal nerve; e, eye; on, optic nerve; von, ventral oesophageal nerve.

3.4. The Suboesophageal Ganglion

Burfield (1927) described an unpaired nerve called the ventral oesophageal nerve arising from the mid line of the oesophageal commissure. We did not confirm Burfield's observation in Sagitta crassa. Electron microscopic observations revealed a single small ganglion beneath the oesophagus from where two nerve bundles (about 15 axons in each bundle) arise (Figs. 10A and 12); one goes downwards and fuses with the oesophageal commissure and the other runs along the oesophagus, forming the ventral oesophageal nerve. The reason why we call this structure a ganglion is the presence of a mass of somata (but not neuropil) at the base of the ventral oesophageal nerve (Fig. 12).

3.5. The Ventral Ganglion

This ganglion forms an elongate rectangle lying between the basement membrane and the outer epithelial layer, similar to the situation of the cerebral ganglion (Figs. 13A and C). Paired main connectives (Fig. 13B) pass obliquely forward and dorsally from the ventral ganglion to the cerebral ganglion. Peripheral nerves arising from here are paired caudal nerves running posteriorly and 12 pairs of nerves extending radially as described by earlier workers. Recently, Bone and Pulsford (1984) clarified the architecture of this ganglion using techniques of methylene blue vital staining and silver impregnation. Figure 14 is the methylene blue stained preparation of Sagitta crassa. Relatively large fibers are stained in the ganglion and the radial nerves.

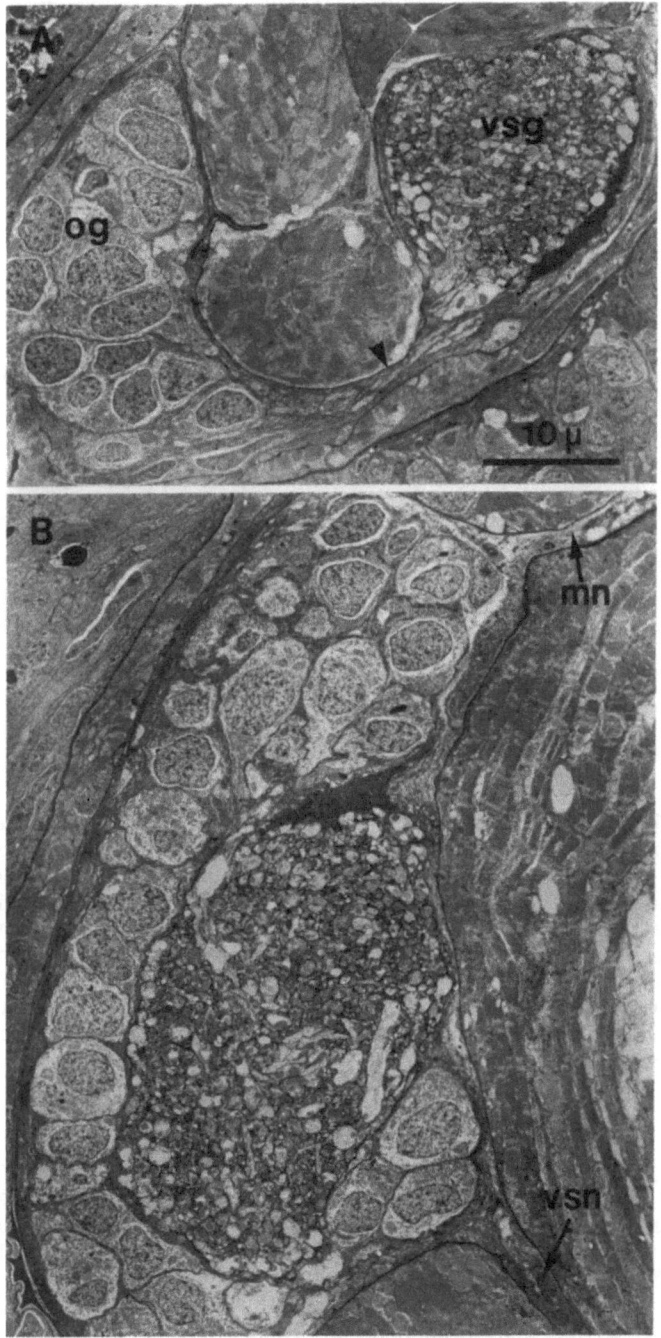

FIG. 11: EM transverse profiles of the vestibular ganglion. A (corresponds to Fig. 9B): The vestibular ganglion (vsg) is connected with the oesophageal ganglion (og) by a connective nerve (arrowhead). B (roughly corresponds to Fig. 4D): The vestibular ganglion giving off the mandibular nerve (mn) and the vestibular nerve (vsn). Dorsal and medial sides are upper and left sides in these figures, respectively.

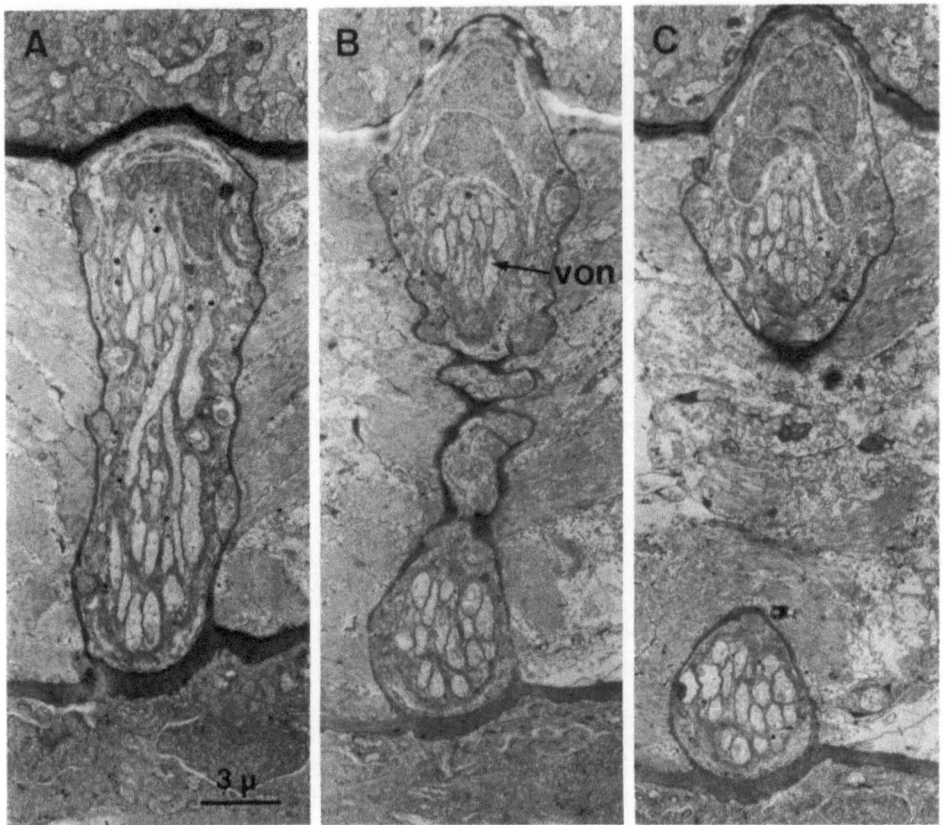

FIG. 12: EM profiles of the suboesophageal ganglion. Transverse sections from anterior (A) to posterior (C). A: A bundle of nerve fibers appears beneath oesophagus (upper side of this figure). B: The bundle begins to separate into two portions. C (corresponds to Fig. 10A): The two portions are completely separated. Lower one is connected to the oesophageal commissure as shown in Fig. 10. Upper one forms the basal region of the ventral oesophageal nerve (von) running along the oesophagus. In B and C, somata provided with a large nucleus are evident at the derivation of the ventral oesophageal nerve.

This ganglion contains an extensive area of central neuropil flanked by rounded somata which are packed in a lateral zone (Fig. 13). The somata have an average diameter of about 5 µm and rarely exceed 10 µm in diameter. Axons run longitudinally, transversely and vertically. Largest axons are 5-6 µm in diameter but most fibers are less than 1 µm. In the mid line of the ventral ganglion just below the ventral surface, there are cell bodies which are regarded as connective tissue cells (Bone and Pulsford 1984). As will be described below, there are also nerve cell bodies which are particularly evident in the posterior portion of this ganglion (Fig. 13C, arrows).

An idea of the layout of this system was given by Bone and Pulsford (1984) (Fig. 15). There are 4 pairs of large neurons which run longitudinally in the ganglion. Three paired fibers named A1 to A3 are derived from neurons within the cerebral ganglion and one pair named C1 arise within the ventral ganglion itself. In the order from near mid line

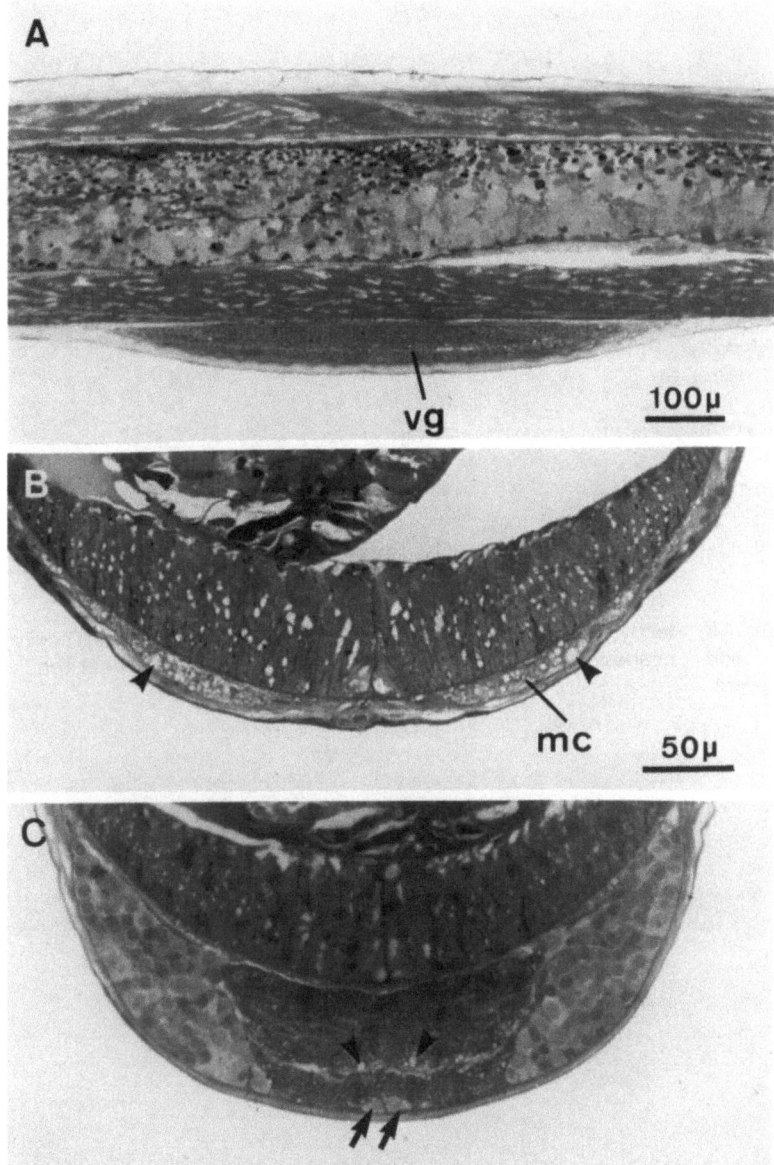

FIG. 13: A: A sagittal profile of the body through the ventral ganglion (vg) which lies beneath the ventral muscle layer. B: A transverse profile through the main connective (mc). Clear fibers (arrowheads) are evident in each nerve. C: A transverse profile of the posterior region of the ventral ganglion. Paired clear fibers (arrowheads) are evident near the mid-line. At most ventral region, two somata are found (arrow). Arrowheads in B and C indicate large fibers.

toward laterally, A1, A2, C1 and A3 are arranged. Only paired A3s pass out to the caudal nerve and others terminate in the posterior end of the ganglion. The C1 pair may be composed of multiple fibers, closely apposed to each other. They are connected by transverse bridges at several points

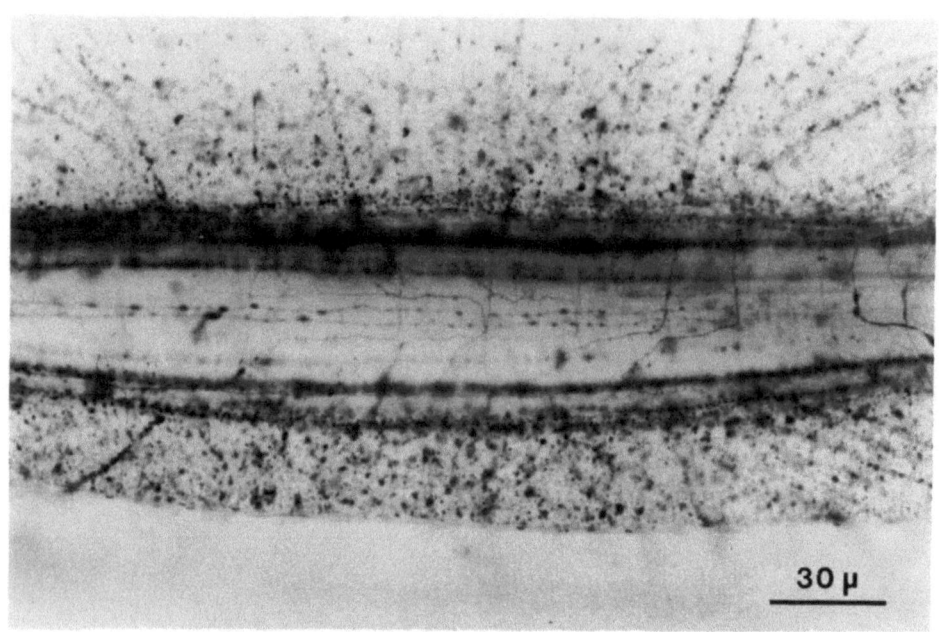

FIG. 14: Methylene blue stained ventral ganglion. Longitudinally and transversely running fibers with varicosities are evident.

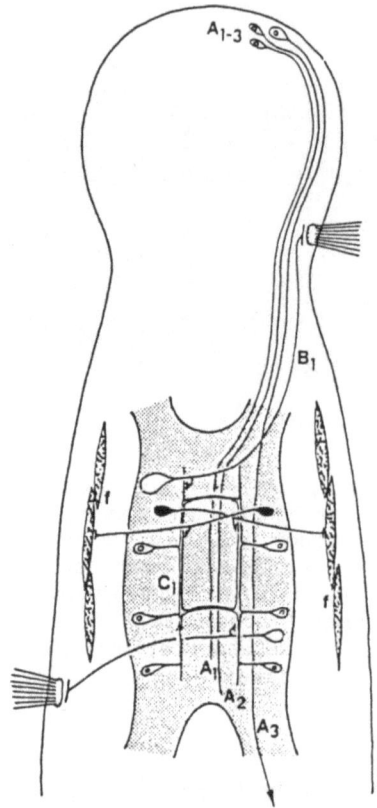

FIG. 15: Schematic diagram of pathways of large fibers (from Bone and Pulsford 1984). For explanation, see text.

along their course and give off lateral branches. Although the somata have not been detected, they are assumed to be situated within the ventral ganglion. There are 10 transversely running fibers, named B1 to B10, which radially pass out to the main connectives, the radial nerves and the caudal nerves. The pathway of this fiber forms a U-shaped curve and the cell body sends its axon to the contralateral side.

We obtained immunohistochemically positive results for three kinds of neuroactive substances (Fig. 16). As found in the cerebral ganglion, there are a pair of 5-HT-like immunoreactive fibers running longitudinally

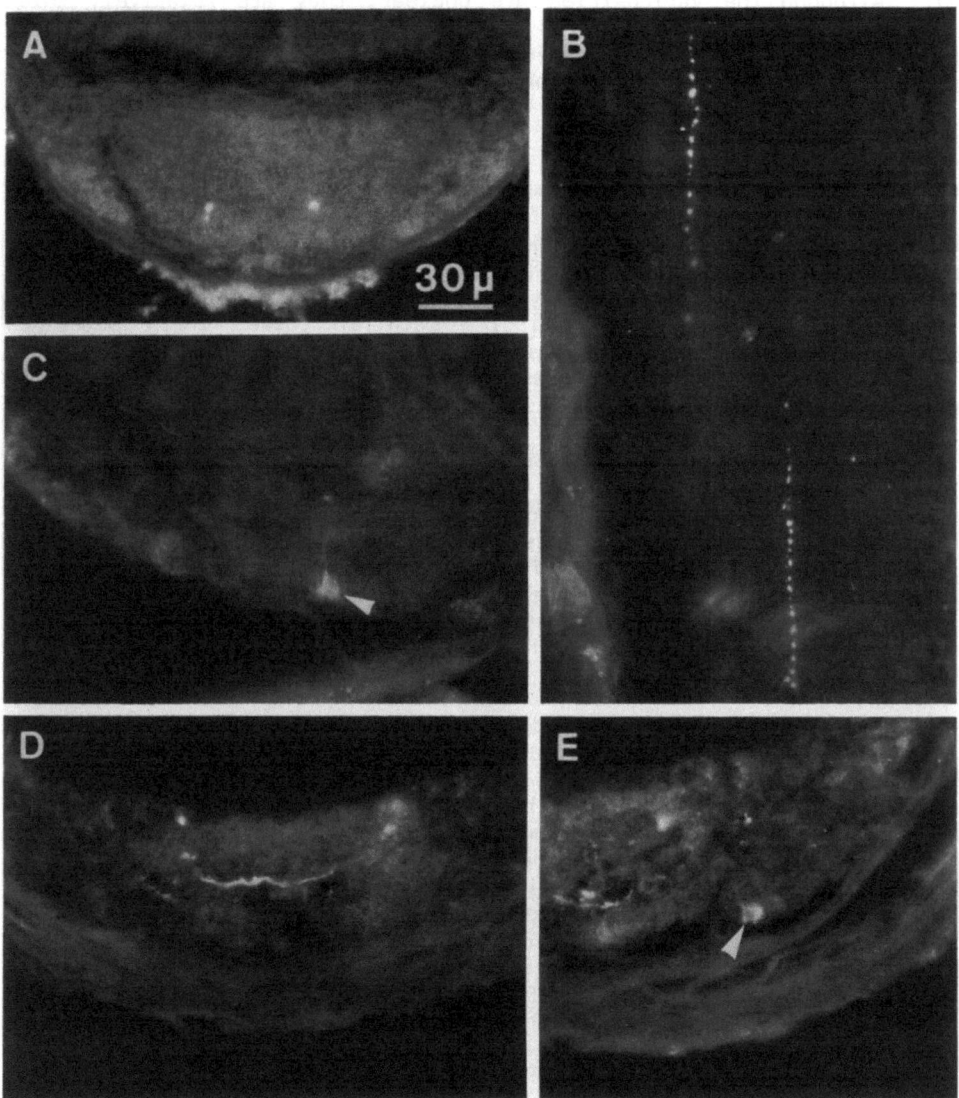

FIG. 16: Immunohistofluorescence of the ventral ganglion. A, B: 5HT-like immunohistofluorescence in transverse (A) and horizontal (B) sections. C: Met-enk-like immunohistofluorescence in a posterior region cut transversely. D, E: NT-immunohistofluorescence of fibers (D) and soma (E) appeared in transverse sections. Arrowheads indicate reacting somata.

within the ventral ganglion (Figs. 16A and B). Varicosities are notable as seen in the preparation stained with methylene blue. However, their somata have not been detected as yet. Considering their pathways and position, these fibers may correspond to C1 neurons. However, each fiber is single, not multiple as noted by Bone and Pulsford (1984). Two other substances which gave positive results are the neuroactive peptides, neurotensine (NT) and methionine-enkephaline (Met-enk). Met-enk-like immunoreactive fibers occur only on rare occasions probably due to their small number (Fig. 16C). The soma is located at the most ventral region near the posterior portion of this ganglion as seen in Fig. 13C (arrowhead). A thin fiber runs dorsally. NT-like immunoreactive fibers run transversely and their somata are located laterally (Figs. 16D and E). Although the varicosities are unclear, they are relatively thick (about 3 μm in diameter). It is not evident whether or not these fibers running transversely are the B neurons stained with methylene blue. These peptide-like immunoreactivities have not been detected in other ganglia.

So far correspondences of these identifiable fibers with their sectional profiles are not clear. Electron microscopic pictures show that many somata extend their axons to the neuropil transversely (Fig. 17A) but it is not sure whether or not the transverse fibers are the B fibers. Bone and Pulsford (1984) found that transversely running fibers are presynaptic to smaller longitudinal fibers. In the transverse sections of the central part of the neuropil, a pair of large fibers (Figs. 13C, 17B, arrowheads) are easily identified. Each of these fibers sometimes contains a large vacuole. They appear to be C1 fibers, considering their size, position and pathway. A pair of large fibers are also found in the main connectives (Fig. 13B). These may correspond to A1.

Peripheral nerves given rise from this ganglion make branches and innervate body muscles (Fig. 15, f) and probably ciliary tufts. Possible neuromuscular junctions at the body muscles show no significant membrane specializations at presumed points of synaptic contacts, but axons contain aggregates of vesicles. Axons and muscles are separated by an acellular basement membrane whose thickness is about 0.7 μm. Duvert and Barets (1983) assumed that this is a type of neuromuscular junction. Similar neuromuscular junctions have been reported in echinoderms (Pentreath and Cobb 1982).

The ventral ganglion is important for movements of arrowworms. Arrowworms swim by repeating quick active upward movements (Goto and Yoshida 1983). Decapitated worms can swim normally but complete deprivation of the ventral ganglion abolishes the worm's movements. The motor center of their active swimming may, therefore, be assumed to exist in this ganglion. When the animal is cut transversely at the center of the ventral ganglion, the posterior portion swims in a normal way though the head portion only moves about over the substrate. There may be several neurons which are related to generation of movements. Bone and Pulsford (1984) recorded electrical activities from the tail tip (Fig. 18A) and the ventral ganglion (Fig. 18B). The firing pattern from the ganglion is similar in its rhythmicity to cyclic electrical activities recorded from locomotor muscles. These results indicate that the neurons in the ventral ganglion drive body muscle contractions. The electrical activity of the ventral ganglion appears to be affected by acetylcholine (Q. Bone, personal communication).

4. PERSPECTIVES

In spite of the interest in chaetognatha from the phylogenetic point of view, they have not received attention from comparative physiologists.

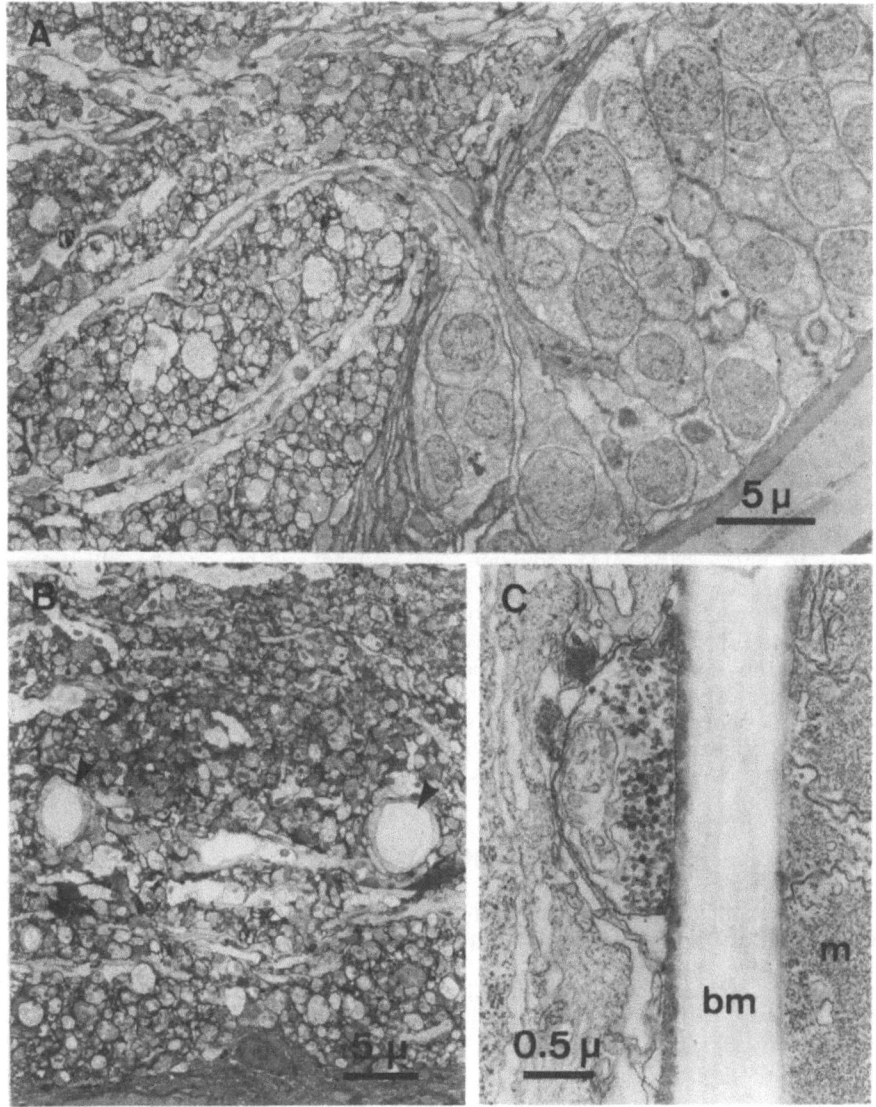

FIG. 17: EM profiles of transverse sections of the ventral ganglion. A: Somata send their axons to neuropil. B: Paired large fibers (indicated by arrowheads in Fig. 13C) containing large vacuoles (arrowheads). C: Presumed neuromuscular junction of peripheral nerves innervating body muscle. bm, basement membrane; m, muscle.

The unpopularity may be due to the difficulty in maintaining arrowworms under laboratory conditions. Fine structural studies on the nervous system and electrophysiological and pharmacological approaches to it have just begun from a few years ago.

Arrowworms have many kinds of putative sensory organs which might be related to their complex behavior. Many speculations have appeared as regard the function of the putative organs. However, we should consider the speculations cautiously. Studies on the role of the sensory organs are much needed to unravel the truth.

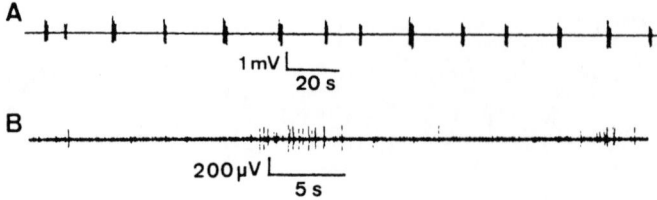

FIG. 18: Rhythmic activities recorded from tail tip (A) and ventral ganglion (B) (from Bone and Pulsford 1984). The species used are <u>Sagitta</u> <u>setosa</u>. A suction electrode was placed on the tail tip of a decapitated worm (A) or on an isolated ganglion (B).

Axonal pathways in the nervous system have been partially clarified by techniques of methylene blue vital staining and silver impregnation (Bone and Pulsford 1984). Our immunohistochemical studies give additional information. Immunohistochemical demonstrations on neuroactive substances present in arrowworms will advance our understanding on the phylogeny of neurotransmitters. The fact that arrowworms contain 5-HT indicates that most of the deuterostomes have this substance except for echinodermata. The presence of two neuroactive peptides, NT and Met-enk, suggests that they are also candidates for a neurotransmitter or a neuromodulator in chaetognatha. Based on these results, the function of histochemically specifiable neurons and the effect of neuroactive substances on the behavior of chaetognatha deserve to be studied.

5. ACKNOWLEDGEMENT

We thank Professor M. Tohyama and Dr. Y. Kumoi of Osaka University, School of Medicine for their kind instructions in immunohistochemistry and supply of the antiserum, Dr. Q. Bone of the Plymouth Laboratory for reading the manuscript, and Mr. M. Isozaki and Mr. W. Godo for their technical help. The work is supported by a Grant-in-Aid from the Ministry of Education, Science and Culture, Japan and by a research fellowship from the JSPS.

6. REFERENCES

Bone Q, Pulsford A (1984) The sense organs and ventral ganglion of <u>Sagitta</u> (Chaetognatha). Acta Zool (Stockh) 65: 209-220.

Bullock TH, Horridge GA (1965) Structure and function in the nervous systems of invertebrates. W.H. Freeman and Company, San Francisco, London.

Burfield ST (1927) <u>Sagitta</u>. Proc Trans Liverpool Biol Soc 41: 1-104.

Duvert M, Barets AL (1983) Ultrastructural studies of neuromuscular junctions in visceral and skeletal muscles of the chaetognath <u>Sagitta</u> <u>setosa</u>. Cell Tissue Res 233: 657-669.

Feigenbaum DL, Reeve MR (1977) Prey detection in the Chaetognatha: response to a vibrating probe and experimental determination of attack distance in large aquaria. Limnol Oceanogr 22: 1052-1058.

Goto T, Yoshida M (1981) Oriented light reactions of the arrowworms *Sagitta crassa* Tokioka. Biol Bull 160: 419-430.

Goto T, Yoshida M (1983) The role of the eye and CNS components in phototaxis of the arrowworm, *Sagitta crassa* Tokioka. Biol Bull 164: 82-92.

Goto T, Yoshida M (1985) The maiting sequence of the benthic arrowworm *Spadella schizoptera*. Biol Bull 169: 328-333.

Grassi GB (1883) Chaetognathi. Fauna Flora Golf. Neapel 5: 1-126.

Hertwig O (1880) Die Chaetognathen: ihre Anatomie, Systematik und Entwicklungsgeschichte, eine Monographie. Gustav Fischer, Jena.

Horridge GA, Boulton PS (1967) Prey detection by Chaetognatha via a vibration sense. Proc R Soc Lond B 168: 413-419.

Hyman LH (1959) Phylum chaetognatha. In: The invertebrates. Vol. 5. McGraw-Hill Book Company Inc., New York. pp. 1-71.

John CC (1933) Habits, structure and development of *Spadella cephaloptera*. Quart J Microsc Sci 75: 625-696.

Ozaki K, Hara R, Hara T (1983) Histochemical localization of retinochrome and rhodopsin studied by fluorescence microscopy. Cell Tissue Res 233: 335-345.

Pentreath VW, Cobb JLS (1982) Echinodermata. In: Shelton GAB (ed) Electrical conduction and behaviour in simple invertebrates. Clarendon Press, Oxford. pp. 440-472.

Rehkamper G, Welsch U (1985) On the fine structure of the cerebral ganglion of *Sagitta* (Chaetognatha). Zoomorphol 105: 83-89.

Scharrer E (1965) The fine structure of the retrocerebral organ of *Sagitta* (Chaetognatha). Life Sci 4: 923-926.

NEUROBIOLOGY OF THE ECHINODERMATA

J.L.S. COBB

Gatty Marine Laboratory

St. Andrews, Fife, Scotland, KY16 8TB

ABSTRACT

There has been little progress since the classical anatomical studies in understanding the form and function of the nervous system of echinoderms until relatively recently. There is, admittedly, a substantial behavioural literature but much of this can be confusing and even contradictory. There were early attempts to record electrical activity with extracellular electrodes but the compound potentials recorded to gross stimulation were clearly highly artifactual. Brehm (1977) showed that it was possible to record single unit activity extracellularly and further that this was possible because the brittlestar preparation that he used contained neurones much larger than average. Since then the large neurones of brittlestars have been exploited to produce a growing amount of information about function in echinoderm nervous systems at the cellular level. It is possible to use multiple recording sites from intact animals and monitor the activity that co-ordinates behaviour. There is data on the sensory perception of a range of environmental parameters. These direct electrophysiological measurements of response to stimuli are invariably consistent and are thus much more valuable than the inconsistent behavioural criteria previously used. It is also now possible to record intracellularly from both ectoneural neurones and hyponeural neurones. Lucifer yellow can be injected iontophoretically into both classes of neurones and preliminary data has now been obtained on the general morphology of individual cells within the nervous system (Cobb 1985). The radial nerve cords consist of connected segmental ganglia. The layout of the large neurones in each segmental ganglia is similar whatever the position of the ganglia within the radial nerve cords. Longitudinally running large neurones pass through at least 4 or 5 segments and show a fine plexus of varicose terminals at each end. These neurones are multimodal in the information they transmit about changes in the environment. The circumoral ring does not show complex structure but appears to act as a connection between the radial nerve cords and does not appear to contain organizing centres. The present evidence suggests that any part of the radial nerve cords when receiving significant local sensory input can act to coordinate whole animal behaviour and thus the echinoderms can be considered "brainless".

The other major advance has been the increasing evidence that the plasticity of connective tissue in echinoderms can be rapidly changed and that this is under direct nervous control. There are two lines of evidence. Extraction procedures have obtained substances that can stiffen or soften connective tissues. Little at present is known of the biochemical nature of these substances. There is also anatomical evidence for a class of juxtaligamental cell (see Wilkie 1984) that sends processes between the collagen bundles of connective tissue. This juxtaligamental tissue is innervated by hyponeural motor nerves. The rapid changes in connective tissue plasticity are known to be Ca^{++} dependent and Emson (1986) sees the evolution of this mechanism as a device for saving energy. In many situations echinoderms require to hold heavy but moveable appendages in fixed positions for long periods of time and he coined the phrase "bone idle" to explain this phenomenon.

I. INTRODUCTION

There is less known about the organization of the echinoderm nervous system than that of any other phylum of metazoan animals. In terms of levels of knowledge about cellular and molecular neurobiological phenomenon that exist for other phyla, the echinoderms simply do not feature. The reason for this is not that they are obscure or that they are not intriguing. It is simply that they are very difficult to work with technically. The cells are very small and in many situations embedded in heavy deposits of hard calcite skeletal material.

Persistence is needed; a crude dissection of a brittlestar reveals a bag of tissue for a gut with one opening, at the right time of year some irregular lobes of sperm or eggs and lots of bits and pieces of hard skeleton. This is not a very encouraging content and confirms the echinoderms as the enigmatic phyla. However, the bits and pieces of hard skeleton do make up a very active and mobile animal and this implies a sophisticated nervous system. The behaviour they perform is intriguing and there is a very large literature but this is not very instructive in trying to come to terms with function in the nervous system. The aim of this review is to indicate what recent techniques have been applied to this phyla and to what purpose. Readers will realize there is no shortage of fundamental problems. Neurobiology in other phyla is so often more of the same - not so in Echinodermata.

There are recent reviews by Hyman (1955), Smith (1965, 1966), Reese (1966), Pentreath and Cobb (1972, 1982) on the nervous system in general. Reviews on more specialized aspects are referred to in the main text. All these reviews should be used to make access to earlier research. Two important aspects of echinoderm neurobiology have become prominent in the last five or so years. First, the phylum has been shown to have a widespread ability to alter the viscosity of connective tissue under the influence of the nervous system. Second, it has been shown that individual cells within the nervous system can be studied structurally and functionally using intracellular techniques. Nevertheless studies on the neurobiology of the echinoderms are twenty years behind! The author would like to dedicate this chapter to the memory of Libbie Hyman whose volume on the phylum will always be the cornerstone of any study and to R. W. Ewer who as his external examiner and a founder of echinoderm physiological studies advised the author to become "a world authority on something obscure".

2. GENERAL MORPHOLOGY OF NERVOUS SYSTEM

2.1. Introduction

There is a detailed description of the general layout of the nervous system for each class of Echinodermata based on the classical and later studies using the light microscope which has been collated and reviewed by Hyman (1955). J.E. Smith (1937, 1945, 1950) carried out an extensive study of the asteroid nervous system using methylene blue, and he himself reviewed the echinoderm system in a chapter in the classic volumes of Bullock and Horridge (1965). Since that time there has been an increasing volume of work describing the ultrastructure of the nervous system particularly by the author (references in detail within the text). The electron microscope work has in general confirmed the histological and vital staining findings of previous workers. The crinoid nervous system is different from that of the other four extant classes in that the animals live oral side up and this has had major repercussions. There are three nervous systems, the ectoneural, the hyponeural and the entoneural but they are not directly comparable to those in the other classes. These nervous systems are enclosed within hard tissues in the class Crinoidea and there is a very sparse literature, probably because of the difficulty of working with it. The account that follows refers in the main to the four other classes but there is great need for a detailed comparative study on crinoid nervous systems. Where recent studies on crinoids are relevant, they are referenced.

The nervous systems of asteroids, echinoids, holothurians and ophiuroids consists of two major parts, the ectoneural nervous system and the hyponeural nervous system. It is possible that there are further sub-divisions and these are discussed below.

The main nervous system is the ectoneural and the most obvious parts of this are the circumoral nerve ring and radial nerve cords. There is a widespread basi-epithelial plexus and this is ganglionated where it is associated with the various organelles of the body surface. The ectoneural nervous system is also continuous with a plexus that underlies the endothelium of all visceral organs. There are sensory, motor and interneurones that make up the ectoneural system.

There is no evidence that the hyponeural nervous system is anything but motor and it is invariably associated with skeletal muscle systems. Cobb (1985b) has recently claimed that it is mesodermal in origin and this is argued below. The muscles of the viscera are also innervated by a nerve plexus that is not ectoneural but there appears to be no direct connection to the hyponeural nervous system. This visceral plexus is discussed more fully in a later section on neuromuscular systems.

Cuenot (1948) describes an entoneural system in starfish and this has been related to the marginal nerves of Smith (1937). At least part of this system is ectoneural (see Cobb and Raymond 1978) but Cuenot's description of this plexus innervating muscles of the body lining needs re-investigating using ultrastructural techniques.

2.2. Ectoneural Nervous System

2.2.1. Basiepithelial Plexus

The epithelium of echinoderms is underlain by a plexus into which sensory axons feed from widespread receptor cells. This plexus has been considered a nerve net (Kinosita 1941) and Bullock (1965) reviewed the evidence for this and decided, at least in echinoids, that it was absent.

The evidence at present available suggests large numbers of sensory neurones feed into a plexus which becomes swollen into a ganglion in the vicinity of an effector organelle. There is direct sensory neurone connection from a region stimulated to all adjacent ganglia. Bullock (1965) discussed whether there was room for such an apparently extravagant system and decided that there was! There are thus typical sensory - interneurone - motor reflex arcs over the whole surface but acting independently. These reflex arcs are also interconnected and connect to the radial nerve cord. Unless fresh evidence can be presented the idea of a nerve net seems outdated, even in asteroids. The separation of sensory, motor and interneurones using any criteria is difficult and the detailed cellular organization of the basiepithelial plexus is unknown except that associated with organelles it becomes of more substantial volume and contains neuropile typified by varicose vesicle filled axons.

2.2.2. Radial Nerve Cords

The ectoneural tissue of the radial nerve cords consists of complex ganglia associated with each segment and interganglionic connectives. There are segmental side branches out to the periphery. Cobb (1970) described the ganglionated nature of echinoid and asteroid nerve cords and showed discrete areas of neuropile were present in predictable locations. Subsequently, Cobb and Stubbs (1981) and Stubbs and Cobb (1981) described the radial nerve cord of an ophiuroid in detail and described the layout of a system of giant fibres. They showed the ganglia of the nerve cord to be regionally organized with neuropile and complex neurone pathways consistently positioned. There were differences in size, but not in complexity and organization, of ganglia from different positions on the nerve cord. The interganglionic regions consist of longitudinally arranged axon bundles.

The cell bodies of the neurones form a layer on the oral surface of the nerve cord and their detailed structure is considered in a later section. It is not possible to distinguish between sensory, motor or interneurone axon processes in the side branches to the periphery and although degeneration manipulations have been used (J.L.S. Cobb unpublished) they are unreliable in echinoderms. It is therefore not clear if sensory axons from the periphery make direct connection with the nerve cord or if there is direct motor output. The ectoneural nerve cord contains a varying percentage of catecholamine containing neurones and there is evidence of a more speculative nature for other transmitter types and neuro-secretory cells (see below). The significance of these in neural function is far from clear except that complex behaviour is generated and the layout of the nervous system is highly organized and sophisticated.

The ectoneural system drives the hyponeural system across a true basement membrane and this has been described in a series of papers (Cobb 1970; Pentreath and Cobb 1972; Cobb and Pentreath 1976; Cobb and Stubbs 1981; Stubbs and Cobb 1981; Cobb 1985a,b).

2.2.3. Circumoral Nerve Ring

It is surprising since many workers have implied that this is the centre of the nervous system in an organizational sense that it has scarcely been studied. Cobb and Laverack (1966a,b) briefly examined it in the echinoid Echinus but the only comprehensive description is that of Cobb and Stubbs (1981) in an ophiuroid. It consists, in the main, of circumferentially running axons connecting the radial nerve cords and local areas of neuropile associated with the visceral innervation, motor output and hyponeural ganglia. There is no evidence for complex intergrative function, only for radial nerve cord connection.

The first ganglion of each radial nerve cord is different from the others since connection is made in both directions round the circumoral ring but there is no evidence from size or layout that it has a predominant controlling function when compared to other segmental ganglia.

2.3. Hyponeural Nervous System

This nervous tissue is associated with large skeletal muscle function and is purely excitatory, motor and apparently cholinergic. It is the hypothesis of the author, since it is always within mesodermal tissue, that it is of mesodermal origin itself and perhaps derived from muscle cells. Ectoneural tissue almost never seems to penetrate echinoderm mesoderm. Hyponeural tissue is found in ten ganglia off the circumoral nerve ring in echinoids associated with large lantern muscles, and probably also associated with the spine muscles (J.L.S. Cobb unpublished) in the radial nerve cord of holothurians and accompanying the radial nerve cords and circumoral nerve ring in asteroids and ophiuroids. Motor axons run from ganglia to innervate muscles. The distribution of hyponeural tissue described by Sloan and Campbell (1982) is in error, as is the description by Bouland et al. (1982) of it in the buccal tentacles in holothurians. There is no real evidence for integrative function in hyponeural ganglia from ultrastructure or physiological studies.

2.4. Muscle Tails

Smith (1950) described 'ribbon axons' in the ampullae of starfish. These were later shown by Cobb (1967) to be modified muscle cells. Narrow processes run from the cells with a uniform diameter for a distance of up to 1 cm. The diameter is often less than 1 μm and the cell processes contain a modified core of both thick and thin myofilaments. More recently, the same type of uniform narrow processes from muscle cells have been described amongst the hyponeural nerves of the brittlestar radial nerve cord (Stubbs and Cobb 1981). These last muscle tails appear to be post-synaptic to varicose, vesicle filled, ectoneural endings on the other side of the basement membrane. It is not clear whether these processes originate from intervertebral muscles (which in the main are innervated by axons from the hyponeural ganglion). They are visible in light microscope sections and are continuous along the whole length of the radial nerve cord. There is no evidence that the muscle tails run for more than one segment. It is likely they are simply involved in innervating intervertebral muscles but the possibility must be considered that they represent a longitudinal conduction system of a unique type. Recent work, using osmium tetroxide and glutaraldehyde mixed as a fixative, has shown that as well as the myofilament core there are large numbers of microtubules arranged longitudinally in the cytoplasm (Fig. 1).

2.5. Ectoneural / Hyponeural Connection

Smith (1950) originally proposed a direct connection between the hyponeural and ectoneural system but this has not been confirmed by an electron microscope study. In most places there is a chemical synapse across a basement membrane (Cobb 1970; Cobb and Pentreath 1976; Cobb and Stubbs 1981; Cobb 1985a,b). Ectoneural nerves and mesodermal muscle are always separated (see below). There are three known exceptions to the separation of the nervous systems. Stubbs and Cobb (1981) using serial light and EM sections showed the basement membrane is not continuous but small fenestrations appear in it through which axon tracts pass. This has also been observed in the circumoral ring (Cobb and Stubbs 1982). It was originally considered that these were hyponeural axons passing orally but

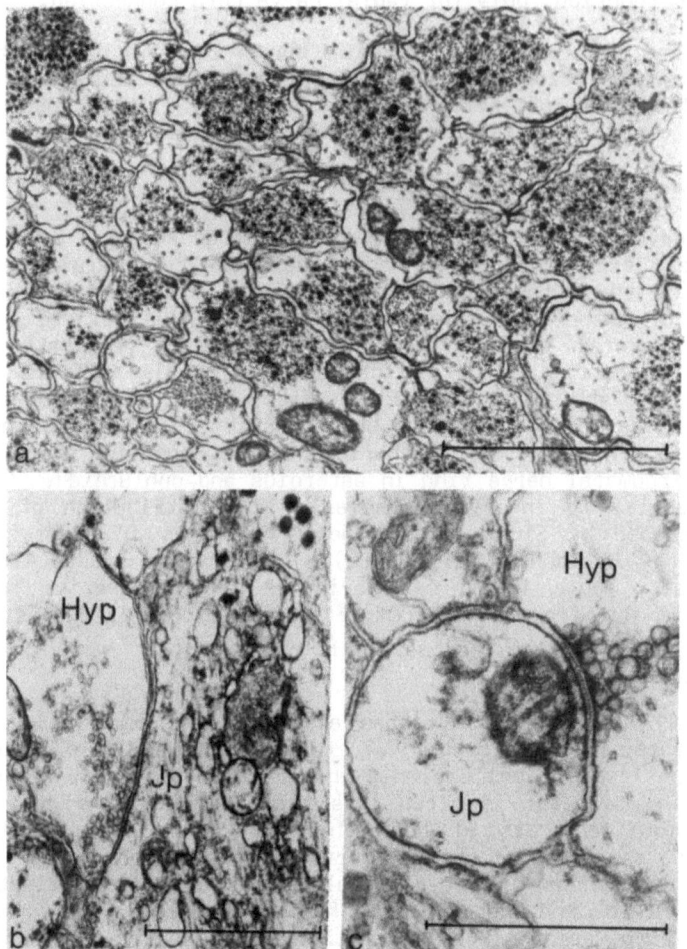

FIG. 1a: Transverse section through muscle tail bundle in the hyponeural ganglion of the radial nerve cord of the brittlestar Ophiura ophiura, (see Cobb and Stubbs 1981). Fixation using a mixture of glutaraldehyde and osmium tetroxide (see Cobb 1985c) shows that substantial numbers of microtubules are present as well as the modified core of myofilaments. Previous work suggests these muscle tails function as motor axons (Cobb 1967). Scale= 2 μm.

1b: Longitudinal section through process (Jp) leaving the soma of a juxtaligamental cell in the node of juxtaligamental tissue associated with the oral plate ligament. A hyponeural motor axon (Hyp) makes an imput synapse onto the process. Scale= 1 μm.

1c: Transverse section through a similar process (Jp) to that in (1d) again with an imput synapse from a hyponeural motor neurone (Hyp). The use of the mixed glutaraldehyde/osmium tetroxide fixation shows considerable specialization of the synapse compared to any others in echinoderms (see Cobb and Pentreath 1976). Material is present in a regularly aligned cleft and there is a fuzz of cytoplasmic material present both pre- and postsynaptically. Examination of other regions of neuropile recent work (Cobb 1985c) has shown these known axon tracts pass round the edge of the radial cord. A current hypothesis is that they represent sensory ectoneural axons from interoreceptors in the intervertebral tissue. Peters (1985) described ectoneural neurones penetrating into mesodermal muscle in the spines of sea urchins where the basement membrane was incomplete. This third exception is considered in a later section.

The greatest enigma in echinoderm neurobiology is this separation of the nervous systems and genuine exceptions to it make it even more unfathomable. This has led Cobb (1985b) to suggest the hyponeural nervous system has developed from mesodermal tissue because ectoderm derived neurones have by some irreversible quirk of evolution not been able to cross basement membranes. It is interesting to speculate that hyponeural nerves have evolved from muscle cells and a comparison of ionic mechanisms between muscle, hyponeural and ectoneural cells, which is now possible, will be interesting. The origin of juxtaligamental tissue apparently responsible in a nerve-like way for the motor innervation of connective tissue (see below) is another fascinating question when considering enigmatic echinoderm cell types.

2.6. Neurone Cell Structure

The ultrastructure of neurones has been described in a wide variety of situations since the original study by Bargmann and Behrens (1963). In particular Kawaguti (1964), Cobb and Laverack (1966b), Cobb (1970), von Hehn (1970), Pentreath and Cottrell (1971), Weber and Grossmann (1977), Burke (1978), Stubbs and Cobb (1981) and Cobb and Stubbs (1981) have described ectoneural and hyponeural cell bodies in detail. One of the most significant findings is the lack of endoplasmic reticulum in the giant neurones of ophiuroids and its replacement by a granular amorphous substance. There is no reason to suppose this is an artefact and because in other phyla neurone somata are areas where substantial membrane based synthesis takes place, this is a considerable enigma. The granular material stains strongly with toludine blue and it would repay investigation with precise histochemical techniques.

2.7. Synapse Ultrastructure

Cobb and Pentreath (1977) and Pentreath and Cobb (1982) have reviewed the literature on synapse ultrastructure. Characteristically neurone-neurone synapses as well as neuromuscular synapses are formed from vesicle containing varicose endings approaching close to the post-synaptic target. There is no evidence of pre- or post-synaptic specializations associated with the membranes either intra- or extracellularly. Cobb and Pentreath (1976, 1977) have reviewed this lack of specialization in a comparative way with those of other phyla. There is always difficulty in describing chemical synapses on morphology alone and the simplicity of these structures in echinoderm makes it even more fraught. Florey and Cahill (1982) using both transmission and scanning electron microscopy show basiepithelial plexus neurones can be varicose even though they

---

using the same technique has not to date shown similar specializations. Recent work (J.L.S. Cobb unpublished) has shown a similar innervation of juxtaligamental nodes of tissue associated with sea urchin spines. The motor neurones in this case also appear to be a class of hyponeural nerve cells which have not previously been described in this situation. These are the same motor nerves as innervate the spine muscles which Peters (1985) described as ectoneural neurones. Scale= 0.5 μm.

contain no vesicles and hence are presumably not sites of transmitter release. Electrophysiological and iontophoretic dye-fills of single neurones (see later section) have confirmed the varicose ending theory for presynaptic terminals. The ectoneural nervous system synapses across the basement membrane that separates it from the hyponeural. Varicose vesicle containing neurones of the ectoneural system end against the basement membrane in places only consisting of basal lamina a few hundred Angstrom thick (Cobb and Pentreath 1976). There is intracellular electrophysiological evidence that this system is functional (Cobb 1985b).

Many invertebrate giant fibres show electrical synapses. These are usually characterized in transmission electron microscopy by a 2 nm gap between the pre- and post-synaptic neurone and a hexagonal array of subunits is visible within the membrane. Freeze fracture studies have shown these junctions are characterized by regular arrays of membrane particles with different structural geometry between the coupled and uncoupled electrical state. Junctions of this type have not been described, although searched for, and some giant fibres have been shown to form chemical synapses physiologically. Coupled cells have been shown to be present in embryological echinoderm tissue using electrophysiological techniques (see Cobb 1982 for review). If they are present in echinoderm adult nervous (or muscular) systems, it will be necessary to demonstrate them physiologically as well as structurally.

3. MOTOR SYSTEMS

3.1. Innervation of Muscle

There are three systems in echinoderms where there are significant amounts of muscle. First, the skeletal system where muscles are functional in structures from as small a pedicellariae to those of the lantern of echinoids. The same class of muscle is also found in the body wall of holothurians. The second situation is the water vascular system which is in part a hydroelastic skeleton with a great deal of muscle associated with it. The third class of muscle is that of the viscera. All these muscles are innervated in quite different ways and will be dealt with below in different sections. The classification outlined above refers to function and not to structure. The cellular structure of contractile tissue is highly varied and ranges from true striated muscle to non muscular epithelial cells thought to be contractile. There is also the question of contractility of non-cellular connective tissue.

Many of the smooth muscles do not resemble the classical vertebrate smooth muscle but contain thick paramyosin-like filaments and have a helical or semi-helical arrangement. Saita et al. (1982) and Carnevali and Saita (1985a,b) have provided careful reviews of these types of muscle in a comatulid crinoid and readers are referred to these accounts for more details.

Cobb and Laverack (1967) reviewed neuromuscular systems in echinoderms but at that time some of the enigmatic structures and functions were still confusing; much good work has been done since.

3.1.1. Non-Muscular Contraction Systems

The question of innervated connective tissue will be dealt with at length in a separate section below. Holland and Grimmer (1981) have however studied the cirri in crinoids and shown that although these are movable there are no typical muscle cells. They have described 5 nm filaments associated with microtubules as very obvious features of the epithelial cells and suggested these could form a contractile system. Neurosecretory cells were also present and resemble the processes of juxtaligamental cells (see Wilkie 1979). These may well be associated with changes in the stiffness of the cirri connective tissue allowing the epithelial cells to move the organelle (see below). Holland and Grimmer quote Jickeli (1884) as having demonstrated that stimulation of the cirri nerve causes movement of the cirri. Harris and Shaw (1984) have however recently used immunological techniques to look at filament bundles surrounded by microtubules in the epithelium of echinoid tubefeet. Admittedly these filaments are 7 to 8 nm and in a totally different situation, but it was shown that they were not actin but more related to vertebrate intermediate filaments. In this situation the evidence suggests that they have a purely cytoskeletal role as opposed to the possibility of a contractile function. Clearly more work is required but innervated non-muscular cellular contraction should be borne in mind as a possibility in this enigmatic phylum.

3.1.2. Visceral Muscle

There are a number of descriptions of visceral muscle, particularly Doyle (1967), Bargmann and Behrens (1968) and Cobb (1969b). The structure of the gut tube and appendages is formed by an endothelial lining with varied numbers of secretory cells and an outer muscle and epithelial layer separated by a basement membrane of varied thickness. There are usually circular and longitudinal smooth muscle layers. Separate nerve plexuses innervate the endothelial layer and the muscle layer and there are no connections between them. There are two types of nerve-like processes between the muscles and these have been described by Cobb (1969b). One contains small agranular vesicles and appears to form 'en passant' synapses throughout the circular and longitudinal muscles. These presumed synapses do not show morphological specializations, but resemble those of the vertebrate autonomic system. The gut has been shown to contain acetylcholine (see Pentreath and Cobb 1972 for review) and is sensitive to this transmitter; there is thus no reason to suppose the innervation of the muscles is not at least in part cholinergic. The second nerve-like process contains large granular vesicles (see Cobb 1969b) following a similar anatomical distribution to the agranular vesicle containing nerve, and this initially suggested dual innervation. These nerves do not contain monoamines. Speculation that these may also be processes from a class of juxtaligamental cell is detailed below but very little further can be said of them. There are three significant questions to answer about the gut muscle plexus. First, are the cholinergic nerves of the outer layer a third totally separate nervous system? They are not derived from, or related to, the ectoneural system and careful serial section studies (J.L.S. Cobb unpublished) have failed to show a connection to the hyponeural system. It cannot be ruled out that such a connection does occur though there is no obvious, or for that obscure, reason why it should, since the hyponeural system is a purely motor/skeletal system. The second question concerns the possibility of a release of transmitter across the basement membrane from the ectoneural plexus either to drive the muscles directly or indirectly via the cholinergic plexus that is present. The function of the ectoneural plexus which contains significant levels of monoamines is far from clear and is discussed in more detail with relation to the innervation of cilia but since cross-basement membrane transmitter release is an established fact elsewhere in echinoderms, it must be considered as a possibility. It seems highly likely therefore that the viscera are innervated by a third, probably mesodermally derived, nervous system quite anatomically separate from the other two. The third question concerns the mechanism by which excitation is spread through the muscle to produce co-ordinated contractures. There are no structures present (see Cobb 1969b) that resemble electrical junctions (gap junctions). They have not been described elsewhere in other neuromuscular systems either. This question is discussed at greater length in a later section. Visceral function is easily the most neglected area of echinoderm neuromuscular function but cellular physiological studies will be very difficult and anything less than that inconclusive.

3.1.3. Muscles of the Water Vascular System

Work over the last decade by a number of authors has made considerable and highly significant progress in understanding this system. It has undoubtedly been hindered by a serious error in the early work of the author (Cobb 1967). The part of the water vascular system most studied consists of the ampulla/tubefoot system. The outside layer of these structures is in contact with either the coelomic cavity or the environment. It consists of a layer of epithelial cells and sensory cells underlain by an ectoneural nerve plexus. There is a basement membrane layer of connective tissue often containing much collagen and of varied thickness. Ectoneural nerves never penetrate this layer. The lumen is lined by muscle and endothelial cells.

Smith (1950) described in the ampullae large flat structures among the muscles with long axon like processes - 'ribbon axons' - arising from them using methylene blue vital staining. Cobb (1967) investigated these using ultrastructural techniques and showed that the 'ribbon axons' of Smith were due to a small proportion of the muscle cells taking up the stain differentially and that all muscles sent long modified processes down to the base of each ampulla. These processes, except for a small atypical central core of myofilaments, resembled axons. The author however misinterpreted the anatomy in this basal region and claimed these muscle processes were directly innervated by ectoneural nerves. Subsequent studies (unpublished) showed the illustrations in the original paper to have been derived from a different region and that the muscles end against the basement membrane but in positions opposite ectoneural nerve endings on the other side. These erroneous findings have misled other authors, notably Cavey and Wood (1981) but at least there is now excellent evidence indicating the unusual way the muscles are actually innervated.

Cobb (1970) realized that hyponeural neurones and muscle cells of the ampullae in a sea-urchin were indirectly innervated by ectoneural nerve endings across a basement membrane. This varied from a few hundred Angstroms of basal lamina to a quite thick layer of collagen sandwiched between two layers of basal lamina. Florey and Cahill (1977, 1980) made a critical examination of the whole problem using scanning and transmission electron microscopy, as well as pharmacological studies on isolated and 'skinned' tubefeet. They used a complex mathematical analysis to show that transmitter release across a wide connective tissue layer could account for the innervation of the underlying muscles by ectoneural nerves. Wood and Cavey (1981) and Cavey and Wood (1981) have also investigated the tubefoot system and shown the inner layer to be a myoepithelium and not two separate layers of cells. These authors provide an excellent description of the detailed morphology of the attachments of the muscles and discuss the difficult problem of whether there is coupling between myo-epithelial cells and how syncronous contraction of the muscles can take place. This problem is discussed further in a section below.

There still remains a major unsolved problem. As with the gut there is present among the muscle cells of the water vascular system a second set of nerve-like cells. They contain large granular vesicles over 100 nm in size and often even larger vesicles in the cell bodies. They do not contain catecholamines. Long processes, often varicose, and of small

(1-2 μm) diameter penetrate among the muscle cells and also into the connective tissue. They are also present in the muscles of the gills (Cobb and Sneddon 1977), the papulae (Cobb 1978), the haemal organ (Doyle 1967) and elsewhere (Von Hehn 1970). They are probably universally present in all non-skeletal muscle. Cobb (1969b) originally described them as containing an unknown class of transmitter and innervating the muscle; they have also been classed as neurosecretory cells (see Pentreath and Cobb 1972 for review). Florey and Cahill (1977) called them mast cells and dismissed the possibility that they innervated the muscle. Cobb (1978) in describing them innervating one type (of the two) of muscles associated with the dermal papulae in asteroids discussed at length their function but without reaching a firm conclusion. A later section in this review speculates further on their function. It is clear however from the work of Florey and Cahill (1980) in particular that the muscles are indirectly innervated by cholinergic nerves across the basement membrane. Studies at the cellular level will again be difficult.

3.1.3. Skeletal Muscle Systems

There are many situations in echinoderms where calcite ossicles form structures articulated by muscles. They may be the intriguingly varied pedicellariae jaw muscles, the massive muscles of the echinoid lantern or intervertebral muscles of ophiuroids. The body wall muscles of holothurians are also of this category. These muscles can be divided into two types; those innervated by ectoneural and those by hyponeural nerves.

3.1.3.1. Ectoneural Innervation

The muscles of the jaws of pedicellariae are innervated in this way. The detail of this system is described in Cobb (1968b). Processes from smooth and striated muscles penetrate between the ossicles and lie adjacent to varicose, vesicle filled, axons of the ectoneural nervous system. A recent re-investigation of this system (J.L.S. Cobb unpublished) shows a continuous basal lamina between the axons and the processes of the muscle cells. The axons in this situation thus do not penetrate into the muscle tissue and are thus confined to ectoneural tissue layers.

3.1.3.2. Hyponeural Innervation

Hyponeural nerves are confined to situations where there are major skeletal structures, with few exceptions. This nervous tissue has been discussed in an earlier section but it is always solely motor and apparently cholinergic. Axons from discrete ganglia penetrate among the muscle cells to innervate by forming non-specialized synapses. The innervation of such muscles has been described in detail for the lantern retractor muscles (Cobb and Laverack 1966c), and the ophiuroid intervertebral muscles by Stubbs and Cobb (1981). The innervation of smaller muscles in the asteroid arm wall has been discussed by Cobb (1978). The only controversial situation is that of the spines of sea urchins. Peters (1985) provides a good and thorough description of this system and shows the outer ring of muscles is innervated by axons which penetrate among them; neurone cell bodies are described as present in this layer. Peters shows there is a basement membrane between the ectoneural ring ganglia around the base of the spine and the muscle layer itself but that this is discontinuous at higher levels up the spine and nerves pass into the muscle layer. The author (J.L.S. Cobb unpublished) has re-examined this system and believes the neurones within the muscle layer are a previously undescribed class of hyponeural motor neurones and that they are innervated across the basement membrane as is found in the radial nerve cords of asteroids and ophiuroids. It must in fairness be said that

Peters (1985) is very positive in his description of direct innervation and that it is difficult to prove this does not exist. The author when beginning to work on this phylum assumed that nerves 'must' directly connect with their post-synaptic targets but with hindsight enigmatic echinoderm function does not require this to be so. Peters' description, if correct, would be the only example of a direct ectoneural connection within muscle tissue.

3.1.4. Coupling Between Muscles

Some muscles, such as the lantern retractor muscles, conduct propagated action potentials (Cobb 1968a; Pentreath and Cobb 1972; Peters and Shelton 1981), these findings being based on intra- and extracellular recordings. Sugi et al. (1985) however, suggest that action potentials in these muscles do not occur, basing their findings on a study using electrical and chemical stimulus of the muscles. It seems likely that the interpretation of their data is not correct but it is possible that there are differences even within the same muscle in different species. Prosser (1954) stated holothurian body wall muscles do not conduct action potentials but in a later study (Prosser and Mackie 1980) suggested they do occur but are too small to record extracellulary and presented evidence that a Ca^{++} spike is involved. Kawaguti (1964) and later Dolder (1972) describe different types of junction within the tubefoot. More recently Cavey and Wood (1981) provide a thorough discussion of the evidence for and against coupling between muscle cells. Possibilities included electrical coupling between the epithelial cells of the myoepithelium of the tubefoot, depolarisation of individual cells by mechanical distortion due to the contracture of neighbouring cells (first proposed by Bargmann and Behrens 1963) or an electrical junction of a different type from that characteristic of other phyla between muscle cells (see Wood and Cavey 1981). Hill et al. (1983) for example, describe large regions of close apposition between muscle cells in a holothurian. Whether there are junctions which allow electrical connection between muscle cells and the spread of contraction is a significant question in understanding the functioning of the innervation. There is as yet no good evidence from freeze-fracture of the characteristic molecular structures in muscle membranes. Green et al. (1979) have described various junctions in endothelial cells and have discussed possible roles, again however physiological evidence for cell coupling is absent. The only unequivocal evidence for electrical coupling between cells is in embryological tissue (see Cobb 1982b for references and review). Whatever the structural evidence that eventually accumulates, it will be of no great value until there is unequivocal physiological evidence at the cellular level that such function occurs. This evidence has not been obtained and it will be difficult to achieve.

There is finally quite a large literature not strictly relevant to this review on the role and storage location of Ca^{++} ions in muscle contracture. Readers are referred to Prosser and Mackie (1980), Suzuki (1982), Hill (1983) and Sugi et al. (1985).

3.2. Innervation of Connective Tissue

There has been a great deal of attention focussed on this problem over the last decade and although the detailed problems have still to be solved the direction which research effort needs to take has been identified. Jordan (1919) and von Uexküll (1926) long ago identified the phenomenon of 'catch' in echinoderm tissues but associated it with a layer of muscle. The major advance was the description by Takahashi (1966, 1967) that the catch 'muscle' is in fact a collagenous connective tissue. There have been suggestions that the connective tissue itself was contractile but there is no evidence for this. This phenomenon is now regarded as being widespread in the phylum and three preparations have been singled out for particular study. These are the body wall dermis of holothurians, the catch mechanisms in sea urchin spines and the intervertebral ligaments of a brittlestar arm. There are excellent recent reviews of the subject by Motokawa (1984b) and Wilkie (1984) and it is not necessary here to detail all the evidence, only summarize the most significant. Two different lines of evidence have led to the current hypothesis on innervated connective tissue. Motokawa in a long series of papers (1981, 1982a,b,c, 1983, 1984a,b) examining visco-elasticity in the holothurian body wall has identified two factors present in the coelomic fluid which respectively 'stiffen' or 'soften' the dermis. One is methanol soluble and the other methanol insoluble.

Byrne (1986) describes the neural control of evisceration in holothurians and shows acetylcholine to be involved in part of the process. She discusses the evidence that an evisceration factor is a low molecular weight neurotransmitter and cites the evidence of Smith and Greenberg (1973). She recognizes and discusses the difficulty of separating the effects of the direct action of the coelomic factor on connective tissue and their interaction via the nervous system. Hidaka (1983), Hidaka and Takahashi (1983) and Smith et al. (1981) have provided physiological and anatomical studies on the catch apparatus of sea urchin spines. Smith et al. also showed very small muscle fibres amongst the catch connective tissue. Hidaka (1983) developed a scheme which suggested the viscosity of the connective tissue is determined by the number of cross links between negatively charged macromolecules. A 'stiffening' nerve and a 'softening' nerve control the charge distribution between macromolecules and the concentration of Ca^{++} ions which in turn reversibly alter the viscosity.

Wilkie in another long series of papers (Wilkie 1978a,b,c, 1979, 1983, 1984; Emson and Wilkie 1980) has provided a detailed description of arm autonomy in brittlestars and produced some of the most conclusive evidence for innervation. This work shows connective tissue changes take place in less than 1 s to a remote stimulus and Wilkie has developed the concept of a juxtaligamental tissue of nerve-like appearance which penetrates between the matrix of the connective tissue. There is now substantial evidence that where the mutable connective tissues (term used by Eylers 1982) occur in echinoderms there are granule-containing cells and nerve-like processes from them. These cells have in the past been called nerves, neurosecretory cells and lately juxtaligamental cells. Until the exact nature and relationship to the nervous system is fully understood it would be best to keep the term justaligamental cells.

There is some evidence that there are two different populations of granule containing cells perhaps representing 'stiffness' and 'softening' nerves (Wilkie 1979; Holland and Grimmer 1981; Biglow 1981; Byrne 1982). Stubbs and Cobb (1981) and Cobb (1985c) have investigated the hyponeural motor innervation of the ophiuroid arm and using serial section techniques have shown axons of hyponeural nerves form endings on juxtaligamental cells in defined nodes of tissue. J.L.S. Cobb (unpublished) has shown a similar situation in nodes of juxtaligamental tissue of echinoid spines where a newly described class of hyponeural motor axons innervate juxtaligamental cells. These nerve endings contain the small agranular vesicles typical of hyponeural motor endings and fulfil the morphological criteria for chemical synapses (Fig. 1). Motor axon bundles of large diameter axons run from the segmental hyponeural ganglion and innervate juxtaligamental nodes of tissue as well as the intervertebral muscles. Bundles of smaller diameter axons run orally round the radial nerve cord and innervate nodes of juxtaligamental tissue associated with the connective tissue of the oral plate ligaments. These are free of muscles (Cobb 1985c). Byrne (1986) suggests two different types of pathways exist for the neural control of holothurian evisceration and ophiuroid autonomy.

The most recent work has suggested that the changes in viscosity of echinoderm mutable connective tissues are brought about enzymatically. Wilkie (1984) reviews suggestions that mutable connective tissues are a device to save energy in echinoderms where heavy organelles and tissues need to be held in fixed positions. Muscle power is used to move organelles but connective tissue holds it fixed in the desired position. This hypothesis has been formalized by Emson (1986).

In summary, connective tissues are mutable because of changes in the binding properties of the proteoglycon matrix allowing bundles of typical collagen filaments to slide past one another. These changes are brought about by the nervous system and involve a class of nerve-like juxtaligamental cells innervated in turn by motor nerves of the hyponeural tissue and perhaps indirectly by the ectoneural system. It is not yet clear how or where the coelomic factors are produced. There is some evidence that two classes of nerves are involved in producing reversible viscosity changes. Previously (see Motokawa 1984b) these changes were not thought to be enzymatic but this view has recently been challenged (Diab and Gilly 1984).

There is however, a more speculative aspect of this innervation of connective tissue. Cells similar in morphological appearance to juxtaligamental cells are present in many tissues not strictly thought of as connective tissues. They are present in gut structures (Doyle 1967; Cobb 1969b), tubefeet (Florey and Cahill 1977), coelomic lining (Cobb and Sneddon 1977, Cobb 1978) and within the epithelium (Markel and Röser 1985). Similar cells were reported in the buccal tenticles of holothurians by Bouland et al. (1982) although they were erroneously described as hyponeural nerves. Cells in these positions have been called neurones with an unknown transmitter, neurosecretory cells or even mast cells. The granular vesicles are distinct from those in monoamine containing nerves in echinoderms. These cells occur between the muscle cells but also penetrate through layers of collagenous connective tissue associated with the basement membrane and are even found in ectodermal tissue (Markel and Röser 1985). This in itself is interesting since echinoderm axons of either the hyponeural or the ectoneural tissue do not penetrate basement membranes. Although based purely on morphological similarity, and hence very speculative, it may be since viscosity changes allow whole collagen bundles to slide past one another that in muscle tissue such as the tubefeet and parts of the gut that need to be highly extensible, the muscle cells themselves are able to move past one another. They would become refixed by reversible changes in the properties of the binding extracellular glycocalyx. The hypothesis at least accounts for the presence of this cell type in some muscle tissues not supposed to be directly innervated.

No doubt progress will be made in the near future in answering a number of questions. What are the natures of the coelomic factor extracted and how are they related to the nervous system and juxtaligamental tissue and what is the detailed molecular basis for the changes in viscosity and how can this be controlled by the nervous system?

3.3. Innervation of Non Muscular Tissues

Cilia are very widely distributed in the echinoderms and although some are associated with receptor function many are clearly motile and involved in movement of fluid. In larval situations there is evidence that ciliary beating is under neural control. Mackie et al. (1969) have recorded electrical activity associated with reversal of ciliary beating but suggested neuroid conduction, Strathmann (1971) used $MgCl_2$ as an anaesthetic and this inhibited synaptic transmission and blocked ciliary reversal in echinoderm larvae. Burke (1978, 1983,a,b) has extensively worked on larval nervous systems and provided a great deal of anatomical evidence that can be related to behaviour. He suggests the organization of the nervous system, and its associated effectors, can be related to the control of feeding. He showed the presence of monoamines in the nervous system and their location. Control of ciliary activity in metazoa is reviewed by Aiello (1972). Cobb and Raymond (1979) suggested that the nerve plexus underlying gut endothelial cells might control ciliary beating. This hypothesis was based on the presence of a monoamine containing nerve plexus beneath the endothelium of the rectal caecae which comprised of a single ciliated cell type. Later studies (J.L.S. Cobb unpublished) with a range of transmitters (dopamine, noradrenaline, adrenaline, etc.) at different dilutions produced no change in direction or rate of beating in a range of echinoderm gut structures. Cobb and Stubbs (1982) described a complex ciliated feeding structure on the outside of the arm of a brittlestar which was also underlain by a monoamine containing plexus. The rate of ciliary beating again was not altered by application of any transmitter. It is of course possible that the failure of these drugs is due to failure to penetrate the receptors on ciliated cells but it must be concluded that there is no evidence for the innervation of ciliated cells in adult echinoderms. This leaves the problem of what the very extensive monoamine containing plexus does where it underlies either the epithelium of the surface of the test or tubefeet and where it underlies the endothelial cells of the gut. There is very little evidence to date but a possibility is that various secretions are released under neural control. These proteinaceous secretions may be mucus or enzymes. Fontaine (1964) suggested the release of mucus by integumentary mucus glands in a brittlestar is under neural control and Kawaguti (1966) has shown that the mucus glands of the body wall of the holothuria <u>Stichopus</u> receive a nerve supply. Bargmann and Behrens (1968) described the innervation of various secretory cells in the gut of <u>Asterias</u> by a nerve plexus. McKenzie (1985) has recently completed an extensive study of types of secretory cells in the tentacles of holothurians. He considers the possibility that secretion is controlled by a neurosecretory mechanism under the control of a particular cell type. Unpublished studies (J.L.S. Cobb) have shown an increased release of proteins when dilute solutions of dopamine are applied. These solutions were washed over the gut surface of various echinoderms or the tubefeet of brittlestars and the washings collected and analysed. This rather crudely obtained evidence is of minor significance until the problem is analysed using much more critical techniques. The function of the extensive ectodermal visceral plexus is still a major enigma.

4. NEUROSECRETION

Welsh (1966) reviewed the early work which chiefly dealt with the gamete-shedding substance (GSS) and Pentreath and Cobb brought the review up-to-date in 1972. There is still argument as to whether described phenomena are neurosecretion in the classical sense but the nervous system is undoubtedly involved in the whole process of gonad maturation and gamete shedding. Great strides have been made in unravelling the complex interactions involved particularly by Kanatani (Kanatani 1964, 1969; Kishimoto and Kanatani 1976; Kishimoto et al. 1982) but see Meijer and Guerrier (1984) for other references to this author. GSS is a thermostable 22 amino acid peptide of about 2100 D (Kanatani et al. 1971) and this in turn stimulates follicle cells to start producing maturation inducing substance (MIS) which has been isolated as 1-methyladenine. There is a recent and substantial review of the considerable progress in this subject by Meijer and Guerrier (1984). The critical question as far as the nervous system is concerned is where the GSS is produced and stored and how it reaches and interacts with the gonads. There are a number of theories from a pheromone-like role by Chaet (1966) to direct transport to the gonads within the nervous system (Schoenmakers et al. 1981). Location of GSS secreting cells has until recently been detailed using rather non specific histological staining techniques for neurosecretory substances. A substantial advance has now been achieved by Caine and Burke (1985) who using a purified extracted GSS have used specific immunological techniques to localize neurones involved. These studies show a concentration of GSS in the perihaemal epithelium adjacent to the radial nerve cords and suggest transportation to the gonads via the radial haemal sinus and the haemal system in general. They also show a correlation between 100 nm dense cord vesicle containing axons and specifically localized GSS.

5. PHARMACOLOGY

Welsh (1966) reviewed the subject of neurotransmitters in depth and work subsequent to that was discussed by Pentreath and Cobb (1972). At that time there was good evidence for a transmitter role for acetylcholine (ACh) and much inconclusive information about other transmitters. The number of transmitters and neuromodulators documented as occurring in the animal kingdom has increased greatly in the last two decades but the amount of hard facts for the phylum Echinodermata is still fairly limited.

Acetylcholinesterase has been demonstrated (Pentreath and Cottrell 1968, Pentreath 1970) in both ectoneural and hyponeural nerves. Florey and Cahill (1980) in a valuable paper examined the evidence for the motor control of the muscles of sea urchin tubefeet. The transmitter involved is ACh and diffusion takes place across a wide connective tissue layer. They show this layer does not contain acetylcholinesterase as would be expected under the circumstances. Evidence is provided that a wide range of drugs potentiates or blocks the effect of ACh when applied directly or released by stimulating the nervous system. Kobzar (1984) looked at transmitter receptors in Cucumaria and showed contraction in the protractor muscle to adrenaline, noradrenaline, dopamine, histamine and GABA as well as ACh but blocking agents in general were ineffective. Protas and Muske (1980) used cleaned and non-cleaned tubefeet to look at transmitter substances. They suggest separate muscarinic and nicotinic ACh receptors with the latter 10 to 100 fold more significant and also GABA receptors. Catecholamines were more effective on uncleaned tubefeet. Noradrenaline and dopamine may cause the lantern muscles in an echinoid to relax (Pentreath and Cobb 1972) and GABA in holothurians (Hill 1970). Florey et al. (1975) showed GABA to be excitatory in tubefoot preparations but showed other transmitters (Florey and Cahill 1980) to be ineffective. Tsuchiya and Ameniya (1977) looked at the contraction of the radial muscle of a soft bodied echinoid and described the mechanical responses to electrical stimulation and drugs. They again showed the importance of ACh as a transmitter but found some responses, such as a reduction in response to continued indirect stimulus difficult to explain. They suggest it necessary to "assume the presence of inhibitory motor nerves or some other kinds of neural interaction, though the present study gives no information about the actual neural mechanism". This statement summarizes the whole problem with the pharmacological studies. It is almost impossible to know where the various transmitters, blockers, modulators, etc. act without a detailed cellular knowledge of the layout of the nervous system and intra-cellular physiological studies of receptor distribution. In many cases failure of drugs may be due to failure to penetrate connective tissue barriers, or between the tightly packed and glial-less axon bundles or simply failure of drugs to dissolve effectively in sea-water. Florey and Cahill (1980) for example showed ACh to be much more effective on 'skinned' tubefeet than when the epithelium was present.

Fluorescent histochemistry has demonstrated the widespread distribution of dopamine and noradrenaline in ectoneural nerves including the epithelial plexus and the endothelial plexus of the gut (see Cobb 1969a and Cottrell and Pentreath 1971). There is a correlation between the distribution of specific fluorescence and the presence of small granular vesicles in nerve endings (Cobb 1969a). Monoamine oxidase has been shown to be present in echinoderm tissues (Blaschko and Hope 1957), Pentreath (1970) used histochemical techniques to demonstrate it within the nervous system.

The organ bath studies, as well as the histochemical localization of pharmacologically significant substances, suggests a wide range of functional specialization in the nervous system. This approach has however yielded little information about the actual roles of these substances other than for ACh.

The pharmacological and neurosecretory data, if nothing else, emphasizes the complexity of the echinoderm nervous system. The ectoneural nervous system contains a range of known transmitter substances but only ACh can be assigned a positive excitatory role in neuromuscular systems. The function of catecholamines is obscure with no evidence for a direct motor role only a probably non-specific effect. Yet there are plenty of clues about possible function since neural inhibition is proposed from a number of behavioural studies and rhythmic behaviour under the control of the nervous system has long been established. There are anatomical clues from electron microscopy with the description of different types of vesicles in neurones implying possible functional differences. There are other snippets of information such as a group of 7-9 cells precisely positioned in each segmental ganglion of the radial cord of _Ophiura ophiura_ which stain with the vital dye neutral red (J.L.S. Cobb unpublished) or the staining of small groups of cells using immunohistochemical procedures and the molluscan small cardioactive peptide (SCP). These last findings are only of a preliminary nature (S. Kempf and J.L.S. Cobb unpublished) but are an indication that the echinoderm nervous system will eventually be shown to have the same sort of complex cellular organization with a wide range of neuromodulators and transmitters associated with specific functions as occurs in other nervous systems. The use of immunolabelling techniques by Caine and Burke (1985) no doubt introduces a new era into the study of echinoderm nervous systems at the cellular level and which combined with the physiological data that will accumulate from the now possible intra-cellular studies will provide for rapid progress in the next decade.

6. SENSORY SYSTEMS

6.1. Introduction

Receptor function in echinoderms is poorly understood but there is a growing ultrastrucural library of presumed receptor structures (see Cobb 1986). There is a substantial literature on sensory perception in this phylum but almost all of it used behavioural criteria to measure a response to stimuli. Sloan and Campbell (1982) have for example reviewed the large literature on chemoreception and shown behavioural studies often produce conflicting results. Recent neurophysiological findings described below make it clear that in many cases the behavioural studies were compromised by investigators introducing stimuli that they were not aware of.

In general receptors on echinoderms would appear to be widely distributed rather than localized into complex sense organs. There are two described exceptions to this, the sensory hillock of globiferous pedicellariae and the ocelli at the tip of the arm in some starfish. Many of the putative receptors described from both scanning and transmission electron microscopy contain a cilium or a modified cilium but it is possible that there are classes of non-ciliated receptors.

6.2. Non-Ciliated Receptors

There are two lines of evidence pertinent to the question of the existence of non-ciliated receptors. Kawaguti and Kamishima (1964) originally proposed that all echinoderm epithelial cells were sensory and Cobb (1968) later proposed that this was so on the pedicellariae of Echinus. Markel and Roser (1983) proposed only a single epithelial cell type on the spines of Eucidaris and this implies either there is no receptor function to the spines or it too is covered only by receptor cells. This idea has been challenged, but Weber and Grossmann (1977) in a thorough survey showed a mixture of sensory cells and epithelial cells, the latter being characterised by the presence of tonofilaments. A different line of evidence is provided by Bouland et al. (1982) and Fankbonner (1978) who have suggested that some of the papillary cells of tubefeet and tentacles which are covered in microvilli but lack ciliary structures may be receptors of the gustatory type. Microvilli are clearly modified and involved in the structure of many of the supposed ciliated receptor cells and it must be considered as possible in some cases the microvilli alone are present in receptor cells.

There is some evidence for non-specialized receptor function in neurones themselves. Yoshida and Millott (1959) and Millott (1968) proposed some neurones were light sensitive and Millott and Coleman (1969) described a structure called the podial pit where a substantial but unspecialized plexus of nerve fibres was present beneath the epithelium and tentatively suggested that these might be photoreceptive. Hendler and Byrne (1985) suggest that glassy tubercles on the surface plates of a brittlestar act as a light guide through the calcite stereom focussing on large nerve fibres within the skeleton. They also suggest that chromatophores in the epithelium which are mobile on a diurnal basis shield or permit light entry to the tubercules. The only unequivocal physiological studies on photoreception are on Ophiura texturata and show that the nerves in the radial cord are not light sensitive but that does not exclude the possibility that peripheral ones are.

Physiological recordings (A. Moore and J.L.S. Cobb, in preparation) show that individual spines are sensitive to water borne vibration and mechanical displacement even with the epithelium removed with bleach. This implies that there are receptors within the ligaments which monitor stretch and this is the conclusion that Lewis (1968) reached after studying the ultrastructure of the echinoid sphaeridia which have a supposed gravity detection function. An examination of such ligaments shows numbers of different types of nerve fibres but no specialized endings. It is therefore also possible that there are morphologically unspecialized neurones that have receptor functions.

6.3. Photoreceptors

Eakin (1966) has reviewed the information available on the ocelli in starfish. There is a great deal of behavioural evidence on the functional significance of these organs. Yoshida and Ohtsuki (1968) carried out a comprehensive study on the starfish Asterias amurensis. They showed the ocelli were most affected by shadow and not by increases in light. They also showed that an extra-ocellar light sense existed but its action was normally masked by the ocelli.

Emerson (1977) reported the structure of larval ocelli and Yamamoto and Yoshida (1978) described the ocelli in a holothurian and described the effects of light and darkness. Eakin and Brandenburger (1979) re-examined the structure of ocelli in three species of starfish in the light of Yamamoto and Yoshida's finding on diurnal changes in the ocellus with varying light/dark regimes. They also described the cyclic structural changes that are associated with the turnover of receptoral membrane. Previously Eakin (1963, 1968) had proposed the receptors were of the ciliary type but in the light of the more recent work now argues they are of the rhadomeric-type. This agrees with Yamamoto and Yoshida's holothurian finding and with recent findings on the starfish Nepanthia by Penn and Alexander (1980) and Takasu and Yoshida (1983) on Asterias. Eakin and Brandenburger (1979) discuss the function of cilia in the formation of the rhadomeric (non-ciliary) type of photoreceptor and reference a longer discussion of this problem (Eakin et al. 1977). Cobb and Moore (1986) in trying to account for the extreme photosensitivity of the brittlestar Ophiura, where the nerve cord itself is non-photoreceptive, described a ciliated receptor in the photically sensitive epithelial. This very modified cilia below the cuticle is associated with an intracellular lamina of membranes. There is however only circumstantial evidence that it is a photoreceptor.

Unusual ciliary structures have been reported in two other situations. Gardiner and Rieger (1980) have shown a modified cilia within a 9+0 structure (interpreted from longitudinal section) on the endothelial cells lining the coelomic cavity of the tubefeet and proposed that these may have a direct stretch receptor function not involving the nervous system. It is interesting to note Mackie et al. (1969) considered the possibility of epithelial conduction to account for larval ciliary beating reversal in echinoderms. Cobb and Stubbs (1981, 1982) also showed cilia are present on the cell bodies of some ectoneural and hyponeural neurones. It is conceivable that these might be associated with some receptor function and from their internal position this is only likely to be proprioception. These cilia would appear to be non-motile.

The role of cilia in receptors thus relies largely on ultrastructural evidence and it must be borne in mind that in even some of the less speculative situations described above they may have nothing to do with transduction of environmental stimuli or even indicate the presence of a receptor cell.

6.4. Interoreceptors

Lewis (1968) examined the sphaeridia of sea urchins in an attempt to describe a receptor responsible for gravity detection. Al echinoderms can orientate with respect to gravity and some brittlestars can 'right' themselves when inverted in less than a second. Lewis found no evidence for neurones within the sphaeridia which could classify it as a sense organ. He proposed drag on the ligaments would be detected by proprioreceptors. This proprioceptive detection of gravitational forces on heavy appendages seems a likely explanation in all classes of echinoderms. A. Moore and J.L.S. Cobb (in preparation) present some direct evidence using physiological techniques on the spines of brittlestars (see below). They showed movement of the spines could still be detected by the animal after the surface epithelium had been removed. The class of neurones responsible for proprioception of this sort has not been identified anatomically.

6.5. Chemoreceptors and Mechanoreceptors

It is not possible to separate these two modalities of perception ultrastructurally. Cobb (1968c) examined the sensory hillock of the globiferous pedicellariae of a sea urchin. He showed that long cilia (capable of movement) with a normal 9+2 configuration were present packed tightly together. Each ciliated cell when cut in transverse section of the surface had a diameter of little more than 1 µm. Campbell (see review in 1983) showed these cilia to be both mechano- and chemoreceptive. Oldfield (1975) has also described sensory cilia in these organelles and further showed that there were swellings at the cilia tip which she did not consider to be artefacts. Frankbonner (1978) has also described the same phenomenon in holothurians.

Whitfield and Emson (1983) and subsequently Byrne and Fontaine (1983) and Cobb and Moore (1986) have described a putative receptor on the surface of brittlestars and crinoids and shown them to be the stabchen of Reichensperger (1908). These are again modified cilia surrounded by a ring of specialized microvilli.

Other ciliated receptors have been described from the tubefeet or buccal tentacles, Martinez (1977), Bouland et al. (1982), Florey and Cahill (1982) and Cobb and Moore (1986). These in general are similar to a single receptor cell of the sensory hillock of pedicellariae and their structure is reviewed in Cobb and Moore (1986). Many of the cilia in the putative receptors are modified from a motile 9+2 arrangement but this is not invariably the case and it is not easy to separate reliably receptor cells from ciliated epithelial cells using solely ultrastrutural criteria.

There are many types of secretory cells associated with the surface epithelia of some echinoderm organs (see Sousa Santos 1966 and Martinez 1977 for review). A number of authors have commented on the close relationship between putative receptors and the secretory cells (Whitfield and Emson 1983; Byrne and Fontaine 1983; De Vos 1985; Cobb and Moore 1986). The significance of this needs to be established.

6.6. Sensory Neurone Structure

The best description of sensory neurones is of the sensory hillock of pedicellariae (Cobb 1968c). In this situation an undoubted group of sensory receptor cells were traced by serial section techniques until bundles of small but characteristic sensory axons arose from the cells. The structure of these sensory neurones was thus a cilium projecting through a narrow collar of microvilli giving rise to a long striated root which ended where the cell was swollen to accommodate the nucleus. Numerous microtubules were associated with the striated rootlet and these passed down to the region where the cell narrowed to form an axon process of uniform diameter. The sensory axons characteristically were filled with microtubules. Cobb (1968c) suggested the ciliated epithelium of other types of pedicellariae were also largely composed of sensory neurones. The work of Weber and Grossmann (1977) examined the structure of the epithelium in the sea urchin and showed using semi-serial techniques that numerous unspecialized sensory cells gave rise to axon processes. They also described other cell classes that were present. Burke (1983c) has also traced axons arising from unspecialized receptor cells in the larvae of a starfish. The evidence at present suggests that the distribution and relative proportion of sensory cells in an epithelium may be very varied according to situation. It is reasonable to assume, until there is evidence to the contrary, that the receptor cells are the sensory neurones themselves and that sensory axons feed directly into the basiepithelial plexus. Direct physiological confirmation of the sensory function of individual cells will be hard to achieve.

7. BEHAVIOURAL STUDIES

There is a substantial, and continually growing, literature on the behaviour of this intriguing phyla. There is no doubt that von Uexküll's image of the echinoderm as a 'republic of reflexes' (1897) without central control was such an out of the ordinary hypothesis as to be constantly challenging. Add to this the fact that echinoderm behaviour is inconsistent and often contradictory, and thus almost any observation is new, and the attraction to behaviourists is even more enticing. This is well illustrated by the account of the distinguished neurobiologist Bullock (1965) sitting down with the Professor Kinosita and coming to opposite conclusions about the experiments they were doing together. Behaviour and the nervous system are synonymous, but because the author is prejudiced and feels that behavioural studies will continue to produce conflicting results until at least a basic understanding of function in the nervous system is achieved at the cellular level, mention will only be made of some major reviews or significant papers.

Hyman (1955) provides a review of the great wealth of early work which so indebts echinoderms biologists to her. Reese (1966) produced a most comprehensive review of the complex behaviour of echinoderms. Smith (1965, 1966) attempted to draw together his own previous studies and show that echinoderm behaviour was controlled by neural centres located within the circumoral nerve ring and immediately adjacent nerve cord. He shows the 'republic of reflexes' concept of von Uexküll (1897) was not a sufficient explanation of observed behaviour. Smith's work is still accepted; embedded in the literature is his concept of a central nervous system. Arshavskii et al. (1976), for example, still describe central controlling centres as interacting to co-ordinate locomotion in ophiuroids although in an earlier paper (1975) they show tubefoot stepping in the same class is under reflex control.

Bullock (1965) discusses at length the evidence for superficial nerve nets in echinoids and asteroids. He decided they were absent in echinoids and present in asteroids. He cautions, however, that behavioural criteria such as he used are not satisfactory without other evidence. Millott in three review papers has again examined a wealth of evidence, much of which is behavioural, on the importance of peripheral and central nervous systems in controlling spine movements (1966), the dermal light sense (1968) and photosensitivity (1975). A great deal of the work reviewed is his own substantial contribution to the literature, particularly in collaboration with Yoshida (see Millott 1975 for references).

A more recent review of one aspect of behaviour is that of Sloan and Campbell (1982) who examined the perception of chemical cues associated with feeding behaviour.

Two recent papers illustrate the dual approaches that are now beginning to be available to study echinoderm behaviour. Valentincic (1985) used behavioural criteria to study the structure-activity relationships of a whole range of stimulatory substances (particularly amino acids). He discusses the short comings of behavioural criteria but shows that soundly planned and well controlled experiments making use of defined biochemical stimuli can produce consistent responses. Moore and Cobb (1985b) used neurophysiological techniques (see later section) to define perceptive ability to a range of chemicals similar to those used by Valentincic and showed consistent and much lower thresholds compared to behaviour experiments run in parallel.

8. LARVAL NERVOUS SYSTEMS

There is a great gulf between larval and adult echinoderm biology. Various embryological stages have been of great significance; much of the work has been on eggs and pre-gastrula stages. There is almost no overlap in the literature as the author found to his cost when trying to review membrane physiology in embryological and adult systems in the phyla. There is much known about some aspects of the former and none about the latter (Cobb 1982). Recently, however, Burke in a series of papers (1978, 1980a,b, 1983a,b,c,d) has used a range of anatomical techniques combined with behavioural observations to describe the importance of the nervous system in both development and metamorphosis. This work shows (Burke 1983d) that cells with the characteristics of nerves first appear in the 60-h larvae. He suggests these cells are derived from the animal plate and develop in association with the ciliary bands. It is suggested that ciliary reversal is carried out by neural intervention (Burke 1983c). This author describes the importance of podial sensory receptors in the induction of metamorphosis (Burke 1980b). Experiments on the neural control of metamorphosis were carried out on Dendraster and described in an important paper (Burke 1983d). He suggests that catecholamine containing neurones are significant in the apical neuropile and the oral ganglion and that these parts of the nervous system are centres that mediate between the perception of the natural cue and the initiation of metamorphosis. It is interesting to reflect Huet and Franquinet (1981) implicated monoamines in regeneration.

9. LEARNING

There have been relatively few studies on learning in echinoderms since the work of Jennings (1907) on the asteroid Sclerasterias using electric shocks. Reese (1966) reviewed the evidence for associative learning. The most recent are the works of Valentincic (1979) on Marthasterias and McClintock and Lawrence (1982) on Luidia. Both studies show associative learning with food stimuli, using an electric shock in the former study and a photic cue in the latter.

There can however be very long term changes in the function of the nervous system due to grossly abnormal stimuli such as the trauma of capture by trawling. Animals treated in this way can show no behavioural or physiological responses for many hours and in some cases the animals eventually die. The reason for this functional failure is not understood and the author feels that if gross stimuli (such as electric shocks) are used neural integratory mechanisms break down and may require many hours to recover. The experiments on the location of the centre for locomotory integration (reviewed by Smith 1965) relied on the massive trauma of cutting off an arm with or without part of the nerve ring. The author feels that the different positions of amputation inflicted different gross stimuli and produced opposite behaviour. It should be remembered that after a period of time that both types of amputation produced the same behaviour (Smith 1950). There is no direct evidence to support these speculations but with progress being made on studying integration at the cellular level a greater understanding should be achieved in the near future.

10. REGENERATION

All the echinoderm classes show varying degrees of regenerative ability. This includes regeneration of the nervous system, not by outgrowths of existing neurons but by the development of totally new populations (J.L.S. Cobb unpublished). Cobb and Stubbs (1981) used degeneration of neurones in an attempt to trace axon pathways. They showed that when a complete transverse cut was made in the radial nerve that severed axons did not degenerate beyond the cut for more than one complete segment. Degeneration was studied over a period of several weeks by which time substantial regeneration was taking place. Recent Lucifer Yellow dye fills of individual neurones (see later section) has shown some axons span at least 5 or 6 segments.

These studies show axons do not degenerate even over long time scales in some parts of the nervous system when severed from cell bodies. The only recent work on regeneration has been by Huet (1979) and Huet and Franquinet (1981) who showed marked changes in the levels of monoamines in the radial nerve cords and circumoral ring between the 2nd and 4th days of regeneration when the blastema is forming. Huet (1975) had previously shown that the arm tip regeneration in the starfish <u>Asterias</u> is controlled by the nervous system. Regenerative mechanism within the nervous system are badly in need of study in this phylum.

11. CELLULAR NEUROBIOLOGY

11.1. Introduction

The final section of this review will deal with recent attempts, particularly by the author, to examine the function and structure of the nervous system at the level of single cells. It will examine the use of extracellular and intracellular recording techniques and the use of dyefills to show the detailed morphology of single, and in some cases, physiologically identified nerve cells. Much of the work described is published but some is new. Sections containing new material will be identified.

11.2. Extracellular Recording Techniques

Attempts were pioneered by Sandeman (1965) and later work was carried out by Millott and Okumara (1968), Binyon and Hasler (1970) and Podolskii (1972). All these studies showed compound potentials decrementally conducted to massive stimuli. There was some evidence that two types of pathway were involved. One has to conclude that this series of papers does not really contribute very much to an understanding of the function of the nervous system. Takahashi (1964) showed unitary spike-like potentials to a photic stimuli in the radial nerve cord of a sea-urchin. He used metal-filled glass micro-electrodes and showed the response to stimulation was delayed for many seconds after it was applied. This work has never been repeated although substantial efforts were made by a number of people over several years.

The breakthrough came when Brehm (1977) made use of a class of giant fibres in an ophiuroid. Giant fibres had been suggested since the work of Christo-Apostolides (1882) and Pentreath and Cottrell (1971) described them ultrastructurally. Brehm (1977) used small suction electrodes to record from the radial nerve cord a through conducted spike associated with a luminescence response to stimulus. This activity was spikes from single neurones. He also recorded compound potentials near the stimulating electrodes. He suggested a small group of large diameter neurones identified during ultrastructural studies were responsible.

Subsequent to the work of Brehm research in this laboratory has used the suction electrode technique on a number of different species of ophiuroid to study neural function. The species _Ophiura ophiura_ is particularly favourable since it is a fast moving active species with many large ectoneural and hyponeural neurones. It is likely that the activity recorded is from a class of ectoneural interneurones but it has proved very valuable in assessing the sensory discriminatory abilities of echinoderms. In all cases thresholds of responses recorded from the nervous system have been compared to behavioural thresholds to the same stimuli.

Stubbs (1982) described a photic response and Moore and Cobb (1985a) have further developed the findings. They have shown great sensitivity to charges in light, particularly to shadow. Behavioural studies showed a shadow induced 'freezing' of movement and it has been suggested this is an anti-predatory device. Sensitivity to wavelength, intensity and rate of change of intensity are all described.

A similar study (Moore and Cobb 1985b) has been carried out on chemoreception. Sensitivity down to 2×10^{-12} has been shown for some amino acids. Behavioural studies show similar stimuli but at higher concentrations and produce 'searching' behaviour. High levels of the same chemicals prove noxious and produce rapid 'avoidance' behaviour.

Mechanical and vibration stimuli have also been classified (A. Moore and J.L.S. Cobb in preparation). Water borne vibrations below 20 Hz produce a 'searching' behaviour and above 20 Hz produce a 'freezing' behaviour. Recorded activity shows different patterns and sizes of spikes for the two behaviours. These studies have also shown by stripping the epithelium off that the mechanoreceptor is likely to be an intero-receptor within the ligaments of the spines. Electrophysiological recordings can be made to the movement of a single spine. Such receptors in ligaments could account for the gravity detecting ability of echinoderms in the manner suggested by Lewis (1968) in his investigation of the proposed function of sphaeridia. Electrodes attached to an animal subsequently inverted showed massive unitary activity associated with the movement (J.L.S. Cobb unpublished).

The use of multiple recording electrodes from different positions on whole animals is allowing an analysis of integrative mechanisms to be undertaken which combined with intracellular studies will give new insight.

Tuft and Gilly (1984) have recently reported another study using suction electrodes but recording mainly compound potentials. This work suffers from the same shortcomings as the early work because of ambiguities in the interpretation of compound potentials. These authors do illustrate spontaneous single unit activity, which our own studies show is often due to the mechanical attachment of electrodes or long term damage response to the initial dissection. The main thrust of their paper relies on conducted compound potentials. They show there are at least two types of these based on a positive going and a negative going compound spike. They offer no explanation for the different polarities. They show using changes in the bathing solution that the two parts of the compound potential can be differentially reduced or eliminated and that they are conducted at different rates. They suggest one is a Na^{++} spike and the other on Ca^{++} spike. It is the opinion of the author that this sort of experiment adds little of value since the anatomical basis for the two classes of fibres is highly equivocal, even assuming the potentials can be related to such classes.

The manipulations of ions solutions and changes used could be affecting receptors and ion channels in all sorts of cells in all sorts of places. They do however show faster conduction velocities than previously described. It is not clear why they measured these between stimulating electrodes and recording electrodes rather than between two recording electrodes. Their technique may give spurious conduction velocities because electronic spread of the large stimulus used was not controlled or measured. They criticised the suggestion of Cobb and Stubbs (1981) who claimed giant longitudinal neurones synapsed in each segment since conduction velocities in their work were too high to involve the synaptic delay in the number of synapses involved. This criticism is valid since more recent work (Cobb unpublished) has shown these neurones to run much longer distances than one segment. This is discussed in detail in the following section.

11.3. Intracellular Techniques

Cobb (1985a,b) has published preliminary intracellular studies on neurones of both the ectoneural and the hyponeural system of the brittlestar <u>Ophiura ophiura</u>. The studies have used the dye Lucifer Yellow which can be iontophoresed into individual cells to illustrate their detailed morphology. At present this can be repeatably and relatively easily achieved for hyponeural cells and with greater difficulty for some ectoneural cells. It is possible to record from single arm preparations or from whole restrained animals and this has allowed a series of very preliminary experiments to be carried out aimed at unravelling the cellular basis for the integration of complex behaviour.

11.3.1. Ionic Basis of the Action Potential

There are suggestions in the literature that one class of spike is a Na^+ spike but that it is tetrodotoxin insensitive. Tuft and Gilly (1984) suggest a second class of axons in ophiuroid nerve cord with a Ca^{++} spike but their evidence and that of Brehm (1977) who proposed only a Ca^{++} spike is not specific enough to be of great value. There is some evidence that muscle spikes are Ca^{++} spikes (Prosser and Mackie 1980). The only intracellular studies on adult echinoderms are those of Cobb (see Pentreath and Cobb 1972) on muscle and those of the same author on neurones (Cobb 1985b). These studies have not looked at the ionic mechanisms and definitive statements can only come from this type of work. A very speculative hypothesis may be that ectoneural spikes are Na^+ and hyponeural and muscle spikes are Ca^{++}. This assumes hyponeural neurones are derived from a line of muscle cells.

11.3.2. Single Cell Morphology (Hyponeural)

The author has tried, without success, to trace neurone pathways using cobalt back-filling or horse raddish peroxidase techniques. Lucifer Yellow (Stewart 1978) has however proved valuable when iontophoretically injected (see Cobb 1985b for techniques). Hyponeural cells can be impaled after either an oral or an aboral dissection with the nerve cord <u>in situ</u>. These cells are either bipolar or unipolar. Bipolar cells send two large diameter motor axons into two separate motor nerves to the intervertebral muscles. There are numerous fine non-varicose branches from the region of the cell body. These form a two-dimensional meshwork against the basement membrane separating the ectoneural nervous system. The area covered by this meshwork occupies up to one half of the total surface area occupied by the hyponeural ganglion on one side of the midline. The anatomy of these cells is illustrated in (Cobb 1985a, b) and subsequent fills have confirmed this. This meshwork consists of post-synaptic dendrites innervated by pre-synaptic varicose terminals of the ectoneural system across the basement membrane. Each of the three motor nerves to the intervertebral muscles contains between 10 and 30 large diameter axons (there are about 50 large hyponeural motor cells in each segmental ganglion on either side of the midline). The axons in the motor nerve then divide close to the muscles (and connective tissue) they innervate into many smaller motor endings. This has been shown (Stubbs and Cobb 1981) using degeneration techniques, which on motor nerves are effective. It is not known if a separate population of motor nerves innervate the juxtaligamental cells of the connective tissue or if these and the muscles are innervated by different terminal branches of the same axon. The nodes of juxtaligamental tissue associated with the muscle-free oral ligaments have separate motor axons. Ultrastructural studies have shown occasional vesicles in the hyponeural dendrites but there is no real evidence for integratory mechanisms in the hyponeural ganglia. There is no evidence from many Lucifer fills that hyponeural nerves interconnect adjacent segmental ganglia as proposed by Pentreath and Cottrell (1971) nor do processes cross the midline.

11.3.3. Single Cell Morphology (Ectoneural)

A large number of partial fills have been achieved but few complete ones. The reasons for this are two-fold. First the cells are smaller and do not present the same defined target as groups of hyponeural cells. Second it is now clear that many cells have axons over 1 cm long and it is difficult to fill with Lucifer to all extremities. There are two classes of cells so far identified (Cobb 1985a, b), one of which runs longitudinally and the other transversely. Most fills have been of the former type and they all follow a characteristic morphological form. The cell body lies orally in the radial nerve cord and a single axons leaves and passes transversely across the nerve cord before running longitudinally following a straight pathway with reference to the midline. In the region of the cell body and where the arising axon runs transversely there are many fine varicose processes given off which cover a substantial area of a single ganglia on both sides of the midline. The axons may run longitudinally either towards the periphery or centrally. The axons terminate in another fine plexus of varicose endings again covering a substantial area of a segmental ganglion. They never make apparent endings while still of large diameter. One neurone was described (Cobb 1985b) that only ran for two segments but subsequent unpublished work has shown that most run at least five or six. The distance is more than 1 cm and at present it has not been possible to inject for sufficient time to fill the whole of these long neurones with high enough levels of dye to identify fine terminal processes at both ends. Many impalements are of axons themselves running longitudinally rather than the cell body regions. A very important and totally consistent finding is that where axons pass right through a segmental ganglia they never branch in any way or diverge from a straight course and thus there is no evidence that they make output except where they terminate in distant segments.

A second class of neurons runs transversely only within one segment of the nerve cord (Cobb 1985a). These neurones have been impaled and filled only rarely. They show substantial varicose branching on both sides of the midline and again occupy a large area of a single ganglion.

There is some evidence for bipolar cells with an axon in each direction longitudinally but few such cells have been even partially filled. A critical point to remember in these studies is that there are neurones which may show cell bodies up to 20 µm and axons to 15 µm down to 3 µm cell bodies and axons less than 1 µm. (Larger neurones reported from anatomical studies [Cobb and Stubbs 1981] may have been virus infected and atypically large). Random impalement will always tend to impale larger cell bodies and axons and this will present an atypical sample of the whole population. Relative proportions of cell types are a fairly meaningless statistic. Only rarely has it been possible to correlate structure of an individual ectoneural cell with known physiological function. Future studies will require this.

11.3.4. Single Cell Function (Hyponeural)

Hyponeural cell bodies can be recognized intracellularly by their failure to spike to depolarizing current injected or on rebound to hyperpolarizing current as well as by continual synaptic potential activity (up to 15 mV) which can be both excitatory or inhibitory (Fig. 2). Penetration of the motor axons, which is more difficult, shows typical spikes are present. Cobb (1985a, b) has shown using dual impalement of two hyponeural cells that there is a varying degree of correlation between the pattern of synaptic potentials recorded and the position of the two electrodes to remote stimulus of the ectoneural nerve cord. At its most extreme this correlation consisted of almost mirror image epsps and ipsps when electrodes were symmetrically placed either side of the midline in a single ganglia. The stimulus used in this experiment was an electrical one and current experiments are using more natural ones. The synaptic potentials also follow remote stimulus to the ectoneural system (even on distant arms in whole animal preparations) and usually cause a summating depolarization (or occasionally hyperpolarization) (Fig. 2).

There is some evidence that the ipsps are caused by local sensory input to the ectoneural system in immediately adjacent segments and epsps to more distant stimuli but much more carefully controlled experimentation is necessary to clarify this. It is certain however that the synaptic potentials are not directly correlated with spike potentials carried in longitudinal ectoneural neurones and there is at least one class of interneurones interposed. Synaptic potentials are recorded to food and noxious chemical stimuli which cause feeding or escape movements of the arm. Similar potentials are recorded less frequently to water disturbance or photic stimuli which normally cause the animal to 'freeze'. Superficially, excitatory synaptic potentials to a 'freeze' behaviour inducing stimuli are an anomaly. It is possible however the 'freeze' response occurs by locking the connective tissue and this too is under excitatory neural control by hyponeural neurones. Dual electrode experiments to carefully defined stimuli will provide a way of progressing towards understanding the integration of motor output.

11.3.5. Single Cell Function (Ectoneural)

At least something is now known of longitudinal cell function and what follows has not been previously published. Most ectoneural cells spike on depolarisation and on rebound spike from hyperpolarisation. Impalements of the cell body or regions close to it (evidence from Lucifer fills) show both epsps and ipsps and spikes. This activity follows remote stimulation of sensory receptors by a range of stimuli but of particular categories (Figs. 3, 4). 'Freeze' stimuli (photic, high frequency vibration or gross water movement) induce spiking in one class of longitudinal neurone. These neurones have different thresholds. The largest neurones usually fire a quick burst to maximal stimuli and smaller ones a large number of spikes over a longer period. These intracellular studies have shown that the extracellularly recorded spike trains to similar stimulus (Moore and Cobb 1985a, b, and in preparation) are caused by a relatively small number of neurones firing repeatedly. They have also shown that the individual events in an extracellular trace do correspond to individual spikes and that the largest and fastest conducting neurones usually produce the biggest extracellular spike.

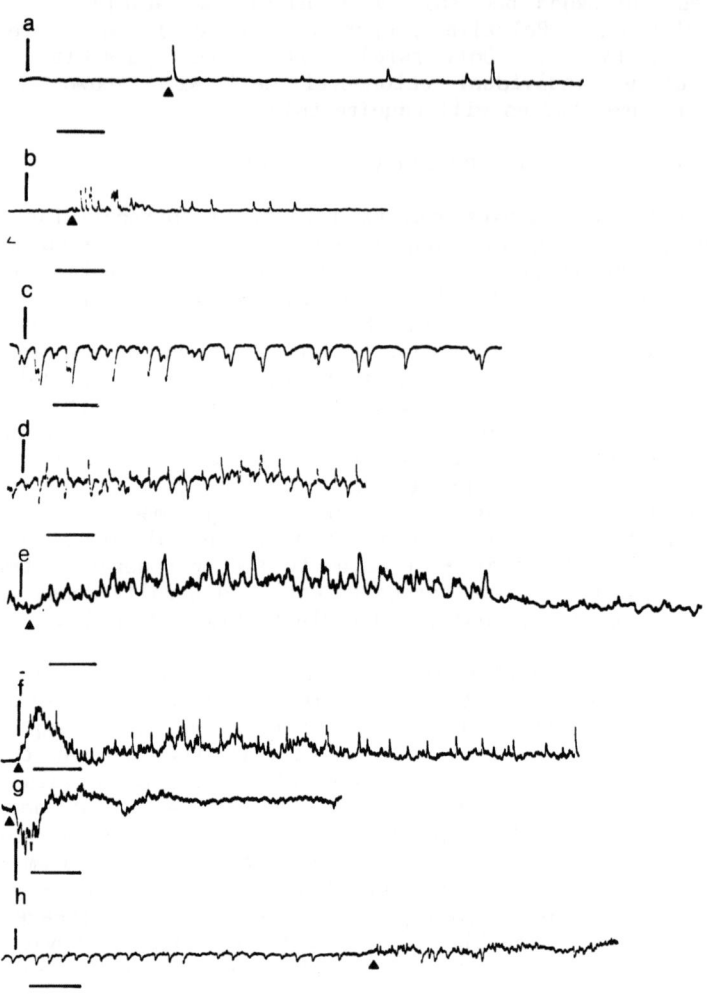

FIG. 2: Intracellular recordings from hyponeural motor neurons.
a. Single epsp (arrow) to the movement of a single spine 15 segments distant. Occasional spontaneous synaptic potential are also present.
b. Epsp to dilute food stimulus applied to tip of arm 25 segments distant; such a stimulus causes arm movements of a food searching nature.
c. Spontaneous ipsps. This impalement was made orally (i.e. through the ectoneural tissue) and presynaptic neurones may have been stimulated to firing by the passage of the electrode.
d. Spontaneous epsps and ipsps. All hyponeural neurones appear to have an excitatory as well as an inhibitory input.
e. Overall depolarization of a hyponeural neurone by summed epsps following a chemical stimulus to the remote armtip.
f. Large and rapid depolarization to noxious chemical stimulus to remote arm tip. Such a stimulus is normally followed by rapid locomotory withdrawal. Epsps persist until withdrawal of stimulus.
g. Large and rapid hyperpolarization to high frequency vibration at remote arm tip. Such a stimulus normally results in the 'freezing' of all movement by the animal, but this stimulus does not always result in ipsps depending on position of cell.

'Flex' stimuli, where behaviourally the animal will flex an arm prior to feeding, are integrated and carried by a separate and perhaps more numerous class of smaller neurones at a lower conduction velocity. 'Flight' stimuli where the animal will rapidly move away from a noxious stimuli produce spikes that are carried by yet another separate class of longitudinal neurones. There are thus separate tracts of neurone for each class of behavioural response. There are other neurones that respond to none of these stimuli types.

Impalements of axons themselves show bursts of spikes to stimuli of the appropriate specific category with no synaptic potentials. Information is conducted in both directions in the nerve cord to the different categories of stimulus and it is just possible that individual axons conduct in both directions. Impalements of single cells have been made and spike trains recorded to bursts of activity travelling in both directions. An alternative, and perhaps more likely, explanation is that these particular cells receive input (and are driven to spike) from different classes of longitudinal nerves which separately conduct in both directions. There is some evidence for this since there are also neurones which show synaptic potentials, both excitatory and inhibitory, as well as spikes to all categories of stimulus. It is surmised that these represent a class of neurone which integrates much of the information entering a particular ganglion from both directions and is the first step in producing a co-ordinated motor output. No Lucifer fills have been achieved of such cells but it is possible they represent the transverse neurones only so far filled on random impalement.

One problem is always going to be smaller neurones, impossible at present to impale, carrying out significant but undetected functions. Tubefoot stepping appears to be under rhythmic control but cannot be related to known neural activity. Okada et al. (1984) have shown rhythmic contractions of the gonads under neural control. Some cells do show bursts of activity of varying types (Fig. 4) but in these preliminary stages injury discharged of various types cannot be excluded when spontaneous activity is encountered in single cells. It is the hope of the author that some types of behaviour involving rapid movement are carried out entirely under the control of large neurones and hence are able to be studied with currently available techniques.

11.3.6. Discussion

This section is necessary since some of the above results have not been previously published. The vital question is how long are the longitudinal giant neurones? The author's hypothesis on the non-centralized nature of the nervous system is not tenable if these neurones are continuous right down the nerve cord and are centrally integrated near or within the nerve ring. The evidence for a centralized giant fibre system is from the conduction velocities which are incompatible with synapses, suggested by Tuft and Gilly (1984); Brehm's (1977) assertion

---

h. Ipsps occuring spontaneously and stimulation (arrowhead) with a dilute food stimulus at remote arm tip. Complex epsp/ipsp response follows. Dual electrode recordings (Cobb 1985b) show a relationship between the position of the neurone within the ganglion and the type of stimulus. The detailed significance of the excitatory and inhibitory input to the motor neurone population to a particular stimulus and related to a particular behavioural response has still to be elucidated.
Vertical scale = 10 mv. Horizontal scale = 0.05 s.

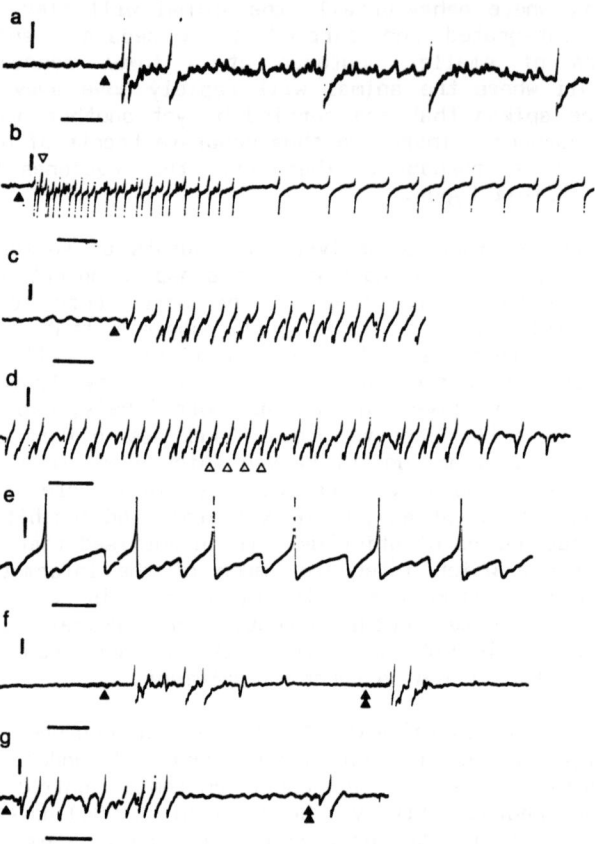

FIG. 3: Intracellular recordings from ectoneural neurones of the radial nerve cord of the brittlestar Ophiura ophiura.
a. Impalement of cell body region of longitudinal neurone showing spikes and junction potentials to a diminution of light intensity (arrowhead).
b. Similar cell to (a) but greater response to identical stimulus. Large epsps are present (open arrowhead) and since these are clearly above threshold presumably this part of the cell is non-excitable. These cells only respond to stimuli that cause 'freeze' behaviour but the threshold of response and the number, burst complexity and frequency of spikes depend on the individual cell when an identical stimulus is given.
c. Neurone showing spikes and both epsps and ipsps to photic stimulation. This cell is probably not a longitudinal neurone since it also responded to other types of stimulus but a second order interneurone of a type not yet fully understood, Lucifer fills of this type of cell have not been achieved.
d. Spontaneous activity from a cell of apparently similar type to that in (c), though there is no direct evidence. This cell shows large ipsps and in many cases spikes follow such events. The functional significance of this type of cell has still to be elucidated but it apparently integrates longitudinally conducted information.
e. Faster timebase recording of a similar cell to (d).
f. Cell, similar to those in (c and d), showing junction potentials and spikes to two different stimuli that cause different behaviour. Light diminution (single arrowhead) causes 'freeze' behaviour and a food stimulus (double arrowhead) causes

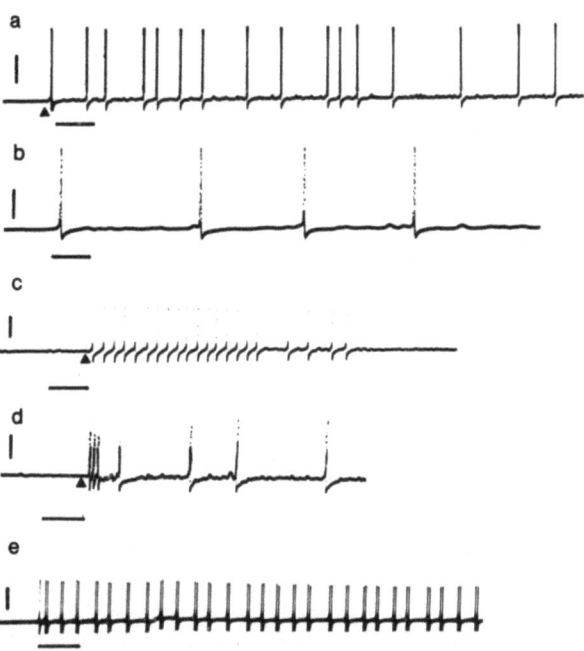

FIG. 4: Intracellular recordings from ectoneural nerves.
a. Longitudinal neurone impaled in the axon showing a long lasting response to diminution of light (arrowhead). Spikes retouched.
b. Faster timebase showing part of a similar response in a longitudinal axon.
c. Rapid initial response in a different neurone to (a), but of the same class, to an identical stimulus.
d. Response to chemical food stimulus to tip of arm (arrowhead). Impalement near terminal region since small psps are present. Spikes retouched.
e. Many neurones show oscillations of the resting potential, spontaneous activity, rhymical bursts of activity or as illustrated double spiking. Further investigations will be needed to show if these various types of activity have functional significance or are artefacts due to cell damage. Vertical scale = 20 mv., time scale = a, c-e = 0.05 s., b = 0.01 s.

'flexure' behaviour anticipatory of feeding. This cell is post-synaptic to the longitudinal neurones and clearly is involved in integrating the response in a single ganglion. No such identified cell has yet been filled with Lucifer Yellow. These cells respond to information travelling longitudinally within the radial nerve cord after stimulation of any arm in a whole animal preparation.
g. Longitudinal neurone (confirmed by Lucifer fill) impaled near cell body. Spikes and both ipsps and epsps to light diminution (single arrowhead) and high frequency vibration (double arrowhead). Both these stimuli give rise to 'freeze' behaviour which although to different modalities is carried by the same neurone. Vertical scale = 10 mv. (Note most spikes are cut off). Time scale a-d, f, g = 0.05 s. e = 0.01 s.

that the same neurone passes right down the nerve cord; and the author's own findings that neurones run at least five or six segments and not the shorter distances proposed from his now discredited earlier degeneration studies (Cobb and Stubbs 1981). The evidence against is that some do end in each segment as is proved physiologically in terms of input synaptic potentials correlating with Lucifer fills showing the cell body had been impaled. Further individual neurones do not make a morphological output (and there is no physiological input) in any segments they pass right through, and yet all segments must receive such output. All these neurones appear to cross over in the region of the cell body before running longitudinally. A longitudinal cut of the nerve cord would thus sever them where they run transversely. As soon as a longitudinal cut down the midline of the nerve cord exceeds six segments extracellularly recorded through conduction fails (J.L.S. Cobb and A. Moore unpublished). This is admittedly crude evidence for neurones about six segments long. If neurones do only run for a proportion of the nerve cord, one then has to ask why? The earlier anatomical studies (Cobb and Stubbs 1981) showed each segmental ganglion to have the same pattern of individual neurone distribution with no evidence for more cells or greater complexity centrally. Clearly this original idea of identical segments communicating only with their immediately adjacent neighbours is too simple. The evidence now points to something inbetween a segmental system and a through conducting centralized system, the whys and wherefores are still to be elucidated but at least this now appears likely to be achieved in the near future. The author now believes the through conducting system consists of numbers of longitudinal neurones carrying different classes of information with over-lapping fields of influence such that each segment can provide input and receive output. The detailed connection may well end up being shown to be far from simple.

12. CONCLUSIONS

The aim of this review has been to show progress towards the goal of understanding the basic cellular mechanisms which underlie the control of behaviour in the echinoderms. The author has pursued this goal for more than twenty years but only relatively recently has a preparation and techniques been developed that have allowed anything but peripheral progress to be made. It has been a guiding conviction that the nervous system with centralized controlling centres advocated by Smith (1965) and deeply embedded in the literature was fundamentally wrong. The evidence available now is still equivocal but the author feels the nervous system is of a uniquely non centralized form. Stated simply, local reflexes of the type described by von Uexküll occur, but the ganglia involved also make output to adjacent ganglia. Summed activity from local ganglia drive segmental ganglia of the radial nerve cords and a group of such ganglia can then dominate and control whole animal behaviour. Control can shift from one group of ganglia (or perhaps an individual ganglion) to another. The circumoral ring connects radii and integrates only central organelles and the viscera. All ganglia are equal, none more dominant than others; function is not centralized. There is one last worry and that is whether the ophiuroid nervous system which will yield an unequivocal answer is typical of the other three Eleutherozoan classes. The anatomical evidence says it is, but physiological evidence will be a literal order of magnitude more difficult to achieve.

To summarize one can say there are widespread receptors able to finely discriminate environmental change and internal state. It is not possible to state how specific individual receptor function is at the single cell level and this will be difficult to achieve. The ectoneural system consists of peripheral ganglia connected to the radial nerve cords which themselves are composed of segmental ganglion. Motor innervation by the ectoneural system does not penetrate the muscle innervated but relies on a number of unusual mechanisms. Large muscles are innervated by a separate hyponeural nervous system driven by the ectoneural system. The hyponeural system is purely motor and non integrative. Echinoderm connective tissue is unique in the animal kingdom in that it is mutable and innervated.

Libbie Hyman was right in "saluting the echinoderms as a noble group especially designed to puzzle zoologists". Many things about their nervous system still do not make sense and it is necessary constantly to be on one's guard against assuming the echinoderms either behave, or are designed like, other animals. Their nervous system is neither simple nor primitive but why it is so enigmatic is not clear. Perhaps it is related to the pentametry, but surely the circumoral ring could have evolved as a 'brain' or the most central radial segmental ganglia evolved to control the behaviour of a single arm? Is it possible that some irreversible genetic change in their development put an evolutionary straight jacket on their future evolution? Perhaps it is related to their powers of autonomy? One cannot however accuse the echinoderms of being unsuccessful since they are dominant in many marine habitats even if somewhat conservative in their life styles. Enigmatic they may be, but undoubtedly they can teach much to the neurobiologist who approaches them with an open mind.

13. REFERENCES

Aiello E (1972) Control of ciliary activity in metazoa. In: Sleigh MA (ed) Cilia and flagella. Academic Press, New York.

Arshavskii Y, Kashin S, Litvinova N, Orlovskii G, Fel'dman A (1975) Coordination of arm movement during locomotion in ophiurians. Neurophysiol 8: 529-537.

Arshavskii Y, Kashin S, litvinova N, Orlovskii G, Fel'dman A (1976) Coordination of movements of the tubefeet and arms of ophiurians during locomotion. Neurophysiol 8: 633-669.

Bargmann W, Behrens B (1963) Uber den Feinbau des Nervensystems des Seesternes (Asterias rubens L.). II. Z Zellforsch 59:746-770.

Bargmann W, Behrens B (1968) Uber des Pylorusanhange des Seesternes (Asterias rubens L.) inbesondere ihre Innervation. Z Zellforsch 84: 536-584.

Bargmann W, von Hehn G (1968) Uber das axialorgan ('mysterious organ') von Asterias rubens. Z Zellforsch 88: 262-277.

Biglow CE (1981) Investigation of variable tensility in echinoderm connective tissue. B Sc thesis, University of Victoria.

Binyon J, Hasler B (1970) Electrophysiology of the starfish radial nerve cord. Comp Biochem Physiol 32: 747-753.

Blaschko H, Hope D (1957) Observation on the distribution of amine oxidases in invertebrates. Arch Biochem Biophys 69: 10-15.

Bouland C, Massin C, Jangoux M (1982) The fine structure of the buccal tentacles of Holothuria forskali. Zoomorphol 101: 133-149.

Brehm P (1977) Electrophysiology and luminescence of an ophiuroid radial nerve. J Exp Biol 71: 213-227.

Bullock TH (1965) Comparative aspects of superficial conduction in echinoids and asteroids. Amer Zool 5: 545-562.

Bullock TH, Horridge A (1965) Structure and function in invertebrate nervous systems. Freeman and Co., San Francisco and London.

Burke RD (1978) The structure of the nervous system of the pluteus larvae of Strongylocentrotus purpuratus Cell Tissue Res 191: 233-247.

Burke RD (1980a) Morphogenesis of the digestive tract of the pluteus larvae of Strongylocentrotus purpuratus. Int J Invertbr Reprod 2: 13-21.

Burke RD (1980b) Podial sensory receptors and the induction of metamorphosis in echinoids. J Exp Mar Biol Ecol 47: 223-234.

Burke RD (1983a) Neural control of metamorphosis in Dendraster excentricus. Biol Bull 164: 176-188.

Burke RD (1983b) The induction of metamorphosis of marine invertebrate larvae: stimulus and response. Can J Zool 61: 1701-1719.

Burke RD (1983c) The structure of the larval nervous system of Pisaster ochraceus. J Morphol 178: 23-35.

Burke RD (1983d) Development of the larval nervous system of the sand dollar, Dendraster excentricus. Cell Tissue Res 229: 145-154.

Byrne M (1982) Functional morphology of a holothurian autotomy plane and its role in evisceration. In: Lawrence J (ed) International echinoderms conference, Tampa. Balkema, Rotterdam.

Byrne M (1986) Ultrastructural changes in the autotomy tissues of Eupentacta quinquesemita during evisceration. In: Keegan B, O'Connor B (eds) Proceedings of 5th international echinoderms conference. Balkema, Rotterdam.

Byrne M, Fontaine AR (1983) Morphology and function of the tubefeet of Florometra serratissima. Zoomorphol 102: 175-187.

Caine GD, Burke RD (1985) Immunohistochemical localization of gonad stimulating substance in the starfish. In: Keegan B, O'Connor B (eds) Proceedings of 5th international echinoderms conference. Balkema, Rotterdam.

Campbell AC (1983) Form and Function of pedicellariae. Echinoderm Studies 1: 139-167.

Carnevali MDC, Saita A (1985a) Muscle system organization in the echinoderms: II Microscopic anatomy and functional significance of the muscle ligament skeletal system in the arm of commatulids. J Morphol 185: 59-74.

Carnevali MDC, Saita A (1985b) Muscle system organization in the echinoderms: Fine structure of the contractile apparatus of the arm flexor muscles of the commatulids. J Morphol 185: 75-87.

Cavey MJ, Wood RL (1981) Specializations for excitation-contracting coupling in the podial retractor cells of the starfish Stylasterias forreri. Cell Tissue Res 218: 475-485.

Chaet AB (1966) Neurochemical control of gamete release in starfish. Biol Bull 130: 43-58.

Christo-Apostolides N (1882) Anatomie et développement des Ophiures. Arch Zool Exp Gen 10: 121-224.

Cobb JLS (1967) The innervation of the ampulla in the starfish Astropecten. Proc R Soc B 168: 91-99.

Cobb JLS (1968a) Observations on electrical activity within the retractor muscles of the lantern of Echinus esculentus using extracellular recording electrodes. Comp Biochem Physiol 24: 311-315.

Cobb JLS (1968b) The fine structure of the pedicellariae of Echinus esculentus. I. The innervation of the muscles. J R Microsc Soc 88: 211-221.

Cobb JLS (1968c) The fine structure of the pedicellariae of Echinus esculentus. II. The sensory system. J R Microsc Soc 88: 223-233.

Cobb JLS (1969a) The distribution of monoamines in the nervous system of echinoderms. Comp Biochem Physiol 28: 967-971.

Cobb JLS (1969b) The innervation of the oesophagus of the sea-urchin Heliocidaris erythrogramma. Z Zellforsch 98: 323-332.

Cobb JLS (1970) The significance of the radial nerve cords in asteroids and echinoids. Z Zellforsch 108: 457-474.

Cobb JLS (1978) An ultrastructural study of the dermal papulae of the starfish, Asterias rubens. Cell Tissue Res 187: 515-523.

Cobb JLS (1982a) The anatomical basis for integratory mechanisms in echinoderms. In: Lawrence J (ed) International echinoderms conference, Tampa. Balkema, Rotterdam.

Cobb JLS (1982b) Membrane physiology of echinoderms. Podesta RB (ed) Membrane physiology of echinoderms. Dekker, New York.

Cobb JLS (1985a) Intracellular studies on the nervous system of an echinoderm. In: Keegan B, O'Connor B (eds) Proceedings of 5th international echinoderms conference. Balkema, Rotterdam.

Cobb JLS (1985b) The neurobiology of the ectoneural/hyponeural synaptic connection of an echinoderm. Biol Bull 168: 432-446.

Cobb JLS (1985c) The motor innervation of the oral plate ligament in the brittlestar Ophiura ophiura. Cell Tissue Res 242: 685-688.

Cobb JLS, Laverack MS (1966a) The lantern of Echinus esculentus, I. Gross anatomy and physiology. Proc R Soc B 164: 624-640.

Cobb JLS, Laverack MS (1966b) The lantern of Echinus esculentus, II. The fine structure of the hyponeural tissue. Proc R Soc B 164: 641-650.

Cobb JLS, Laverack MS (1966c) The lantern of Echinus esculentus, III. The fine structure of the lantern retractor muscle and its innervation. Proc R Soc B 164: 651-658.

Cobb JLS, Laverack MS (1967) Neuromuscular systems in echinoderms. Symp Zool Soc Lond 20: 25-51.

Cobb JLS, Moore A (1986) Comparative studies on receptor structure in the brittlestar Ophiura ophiura. J Neurocytol 15: 97-108.

Cobb JLS, Pentreath VW (1976) The identification of chemical synapses in echinoderm nervous systems. Thalass. Jugoslavica 12: 81-85.

Cobb JLS, Pentreath VW (1977) Anatomical studies of simple invertebrate synapses using stage rotation electron microscopy and densitometry. Tissue Cell 9: 125-135.

Cobb JLS, Raymond AM (1979) The basiepithelial nerve plexus of the viscera and coelom of eleutherozoan Echinodermata. Cell Tissue Res 202: 155-163.

Cobb JLS, Sneddon E (1977) An ultrastructural study of the gills of Echinus esculentus. Cell Tissue Res 182: 265-274.

Cobb JLS, Stubbs T (1981) The giant neurone system in ophiuroids. I. The general morphology of the radial nerve cords and circumoral ring. Cell Tissue Res 219: 197-207.

Cobb JLS, Stubbs T (1982) The giant neurone system in ophiuroids. III. The detailed connections of the circumoral ring. Cell Tissue Res 226: 675-687.

Cottrell GA, Pentreath VW (1970) Localization of catecholamines in the nervous system of a starfish Asterias rubens. Comp Gen Pharmac 1: 73-81.

Cuenot L (1948) Anatomie, Ethologie et Systématique des Echinoderms. In: Grosse P (ed) Traité de Zoologie. Masson et Cie, Paris.

De Vos L (1985) Ultrastructure of the tubefeet of the ophiuroid Amphipholis squamata. Proc. 5th Int. Echinoderm Conference, Galway. Keegan B, O'Connor B (eds). Balkema, Rotterdam.

Diab M, Gilly WM (1984) Mechanical properties and control of non-muscular catch in spine ligaments of the sea-urchin Strongylocentrotus. J Exp

Dolder H (1972) Ultrastructural study of the smooth muscle in the tubefeet of echinoderms. J Submicr Cytol 4: 221-232.

Doyle WL (1967) Vesiculated axons in haemal vessels of an holothurian, Cucumaria frondosa. Biol Bull 132: 329-336.

Eakin RM (1963) Lines of evolution of photoreceptors. In: Mazia D (ed) General physiology of cell specialization. McGraw-Hill, New York.

Eakin RM (1966) Evolution of photoreceptors. Cold Spring Harbor Symposia on Quantitative Biology. 30: 367-370.

Eakin RM (1968) Evolution of photoreceptors. In: Dobzhansky T (ed) Evolutionary biology. Appleton-Crofts, New York.

Eakin RM, Martin GG, Reed CT (1977) Evolutionary significance of fine structure of archiannelid eyes. Zoomorphol 88: 1-18.

Eakin RM, Brandenburger JL (1979) Effects of light on ocelli of seastars. Zoormorphol 92: 191-200.

Emerson CJ (1977) Larval development of the seastar with particular reference to the optic cushion. In: Scanning electronmicroscopy, vol II. Proceedings of workshop on SEM/STEM. IIT Research Institute, Chicago.

Emson RH (1986) Bone idle - a recipe for success. In: Keegan B, O'Connor B (eds) Proceedings of 5th international echinoderms conference. Balkema, Rotterdam.

Emson RH, Wilkie IC (1980) Fission and autonomy in echinoderms. Oceanogr Mar Biol Ann Rev 18: 155-250.

Eylers JP (1982) Ion-dependent viscosity of holothurian body wall and its implications for the functional morphology of echinoderms. J Exp Biol 99: 1-8.

Fankbonner PV (1978) Suspension feeding mechanisms of the armoured sea cucumber Psolus chitinoides. J Exp Mar Biol Ecol 31: 11-25.

Florey E, Cahill MA (1977) Ultrastructure of sea urchin tubefeet. Cell Tissue Res 177: 195-214.

Florey E, Cahill MA (1980) Cholinergic motor control of sea urchin tubefeet: Evidence for chemical transmission without synapses. J Exp Biol 88: 281-292.

Florey E, Cahill MA (1982) Scanning electron microscopy of echinoid podia. Cell Tissue Res 224: 543-551.

Florey E, Cahill MA, Rathmayer M (1975) Excitatory actions of GABA and of acetylcholine in sea urchin tubefeet. Comp Biochem Physiol 51C: 5-12.

Fontaine AR (1964) The integumentary secretions of the ophiuroid Ophiocomina nigra. J Mar Biol Ass UK 44: 145-162.

Gardiner SL, Rieger RM (1980) Rudimentary cilia in muscle cells of annelids and echinoderms. Cell Tissue Res 213: 247-252.

Green CR, Bergquist PR, Bullivant S (1979) An anastomosing septate junction in endothelial cells of the phylum Echinodermata. J Ultrastruct Res 67: 72-80.

Harris P, Shaw G (1984) Intermediate filaments, microtubules and microfilaments in epidermis of sea-urchin Strongylocentrotus tubefeet. Cell Tissue Res 236: 27-34.

Hehn G von (1970) Uber den Feinbau des hyponeuralen Nervensystems des Seesternes. Z Zellforsch 105: 137-154.

Hendler G, Byrne M (1985) First description of a brittlestar photoreceptor system. Amer Zool 25: 143A.

Hidaka M (1983) Effects of certain physico-chemical agents on the mechanical properties of the catch apparatus of the sea-urchin spine. J Exp Biol 103: 15-29.

Hidaka M, Takahashi K (1983) Fine structure and mechanical properties of the catch apparatus of a sea urchin spine. J Exp Biol 103: 1-14.

Hill RB (1970) Effects of some postulated neurohumors on rhythmicity of the isolated cloaca of a holothurian. Physiol Zool 43: 109-123.

Hill RB (1983) Restoration of contractility by depolarizing agents and by calcium after caffeine treatment of holothurian muscle. Comp Biochem Physiol 75C: 5-15.

Hill RB, Sanger JW, Chen C (1983) Close apposition of muscle cells in the longitudinal bands of the body wall of a holothurian Isostichopus. Cell Tissue Res 23: 467-473.

Holland ND, Grimmer JC (1981) Fine structure of the cirri and a possible mechanism for their motility in stalkless crinoids. Cell Tissue Res 214: 207-217.

Huet M (1975) Le rôle du système nerveux au cours de la regénération du bras chez une étoile de mer: Asterina gibbosa. J Embryol Exp Morphol 33: 535-552.

Huet M (1979) Système nerveux et aptitude à la regénération du bras de l'étoile de mer Asterina gibbosa. Actes du Colloque européen sur les Echinodermes. Balkema, Rotterdam.

Huet M, Franquinet R (1981) Histofluorescence study and biochemical assay of catecholamines during the course of arm tip regeneration in the starfish. Histochem 72: 149-154.

Hyman LH (1955) The Invertebrates vol. IV Echinodermata. McGraw-Hill, New York.

Jennings HS (1907) Behaviour of the starfish, Asterias forreri. Univ Calif Publ Zool 4: 53-185.

Jickeli CF (1884) Vorlaufige Mitteilungen uber den Bau der Echinoderm. Zool Anz 7: 346-370.

Jordan H (1919) Uber 'reflexarme' Tiere. IV Die Holothurien Zool Jb (allg Zool) 36: 109-156.

Kanatani H (1964) Spawning of the starfish: action of gamete shedding substance shed from radial nerves. Science 146: 1177-1179.

Kanatani H (1969) Mechanisms of starfish spawning: action of the neural substance on the isolated ovary. Gen Comp Endocrinol 2: 582-589.

Kanatani H, Shirai H (1967) In vitro production of mitosis inducing substance by nerve extract in the ovary of the starfish. Nature 216: 284-286.

Kanatani H, Ikegami S, Shirai H, Oide H, Tamura S (1971) Purification of gonad stimulating substance obtained from the radial nerve cord of a starfish. Devel Growth Differet 13: 151-164.

Kawaguti S (1964) Electron microscopic structure of the podial wall of an echinoid with special reference to the nerve plexus and muscle. Biol J Okayama Univ 11: 41-52.

Kawaguti S (1966) Electron microscopy on the body wall of a sea-cucumber with special attentions to its mucous cells. Biol J Okayama Univ 12: 35-45.

Kawaguti S, Kamishima Y (1964) Electron microscopic study on the integument of an echinoid, Diadema setosum. Annotnes Zool Jpn 37: 147-152.

Kinosita H (1941) Conduction of impulses in superficial nervous system of sea urchin. Jpn J Zool 9: 221-245.

Kishimoto T, Kanatani H (1976) Cytoplasmic factor responsible for germinal vesicle breakdown and meiotic maturation in starfish oocytes. Nature 260: 321-332.

Kishimoto T, Cayer MI, Kanatani H (1982) Starfish oocyte maturation and reeuction of disulphide-bond on oocyte surface. Exp Cell Res 101: 104-111.

Kobzar GT (1984) Muscle chemoreceptors in the holothurian Cucumaria japonica. Zh Evol Biokhim Fixiol 20: 419-422.

Lewis JB (1968) The function of the sphaeridia of sea-urchins. Can J Zool 46: 1135-1138.

Mackie GO, Spencer AN, Strathmann RR (1969) Electrial activity associated with ciliary reversal in echinoderm larvae. Nature 223: 1384-1385.

Markel K, Roser U (1983) The spine tissues in the echinoid Eucidaris tribuloides. Zoomorphol 103: 25-41.

Markel K, Roser U (1985) Comparative morphology of echinoderm calcified tissue: Histology and ultrastructure of ophiuroid scales. Zoomorphol 105: 197-207.

Martinez JL (1977) Ultraestructura del tejido nervioso podial de Ophiothrix fragilis, Bol Ro Sox Espanola Hist Nat (biol) 75: 315-333.

McClintock HB, Lawrence JM (1982) Photo response and associative learning in Luidia clathrata. Mar Behav Physiol 9: 13-21.

McKenzie JD (1985) The tentacular ultrastructure of dendrochirote holothurians, A comparative SEM study. In: Keegan BF, O'Connor B (eds) Proceedings of 5th international echinoderms conference. Balkema, Rotterdam.

Meijer L, Guerrier P (1984) Maturation and fertilization in starfish oocytes. Int rev Cytol 86: 130-192.

Millott N (1966) Co-ordination of spine movements in echinoids. In: Boolootian RA (ed) Physiology of the Echinodermata. Interscience, New York.

Millott N (1968) The dermal light sense. In: Carthy J, Newell G (eds) Invertebrate photoreceptors. Symp Zool Soc Lond Academic Press, London.

Millott N (1975) The photosensitivity of echinoids. Adv Mar Biol 13: 1-52.

Millott N, Coleman R (1969) The podial pit - a new structure in the echinoid Diadema antillarum. Z Zellforsch 95: 187-197.

Millott N, Okumura H (1968) The electrical actvity of the radial nerve cord in Diadema antillarum. J Exp Biol 48: 279-287.

Moore A, Cobb JLS (1985a) Neurophysiological studies on photic responses in Ophiura ophiura. Comp Biochem Physiol 80A: 11-16.

Moore A, Cobb JLS (1985b) Neurophysiological studies on the detection of amino acids by Ophiura ophiura. Comp Biochem Physiol 82A: 395-399.

Motokawa T (1981) The stiffness change of the holothurian dermis caused by chemical and electrical stimulation. Comp Biochem Physiol 70C: 41-48.

Motokawa T (1982a) Fine structure of the dermis of the body wall of the sea cucumber Stichopus. Galaxea 1: 55-64.

Motokawa T (1982b) Rapid change in the properties of echinoderm connective tissue caused by coelomic fluid. Comp Biochem Physiol 73C: 223-229.

Motokawa T (1982c) Factors regulating the properties of holothurian dermis. J Exp Biol 99: 29-41.

Motokawa T (1983) Mechanical properties and structure of the spine joint ligament of the sea-urchin. J Zool 201: 223-235.

Motokawa T (1984a) Viscoelasticity of the holothurian body wall. J Exp Biol 109: 63-75.

Motokawa T (1984b) Connective tissue catch in echinoderms. Biol Rev 59: 255-270.

Okada Y, Iwata KS, Yanagihara M (1984) Synchronized rhythmic contractions among five gonadal lobes in the shedding sea-urchins: Coordinative function of the aboral nerve ring. Biol Bull 166: 228-236.

Oldfield SC (1975) Surface fine structure of the globiferous pedicellariae of the regular echinoid, Psammechinus miliaris. Cell Tissue Res 162: 377-385.

Penn PE, Alexander CG (1980) Fine structure of the optic cushion in the asteroid Nepantia belcheri. Mar Biol 58: 251-256.

Pentreath VW (1970) A study of neurotransmitters in the Asteroidea, Crinoidea and Ophiuroidea. M Sc thesis Univ St Andrews.

Pentreath VW, Cobb JLS (1972) Neurobiology of Echinodermata. Biol Rev 47: 363-392.

Pentreath VW, Cobb JLS (1982) Echinodermata. In: Shelton G (ed) Electrical conduction and behaviour in 'simple' invertebrates. Clarendon, Oxford.

Pentreath VW, Cottrell GA (1968) Acetylcholine and cholinesterase in the radial nerve cord of Asterias rubens. Comp Biochem Physiol 27: 775-785.

Pentreath VW, Cottrell GA (1971) 'Giant' neurones and neurosecretion in the hyponeural tissue of Ophiothrix fragilis. J Exp Mar Biol Ecol 6: 249-264.

Peters BH (1985) The innervation of spines in the sea-urchin Echinus esculentus. An electron microscope study. Cell Tissue Res 239: 219-228.

Peters BH, Shelton GAB (1981) Electrical activity during a simple behaviour: spine-pointing in a sea-urchin. Comp Biochem Physiol 70A: 397-403.

Podol'skii OG (1972) Responses of the radial nerve cord of the starfish Asterias rubens to single and rhythmical electric shocks. Zhurnal Evolyutsionnoi Biokhimmi i Fiziologii 8: 517-522.

Prosser CL (1954) Activation of a non-propagating muscle in Thyone. J Cell Comp Physiol 44: 247-254.

Prosser CL, Mackie GO (1980) Contractions of holothurian muscle. I Comp Physiol 136: 103-112.

Protas LL, Muske GA (1980) Effects of some transmitter substances on the tube foot muscles of the starfish Asterias amurensis. Gen Pharmacol 11: 113-118.

Reese ES (1966) The complex behaviour of echinoderms. In: Boolootian RA (ed) Physiology of Echinodermata. Wiley, New York.

Reichensperger A (1908) Die Drusengebilde der Ophiuren. Z Wiss Zool 91: 304-350.

Saita A, Carnevali MDC, Canonaco M (1982) Muscle system organization in the echinoderms. J Submicrosc Cytol 14: 291-304.

Sandeman DC (1965) Electrical activity in the radial nerve cords and ampullae of sea-urchins. J Exp Biol 43: 247-256.

Schoenmakers HJN, Colebrander PHJM, Peute J, Oordt PGWJ (1981) Jangoux M, Lawrence JL (eds) Anatomy of the ovaries of a starfish Asterias rubens. Cell Tissue Res 217: 577-597.

Sloan NA, Campbell AC (1982) Perception of food. In: Jangoux M, Lawrence JL (eds) Echinoderm nutrition. Balkema, Rotterdam.

Smith DS, Wainwright SA, Baker J, Cayer ML (1981) Structural features associated with movement and 'catch' of sea-urchin spines. Tissue Cell 13: 299-320.

Smith GN, Greenberg MJ (1973) Chemical control of the evisceration process in Thyone briareus. Biol Bull 144: 421-436.

Smith JE (1937) On the nervous system of the starfish Marthasterias glacialis. Phil Trans R Soc B 227: 111-173.

Smith JE (1945) THe role of the nervous system in some activities of starfishes. Biol Rev 20: 29-43.

Smith JE (1950) The motor nervous system of the starfish, Astropecten irregularis with special reference to the innervation of the ampullae and tubefeet. Phil Trans R Soc B 234: 521-558.

Smith JE (1965) Echinodermata. In: Bullock TH, Horridge GA (eds) Structure and function of the nervous systems of invertebrates, vol. II. Freeman and Co, San Francisco.

Smith JE (1966) The form and functions of the nervous system. In: Boolootian RA (ed) Echinoderm physiology. Wiley, New York.

Sousa Santos H (1966) The ultrastructure of the mucous granules from starfish tubefeet. J Ultrastruct Res 16: 41-51.

Stewart WW (1978) Functional connections between cells as revealed by dye-coupling with a high fluorescent naphalimide tracer. Cell 14: 741-759.

Strathmann RR (1971) The feeding behaviour of planktotrophic echinoderm larvae: mechanisms, regulation and rates of suspension feeding. J Exp Mar Biol Ecol 6: 109-160.

Stubbs T (1982) The neurophysiology of photosensitivity in ophiuroids. In: Lawrence JM (ed) Echinoderms: Proceedings of the International Conference, Tampa. Balkema, Rotterdam.

Stubbs T, Cobb JLS (1981) The giant neurone system in ophiuroids, II. the hyponeural motor tracts. Cell Tissue Res 220: 373-385.

Stubbs T, Cobb JLS (1982) A new ciliary feeding structure in an echinoderm. Tissue Cell 14: 573-583.

Sugi H, Gomi S, Toride M, Emura A, Tsuchiya T, Takei N (1985) Mechanical activity in the lantern retractor muscle of a sea-urchin Anthocidaris crassipina. Comp Biochem Physiol 81A: 397-401.

Suzuki S (1982) Physiological and cytochemical studies on activator calcium in contraction by smooth muscle of a sea cucumber, Isostichopus badionotus. Cell Tissue Res 222: 11-24.

Takahashi K (1964) Electrical responses to light stimuli in the isolated radial nerve of the sea-urchin, Diadema setosum. Nature 201: 1343-1344.

Takahashi K (1966) Muscle physiology. In: Boolootian RA (ed) Physiology of Echinodermata. Wiley, New York.

Takahashi K (1967) The catch apparatus of the sea-urchin spine, II. Responses to stimuli. J Fac Sci Tokyo Univ, Sec IV 11: 109-120.

Takasu N, Yoshida M (1983) Photic effects on photo sensory microvilli in the seastar Asterias amurensis. Zoomorphol 103: 135-148.

Tsuchiya T, Amemiya S (1977) Studies on the radial muscle of an echinothuriid sea-urchin, Asthenosoma. I Mechanical responses to electrical stimulation and drugs. Comp Biochem Physiol 57C: 69-73.

Tuft PJ, Gilly WF (1984) Ionic basis of action potential propagation along two classes of 'giant' axons in the ophiuroid Ophiopteris papillosa. J Exp Biol 113: 337-350.

Uexkull J von (1897) Uber reflexe bei den Seeigeln. Z Biol 37: 334-403.

Uexkull J von (1926) Die Sperrmuskulatur der Holothurien. Pflugers Archiv 212: 1-14.

Valentincic T (1979) Associative learning in the starfish Marthasterias glacialis, a simple model for the study of learning. Proceedings of European colloquium on Echinoderms. Balkema, Rotterdam.

Valentincic T (1985) Behavioural study of chemoreception in the seastar Marthasterias glacialis: Structure-activity relationships of lactic acid, amino acids, and acetylcholine. J Comp Physiol A 157: 537-545.

Weber W, Grossmann M (1977) Ultrastructure of the basiepithelial plexus of the sea-urchin, Centrostephanus. Cell Tissue Res 175: 551-562.

Welsh JH (1966) Neurohumors and neurosecretion. In: Boolootian RA (ed) Physiology of Echinodermata. Wiley, New York.

Whitfield PJ, Emson RH (1983) Presumptive ciliated receptors associated with the fibrillar glands of the spines of the echinoderm Amphipholis squamata. Cell Tissue Res 232: 609-624.

Wilkie IC (1978a) Arm autonomy in brittlestars (Echinodermata, Ophiuroidea) J Zool (Lond) 186: 311-330.

Wilkie IC (1978b) Nervously mediated changes in the mechanical properties of a brittlestar ligament. Mar Behav Physiol 5: 289-306.

Wilkie IC (1978c) Functional morphology of the autotomy plane of the brittlestar Ophiocomina nigra. Zoomorphol 91: 289-305.

Wilkie IC (1979) The juxtaligamental cells of Ophiocomina nigra, and their possible role in mechano-effector function of collagenous tissue. Cell Tissue Res 197: 515-530.

Wilkie IC (1983) Nervously mediated change in the mechanical properties of the cirral ligaments of a crinoid. Mar Behav Physiol 9: 229-248.

Wilkie IC (1984) Variable tensility in echinoderm collagenous tissues: a review. Mar Behav Physiol 11: 1-34.

Wood RL, Cavey MJ (1981) Ultrastructure of the coelomic lining in the podium of the starfish Stylasterias forreri. Cell Tissue Res 218: 449-473.

Yamamoto M, Yoshida M (1978) Fine structure of the ocelli of a synaptid holothurian, Opheodesoma spectabilis, and the effects of light and darkness. Zoormophol 90: 1-17.

Yoshida M, Millott N (1959) Light sensitive nerve in an echinoid. Experientia 15: 13-14.

Yoshida M, Ohtsuki H (1968) The phototactic behaviour of the starfish, Asterias amurensis. Biol Bull 134: 516-532.

TUNICATES

Q. BONE

The Laboratory, Marine Biological Association

Citadel Hill, Plymouth PL1 2PB - United Kingdom

ABSTRACT

Tunicates are diverse: most is known of the pelagic groups, although even here, the CNS is poorly known. In most, mechanosensitive excitable epithelia extend the field of a small number of sensory cells, and interactions between these epithelial sheets and the CNS are a feature underlying tunicate behaviour. Control of the locomotor muscles differs in different groups and may involve dual innervation and coupled muscle cells, to give a wide repertoire of activity.

1. GENERAL FEATURES OF TUNICATES

After the recognition of the chordate affinities of tunicates in the last century following Kowalewsky's work on the tadpole larva of Ciona (Kowalewsky 1867, 1871) tunicates were looked at in some detail by histologists and embryologists, most of whom worked on the readily available sessile forms, although the planktonic larvaceans also attracted attention because of their "chordate-like" appearance (e.g. Fol 1872; Martini 1909). More recently, physiologists have found the group of interest, and have on the whole paid most attention to the pelagic forms whose rhythmic activity and transparency more or less "pre-adapt" them for electrophysiological recording. Sessile tunicates have been relatively neglected neurophysiologically but as a very recent contribution by Nevitt and Gilly (1986) on the muscle fibres of Ciona shows, there is much of interest to be discovered here. The group is divided into three classes (Fig. 1): Ascidiacea, Thaliacea and Larvacea. All but the first are pelagic. Good accounts of the anatomy and general organisation of the three classes are given by Berrill (1950), and by Brien (1948) in vol. XI of the Traité de Zoologie edited by P.-P. Grassé. For the present purpose a brief outline of each class will suffice, but it should be supplemented by reference to Berrill (1950) and Brien (1948). The Ascidiacea are generally agreed to be the most primitive class from which the others have been derived, and consist essentially of a large branchial chamber with multiple ciliated gill slits enclosed in an atrium protected by the outer cellulose test (Fig. 1a). Water flows into the branchial chamber via an oral siphon, and out via the atrial siphon, both are provided with circular muscles, and the ascidian can vary the size of both apertures. Bundles of smooth muscle fibres run circumferentially and longitudinally

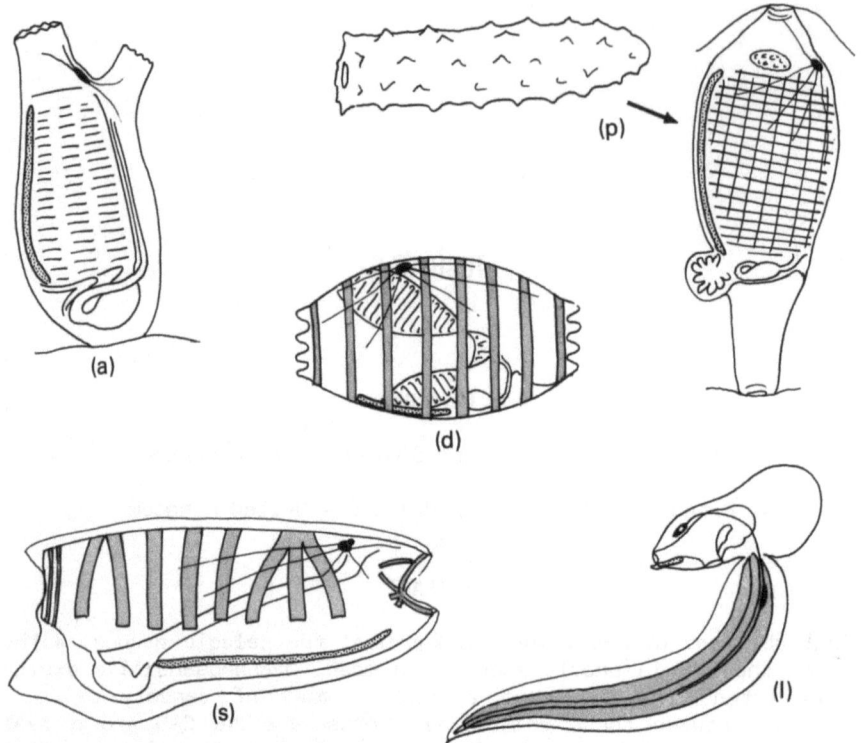

FIG. 1: General appearance of different tunicates. (a) Solitary ascidian; (p) Pyrosoma, single zooid on right from colony on left; (d) Doliolum; (s) Salpa solitary oozooid; (l) Oikopleura. Not to same scale. In each, striated muscle: fine stipple; endostyle: coarse stipple. All have a centralised brain from which mixed nerves radiate, in larvaceans there is also a caudal ganglion near the base of the tail. From Bone and Mackie (1982).

in the body wall and those species like Ciona with a thin flexible test, can change body shape considerably, contracting it down to a wrinkled lump.

In others, where the test is thicker and tougher, such as Phallusia, the body does not change much in shape; when stimulated the animal closes the siphonal apertures, and contracts very little, the muscles are opposed by test elasticity. With a few exceptions known from deep water (Monniot and Monniot 1975; 1978) which seem to be able to catch large prey, ascidians are filter feeders as are all members of the other two classes. They may be solitary or colonial, and in colonial forms, there is some co-ordination between the activities of the different zooids (Mackie 1974). The brain in all ascidians lies between the bases of the two siphons, and from it mixed nerves pass out to the body wall, branchial sac and (probably) to the viscera. In close proximity to the brain there is a neural gland, opening to the branchial cavity by a ciliated funnel, which some workers have considered to be homologous with the chordate hypophysis. The ascidian larval stage, the tadpole larva studied by Kowalewsky, is totally unlike the adult and since it swims by oscillating a muscular tail supported by a notochord above which is a dorsal nerve cord, manifests the chordate affinities of the group.

The Thaliacea are divided into three orders, rather unlike each other. The Pyrosomida are close to the Ascidiacea (Fig. 1p) consisting of large numbers of ascidian-like zooids lying in a common test shaped like a test-tube. Water entering by the siphons of each zooid passes into the common cavity of the "test-tube" and flows out of the opening at the hinder end of the colony, slowly propelling it forwards. Unfortunately, although (as Berrill remarked) the anatomy and development of Pyrosoma has been well-studied, and although it is very strikingly luminous, the living animal and its general physiology are little known.

The other two Thaliacean orders, the Salpida and Doliolida (Fig. 1, s and d) are better known physiologically, and contain much more active animals, which swim by jet-propulsion driven by bands of striated muscle fibres. The muscle bands encircle the body (only partially in most salps), and deform the thin test to drive water out of anterior or posterior valved apertures, according to the direction of swimming. Both salps and doliolids have circular brains from which mixed nerves radiate over the body; only in salps of all tunicates has any progress been made in electrical recording from central neurones. Doliolids filter feed using delicate gills with multiple ciliated openings, but in salps, the gills are reduced to a single bar, and the feeding current is produced by the rhythmic swimming activity. Life cycles in the two groups are complex, particularly so in doliolids (Braconnot 1971), where there is a vestigial tadpole type larva superficially resembling that of ascidians; in both, sexual and asexual generations alternate.

Finally, the larvaceans or appendicularians (usually much smaller than members of the other classes) are highly specialised filter feeders which are not unlike ascidian tadpole larvae (Fig. 1 l) and use their muscular tails both for swimming, and to drive water through the extraordinarily complex filtering house which they secrete. The larva is very similar to the adult, except that it possesses a very thin test which is lost during development, so that the adult larvacean is the only tunicate without a cellulose test.

There are therefore considerable differences in design between the different kinds of tunicate, which make their behaviour and physiology very different, and make it difficult to provide a synoptic view of the structure and operation of the nervous system in the phylum as a whole. But there are other difficulties lying in the path of such an attempt. First, as we shall soon see in what follows, in comparison with the other groups that have been discussed in this volume, neurones and nerve fibres are small (the largest tunicate nerve fibre is only 5 µm in diameter!), and little progress has been made in establishing their form and synaptic connexions. We have to admit, in fact, that not much is known of nervous system histology in any tunicate class; only in some of the pelagic groups is the anatomy of the peripheral nervous system reasonably well understood.

A second and more interesting obstacle is provided by the fact that even where tunicates of different classes seem at first sight to have a rather similar organisation and pattern of behaviour, the underlying structures are actually rather different, so that even here, there is not really a typical arrangement. For example, ascidian tadpole larvae, doliolid tadpole larvae, and larvaceans swim in essentially the same way, by oscillating a flattened tail supported by a notochord and driven by striated muscles, but they do not control or move the tail in the same way. In one case, the muscles are not innervated, and movements are myogenic, in another only a proportion of the muscle cells are innervated, whilst in the third, the muscle cells receive a dual pattern of innervation. Although they may look alike therefore, the locomotor systems of these tunicates are really very different, as we shall see later on.

Despite these different kinds of difficulties in approaching the nervous system of tunicates, there are certain features common to the different classes, for example, the presence and operation of excitable conducting epithelial sheets, and it is possible to try and make some tentative phylogenetic conclusions about the development of the nervous system within the group.

A fairly comprehensive review of the tunicate nervous, muscular, and ciliary systems was given a few years ago (Bone and Mackie 1982), and this brief review will therefore deal mainly with topics with which some recent progress has been made; for a more complete account (including the nervous control of ciliary activity) the reader is referred to the earlier review.

First, sensory systems will be briefly considered, including the important role of mechanosensitive epithelia, and then neuromuscular co-ordination will be discussed, before examining what little is known of the central nervous system.

2. SENSORY RECEPTORS

2.1. Mechanoreceptors

All classes of tunicate except the larvacea possess mechanoreceptors, which are typically epidermal cells bearing an elongate single cilium arising from amongst an apical cluster of microvilli (Fig. 2). The cilium passes into the test, and may be 150 μm or more long, according to test thickness. Such cells are most abundant around the margins of siphonal apertures and around the anterior and posterior apertures of salps and doliolids, but they are also scattered less abundantly over the body surface. In salps, and in the atrial siphons of ascidians, sensory cells of this kind are grouped to form receptor organs, in which the ciliary processes pass into a flag or cupula of test material that projects into the water (Fig. 3). It seems clear that such organs respond not only to direct mechanical stimuli, but also to nearfield vibrations in the surrounding water, (as do lateralis receptors); preliminary experiments on the cupular organs of Ciona indicate that they respond to probes 4 cm away from the atrial siphon vibrating between 25-400 Hz (Bone and Ryan 1978). Sensory organs of this kind perhaps provide (for those with phylogenetical leanings) a reasonable starting point for the vertebrate acustico-lateralis system, particularly since it is only the sensory cells within these organs that have an apical electron lucent zone which in some ways resembles the apical plate of vertebrate acustico-lateralis receptors (Bone and Ryan 1978).

The evidence that cells of the kind shown in Fig. 3 are mechanoreceptors, rather than responding to other kinds of stimuli, is simply that they are concentrated in areas on the tunicate body which are most sensitive to mechanical stimulation. No central or peripheral records have been obtained in any ascidian or thaliacean, so far as I am aware, which would prove directly that they are indeed mechanoreceptors, though of course this seems very likely. There are other sensory cells of similar morphology whose function is less evident, for example, the sensory cells whose processes emerge from the epithelial layer of the tail in ascidian tadpole larvae (Torrence and Cloney 1982) to pass into the test of the caudal fin. Although these are probably mechanoreceptors, there remains the possibility that they might be stimulated by tail movements and hence function as proprioceptors. However, since their

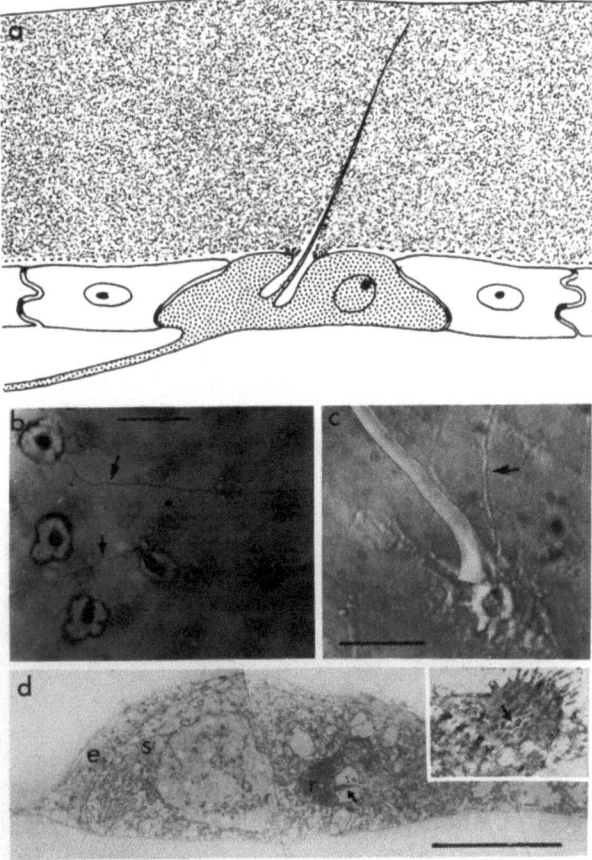

FIG. 2: a. Schematic diagram of salp mechanoreceptor cell. The cilium arises deep inside the cell and passes out of the cell apex (surrounded by microvilli) into the test. The sensory cell sends its own axon to the brain, and is probably coupled by gap junctions to the cells of the outer excitable epithelium amongst which it lies.

b. and c. Similar cells seen by Nomarski interference microscopy in the salp outer epithelium. Axons of sensory cells arrowed. b: fixed with Flemming, C: living. Scale bars: 90 μm.

d. Survey view of section of similar cell, with (inset at right) adjacent section showing apical microvilli and cilium (arrowed) near exit from cell. e: epithelial cell; s: sensory cell; r: ciliary rootlet area. Base of cilium arrowed. Scale bar: 5 μm.

axons do not apparently connect with the motor axons of the dorsal cord within the tail, and isolated tails will swim, input from this system is not essential for swimming. Again, in doliolid oozooids there is a group of ciliated sensory cells at the base of the stalk which bears the buds of the next generation. This cadophore organ seems likely to be mechanoreceptive, but it is unclear how it may be stimulated.

All of these types of assumed mechanoreceptor cells are primary sensory cells, sending their own axons to the central nervous system.

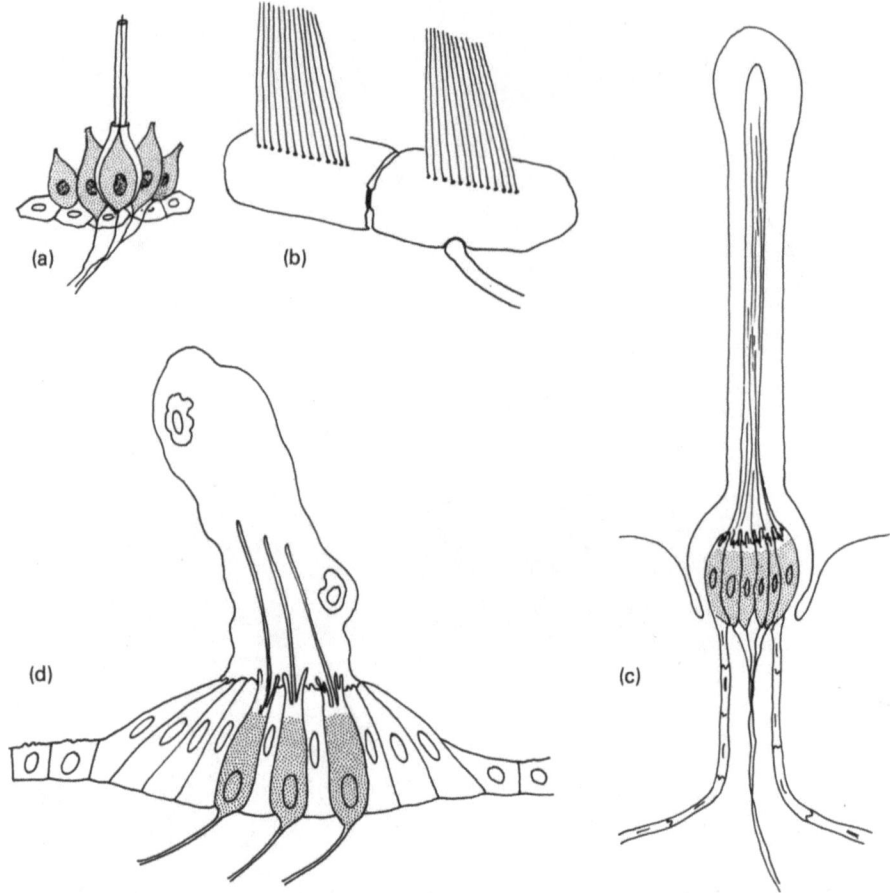

FIG. 3: Various sensory cells from different tunicates. (a): Group of mechano - or chemosensory cells from front lips of Salpa maxima; (b): Ciliary fence mechanoreceptors from the lower lips of the larvacean Fritillaria; (c): Flag or cupular organ from Salpa; (d): Cupular organ from atrial siphon of Ciona.

Unlike the scattered mechanoreceptors in salps and ascidians, those of the cupular organs have a specialised fibrous apical zone. From Bone and Mackie (1982).

Nothing is known of their central connexions. It is remarkable that primary sensory cells are not found in Larvaceans, which also possess mechanoreceptors; these are secondary sensory cells, linked to the central nervous system by the processes of central cells which are coupled to the base of the sensory cells by gap junctions.

The best-known of these receptors are the paired Langerhans bristle receptors on either side of the trunk in Oikopleura (Fig. 4). Each receptor cell possesses a stout bristle derived from a cilium (there are multiple tubules within it, and a rootlet system at its base) which projects into the water, since there is no test. When the animal is inside its filtering house, it is likely that the bristle is in contact with the wall of the house, since even slight vibration or touch to the house causes the animal to exit rapidly and swim around frenetically. The

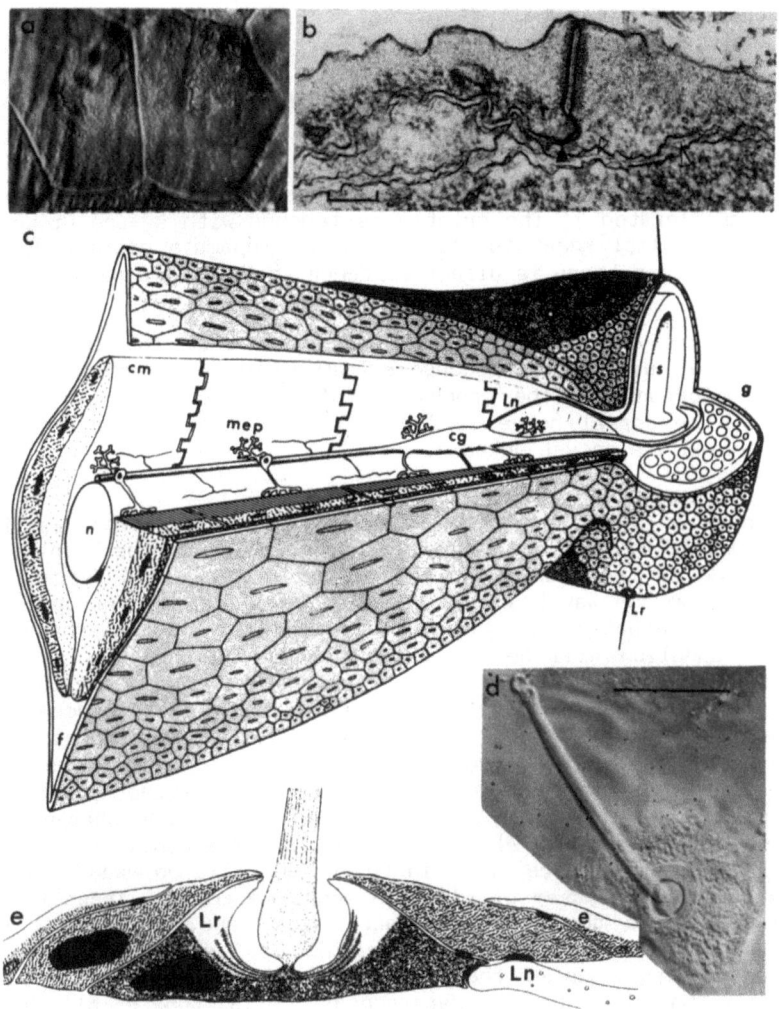

FIG. 4: The Langerhans receptor of <u>Oikopleura</u>.

(a) and (b): The outer epithelial cells, (a) surface view (scale bar: 50 μm), (b) section showing remnants of glycocalyx at upper right, and gap junction (solid arrow) between two cells. Open arrows show basement membrane between outer epithelial cells and outer border of caudal muscle cell. Scale bar: 0.2 μm. (c): Stereogram of general arrangement of receptors. cg: caudal ganglion; cm: caudal muscle; f: fin; g: gonad; Ln: Langerhans "axon"; Lr: Langerhans receptor cell; mep: motor end plate; o: oikoplastic epithelium (non-excitable). (d): Surface view of receptor cell and its bristle. Scale bar: 50 μm. (e): Diagram of structure and innervation of receptor cell (Lr). Note that it is coupled to the excitable outer epithelial cell (e) via gap junctions with an accessory cell, and that the Langerhans "axon" (Ln) makes gap junctions with the receptor cell and with the accessory cell. Combined from Bone and Mackie (1975); Bone et al. (1977); and Bone and Ryan (1979).

nerve fibre (the Langerhans nerve) passing to the receptor cell forms a gap junction with it at its base, and also is coupled by a gap junction to the adjacent epithelial cell (Fig. 4e). Since the outer epithelium is mechanosensitive, the nerve fibre may be fired either by direct mechanical stimulation of the bristle, or by mechanical stimulation at any point on the surface of the epithelium, which thus greatly extends the receptive field of the receptor cell.

Oikopleura shows a characteristic rhythmic pattern of tail movement (Fig. 6), accelerated if the bristle is touched with a fine probe (as it is if the epithelial sheet is stimulated to propagate action potentials) and so in this case there is direct evidence that it is a mechanoreceptor.

Very recently, Holmberg (1986) has shown that the two Langerhans nerves (Fig. 4g) are in fact processes from a single neurone in the caudal ganglion; unfortunately the connexions of this neurone with the motor neurones of the dorsal nerve cord or with possible interneurones in the caudal ganglion are as yet unknown.

The obvious implication from this interesting discovery is that the animal is unable to distinguish between stimulation of the bristle on one or other side of the trunk, and indeed, we might have inferred such an arrangement from the way that the receptor cell field is extended by the epithelial action potential system, since input to the Langerhans cell in the caudal ganglion will be the same wherever on the body surface the animal is stimulated. We have here a good example of the notable economy of means that tunicates utilise, most particularly the larvaceans, where cell numbers are small, and there are only two mechanoreceptor sensory cells on then outside of the animal. There are, however, other ciliated cells on the lower lips, which are probably mechanoreceptors, responding to particles within the inhalent water flowing into the pharynx. These were first seen by Fol (1872) in fritillariid larvaceans, and consists of large cells bearing a fence of cilia (over 100 cilia on each cell) linked by an extracellular system of electron dense material. These fences lie at right angles to the incurrent water flow, and are motile, beating towards the pharyngeal opening at intervals of 1 s or so. Nerve fibres are linked to the bases of these curious cells by gap junctions and it seems likely that they are secondary sensory cells (Bone et al. 1979).

It is noteworthy therefore that the only known sensory exteroreceptor cells in larvaceans are secondary sensory cells unlike those of the other tunicate groups.

2.2. Other Receptors

Larvaceans possess statocytes in the anterior brain vesicle, whose structure has been examined by Holmberg (1984); it is not known whether these are gravity receptors or whether they may respond to vibration or act as accelerometers. Since Fritillaria possesses a statocyte, but appears to adopt any position in the water (Bone et al. 1979) it may be that they are not gravity receptors, as the analogous statocytes of ascidian tadpole larvae seem to be, for ascidian tadpole larvae respond to gravity (e.g. Grave and Woodbridge 1924). So far as I am aware, no recent work has been done on tunicate photoreceptors (particularly prominent in Pyrosoma and salps, but found also in many ascidian tadpole larvae and possibly in some adult ascidians); a summary of their structure and operation is given by Bone and Mackie (1982).

3. EXCITABLE EPITHELIA

Epithelial sheets across which action potentials are propagated as a result of mechanical stimulation are found in a variety of animals, e.g. they form an important part of the response system in hydrozoa (see Andersen 1980), but in some respects at least, the epithelial conducting systems of tunicates are better understood than those of other groups, and show some interesting features not found elsewhere.

In both salps and oikopleurid larvaceans, the outer epithelium is excitable and plays a significant role in their locomotor behaviour, as it has been shown to do in a single ascidian tadpole larva. The inner epithelium of salps, and of the old nurse stage of the doliolid Dolioletta also propagate action potentials, but the role of these excitable systems is not yet understood. Table 1 summarises the occurrence and main features of the excitable epithelia of tunicates. The action potentials in all (Fig. 5) are similar rapid events with little or no undershoot, apart from those of the Dendrodoa tadpole larvae and those produced by repetitive stimulation in the embryonic salp stolon (Anderson 1979). The

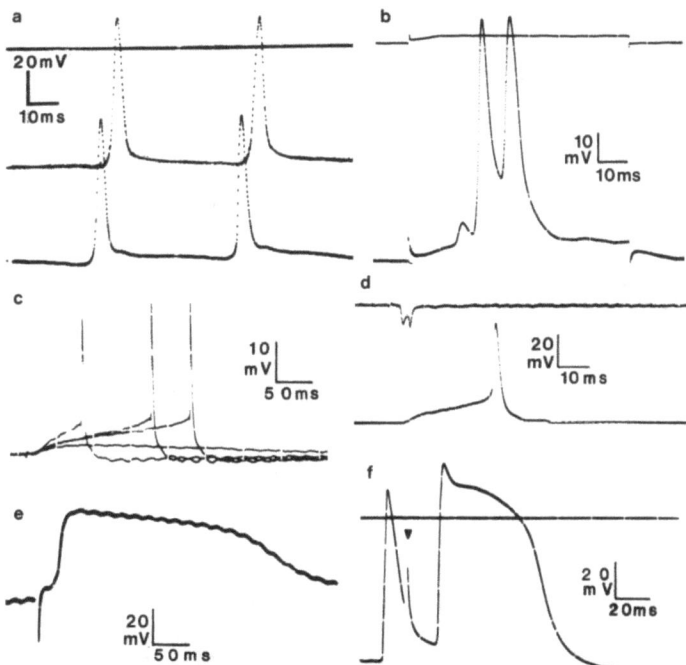

FIG. 5: Action potentials from tunicate epithelia. (a) Action potentials evoked by mechanical stimulation recorded at two electrodes 3.2 mm apart in Oikopleura albicans. (b) Typical "paired" action potentials evoked by current injection in O. dioica after treatment with 25 mM Co^{++}. (c) and (d) Action potentials evoked by mechanical stimulation in O. dioica preceded by receptor potentials, in (c) stronger stimuli increase rate of rise of receptor potentials and action potentials arise earlier, in (d) stimulus marked on zero potential line. (e) Dendrodoa tadpole larva. (f) Salp stolon, second stimulus (arrow) evokes plateaued potential. From Anderson (1979); Mackie and Bone (1976); Bone (1985) and unpublished.

TABLE 1. CONDUCTING EPITHELIA IN TUNICATES

Tissue	Reference	Resting potential (mV negative)	Amplitude (mV)	Duration (ms)	Conduction velocity (cm s^{-1})	Refractory period (ms)	Behavioural role
Larvacea Oikopleura tail skin	Bone (1985)	80	98	15	15-21	8-12	Excites locomotor pacemaker
Ascidiacea Dendrodoa tail skin	Mackie and Bone (1976)	42	48	200	6-8	200-250	Inhibits locomotor pacemaker
Salpida Salpa outer epithelium	Mackie and Bone (1977); Anderson et al. (1979)	80	90	20	17-38	7	Locomotor co-ordination
Salpa stolon epithelium	Anderson (1979)	75-96	84-104	120	4-8	7	-

Duration of action potentials taken as time to repolarization, ignoring after hyperpolarization. Modified from Bone and Mackie 1982.

mature salp shows similar rapid potentials to those of larvaceans (Fig. 9) and it seems likely that the elongate plateaued action potentials are typical of embryonic cells, as Roberts (1975) suggested.

3.1. Oikopleurid Larvaceans

Action potentials from tunicate excitable epithelia were first found in Oikopleura by Galt and Mackie (1977), who recorded small potentials from the trunk region. Subsequently, (Bone and Mackie 1975) it was shown that these potentials propagated across from the entire outer surface of the animal (apart from the oikoplastic epithelium secreting the house), and that they evoked or accelerated bursts of locomotor activity, being linked to the caudal ganglion driving the tail by the pair of Langerhans "axons" in connection with the bristle receptors of the trunk.

The outer epithelium is thin (0.5 - 3.5 μm) but despite this, it is not too difficult to obtain stable intracellular penetrations, possible because of the curious fibrous nature of the outer region of the cells (Fig. 4b) and hence the passive membrane properties of the system have been determined in two species (Bone 1985). Like the excitable epithelium of the hydrozoan jellyfish Euphysa (Josephson and Schwab 1979), the internal resistance of the cells (including the gap junctions coupling them) is very low (Table 2), but unfortunately except in Oikopleura, the technical difficulty of a 2-electrode study on such thin cells, as all conducting epithelia are, has prevented studies of other examples, so it is not known if this is a general feature of excitable epithelia.

In Oikopleura, it has also been possible to evoke graded depolarisations by applying graded mechanical stimuli from a fine probe (Fig. 5); these led to action potentials if the stimulus is sufficient. There seems little doubt that these are receptor potentials of the same kind as are found in other mechanoreceptors, and suggest that Oikopleura might offer a useful preparation (albeit rather delicate!) in which to examine the properties of mechanosensitive ion channels.

It was early observed (Bone and Mackie 1975) with extracellular recording techniques that the outer epithelium of Oikopleura was strikingly impermeable to such agents as curare or Co^{++} in the water surrounding intact preparations, and more recent investigations underway at present have confirmed this, for normal action potentials can be recorded intracellularly even after the preparation has long sojourned in seawater containing 25 mm Co^{++} or in artificial seawater (ASW) where Na^+ has been replaced by choline! These somewhat baffling results afford clear evidence of a permeability barrier, and it has now been found that digestion for a short period in dilute pronase solutions has no effect on the action potentials, but permits access of external ions to the cell membrane. Presumably the impermeability barrier is provided under normal circumstances by the conspicuous glycocalyx, removed by pronase. Investigations of the ionic basis of the action potentials are incomplete, but since action potentials are reversibly abolished by $0Na^+$ ASW (Na^+ substituted by choline) and tetrodotoxin (TTX) 10^{-5}; and are not at first affected by such Ca^{++} - blockers as Cd^{++}, Mn^{++}, or Co^{++}, it seems clear that they are largely if not entirely carried by Na^+. The subsequent effects of Ca^{++} - blockers and of tetraethyl ammonium or 4- aminopyridine indicate that there are probably two types of K^+ channel involved in repolarisation; Ca^{++} - blocking agents lead to doubling of potentials (Fig. 5b).

Since the preliminary studies of Anderson (1979) on the ionic basis of salp epithelial action potentials also indicate that they are mainly carried by Na^+ it seems probable that this is general in tunicates. The

TABLE 2. SOME ELECTRICAL PROPERTIES OF CONDUCTING EPITHELIA

Tissue	Space constant, λ (μm)	Membrane resistivity, R_m KΩcm^2	Specific intraepithelial resistivity, R_i Ωcm	Mean thickness (μm)	Mean resting potential (mV)
O. dioica outer epithelium (a)	922	4.3	82.7	1.64	82.6
O. longicauda outer epithelium (a)	3350	35.6	104.5	3.3	80.2
Euphysa exumbrellar ectoderm (b)	1300	23	196	1.4	46

(a) Bone (1985)

(b) Josephson and Schwab (1979)

ionic basis of the receptor potentials has not yet been examined but action potentials may be evoked by mechanical stimulation in 0Ca^{++} ASW, so it is possible that like the action potentials they are mainly carried by Na$^+$.

Oikopleurid larvaceans certainly offer the most promising preparation for the study of epithelial conduction in any animal, for the outer epithelium is easily penetrated by electrodes and is directly accessible. An approach now in progress which should yield results of general interest is to examine the mechanoreceptor channels using patch electrodes; accessible mechanoreceptor membranes are not too easy to come by, and it will be interesting to determine the manner in which receptor potentials arise.

Curiously enough, the other larvacean group, the fritillariids, do not have epithelial action potentials, although they have Langerhans "axons". This is simply because they do not possess an outer epithelium! Fol (1872) suggested that the outer epithelium of the larvae was lost in the adult and the outer surface of the adult represented the basal membrane of the larval epithelial cells, but it is nearly 2 μm thick and has a fibrous structure resembling that of oikopleurid outer epithelial cells, so it is also possible that it is formed by the breakdown of cell boundaries as development proceeds (Bone et al. 1977).

In oikopleurid larvaceans, action potentials in the outer epithelium activate the Langerhans "axons" of the two trunk bristle receptors, and either evoke bursts of rapid swimming, or if they occur during a swimming burst, increase the frequency of swimming (Fig. 6). Similar changes occur (in the absence of epithelial action potentials) if the bristle receptors are stimulated directly, and it is obvious that the epithelial system functions to extend the sensory field of these two receptors.

Presumably when the animal is feeding normally in its house, oscillating its tail in short bursts to draw water through the filters, the outer epithelium is unlikely to be mechanically stimulated, and escape from the house when it is touched is mediated by the bristle receptors. Paradoxically therefore, the epithelial system seems only to be of use to the animal when it has escaped from the house, during the interval before it expands and enters a new one. In this short interval, it is exposed to predators (such as fish larvae), and so requires a rapid escape system. In the large Oikopleura labradoriensis, extracellular records of epithelial potentials and muscular activity show that muscular activity begins some 20 ms after an epithelial action potential is recorded 1 mm from the base of the tail. Conduction velocity is around 17.8 cm s^{-1} and the recording electrode lay some 2 mm from the bristle receptor. Thus about half of the response time is occupied by epithelial conduction, the remainder by conduction along the Langerhans "axons"; synaptic delays in the caudal ganglion and cord; and by delays along the cord and at the neuromuscular junction. In a specimen touched at the tip of the tail, the response time would be around 50-60 ms. Similar rapid responses are seen in the smaller O. dioica; here mechanical stimulus near the tail tip evoked an epithelial action potential in a cell in the mid-region of the tail 10 ms later and movement began around 20 ms later. Swimming speed in O. labradoriensis during evoked swimming bursts is not known; in the smaller O. dioica however, it is around 3 cm s^{-1}, which corresponds to over 15 $1s^{-1}$: the escape system is obviously an effective one.

As well as evoking rapid swimming and accelerating swimming activity, intracellular records from caudal muscle cells show that during these changes in the pattern of caudal muscle activity, the spike-like muscle potentials increase in amplitude (Fig. 6). During the regular rhythmic swimming bursts, muscle potentials are around 50 mV, and do not overshoot

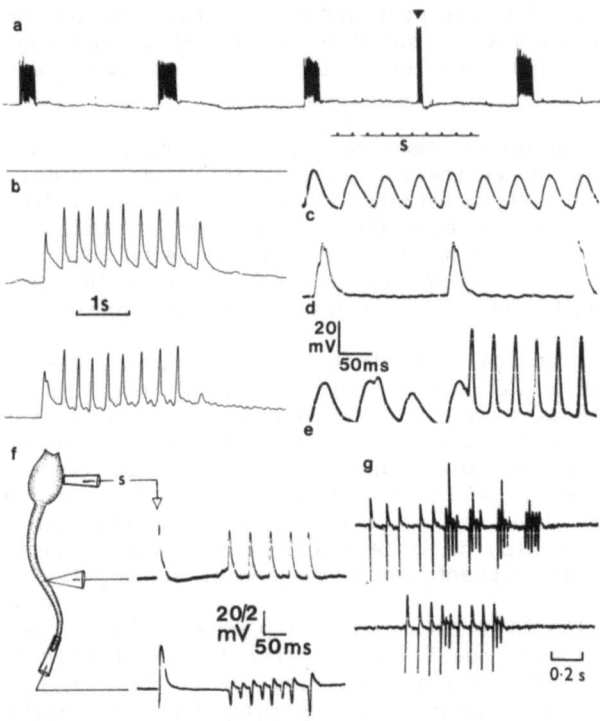

FIG. 6: Locomotor activity in Oikopleura and Dendrodoa larvae.
(a) Intracellular records from Oikopleura caudal muscle cell showing regular swimming interrupted by higher amplitude short burst (triangle) evoked by mechanical stimulation of epithelium. (b) Simultaneous records from two caudal muscle cells in Oikopleura (anterior cell below), note final potential smaller in anterior cell. (c) - (e) Records of caudal muscle cell activity in Oikopleura showing continuous activity in (c), compound nature of potentials in (d) and change in potential form in (e) evoked by epithelial action potentials. (f) Suction electrode and intracellular electrode records from Dendrodoa larval showing four potentials from either side (lower record) followed by single contraction on side monitored by microelectrode (upper). (g) Suction electrode records from Dendrodoa larva tail tip showing irregular coupling of activity on either side of tail (recorded at different amplitudes). From Mackie and Bone (1976); Bone (1985) and unpublished results.

resting potential (around 65 mV), but during the rapid bursts evoked by epithelial action potentials, they may overshoot resting potential. It is not clear how this difference arises.

3.2. Salps

In contrast to the relatively simple outer epithelial system found in oikopleurid larvaceans, salps are more complex, for they not only have endodermal and ectodermal conducting epithelia (the former divided into three separate regions), but the ectodermal outer epithelial system is innervated by epithelio-motor nerves which generate "driven" epithelial action potentials in addition to those evoked by mechanical stimulation. In fact, salps provide the most complicated example of epithelial conduction known in any animal, and the role that the system plays in the behaviour of the animal is not yet fully understood.

The inner endodermal epithelium that lines the large branchial cavity is excitable in three regions (Fig. 7), which are electrically independent of each other; inner skin impulses in one region are not conducted to the other regions. Anteriorly, the inner epithelium bordering the front lips is excitable over its whole extent, separated from the inner epithelium posteriorly by the peripheryngeal ciliated bands. Only part of the epithelium lining the remainder of the branchial cavity is excitable (excitability diminishing dorsally and posteriorly) (Fig. 7); this is divided into two halves by the endostyle, and although the epithelium covering the gill bar is excitable, impulses in the right half of the system do not pass to the left half, probably because the epithelium is very thin at the top and bottom of the gill bar (only 55-80 nm) so that

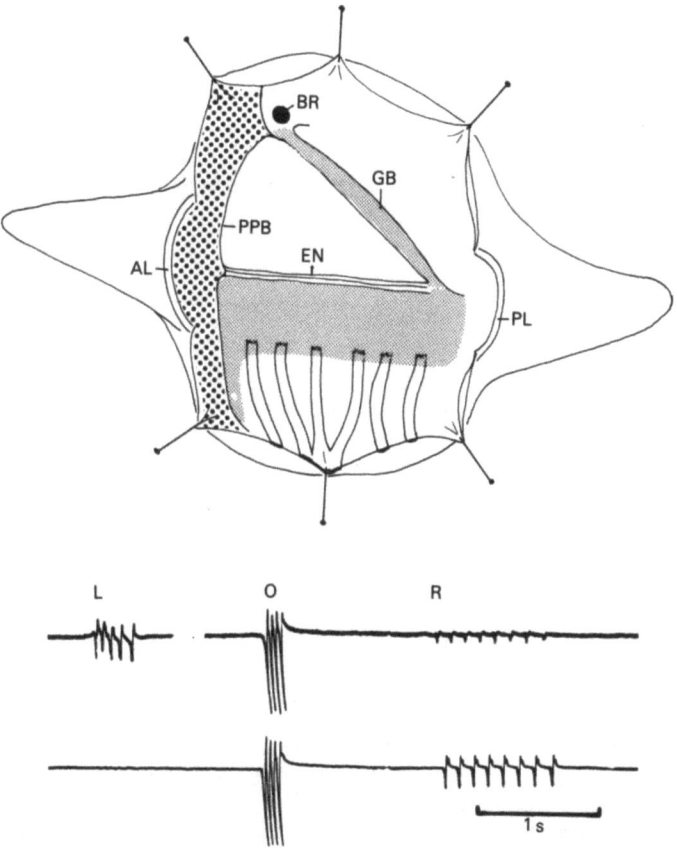

FIG. 7: Endodermal inner epithelial action potentials in Salpa fusiformis blastozooid. Above: animal opened by mid-dorsal cut and pinned out, showing anterior excitable region (coarse stipple) between anterior lip (AL) and the peripharyngeal bands (PPB). The left inner epithelial excitable region (fine stipple) extends up the left side of the gill bar (GB) towards the brain (B) and is separated from the right region ventrally by the endostyle (EN). PL: posterior lips. The suction electrode records below show action potentials in the inner endodermal epithelium recorded directly from left and right sides (L and R), and outer ectodermal epithelial action potentials (O) picked up electrotonically, all evoked by mechanical stimulation. From Mackie and Bone (1977).

voltage attenuation and conduction block occurs there. The inner epithelium is thicker elsewhere, but in young Salpa fusiformis blastozooids it is still less than 1 μm thick, and intracellular records have not yet been obtained.

Extracellular records show that conduction velocity along the gill bar is around 8.5 cm s^{-1}; such records are of the same form as those from epithelia where intracellular records have been obtained. Mechanical stimulation of the inner surface of the lips evokes inner skin pulses from the anterior region, and these lead to a brief interruption of the regular swimming rhythm (Fig. 8). It is not clear how such epithelial action potentials "enter" the brain to evoke changes in locomotor behaviour, but

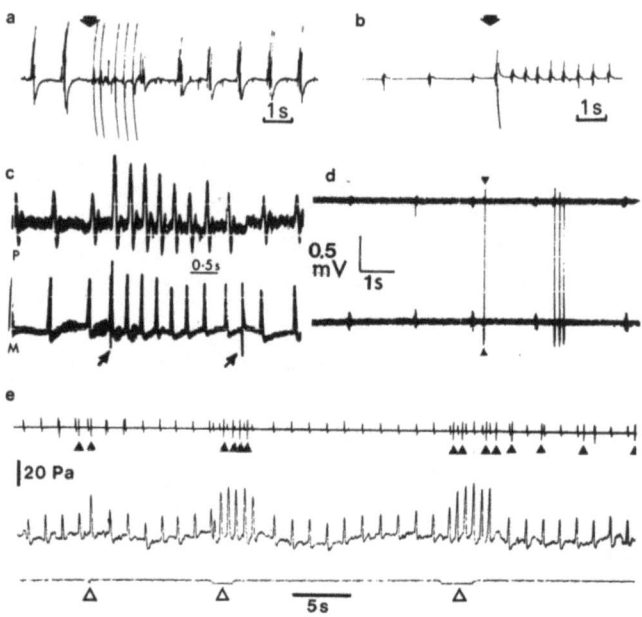

FIG. 8: The effects of epithelial action potentials in S. fusiformis. (a) A series of inner epithelial pulses evoked by mechanical stimulation inside the front lips (arrow) inhibits forward locomotion of a blastozooid (regular potentials from body muscle band). (b) A single outer epithelial action potential evoked by mechanical stimulation of rear lips (arrow) accelerates the regular swimming of a young oozooid. (c) The first of two outer epithelial action potentials (arrows on lower line) increases amplitude of locomotor muscle potentials in a blastozooid as well as their frequency, and much increases chamber pressures (upper line). (d) A single outer epithelial potential in an oozooid (triangles) has no effect on the regular activity of anterior (upper) or posterior lip muscles, but three epithelial potentials subsequently slow the swimming rhythm. (e) Upper line, potentials from anterior lip muscle and outer epithelial potentials (triangles); lower line chamber pressures. Mechanical stimulation of the anterior lips for longer or shorter periods (indicated by event marker on bottom line) evokes epithelial action potentials, and reduces muscle activity as this small blastozooid attempts to reverse keeping the anterior aperture open, and increases chamber pressures. From Bone and Trueman (1983); Mackie and Bone (1977) and Bone (1982).

it seems most probable that they pass via gap junctions to sensory cells in the inner epithelium, and thence to the brain via sensory cell axons. As in the outer epithelium, however, gap junctions have yet to be demonstrated between sensory cells and epithelial cells.

Presumably both the anterior region and the two posterior halves of the inner system are in life involved in food rejection reflexes, e.g. when a large particle enters the pharynx, but the effects of inner skin pulses from the posterior regions on locomotor activity have not been investigated.

One problem with pinned out preparations, where these two zones are accessible for mechanical or electrical stimulation, is that in such preparations, as Mackie and Bone (1977) observed, a great deal of skin pulse activity can occur without any visible effect on locomotion. Presumably adaptation or fatigue occurs quickly at some step or steps along the sensory pathway, with repeated stimulation, but of course, in nature, skin pulses are likely to be rare events, and single skin pulses probably affect the locomotor rhythm.

The outer skin pulse (OSP) system is easier to investigate, for mV potentials in the outer epithelium can easily be recorded extracellularly (through the test). They are conducted across the whole of the outer epithelium at velocities around 17 cm s^{-1} (exceptionally up to 38 cm s^{-1}) and evoke interruptions or reversals of the locomotor rhythm if generated anteriorly, accelerations if generated posteriorly (Fig. 8). During such changes in the locomotor rhythm, not only does the rhythm alter but especially in reverse locomotion, the muscles contract more strongly, as evidenced by large increases in chamber pressure (Fig. 8). In intact animals collected in plankton nets, single outer skin pulses often evoke no locomotor changes, whilst a series of two or three in succession do so (Fig. 8), but (as for the inner epithelial system) it is probable that in nature, single OSPs suffice, as in Fig. 8 (6).

In both oozooids and blastozooids, a most striking and remarkable effect is often observed when two recording electrodes are placed on the surface of the test, and the outer epithelium stimulated via a third electrode or with a fine probe. Sometimes, as expected, OSPs are recorded first at the electrode nearest the site of stimulation, and then after an interval of conduction, at the second recording electrode, as seen in Fig. 9a. But in Figs. 9c and d whilst the first (c) or first pair (b) of skin impulses are recorded by the two electrodes in this sequence, the last skin potential in each case is reversed, being recorded first at the anterior electrode farthest away from the stimulus site. In other words, OSPs travel first in one direction, and then in the opposite direction. In Fig. 9d, a mechanical stimulus at the rear of the animal evoked an epithelial action potential which began anteriorly and passed posteriorly, without a potential first arising in the opposite sense. By varying the placement of the recording electrodes it is evident that the first OSPs in Figs. 9b and c arise at the stimulus site, and the latter from a site close to the brain, which is near the anterior end of the animal. The outer epithelial cells in a limited region around the brain are thicker than elsewhere. Anderson and Bone (1980) were able to record from them intracellularly, (Fig. 9) showing that the epithelial potentials arising near the brain differ from those propagating into the cells from elsewhere (the first potential in Fig. 9c) for they arise from large depolarisations and are preceded by excitatory post synaptic potentials (the second and the third potentials). Evidently these cells are innervated by epitheliomotor nerves, and as Anderson et al. (1979) showed, nerve terminals are found upon them (Fig. 10), in a restricted area around the brain in both oozoids and gonozooids. In oozooids, double stimulus pulses delivered to

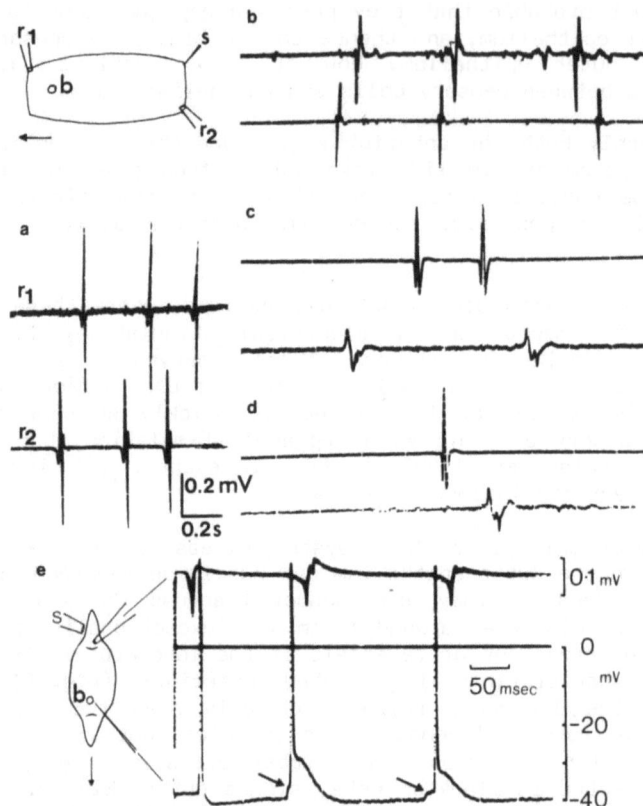

FIG. 9: "Driven" epithelial action potentials in S. fusiformis and Pegea confederata. (a) - (d) oozooid; (e) blastozooid. (a) Outer epithelial action potentials recorded by suction electrodes r1 and r2 placed as in diagram top left when oozooid is stimulated mechanically at s. (b) and (c) Similar records to (a) but showing potentials arising first near anterior electrode. (d) Delicate mechanical stimulation of rear lips give rise to "driven" potential appearing first at anterior electrode without preceding postero-anterior epithelial pulse. In this case, stimulation presumably activated a sensory cell of the rear lips. (f) Simultaneous extracellular (upper) and intracellular records of two "driven" potentials following a postero-anterior epithelial potential. Arrows indicate e.p.s.p's preceding "driven" potentials. All from S. fusiformis except (f) from P. confederata. From Anderson and Bone (1980) and Bone (1982).

a site near the front lips produce trains of driven OSPs; similar trains are evoked by injection of small drops of acetylcholine solution (10^{-4}M) into the space between the inner and outer epithelia around the brain. D-tubocurarine abolishes the production of driven OSPs, but OSPs evoked by stimulation are unaffected. It seems possible, therefore, that acetylcholine may be the neuroepithelial transmitter, as it probably is at the neuromuscular junctions in all tunicates. The presence of neuro-epithelial motor terminals, which can drive OSPs in the excitable epithelium from a site around the brain is unique to salps, and it certainly seems rather curious that the effect of stimulating a mechano-sensory system may be to evoke potentials in the system driven by the CNS. Clear evidence for the role of driven OSPs has, however, been obtained in the aggregated

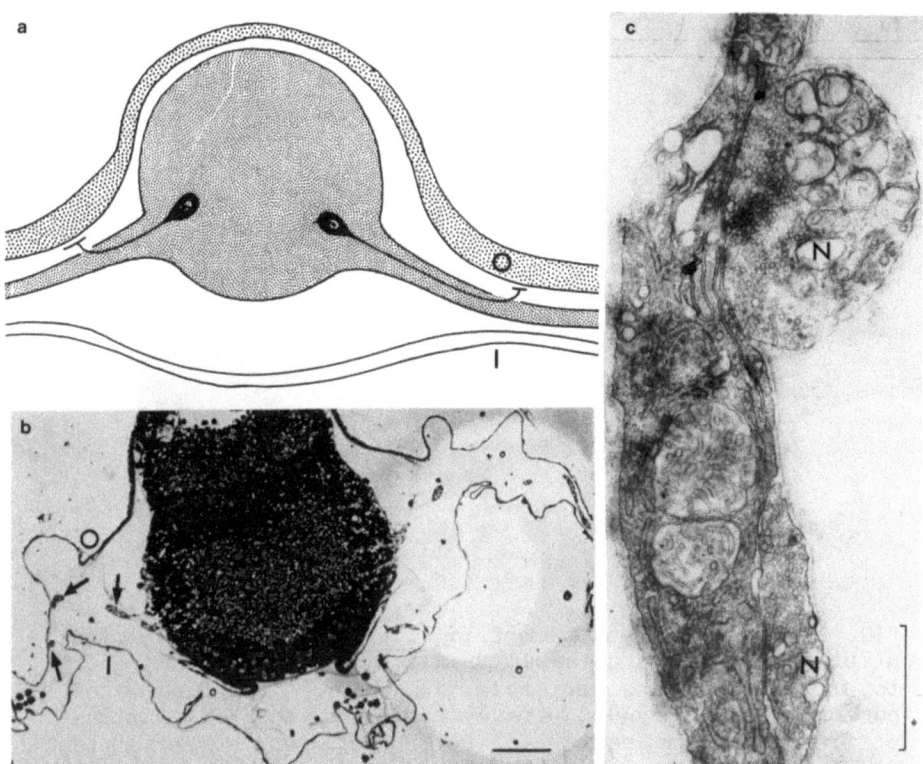

FIG. 10: Histological basis for "driven" epithelial action potentials in S. fusiformis. (a) Diagram of brain showing inner (I) and outer (O) epithelia and position of neuro-epithelial synapses. (b) Section of brain showing inner and outer epithelia and (arrowed) nerve bundles close to outer epithelium. Scale bar= 50 μm. (c) Neuro-epithelial synapses (N) on outer epithelial cells near brain of oozooid. Scale bar= 1 μm.

blastozooid generation, and it is probable that this unique system has arisen as a consequence of the alternation of generations and manner of budding in salps.

The blastozooids are produced as a long chain that arises as a stolon from the hinder end of the solitary oozooid, and eventually breaks off forming a chain of independent linked zooids. Each is attached to its neighbours by oval attachment plaques (Fig. 11) in such a way that each zooid is attached to another by two plaques.

Although the locomotor activities of the zooids in the chain are normally asynchronous (Fig. 12a) if the anterior or posterior zooid is stimulated, the entire chain reverses or accelerates (Fedele 1923), and the zooids contract in synchrony as seen in Fig. 12b. Since OSPs are conducted along the chain from one zooid to the next (Fig. 14), epithelial conduction seems an obvious co-ordinating mechanism. Yet at the attachment plaques, the epithelia of the two different zooids are not in connexion, being separated over most of their area by some 15 μm of test material. Detailed investigations of the plaques (Bone et al. 1980), have shown that the two linking one zooid to a neighbour are asymmetrical, and that one side of one plaque is innervated by a small group of sensory cells, at the other plaque, the other side is innervated. By separating

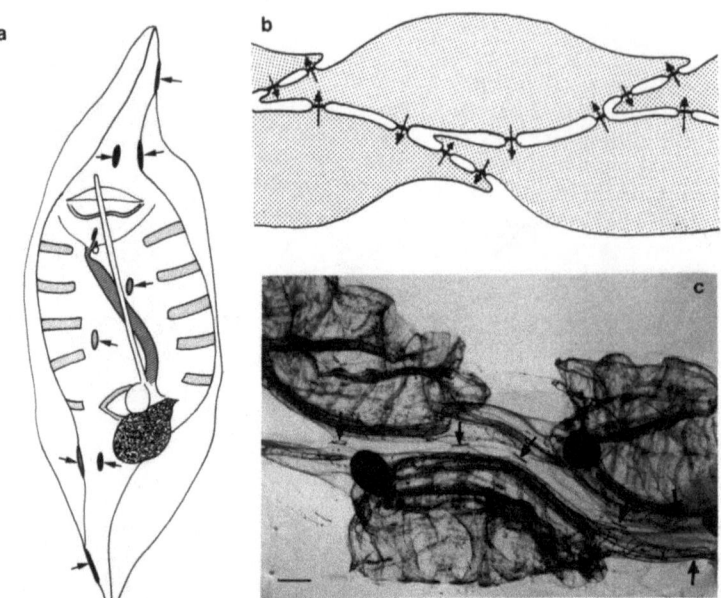

FIG. 11: Blastozooid attachment in S. fusiformis. (a) Diagram of blastozooid showing the eight attachment plaques linking it to four neighbouring zooids in the chain. (b) Diagram of portion of chain showing attachment plaques, and the direction of transmission at each. (c) Portion of chain (anterior to right) showing some attachment plaques. Scale bar= 2 mm. Partly from Bone et al. (1980).

one or other plaque and stimulating a linked pair of zooids, it was found that OSPs only passed from the non-innervated to the innervated side of the plaque. Ultrastructural investigations revealed a bizarre kind of coupling between the cilia of the sensory cells, and the outer epithelium of the neighbouring zooid on the other side of the plaque (Fig. 13). After emerging from the sensory cell, the cilium expands and produces many branches containing irregular numbers of axonemal tubules; these branches terminate against the membrane of the outer epithelium of the adjacent zooid, often in small "Synaptic"-like depressions in the membrane. The sensory cells themselves are not coupled by gap junctions to the epithelial cells amongst which they lie. It seems therefore that conduction of OSPs along the chain proceeds in the following manner. Stimuli applied to one zooid evoke an OSP or several OSPs which propagate into the plaque regions. Here, they activate the branches of the sensory cilium in contact with them and fire the cell or cells on the innervated side of the plaque in the second zooid, which in turn send action potentials along their axons to the brain. Here, neuro-epithelial neurons are activated, and driven OSPs pass around the outer epithelium or the second zooid to reach a non-innervated plaque, where they activate the sensory cells at that plaque belonging to the third zooid, and so on. It would certainly be possible to imagine a simpler system, which did not involve driven OSPs (for example at the plaques, the outer epithelia of adjacent zooids might be approximated without intervening test material, and linked by gap junctions), but it is important for the blastozooids to be able to separate from their neighbours in the chain, and it may be for this reason that the complicated system actually found has been evolved.

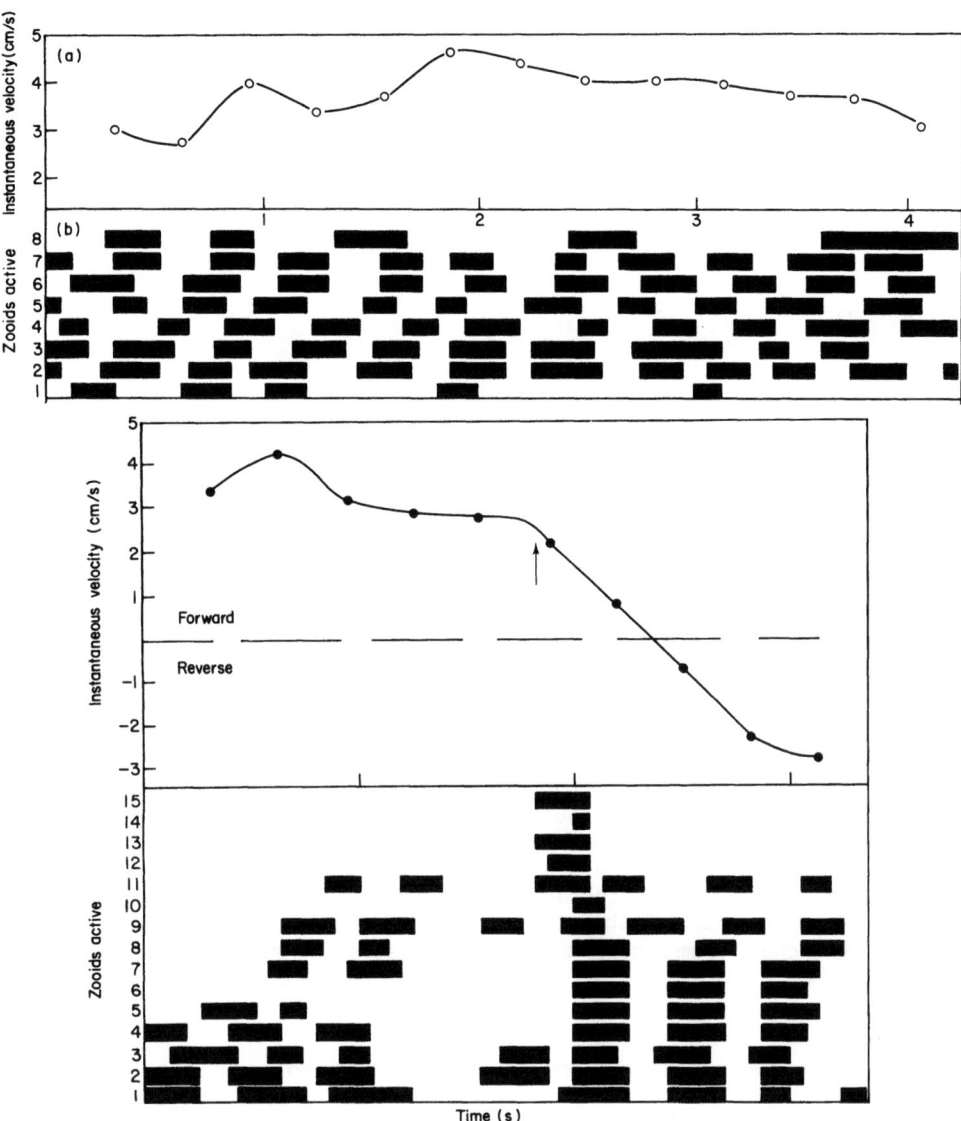

FIG. 12: Swimming behaviour of chains of *S. fusiformis* blastozooids. Upper: (a) Instantaneous velocities of 8-member chain. (b) Exhalent phases of jet cycle for each member of chain (solid bars), note lack of simultaneous activity. Lower: similar diagram showing co-ordination of activity of 15 zooids in another chain when first member is stimulated anteriorly to evoke reverse locomotion (first reverse pulse arrowed). Reversal involves co-ordination of all zooids in chain, including those previously inactive. From Bone and Trueman (1983).

In chains of blastozooids that have been anaesthetised with MS222 the zooids are firmly attached to each other, and remain linked even if strongly stimulated mechanically. But in un-anaesthetised chains, similar stimuli lead to immediate separation, and all the zooids swim off independently. This behaviour has obvious survival value, for if they could

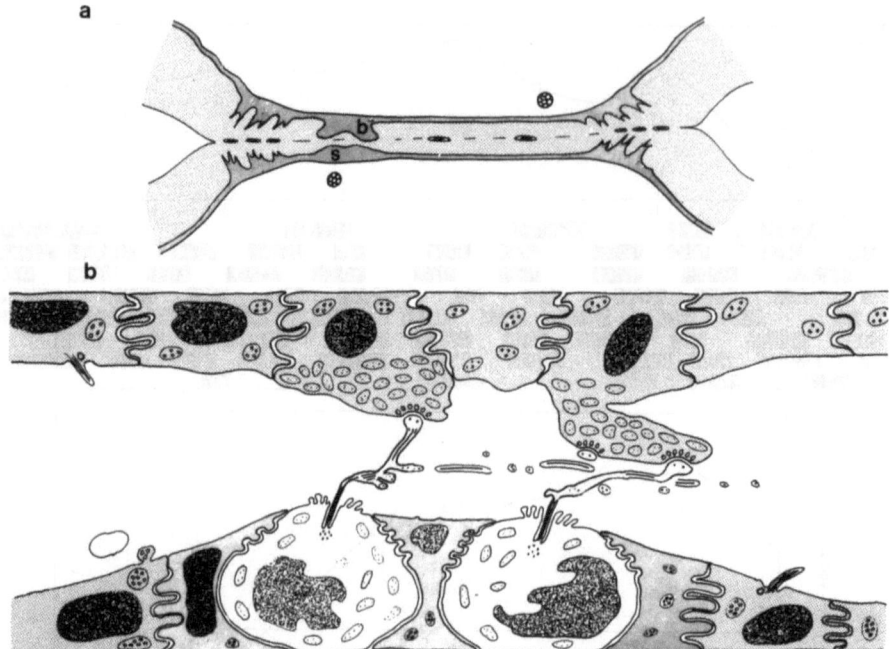

FIG. 13: The arrangement of the attachment plaques between S. fusiformis blastozooids. (a) Median section through plaque showing test material (light stipple) between the outer epithelium of the two zooids, and the position of a sensory cell group (s) in the lower zooid, opposite to a thickened region of the epithelium (b) in the upper zooid. (b) Diagram of ultrastructure of sensory cell region of plaque. The sensory cells in the epithelium of the lower zooid are in connection with the thickened region of the epithelium of the upper zooid by means of branches from their ciliary processes. From Bone et al. (1980).

not separate, a predator such as the sunfish Mola or the stromateid, Tetragonurus (both of which feed on salps) would simply begin at one end of a chain and munch its way along till all were eaten.

Separation involves bursts of OSPs, but it is not clear how these bring about separation; it is not due to mechanical activity by neighbouring zooids.

Since the alternating system of epithelio-neural and neuro-epithelial synapses described above evidently cannot obtain in the solitary oozooid stage, it is unexpected to find driven OSPs in this stage of the life cycle also. If OSPs evoked by mechanical stimulation "enter" the CNS via the axons of sensory cells, as is probably the case, it is hard to see how driven OSPs would not also do so, and puzzle the animal about the direction in which it should swim to avoid the stimulus.

One possibility is that bursts of driven OSPs evoked by strong mechanical stimulation may propagate into the stolon and cause detachment of the attached chain of young blastozooids, much as they seem to cause separation in the older blastozooid chain, but this possibility remains to be tested. Figure 14 compares the suggested operation of the OSP system in oozooids and blastozooids.

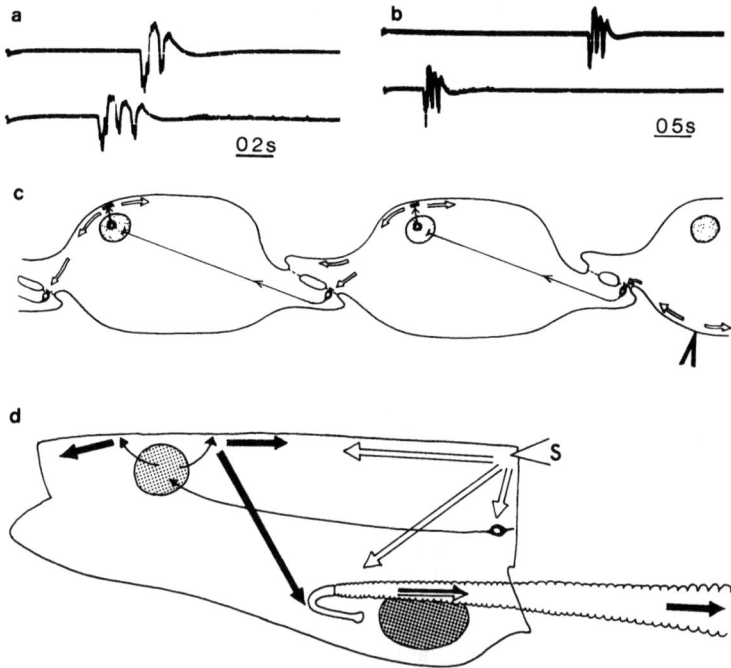

FIG. 14: Comparison of the epithelial conduction system in S. fusiformis oozooid and blastozooids. (a) and (b) Transmission of epithelial action potentials along chains of blastozooids. (a) Recorded from two zooids separated by single zooid (note different form of action potential burst in each). (b) Recorded from two zooids with 20 zooids intervening. (c) Transmission along blastozooid chain after stimulation of right zooid at arrow. Epithelial potentials open arrows, nerve impulses solid arrows. (d) Suggested operation of system in oozooid. Open arrows: epithelial action potentials evoked by mechanical stimulation, solid thicker arrows: "driven" epithelial potentials, thin arrows: nerve impulses. From Mackie and Bone (1977), Anderson and Bone (1980) and Bone (1982).

3.3. Doliolids

An excitable inner endodermal epithelium lining the body cavity is found in the "old nurse" stage of Dolioletta (Bone and Mackie 1977), which conducts potentials at 11-12 cm s^{-1} in the longitude direction and at 8 cm s^{-1} circumferentially. Such inner skin pulses have no effect on the regular locomotor activity of the animal, and their function is unknown. All other doliolid species and stages examined do not possess excitable epithelia.

3.4. Ascidians

Adult ascidians, as far as is known, do not have excitable sheets of epithelia, but the outer epithelium of a single tadpole larva, that of Dendrodoa, propagates long plateaued action potentials (Fig. 5e) in the outer epithelium, which block the swimming rhythm. The adults live attached to stones, and this inhibitory response (which is different to that of other tunicates where epithelial action potentials evoke escape responses) seems adapted for the larvae to halt and investigate any solid objects that they may encounter, as a possible site for settlement. So far, other species of ascidian tadpoles with an excitable outer epithelium have not been found, although in those investigated, the outer epithelial cells are linked by gap junctions. Dendrodoa has a relatively large larvae (2 mm long) and lack of success in demonstrating epithelial conduction in smaller larvae may simply reflect technical difficulties.

4. NEUROMUSCULAR SYSTEM

Muscle ultrastructure has been investigated in ascidian tadpole larvae, and in the adults of all other groups apart from the Pyrosomida, and as we should expect, differs according to the habits of the animal. In doliolids, for example, which swim by jet propulsion, and where there are large length changes in the hoop-like muscle bands encircling the body, the muscle fibres are obliquely-striated, as they are in other animals where muscle fibres contract around a body cavity and there is an hydrostatic skeleton (e.g. squid or annelid worms). In larvaceans, and in ascidian tadpoles and salps, they are cross-striated, whilst in adult ascidians they are unstriated. In all tunicates, the somatic musculature appears to be innervated by cholinergic nerves, although in none have more than one or two of the different criteria required to prove acetylcholine is the transmitter been satisfied.

Figure 15 shows the innervation patterns of somatic musculature in different tunicates. In all except larvaceans and possibly ascidian tadpole larvae, striated muscle fibres are multiply-innervated, and compound spike-like potentials are found during spontaneous activity. In salps for example, multiple small potentials summate to give compound muscle potentials, and recordings from motoneurones and motor nerves show bursts of action potentials, whilst stimulation of the motor nerves with pulse trains evokes muscle potentials similar to those seen during

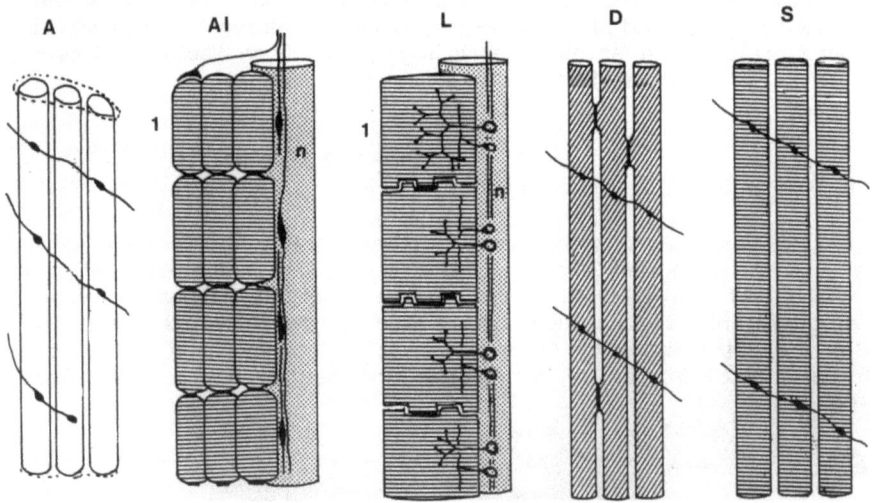

FIG. 15: Pattern of innervation of somatic muscles in different tunicates. In the diagrams of the ascidian tadpole larvae and the larvacean, 1: anterior muscle cell; n: notochord. Note different pattern of innervation of anterior cell or cell group to those following. A: adult ascidian; AL: ascidian larva; L: larvacean; D: Doliodum; S: salp.

spontaneous activity. In salps as in doliolids (Fig. 17) simultaneous intracellular records from two sites show that the largest potentials are recorded near the sites of motor terminals, and it is clear that muscle potentials are decremental. Remarkably enough, recent work by Nevitt and Gilly (1986) has shown that the non-striated somatic muscles of Ciona produce overshooting muscle spikes on stimulation (Fig. 16) and these apparently propagate along the fibres. In this case, as in the case of Doliolum (Fig. 17) external Ca^{++} is required for contraction and for spike production and it seems probable that in both cases Ca^{++} to activate the contractile mechanism enters during the muscle spike.

In larvaceans and ascidian tadpole larvae, the innervation pattern is rather complicated and not fully understood. In both, anterior muscle cells at the base of the tail are innervated differently to those elsewhere (in larvaceans by a much larger end-plate; in ascidian tadpole, larvae by an extra ventral end-plate) and in both, all muscle cells are electrically coupled by gap junctions. More complex still, in larvaceans, each muscle cell appears to be innervated by two motor endings at only one of which is acetylcholinesterase found (Flood, personal communication).

It is hardly surprising that in larvaceans there is a rather varied repertoire of muscle movement (Bone and Mackie 1975) and that intracellular records of caudal muscle activity show single potentials of varying amplitude; compound potentials; and smaller slower potentials (Fig. 7), some of which may possibly be of myogenic origin. Epithelial action potentials evoked by mechanical stimulation increase muscle potential amplitude and change their frequency (Fig. 7).

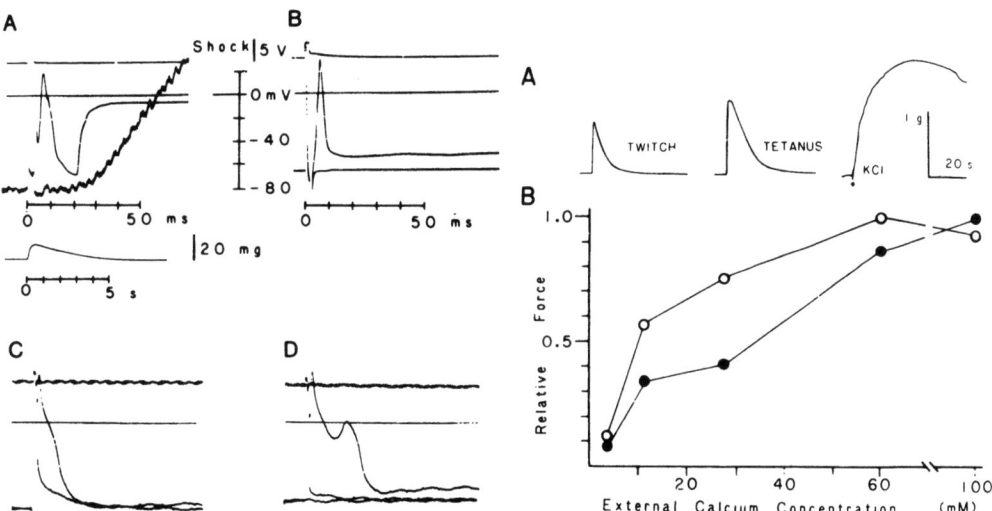

FIG. 16: Action potentials from Ciona unstriated somatic muscle cells. On left: A: Action potential and twitch (contraction displaces electrode, mechanical record heavy line); B: Sub- and supra-threshold shocks; all-or-none action potential evoked by suprathreshold stimulus. C and D action potentials evoked by intracellular current injection. Note hump on falling phase in C and doublet-type spike in D. On right: Dependance of force generated on external Ca^{++} concentration (open circles tetani, closed circles, twitches. Examples of twitch, tetanic, and KC contractions are shown above. From Nevitt and Gilly (1986).

Manifestly, much more work is needed before the operation of this complex system is understood, but unfortunately, the preparation is not an easy one, and it is in fact quite difficult to record from two or more muscle cells simultaneously in such small active animals.

The tadpole larvae of the ascidian Dendrodoa is easier to record from, but simultaneous intracellular records from two muscle cells have not yet been attempted. A striking point about muscle activation in this larvae, is that the activity of the two sides of the animal is not always symmetrical, for unilateral flexions may follow (Fig. 6f) or precede (Fig. 6g) bursts of symmetrical swimming activity. Symmetrical swimming activity is found in newly released larvae, but as the larvae ages, unilateral flexions become more and more prominent, i.e. the activity of the two sides becomes more and more independent. The pacemakers driving rhythmical swimming must lie in the cord, because swimming continues after removal of the trunk and motor neuron somata. As in the case of Oikopleura, it has not so far been possible to distinguish between electrical activity generated in a particular muscle cell by the action of motor nerves, and that propagated into the cell via gap junctions from an innervated cell elsewhere in the system, as for example from the anterior ventral muscle cell in ascidial larvae, or the anterior cell with the large end-plate in larvaceans.

5. THE CENTRAL NERVOUS SYSTEM

Comparatively little is known of the central nervous system in any tunicate, and this is unfortunate, for several different features suggest that there are really interesting discoveries to be made. For example, the reversal of swimming in salps and doliolids involves switching the

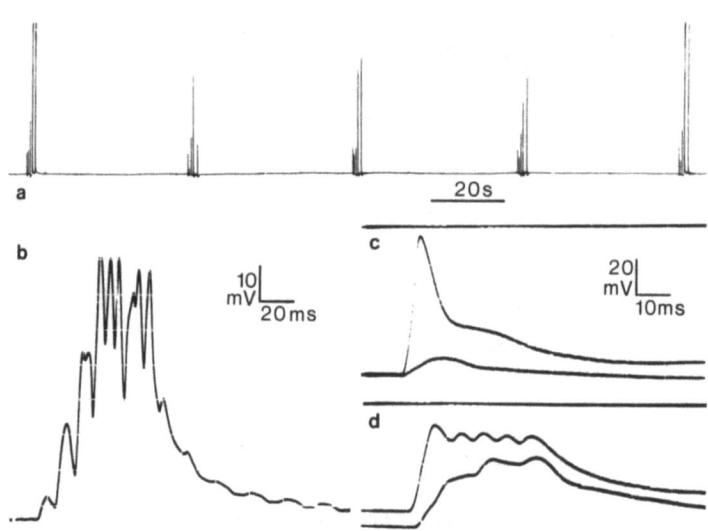

FIG. 17: Muscle potentials in doliolids. (a) and (b) Intracellular records from locomotor muscle of Dolioletta gonozooid showing regular slow rhythmic activity and multiple events which may overshoot resting potential. (c) and (d) Muscle potentials from Doliolum. In (c) simultaneous records from muscle band close to and distant from innervated region: in (d) compound potentials associated with longer contraction. From Bone and Trueman (1984) and unpublished.

delays between the contraction of the muscles controlling the lips and the main locomotor muscle bands; again, in ascidians that brain removal merely slows reflexes rather than abolishing them (Florey 1951) suggests that the central nervous system operates rather differently to that in other tunicates (or other animals for that matter).

Only in salps have any records been made from brain neurones (Fig. 18). These have included motoneurones and more or less regularly firing neurones whose activity is at a higher frequency than the swimming rhythm, but which are evidently linked to the rhythm, since OSPs and changes in illumination which alter the swimming rhythm also change their firing pattern (Mackie and Bone 1977; Anderson et al. 1979). Nothing has been discovered about the "wiring diagram" of the salp brain (Fig. 18) since Fedele's pioneering work on S. maxima (Fedele 1933) but investigations have begun using antisera to various peptides (M. Pestarino, personal communication) which may help to resolve some of the circuitry.

In ascidians, the immunocytochemical approach has been used by several workers (e.g. Fritsch et al. 1982; Thorndyke 1982; Pestarino 1985) to localise neurone somata containing different peptides in the brain cortex, and in some cases also, cells in the neural gland complex. These studies have demonstrated a considerable variety of peptides (or rather, have demonstrated reactivity to antisera raised against a variety of mammalian neuropeptides) in brain neurons, but have not so far been of much assistance in deciding what their functional role may be; whether fibres containing such peptides play any part in the peripheral innervation of effectors such as muscle fibres or ciliated cells; nor in unravelling the connexions of neurones in the brain.

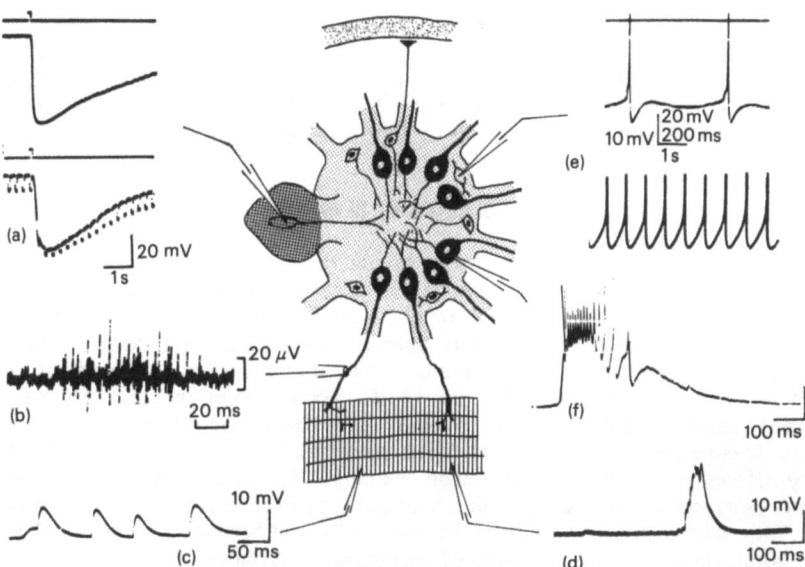

FIG. 18: Organisation and activity of salp brain. (a) Hyperpolarising potentials from visual cells; (b) extracellular record of motor nerve burst from a mixed nerve. (c) Junction potentials (jp) from locomotor muscle fibre; (d) summated jp's during muscle contraction; (e) records from regularly - firing cells (upper record showing synaptic potentials). (f) Motoneurone burst. From Bone and Mackie (1982).

Nevertheless, there is real hope that further immunocytochemical work may provide valuable information about ascidian brain connexions, which have remained obscure in part because the usual neurological methods (i.e. reduced silver and methylene blue staining) have proven difficult to adapt to ascidians.

Dilly (1969) showed that the chemical synapses in the brain contain vesicles of two kinds, those with electron dense contents ranging up to 100 nm in diameter, so that it seems highly probable that at least some of the peptides reactivity found by immunocytochemical techniques may in fact be from peptide neurotransmitters. It is a relatively simple matter to record neuronal activity from the brain with extracellular suction electrodes (Brace, personal communication) hence the activity of the different peptides found in brain neurones might initially be tested in this way. Osborne et al. (1979) assayed the brain of Ciona for possible neurotransmitters, finding a spectrum of putative neurotransmitters such as gamma aminobutyric acid (GABA), glutamate, noradrenalin and dopamine, but once again, it should be possible by relatively simple experiments to see whether any of these have effects upon brain neurones.

An intriguing feature of the ascidian brain is that if it is ablated, the animal is capable of regenerating a new brain and neural gland, as Schultze (1899) discovered. It is obvious that this offers interesting opportunities for the study of the re-appearance of different neurone types (e.g. by immunocytochemical means) but as yet advantage has not been taken of this unique preparation.

In the other tunicate groups, brain removal results in immediate paralysis of muscular activity, and rapid death, but in larvaceans, the anterior portion of the central nervous system containing the statocyte (sometimes termed the brain), may be removed without affecting locomotor activity, which is controlled by the caudal ganglion near the base of the tail. In both larvaceans and ascidian tadpole larvae, cell numbers in the caudal ganglion and brain are small, and it will certainly be possible by EM serial sectioning techniques to work out the connexions of most if not all neurones. So far this approach has not been used, nor have the alternative techniques of dye injection or uptake.

6. CONCLUDING REMARKS

All students of tunicates have borne in mind that they are dealing with a group in some ways intermediate between invertebrates and vertebrates, though with obvious chordate affinities, and not only have relationships between classes been speculated upon, but also, the relationship of tunicates to vertebrates. The structure and operation of the tunicate nervous system, insofar as it is known does not, in my opinion, throw very much light yet on either of these problems. Within the tunicates themselves, however, it does seem clear that the larvaceans are not very closely allied to ascidian tadpole larvae, and are unlikely to have been derived from them by any neotenic transformation. The organisation of the neuromuscular system is entirely different in the two groups, and the appearance in larvaceans of secondary sensory cells suggests that larvaceans have been much modified from an ancestral form and have "invented" a number of entirely new features, such as this type of sensory cell, or the dual pattern of motor innervation. It seems more likely that they could have been derived, as Garstang (1928) suggested, from a doliolid type of larvae by a process of simplification and reduction in cell number and subsequent new developments of the nervous system.

Doliolids and salps are evidently fairly closely allied by morphological features as well as by their life-histories; the loss of multiple ciliated gill bars in salps and the development of epithelial conduction including neuro-epithelial motor innervation suggests that salps are farthest removed from the ancestral jet-propelled form. As regards the relationships of the group with vertebrates, whatever they may be, the most interesting light that the nervous system may throw upon this seems likely to come from further immunocytochemical studies of brain peptides. So far, despite clear evidence of the existence of a variety of peptides in the adult ascidian brain (though their true nature is unknown), nothing is known of their existence in ascidian larvae, nor in larvaceans, and it would be interesting indeed to know whether in these forms, which operate in an essentially vertebrate-like manner, there are peptide systems that may be homologous with those found in higher chordates, such as amphioxus. There is evidently much that still needs to be done with the tunicate nervous system, and even in this limited review, several lines have been mentioned which should yield results of general interest.

7. REFERENCES

Anderson PAV (1979) Epithelial conduction in salps. I. Properties of the outer skin pulse system of the stolon. J Exp Biol 80: 231-239.

Anderson PAV (1980) Epithelial conduction: its properties and functions. Prog Neurobiol 15: 161-203.

Anderson PAV, Bone Q (1980) Communication between individuals in salp chains II. Physiol Proc Soc B 210: 559-574.

Anderson PAV, Bone Q, Mackie GO (1979) Epithelial conduction in salps II. The role of nervous and non-nervous conduction system interactions in the control of locomotion. J Exp Biol 80: 241-250.

Berrill JN (1950) The Tunicata, with an account of the British species. Ray Soc Monogr 133: 1-354. Ray Society, London.

Bone Q (1982) The role of the outer conducting epithelium in the behaviour of salp oozooids. J Mar Biol Assoc UK 62: 125-132.

Bone Q (1985) Epithelial action potentials in Oikopleura (Tunicata: Larvacea). J Comp Physiol A 156: 117-123.

Bone Q, Mackie GO (1975) Skin impulses and locomotion in Oikopleura (Tunicata: larvacea). Biol Bull Mar Lab Woods Hole 149: 267-286.

Bone Q, Mackie GO (1977) Ciliary arrest potentials, locomotion and skin impulses in Doliolum (Tunicata: Thaliacea). Riv Biol Norm Pathol 3 (5): 181-191.

Bone Q, Mackie GO (1982) Urochordata. In: Shelton GAB (ed) Electrical conduction and behaviour in "Simple" invertebrates. Clarendon Press, Oxford, pp 473-535.

Bone Q, Ryan KP (1978) Cupular sense organs in Ciona (Tunicata: Ascidiacea). J Zool Lond 196: 417-429.

Bone Q, Ryan KP (1979) The Langerhans receptor of Oikopleura (Tunicata: larvacea). J Mar Biol Assoc UK 59: 69-75.

Bone Q, Trueman ER (1983) Jet propulsion in salps (Tunicata: Thaliacea). J Zool Lond 201: 481-506.

Bone Q, Trueman ER (1984) Jet propulsion in Doliolum (Tunicata: Thaliacea). J Exp Mar Biol Ecol 76: 105-118.

Bone Q, Fenaux R, Mackie GO (1977) On the external surface in Appendicularia. Ann Inst Oceanogr 53: 237-244.

Bone Q, Gorsky G, Pulsford A (1979) On the structure and behaviour of Fritillaria (Tunicata: larvacea). J Mar Biol Assoc UK 59: 399-411.

Bone Q, Anderson PAV, Pulsford A (1980) The communication between individuals in salp chains. I. Morphology of the system. Proc R Soc B 210: 549-558.

Braconnot J-C (1971) Contribution à l'étude des stades successifs dans le cycle des Tuniciers pelagiques Doliolides. II. Les stades phorozoid et gonozoide des Doliolides. Arch Zool Exp Gen 112: 5-32.

Brien P (1948) Tuniciers. In: Grassé P-P (ed) Traité de zoologie, XI. Masson et Cie, Paris. pp 553-930.

Dilly N (1969) Synapses in the cerebral ganglion of adult Ciona intestinalis. Z f Zellf 93: 142-150.

Fedele M (1923) Simmetria ed unita dinamica nelle catene di Salpa. Boll Soc Nat Napoli 36: 20-32.

Fedele M (1933) Sul ritmo muscolare somatico delle salpe. Boll Soc Ital Biol Sper 8: 475-478.

Florey E (1951) Reizphysiologische Untersuchungen an der Ascidia Ciona intestinalis L. Biol Zentralbl 70: 523-570.

Fol H (1872) Études sur les appendiculaires du détroit de Messine. Mém Soc Phys Hist Nat Genève 21: 445-499.

Fritsch HAR, van Noorden S, Pearse AGE (1982) Gastro-intestinal and neurohormonal peptides in the alimentary tract and cerebral complex of Ciona intestinalis (Ascidiaceae). Cell Tissue Res 223: 369-402.

Garstang W (1928) The morphology of the Tunicata, and its bearings on the phylogeny of the Chordata. Quart J Micro Sci 72: 51-187.

Grave C, Woodbridge H (1924) Botryllus schlosseri (Pallas); the behaviour and morphology of the free-swimming larvae. J Morphol Physiol 39: 207-247.

Holmberg K (1984) A transmission electron microscopic investigation of the sensory vesicle in the brain of Oikopleura dioica (Appendicularia). Zoomorphol 104: 298-303.

Holmberg K (1986) The neural connection between the Langerhans receptor cells and the central nervous system in Oikopleura dioica (appendicularia). Zoomorphol 106: 31-34.

Josephson RK, Schwab WE (1979) Electrical properties of an excitable epithelium. J Gen Physiol 74: 213-236.

Kowalewsky AO (1867) Entwicklungsgeschichte der einfachen Ascidien. Mem Acad Sc St Pétersb 10 (7): 1-19.

Kowalewsky AO (1871) Weitere Studien uber die Entwicklung der einfachen Ascidien. Arch mikr anat 7: 101-130.

Mackie GO (1974) Behaviour of a compound ascidian. Can J Zool 52: 23-27.

Mackie GO, Bone Q (1976) Skin impulses and locomotion in an ascidian tadpole. J Mar Biol Assoc UK 56: 751-768.

Mackie GO, Bone Q (1977) Locomotion and propagated skin impulses in salps (Tunicata: Thaliacea). Biol Bull Mar Lab Woods Hole 153: 180-197.

Martini E (1909) Studien uber die Konstanz histologischer Element. I. Oikopleura longicauda. Z wiss zool 92: 563-626.

Monniot C, Monniot F (1975) Abyssal Tunicates: an ecological paradox. Ann Inst Oceanogr Paris 51: 99-129.

Monniot C, Monniot F (1978) Recent work on the deepsea tunicates. Ann Rev Oceanogr Mar Biol 16: 181-228.

Nevitt G, Gilly WF (1986) Morphological and physiological properties of nonstriated muscle from the tunicate, Ciona intestinalis: parallels with vertebrate skeletal muscle. Tissue Cell 18: 341-360.

Osborne NN, Neuhoff V, Ewers E, Robertson HA (1979) Putative neurotransmitters in the cerebral ganglia of the tunicate Ciona intestinalis. Comp Biochem Physiol 63C: 209-213.

Pestarino M (1985) Occurrence of β-endorphin-like immunoreactive cells in the neural complex of a protochordate. Cell Molec Biol 31: 27-31.

Roberts A (1975) Some aspects of the development of membrane excitability, the nervous system and behaviour in embryos. In: Usherwood PNR, Newth DR (eds) Simple nervous systems. Arnold, London, pp 27-65.

Schultze LS (1899) Die Regeneration des Ganglions von Ciona intestinalis L und über das Verhaltniss der Regeneration und Knospung zur Kiemenbatterlehne. Jean Zeit f Naturw 33: 263-344.

Thorndyke MC (1982) Cholecystokinin (CCK) gastrin-like immunoreactive neurones in the cerebral ganglion of the protochordate ascidians Styela clava and Ascidiella aspersa. Regulatory Peptides 3: 281-288.

Torrence SA, Cloney RA (1982) Nervous system of ascidian larvae: caudal primary sensory neurons. Zoomorphol 99: 103-115.

NERVOUS MECHANISMS OF SPAWNING IN REGULAR ECHINOIDS

M. YOSHIDA, H. NOGI and Y. TANI

Ushimado Marine Laboratory

130-17 Kashino, Ushimado

Okayama 701-43, Japan

ABSTRACT

Spawning is chosen as a criterion response to elucidate nervous mechanisms controlling co-ordination in effector systems in regular echinoids. A few drops of isotonic KCl applied to the test cavity of the upper hemisphere induce a long lasting contraction (arrhythmic, ArC) of gonads. Superposed on it, small and repetitive contractions (rhythmic, RC) often occur synchronously in all five gonads. Electrical stimulation and acetylcholine application also induce ArC and these three kinds of stimulants are shown to affect directly gonadal muscles. To induce RCs, the presence of the gonoduct which connects the gonad with the aboral ring system is needed. The aboral ring system is composed of an inner and an outer circular nerves, the latter being connected to the gonad via the gonoduct. It is assumed that a pacemaker residing within the anal region is responsible for rhythm generation and that the pacemaker activities are facilitated by gamma-aminobutyric acid (GABA) and inhibited by 1-glutamic acid and 1-aspartic acid.

1. INTRODUCTION

The nervous system of echinoderms consists of circular and radial cords. Such a basic plan led workers in the middle of the century to conceive an idea of a central nervous system or a motor center to account for a variety of beautifully co-ordinated movements such as stepping of tube feet, righting the overturned body, etc (Smith 1945; Kerkut 1954). Investigations in the past two decades, however, have forced us to discard such a notion as centralization of the nervous system and an old phrase "Reflex Republik" by von Uexkull (1897) has now been modified to "republic of ganglia" (for reviews, see Pentreath and Cobb 1982). In this article, we approach the enigma in this phylum by focussing our attention on a single phenomenon, which is relatively simple, yet nicely co-ordinated and precisely recordable. As a criterion response to meet the above requirements, we choose the spawning activities in regular echinoids.

2. SPAWNING PROCESS

2.1 Methods of Spawning Induction

Methods commonly used by cell biologists to obtain gametes are either application of isotonic KCl into the test cavity or AC stimulation, about 20 V between the oral and the aboral poles (Iwata 1950; Harvey 1952). Other means such as raising the temperature and rubbing the test surface are reported to be effective. Apart from these non-physiological methods, a heat-stable oligopeptide extractable from the radial nerve (Cochran and Engelmann 1972) and γ aminobutyric acid (GABA) (Takahashi 1986) which is actually present in various tissues (Osborne 1971) are known to be effective inducers. In line with the general notion that neuromuscular junctions in the echinoderms are cholinergic, acetylcholine is shown to induce spawning (Iwata and Fukase 1964).

2.2 Recording Method

The spawning process may be followed photo-electrically (Fig. 1. I-J system). Here, released gametes interrupt the beam passing from the light source to a detector (CdS), from which the signals can be recorded after amplification. However, the spawning itself is merely the final result of gonadal activities. Detailed processes occurring in gonads were directly followed by two methods; one by measuring internal pressure changes (Iwata

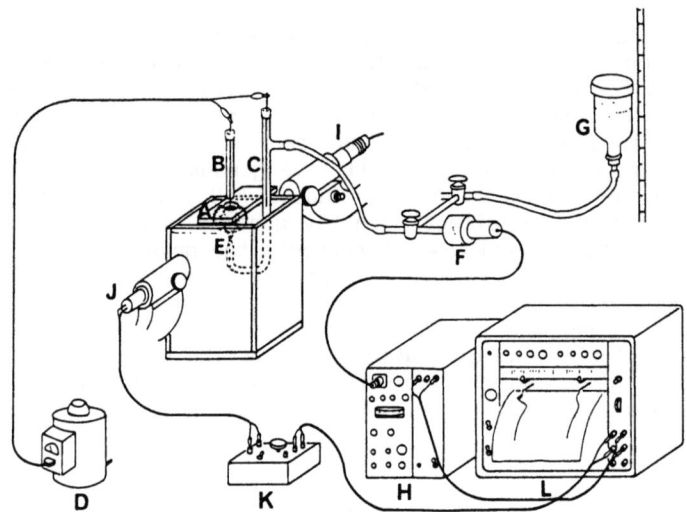

FIG. 1: A method of simultaneous recording of the spawning process and development of internal pressure of a gonad (Iwata 1976). A sea urchin (A) is placed anal side down and a U-shaped glass tubing (E) is firmly attached to a gonopore via a short rubber tubing. Gonads are induced to shed gametes by electrical stimulation (electrodes B and C, and a variac D). Changes in intensity of inspection light (I) caused by gametes being shed are detected by CdS (J), the signals of which are recorded via a bridge circuit (K) on one channel of a two-pen recorder (L). Changes in internal pressure of a gonad is detected by an electric manometer (F), the signals of which are also recorded on the other channel of the recorder (L) after amplification (H). G is a sea water reservoir used for pressure application and also for pressure calibration.

1976) and the other by recording the degree of displacements of the gonadal wall (Okada et al. 1984). In the former case (Fig. 1), a glass tube (E), 2 mm in internal diameter, was filled with sea water and a rubber tubing attached to it was placed tightly onto one of the five genital pores. The other end of the glass tube was connected to a high-gain pressure transducer (F). In the latter case (Fig. 2), a straw was placed lightly on the gonadal surface and the upper end of the straw was attached to a high-gain force displacement transducer. With this set-up, it becomes possible to record simultaneously movements of more than one lobe. A comparison of the three kinds of recordings (Fig. 3) reveals that they all coincide well in amplitude as well as in timing. In other words, recordings of gonadal displacements can be regarded as reflecting faithfully the spawning activities. The coincidence also indicates that the recorded displacements are caused by actual contraction of gonadal lobes which must have raised the internal pressure.

2.3 Contraction Pattern

The pattern of gonadal contraction differs according to the means of artificial stimulation. Five typical examples are shown in Fig. 4. When a gonad is connected to the anal system via the gonoduct, electrical stimulation induces a tetanic contraction with small repetitive contractions superposed on it (A). When the gonoduct is cut across, the pattern becomes smooth and biphasic so that an initial twitch-like contraction is followed by a slow, small and long-lasting one (B). Application of acetylcholine (10^{-5} M) induces only a smooth monophasic contraction (C), upon which small repetitive contractions become superposed when electrical stimulation is added (D). The KCl induced pattern (E) is similar to that induced electrically but lasts longer. Application of GABA also induces long lasting contractions and repetitive

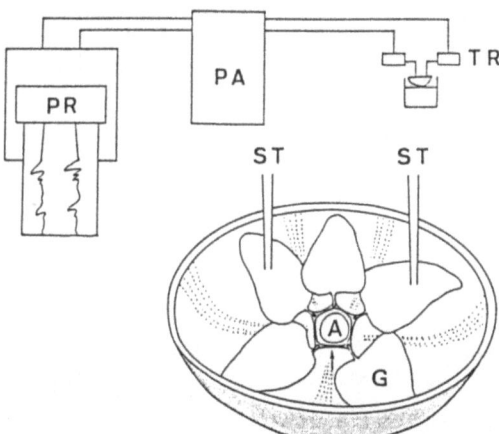

FIG. 2: A method of simultaneous recording of movements occurring in more than one gonad. After removing all spines by a wire brush, the test animal is cut along the equator and alimentary canals are carefully removed. The aboral half is placed anal side (A) down on a plate with a hole at the center. Two or three straws (ST) are placed separately on each surface of gonads (G). The upper end of each straw is connected to a high gain force displacement transducer (TR), the potential changes of which are recorded on a multichannel pen recorder (PR) after amplification (PA).

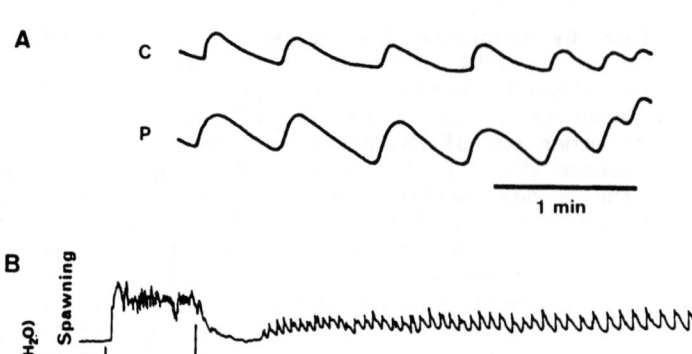
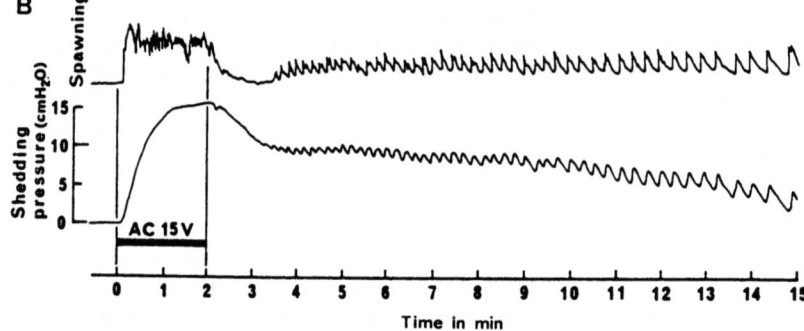

FIG. 3: Examples of recordings. A: Movements of a gonad (C) coincide in time and amplitude with changes in internal pressure (P) of the same gonad (Okada et al. 1984). B: Changes in internal pressure of a gonad (upper trace) coincide with gamete (eggs) shedding from the same gonad (lower trace). During AC stimulation, the internal pressure increases sigmoidally and the photo-electric recording keeps a high level. A few minutes after the end of stimulation, a small mount of eggs are shed intermittently and the recording of the internal pressure follows faithfully the intermittent shedding activity (Iwata 1976).

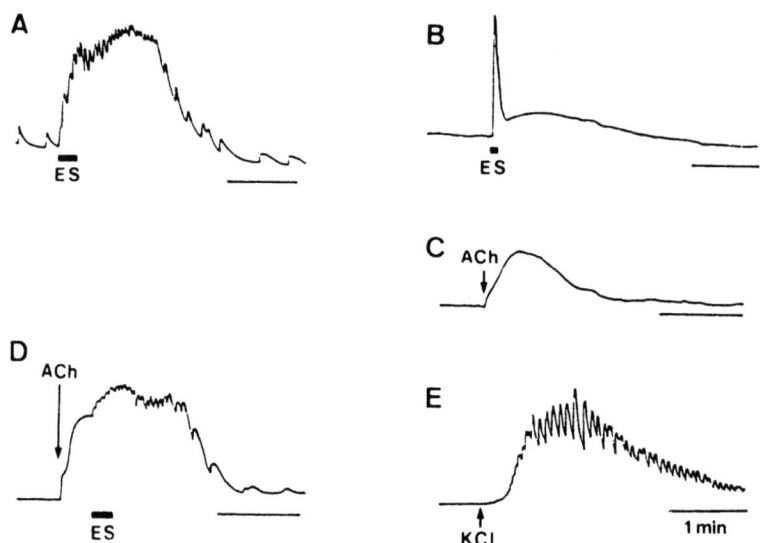

FIG. 4: Difference in contraction pattern depending on the type of stimulation. All the recordings from this onwards are made with the force displacement detecting technique. A: Electrical stimulation (ES) is given to the aboral region of the test. B: The gonoduct of the recorded lobe is cut across. C: 10^{-6} M ACh is applied into the test cavity. D: An electrical stimulation is followed 0.5 min after application of ACh. E: A few drops of 0.5 M KCl is applied into the sea water filled test cavity. The gonoducts are intact in all but B (Okada et al. 1984).

contractions (Fig. 5B). Following the terminology of Okada et al. (1984), we call the large long lasting contraction "Arrhythmic contraction" (ArC) and the small repetitive contraction "Rhythmic contraction" (RC). It must be noted that RCs seldom occur out of the breeding season.

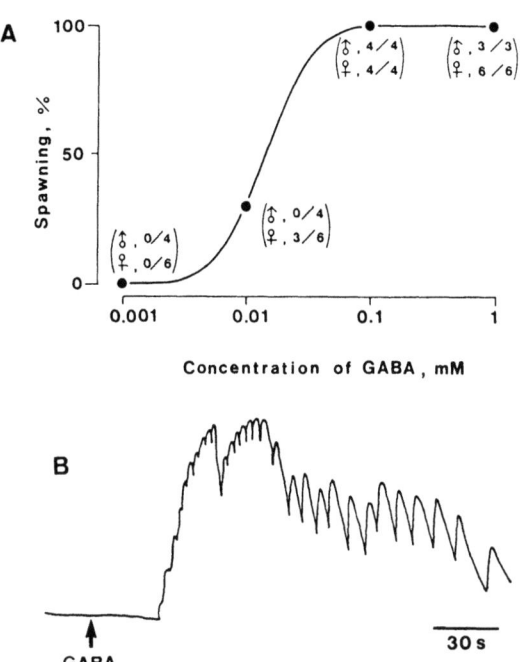

FIG. 5: Effects of GABA. A: Dose-response curve. GABA at the concentration indicated on abscissa is applied into the test cavity. In this figure only, the material used is <u>Strongylocentrotus</u> <u>intermedius</u> (By courtesy of Dr. Takahashi, Sapporo Medical University). B: Contraction pattern. From this onwards, concentrations of each chemical mentioned in Figures are final concentrations calculated after application of test solutions as a drop of 0.04 ml into the sea-water filled test cavity (5-10 ml).

3. EFFECTIVE SITE OF STIMULATION

3.1 Arrhythmic Contraction

By anesthetization with isotonic magnesium salts (SO_4^{2-}, Iwata and Fukase 1964; Cl^-, Okada et al. 1984), the ArCs in response to either electrical stimulation or KCl application are not affected but the RCs are completely abolished (Fig. 6). This result suggests that stimulants

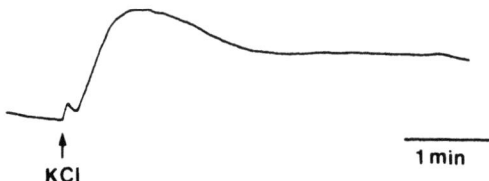

FIG. 6: Effect of anesthetization. The gonad is pre-treated for 15 min with 0.5 M $MgCl_2$. No RCs are induced (Okada et al. 1984).

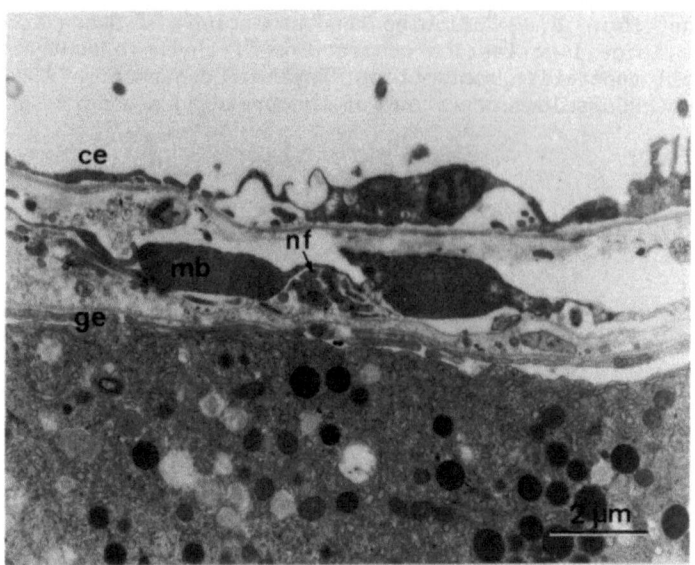

FIG. 7: Electron micrograph of the gonadal wall. ce, coelomic epithelium; ge, germinal epithelium; mb, bundle of myofilaments; nf, nerve fiber.

inducing the ArC must affect directly the gonadal muscles, which are distributed in bundles along the gonadal wall (Fig. 7). This suggestion is supported by the following two experiments.

3.1.1. Effects of Acetylcholine

As shown in Fig. 4, acetylcholine at a concentration of 10^{-6} M induces only ArC but not RC. The effect is potentiated by pre-treatment with eserine (Fig. 8A) and inhibited by hexamethonium (Fig. 8B) (confirmation of Iwata and Fukase 1964). Thus, gonadal muscles may be assumed to be cholinergic as other muscular systems of echinoderms (holothurian longitudinal retractor muscles, Prosser and Mackie 1980; Sugi et al. 1982: smooth muscles of sea urchin spine, Peters and Shelton 1981: sea urchin tube feet, Florey et al. 1975; Florey and Cahill 1980). However, varicosities or terminals of the nerve fibers in the muscle layer of the gonadal wall (Fig. 9) contain dense cored vesicles of various sizes and no cholinergic nerve fibers characterized by small clear vesicles (Pentreath and Cobb 1982) are encountered.

3.1.2. ArC Occurs in Isolated Gonads

As mentioned above (Fig. 4B), transection of the gonoduct abolishes the RCs, leaving only ArC. These results suggest, in addition, that an RC controlling factor should reside outside the gonadal lobe (see below).

3.2. Rhythmic Contraction

3.2.1. Nervous Involvement

As noted above, a candidate for the rhythm generator appears to be located proximal to the gonoduct which runs out of the aboral nerve ring (ANR). Here, a brief description of the nervous system in the aboral

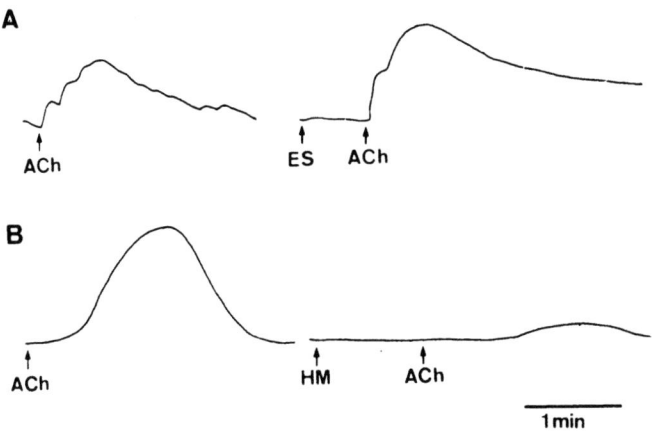

FIG. 8: Effects of eserine (A) and hexamethonium (B).

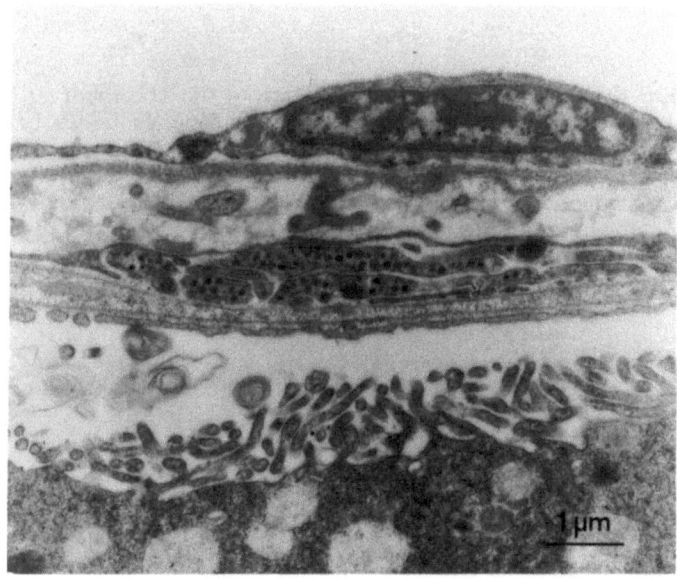

FIG. 9: Electron micrograph of the gonadal wall. The varicosities of nerve fibers covered by thin sheets of gonadal muscle cells contain dense cored vesicles.

region (Fig. 10) may be needed. Each gonoduct is in contact with the aboral sinus just before forming a gonopore (gp) at the end. Two nerve cords run separately along the inner and the outer margins of the aboral sinus. The outer circumanal nerve cord (oc) contains five neuropil regions (np) and protrudes five short branches (gn). The neuropils are present in the region where the gonoducts are in contact with the aboral sinus. In this region nerve fibers of varying diameters interweave and form varicosities containing many small vesicles (Fig. 11). The neuropil regions are connected with tracts consisting of regularly and longitudinally arranged thin fibers (Fig. 12), about 6000-8000 in number in a cross section. Each of the short branches consisting of about 4000

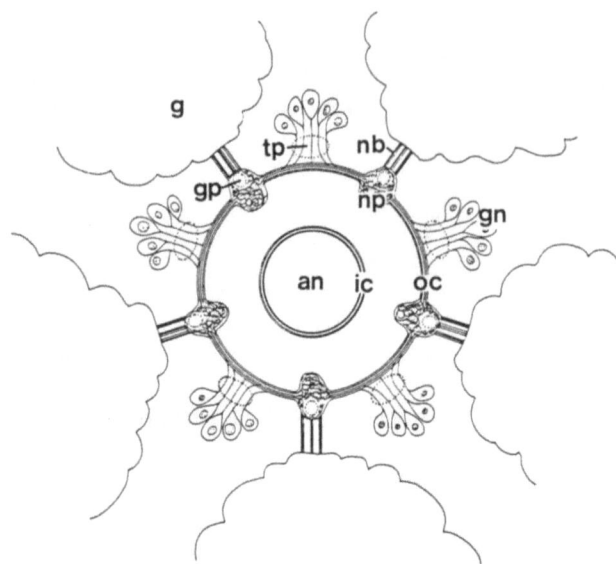

FIG. 10: A diagrammatic representation of the aboral nervous system. an, anus; g. gonad; gn, group of nerve cells branching out of the outer circumanal nerve cord; gp, gonopore; ic, inner circumanal nerve cord; nb, nerve bundles in the flattened sinus of the gonoduct wall; np, neuropil region; oc, outer circumanal nerve cord; tp, terminal pore.

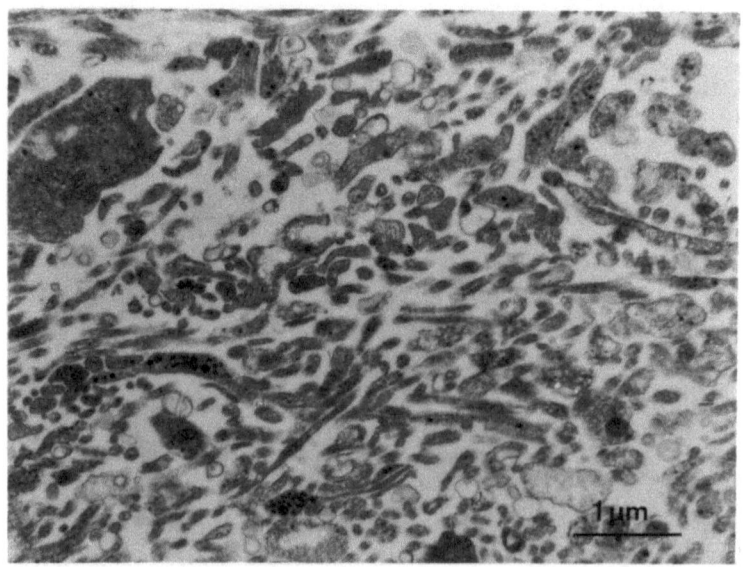

FIG. 11: Electron micrograph of a region of the neuropil (np in Fig. 10), showing interweaving nerve fibers which form varicosities containing many vesicles.

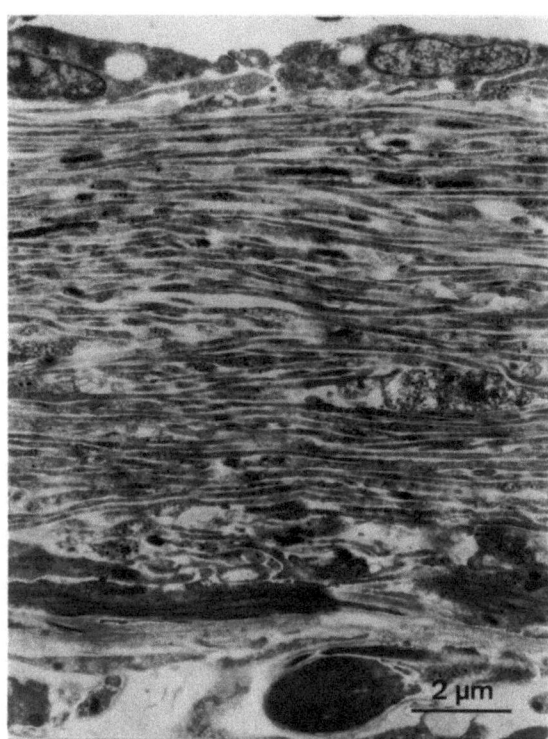

FIG. 12: Longitudinal section of the outer circumanal nerve cord.

nerve fibers arises from midway between two neuropil regions and ends near the terminal pore (tp), the apical end of the radial nerve cord. Near the terminal part of the short branch a group of neuronal cell bodies (gn) with small dense cored vesicles is present. These cell bodies are separated from the radial nerve cord by a basement membrane and give off axons toward the outer circumanal nerve cord (Fig. 13). Nerve fibers from the outer circumanal nerve cord seem to innervate the muscle layer of the gonadal wall via bundles of nerve fibers (nb) running within the inner wall of the flattened sinus in the gonoduct wall.

Okada et al. (1984) have shown that when the ANR is intact, the RCs are strikingly synchronized in phase as well as in relative magnitude (Fig. 14) although supernumerary (open circle) or lack (arrowhead) of contractions does occur on rare occasions. When a cut is made in the ANR, out of phase contractions become more frequent (B). When a gonoduct is cut across, no RC appears distal to the cut (C). If, however, the junction of the gonoduct with the ANR is left intact, the lobe belonging to this sector is capable of contracting rhythmically (D). However, as this sector is isolated from neighboring lobes by cuts made on either side of the gonoduct, synchronization with lobes in other sectors no longer appears. An obvious conclusion obtained from these experiments is that the working hypothesis put forward at the beginning of this section is fully substantiated.

3.2.2. Pharmacology

Okada and Iwata (1985) have shown that quiescent gonads can sometimes be induced to contract rhythmically when washed with fresh sea water

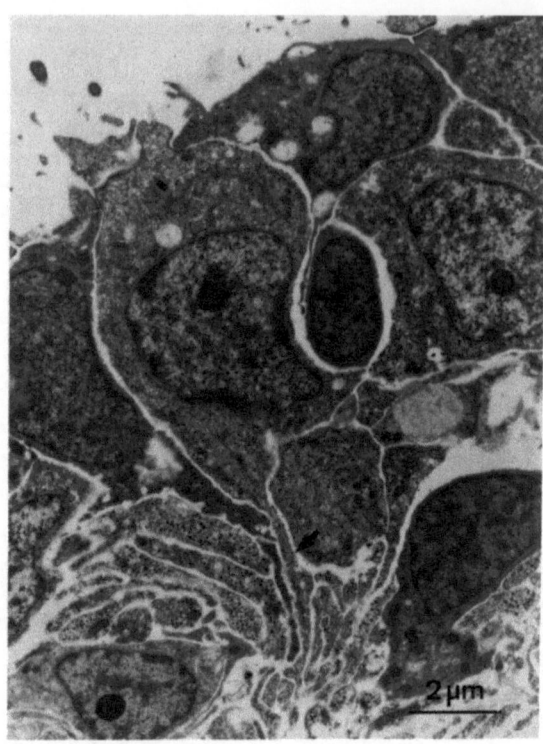

FIG. 13: Nerve cells occurring at the end of a short branch of the outer circumanal nerve cord. The nerve cell bodies contain a large nucleus, rough endoplasmic reticulum, mitochondria, Golgi apparatus, and many small dense cored vesicles. The cell gives off an axon (arrow).

(Fig. 15A) and that the rhythm is abolished by application of coelomic fluid (Fig. 15B) or by a heat-stable principle (Fig. 15C) found in extracts of various tissues (gonad, gill, muscle) of widely varied animals such as sea urchins, seastars, a sea cucumber, an oyster, a clam, and fishes.

Chloralhydrate known to anesthetize the nervous system of tube feet of sea urchins (Florey and Cahill 1980) blocked RCs but not ArCs in response to ACh. The implication of the nervous elements being involved in eliciting the RCs led us to test effects of substances known as neurotransmitters. Dopamine and serotonin were found to be ineffective at a concentration of 10^{-3} M. Among 20 amino acids tested, 1-glutamic and 1-aspartic acids are by far the most potent inhibitors of the rhythm. As shown in Fig. 16, these two amino acids are fully effective at a concentration of 10^{-5} M.

It is generally held that glutamic and aspartic acids are excitatory and γ-aminobutyric acid (GABA), inhibitory. An entirely opposite situation occurs in their effects on the RC so that application of GABA leads quiescent gonad to show RCs superposed on an ArC (Fig. 5B). Picrotoxin, a GABA antagonist, reduces or abolishes RCs at a concentration of 10^{-5} M. Excitatory action of GABA is not unusual in echinoids, however, GABA evoked contractions have been shown for tube feet (Florey et al. 1975). Also the occurrence of GABA in the central nervous system has been reported in some echinoderms (Osborne 1971). Glutamic acid is

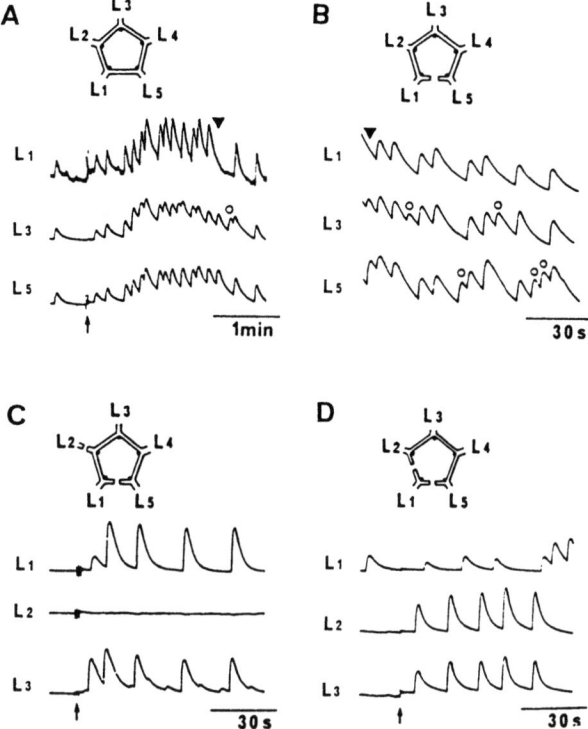

FIG. 14: Coordination of RCs in different gonadal lobes. The drawing above recordings shows schematically the aboral nerve ring (ANR). L's denote gonadal lobes and suffixes to them, the lobes from which recordings are made. A: ANR is intact. B: ANR is transected between L_1 and L_5. C: ANR is transected between L_1 and L_5, also, the gonoduct of L_2 is transected. D: ANR is transected on both sides of the sector L_1.

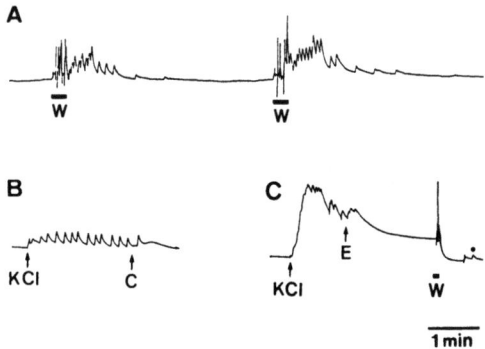

FIG. 15: RC induction and inhibition. A: The quiescent gonad is induced to contract repetitively by washing the test cavity with fresh sea water. B: The RC-showing gonad stops the rhythm when coelomic fluid is applied into the test cavity. C: The KCl-induced rhythm is abolished by application of a testes extract and resumed after washing (a dot after w) (Okada and Iwata 1985).

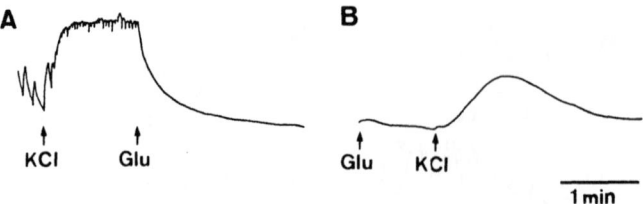

FIG. 16: RC-inhibiting effect of glutamic acid.

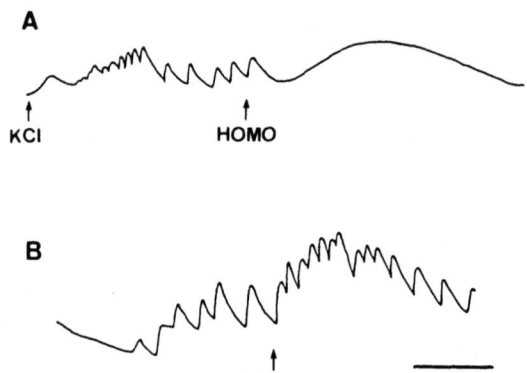

FIG. 17: Effect of glutamic acid decarboxylase (GAD) on the inhibitory action of gonadal homogenate. Homogenates were prepared distilled water and centrifuged for 1 h at 12000 g at 4°C. The supernatant was then heated at 100°C for 20 min and again centrifuged for 1 h at 30000 g at 4°C and lyophilized. The lyophilized sample of testes was dissolved in distilled water (75 mg/ml), filtered and bioassayed (A). The same sample was acidified to pH 4.5 and incubated with 70 μg/ml of 1-glutamic acid decarboxylase (GAD; Kyowa Hakko Industrial Co.) for 2 h at 37°C. After heating at 100°C for 10 min, the incubated mixture was neutralized to pH 7.8 and then bioassayed (B).

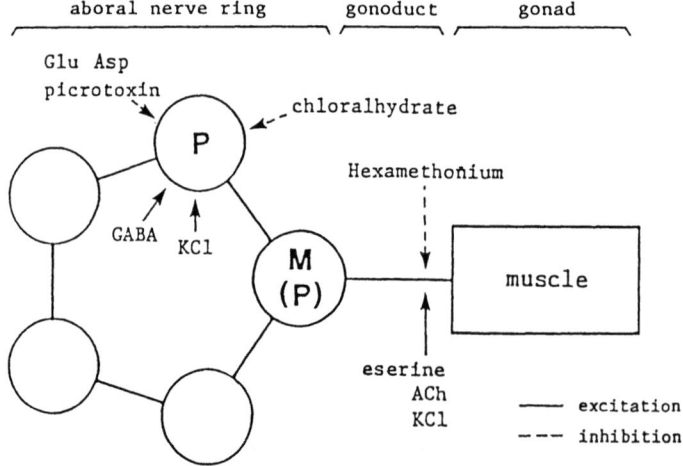

FIG. 18: Summary diagram of the spawning mechanism.

converted to GABA by glutamic acid decarboxylase (GAD). As shown in Fig. 17, gonadal homogenate (A) inhibits KCl-induced RCs, whereas GAD-treated homogenate appears even to enhance spontaneously occurring RCs (B). The enhancement is understandable if the enzyme converted inhibitory glutamic acid into facilitatory GABA.

Figure 18 shows a summary diagram to explain results so far obtained.

4. REFERENCES

Cochran LS, Engelmann F (1972) Echinoid spawning induced by a radial nerve factor. Science 178: 423-424.

Florey E, Cahill MA (1980) Cholinergic motor control of sea urchin tube feet: Evidence for chemical transmission without synapses. J Exp Biol 88: 281-292.

Florey E, Cahill MA, Rathmayer M (1975) Excitatory actions of GABA and acetylcholine in sea urchin tube feet. Comp Biochem Physiol 51C: 5-12.

Harvey EB (1952) Electrical method of "sexing" Arbacia, and obtaining small quantities of eggs. Biol Bull 102: 284.

Iwata KS (1950) A method of determining the sex of sea urchins and of obtaining eggs by electrical stimulation. Annot Zool Jpn 23: 39-42.

Iwata KS (1976) Gonadal pressure in the sea urchin. Zool Mag (Tokyo) 85: 270-272 (In Japanese).

Iwata KS, Fukase H (1964) Artificial spawning in sea urchins by acetylcholine. Biol J Okayama Univ 10: 51-56.

Kerkut GA (1954) The mechanics of co-ordination of the starfish tube feet. Behaviour 6: 206-232.

Okada Y, Iwata KS (1985) A substance inhibiting rhythmic contraction of the gonad in the shedding sea urchin. Zool Sci 2: 805-808.

Okada Y, Iwata KS, Yanagihara M (1984) Synchronized rhythmic contractions among five gonadal lobes in the shedding sea urchins: coordinative function of the aboral nerve ring. Biol Bull 166: 228-236.

Osborne NN (1971) Occurrence of GABA and taurine in the nervous systems of the dogfish and some invertebrates. Comp Gen Pharmacol 2: 433-438.

Pentreath VW, Cobb JLS (1982) Echinodermata. In: Shelton GAB (ed) Electrical conduction and behaviour in "simple" invertebrates. Clarendon Press, Oxford, p 440-472.

Peters BH, Shelton GAB (1981) Electrical activity during a "simple" behaviour: Spine-pointing in a sea urchin. Comp Biochem Physiol 70A: 329-403.

Prosser CL, Mackie GO (1980) Contractions of holothurian muscles. J Comp Physiol 136: 103-112.

Smith JE (1945) The role of the nervous system in some activities of starfish. Biol Rev 20: 29-43.

Sugi H, Suzuki S, Tsuchiya T, Gomi S, Fujieda N (1982) Physiological and ultrastructural studies on the longitudinal retractor muscle of a sea cucumber Stichopus japonicus. I. Factors influencing the mechanical response. J Exp Biol 97: 101-111.

Takahashi N (1986) The spawning of the sea urchin, Strongylocentrotus intermedius, by γ-aminobutyric acid. Bull Jpn Soc Sci Fish (in press).

Uexkull J von (1897) Über Reflexe bei den Seeigeln. Z Biol 34: 289-318.

NEURAL CONTROL MECHANISMS IN BIOLUMINESCENCE

MICHEL ANCTIL

Département de sciences biologiques et

Centre de recherche en sciences neurologiques

Université de Montréal

Montréal, Québec

H3C 3J7, Canada

"And the nervous system commanded: Let there be light."

Neurogenesis, Verse hν.

ABSTRACT

Light emissions in invertebrates are expressed through many activities in a variety of epithelial and other structures. This contribution reviews our current understanding of the means by which invertebrate nervous systems shape these behaviours. The discussion focusses on the firefly lantern, the bioluminescent scales of polynoid worms, the luminescent epithelia of chaetopterid tube-worms, and some coelenterate systems. The range of nerve circuitries and neural control mechanisms involved in luminescence regulation is a fair reflection of the increasing complexity and diversity of nervous system functions from coelenterates to insects. Future challenges will be to specify the cellular basis of the circuits associated with luminescent behaviours, and the role of neural mechanisms in the several luminescent systems which use epithelial conduction.

1. INTRODUCTION

As a topic of comparative neurobiology, bioluminescence has been denied the attention it enjoyed from biochemists and general physiologists in the past (see Hogben 1926; Harvey 1952; Prosser and Brown 1961). During the last two decades, physiological control mechanisms in luminescent systems have attracted more interest (see Herring 1978). It is thus appropriate at this time to attempt a critical examination of the contribution of the invertebrate nervous systems to the expression of luminescent patterns.

Light emission is a relatively widespread effector function among invertebrates, especially those inhabiting marine environments (Morin 1983; Young 1983). It is, as the final expression of instructions transmitted by the nervous system of these organisms, a potentially more convenient and effective methodological window than muscle or glands to peep at neural control mechanisms whenever electrophysiological approaches are lacking, inappropriate or impossible to achieve. Indeed, while invertebrate bioluminescence in nature assumes numerous behavioural functions, to the investigator it serves as a public advertisement of the workings of some of the invertebrate nerve circuitries. Although it is often easier to record luminescence accurately than other effector activities, the luminescent effectors themselves display a wide range of organizational complexity. This structural diversity is well matched by the range of innervation circuitries associated with these effectors (see Case and Strause 1978; Anctil 1979b).

In the following account, we shall review (1) patterns of neural circuitry and activity associated with the luminescent output, (2) modes of neuro-photocyte transmission, and (3) tentative identification of putative neurotransmitter mechanisms involved in these activities. Because only species from a few invertebrate phyla - Cnidaria, Ctenophora, Annelida and Arthropoda - were investigated in sufficient detail to secure glimpses of the workings of neurally controlled luminescence, this chapter will deal essentially with them. Mentions of less well understood luminescent systems from other phyla will occasionally surface for the sake of comparison (Table 1), and further details concerning these systems may also be found in accounts by Case and Strause (1978) and Anctil (1979b).

2. COELENTERATES

The phyla Cnidaria and Ctenophora are included under this heading, although the closeness of the phylogenetic relationship between them is the subject of much debate. Morin (1974) provided a general review of bioluminescence in these two phyla. Two themes have permeated investigations of the coordinating mechanisms of coelenterate bioluminescence. First, the opportunity to analyse the properties of the most primitive nervous systems known to exist in multicellular organisms, using bioluminescence as a convenient probe of neural activity. This was the approach used by Parker (1920a, b), a pioneer in the study of primitive nervous systems. And secondly, the realization that luminescence in some coelenterates is at least partly mediated by nerve-free, epithelial propagation of excitability.

2.1. Cnidaria

2.1.1. Hydrozoa

Among the cnidarians, only studies on Hydrozoa and Anthozoa have turned up physiologically meaningful information. The hydromedusan genera Aequorea, Halistaura (Davenport and Nicol 1955) and Euphysa (Mackie and Mackie 1963) have been investigated to some extent. Luminescence is localized at the bell margin in the first two genera and over the bell surface in the last. In all cases, luminescence is intracellular, and the photocytes are endodermal cells. Flash activity is elicited by mechanical or electrical stimulation. The luminescent response appears to be local in Aequorea (Davenport and Nicol 1955), although this has been questioned (Herring 1978). In Ephysa it spreads all over the bell, possibly by epithelial conduction (Mackie and Mackie 1963, 1967). The possible role of neural pathways in initiating luminescence excitation is still unexplored in this group despite the proved feasibility of intracellular electrophysiology (Satterlie 1985).

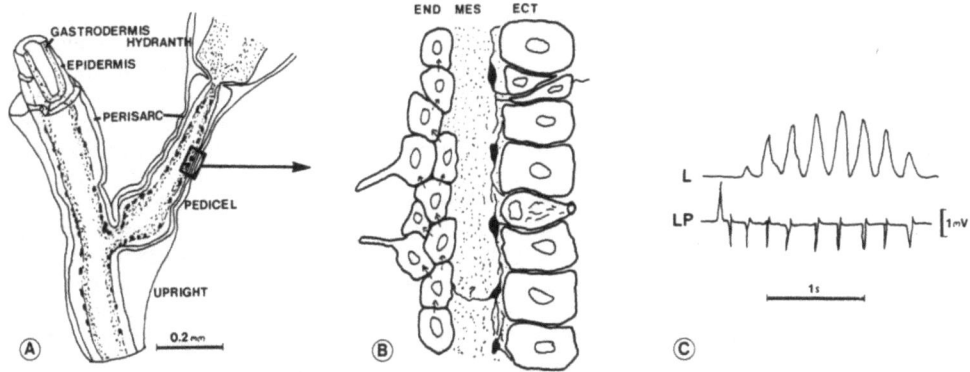

FIG. 1: The luminescent system of Obelia geniculata. (A) Organization of a portion of the colony in which the distribution of photocytes is indicated by large dots. (B) Reconstruction of pedicel tissue in the region delineated by rectangle in A. Photocytes are implicitly present in the endoderm (END, especially the cells with a process) adjacent to the mesoglea (MES) where epithelial conduction of excitation is presumably occurring (arrows). The nerve net underlying the ectoderm (ECT) is assumed here to give off occasional projections through the mesoglea toward the endoderm (?). (C) Luminescence (L) and luminescent potentials (LP) simultaneously elicited by a train of electrical stimuli in Obelia. The first upward deflection in LP is a contraction potential. A and C redrawn after Morin and Cooke (1971a, b).

In the colonial hydroid Obelia and other calyptoblastoid hydroids, luminescent flashes originating from discrete locations in the stolons and pedicels (Morin and Reynolds 1974; Fig. 1A, B) are elicited by mechanical, electrical and chemical (KCl) stimulation (Morin and Cooke 1971a). Extracellularly recorded luminescent potentials (LPs) of relatively large amplitude and duration are also elicited by electrical stimulation, precede each luminescent flash in a burst following a single stimulus pulse (Fig. 1C), and can be conducted through the colony in a non-decremental fashion at a rate of 22 cm/s at 12°C (Morin and Cooke 1971a, b). The suggestion by Morin and Cooke (1971a) that the LPs are epithelial in nature, and therefore that conduction of luminescence excitation is epithelial, is supported by recent electrophysiological recordings from isolated photocytes and clumps of epithelial cells (Brehm et al. 1985). Brehm et al. (1985) used whole-cell clamp techniques to demonstrate that the photocytes in isolation fail to luminesce in response to stimulation and lack calcium channels, whereas the latter are present in support cells which, when present in clumps including photocytes, make luminescent responses possible. Whether the source of the spreading excitation is neuronal or not remains to be investigated. A neuronal involvement is probable because flash-inducing tactile excitation on the ectoderm must somehow be transmitted through the mesoglea to reach the endoderm (Fig. 1B).

As in Obelia, several consecutive luminescent flashes in the siphonophore Hippopodius can be elicited by single electrical shocks, and this response is subjected to facilitation, summation and fatigue (Bassot et al. 1978). "Frenzy" (sensu Buck 1973) responses such as bursts of flashes of increasing brightness and flicker frequency, are elicited after a strong stimulation regime. Contrary to other hydrozoans, the ectoderm, not the endoderm, appears to be the source of the luminescence (Bassot et al. 1978). The nerve-free epithelial cells of the medusa-like nectophores, in which luminescence originates, are interconnected by gap junctions. They produce potentials which propagate through the epithelium and trigger flash activity on a one-to-one basis following an initial priming effect (Bassot et al. 1978). Hence the evidence is compelling in favour of epithelial conduction of flashing. This seems to be part of an overall protective strategy since other activities such as blanching (an unusual opacity response of the nectophores), muscular involution and exocrine secretion are similarly spread by epithelial conduction upon stimulation (Bassot et al. 1978). Again, the Hippopodius and other hydrozoan luminescent systems may well rely for coordination of these activities on an interplay of neural patterning and epithelial conduction. However, no hard evidence of neural involvement has yet been produced.

2.1.2 Anthozoa

Contrary to hydrozoans, epithelial conduction and gap junctions have not been convincingly demonstrated in anthozoans despite considerable efforts in that direction (Anderson 1980; Mackie et al. 1985). Luminescence is restricted to a few colonial forms among whom the Pennatulacea received the greater share of attention. In the latter group, the luminescence of the sea pansy Renilla (Fig. 2A) has been investigated for over six decades as a means to "illuminate" the properties of the anthozoan nerve net (Parker 1920b; Nicol 1955a, b; Buck 1973; Anderson and Case 1975). In the process, it became increasingly clear that luminescence excitation and propagation through the colony were controlled by a colonial nerve net system.

Clumps of endodermal photocytes are restricted in distribution to the base and crown of the autozooids (feeding polyps) and to specific chambers of the water-pumping siphonozooids as revealed principally by fluorescence microscopic observations (Morin 1974; Fig. 2B-D). Luminescence from these sources arises as a wave advancing in all directions from the site of mechanical or electrical stimulation in the colony (Nicol 1955a; Buck 1973). Although seasonal differences may exist in luminescence excitability of the two types of polyps (but see Anderson and Case 1975), the siphonozooids appear to be the major contributors to this wave-like response (Buck 1973). The autozooid crowns are capable of local luminescent responses of longer duration (glows), but the control mechanisms of the photocyte clumps associated with each of the eight tentacle bases of these polyps are unknown.

Once the wave has been "primed" to appear after the first two or three stimulating pulses, it spreads non-decrementally to the edge of the colony, thus reflecting a non-facilitated through-conduction process (Parker 1920b; Nicol 1955a, b). With exhaustive stimulation it is possible to induce frenzies reminiscent of those recorded in Hippopodius. In Renilla luminescent waves of increasing brightness and frequency reverberate from the edge of the colony back in the opposite direction (Buck 1973). Another impressive display in Renilla is the cancellation of colliding luminescent waves travelling from distinct sources, a phenomenon attributed to the refractory period of the underlying excitable nerve net (Moore 1926). Similar observations were made with the European pennatulacean Veretillum (Bilbaut 1975a, b).

Anderson and Case (1975) were able to record nerve net pulses (NN)

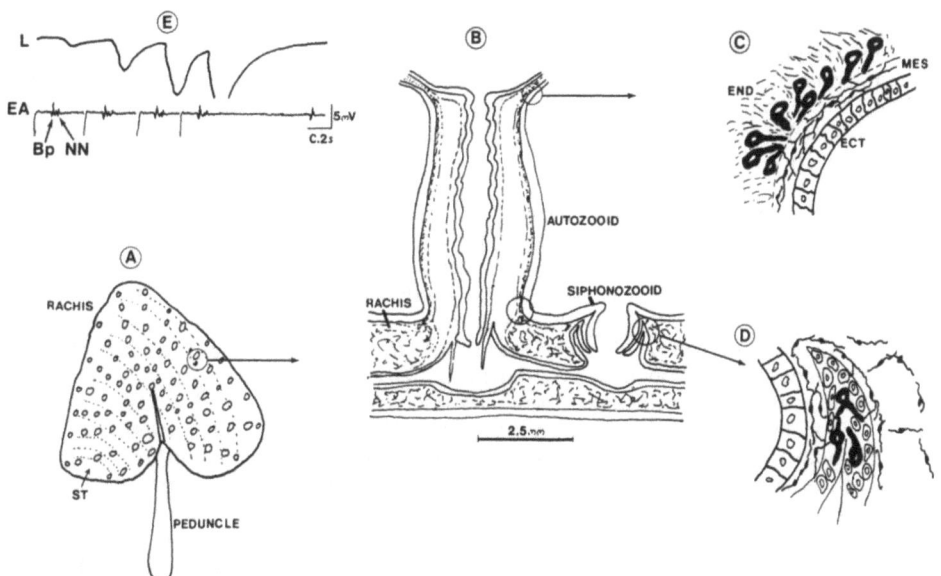

FIG. 2: The luminescent system of Renilla kollikeri. (A) The colony comprises a rachis whose upper surface is covered with polyps (autozooids and siphonozooids) and a muscular peduncle which anchors the colony in sandy bottoms. Mechanical or electrical stimulation of the rachis (ST) evokes through-conducted wavefronts of light (dotted lines). (B) Cross-section through a small portion of colony, showing rachidial and polyp tissues as well as the regions where photocyte clumps are located (circled). (C) and (D) are close-up views of luminescent areas in autozooid crown and siphonozooid, respectively. Note the endodermal (END) location of photocytes (large black cells) and the neighbouring neurones of the mesogleal (MES) nerve net. ECT, ectoderm. (E) Luminescence (L) and electrical activity (EA) recorded simultaneously from the exhalent siphonozooid following repetitive electrical stimulation (large downward deflections in lower trace). Note the larger biphasic pulse (Bp), followed by the smaller nerve net (NN) pulse associated with the facilitated luminescence recorded. E redrawn from Anderson and Case (1975).

extracellularly after electrical stimulation of Renilla autozooids. The NN pulses propagated through the colony at speeds similar to those of the luminescent waves (6-7 cm/s at 16°C), and were invariably accompanied by luminescent episodes after a delay of 20-50 ms (Fig. 2E). The coupling between the two events is a facilitated process, quite independent of NN pulse parameters. Anderson and Case also established that the abovementioned luminescent frenzy is associated with post-stimulatory hyperactivity of the through-conducting nerve net. Thus there is convincing evidence that luminescence in Renilla is an effector activity coordinated by the colonial nerve net and, in the autozooids, regulated also by a presumably semi-autonomous zooid nerve net (ZNN: Anderson and Case 1975). Other activities of Renilla, such as autozooid withdrawal and rachidial contraction, are also controlled by the colonial nerve net in an integrated manner. These conclusions were found to be widely applicable to other luminescent pennatulaceans, some of which displaying faster conduction velocities of NN pulses and luminescent waves (70-140 cm/s) than those of Renilla (Satterlie et al. 1976, 1980). The latter studies also provided glimpses of a possible neuroanatomical substrate for the through-conducting nerve net, a plexus of nerve tracts associated with amoebocytes and spanning the entire mesogleal layer of rachis and autozooids (Satterlie et al. 1980; C.J.P. Grimmelikhuijzen and M. Anctil, in preparation).

The neuropharmacological basis of luminescence control in cnidarians has eluded resolution despite its potential implications for an understanding of the evolutionary origin of neurotransmitters in nervous systems. A cautionary optimism can be contemplated regarding this problem because the anthozoan Renilla has recently proved to be a promising model for such investigations. On the basis of a pharmacological dissection of the luminescent responses of Renilla köllikeri, the involvement of excitatory catecholamine-like mechanisms was proposed (Anctil et al. 1982). Adrenaline elicits intermittent flashes at low concentrations, thus suggesting that the potentially significant adrenaline effect, flashing, is mediated through enhanced excitability in the nerve net. The latter interpretation is supported by evidence that isolated Renilla photocytes respond only to high concentrations (0.1-10 mM) of adrenaline in the form of glows and superimposed flashes (G. Germain and M. Anctil, unpublished). Renilla also possesses high affinity uptake processes for noradrenaline and adrenaline (Anctil et al. 1984), although their localization in nerve cells remains to be unequivocally demonstrated. Dopamine, noradrenaline and adrenaline were radioenzymatically detected in significant amounts in tissues of Renilla köllikeri (DeWaele et al. 1987) and discrete populations of ectodermal and endodermal dopamine- and serotonin-immunoreactive neurones were recently detected in the same species (D. Umbriaco and M. Anctil, unpublished). Furthermore, the adrenaline response is selectively sensitive to propranolol and can be mimicked by isoproterenol, thus suggesting the mediation of a beta-adrenergic type of receptor mechanism (Anctil et al. 1982). In sum, although evidence is mounting in favour of an indirect catecholamine-like regulation of the photogenic tissues, an unequivocal demonstration is still lacking. Hence the neurotransmitter system involved in direct neuro-photocyte transmission remains to be identified. It is interesting in this regard that a recent immunocytochemical mapping of the overwhelmingly prominent neurones reactive to an Arg-Phe-amide antibody in Renilla, including many in the mesogleal nerve net, failed to demonstrate them in the vicinity of clumps of photocytes (C.J.P. Grimmelikhuijzen and M. Anctil, in preparation). In addition, the peptides Arg-Phe-amide and Phe-Met-Arg-Phe-amide had no effect on Renilla luminescence (Anctil 1987). Hence it is apparent that a select, relatively minor neurotransmitter pathway must be directly associated with the photocytes.

2.2. Ctenophora

Luminescence in this phylum has its intracellular source in the endoderm (Freeman and Reynolds 1973; Labas and Mashanskii 1976; Anctil 1985a), except for the extracellular luminescent secretion of Eurhamphea (Hamner et al. 1975). As in anthozoans, the luminescent displays appear to be coordinated by a nerve net. However, much less is known about the neurobiology of luminescence in this group than in Cnidaria.

Single flashes originating in the endodermal wall of the meridional canals (Fig. 3B, C) and lasting approximately 300 ms can be elicited by mechanical or electrical stimulation in the lobate Mnemiopsis (Chang 1954; Fig. 3A). Decay of the flash response is temperature-dependent (Chang 1954), whereas flash intensity is sensitive to pressure (Chang and Johnson 1959). Repetitive stimulation regimes revealed the familiar facilitation,

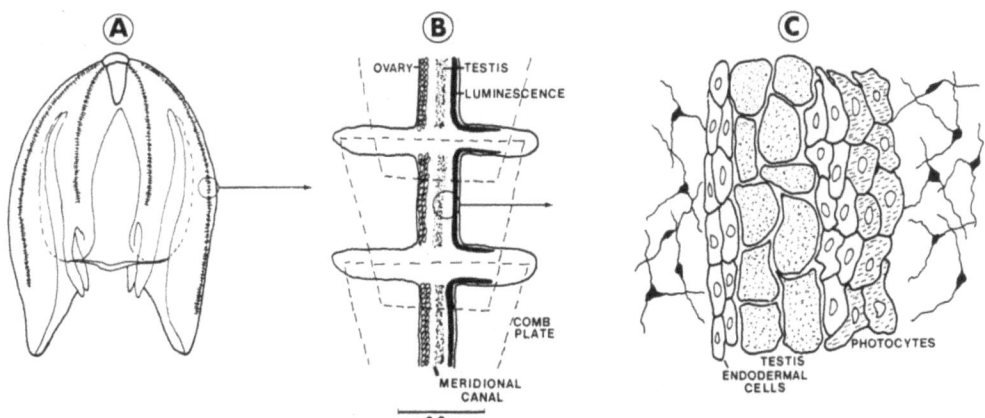

FIG. 3: The luminescent system of Mnemiopsis leidyi. (A) View of the animal with four of its eight ciliary comb plate rows, portion of one of which (circled) is enlarged in B. (B) The endodermal meridional canal, in whose wall the photocytes are located, is situated underneath the ectodermal comb plate row (profiled with dashed lines). Note lateral extensions of meridional canal where the photocyte lining is interrupted. (C) Reconstituted cellular organization of a tangential section through the meridional canal wall (circled in B). Note that photocytes (hatched) are associated with other endodermal cells (clear) and these are all interconnected by gap and septate junctions. Note neurones of the mesogleal nerve net which are in the vicinity of photocytes from the outside of the wall.

summation and fatigue of other coelenterates. The conduction velocity of the flash response travelling along the meridional canals was estimated at about 13 cm/s (Chang 1954), thus well within the velocity range of luminescence conduction in pennatulaceans where nerve nets have been implicated. However, epithelial conduction of flashes along the meridional canals is also a strong contender because the photocytes of Mnemiopsis are interconnected by gap junctions (Anctil 1985a). Assuming such is the case, neurally mediated conduction of flashing may still be involved in propagating excitation between successive comb plate rows where the continuity of the luminescent epithelium is broken (Freeman and Reynolds 1973; Fig. 3B). Despite unconfirmed claims of extracellularly recorded action potentials accompanying flash activity in Mnemiopsis (Chang 1954), the electrophysiology of flashing activity remains open to investigation.

In the ctenophore Bolinopsis, as in Renilla, luminescence is one of several coordinated activities aimed at producing a stereotyped defensive reaction (Labas 1977a). Electrical or mechanical stimulation invariably and concurrently caused flashing, ciliary arrest and muscle contraction. Labas (1977a) further showed that exposure of Bolinopsis to single, brief pulses (1 s) of artificial light induced flashing and ciliary arrest, pulses of longer duration being increasingly less effective. Increasing calcium and decreasing magnesium concentrations suppressed ciliary beating while exciting luminescence and muscle contraction (Labas 1977b). Reversing the above-mentioned order of concentration reversed the sign of the responses. Hence the ciliary apparatus receives an inhibitory input, and the luminescent and contractile effectors an excitatory one, apparently under the command of a neural coordinating system potentially influenced by a sensory input (photoreceptive?). The most likely pathways to account for these observations are the ectodermal and mesogleal nerve nets of these ctenophores (Hernandez-Nicaise 1973a, b), provided they are interconnected. In Pleurobrachia, electrophysiologically identified conducting systems of neural origin were suspected to mediate inhibitory and excitatory pathways to ciliary and contractile effectors (Satterlie 1978).

The neurochemical basis of neuro-photocyte transmission is unknown in ctenophores. A few nerve terminals containing synaptic vesicles are present in the vicinity of photocytes in Mnemiopsis (Anctil 1985a). In a recent pharmacological analysis of flashing in Mnemiopsis, a beta-like adrenergic system was detected, with adrenaline as the most potent excitatory catecholamine (Anctil 1985b). Furthermore, a curare-sensitive excitatory cholinergic system also appeared to be involved in this luminescent system. However, both neurochemical systems are unlikely to operate directly on photocytes because adrenergic and cholinergic agonists induced flashing, not the expected glow after sustained exposure to these drugs (Anctil 1985b). Further evidence that the cholinergic system influences light emission indirectly is that acetylcholine-induced flashing was blocked by an adrenergic antogonist whereas adrenaline-induced flashing was unaffected by curare (Anctil 1985b). This analysis is a promising starting point in the identification of transmitter-specific conduction pathways involved in the coordination of luminescent and other effectors in coelenterates. However, the neurotransmitter system(s) acting on the photocytes themselves remains to be explored by pharmacological and neurochemical means.

3. POLYCHAETE WORMS

Although there are a few bioluminescent oligochaete worms, so little is known of their physiology and even less of their neurobiology as to make any attempt to review them futile. In contrast, luminescence is widespread among the polychaetes, and its expression betrays the active involvement of the nervous system. Only the two most thoroughly investigated groups will be discussed. Although both are epithelial luminescent systems, one involves intracellular light emission in tightly coupled photocytes (polynoid worms) while the other involves extracellular luminescence from uncoupled gland cells (chaetopterid worms).

3.1. Polynoid worms

The intracellular luminescence of polynoid worms (Fig. 4A) is the best known annelid luminescent system from the photometric, electrophysiological, biochemical and subcellular viewpoints. A photoprotein, polynoidin, present in the membranes of the photocyte granules, photosomes, is responsible for light emission when reacting with a peroxide and ferrous ions (Nicolas et al. 1981, 1982). Repetitive flashes of short duration (approx. 100 ms) are elicited from deciduous scales (elytra) upon mechanical stimulation of the body or electrical stimulation of the ventral nerve cord (Nicol 1954; 1957a, b). Flash intensity can be facilitated and amenable to summation and fatigue upon repetitive stimulation. This facilitation is achieved by recruitment of additional luminescent sources. Autotomy or electrical stimulation of carefully excised elytra (Fig. 4B) can induce long trains of rhythmic flashes of remarkable regularity (Bilbaut and Bassot 1977; Fig. 5). This automomous activity goes on unimpeded in the absence of the elytral ganglion, a local neuropile which is connected to the second segmental nerve through a

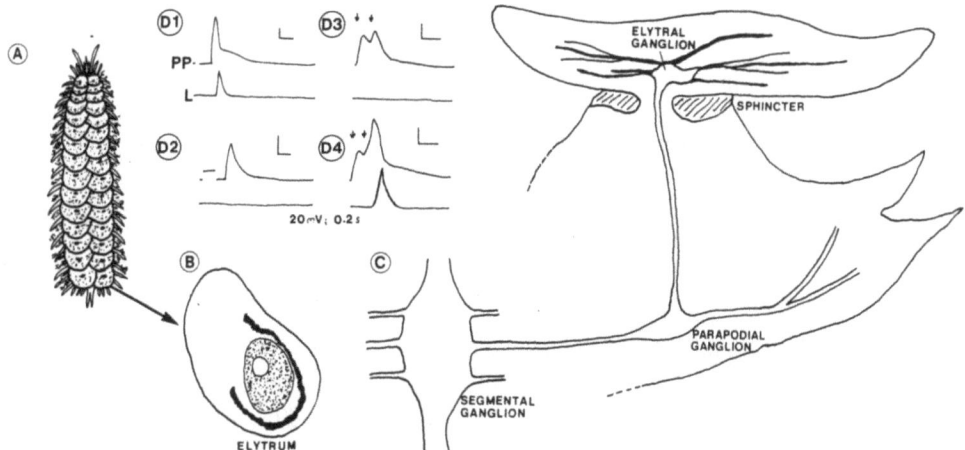

FIG. 4: The luminescent system of polynoid worms. (A) <u>Harmothoe imbricata</u> with its overlapped dorsal scales (elytra). (B) Elytrum of <u>H. lunulata</u> viewed from the upper surface. Note the heavily pigmented band partly encircling the photogenic zone (dotted). The white circle within the photogenic zone represents the insertion of the elytral stalk (elytrophore). (C) Representation of the neural pathway to the elytra. The second segmental nerve of the ganglionic chain ends in a parapodial ganglion, one root of which gives off the parapodial nerve entering the elytrophore to form the elytral ganglion. From the latter irradiate the elytral nerves. Note position of sphincter muscle responsible for the autotomy response. (D) Simultaneous recordings of photocyte potential (PP) and luminescence (L) in <u>Hesperonoe complanata</u>. In D1 a calcium spike and a single flash are elicited by electrical stimulation of the second segmental nerve. IN D2-3 electrical stimulation of this nerve (arrows) causes the appearance of postsynaptic potentials (PSPs) in photocytes (D2) which summate (D3) without concomittent flashing activity. In D4 summation of PSPs leads to a calcium spike and flashing. A and B redrawn after Nicol (1954) and D after Herrera (1979).

parapodial ganglion (Fig. 4C). Spontaneous muscular activity has been reported to precede the emergence of flash activity (Pavans de Ceccaty et al. 1972).

A study by Nicolas (1977) of the regeneration of elytra following autotomy showed that these organs derive from strictly ectodermal outgrowths whose middle compartment becomes filled with nerve tissue. The photocytes of <u>Acholoe</u> are localized in the lower epithelium of the elytra, in a photogenic area concentric to the site of insertion of the elytral stalk. They are cuboidal cells with long fibrous processes (pillars) extending far into the middle compartment of the elytra (Pavans de Ceccaty et al. 1977; Fig. 6). Some of these pillars are contacted by processes or terminals from a neuronal plexus which sprawls the middle compartment (Pavans de Ceccaty et al. 1977). Although all photocytes appear to be interconnected by gap junctions, an image intensification analysis of the spread of luminescent sources over the elytra suggests that coupling is limited by sectorial "barriers" whose nature remains to be determined (Bassot and Bilbaut 1977). Yet, coupling of excitation among photocytes

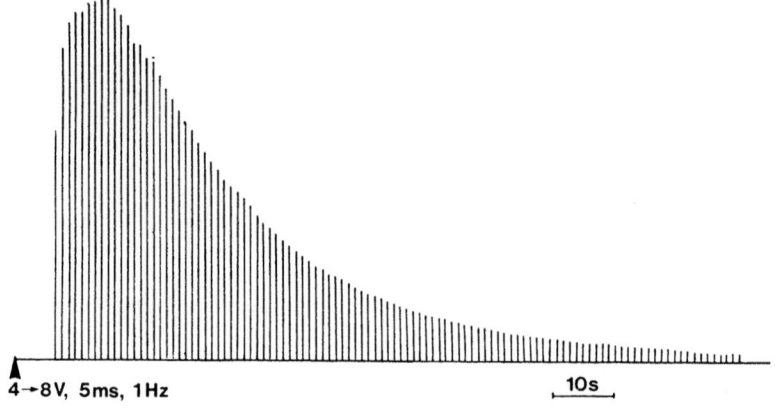

FIG. 5: Flashing activity of an isolated elytrum of H. lunulata in response to threshold, repetitive field stimulation starting at arrowhead. Note the striking regularity of the emissions, demonstrating facilitation and fatigue.

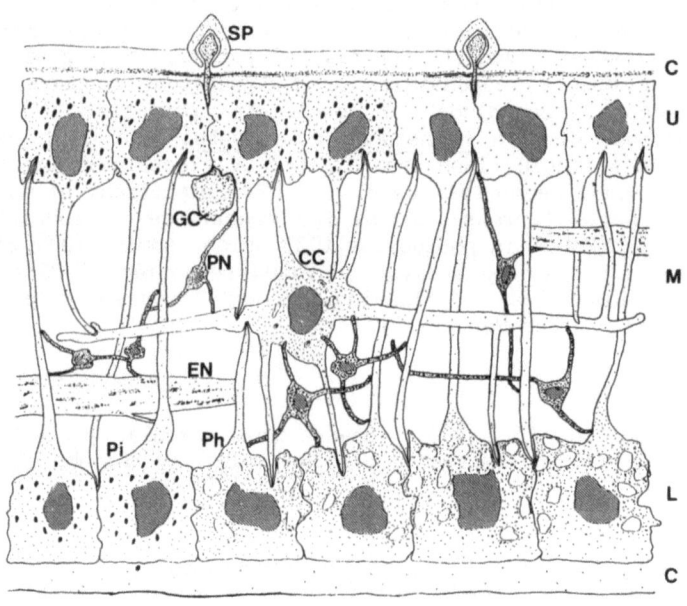

FIG. 6: Schematic reconstruction of the cellular organization of an elytrum from H. imbricata. Labelling (small dots) represents the distribution of immunohistochemical and radio-autographic reactions for serotonin. Note the epithelial cell processes (pillars) which extend across the middle compartment (M) and anchor to cells of the opposite epithelium. Drawing is not to scale. C, cuticle; CC, clear cell; EN, elytral nerve; GC, ganglion cell; L, lower epithelium; Ph, photocyte; Pi, pigmented cell; PN, plexus neuron; SP, sensory papilla; U, upper epithelium. After Miron et al. (1987).

is now well documented in Acholoe (Bilbaut 1980a). Electrophysiological investigations of the luminescent epithelium of Acholoe (Bilbaut 1978, 1980a, b) and Hesperonoe (Herrera 1979) have gone a long way to unravel control mechanisms in this effector system. Overshooting, calcium-dependent action potentials recorded in the photocytes precede light emission. The spiking activity of the photocytes is complex in that a sodium-dependent action potential can also be evoked in the absence of calcium, but no luminescence is temporally linked to such spikes. Herrera (1979) demonstrated neuro-photocyte transmission by electrically stimulating an exposed segmental nerve and recording postsynaptic potentials leading to a calcium spike and luminescence in photocytes from an elytra of the corresponding segment (Fig. 4D). Hence it is reasonably certain that at least some of the photocytes are directly innervated, excitation being propagated through fields of coupled photocytes, and excitation-luminescence coupling being most likely achieved by calcium entry into photocytes.

There is a growing body of neuropharmacological evidence for the involvement of classical neurotransmitter systems in polynoid luminescence. Nicolas et al. (1978) presented some evidence pointing to a muscarinic cholinergic mechanism as part of the excitatory pathway of elytral luminescence in Harmothoe lunulata. Pavans de Ceccaty et al. (1977) reported a strong acetylcholinesterase histochemical reaction in the elytral ganglion and its arborescence of elytral nerves. A recent pharmacological analysis in another species (H. imbricata) indicates that monoamines at high concentrations (0.1-10 mM) elicit luminescence (M. Anctil and J.-M Bassot, unpublished). The elytral luminescent system was particularly activated by serotonin, a response blocked by the serotonin antagonist methysergide. This response was calcium-dependent and unaffected in the presence of tetrodotoxin or absence of external sodium, thus suggesting that the monoamine acted directly on photocytes or on other epithelial cells directly coupled to photocytes.

Although the high monoamine concentrations necessary to activate luminescence cast doubt on the physiological relevance of monoamines as putative neurotransmitters for luminescence, there is coincident immuno-histochemical and radioautographic evidence that part of the neuronal population and all the epithelial cells in the elytra of H. imbricata accumulate biogenic amines (Miron et al. 1987; Fig. 6). Numerous plexus neurones heavily labelled with tritiated serotonin were seen in the vicinity of as heavily labelled photocytes. The low sensitivity of these elytra to monoamines may be due to the potential diffusion barrier erected by the tightly coupled epithelium of these organs. An alternative explanation is that these monoamines (especially serotonin) may act intracellularly as suggested by histochemical evidence of their presence and transport within photocytes (Miron et al. 1987). Assuming the presence of intracellular monoaminergic receptors, the apparent relative insensitivity of this preparation to monoamines would be expected. It is interesting in this regard to note that the tritiated serotonin accumulated inside photocytes is distributed mostly in compartments other than the photosomes, the light-emitting granules (Miron et al. 1987; Fig. 6). Whatever the outcome of further analyses of this luminescent system, it is already apparent that these elytra are neuroectodermal organs in which monoamines must play important roles.

3.2. Chaetopterid worms

In tube-worms of the genus Chaetopterus (Fig. 7A), luminescence is also epithelial and likewise involves a photoprotein, this one activated by a peroxide and ferrous ions (Shimomura and Johnson, 1966). However, the divergences between this and the polynoid system are more striking than their similarities as will become clear in the following account.

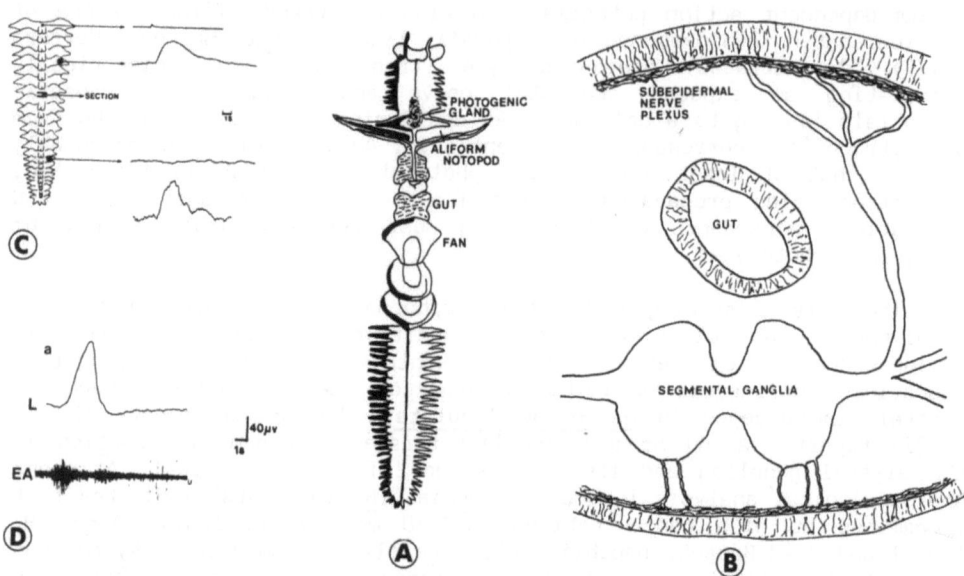

FIG. 7: The chaetopterid luminescent system. (A) Dorsal view of the worm, with the light-producing regions shown in black on the left side. Note also the photogenic gland (stippled) spanning segments 10-12. (B) Representation of the neural pathway to luminescent epithelia. From the ventral ganglionic chain the first segmental nerve courses dorsally and some of its branches merge into the subepidermal plexus. (C) Effect of sectioning the ventral nerve cord on propagation of luminescence. Upper trace, electrical stimulation; all other traces, recordings of light emission. Note failure of luminescence to spread posterior to section, compared with control response before section (lower trace). (D) Recording of spontaneous impulse discharges in the nerve cord (EA) preceding a spontaneous luminescent event (L) in the same posterior segment. A adapted from Nicol (1952a), C and D after Martin and Anctil (1984b).

Luminescence is extracellular in Chaetopterus and originates in epithelial patches of all the dorsal segmental appendages (Nicol 1952a). The luminescent mucus is expelled from one of two types of resident gland cells, known as the eosinophil (Nicol 1952a) or orthochromatic cells (Anctil 1979a) according to their staining properties. Glows of variable amplitude and duration are elicited by mechanical or electrical stimulation of the worm (Nicol 1952b, c). Facilitation, summation and fatigue of luminescence conducted between consecutive body segments were demonstrated in Nicol's studies. These were also properties of luminescent responses of isolated notopods, facilitation being the result of recruitment of active gland cells (Anctil 1981).

In the absence of evidence of epithelial conduction, the luminescent epithelia appear to be controlled by the nervous system. The general organization of the latter was described by Martin and Anctil (1984a). The key role of the cerebral ganglia and ventral nerve cord in the coordination of luminescence propagation was examined by Martin and Anctil (1984b). The latter showed that spike volleys recorded in the nerve cord invariably preceded luminescent glows (Fig. 7B, D) and were conducted along the body at the same velocity (5-8 cm/s) as luminescence. The conduction speed here is almost identical with that of Renilla luminescence, presumably because synaptic nets of similar properties are involved in both cnidarians and polychaetes. Even the small spontaneous puffs of non-conducted luminescent secretions were clearly associated with nerve cord activity. Both nerve cord volleys and luminescence spread more readily in the antero-posterior direction of the body axis than in the reverse direction (Martin and Anctil 1984b), as did luminescence in polynoids (Nicol 1954). Although there is an extensive subepidermal nerve plexus throughout the body of Chaetopterus (Anctil 1979a; Martin and Anctil 1984a), it apparently cannot serve as a pathway for intersegmental luminescence propagation as shown by the failure of luminescence conduction after sectioning the ventral nerve cord (Martin and Anctil 1984b; Fig. 7B, C). It is noteworthy that the aliform notopods of the 12th segment, which contribute the most intense luminescent display and are the most richly innervated segmental appendages of the worm, are also the only ones that could be field stimulated to elicit a decremental propagation of luminescence along the entire body (Martin and Anctil 1984b). Hence recruitment of neural units and increased ability to overcome synaptic barriers may account for this phenomenon.

A microscopical study of the luminescent epithelium of Chaetopterus provided clues as to the means by which neurally controlled luminescence is effected at the cellular level (Anctil 1979a; Fig. 8). In vivo examination of the luminescent epithelia by fluorescence microscopy revealed a constellation of fluorescent spots which coincided in size and

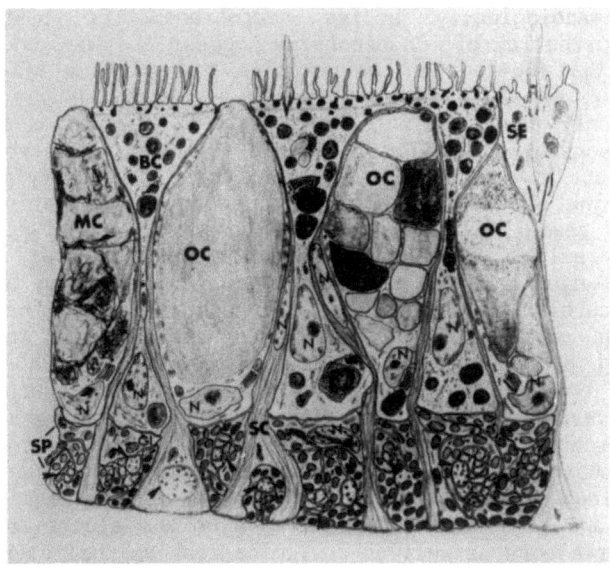

FIG. 8: The organization of the luminescent epithelium in Chaetopterus. BC, biconcave cell; GI, gliointerstitial process; MC, mucus cell; N nuclei; OC, orthochromatic cells; SC, supportive (musculo-epithelial) cell; SC, sensory cell; SP, subepidermal nerve plexus. Neurites are indicated by arrowheads. After Anctil (1979a).

shape with the apical plug of the equally fluorescent orthochromatic cells of the histologically processed epithelia. As shown by electron microscopy, these are exocrine goblet cells the greatest portion of which is occupied by immature or mature secretory products released through the apical pore during luminescent episodes (Anctil 1979a). Since no intracellular element suggestive of a contractile function was detected in these large cells, it was assumed that the epithelio-muscular cells ensheathing the base of the gland cells could provide the contractile mechanism necessary to push out the secretory mass during sudden, phasic episodes of luminescence. This assumption was supported by the presence of nerve endings arising from the subepidermal plexus and making contacts with the epithelio-muscular, not the gland cells (Fig. 8). Hence it was proposed that the gland cells secreting the photoprotein are indirectly controlled by the nervous system, through the mediation of a motor system.

Recent evidence of the dual character (neurogenic and aneurogenic) of luminescent displays in Chaetopterus was secured by observations on a hypertrophied mass of gland cells causing a bulge along the midline of segments 10 to 12 (Martin and Anctil 1984b; Fig. 7A). This photogenic gland, quite unlike the regular luminescent epithelia, was completely refractory to mild mechanical or to electrical stimulation. Only physical treatment causing the rupture of the membrane overlying the gland could induce the forceful release of its large store of luminescent mucus. Martin and Anctil (1984b) hypothesized from various lines of evidence that the release of luminescent clouds from the tube in which the worm dwells, presumably a defensive response, originates largely from the photogenic gland as the result of sudden folding and twisting movements of the worm inside the tube. Although the biological role of the neurally controlled luminescent epithelium, which is extremely sensitive to tactile stimulation (Martin and Anctil 1984b), is not understood at present, it seems clear that both types of effectors require mechanical forces to use luminescence to maximum effect.

The pharmacologically active neurotransmitter systems of the luminescent epithelia of Chaetopterus appear to be cholinergic and GABAergic (Anctil 1981). Of the putative transmitters effective in the luminescent system of polynoid worms - acetylcholine, serotonin, adrenaline - only acetylcholine was active in Chaetopterus albeit with much greater sensitivity. The threshold for the cholinergic excitation was 10 μM, but it was increased sharply by the acetylcholinesterase inhibitor eserine and significantly reduced by the muscarinic antagonist atropine, thus revealing the presence of a potent inactivating mechanism as well as a selective muscarinic-type receptor system. A separate inhibitory GABAergic pathway was postulated from evidence that the amino acid GABA acutely depressed glow responses to either electrical stimulation or acetylcholine, and this effect was antagonized by picrotoxin and bicuculline. It is unclear whether GABA acts as an interruptor of triggered luminescence to limit the glow duration, or as a modulator of overall excitability. The neurotransmitter-specific pathways of the luminescent epithelia have yet to be traced histochemically to provide more substantive support for their involvement in neurotransmission of luminescent signals. It should be stressed that both acetylcholine and GABA have potent effects on the electrical activity of the ventral nerve cord as well as on luminescent epithelia (Martin 1984).

4. LAMPYRID FIREFLIES

Fireflies are the best known of the bioluminescent organisms, and this is reflected by the large amount of physiological work as well as by the depth of understanding of luminescence control that was achieved in this group (see Carlson 1969; Case and Strause 1978 for reviews). Bioluminescence is an essential ingredient of the mating behaviour of fireflies and the study of its physiological regulation has been subordinated to that realization. However, there still are nagging problems in need of resolution as will be made manifest in the following account. The latter will address such issues as the complex central control mechanisms underlying the sometimes intricate luminescent behaviour of fireflies, and the controversy around the respective roles of peripheral neurotransmission and oxygen access in the output of photocytes. Whenever possible, adult and larval fireflies will be treated separately, since the structural organization of the light organ (lantern) and the phenomenology of the luminescent output differ markedly between these two life stages.

4.1. Central control

Our basic understanding of central mechanisms of adult firefly luminescence stems from studies by Buck and collaborators in the sixties. Buck and Case (1961) showed that the firefly lantern (Fig. 9A, C) responds to electrical stimulation essentially as do other neurally controlled effectors such as muscle. Flashes elicited in this fashion displayed

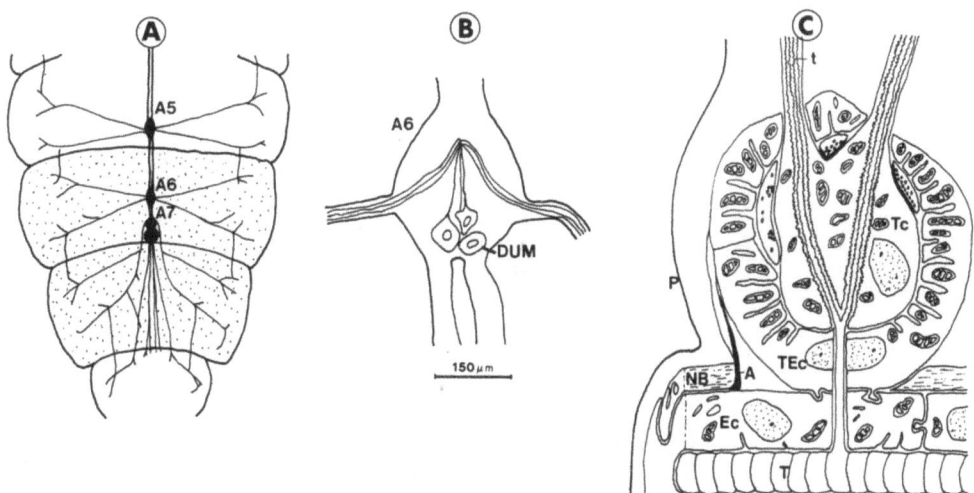

FIG. 9: The luminescent system of the adult firefly Photuris. (A) Innervation of the male lantern. The lantern covers most of the ventral face of the sixth and seventh abdominal segments (stippled), and are innervated by nerves projecting from three abdominal ganglia (A5-A7). (B) Drawing of the three dorsal unpaired median (DUM) neurones of the sixth abdominal ganglion (A6) which send bilateral processes to the lantern nerve roots. (C) Drawing illustrating the relationship between nerve tissue, the tracheal system and photocytes in the lantern. A, axon; Ec, tracheal epithelial cells; NB, peripheral nerve branch; P, photocyte margins; T, trachea; t, tracheoles; Tc, tracheolar cell; TEc, tracheal end cell. Note the three nerve endings with their vesicles, which are sandwiched between tracheolar and tracheal end cells. A redrawn from Hanson (1962), B from Christensen and Carlson (1981) and C from Smith (1963).

kinetic properties similar to those of spontaneous flashes, while facilitation, summation and fatigue were also evident. Case and Buck (1963) demonstrated the necessary role of the brain in producing the rhythmic flashes that are eminently characteristic of males from each firefly species. Electrophysiological recordings revealed that volleys of action potentials in the abdominal nerve cord and lantern precede spontaneous flashing (Fig. 10B, C). The volleys were conducted at 15-50 cm/s.

Although it is clear that a pacemaker in the brain is responsible for the rhythmic flashing of male fireflies, the work of Buck and collaborators did not address the dual problem of its precise location and mechanism. Using techniques of localized electrolytic lesions and of electrical stimulation of brain and related structures, Bagnoli et al. (1976) were able to show that "photoneurones" associated with the deep protocerebral neuropile were responsible for producing tonic and asynchronous spike activity driving the lantern nerves of Luciola (Fig. 10A). The activity of these photomotor neurones, according to Bagnoli et al. (1976), is driven by a phasic oscillator, presumably located in the optic lobes as deduced from the loss of rhythmic activity after sectioning tracts between the protocerebrum and lobula. Under such conditions, rhythmic flashing is replaced by a continuous glow and asynchronous lantern spike discharges. Flashing is only one of several rhythmic activities in insects which are controlled by an oscillator in the optic lobes (Saunders 1976), and in the case of Luciola Bagnoli et al. (1976)

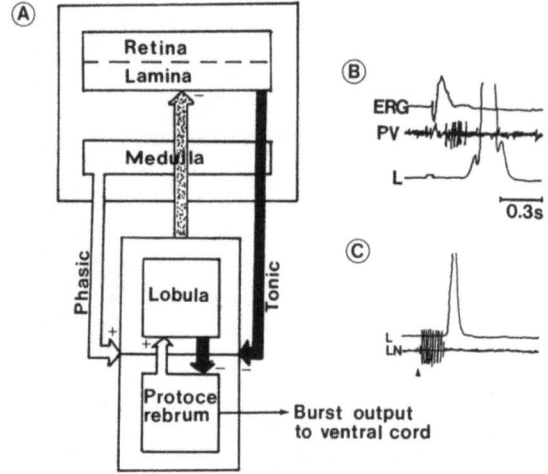

FIG. 10: Central control of flashing in the firefly Luciola. (A) Model of the postulated organization of neural mechanisms underlying flashing, based on localized lesion and stimulation experiments. (B) Simultaneous recording of the electroretinogram (ERG), impulse activity in the lantern (photogenic volley, PV) and flashing (L) in response to photic stimulation (small square signal in L). (C) Simultaneous recording of impulse activity in the lanternal nerve (LN) and of flash activity (L) in the lantern following repetitive electrical stimulation of the deep protocerebrum through a microelectrode (arrowhead). A and C redrawn from Bagnoli et al. (1976), B from Brunelli et al. (1977).

proposed that the rhythm thus generated is probably endogenous since blindfolding or retinal ablation failed to disrupt flashing rhythmicity.

The activity of the pattern generator in the optic lobes and of the photomotor neurones can be influenced by several types of input. Case and Buck (1963) reported that electrical stimulation of the eye of Photuris can suppress or enhance flash activity depending on stimulus strength. It was soon realized that photic modulation mimicked these effects in either Photuris or Luciola, low intensity light evoking enhancement and high intensity light inhibition of flash activity (Magni 1967; Case and Trinkle 1968). Magni (1967) suggested that photic inhibition in Luciola was caused by two distinct mechanisms, namely a central effect on the pacemaker and a peripheral effect on the lantern itself. Several lines of experimental evidence pointed to a hormonal factor released by cortical cells of the male gonad of Luciola (possibly noradrenaline) as responsible for the inhibitory effect of light on spontaneous flashing or on electrically stimulated lanterns (Brunelli et al. 1968, 1970; Bagnoli et al. 1970, 1972, 1973). Peripheral inhibition of the type depicted in Luciola was not reproducible in Photuris (Case and Trinkle 1968). It is thus possible that fundamentally different inhibitory mechanisms exist in these two genera of fireflies.

The central inhibitory mechanism, which is present in both Luciola and Photuris, was investigated by Bagnoli et al. (1976). Ablation and stimulation experiments indicated that selective inputs from the retina-lamina complex of the visual system mediate the inhibitory influence of light on flashing. The enhancement effect, a low-threshold phasic response to photic stimulation, was attributed to activity arising in the medulla neuropile of the CNS, on the basis of a similar experimental approach (Fig. 10A). The biological significance of these mechanisms has been examined with regard to courtship flashing (Brunelli et al. 1977; Carlson 1981) or pacemakers of synchronous fireflies (Hanson et al. 1971; Hanson 1978; Buck et al. 1981a, b).

4.2. Peripheral control

Peripheral control mechanisms of luminescence in adult and larval fireflies have become increasingly better understood since the sixties. In an attempt to account for the relatively large excitation-response delay of the adult firefly lantern when compared with neurogenic conduction delays, Buck et al. (1963) conducted a series of experiments which led to the distinction between slow, fast and ultrafast flashes depending on the duration and voltage of stimulus pulses. They concluded that (1) the ultrafast response involved direct excitation of photocytes proximal to electrode location, (2) the fast flash resulted from the direct excitation of the tracheal end-organ which is located near the photocytes (Smith 1963; Case and Strause 1978), and (3) the slow flash was due to the excitation of the lantern nerves ending on the tracheal end-organ. Hence a three-step control system seems operative in adult lanterns. Experimental evidence suggests that flashing activity of the European firefly Luciola is subjected to similar peripheral mechanisms (Buonamici and Magni 1967).

The adult lantern of Photuris is innervated by the two most posterior abdominal ganglia (Hanson 1962; Fig. 9A). Backfilling neuroanatomical methods demonstrated that 3-4 large neurones in each of the two ganglia innervate the lantern of male Photuris (Christensen and Carlson 1981; Fig. 9B). These belong to a class of well identified insect neurones (Hoyle et al. 1974), the dorsal unpaired median (DUM) neurones. Their axon bifurcates in a manner such that it is distributed bilaterally in the left and right lantern nerves, an arrangement accounting for the decussating path of flash production postulated by Hanson (1962) from experimental evidence. Each bilateral axon branches profusely in the lantern, thus allowing for a very precise and synchronous activation of luminescence and especially the production of very short-lived flashes (Christensen and Carlson, 1981). Whether these DUM neurones are synaptically driven by the output from the brain oscillator remains to be determined.

Four DUM neurones in the last abdominal ganglion similarly innervate the paired lanterns of the larval *Photuris*. The intrasomatically recorded spike activity of these neurones coincided well temporally with glow activity such that these neurones seem to represent the only photomotor output from the nerve cord to the lanterns (Christensen and Carlson 1982).

4.3. Neuro-photocyte transmission

Intracellular recordings of photocytes from larval lanterns of *Photuris* indicated that a slow, graded depolarization of small amplitude, accompanied by a reduction of photocyte membrane resistance, always followed stimulation of the lantern nerve (and presumably of the DUM axons) and preceded the onset of the typical larval glow (Oertel and Case 1976). However, the significance of this electrical activity for photocytic excitation-luminescence coupling is unclear because the same authors reported that the larval glow can be elicited by noradrenaline, adrenaline and synephrine without the concomittent photocyte depolarization. It is possible that lantern nerve excitation causes the release of a synaptic co-transmitter or modulator responsible for the depolarizing effect. The excitatory neurotransmitter has been identified as octopamine on the basis of assays of whole lantern extracts (Robertson and Carlson 1976) or microassays of DUM neurone somata (Christensen et al. 1983). Supporting evidence include also the presence of a high affinity uptake system for octopamine in larval firefly lanterns (Carlson and Evans 1986) and calcium-dependent octopamine release in lanterns exposed to high potassium saline (Carlson and Jalenak 1986). A suggestion by Oertel and Case (1976) that the aminergic transmitter action is mediated by a cyclic nucleotide was later supported by evidence of a sharp rise in adenylate cyclase activity elicited by octopamine in the larval (Nathanson and Hunnicutt 1979) and adult firefly lantern (Nathanson 1979). Formamimide pesticides also activated adenylate cyclase activity in adult firefly lanterns and thereby induced a bright glow (Hollingworth and Murdock 1980). This octopamine response, like that of other insect effectors, is more sensitive to alpha- than to beta-adrenergic antagonists (Oertel and Case 1976; Nathanson 1979). Strause and Case (1981) reported that the octopaminergic receptors of developing adult lanterns are functional before the innervation becomes apparent.

The means by which the intracellular constituents of the photocytes are activated to produce light are poorly understood, and the problem is compounded by the realization that "excitational latencies of adult and larval firefly lanterns are quite the opposite of what might be expected from what is known about innervation patterns in adult and larval light organs" (Oertel et al. 1975). Indeed, the nerve terminals are closely apposed to, and even make synapses with photocytes of the larval lantern which responds sluggishly to stimulation (Oertel et al. 1975). In contrast, adult lanterns respond quickly to stimulation by emitting short-lived flashes, even though the nerve terminals make synaptic contacts with tracheal end-organs and not photocytes (Case and Strause 1978). This arrangement has led to speculations that light emission of the adult lantern is oxygen-limited. This is substantiated by the peripheral distribution of mitochondria in adult photocytes, thus possibly forming a diffusion barrier for oxygen at rest. This mitochondrial compartmentation is absent in photocytes of developing adult lanterns of late pupal stages (Strause and Case 1981). Lantern nerve stimulation is believed to cause the release of the transmitter octopamine which in turn changes the oxygen permeability of the tracheal end cells, thus resulting in a large and sudden influx of oxygen in the photocytes which is then able to overcome the mitochondrial barrier (Ghiradella 1977). In support of the oxygen theory of flash control, Ghiradella (1977, 1978) reported the presence of reinforced tracheoles in adult, but not larval lanterns, presumably to counteract osmotic forces associated with the maintenance of high oxygen tensions.

5. NEURONAL CIRCUITRY, NEUROTRANSMISSION AND BIOLUMINESCENCE

Given the large number of invertebrate taxa which include luminescent species, surprisingly few of the latter have been investigated with respect to nervous control. Furthermore, of the taxa investigated (Table 1), only the neurobiology of polychaete and firefly luminescent systems is reasonably well understood. In many of the groups listed in Table 1, not only are photocytes epithelial constituents but they are also derived from pre-existing epithelial effectors (epithelio-muscular cells, gastrodermal cells, mucous cells), and therefore are subjected to the modes of neural control already in place for such effectors. Hence there is no need to propose the evolutionary implantation of novel and/or distinct nerve pathways to account for luminescence control. The neural substrates alluded to here are elementary nerve nets or plexuses which integrate several types of activity to orchestrate simple behaviours (attracting preys, escaping or discouraging predators, etc.). In some taxa, epithelial conduction appears to at least partly substitute for that role, although neuronal elements may be (Hydrozoa, Ctenophora) or are certainly involved (polynoid worms) as well in these systems.

In invertebrates possessing "simple" nervous systems (coelenterates, hemichordates, echinoderms), the command units for luminescence are integrated components of the diffuse nerve nets themselves. In luminescent animals with a more complex nervous system (bivalve molluscs, polychaete worms), the epithelial luminescent systems with their underlying nerve plexus are still in place, but the centralized nervous systems of these groups (ganglionic chains) now provide the command neurones responding to some external input. Finally, in invertebrates possessing complex luminescent organs (photophores), the involvement of complex CNS patterning of the luminescent output is either suspected (cephalopods) or largely substantiated (fireflies). These organizational principles are not unique to luminescence, but they are often manifested vividly in luminescent systems.

The electrophysiology and pharmacology of neuro-photocyte transmission are even less well understood than the overall integration mechanisms underlying the expression of luminescent behaviour (Table 1). Only in fireflies can it be confidently said that the neurotransmitter (octopamine) responsible for luminescence excitation is identified with reasonable certainty, although its effects on postsynaptic membranes have not been documented yet. It is interesting that octopamine and the neurones which contain it (DUM) are also involved in the peripheral control of several other activities (muscular, metabolic) in insects (see David and Coulon 1985). Although classical monoaminergic and cholinergic mechanisms influence the luminescent systems of several invertebrate groups, including the most primitive ones, corresponding innervations have not yet been demonstrated except for the dense monoaminergic innervation

TABLE 1. SYNOPSIS OF SELECTED PHYSIOLOGICAL FEATURES OF INVERTEBRATE LUMINESCENT SYSTEMS

	LUMINESCENT SOURCES	TYPES OF ACTIVITY	MODES OF COORDINATION	PHARMACOLOGY
CNIDARIA				
Hydrozoa	Scattered photocytes	Spreading or local	Epithelial conduction (?)	Unknown (1)
Anthozoa	Clumps of photocytes	flashes	Mesogleal nerve net	Adrenergic-like (1)
CTENOPHORA	Epithelium (intracellular) Endodermal	Spreading or local flashes	Mesogleal nerve net (?) (+ epithelial conduction?)	Cholinergic and (1) adrenergic-like
MOLLUSCA				
Bivalvia	Epithelium (secretion)	Luminescent cloud	Subepidermal nerve plexus	Unknown (2)
Cephalopoda	Epithelium (secretion) Symbiotic (bacteria) Photophores	Luminescent cloud Glows Flashes	Unknown Photophores innervated	Unknown (3)
ANNELIDA				
Polychaeta	Epithelium (extra- or intracellular)	Glows, spreading flashes, luminescent cloud	Subepidermal nerve plexus Ventral nerve cord	Cholinergic and (1) monoaminergic-like
Oligochaeta	Coelomic cells	Pulsing glows	Unknown (non-nervous?)	Unknown (4)

(1) See text for references; (2) Bassot 1966, Nicol 1960; (3) Arnold and Young 1974, Herring 1978; (4) Wampler and Jamieson 1980.

592

	LUMINESCENT SOURCES	TYPES OF ACTIVITY	MODES OF COORDINATION	PHARMACOLOGY
ARTHROPODA				
Crustacea	Hepatopancreas-derived Epithelium (secretion) Photophores	Luminescent cloud Glows, steady emissions	Some photophores innervated Neuroendocrine pathways?	Serotonergic-like (5)
Insecta	Photophores (lanterns) Clumps of photocytes	Flashes Glows	Optic and photomotor centers in firefly brain; DUM neurons in firefly nerve cord	Octopaminergic (1,6) (firefly)
PROTOCHORDATES	Epithelium (secretion and intracellular)	Luminescent slime Flashes	Subepithelial nerve net or no apparent innervation	Unknown (7)
ECHINODERMATA				
Asteroidea	Photocytes in ovary Epithelium (secretion)	Luminescent cloud Glows	Unknown	Unknown (8)
Ophiuroidea	Photocytes (radial nerve)	Spreading flashes	Radial nervous system	Unknown (9)

(5) Herring 1978, Latz and Case 1982; (6) Halverston et al. 1973; (7) Baxter and Pickens 1964, Mackie and Bone 1978; (8) Herring 1974; (9) Brehm 1977, Brehm and Morin 1977.

of the bioluminescent scales of polynoid worms. All the active putative transmitters thus belong to aminergic categories which have long been exploited by nervous systems, as further shown by recent immunohistochemical evidence of serotonergic-like neurones in the coelenterate <u>Renilla</u> (D. Umbriaco and M. Anctil, unpublished).

6. PERSPECTIVES

Despite recent advances in our understanding of neurally controlled luminescent systems, there are still major deficiencies in the neurobiological schemes constructed from our knowledge of the most investigated taxa and in the efforts to unravel control mechanisms of the less known taxa.

A major obstacle to the improvement of our understanding of these systems has been the lack of focus on central nervous mechanisms. Even in luminescent organisms possessing decentralized, diffuse nerve nets, the nuts and bolts of information processing in these networks have not come to light. Anderson's (1985) intracellular study of such neurones in the coelenterate <u>Cyanea</u> is a promising step in the right direction. The need to specify the neuronal circuits of luminescent systems in many of these invertebrates is evident, although this may prove difficult because of the less than ideal size and layout of these circuits. The best known system in this regard, that of the adult firefly, is understood more in terms of overall integrated pathways than in terms of cellular mechanisms.

Because light emission is often used as an intra- or interspecific communication signal, its role in modulating the light output of congeners or other luminescent species should be investigated more strenuously. Light has direct inhibitory effects on the intracellular control mechanism of photocytes (<u>Renilla</u>: Kreiss and Cormier 1967; G. Germain and M. Anctil, unpublished) or on the chemiluminescent reactants themselves (<u>Mnemiopsis</u>: Anctil and Shimomura 1984). There is also experimental evidence that certain parameters of light exposure can alter the luminescent behaviour of cephalopods (Young et al. 1979; Young and Mencher 1980), polynoid worms (Pavans de Ceccaty et al 1972), crustaceans (Warner et al. 1979) and fireflies (see above), probably through the mediation of neural pathways. The ultimate goal here is to identify sensory modes of significance for a particular luminescent behaviour, and to trace the pathways between the sensory input and photomotor output as well as their cellular basis. The sensory mode need not be photoreceptive, mechanoreception being an alternative mode in luminescent coelenterates and chaetopterid worms, in which photoreception has not been clearly demonstrated.

The search for neurotransmitter-specific pathways involved in luminescence control is well under way. The progress is slow because there are a number of rather stringent criteria to be met for the secure identification of such pathways. Few luminescent systems are good models for the interdisciplinary approach (pharmacology, neurochemistry, histochemistry, cellular electrophysiology) required to satisfy all the criteria. It is important for the search to go on, because many of the luminescent invertebrates occupy key positions in the phylogenetic tree in terms of providing useful information on the origin and evolution of neurotransmitters in multicellular organisms.

The matter just raised addresses an issue of wider import, namely the ongoing search for adequate neurobiological models to shed light on neural mechanisms and functions at different levels of the integrated organism. Can and should luminescent systems provide models whose adequacy and expository power compete favourably with the existing ones? Except for the firefly system, no luminescent system is controlled by a network of a few large nerve cells easily amenable to intracellular electrophysiological work. The use of luminescent effectors to guide our search for the earliest known neurotransmitter systems in the evolutionary context is the best that can be achieved at this time. Isolated luminescent cells (Obelia: Brehm et al. 1985; Renilla: G. Germain and M. Anctil, unpublished) are also potentially useful probes for the study of membrane ionic channels associated with an easily monitored calcium-dependent activity. Beyond modelization, the study of animals which use their nervous system to command the appearance of light is a fascinating pursuit in itself.

7. REFERENCES

Anctil M (1979a) The epithelial luminescent system of Chaetopterus variopedatus. Can J Zool 57: 1290-1310.

Anctil M (1979b) Physiological control of bioluminescence. Photochem Photobiol 30: 777-780.

Anctil M (1981) Luminescence control in isolated notopods of the tube-worm Chaetopterus variopedatus: effects of cholinergic and GABAergic drugs. Comp Biochem Physiol 68C: 187-194.

Anctil M (1985a) Ultrastructure of the luminescent system of the ctenophore Mnemiopsis leidyi. Cell Tissue Res 242: 333-340.

Anctil M (1985b) Cholinergic and monoaminergic mechanisms associated with control of bioluminescence in the ctenophore Mnemiopsis leidyi. J Exp Biol 119: 225-238.

Anctil M (1987) Bioactivity of FMRFamide and related peptides on a contractile system of the coelenterate Renilla köllikeri. J Comp Physiol 157: 31-38.

Anctil M, Shimomura O (1984) Mechanism of photoinactivation and reactivation in the bioluminescence system of the ctenophore Mnemiopsis. Biochem J 221: 269-272.

Anctil M, Boulay D, LaRivière L (1982) Monoaminergic mechanisms associated with control of luminescence and contractile activities in the coelenterate, Renilla köllikeri. J Exp Zool 223: 11-24.

Anctil M, Germain G, LaRivière L (1984) Catecholamines in the coelenterate Renilla köllikeri. Uptake and radioautographic localization. Cell Tissue Res 238: 69-80.

Anderson PAV (1980) Epithelial conduction: its properties and functions. Prog Neurobiol 15: 161-203.

Anderson PAV (1985) Physiology of a bidirectional, excitatory, chemical synapse. J Neurophysiol 53: 821-835.

Anderson PAV, Case JF (1975) Electrical activity associated with luminescence and other colonial behavior in the pennatulid Renilla köllikeri. Biol Bull 149: 80-95.

Arnold JM, Young RE (1974) Ultrastructure of a cephalopod photophore. I. Structure of the photogenic tissue. Biol Bull 147: 507-521.

Bagnoli P, Brunelli M, D'Ajello V, Magni F (1970) Further evidence for peripheral inhibition of flashing and for the role of the male gonads in Luciola lusitanica (Charp.). Arch Ital Biol 108: 181-206.

Bagnoli P, Brunelli M, Magni F, Viola MP (1972) The identification of a flash inhibiting substance from the male gonads of Luciola lusitanica (Charp.). Arch Ital Biol 110: 16-34.

Bagnoli P, Brunelli M, Magni A, Pellegrino M (1973) Central and peripheral mechanisms in the control of the diurnal rhythm of flashing in Luciola lusitanica (Charp.). Arch Ital Biol 111: 170-186.

Bagnoli P, Brunelli M, Magni F, Musumeci D (1976) Neural mechanisms underlying spontaneous flashing and its modulation in the firefly Luciola lusitanica. J Comp Physiol 108: 133-156.

Bassot J-M (1966) Données histologiques et ultrastructurales sur les organes lumineux du siphon de la pholade. Z Zellforsch 74: 474-504.

Bassot J-M, Bilbaut A (1977) Bioluminescence des élytres d'Acholoe. III. Déplacement des sites d'origine au cours des émissions. Biol Cell 28: 155-162.

Bassot J-M, Bilbaut A, Mackie GD, Passano LM, Pavans de Ceccaty M (1978) Bioluminescence and other responses spread by epithelial conduction in the siphonophore Hippopodius. Biol Bull 155: 473-498.

Bassot CH, Pickens PE (1964) Control of luminescence in hemichordates and some properties of a nerve net. J Exp Biol 41: 1-14.

Bilbaut A (1975a) Etude de la bioluminescence chez l'octocoralliaire Veretillum cynomorium (Pall.). I. Les réponses lumineuses des autozoïdes isolés de la colonie. Arch Zool Exp Gén 116: 27-42.

Bilbaut A (1975b) Etude de la bioluminescence chez l'octocoralliaire Veretillum cynomorium Pall. II. Les réponses lumineuses de la colonie. Arch Zool Exp Gén 116: 321-341.

Bilbaut A (1978) Excitabilité et conductibilité des cellules épithéliales photogènes et non photogènes de l'élytre chez Archoloe astericola (Annelide, Polynoinae). CR Acad Sci Paris D286: 985-988.

Bilbaut A (1980a) Cell junctions in the excitable epithelium of bioluminescent scales on a polynoid worm: a freeze-fracture and electrophysiological study. J Cell Sci 41: 341-368.

Bilbaut A (1980b) Excitable epithelial cells in the bioluminescent scales of a polynoid worm: effects of various ions on the action potentials and on excitation-luminescence coupling. J Exp Biol 88: 219-238.

Bilbaut A, Bassot J-M (1977) Bioluminescence des élytres d'Acholoe. II. Données photométriques. Biol Cell 28: 145-154.

Brehm P (1977) Electrophysiology and luminescence of an ophiuroid radial nerve. J Exp Biol 71: 213-227.

Brehm P, Morin JG (1977) Localization and characterization of luminescent cells in Ophiopsila californica and Amphipholis squamata (Echinodermata: Ophiuroidea). Biol Bull 152: 12-25.

Brehm P, Takeda T, Dunlap K (1985) Control of light emission from cells containing endogenous Ca-activated photoprotein: roles of Ca current and gap junctions. Biol Bull 169: 548 (abstract).

Brunelli M, Buonamici M, Magni F (1968) Mechanisms for photic inhibition of flashing in fireflies. Arch Ital Biol 106: 85-99.

Brunelli M, Buonamici M, Magni F, Viola MP (1970) The role of the male gonads in the peripheral inhibition of flashing in Luciola lusitanica (Charp.). Arch Ital Biol 108: 1-20.

Brunelli M, Magni F, Pellegrino M (1977) Excitatory and inhibitory events elicited by brief photic stimuli on flashing of the firefly Luciola lusitanica (Charp.). J Comp Physiol 119: 15-35.

Buck J (1973) Bioluminescent behavior in Renilla. I. Colonial responses. Biol Bull 144: 19-42.

Buck J, Case JF (1961) Control of flashing in fireflies. I. The lantern as a neuroeffector organ. Biol Bull 121: 234-256.

Buck J, Case JF, Hanson FE Jr. (1963) Control of flashing in fireflies. III. Peripheral excitation. Biol Bull 125: 251-269.

Buck J, Buck E, Hanson FE, Case JF, Mets L, Atta GJ (1981a) Control of flashing in fireflies. IV. Free run pacemaking in a synchronic Pteroptyx. J Comp Physiol 144: 277-286.

Buck J, Buck E, Case JF, Hanson FE (1981b) Control of flashing in fireflies. V. Pacemaker synchronization in Pteroptyx cribellata. J Comp Physiol 144: 287-298.

Buonamici M, Magni F (1967) Nervous control of flashing in the firefly Luciola italica L. Arch Ital Biol 105: 323-338.

Carlson AD (1969) Neural control of firefly luminescence. Adv Insect Physiol 6: 51-96.

Carlson AD (1981) Neural control of the male Photuris versicolor firefly flash. J Exp Biol 92: 165-172.

Carlson AD, Evans PD (1986) Inactivation of octopamine in larval firefly light organs by a high affinity uptake mechanism. J Exp Biol 122: 369-385.

Carlson AD, Jalenak M (1986) Release of octopamine from the photomotor neurones of the larval firefly lanterns. J Exp Biol 122: 453-457.

Case JF, Buck J (1963) Control fo flashing in fireflies. II. Role of central nervous system. Biol Bull 125: 234-250.

Case JF, Strause LG (1978) Neurally controlled luminescent systems. In: Herring PJ (ed) Bioluminescence in action. Academic Press, London, New York, p 331-366.

Case JF, Trinkle MS (1968) Light-inhibition of flashing in the firefly Photuris missouriensis. Biol Bull 135: 476-485.

Chang JJ (1954) Analysis of the luminescent response of the ctenophore, Mnemiopsis leidyi, to stimulation. J Cell Comp Physiol 44: 365-394.

Chang JJ, Johnson FH (1959) The influence of pressure, temperature and urethane on the luminescent flash of Mnemiposis leidyi. Biol Bull 116: 1-14.

Christensen TA, Carlson AD (1981) Symmetrically organized dorsal unpaired median (DUM) neurones and flash control in the male firefly, Photuris versicolor. J Exp Biol 93: 133-147.

Christensen TA, Carlson AD (1982) The neurophysiology of larval firefly luminescence: direct activation through four bifurcating (DUM) neurons. J Comp Physiol 148: 503-514.

Chrisensen TA, Sherman TG, McCaman RE, Carlson AD (1983) Presence of octopamine in firefly photomotor neurons. Neurosci 9: 183-189.

Davenport D, Nicol JAC (1955) Luminescence in hydromedusae. Proc R Soc Lond B 144: 399-412.

David J-C, Coulon J-F (1985) Octopamine in invertebrates and vertebrates. A review. Prog Neurobiol 24: 141-185.

De Waele J-P, Anctil M, Carlberg M (1987) Biogenic catecholamines in the cnidarian Renilla köllikeri: radioenzymatic and chromatographic detection. Can J Zool 65 (in press).

Freeman G, Reynolds GT (1973) The development of bioluminescence in the ctenophore Mnemiopsis leidyi. Develop Biol 31: 61-100.

Ghiradella H (1977) Fine structure of the tracheoles of the lantern of a photurid firefly. J Morphol 153: 187-204.

Ghiradella H (1978) Reinforced tracheoles in three firefly lanterns: further reflections on specialized tracheoles. J Morphol 157: 281-300.

Halverson RC, Case JF, Tiemann D (1973) Control of luminescence in phengodid beetles. J Insect Physiol 19: 1327-1339.

Hamner WM, Madin LP, Alldredge AL, Gilmer RW, Hamner PO (1975) Underwater observations of gelatinous zooplankton: sampling problems, feeding biology and behavior. Limnol Oceanogr 20: 907-917.

Hanson FE (1962) Observations on the gross innervation of the firefly light organ. J Insect Physiol 8: 105-111.

Hanson FE (1978) Comparative studies of firefly pacemakers. Fed Proc 37: 2158-2164.

Hanson FE, Case JF, Buck E, Buck J (1971) Synchrony and flash entrainment in a New Guinea firefly. Science 174: 161-164.

Harvey EN (1952) Bioluminescence. Academic Press, New York.

Hernandez-Nicaise M-L (1973a) Le système nerveux des Cténaires. I. Structure et ultrastructure des réseaux épitheliaux. Z Zellforsch 137: 223-250.

Hernandez-Nicaise M-L (1973b) Le système nerveux des Cténaires. II. Les éléments nerveux intra-mésogleens des béroidés et des cydippidés. Z Zellforsch 143: 117-133.

Herrera AA (1979) Electrophysiology of bioluminescent excitable epithelial cells in a polynoid polychaete worm. J Comp Physiol 129: 67-78.

Herring PJ (1874) New observations on the bioluminescence of echinoderms. J Zool 172: 401-418.

Herring PJ (ed) (1978) Bioluminescence in action. Academic Press London, New York.

Hogben LT (1926) Comparative physiology. Sidgwick and Jackson, London.

Hollingworth RM, Murdock LL (1980) Formamidine pesticides: octopaminelike actions in a firefly. Science 208: 74-76.

Hoyle G, Dagan D, Moberly B, Colquhoun W (1974) Dorsal unpaired medial insect neurons make neurosecretory endings on skeletal muscle. J Exp Zool 187: 159-165.

Kreiss P, Cormier MJ (1967) Inhibition of Renilla reniformis bioluminescence by light: effects on luciferase and its substrates. Biochim Biophys Acta 141: 181-183.

Labas YA (1977a) Triggering and regulatory mechanisms of ciliary beating in Ctenophora. I. Coordination of ciliary beating with intracellular bioluminescence and muscle contractions. Tsitologiya 19: 514-521.

Labas YA (1977b) Triggering and regulatory mechanisms of ciliary motion in the ctenophore Bolinopsis. II. Effect of ionic changes on the ciliary apparatus and bioluminescence system of a ctenophore. Tsitologyia 19: 644-654.

Labas YA, Mashanskii VF (1976) Structural basis of luminescence in ctenophores. Biol Morya 1: 57-66.

Latz MI, Case JF (1982) Light organ and eyestalk compensation to body tilt in the luminescent midwater shrimp, Sergestes similis. J Exp Biol 98: 83-104.

Mackie GO, Bone Q (1978) Luminescence and associated effector activity in Pyrosoma (Tunicata: Pyrosomida). Proc R Soc Lond B 202: 483-495.

Mackie GO, Mackie GV (1963) Systematics and biological notes on living hydromedusae from Puget Sound. Contr Zool Nat Mus Can Bull 199: 63-84.

Mackie GO, Mackie GV (1967) Mesogleal ultrrastructure and reversible opacity in a transparent siphonophore. Vie Milieu 18: 47-71.

Mackie GO, Anderson PAV, Singla CL (1985) Apparent absence of gap junctions in two classes of Cnidaria. Biol Bull 167: 120-123.

Magni F (1967) Central and peripheral mechanisms in the modulation of flashing in the firefly, Luciola italica L. Arch Ital Biol 105: 339-360.

Martin N (1984) Le système nerveux de Chaetopterus variopedatus (Polychète tubicole): morphologie, neuropharmacologie et rôle dans le contrôle de l'activité lumineuse. M.Sc thesis, University of Montreal.

Martin N, Anctil M (1984a) The nervous system of the tube-worm Chaetopterus variopedatus (Polyshaeta). J Morphol 181: 161-173.

Martin N, Anctil M (1984b) Luminescence control in the tube-worm *Chaetopterus variopedatus*: role of nerve cord and photogenic gland. Biol Bull 166: 583-593.

Miron M-J, LaRivière L, Bassot J-M, Anctil M (1987) Immunohistochemical and radioautographic evidence of monoamine-containing cells in bioluminescent elytra of *Harmothoe imbricata* (Polychaeta). Cell Tissue Res (in press).

Moore AR (1926) On the nature of inhibition in *Pennatula*. Amer J Physiol 76: 112-115.

Morin JG (1974) Coelenterate bioluminescence. In: Muscatine L, Lenhoff HM (eds) Coelenterate biology: Reviews and new perspectives. Academic Press, New York, p 397-438.

Morin JG (1983) Coastal bioluminescence: patterns and functions. Bull Mar Sci 33: 787-817.

Morin JG, Cooke IM (1971a) Behavioural physiology of the colonial hydroid *Obelia*. II. Stimulus-initiated electrical activity and bioluminescence. J Exp Biol 54: 707-721.

Morin JG, Cooke IM (1971b) Behavioural physiology of the colonial hydroid *Obelia*. III. Characteristics of the bioluminescent system. J Exp Biol 54: 723-735.

Morin JG, Reynolds GT (1974) The cellular origin of bioluminescnece in the colonial hydroid *Obelia*. Biol Bull 147: 397-410.

Nathanson JA (1979) Octopamine receptors, ademosine 3', 5'-monophosphate, and neural control of firefly flashing. Science 203: 65-68.

Nathanson JA, Hunnicutt EJ (1979) Neural control of light emission in *Photuris* larvae: identification of octopamine-sensitive adenylate cyclase. J Exp Zool 208: 255-262.

Nicol JAC (1952a) Studies on *Chaetopterus variopedatus* (Renier). I. The light-producing glands. J Mar Biol Assn UK 30: 417-431.

Nicol JAC (1952b) Studies on *Chaetopterus variopedatus* (Renier). II. Nervous control of light production. J Mar Biol Assn UK 30: 433-452.

Nicol JAC (1952c) Studies on *Chaetopterus variopedatus* (Renier). III. Factors affecting the light response. J Mar Biol Assn UK 31: 113-144.

Nicol JAC (1954) The nervous control of luminescent responses in polynoid worms. J Mar Biol Assn UK 33: 225-255.

Nicol JAC (1955a) Observations on luminescence in *Renilla* (Pennatulacea). J Exp Biol 32: 299-320.

Nicol JAC (1955b) Nervous regulation of luminescence in the sea pansy *Renilla kollikeri*. J Exp Biol 32: 619-635.

Nicol JAC (1957a) Luminescence in polynoids. II. Different modes of response in the elytra. J Mar Biol Assn UK 36: 261-269.

Nicol JAC (1957b) Luminescence in polynoids. III. Propagation of excitation through the nerve cord. J Mar Biol Assn UK 36: 271-273.

Nicol JAC (1960) Histology of the light organs of Pholas dactylus (Lamellibranchia). J Mar Biol Assn UK 39: 109-114.

Nicolas M-T (1977) Bioluminescence des élytres d'Acholoe. V. Les principales étapes de la regénération. Arch Zool Gén Exp 118: 103-120.

Nicolas M-T, Moreau M, Guerrier P (1978) Indirect nervous control of luminescence in the polynoid worm Harmothoe lunulata. J Exp Zool 206: 427-433.

Nicolas M-T, Bassot J-M, Shimomura O (1981) Caractérisation d'une photoprotéine nouvelle dans le système bioluminescent des annélides polynoinae. CR Acad Sci Paris 293: 777-780.

Nicolas M-T, Bassot J-M Shimomura O (1982) Polynoidin: a membrane photoprotein isolated from the bioluminescent system of scale-worms. Photochem Photobiol 35: 201-207.

Oertel D, Case JF (1976) Neural excitation of the larval firefly photocyte: slow depolarization possibly mediated by a cyclic nucleotide. J Exp Biol 65: 213-227.

Oertel D, Linberg KA, Case JF (1975) Ultrastructure of the larval firefly light organ as related to control of light emission. Cell Tissue Res 164: 27-44.

Parker GH (1920a) The phosphorescence in Renilla. Proc Amer Phil Soc 59: 171-175.

Parker GH (1920b) Activities of colonial animals. II. Neuromuscular movements and phosphorescence in Renilla. J Exp Zool 31: 475-515.

Pavans de Ceccaty M, Bilbaut A, Bassot J-M (1972) Corrélations entre les signaux électriques spontanés, la motricité et la luminescence chez Acholoe astericola Delle Ch. CR Acad Sci Paris 275: 2523-2526.

Pavans de Caccaty M, Bassot J-M, Bilbaut A, Nicolas M-T (1977) Bioluminescence des élytres d'Acholoe. I. Morphologie des supports structuraux. Biol Cell 28: 57-64.

Prosser CL, Brown FA Jr. (1961) Comparative animal physiology, 2nd. ed. Saunders, Philadelphia, p 489-501.

Robertson HA, Carlson AD (1976) Octopamine: presence in firefly lantern suggests a transmitter role. J Exp Zool 195: 159-164.

Satterlie RA (1978) Feeding mechanisms in the ctenophore Pleurobrachia pileus. Biol Bull 155: 464 (abstract).

Satterlie RA (1985) Control of swimming in the hydrozoan jellyfish Aequorea aequorea: direct activation of the subumbrella. J Neurobiol 16: 211-226.

Satterlie RA, Anderson PAV, Case JF (1976) Morphology and electrophysiology of the through-conducting systems in pennatulid coelenterates. In: Mackie GO (ed) Coelenterate ecology and behavior. Plenum Press, New York, p 619-627.

Satterlie RA, Anderson PAV, Case JF (1980) Colonial coordination in anthozoans: Pennatulacea. Mar Behav Physiol 7: 25-46.

Saunders DS (1976) Insect clocks. Pergamon Press, Oxford.

Shimomura O, Johnson FH (1966) Partial purification and properties of the Chaetopterus luminescence system. In: Johnson FH, Haneda Y (eds) Bioluminescence in progress. Princeton University Press, Princeton, p 495-521.

Smith DS (1963) The organization and innervation of the luminescent organ in a firefly, Photuris pennsylvanica (Coleoptera). J Cell Biol 16: 323-359.

Strause LG, Case JF (1981) Neuro-pharmacological studies on firefly light organs during metamorphosis. J Insect Physiol 27: 5-15.

Wampler JE, Jamieson BGM (1980) Earthworm bioluminescence: comparative physiology and biochemistry. Comp Biochem Physiol 66b: 43-50.

Warner JA, Latz MI, Case JF (1979) Cryptic bioluminescence in a midwater shrimp. Science 203: 1109-1110.

Young RE (1983) Oceanic bioluminescence: an overview of general functions. Bull Mar Sci 33: 829-845.

Young RE, Mencher FM (1980) Bioluminescence in mesopelagic squid: diel color change during counterillumination. Science 208: 1286-1288.

Young RE, Roper CFE, Walters JF (1979) Eyes and extraocular photoreceptors in midwater cephalopods and fishes: their roles in detecting downwelling light for counterillumination. Mar Biol 51: 371-380.

ACOUSTIC COMMUNICATION IN CRICKETS:
BEHAVIORAL AND NEURONAL MECHANISMS OF SONG RECOGNITION AND LOCALIZATION

KLAUS SCHILDBERGER

Max-Planck-Institut f. Verhaltensphysiologie

8131 Seewiesen FRG

ABSTRACT

The acoustic communication of crickets provides a model system for studying the mechanisms underlying sexual behavior. Female crickets walk toward a singing male. Two basic problems occur. Such phonotactic behavior is dependent upon recognition of the conspecific song and the localization of the sound source. Recognition is based upon the behavioral selectivity for the temporal pattern; neuronal correlates for this selectivity were found in the cricket brain. Localization is believed to be based on the excitatory difference between the two sides of the auditory pathway, even in animals with only one ear. By monitoring phonotaxis and simultaneously recording from identified auditory neurons one can clarify some causal relations between phonotaxis and neurons of the auditory pathway.

INTRODUCTION

Male crickets attract their females by the calling song. If a female is sexually responsive and within the acoustic range she will approach the male. As Johannes Regen (1913) showed, sound alone can elicit phonotactic behavior and other cues are not necessary. If the female arrives at the singing male the behavior will continue with antennal contacts, male courtship song, copulation and male guarding. Up to now we studied only the initial steps of the behavioral sequence, e.g. the singing of the male and the phonotactic response of the female. Only the latter aspect will be described here from a behavioral and the neurobiological points of view. As far as our knowledge allows we will deal with two basic concerns of the female: the recognition of the calling song and the localization of the sender by the receiver.

RECOGNITION OF THE CONSPECIFIC SONG

To determine which crucial parameters of the male calling song are necessary and sufficient to attract females, phonotaxis studies have been done in our lab on a walking compensator developed by E. Kramer in Seewiesen. An animal is placed on top of a sphere (50 cm in diameter). A reflecting foil is mounted on the thorax and an overhead infrared camera is focused down on the foil. If the animal now departs from the center the camera senses the deviation from the middle position. The camera output is used to drive two motors that counterturn the sphere. The animal is free to move but kept close to the center. As a result the angle and the velocity of the walking path can be measured for a long time. The treadmill is placed in an anechoic chamber containing two loudspeakers placed 135 degrees apart from each other. When an attractive auditory stimulus is presented, the female walks in the direction of the active loudspeaker meandering around the midline. She immediately orients to each change of sound direction (Fig. 1).

Based on a critical analysis of the calling song, J. Thorson and T. Weber developed various sound models and tested them for attractiveness in phonotaxis (Weber et al. 1981; Thorson et al. 1982). A typical cricket calling song consists of a series of chirps, each chirp divided into single sound groups called syllables with a carrier frequency of 4.8 kHz. Each syllable is about 18 ms long and is repeated in a chirp 4-5 times at an interval of about 35-40 ms. By systematically changing all parameters Thorson and Weber found that in Gryllus campestris the number of syllables in a chirp does not influence the phonotactic response, even when the chirp becomes a continuous trill. Likewise, the length of the syllable and the length of the pause between syllables do not play a role unless there is no pause; this constitutes a burst or unmodulated 5 kHz tone. Such burst elicits no tracking. The most critical parameter is the syllable repetition interval which is phonotactically effective only between about 25-55 ms. From these results they conclude that the only

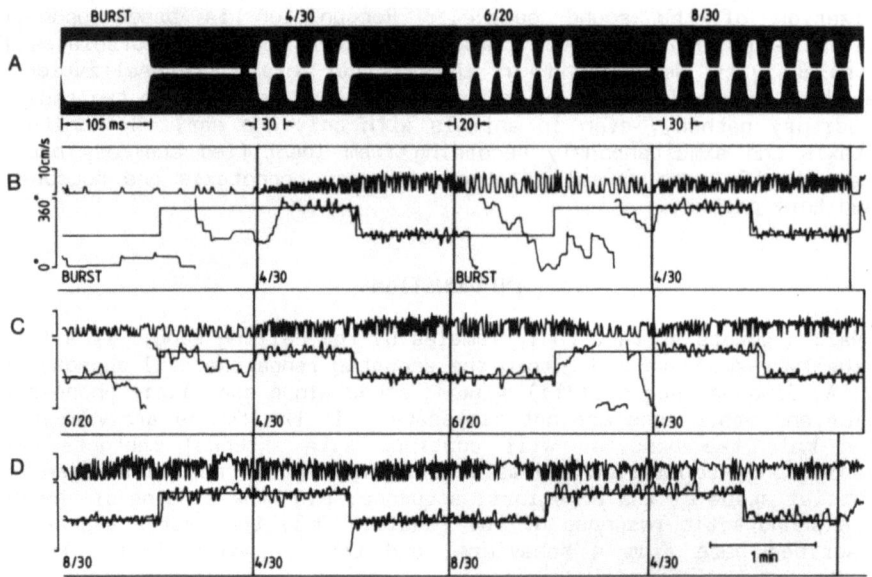

FIG 1: Tracking performance of a cricket on the Kramer treadmill for different sound patterns. The four simulated songs (A one conspecific song - 4/30 - and 3 'wrong' songs), all modulations of a 5-kHz carrier, are defined in A and presented as shown in B-D, at 70 dB in all cases, with chirp repetition period 350 ms. In each pair, the upper trace is the animal velocity and the lower trace the direction of the animal's walking course. The horizontal lines within the direction fields denote the direction of the loudspeaker. The vertical lines mark changes of conditions, the rightmost ones in B and D denoting 'sound off' (from Weber et al. 1981).

necessary and sufficient parameter for eliciting phonotaxis is the syllable repetition interval (SRI). Therefore, the recognition of the conspecific signal may be a simple process and can be tested using electrophysiological methods.

NEURONAL CORRELATES FOR RECOGNITION

Sensory fibers originate within the tympanal organ in the tibia of the foreleg and project into the prothoracic ganglion where they end ipsilaterally in an area called auditory neuropil. Some receptor cells are tuned to the carrier frequency of the calling song and copy the temporal structure of auditory stimuli (Esch et al. 1980). In the ganglion D. Wohlers and other authors (Casaday and Hoy 1977; Popov et al. 1978; Wohlers and Huber 1978, 1982, 1985) identified six types of auditory interneurons (Fig. 2): two bilaterally organized local "omega" neurons

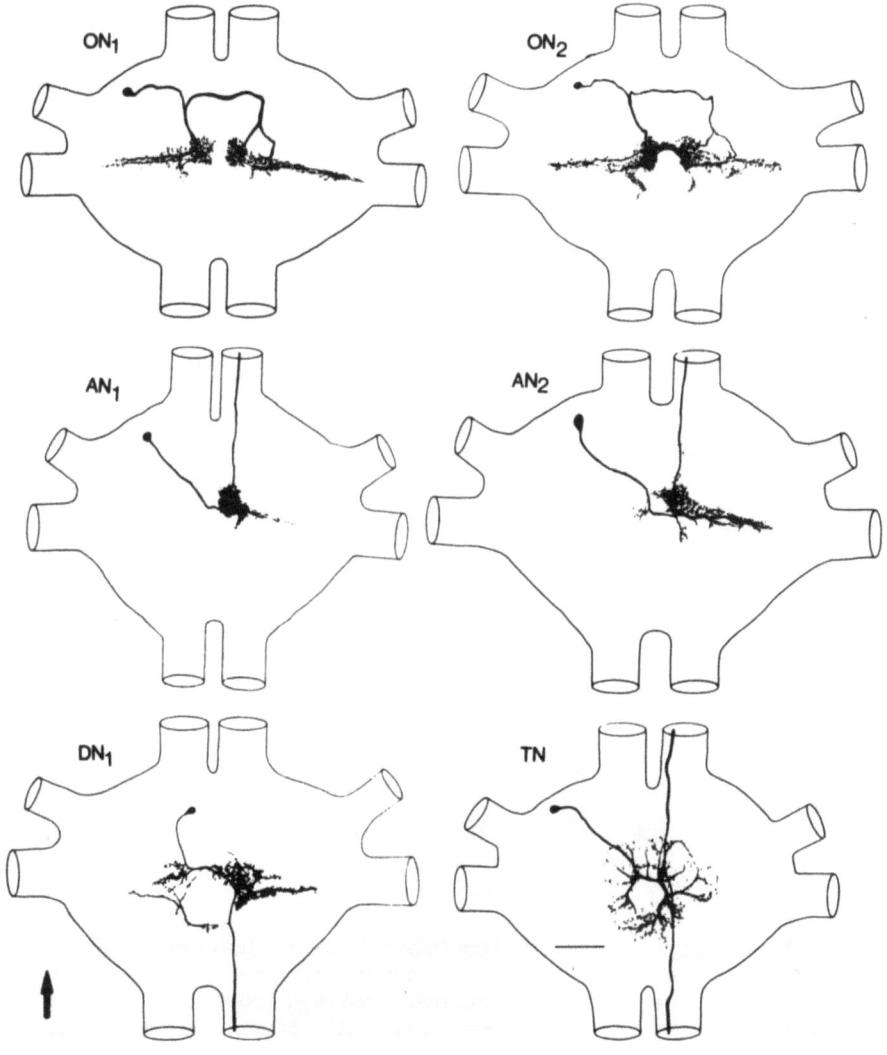

FIG. 2: The six auditory interneuron types reconstructed from color slides of serial sections of ganglia containing a Lucifer Yellow filled interneuron. Each type has a mirror-image partner; scale bar: 200 μm (from Wohlers and Huber 1982).

(ON1, ON2), two neurons with an ascending axon (AN1, AN2), a through conducting fiber (TN1) and a descending neuron (DN1). Some of them are sharply tuned to the carrier frequency of the calling song (ON1 and AN1) while the others exhibit more broadly banded tuning. When testing these cells with temporal patterns similar to those used in the phonotactic experiments all the cells respond to all patterns; some of them (AN1, ON1) precisely copy patterns throughout a wide range of SRI, but no cell selectively responds to or copies only the pattern of the calling song.

The cells AN1 and AN2 ascend to the brain and end there in specific areas (Boyan and Williams 1982; Schildberger 1984). Several types of auditory neurons could be identified in the brain (Boyan 1980; Schildberger 1984). Based on such anatomical criteria as arborization areas and overlapping projection fields one may classify the cells and formulate a first hypothesis on the auditory pathway in the brain (Fig. 3). The auditory information may travel from the ascending neurons via a

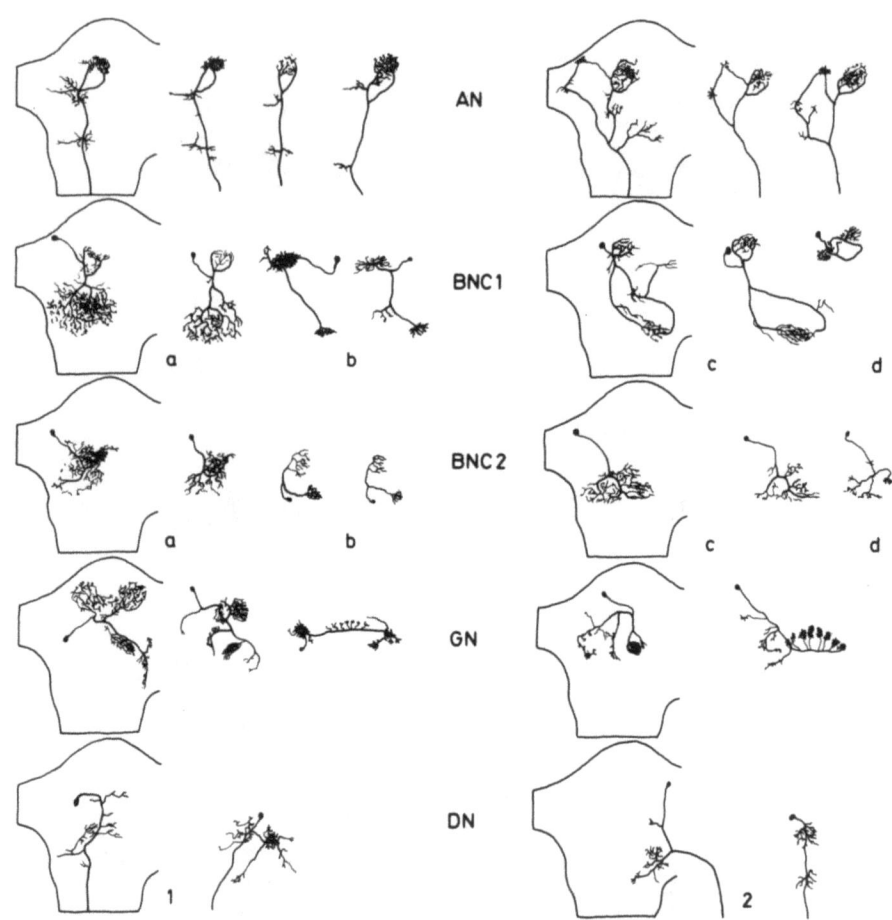

FIG. 3: Auditory neurons in the brain. Left: low-frequency (5 kHz) tuned auditory neurons; reconstructions from serial sections of Lucifer Yellow stained cells; frontal view; GN neurons are from Acheta Domesticus, all other neurons from Gryllus bimaculatus; abbreviations (as in all other figures): AN= ascending neuron, BNC1= brain neuron class 1, BNC2= brain neuron class 2, GN= extrinsic glomerular neuron, DN= descending brain neuron. Right: high-frequency and broad-band tuned auditory neurons (from Schildberger 1985).

class of overlapping brain neurons (BNC1) to a second class of brain neurons (BNC2) which overlap BNC1 neurons but do not overlap the ascending neurons. From here the information may be transferred to descending cells. The hypothesis is supported by the latencies of the respective cell types: ascending neurons having the shortest, followed by the increasingly longer latencies of BNC1, BNC2 and descending cells respectively.

Brain neurons were tested for pattern selectivity with model songs that were identical in temporal pattern, carrier frequency and sound intensity to those of the models tested in the behavioral experiments. Brain cells show a variety of different response types to the patterns tested (Fig. 4). Two response properties were analyzed in detail: the ability of the cells to copy the temporal pattern and the response magnitude, e.g. the number of action potentials per chirp. AN1 cells precisely copy the temporal structure of songs that elicit phonotaxis. When testing different patterns the acuity of copying increases with increasing syllable repetition interval (SRI), the parameter most relevant for recognition, and there is significant copying of temporal patterns that are not attractive as well. BNC1 cells copy less precisely and significant copying is achieved only at long SRIs, BNC2 neurons do not copy the chirp structure. So no evidence was found that conspecific temporal patterns produce higher degrees of synchronization than unattractive ones.

As far as response magnitude is concerned varying the SRI in chirps with constant-energy and constant length has no influence on the response magnitude of ascending and most BNC1 neurons. But in BNC2 neurons only intermediate SRIs elicit a response while longer and shorter ones fail to do so. This temporal selectivity in the neurons correlates precisely with the behavioral selectivity found in phonotactic experiments; the neurons

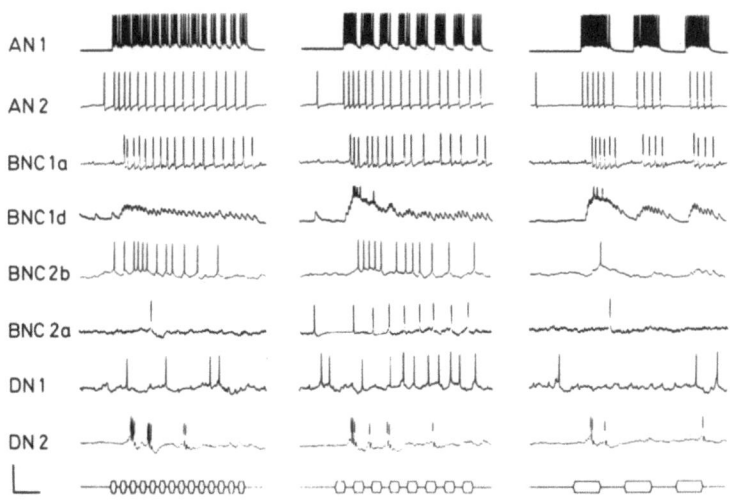

FIG. 4: Responses of different brain neurons to chirps with varying syllable repetition interval. Only AN1 copies all patterns; local and descending brain neurons show specific filter properties. SRIs are 18 ms (left), 34 ms (middle) and 98 ms (right) at 5 kHz and 80 dB SPL; calibration: abscissa= 80 ms, ordinate= 50 mV (traces 1-3) and 25 mV (traces 4-8); identification labels correspond to those in Fig. 2 (From Schildberger 1985).

could be considered band-pass cells (Fig. 5). Additionally, some BNC1 neurons respond to intermediate and long SRIs but not to short ones (low-pass) and some BNC2 neurons respond to intermediate and short but not to long SRIs (high-pass).

On the basis of these findings one may speculate about different models that could be responsible for recognizing the temporal pattern of the conspecific song (see Huber and Thorson 1985). The first model - discussed for about 20 years for birds, frogs and crickets - suggests that the female has an inbuilt template or pattern generator to compare with the temporal pattern of the incoming male song through some kind of cross-correlation. One possible realization for that kind of operation could be to synchronize the activity of the presumed internal pattern generator with the incoming activity. As a result responses to attractive temporal patterns should be better synchronized with the stimulus structure than those which are ineffective. Though this can not be supported by the current data in brain neurons, it remains a possible solution. A second model for recognition of a particular temporal rate was discussed by R. Reiss. It invokes two pathways by which the signal reaches the recognizer. One pathway involves a specific delay and if the recognizer requires temporal coincidence of events from both pathways then it responds only to certain temporal patterns. The model should recognize multiples of the designed rate, for coincident events from the two pathways occur for such multiples as well. But this multiples have not been found neither in the behavior nor in the responses of the brain cells.

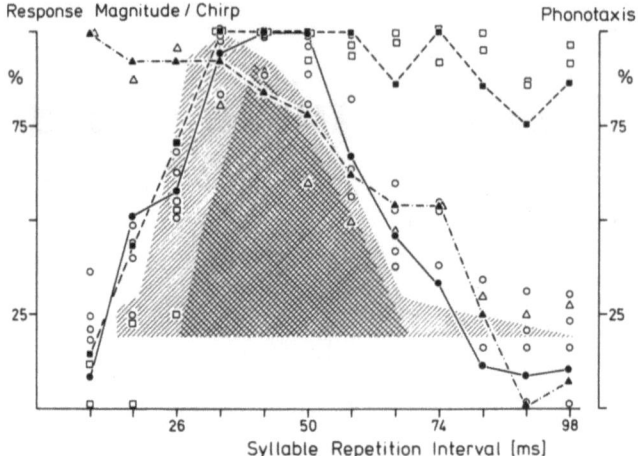

FIG. 5: Relative response magnitude of auditory brain neurons with specific filter characteristics to chirps varying in syllable repetition interval. Data points with squares are from the neuron BNC1d, triangles from BNC2b and circles from BNC2a. Closed symbols for three such cells are connected by lines. The open symbols show the responses of other examples of these identified neurons in different animals to indicate variability. The hatched areas show the relative effectiveness (right ordinate) of the syllable repetition intervals in eliciting phonotactic tracking in <u>Gryllus campestris</u> and <u>Gryllus bimaculatus</u> (cross-hatched); sound frequency, 5 kHz, intensity, 80 dB SPL (from Schildberger 1985).

Still another model of recognition – as proposed by Capranica and Rose (1983) – results from the discovery in crickets and frogs that band-pass cells are found in close association with high-pass and low-pass cells. Perhaps the nervous system resorts to two steps. First, two sets of intermediate cells decide whether a rate is above a given rate (high-pass) or below (low-pass). In the second step, recognizer cells respond only if both the high-pass and the low-pass cells are active. So this model describes a logical AND operation. It is the only one of the three where we have evidence from behavioral and neurophysiological data, at least in the cricket.

SOUND LOCALIZATION

In behavioral experiments J. Rheinlaender and G. Blaetgen (1982) showed that crickets always correctly turn to the side of an active loudspeaker when the sound source is at least 20 degrees aside from the body axis. At smaller angles, significant error in turning direction occurs. So there is a frontal sector of ambiguity of about 30-40 degrees. This may explain the meandering walking style of a freely moving animal. Because the ears possess a directional sensitivity, the animal turns toward the ear most strongly stimulated, thereby crossing the ambiguity sector. At a certain point the other ear becomes more strongly stimulated and the animal turns back to the former side and so on. Measuring the angular velocity for a fixed animal in an open loop situation, J. Stabel and G. Wendler (in preparation, see Fig. 6) found that the turning tendency

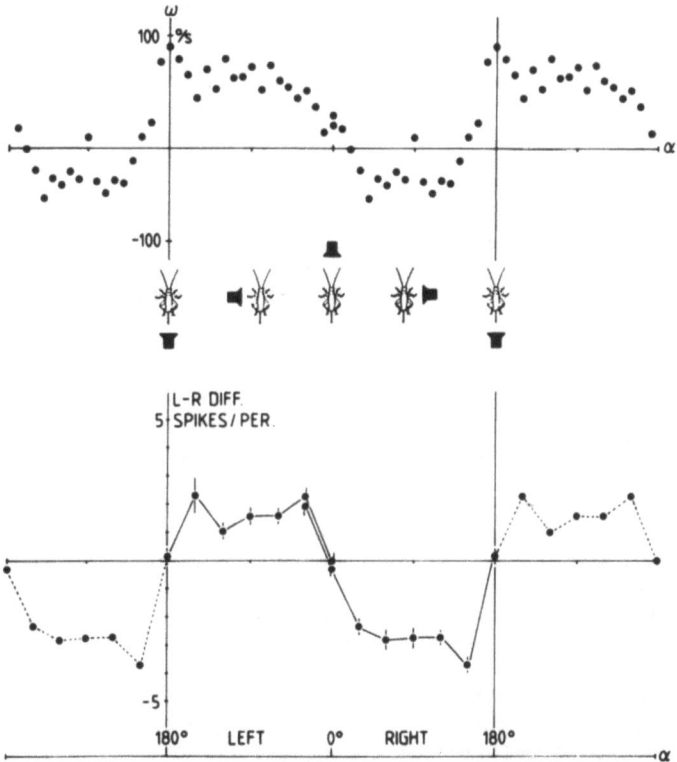

FIG. 6: Top: Angular velocity of a cricket as a function of stimulus angle determined in an open loop situation. Bottom: Differences of spike activity of neurons recorded in the left and right cervical connective (probably AN2) at different stimulus angles. Calling song intensity in both cases 70 dB; behavioral and neuronal data are not from one individual animal (Courtesy of J. Stabel, J. Stabel and G. Wendler in prep.).

toward the active loudspeaker increases linearly with increasing stimulus angles up to about 30-40 degrees. The hearing system with its two ears therefore has to provide sufficient information about the direction of sound for these stimulus angles.

The peripheral sound detection system of the cricket including tympanal membranes transduces incident sound to excite an array of auditory receptor cells and in turn central nerve cells. The ears are located in the tibiae of the forelegs just below the 'knees'. The two ears are coupled by an air-filled tube and major branches of the tube end in openings on the two sides of the body. Sound pressure reaches the outsides of the tympana directly and the insides via these openings and tubes. The H-shaped tube is an acoustically specialized part of the tracheal system that no longer serves only for respiration. Externally the four ends of the tube system are marked by spiracular openings on either side of the prothorax and by a pair of tympana on the surface of each foreleg tibia. Each auditory organ includes 55-60 sensory cells lined up in a row against the wall of an adjacent smaller branch of the main acoustic trachea. The axons of these cells course centrally through the leg, project into the prothoracic ganglion and end in an area called auditory neuropil.

This system which includes four tympana and the tube provides a directional characteristic. When measuring the directional sensitivity at the level of the auditory nerve Boyd and Lewis (1983) discovered a cardioid with a maximal sensitivity ipsilateral and a minimal one contralateral to the stimulated ear. The directionality critically depends on the different sound entrances in various ways. Whereas blocking of the contralateral tympanum has no effect on the directionality of the ipsilateral ear, blocking of the ipsilateral and/or contralateral spiracle openings destroys the directionality. Directional sensitivity also occurs in central interneurons (Boyan 1979; Wiese and Eilts 1985). A pair of mirror-image local interneurons, the Omega cells, may serve to sharpen the binaural intensity difference. Each of these cells receives excitatory input from the ear ipsilateral to the cell body and inhibitory input from the contralateral ear (Wohlers and Huber 1982). Selverston et al. (1985) selectively killed one of the Omega cells. In this case the inhibition in the other remaining Omega cell is no longer detectable (Fig. 7). So these two cells are coupled by reciprocal inhibition. The Omega neuron also inhibits an ascending neuron AN2 on the contralateral side.

The neuron pair AN2 may provide information on sound direction - sharpened by the Omega neurons - because of the difference of excitation between the left and the right AN2. When measured under identical acoustic conditions the dependence of the angular velocity of a cricket from the stimulus angle follows a course which, when plotted, is similar to the difference of excitation in ascending neurons of the left and right side (J. Stabel and G. Wendler in preparation, Fig. 6). The most prominent changes in angular velocity and differences in spike numbers occur at stimulus angles up to 30-40 degrees from the body axis. So, the turning of the animal is closely correlated to the neuronal responses.

So far we have assumed that the identified neurons described earlier are really involved in phonotactic behavior. However, this assumption is based only on the correlation of the behavioral and neuronal characteristics. With a new experimental design we can now study the causal relationships of behavior and neurons (Fig. 8). Animals are fixed on a holder in such a way that they cannot move their body but are free to move their legs. Under this condition the cricket can turn an air supported ball. A specific camera system measures the movement of the ball and therefore the intended translation and rotation of the animal. Switching on a loudspeaker placed 50 degrees left from the body axis causes the animal to

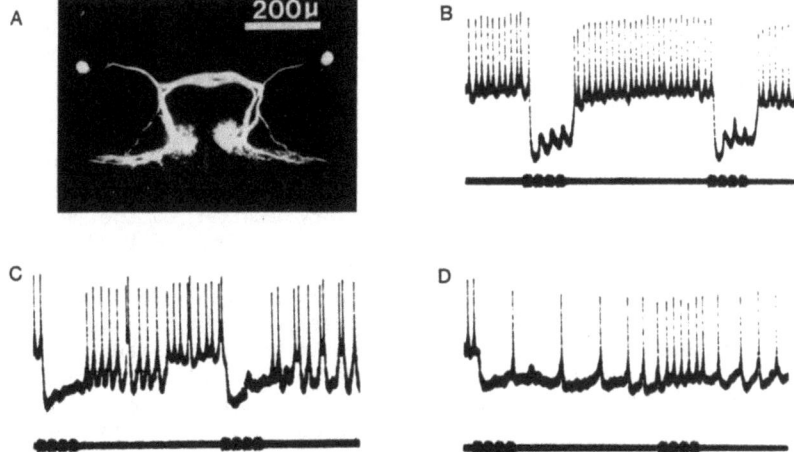

FIG. 7: Photoinactivation of one ON1 neuron while recording from the other (pairs shown in A) removes the inhibitory response to contralateral stimulation (5 kHz 85 dB). B to D show the graded reduction in the summed IPSP response during irradiation with blue light. D, in particular, shows the transition between the last inhibitory response and the weak excitatory response unmasked by the killing procedure (From Selverston et al. 1985, modified).

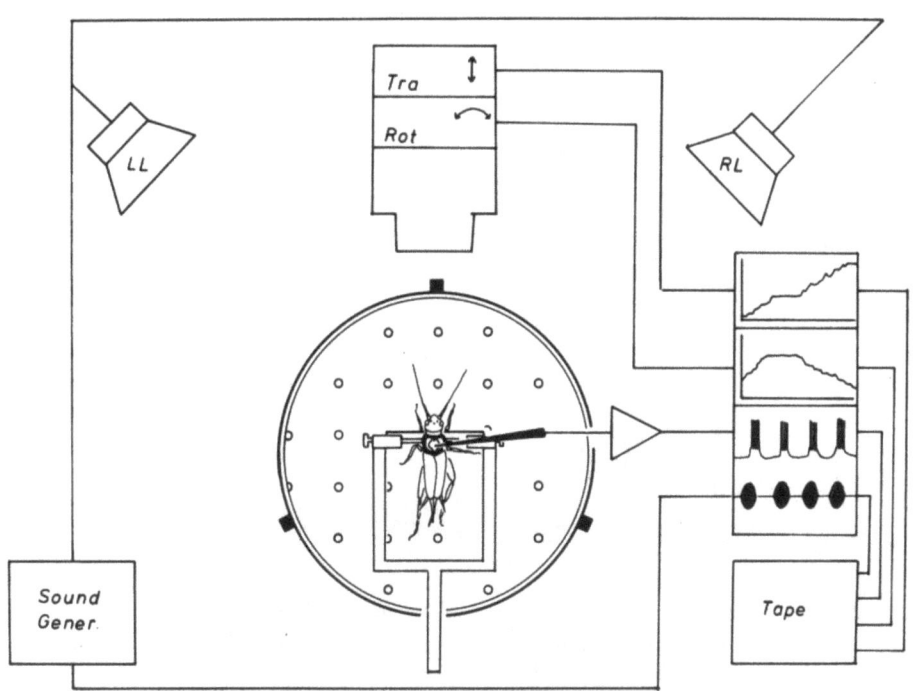

FIG. 8: Experimental setup for intracellular recording during walking, seen from above. The animal is fixed on a holder but the legs can turn an air supported styrofoam ball. The camera senses two components of the ball rotation and therefore of the animal's intended rotation and translation. Sound is delivered by two loudspeakers, each 50 degrees apart from the longitudinal axis of the animal; not to scale.

turn the ball clockwise and backwards so the intended movement is forward and toward the loudspeaker side. Switching to a second loudspeaker placed symmetrically on the right side causes an intended movement of the animal to the right. So the animal tries to orientate in this open-loop situation.

Under the conditions described we are able to record intracellularly from identified auditory interneurons. As long as the animal is not walking, the responses of all cells are very similar to the responses of the same identified cells in totally fixed animals. But at the moment when the animal starts to walk and during sustained walking the responses to sound are suppressed in all prothoracic auditory neurons thus far identified (Fig. 9). The amount of suppression during walking depends on the sound pressure level. At low intensities no excitation to sound stimulation is observed; with moderate and higher intensities the responses appear at a lower magnitude. The suppression is caused by inhibition, because the membrane potential becomes more negative when the animal starts to walk. We have not yet examined the origin of the inhibition but in principle it could arise via a peripheral input from leg receptors that are activated during walking or by a central mechanism that generally switches off the auditory pathway during movement. The consequence for phonotactic behavior could be that an animal far away from the sender localizes only when standing, but when closer to the sound source could also localize during walking.

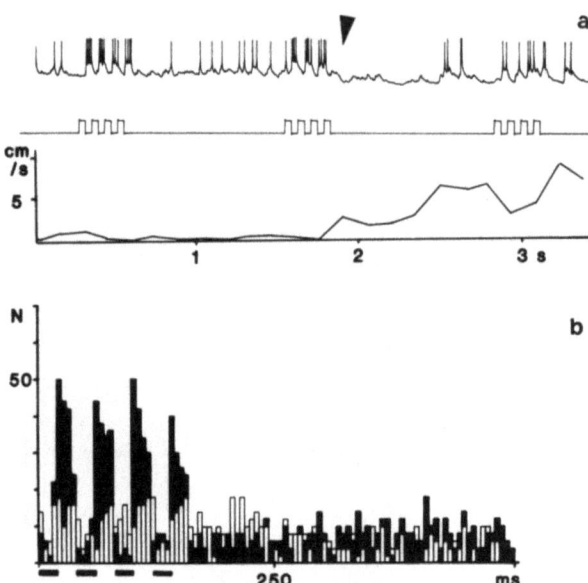

FIG. 9: Auditory responses of ON1 recorded in the situation shown in Fig. 8. a: upper trace is the intracellular recording, middle trace the stimulus marker (5 kHz 80 dB) and lower trace the simultaneous walking speed of the animal. Arrow marks the absence of a neuronal response to the fourth syllable of the chirp just after the animal starts walking. b: pst histogram of the responses of the ON1 to chirps (marked below) at 5 kHz and 60 dB. Black histogram shows the responses to 50 chirps when the walking speed was below 1 cm/s, i.e. the animal was not walking. White histogram shows the responses to 50 chirps when the walking speed was above 2 cm/s. Note suppression of the response nearly to noise level when the animal walks.

Switching on one loudspeaker causes a turning tendency to that side, as described before. When the AN1 cell of that side is now hyperpolarized to the threshold response level the animal now turns to the other side away from the active loudspeaker (Fig. 10). Switching off the hyperpolarizing current while that loudspeaker is still active results in a return of the excitation in AN1 to the pre-hyperpolarized level and consequently a turning tendency back to that side. One can interpret these results in the following way: Hyperpolarizing one AN1 changes the balance of the two mirror-image AN1 cells such that the non-hyperpolarized AN1 is always more strongly excited than the other independent of the loudspeaker direction. Two conclusions may be drawn: Firstly, even if one AN1 cell is desactivated the recognizer in the brain is still operational because the female shows phonotaxis. Secondly, it is indeed the balance of these two cells that determines the turning direction and therefore the acoustic orientation.

If the same experiment is carried out by hyperpolarizing another ascending neuron, AN2, different effects occur. In most animals, no change of the turning direction or the angular velocity was visible upon hyperpolarization. But in some very sensitive animals hyperpolarization of AN2 causes a decrease of angular velocity, but not a switch to the other side. The effect is only significant at intensities higher than 75 dB and in animals with a low behavioral threshold. So AN2 may also play a role in sound localization, but in contrast to AN1 the effects are weaker and occur, not at low, but at high sound intensities. Surprisingly, we could not detect any influence of the omega neuron on the turning tendency using hyperpolarizing methods.

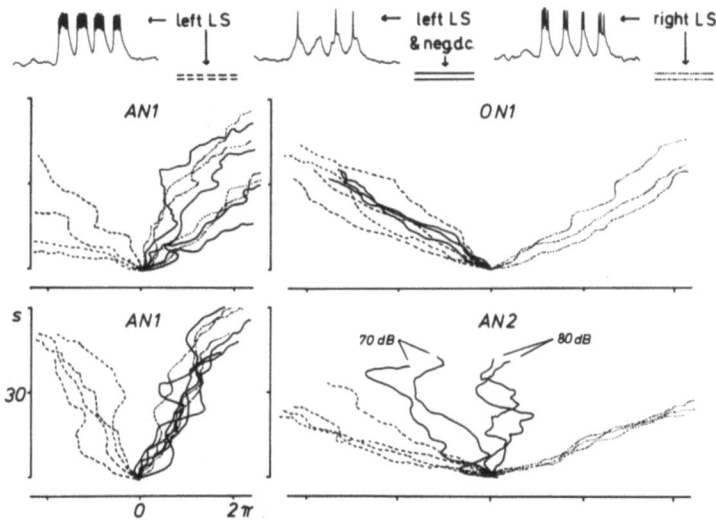

FIG. 10: Turning tendencies of female crickets during calling song determined in the situation shown in Fig. 8. Simultaneous recordings were made from the indicated neurons of the left side (the neurons which are excited by the left ear). The angle of the ball rotation is plotted over time for different conditions. For each conditional change the trace is reset to the starting point. Stipled line - left loudspeaker active. Dotted line - right loudspeaker active. Solid line - left loudspeaker active and respective neuron hyperpolarized to threshold level. A recording example from the left AN1 is shown above for the different conditions. If not indicated, sound intensity was 70 dB.

PHONOTAXIS WITH ONE EAR

As mentioned above, sound localization requires two ears and for a long time it was believed that crickets need two ears for orientation. But it was forgotten that J. Regen, the discoverer of orthopteran phonotaxis, reported in the beginning of this century that crickets with only one ear can reach a sound source (see Huber et al. 1984). Huber et al. (1984) reexamined his statement by cutting one foreleg in an adult. The animal circles towards the side of the remaining ear. But there is a considerable amount of walking in a loudspeaker dependent direction. When the amputation is done in a larval instar, the leg will regenerate but lacks an ear. Some of these animals, tested as adults, tracked without any circling; others circled all the time toward the side of the remaining ear, and still others walked in cycloids (Fig. 11, B. Schmitz in press). So the loss of an ear can be compensated in some animals during larval development.

This finding raises the question of how an animal with only one ear can detect sound direction. If, in such one-eared animals, the auditory pathway in the central nervous system were functional only on the side of the remaining ear, orientation would only be possible if the animal measures sound intensity successively to determine the direction of the sound source. But, if in one eared animals, both sides of the central auditory pathway are functional, other mechanisms for localization would

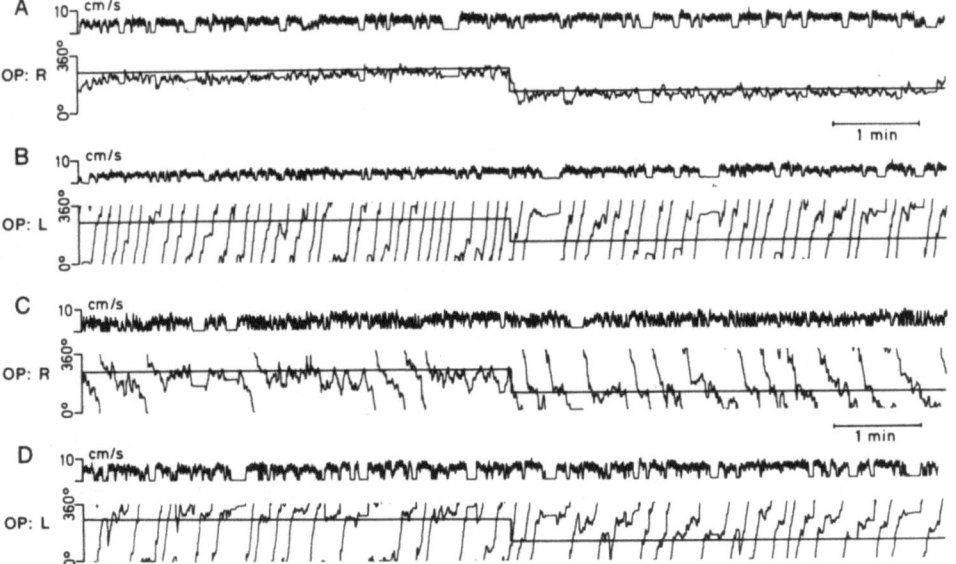

FIG. 11: Walking behavior (with calling song presentation at 80 dB) of 4 adult, female crickets (A-D), one foreleg of which was amputated in the 8./9. larval instar; upper traces: time courses of walking velocity, lower traces: time courses of walking direction; horizontal lines indicate sound direction changed after 5 min by 135 degrees. A: cricket tracking a stable course deviating from sound direction toward the left (intact) side; B: cricket, permanently circling toward the right (intact) side; C, D: crickets walking in cycloids, i.e. circling toward the intact side, but at times also preferring a walking direction deviating from sound direction toward this side; operated leg marked by R or L (Courtesy of B. Schmitz).

be possible. In animals operated in a larval instar we recorded from central auditory neurons that would normally receive excitatory input from the now deafferented side and stained the neurons by intracellular dye injection. Most of the neurons were changed in their morphology (Fig. 12). In contrast to the situation in the intact animal, the cells of the operated side send dendritic branches over the midline and reach the auditory neuropil of the intact side. ON1s that show this dendritic sprouting over the midline are excited by the remaining ear, but in those ON1s that do not exhibit these new branches any perceivable excitation is nearly overridden by inhibition (Fig. 12a, b). AN1s and AN2s exhibiting dendritic sprouting were also excited by the remaining ear. Threshold and intensity/response functions are very similar to those of intact animals, with the exception of those from the ON1 that did not sprout. So these new connections are functional. The consequence is that the responses of auditory interneurons from the operated side are only different from the responses of those of intact animals in the sense that their excitation arises from the 'wrong' ear.

These one-eared animals have functional central pathways on both sides and can localize the sound source. If this localization results from the comparison of excitation from the left and right side of the CNS, then a different threshold and different slopes of the intensity/response functions of the left and right side would be sufficient to explain orientation. If the two intensity/response functions do not cross under these conditions, the result would be a permanent turning to the side of the more sensitive neuron. If they cross then an intensity range should exist whereby excitation is balanced and a stable walking course could be maintained. Recordings from the left and right AN2 neuron of such an animal show indeed that the neurons of the intact and of the deprived side have different thresholds and that the intensity/response functions cross. But further behavioral and neuronal data from the same individual are necessary to support the hypothesis.

Under experimental conditions differing from those described above, in many cases an ear can regenerate on a regenerated leg (Fig. 13). The behavioral test shows that the animals can orient. Cutting away the regenerated leg leads to circling. So the regenerated ear is functional. The omega neurons in these animals exhibit a totally normal morphology and physiology, in a sense that dendritic sprouting is not visible and excitatory input arises from the regenerated ear, inhibitory input from the intact ear. This raises the question over the mechanisms and the time scale of morphological and functional plasticity in the auditory system of the cricket.

ACKNOWLEDGEMENTS

I thank B. Schmitz, J. Stabel and D. W. Wohlers for providing unpublished data, M. Horner and M.-L. Obermayer for help with some experiments, H.-U. Kleindienst for computer programs and D. W. Wohlers for criticism of the manuscript.

REFERENCES

Boyd P, Lewis B (1983) Peripheral auditory directionality in the cricket. J Comp Physiol 153: 523-532.

Boyan GS (1979) Directional responses to sound in the central nervous system of the cricket Teleogryllus commodus. I. Ascending interneurons. J Comp Physiol 130: 137-150.

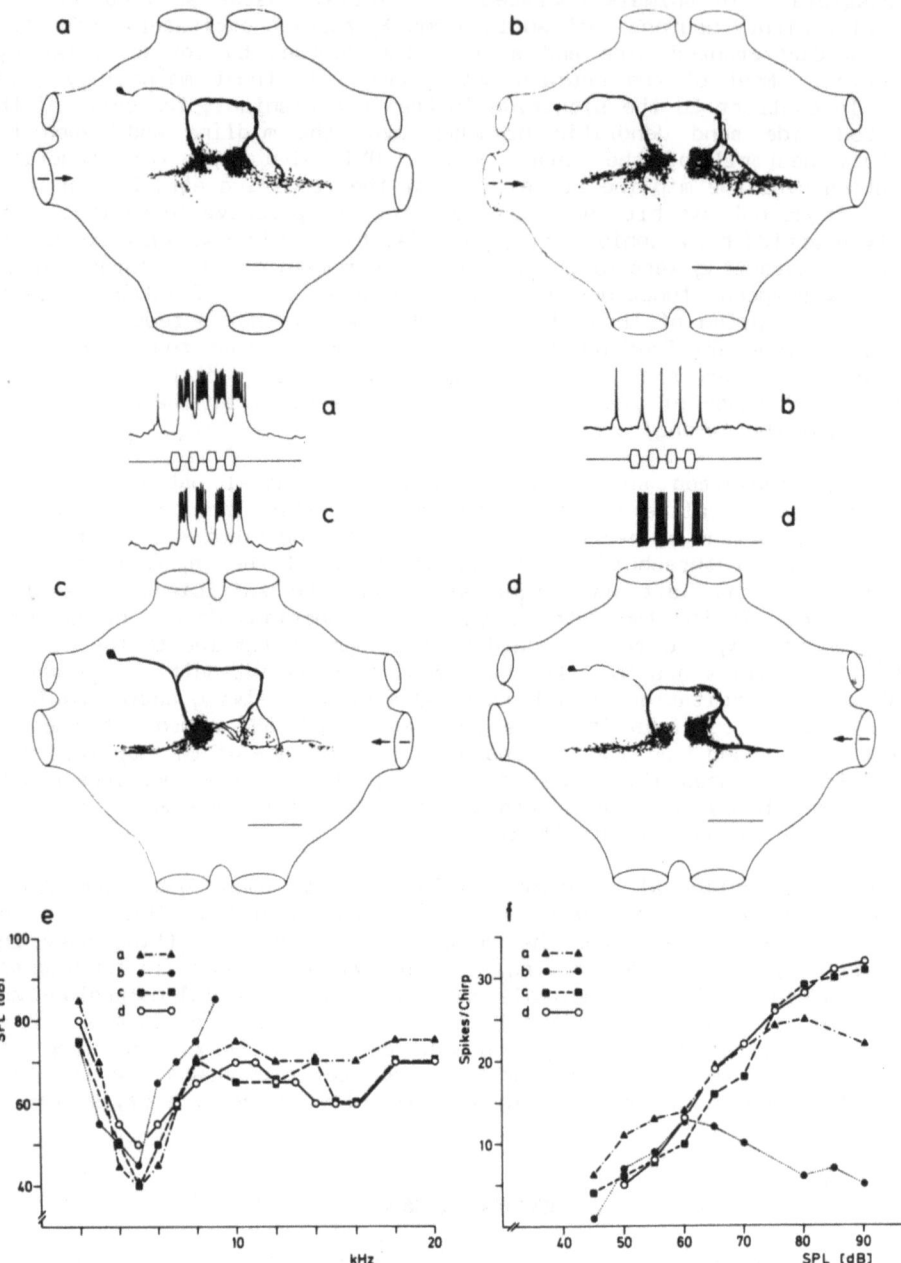

FIG. 12: Morphology and physiology of Omega Neurons (ON1s) in unilaterally deafferented animals. Reconstructions of cells from the deafferented (a,b) and the intact (c,d) side. Arrows denote the deafferented side; scale 200 μm. The response of each cell to one chirp of the simulated song (5 kHz, 80 dB) is shown below (a,b) or above (c,d) the reconstruction. The recordings in a-c were made from the soma-ipsilateral (left) side, in d from the soma-contralateral (right) side. Note the different responses in a and b and compare with the different morphology. e: threshold and f: 5 kHz intensity/response functions of the drawn neurons (the letters a-d correspond to the lettered drawings) (From Schildberger et al. 1986).

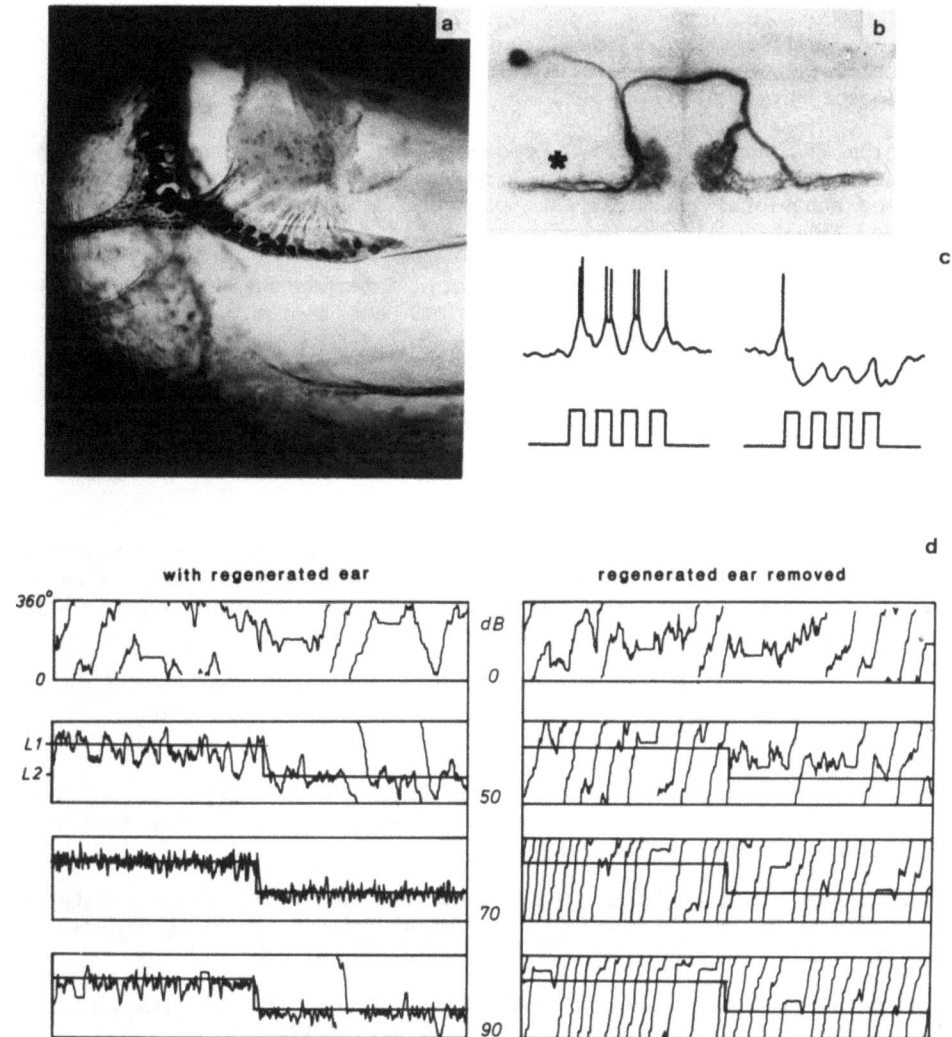

FIG. 13: a: regenerated tympanal and subgenualorgan in a regenerated foreleg of an adult cricket which was amputated at the femur/tibia joint in the 6th larval instar; silver intensified cobalt back-fill through the auditory nerve. b: Omega neuron ON1 of an adult cricket with a regenerated left ear (regenerated side denoted by *). The left foreleg was amputated at the femur/tibia joint in the 6th larval instar. There are no obvious morphological differences between this neuron and ON1s in unoperated animals (see Fig. 6). c: responses of the neuron shown in b to calling song (5 kHz 90 dB) presented in a closed sound field to the left (regenerated) ear and to the right (normal) ear. Excitation arises from the regenerated ear, inhibition from the contralateral (normal) ear. d: Walking behavior of an adult cricket with a regenerated ear during calling song at various sound intensities, amputation as in a, b. Traces show the walking direction, horizontal lines the loudspeaker direction; each loudspeaker condition was maintained for 2 min. The animal was tested, then one day later the regenerated leg was removed and the animal retested. With the regenerated ear the animal tracks over the total range of tested intensities. After removal of the ear, the animal circles towards the side of the remaining ear.

Boyan GS (1980) Auditory neurons in the brain of the cricket *Gryllus bimaculatus*. J Comp Physiol 140: 81-93.

Boyan GS, Williams JLD (1982) Auditory neurones in the brain of the cricket *Gryllus bimaculatus* (De Geer): Ascending interneurones. J Insect Physiol 28: 493-501.

Capranica RR, Rose G (1983) Frequency and temporal processing in the auditory system of anurans. In: Huber F, Markl H (eds) Neuroethology and behavioral physiology. Springer, Berlin, Heidelberg, New York. pp. 136-152.

Casaday GB, Hoy RR (1977) Auditory interneurons in the cricket *Teleogryllus oceanicus*: physiological and anatomical properties. J Comp Physiol 121: 1-13.

Esch H, Huber F, Wohlers DW (1980) Primary auditory neurons in crickets: Physiology and central projections. J Comp Physiol 137: 27-38.

Huber F, Thorson J (1985) Cricket auditory communication. Sci Amer 253: 60-68.

Huber F, Kleindienst H-U, Weber T, Thorson J (1984) Auditory behavior of the cricket. III. Tracking of male calling song by surgically and developmentally one-eared females, and the curious role of the anterior tympanum. J Comp Physiol A 155: 725-738.

Popov AV, Markovich AM, Andjan AS (1978) Auditory interneurons in the prothoracic ganglion of the cricket, *Gryllus bimaculatus*. I. The large segmental auditory neuron (LSAN). J Comp Physiol 126: 183-192.

Regen J (1913) Uber die Anlockung des Weibchens von *Gryllus campestris* L. durch telephonisch ubertragene Stridulations-laute des Mannchens. Pflugers Arch 155: 193-200.

Rheinlaender J, Blaetgen G (1982) The precision of auditory lateralization in the cricket *Gryllus bimaculatus*. Physiol Entomol 7: 209-218.

Schildberger K (1984) Temporal selectivity of identified auditory neurons in the cricket brain. J Comp Physiol A 155: 171-185.

Schildberger K (1985) Recognition of temporal patterns by identified auditory neurons in the cricket brain. In: Kalmring K, Elsner N (eds) Acoustic and vibrational communication in insects. Paul Parey, Berlin. pp. 41-49.

Schildberger K, Wohlers D, Schmitz B, Kleindienst H, Huber F (1986) Morphological and physiological changes in central auditory neurons following unilateral foreleg amputation in larval crickets. J Comp Physiol A 158: 291-300.

Selverston AL, Kleindienst H-U, Huber F (1985) Synaptic connectivity between cricket auditory interneurons as studied by selective photoinactivation. J Neurosci 5: 1283-1292.

Thorson J, Weber T, Huber F (1982) Auditory behavior of the cricket II. Simplicity of calling song recognition in *Gryllus* and anomalous phonotaxis at abnormal carrier frequencies. J Comp Physiol 146: 361-378.

Weber T, Thorson J, Huber F (1981) Auditory behavior of the cricket. I. Dynamics of compensated walking and discrimination paradigms on the Kramer treadmill. J Comp Physiol 141: 215-232.

Wiese K, Eilts K (1985) Evidence for matched frequency dependence of bilateral inhibition in the auditory pathway of _Gryllus bimaculatus_. Zool Jb Physiol 89: 181-201.

Wohlers DW, Huber F (1978) Intracellular recording and staining of cricket auditory interneurons (_Gryllus campestris_ L., _Gryllus bimaculatus_ De Geer). J Comp Physiol 127: 11-28.

Wohlers DW, Huber F (1982) Processing of sound signals by six types of neurons in the prothoracic ganglion of the cricket, _Gryllus campestris_ L. J Comp Physiol 146: 161-173.

Wohlers DW, Huber F (1985) Topographical organization of the auditory pathway within the prothoracic ganglion of the cricket _Gryllus campestris_ L. Cell Tissue Res 239: 555-565.

A MODEL FOR DECISION MAKING IN THE INSECT NERVOUS SYSTEM

JENNIFER S. ALTMAN and JENNY KIEN

Institut für Zoologie,

Universität Regensburg,

Fed. Rep. Germany

ABSTRACT

Little is known about the neuronal mechanisms for selecting behavioural outputs appropriate to ongoing conditions. We present a model in which decisions are made by a concensus between the inputs at each stage in the system, not by a few neurones in a single centre. The stages are interconnected by loops of varying lengths, each with specific control functions. Neuromodulators and hormones contribute to the overall output by altering excitability but no single input is necessary and sufficient for producing any output.

1. INTRODUCTION

Among the most challenging problems in neuroscience are perception, memory and decision making. Whereas the first two are subjects of intensive research and debate, little attention has been paid to the mechanisms by which decisions are made. Here we present a model of the neuronal processes that result in changes in an animal's behaviour, which we hope will stimulate both the way that we think about nervous systems and the design of experiments to test the model.

Deciding what to do next is a central and continuous part of the neural control of behaviour, whether it is continuing to do the same thing, making small changes such as turning a few degrees to the left instead of walking straight ahead, or switching from one programme to another, from grooming to feeding, for example. Equally important, although less obvious, are decisions not to respond to a stimulus on a particular occasion, which amounts to deciding not to change programmes. These problems have been studied on the behavioural level by ethologists and phychologists, who talk in terms of motivation, arousal and hierarchies of behaviours, but there have been few attempts to determine what is going on in the nervous system to direct such changes. Some of the neuronal networks controlling particular motor outputs, such as flight and kicking in locusts, are being intensively studied and are at least partially described but we still know very little about the mechanism for selecting the output appropriate to the conditions of the moment.

Over the past decades the way we think about nervous systems has been dominated by concepts which we consider are now restrictive and hindering progress towards understanding the more complex integrative functions performed by the nervous system. First, the use of intracellular recording and staining methods have led, especially in invertebrates, to the idea of the single identifiable neurone (eg. Hoyle 1977), which is usually studied on its own without considering its normal biological context; its inputs are examined and its outputs characterised under rigourously controlled and restricted conditions. This approach has provided much valuable information on neuronal interaction but at the same time has led to the view of individual neurones as independent information channels and tends to make us think of the output of the single neurone under study as the important input to the next neurone in a linear circuit.

Second, the concept of a linear hierarchy of information processing and instructional levels still has considerable force (see also Davis 1976). In this scheme, the higher centres of the brain are seen as being in command, dictating what should be done by the lower centres of the nerve cord. Integral to this approach is the idea that there is a decision centre in the brain, with a watershed between the inflowing sensory information and the generation of motor commands.

Data from both vertebrates and invertebrates that do not fit a linear system containing a command centre have gradually been accumulating (Eaton and DiDomenico 1987) and the model we present here provides a more appropriate framework. Central to our concept is the fact that information at all stages of processing, including feedback on the state of the networks at each level, is fed to every other level of the system, so that the levels are all interconnected in a number of loops (see also Davis 1985). We propose that there is no single decision-making centre but that decision making is a distributed function of the whole nervous system: particular behaviours occur as a result of the balance of activities in different parts of the nervous system at any instant.

In this context we define a behaviour as a routine dominated by a single motor output. Cricket courtship, for example, which is a behaviour in ethologists' terms, consists of a number of routines such as walking, singing the calling song, orientation, singing the courtship song and copulation (Huber 1980). Each of these routines we term a behaviour - and doing nothing as also a behaviour! Every behaviour involves the whole animal: flying is not just wing beats, walking is more than a sequence of leg movements, and the control of any action involves not only the CNS but the peripheral nervous system, the effectors and the skeleton as well.

Behaviour must also be seen as a continuum, not as a series of separate, step-function movements with sudden switches between them. The result is a continuous interaction with the environment that produces a continual change both outside and inside the animal. Clearly it is important to think about the system as a whole in order to assess the contribution of the parts.

2. THE APPROACH

Any central nervous system can be considered as a number of processing centres, which we term stations, joined by transmission lines or connexions. A station, its inputs and its outputs form a stage (Fig. 1). Our model concerns the generation of outputs by a station in response to information received from other stations in the system. We do not deal with the local networks within each station that process the inputs (see Robertson 1987) but deal solely with the information that is transmitted between stations. <u>We propose that the same operating principles apply at</u>

all stations, so that any one can be taken as a model for all. In practice, this means that the ouptuts of the motor neurones can be treated in exactly the same way as those of the descending interneurones from the brain.

Each station contains the targets for the outputs of other stages in the system. A target can be a single neurone or muscle, a group or groups of neurones or muscles, or a whole local network. It is a collector and integrator of information and the information it receives at any moment determines the output it generates at that time. To appreciate the way the system works, it is helpful to look backwards at the sources of incoming information rather than concentrating solely on the outputs of a station. This output is the result of combining what comes into a station with what is already going on in it and so it depends on the relative strengths of the various inputs at any time - it is a concensus rather than the result of a command from elsewhere (Kien 1983).

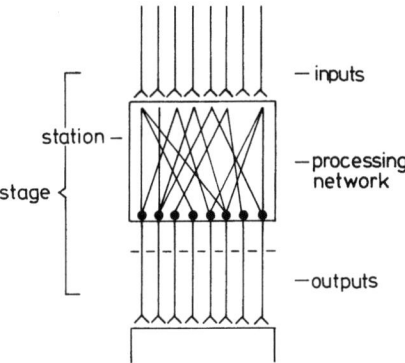

FIG. 1: Definitions of the terms used in this chapter. To get away from the linear hierarchy concept embedded in terms such as neural centre or level, neutral terms have been used: a <u>station</u> is a structure such as a ganglion, that contains <u>processing networks</u> of interneurones. Stations receive <u>inputs</u> from other stations; these feed either directly or through the processing networks onto the neurones or muscles that form the <u>outputs</u> of the station. A station with its inputs and outputs form a <u>stage</u>. The dotted line indicates the across fibre pattern (AFP) which can be read across the outputs of a station.

3. THE PRINCIPLES

3.1. Principle 1

Each target neurone or muscle integrates a large number of inputs (Fig. 2). When and how a target neurone or muscle fires depends on the pattern of excitation at any moment across the neurones that provide its inputs, irrespective of the source of the input (ie. the "across fibre pattern", AFP; Erikson 1963). Not all the inputs will be active at once and not all will be excitatory. Activity in the inputs may also vary with time in their frequency and patterning. The sources of the inputs, dealt with in detail in section 4, include local and distant primary sensory neurones, the outputs of local signal detection and pattern generator networks, and the outputs of other stages.

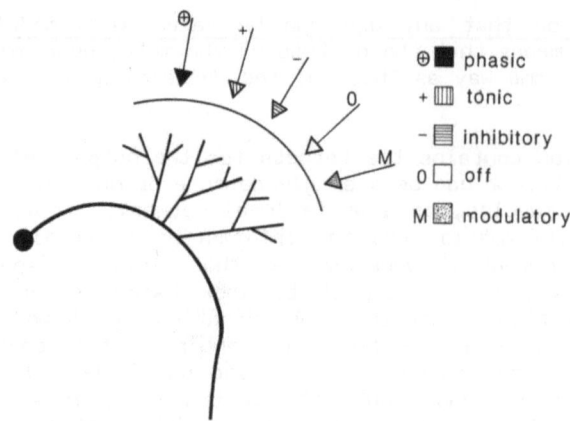

FIG. 2: Each target neurones of muscle collects and integrates four main types of inputs: phasic excitatory; tonic excitatory; inhibitory and modulatory. Not all the inputs are necessarily active at the same time, so a fifth class of silent or off neurones is included - these can of course be any of the other four types. These symbols are used throughout the figures describing AFPs.

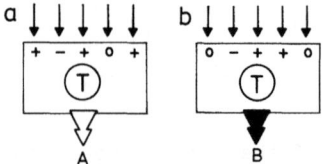

FIG. 3: The input to a target (T) is the pattern of activity across all the individual inputs at any moment. This is the AFP (across fibre pattern, Erikson 1963). It can change from moment to moment and each AFP results in a different output (a, b). The AFP includes inputs that are silent and the output resulting from some AFPs may be zero (see also Fig. 6). In a) and b) the second and third units from the left have the same activity but the overall combination of active units in the two examples is different and so the outputs are not the same. A target for inputs may be a single neurone or muscle, groups of neurones or muscles, or a whole station, depending on the level of analysis.

3.2. Principle 2

The input to a station is not the firing of individual neurones but the across fibre pattern (AFP) of activity in all the inputs it receives; this includes fibres that are silent (Fig. 3). Qualitative or quantitative changes in the AFP lead to changes in the activity of the target, changes in its output and ultimately to changes in behaviour. The inputs do not necessarily all synapse with the same target within a station, nor does each target necessarily receive all the inputs.

3.3. Principle 3

The AFPs for different behaviours are not mutually exclusive but units may be active in more than one (Figs. 3, 4, 5, 7), an arrangement that is both economical and flexible. Each AFP is the outcome of a particular pattern of distant and local inputs onto the arborisations of the output neurones of a stage. As the inputs to the individual output neurones vary and the weighting of distant and local inputs shifts, so the output neurones that are activated change and hence the AFP changes.

3.4. Principle 4

Each station receives inputs from all other stages in the system. A station also sends outputs to all other stages. The whole system is thus formed of loops of varying lengths, each of which plays a different role in the organisation of motor outputs. All the loops are essential for accurate performance of any motor programme and so none can be said to be more important or to control the others in a linear hierarchical sense.

Principles 1 - 3 concern the generation of the output at any stage in the system, whereas Principle 4 deals with the sources of the inputs to a stage and the interactions between stages. The first three principles are therefore treated together in the examples detailed below, followed by a separate discussion of the fourth principle. As it is our contention that these principles operate at all stages in the system, we take the simplest, most accessible stage, the input from motor neurones to muscles as a good model for all other stages, and then extend the description to the inputs to the motor neurones and to the inputs and outputs of the descending interneurones from the brain to the thoracic ganglia.

4. EXAMPLES

4.1. The control of muscle contraction by motor neurones

The way an effector such as a leg or a wing is used depends on which behaviour is being performed: the amplitude and speed of movement, the angle with respect to the body and the pattern of coordination with other effectors are all characteristic for each behaviour. Changes in these parameters result from changes in the pattern of excitation, or AFP, in the motor neurons that innervate the muscles controlling the effectors and so can be monitored by recording from the motor nerves.

The large hind (metathoracic) legs of the locust are a good example. They are rapidly extended as a pair during jumping, singly in defensive kicking; they are extended and flexed relatively slowly during walking, when they mostly alternate, and in grooming, when only one is used; and they maintain various positions for different static postures. During flight they are tucked up in a flexed position against the body for streamlining or extended as balancers or rudders at slow flight speeds, in turbulent air or after take off and before landing. All this is achieved by a group of muscles in the leg and thorax wall, whose contractions are regulated by a relatively small group of motor neurones in the metathoracic ganglion.

Unlike mammalian muscles, insect muscle fibres may be innervated by up to four types of motor neurone: fast, slow, inhibitory and modulatory. As well as monoamines or peptides released from modulatory neurones (Evans and O'Shea 1977; O'Shea and Adams 1981), the responsiveness of some muscles may also be altered by blood-borne modulator substances (Kravitz et al. 1980). The surface of the muscle fibre thus acts as an integrator of information and the fibre contracts if the sum of the inputs brings its membrane above threshold (Principle 1, Fig. 2). The type of contraction produced by a muscle therefore depends on the AFP in the neurones that innervate it (Principle 2, Fig. 3). If none of the inputs is active or the inhibitory input is stronger than the excitatory, the muscle does not contract. One motor neurone may be active in more than one AFP (Principle 3, Fig. 4), so that the behaviour of the muscle cannot be described adequately by recording only from one motor neurone. For example, when fast and slow motor neurones are active together, the contraction will differ in amplitude and duration from that produced by a slow motor neurone and a modulator.

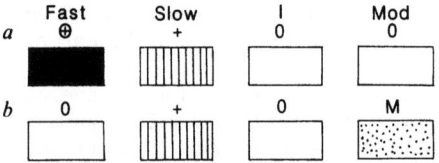

FIG. 4: A simple example of AFPs in motor neurones innervating a muscle. Each muscle fibre receives inputs from four motor neurones: fast (equivalent to phasic), slow (equivalent to tonic), inhibitory and modulatory. The type of contraction depends on the combination of motor neurones active at one time, that is on the AFP of motor neurone activity. The AFP in a) would produce a fast-onset, maintained contraction; that in b) one with a slower onset and longer duration. Note that the slow motor neurone is active in both cases but it is the AFP i.e. the total input to the muscle that determines the type of contraction.

The inputs may vary not only in whether or not they fire but also in how they fire, that is in their frequency and burst pattern. This is best seen in the most extreme case, where muscle fibres are innervated by a a single excitatory motor neurone, as in direct flight muscles of locust or mammalian skeletal muscle. Here the AFP of the input to the muscle fibre consists of activity/no activity in a single motor neurone, but amplitude and duration of contraction are determined by the frequency and patterning of the firing in this motor neurone. A very clear example is the dual function wing/leg muscles of the locust thorax (Wilson 1962): the muscles that are synergists for flight are antagonists for walking and the motor neurones fire continuously and rhythmically at about 20 Hz during flight but in discontinuous high frequency bursts during leg movements (Wilson 1962). That is, the muscle membrane integrates the motor neurone output to produce two very different patterns of contraction, repeated fast twitches during flight and periodic sustained contractions for walking.

These principles also apply to the coordination of the angles of all the joints in a single leg, which will depend on the AFP in the motor neurones of all the muscles in the leg, and to the coordination of one leg with all the others, resulting from the AFP in the whole population of motor neurones innervating all the legs. Changes in the movements at a single joint or of a whole limb, or in the coordination between limbs result from changes in the relevant AFP. It is, however, the same units that are active in different combinations that produce the different patterns of coordination.

The control of the femur-tibia joint of the metathoracic leg of the locust provides another good example of how different AFPs contain different combinations of the same active elements (Fig. 5). To produce a kick (Heitler and Burrows 1977) the tibia is first flexed against the femur and then rapidly extended. The flexion is produced by contraction of the flexor muscle in the femur, which is driven by fast and slow motor neurones; the extensor muscle is relaxed and presumably the extensor motor neurones are silent or firing at very low frequency (Fig. 5, top row). It is not known whether the modulator(s) is active or whether the inhibitor to the extensor fires. Next, in preparation for kicking the extensor muscle contracts as well, to produce a state known as co-contraction (Heitler and Burrows 1977) - the excitatory motor neurones of both flexor and extensor are active simultaneously (Fig. 5, row 2). The AFP has changed but the flexor motor neurones are still firing. In the final phase, the tension in the femur-tibia joint produced by co-contraction is released by sudden inhibition of the flexor motor neurones and relaxation of the flexor muscle, so that the tibia is rapidly extended. The AFP has changed again (Fig. 5, row 3), although the excitatory neurones of the extensor continue to fire.

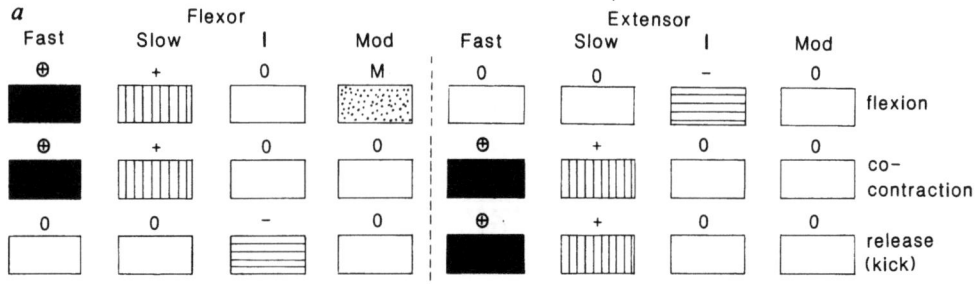

FIG. 5: Partly hypothetical schemes of AFPs in the motor neurones innervating the tibial extensor and flexor muscles in the locust metathoracic leg. This emphasises that the concept of AFPs can also be applied to all motor neurones controlling a single joint or a whole limb. Three phases in generation of a kick are shown. Top row: flexing the tibia against the femur by contracting the flexor muscle. Middle row: flexor and extensor muscles contact simultaneously to tense the femur-tiba joint. Bottom row: tibia is rapidly extended by release of this tension as the flexor rapidly relaxes. These schemes illustrate the principle that the same unit can be active in more than one AFP and the resultant output depends on the AFP rather that on individual inputs.

Although the innervation of muscles is relatively simple and the neuromuscular system provides easy access for analysis, there appears to be no complete description of the control of motor output in terms of changing constellations of active neurones, even in the well-studied tibial flexor and extensor muscles of the locust. Most experiments have focussed on the activity of a single motor neurone, usually either fast or slow excitor, and the whole output pattern has not been considered. Consequently, the roles of some of the units in Fig. 5 remain guesswork.

4.2. Inputs to motor neurones

The principles that govern the control of muscle contraction by motor neurones are clearly reiterated at the preceding stage, the inputs to the motor neurones and the generation of the motor output pattern (Fig. 6). Each motor neurone can be considered in exactly the same way as each muscle, as an integrator of information from a number of sources (Principle 1, Fig. 2). The output of a motor neurone will depend on which inputs are active at a particular time (Principle 2, Fig. 3), that is on the AFP of its inputs. The motor neurone will fire if any combination of inputs raises its membrane potential above threshold - a single input driven hard enough, as in some experimemtal paradigms, may suffice but this does not necessarily mean that this input alone drives the motor neurone in a normal animal. The response of a motor neurone to stimulation of any one of its inputs is therefore not strictly predictable but is context dependent, as it depends on which other inputs are active at the same time.

Changes in the AFP in the inputs to a single motor neurone may change the frequency and pattern of its firing, although the AFPs for different output patterns may share some of the same active units (Principle 3, Fig. 3). Motor neurones may share some inputs (Fig. 6) but this does not imply they necessarily fire together because the AFP of the inputs to each will differ.

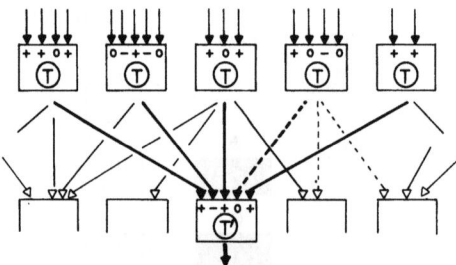

FIG. 6: The principle that output depends on the pattern of inputs, or AFPs, in the preceding stage is reiterated through the whole system. Here, several output neurones (T) provide inputs to a target (T') in the next station. Each T receives a particular pattern of inputs (which change with time, see Fig. 3) which determines its output; the output of T' thus depends on the inputs it receives from the Ts, which in turn depend on the inputs they receive from preceding stages. Note that each T can provide inputs to more than one T' in other stations, with the exception of the motor neurones, most of which innervate only one muscle.

The total output of AFP of a pool of motor neurones innervating a group of muscles, for example those of the metathoracic leg, thus depends on the total AFP in the inputs to these motor neurones. Changes in this pattern lead to changes in the pattern of activation of motor neurones, thence to changes in the pattern of muscle contraction and so to changes in the performance of the limb.

The inputs to the motor neurones come from all stages in the system, not just from the premotor interneurone networks but also include direct sensory inputs and highly processed information from distant parts of the system (Principle 4). These will be discussed in section 5.

4.3. Descending interneurones originating in the brain

An important input to the motor neurones is information about the changes in the world around the animal that are monitored by sense organs on the head - in an insect chiefly the eyes, ocelli, antennae and head hairs - and transmitted to the thoracic ganglia by descending interneurones (DINs). These DINs have their cell bodies and input arborisations in the head ganglia (brain and suboesophageal ganglion) and their axons terminate in the ganglia of the ventral nerve cord, where they synapse directly with the motor neurones or with premotor interneurones or both (Burrows and Rowell 1973; Bacon and Tyrer 1979; Kien 1979; Pearson et al. 1980; Rowell and Pearson 1983).

The general form of these DINs is similar to that of a motor neurone, with a collecting arborisation in one area and an axon projecting to a distant target. We consider that the inputs and outputs of the DINs can be treated in exactly the same way as those of the motor neurones - the four principles also operate at this level. Each DIN integrates a number of inputs across its arborisation; its output depends on the AFP of these inputs; a change in the AFP of the inputs may lead to a change in the output of the DIN but AFPs producing different outputs can have active units in common; DINs receive inputs from all stages of the system not just from the sense organs of the head. The only organisational difference between the output of the DINs and of the motor neurones is that one DIN may provide inputs to several or many target neurones (Fig. 6), whereas motor neurones usually innervate only one muscle. As each target neurone receives a number of other inputs (Fig. 6) not all the neurones receiving inputs from one DIN will necessarily be activated together.

A really detailed anatomical and physiological description of the inputs to DINs with arborisation in the brain is available only for the giant descending neurones of the fly (see Nässel 1987). This neurone receives inputs from the eye and antenna, as well as ascending inputs from the thoracic ganglia, each of which terminates on a different part of the arborisation (Bacon and Strausfeld 1986). The different inputs to the tritocerebral commissure giant (TCG) neurone also appear to be morphologically separated (Bacon and Möhl 1983), suggesting that segregation of inputs to DIN arborisations in the brain may be common. If this spatial segregation has a functional significance in response generation, rather than merely reflecting the spatial organisation in the brain, then our first principle is too simplistic in suggesting there is a simple summation of inputs across the arborisation of a neurone. Suggestion of functional subdivisions within the arborisations of motor neurones (Altman 1981; Robertson 1987) and interneurones (Siegler and Burrows 1979) have been made previously.

Looking at the outputs of all the DINs leaving the brain, the AFP will depend on which are activated or inhibited by a particular set of inputs at any given moment (Fig. 6). The directional sensitivities of optomotor DINs in the bee illustrate how precise information on direction of motion can be given by the AFP of a group of DIN axons (W.A. Fletcher, R.G. Guy, P.G. Mobbs and L.J. Goodman in preparation). There are ten DINs in each cervical connective that are sensitive to wide field stimuli moving in particular directions. They are broadly tuned but each has a characteristic response curve with a peak in a certain direction. The peak responses are not evenly distributed but are all close to either horizontal or near vertical. The result is that in each connective there are four units responding preferentially to downward, two to upward and four to horizontal movements with a bias to left in the left connective and to right in the right connective. Intermediate directions are presumably coded by different combinations of excitation and inhibition in the various horizontal and vertical units, the exact direction being read out as a vectorial sum of the activities of all the neurones. That is, directions of stimulus movement are coded by the AFPs in the twenty neurones.

A change in the AFP in the DIN outputs will produce a change in the inputs to the target neurones of the motor networks and so can lead to a change in behaviour, always provided that the other inputs to the target are in a permissive state. An example of this is the gating of the input from descending ocellar interneurones to flight motor neurones (Reichert et al. 1985). This input is effective in changing the firing of the motor neurones only if the flight pattern generator is active and in the correct phase. That is, there is a local input that determines whether or not the descending information is effective.

So far, little work has been done to correlate the DIN AFP with particular behaviours - experimentally this is at present very difficult as it requires recording from a large number of units simultaneously in moving animals. In theory, however, it is possible to see that a slight change in the DIN AFP (Fig. 7a) could lead to a subtle change in the expression of an ongoing behaviour, for example an increase in the power output of the wing depressor muscles on one side initiating a change in flight direction. It is possible that a number of DINs remain active throughout, providing the basic input to keep the animal flying (those enclosed by dotted lines in Fig. 7a), while the subtle changes in detail are provided by changes in other units contributing to the AFP. A change in type of behaviour, from flying to perching on a grass stem for instance, would involve a radical change in the AFP (Fig. 7b), though some of the same units may still be active. Again the basic behaviour may

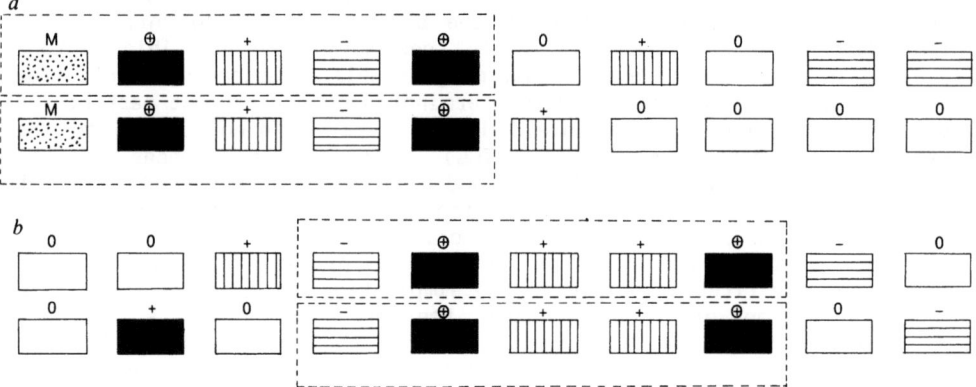

Fig. 7: A hypothetical example illustrating how a single group of descending interneurones (DINs) can be involved in regulating different behaviours by changes in the AFP. Suppose the two rows in a) represent two AFPs during flight, those in b) two AFPs during perching on a grass stem. The units enclosed by dotted lines are active throughout each behaviour and are involved in expression of the basic output for the behaviour. Changes in activity in other units in the row (compare a) top and bottom; b) top and bottom) then regulate finer details of the output, such as wing angle or flexor tension. Note that some units belong to the 'basic' group for both behaviours - compare units in the boxes in a) and b). In practice it is easier to speak of <u>assemblies</u> of active DINs: the shaded boxes in each row represent an assembly and there is a different assembly for every aspect of behaviour (see text for further discussion).

result from a specific group of active units and subtle changes in posture by changes in the activity of other units.

A further practical difficulty in defining the AFP in DINs derives from the impossibility of identifying silent units when recording from a few out of several hundred units in a connective. From an experimental point of view it is more useful to speak in terms of the units active at one time as an assembly. That is, in the case of the optomotor interneurones above, each direction of stimulus motion would be represented by a different assembly of active DINs. Unlike an AFP, the composition of the assembly for any behaviour is fixed but a unit may participate in more than one assembly at different times.

Even using the concept of an assembly, problems of definition still remain because a unit that changes its firing characteristics cannot always be identified as the same unit under different circumstances. Furthermore, inhibitory and modulatory units cannot be identified as such from the activity in their axons, and the sign of an input may be changed locally by an inverting interneurone, so that one DIN could be responsible for exciting one target neurone and inhibiting another (Ramirez 1986). These problems can be resolved only by recording from post-synaptic neurones as well as from DINs, and to do this requires a thorough study of the connexions identified DINs make with their target neurones, to provide information that is so far woefully lacking.

5. INTERCONNEXIONS BETWEEN STAGES

Each stage in the system receives inputs from all other stages. This is crucial evidence that the system is not organised as a linear hierarchy but in several parallel loops of different lengths. The information that converges on a single output neurone in any station may include direct sensory inputs as well as the outputs of complex interneuronal networks such as the mushroom and central bodies in the brain or the premotor networks in the ventral nerve cord ganglia. Diagrams of the inputs to a motor neurone and a DIN (Fig. 8) therefore look very similar, although the actual sources of the inputs are different.

Essentially, inputs are of two types: self-generated changes resulting from the action being performed, and novel stimuli produced by events in the environment not generated directly by the animal's movement (Fig. 9). As well as proprioceptive feedback, self-generated stimuli include all changes in visual field, sound field and air flow round the body, as well as sounds resulting from movement. These are propagated externally and are detected by sense organs on other segments, chiefly those on the head and the cerci on the last abdominal segments but also to some extent on every segment.

A third type of input, indicated by the sources labelled M in Fig. 8, comes from the internal milieu of the animal and includes neuromodulation, hormonal levels indicating the physiological state of the animal and the output from memory stores. These will be dealt with in the following section.

Both novel and self-generated signals converge onto the intersegmental interneurones that transmit information between stations: by ascending interneurones (AINs) from the body segments to the brain and by DINs from the brain to the ventral nerve cord ganglia (Fig. 9). Intersegmental interneurones also provide links between the ganglia (Fig. 10a). Self-generated stimuli are thus doubly represented at each station, directly through inputs from local sense organs and indirectly through interneuronal pathways.

In addition to direct loops between segmental ganglia and between the segmental ganglia and the brain, another set of loops has recently been discovered between the brain, suboesophageal ganglion[1] and segmental ganglia of the body (Figs. 9, 10a; see Altman and Kien 1986 for review). Lesion and stimulation studies in crickets, locusts and stick insects where the suboesophageal ganglion and brain are separate, have shown that the brain and suboesophageal ganglion both have important, though rather different, roles in the control of locomotion (Huber 1955; Kien 1983; Bässler et al. 1985). The brain is concerned with the initiation of a behaviour and its directional regulation, whereas the suboesophageal ganglion, also involved in initiation, is essential for the maintenance and correct performance of the behaviour.

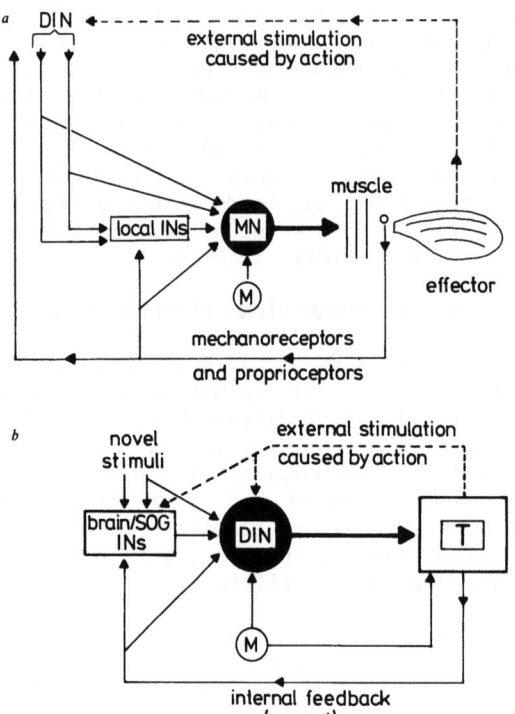

Fig. 8: The inputs to the output neurones in any station come from a variety of sources, including those carrying information on novel stimuli, external and internal feedback, and a range of modularoty inputs (M). Although the actual sources of inputs to a) motor neurones (MN) and b) DINs differ, the types of inputs are very similar for both. The inputs shown as direct to the output neurones include those that have passed through several layers of feature abstracting neurones (not shown; for example in the optic lobes); the signals from the same sources may also reach the output neurones indirectly through the motor pattern generating networks of local interneurones (a), or the interneurones of the mushroom bodies and central body in the brain (b).

---

[1] In insect, the term 'brain' strictly applies to the supraoesophageal ganglionic mass; the suboesophageal ganglion, lying in the head directly below the oesophagus, is three fused segmental ganglia innervating the mouthparts. In more advanced insects, such as flies and bees, these fuse with the supraoesophageal ganglia.

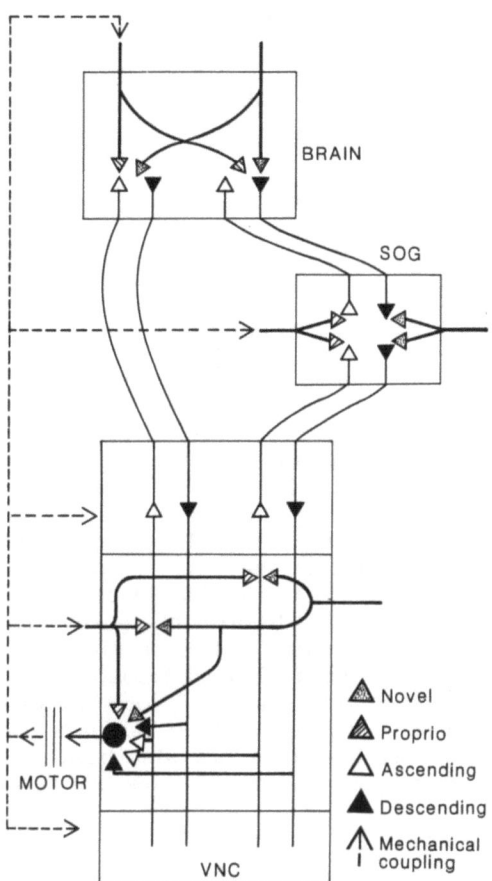

FIG. 9: Stations are connected with other stations by <u>loops</u> of various lengths here shown schematically for one station in the ventral nerve cord (VNC). Inputs signalling both novel inputs and the self-generated results of action ('proprio') feed directly to motor neurones as well as on to ascending interneurones that carry information to other stations in the VNC and in the head (brain and suboesophageal ganglion – SOG). Proprio signals also feed, by external routes (mechanical coupling) to other stations. In the brain and SOG they combine with signals from the ascending interneurones and other types of novel stimuli to provide inputs to the descending interneurons that return to the VNC stations. The brain and SOG form two distinct loops with the VNC stations (see Fig. 10 for details). The lines represent information channels, not identified neurones.

The anatomical and physiological substrate for these different functions is provided by separate sets of DINs, one originating in the brain and the other, newly described (Kien and Altman 1984; Boyan and Altman 1985; Ramirez 1986; J. Kien, W.A. Fletcher, J.S. Altman, J.M. Ramirez and U. Roth, in preparation), that originates in the suboesophageal ganglion. Physically, of course, the brain DINs pass through the suboesophageal ganglion, so the cervical connectives, joining the suboesophageal ganglion and the first thoracic ganglion, contain both

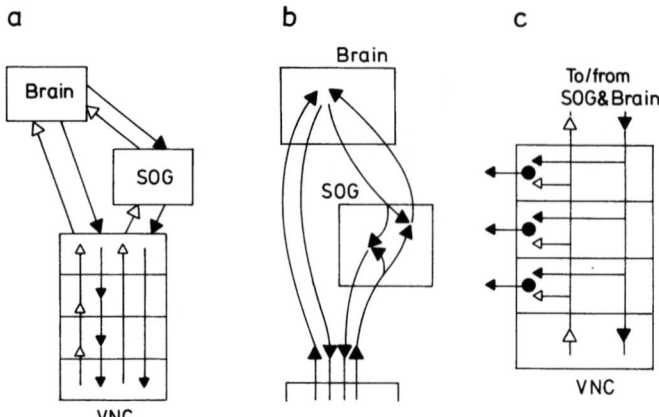

FIG. 10: Schematic diagrams of the loops between various stations in the insect CNS. (The peripheral pathways are omitted for clarity, but see Figs. 9, 11). a) Simplified scheme of the major loops between brain and VNC, SOG and VNC, brain and SOG, and more local loops between VNC stations. Note that there are local loops between adjacent (on left) and subadjacent VNC stations (on right). b) Detail of brain-VNC, SOG-VNC and brain-SOG loops. c) Ascending and descending interneurones that are part of the major loops also contribute to the local loops between VNC stations. The brain-VNC loop is shown outside the SOG for convenience - physically both pathways pass through the SOG. Not shown here, the loops are also bilaterally organised, with mirror image components on each side of the nervous system.

sets of DINs. For convenience, these have been separated out in the figures (Figs. 9, 10b).

Some of the suboesophageal DINs have been characterised by intracellular recording and dye marking in the suboesophageal ganglion or in the cervical connectives in both free moving and fixed preparations (Kien and Altman 1984; Boyan and Altman 1985; Ramirez 1986). They are generally multimodal, receiving inputs from eyes, antennae and hairs on the head as well as from mechanoreceptors, proprioceptors and auditory receptors on the body. Most of the inputs from other segments appear to be interneuronal, many apparently from collaterals of the brain DINs and interneurones ascending from segmental ganglia to the brain. Again for simplicity, these pathways are shown separately in Figs. 9 and 10.

The suboesophageal DINs thus seem to combine quite detailed information about the body with inputs received through sense organs on the head (Ramirez 1983; Kien and Altman 1984; Boyan and Altman 1985) and they signal directly back to the motor networks of the ventral nerve cord ganglia, in a loop that is separate from and parallel to the direct loop between the brain and the ventral nerve cord ganglia (Figs. 9, 10a, b). Not only do the two loops seem to carry messages containing different mixes of information but there will also be timing differences for information from the same sources travelling in the two loops to the same destination.

It is important to note that we are speaking of the functions of the loops and not of the stations. Because both brain and suboesophageal DINs receive large inputs signalling the results of action and because the inputs from the brain and suboesophageal ganglion both influence the target neurones only when the local context is correct, we consider it is more accurate to ascribe a particular function to the loop that joins two stations, rather than to one of the stations. That is, the neurones that form the loops between the brain and the ventral nerve cord determine which behaviour is initiated and regulate its spatial direction; they determine what is done. Similarly, the neurones in the loops between the suboesophageal ganglion (SOG) and the ventral nerve cord ganglia, maintain and regulate the performance of the behaviour; they determine how it is done.

Of course, these two loops cannot operate completely independently. Not only do they receive information from the same sources but there is also a loop between the brain and the suboesophageal ganglion. AINs originating in the suboesophageal ganglion, that branches in the same regions as the DINs (Boyan and Altman 1985; Kien et al. in prep), project to the brain, where some terminate in the same zones as the neurones directly ascending from the ventral nerve cord ganglia, or in the areas containing branches of the DINs (Boyan and Altman 1985). The only inputs from the brain to the SOG described so far are from the brain DINs that have collaterals in the suboesophageal ganglion. Cobalt fills from cervical and circum-oesophageal connectives to the brain indicate that a number of DINs in the brain do terminate in the suboesophageal ganglion but their morphology and function is unknown (J.S. Altman, unpublished).

The functions of the brain-suboesophageal loop are not yet defined. Apart from coordinating the functions of the two loops and delivering information from the head sense organs to the suboesophageal DINs, it may provide a delay line. Signals from the tympanic organs on the locust abdomen reach the brain through large, direct AINs that have collaterals in the suboesophageal ganglion. AINs originating in the suboesophageal ganglion relay these signals to the brain, where they arrive at least 42 ms after those in the direct AINs (Boyan and Altman 1985). The role of such a delay in processing auditory signals is not yet known.

There are also loops of various lengths between the various ventral nerve cord ganglia (Figs. 9c, 10a, c), made both by interneurones projecting from one ganglion to the next and by collaterals of AINs and DINs. Segmental sensory neurones provide inputs to AINs as well as to the segmental interneurones, and external self-stimulation is also an important component of the loops between the segmental ganglia (Fig. 9).

The segmental loops are essential for the local coordination of effectors. The brain and suboesophageal loops superimposed on them ensure that all segments respond in the same way to stimuli that may not effect them directly, and prevent different parts of the body from expressing different responses simultaneously.

Although the same types of sensory information are received in every stage of the system (Fig. 10), the output of each stage will be unique because it depends on the combination of these inputs with the processed outputs of other stages (Fig. 11). The output of the suboesophageal DINs will combine local inputs, feedback from body segments and information from the brain, whereas the thoracic motor neurones will combine the outputs from both the brain and suboesophageal DINs with local inputs and with feedback from other segments, which includes the outputs of other stages.

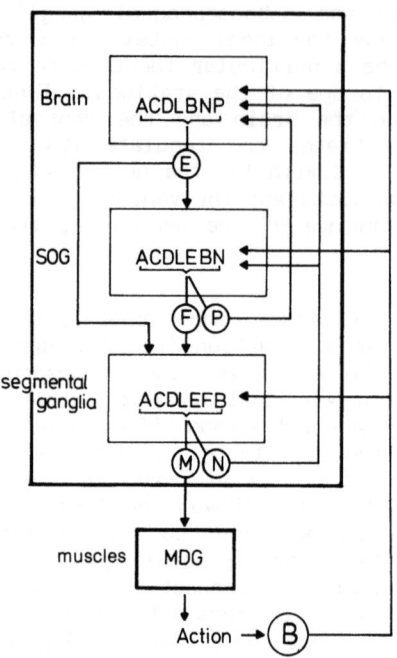

FIG. 11: Each station contains a unique set of information and produces a unique output, which forms an element of the input to other stations. Information flows in both directions, so all stations are kept informed about the state of the others. Types of information are represented by a letter code: A - novel information; B - results of action ('proprio' in Fig. 9); C - information from other segments; D - internal state, including modulation memory/arousal/hormonal; E - brain output; F - SOG output; G - mechanical constraints; L - local network computations; M - motor output; N - report back from VNC stations; P - report back from SOG to brain.

6. MODULATION

So far the discussion has treated the connexions between neurones in different stations in terms of classical neural and synaptic transmission, where neurones fire action potentials when the sum of their inputs raises the membrane potential above threshold. In the past few years, it has become increasingly clear in a wide range of species that the performance of neuronal circuits can be influenced by modulator neurones that are not directly involved in the generation of the coordinated behavioural output. These are tuning neurones: by changing the responsiveness of many neurones and muscles simultaneously, they facilitate the expression of one behaviour over others (for example Siegelbaum et al. 1982; Harris-Warrick and Kravitz 1984; Harris-Warrick and Flamm 1986). They may also be involved in the long-term changes in membrane properties or synaptic transmission that may underlie learning (Goelet et al. 1986).

The neurones broadly classed as modulators generally use a catecholamine or a peptide as their transmitter, though acetylcholine acts as a modulator in some systems. They do not usually make specific contacts with a small group of target neurones but have very widespread branching, often in several different stations. In gastropod molluscs, neurones such as the metacerebral cell (see Croll 1987) have branches in the periphery as well as in the CNS but so far there are no reports of similar neurones in insects. There are however motor-neurone-like cells containing octopamine (Evans and O'Shea 1977), proctolin (O'Shea and Adams 1981) and serotonin (Tyrer et al. 1984; Davies 1987; P. Bräunig and N.M. Tyrer, personal communication) that arborise in motor neuropiles and either terminate directly on the muscles or are presynaptic to 'classical' motor neurones. Others, such as the paired oxytocin/vasopressin-reactive neurones in the locust suboesophageal ganglion (Rémy and Girardie 1980; Altman and Kien 1985) have branching throughout the CNS but restricted to particular neuropiles in each ganglion, with projections into the bases of some peripheral nerves that seem to terminate among the axons (N.M. Tyrer, personal communication). From the little evidence presently available (Hoyle et al. 1980) and by analogy with vertebrates, it is probable that many of these neurones do not make classical synaptic contacts but release transmitter in a more diffuse, less directed way that could influence many neurones and/or muscles in the vicinity.

There is increasing evidence (see Menzel and Bicker 1987 - for review) that modulator neurones are involved in the selection and coordination of behavioural outputs. In lobster, injection of serotonin into the haemolymph elicits tonic flexion of all limbs to produce a posture similar to the submissive stance, whereas octopamine produces tonic extension and an aggressive-like posture (Kravitz et al. 1980). Similarly in the lobster stomatogastric ganglion 14 neurones generate the motor pattern driving the pyloric muscles but the type of coordination pattern is determined by several modulators using acetylcholine, catecholamines and a variety of peptides as transmitters (Harris-Warrick and Flamm 1986). These influence a range of parameters from synaptic efficacy to the form of action potential and recruitment of particular neurones, so that each moulds the basic pattern produced by the circuit in a different way. The pentapeptide proctolin, for example, modifies the firing pattern of some motor neurones and works directly on the muscles. Its presence or absence determines the type of coordination between the lateral and medial teeth in the stomach (H.G. Heinzel and A. Selverston, personal communication). By changing integrative properties simultaneoulsy at several points in a network, it seem that modulatory neurones can coordinate or orchestrate (Sombati and Hoyle 1984) the activation of an assembly of neurones that will produce the output appropriate for a particular behaviour.

This might suggest that the modulatory neurones control the selection of behaviour patterns - that they are the decision makers. We believe that this is not so. Research on neuromodulators is at an early stage and has concentrated on defining their actions, either in isolated nerve cord preparations (for example Lent 1985) or in reduced and restrained animals. So far, little attention has been paid to the inputs to modulatory neurones and their activation has not been examined in anything approaching a normal behavioural situation. Indeed, the same rules most probably apply to the modulatory neurones as to the DINs and motor neurones. It is likely that they receive a similar range of inputs - sensory stimuli, neural feedback and signals concerning the internal state of the animal - and that their activity is determined by the AFP of these inputs. Their effects on the output of any station will depend on which other inputs to that station are active and on the state of the local networks. That is, neuromodulatory inputs should be seen as one element of the AFP of the inputs to a station.

The modulatory neurones should perhaps be seen as part of another loop, although one that is both anatomically and physiologically more diffuse than the loops of the pattern-generating circuits, because modulators can have their effects at several points in the system simultaneously. They do, however, constitute just one of the inputs to the targets in any station and for this reason, we have included them on an equal basis to other inputs in Figs. 2, 4, 5, 7 and 11.

Two other types of inputs have been referred to briefly (Fig. 11): memory and internal physiological state. The latter can be considered as an extension of neuromodulation - hormones, acting on a longer time scale and more globally, determine the responsiveness of the nervous system to external stimuli according to the state of hunger, thirst, fatigue, development or reproductive readiness. For example, the cercal startle response is supressed in crickets ready to copulate (Huber 1965); female crickets that have recently deposited eggs do not respond to the male's calling song (J.S. Altman, unpublished).

Hormones can also determine which motor programme is expressed, as in the triggering of eclosion behaviour by eclosion hormone in the moth, Manduca sexta (Tublitz et al. 1986). This occurs towards the end of the pupal instar when eclosion hormone is released into the haemolymph from a pair of neurone clusters in the brain. This release occurs only when the correct pattern of sensory inputs is combined with appropriate internal physiological signals, so development must be complete and the environmental conditions correct before eclosion can start.

Yet another level of complexity must increasingly be taken into account when considering neuronal integration and the generation of behaviour: modulatory transmitters and hormones regulate neuronal responsiveness by changes in intracellular biochemistry that can far outlast the signal and may be permanent. The nervous system in Manduca is unresponsive to eclosion hormone until shortly before ecdysis, when neurones produce new proteins that are essential to the intracellular signal-transduction pathway linking the eclosion hormone receptor to membrane responses (Morton and Truman 1986). It seems these proteins are generated in response to another hormone signal. When the signal molecules bind to their receptors in the cell membrane they trigger intracellular enzyme cascades that may result indirectly in changes in the functioning of a variety of ion channels, which can lead to long-lasting changes in membrane properties or synaptic vesicle release (Siegelbaum et al. 1982).

The results of past actions and experience provide a further input influencing the generation of behaviour although the mechanisms by which they are expressed are still unclear. Most probably, changes in synaptic transmission are involved through intracellular modifications of membrane proteins, either through long-lasting changes in the intracellular signalling cascades or by changes in gene expression, possibly as an indirect action of these cascades (Goelet et al. 1986). The synaptic modification could occur at synapses between neurones of the pattern generating circuits, or in ancillary loops, perhaps involving structures such as the mushroom bodies and central body in the brain (see Nässel 1987).

The overall responsiveness of any neurone thus seems to be the result of complex interactions between a range of signals with different time courses, all of which are components of the AFP of the neurone's inputs. That is, whether and how a neurone responds to any input depends on a number of chemical and electrical factors, all more or less necessary but none sufficient on its own. It is only when several of these factors coincide that a particular output is produced and a change in one factor may result in another output or no response at all.

7. CONCLUSIONS

In presenting this model we have emphasised the need to consider the whole animal analysing the neural organisation of behaviour. This 'holistic' approach is not intended to exclude or replace reductionist methods for dissecting the operations of the system but rather to provide a conceptual frame of reference for such dissections. Intracellular recording from single identified neurones is still one of the most powerful analytical tools available but the results can be very misleading. To put them in context and to maintain a more accurate perspective, we need to combine the single unit approach with other, less precise methods, ranging from intracellular recording in moving animals to multiunit recording and focal stimulation, as well as more exact descriptions of behaviours. The introduction of voltage-sensitive dyes and optical recording methods (Blasdel and Salama 1986; Grinvald et al. 1986) may with further refinement provide a better window to spatio-temporal patterns of activity. It is also necessary to work on longer time scales which is essential if the slower effects of neuromodulators and hormones are to be taken fully into account.

For many years, the command neurone hypothesis has governed thinking about the selection of appropriate behaviours in invertebrates. However, no neurone that meets the criterion of being both "necessary and sufficient" (Kupferman and Weiss 1978) to elicit an organised output in a freely behaving animal has yet been identified. The model proposed here does not exclude the possibility of a single neurone driving the expression of a behaviour but suggests it is an extreme case - a neuronal assembly could in theory consist of a single neurone. But such simple all-or-nothing triggers are unsatisfactory to explain the subtle and complex regulation of the ongoing behaviour patterns that are the basis of all day and everyday life. Instead the expression of most behaviours is much more likely to be the result of the balance of activities in a number of loops, such as those we have described above. Decision making in the nervous system should be seen as the outcome of a concensus among many neurones rather than command by a few.

Although the discussion here has mainly been limited to insects, the principles we have defined may well apply to other groups. Indeed, even the echinoderms seem to display characteristics that comply with these principles (see Cobb 1987). On a wider scale, the concept of neuronal assemblies consisting of different combinations of the same units, each activated by a different group of input conditions and reinforced by use have been proposed in models of mammalian memory systems (Dudai et al. 1987) and in the theoretical networks proposed by the connexionist school of computer modellers (Rummelhart et al. 1986). Perhaps the insect nervous system will prove to be a good experimemtal system for testing these more generally applicable ideas.

8. ACKNOWLEDGEMENTS

This work was supported by project H2 from SFB-4 of the Deutscheforschungsgemeinschaft. We thank all our colleagues for many stimulating discussions over the years and especially those who have allowed us to quote from unpublished work.

9. REFERENCES

Altman JS (1981) Functional organisation of insect ganglia. Adv Physiol Sci 23: 537-555.

Altman JS, Kien J (1985) The anatomical basis for intersegmental and bilateral coordination in locusts. In: Bush B, Clarac F (eds) S.E.B. Seminar Series, 24, Cambridge Univ Press, Cambridge, pp 91-119.

Altman JS, Kien J (1986) Functional organisation of the suboesophageal ganglion in insects and other arthropods. In: Gupta AP (ed) Arthropod brain: its evolution, development, structure and functions. John Wiley and Sons, NY (in press).

Bacon JP, Möhl B (1983) The tritocerebral commissure giant (TCG) wind-sensitive interneuron in the locust. I. Its activity in straight flight. J Comp Physiol 150: 439-452.

Bacon JP, Strausfeld NJ (1986) The dipteran 'giant fibre' pathway: neurons and signals. J Comp Physiol 158: 527-548.

Bacon JP, Tyrer M (1979) Wind interneurone input to flight motorneurones in the locust, Schistocerca gregaria. Naturwiss 66: 116.

Bässler U, Foth E, Breutel C (1985) The inherent walking directions differ for the prothoracic and mesothoracic legs of stick insects. J Exp Biol 116: 301-311.

Blasdel GG, Salama G (1986) Voltage-sensitive dyes reveal a modular organisation in monkey striate cortex. Nature 321: 579-585.

Boyan GS, Altman JS (1985) The suboesophageal ganglion: A "missing link" in the auditory pathway of the locust. J Comp Physiol 156: 413-428.

Burrows M, Rowell CHF (1973) Connections between descending visual interneurons and metathoracic motoneurons in the locust. J Comp Physiol 85: 221-234.

Cobb JLS (1987) Neurobiology of the Echinodermata. (This volume)

Croll RP (1987) Identified neurons and cellular homologies. (This volume)

Davies NT (1987) Neurosecretory neurons and their projections to the serotonin neurosecretory system of the cockroach Periplaneta americana (L.), identification of mandibular and maxillary motor neurons associated with this system. J Comp Neurol (in press).

Davis WJ (1976) Organisational concepts in the central motor networks of invertebrates. Adv Behav Biol 18: 265-292.

Davis WJ (1985) Central feedback loops and some implications for motor control. In: Barnes WJP, Gladden MM (eds) Feedback in motor control. Croon-Helm, London, pp 13-34.

Dudai Y, Amari SI, Bienenstock E, Dehaene S, Fuster J, Goddard GV, Konishi M, Menzel R, Mishkin M, Müller CM, Rolls ET, Schwegler HH, von der Malsburg C (1987) On neural assemblies and memories. In: Changeux J-P, Konishi M (eds) Neural and molecular bases of learning. Dahlem Conference, Springer Verlag, Berlin (in press).

Eaton RC, DiDomenico R (1987) Command and the neural causation of behaviour: a theoretical analysis of the necessity and sufficiency paradigm. Brain Behav Evol (in press).

Erickson RP (1963) Sensory neural patterns and gustation. In: Zotterman Y (ed) Olfaction and taste. Pergamon Press, New York, pp 205-213.

Evans PD, O'Shea M (1977) An octopaminergic neurone modulates neuromuscular transmission in the locust. Nature 270: 257-259.

Goelet P, Castellucci VF, Schacher S, Kandel ER (1986) The long and the short of lont-term memory: a molecular framework. Nature 322: 419-421.

Grinvald A, Licke E, Frostig RD, Gilbert C, Wiesel TN (1986) Functional architecture of cortex revealed by optical imaging of intrinsic signals. Nature 324: 361-364.

Harris-Warrick RM, Flamm RE (1986) Chemical modulation of a small central pattern generator circuit. Trends Neurosci 9: 432-437.

Harris-Warrick RM, Kravitz EA (1984) Cellular mechanisms for modulation of posture by octopamine and serotonin in the lobster. J Neurosci 4: 1976-1993.

Heitler WJ, Burrows M (1977) The locust jump. I. The motor programme. J Exp Biol 66: 203-219.

Hoyle G (1977) Identified neurons and behavior of arthropods. Plenum, New York.

Hoyle G, Colquhoun W, Williams M (1980) Fine structure of an octopaminergic neuron and its terminals. J Neurobiol 11L 103-126.

Huber F (1955) Sitz und Bedeutung nervöser Zentren für Instinkthandlung beim Männchen von <u>Gryllus campestris</u> L. Z Tierpsychol 12: 12-48.

Huber F (1965) Brain controlled behaviour in orthopterans. In: Treherne JE, Beament JWR (eds) The physiology of the insect central nervous system. Academic Press, London, pp 233-246.

Huber F (1980) Zoologische Grundlagenforschung aus der Sicht eines Insektenbiologen. Verh Dtsch Zool Ges, 73 Jahresversammlung, pp 12-37.

Kien J (1979) Variability of locust motorneuron responses to sensory stimulation: A possible substrate for motor flexibiltiy. J Comp Physiol 134: 55-68.

Kien J (1983) The initiation and maintenance of walking in the locust. An alternative to the command concept. Proc R Soc Lond B 219: 137-174.

Kien J, Altman JS (1984) Descending interneurones from the brain and suboesophageal ganglia and their role in the control of locust behaviour. J Insect Physiol 30: 59-72.

Kravitz EA, Glusman S, Harris-Warrick RM, Livingstone MS, Schwarz T, Goy MF (1980) Amines and a peptide as neurohormones in lobsters: Actions on neuro-muscular preparations and preliminary behavioural studies. J Exp Biol 89: 159-175.

Kupfermann I, Weiss KR (1978) The command neuron concept. Behav Brain Sci 1: 1-10.

Lent CM (1985) Serotonic modulation of the feeding behaviour of the medicianl leech. Brain Res Bull 14: 643-655.

Menzel R, Bicker G (1987) Plasticity in neuronal circuits and assemblies of invertebrates. In: Changeux J-P, Konishi M (eds) Neural and molecular bases of learning. Dahlem Conference. Springer Verlag, Berlin (in press).

Morton DB, Truman JW (1986) Substrate protein availability regulates eclosion hormone sensitivity in an insect CNS. Nature 323: 264-267.

Nässel DR (1987) Aspects of the functional and chemical anatomy of the insect brain. (This volume)

O'Shea M, Adams M (1981) Pentapeptide (Proctolin: Arg - Tyr - Len - Pro - Thr) associated with an identified neuron. Science 213: 567-569.

Pearson KG, Heitler WJ, Steeves JD (1980) Triggering of locust jump by multimodal inhibitory interneurons. J Neurophysiol 43: 257-278.

Ramirez J-M (1983) Untersuchung der sensorischen Eingänge von multimodalen Neuronen im Unterschlundganglion der Heuschrecke (Schistocerca gregaria). Diplom-Thesis, Universität Regensburg, Regensburg.

Ramirez J-M (1986) Interneuronal control of flight. Doctoral Thesis. University of Regensburg, Regensburg.

Reichert H, Rowell CHF, Griss C (1985) Course correction circuitry translates feature detection into behavioural action in locusts. Nature 315: 142-144.

Rémy C, Girardie J (1980) Anatomical organization of two vasopressin-neurophysin-like neurosecretory cell throughout the central nervous system of the migratory locust. Gen Comp Endocrinol 40: 27-35.

Robertson RM (1987) Insect neurons: synaptic interactions, circuits and the control of behavior. (This volume)

Rowell CHF, Pearson KG (1983) Ocellar input to the flight motor system of the locust: structure and function. J Exp Biol 103: 265-288.

Rummelhart DE, Hinton GE, Williams RJ (1986) Learning representations by back-propagating errors. Nature 323:533-536.

Siegelbaum SA, Camardo JS, Kandel ER (1982) Serotonin and cyclic AMP close single K^+ channels in Aplysia sensory neurones. Nature 299: 413-417.

Siegler MVS, Burrows M (1979) The morphology of local non-spiking interneurons in the metathoracic ganglion of the locust. J Comp Neurol 183: 121-147.

Sombati S, Hoyle G (1984) Generation of specific behaviors in a locust by local release into neuropil of the natural neuromodulator ocotpamine. J Neurobiol 15: 481-506.

Tublitz NJ, Copenhauer PF, Taghert PM, Truman JW (1986) Peptidergic regulation of behavior: an identified neuron approach. Trends Neurosci 9: 359-363.

Tyrer NM, Turner J, Altman JS (1984) Identifable neurons in the locust central nervous system that react with antibodies to serotonin. J Comp Neurol 227: 313-330.

Wilson DM (1962) Bifunctional muscles in the thorax of grasshoppers. J Exp Biol 39: 669-677.

GENERAL CONCLUSIONS

M.A. ALI

Département de biologie

Université de Montréal

Montréal, Québec, Canada H3C 3J7

This chapter is a gist of the discussions which took place on the last working day of the ASI. As in previous ASIs, eight persons were asked to enumerate and discuss topics which they felt would add to make the conference more complete. Comments and/or questions from the floor were encouraged during and after these presentations.

The aims of those present at this ASI varied. There were those who looked at the nervous systems of invertebrates from morphological, ecological, physiological and biochemical view points. But no matter what approach was used it was agreed that: terminology/taxomomy must follow a strict set of rules (e.g. those set by the International Commission for Zoological Nomenclature), material had to be correctly identified and anatomical terms specifically defined.

Invertebrate biologists (including neurobiologists) have always felt that they had to justify their interest in invertebrates. Thus there are those who maintain that their interests lie in the convenience (large cells, ease of handling, etc.) which invertebrates offer in order to understand the diversity and complexity of the mammalian nervous system. However, is it appropriate to use information derived from invertebrates to relate to vertebrates (especially mammals) where development is so totally different and where the nerve cells are themselves so fundamentally different in organisation? Invertebrate biologists (including neurobiologists) should stop looking for excuses and stop thinking that work on vertebrates is superior. The elegant preparations of invertebrates are taking prominence in the development of neurobiology. Insect preparations are ideally suited to at least two specific fields: the role and extent of local interactions in the central nervous system, and the central processing and role of sensory input in generating and collecting motor output patterns. There will come a time in the near future when it is possible to take the techniques developed and used by the people at this ASI and to apply them not only to man but also to ancestral or closely related forms still available and to obtain by this measure a better understanding of the evolution of very specific and homologous functions (e.g. colour vision has been worked on by systematic neurobiological study). Thus if we must really find some justifable cause for the study of invertebrates we can talk of man in an ecosystem which includes other animals or we can talk of man as the outcome of an evolutionary process which is related to other aminals. However, if we are honest we will admit that we study invertebrates because we enjoy them and are attracted to them albeit the other arguments hold.

Two broad approaches have been employed for the study of invertebrate neurobiology. The first involved the formation of a question and then looking around in the animal kingdom for an answer (preparation). The second starts with a preparation, followed by the collection and build up of data and finally the decision of what to do with the accumulated data. Fundamentally, however, observations should come first, then questions on the observations which we do not understand.

Invertebrate neurobiology has given us new insights. It has allowed us to: 1) form a different philosophical basis about ourselves and about the idea of how special our nervous system is; 2) question that we are the only ones who learn (obviously this is no longer true); 3) question that we are the only ones who are fearful, who are conditioned to fear and aphasia. We get an idea of what we think is special when we begin to identify that same phenomenon in nervous systems which we consider simple. This gives a view of how special we are from a philosophical point. It unables us to argue that we are not mechanistic because our behaviour is so complex. But when we look at the behaviours and nervous systems of many invertebrates we begin to realise that they too are awfully complex. In fact some instinctive and learned behaviour patterns in invertebrates (especially arachnids and insects) are very intricate. This realisation of complexities around us is in itself a significant outcome of invertebrate neurobiology.

Another philosophical point put forward is that the human brain represents the ultimate problem in neurobiology and it seems logical to start with something "simple" which will throw light on the human brain. However, it can also be argued that the human brain is widely studied because most information (function) available is on it. Not all present at the ASI upheld the latter view. Still another philosophical approach is that the invertebrates are the result of several experiments in the building of an animal and its nervous system. Each phylum tried a different solution to the problem and came out with a different answer. Thus suggesting that a comparative study of the different answers will reveal some fundamental principle which will apply just as well to echinoderms, to insects and even perhaps to man. There is evidence from growth at the cellular level that neurones are similar throughout the animal kingdom and that they function in the same way regardless of their origin.

The distinction between vertebrate and invertebrate is a barrier which is totally artificial and depends on a few old fashioned ideas of separating one group of animals from another. This idea is destructive and outweighs the little beneficial effects it may have. Take insulin for example, it was first identified in vertebrates (specifically mammals in this context) and became firmly recognised as a molecule secreted by the pancreas and having very specific functions in metalolism. As such it is an unlikely candidate molecule in invertebrates, but insulin-like molecules have since been identified in invertebrates. Thus other factors are equally important in defining animals e.g. morphology, behaviour, biochemistry, etc.

Neurones can be identified by chemical means. Thus chemical identification of neurones represents the expression of a substance contained within the neurone and in no way does it represent the formal expression of the neurone. The neurone may well contain other chemicals which are of greater importance than the one identified and used as reference for that neurone. Furthermmore, the fact that the same molecule is found in diverse organisms does not mean that it is functioning in the same capacity. The surprise caused by the discovery of insulin-like molecule in the nervous systems of invertebrates is a clear example of "boxed" thinking - our narrow thinking of the endocrine and nervous systems as complex subsystems. Thus it is important not to become preoccupied with the molecule, other factors (e.g. receptor/acceptor) are just as important. It is quite probable that the difference in physiological functions of the same molecule is dependent on the receptor/acceptor. Nevertheless the molecule must be identified first, then both molecule and receptor/acceptor must be considered as it is impossible to have the effect of one without the other; and possibly both the molecule and its receptor/acceptor evolved simultaneously. This presents extremely difficult problems. For example, calmodulin is a substance with a long chemical history. It is a multifunctional protein which has no intrinsic enzymatic properties but which regulates a wide spectrum of cellular functions. Furthermore calmodulin in ubiquitous in the eukaryotes and constitutes between 0.1% and 1.0% of the total protein from the lowest- to highest-order plants and animals; and as expected it is remarkably well conserved phylogenetically. This molecule has been sequenced from representatives of mammalian, fish, cnidarian, protozoan and plant phyla, and of the 148 amino-acids no more that seven have been altered between the species. In addition to its primary amino-acid sequence calmodulins appear to be immunologically identical. What is bothersome with these studies is that we are separating the molecule from the very cell which it is dependent upon for its expression and still expect it to retain its biological functions. The chemical environment (cations, anions) of the molecule is important. There are intracellular controls for the release of different substances at differrent times and under different conditions.

To further complicate matters it is extremely difficult to conceive the evolution of a molecule. For example, insulin starts off as a small molecule to which more and more amino acid sequences are added until finally the modecule insulin results. However a molecule is either "insulin" in the terms of the method used in its definition or it is not insulin - there is no transient molecule; it must be a complex tertiary polypeptide! In the same way one is either a vertebrate or an invertebrate by definition - there are no intermediates. Thus coming back to the point of chemical identification we can realise that it is a very narrow means of defining cells as it does not take into account the "building blocks" or enzymes which manufacture them. Immunopeptide cytochemistry is important because is shows us what neurones we have, what neurones are likely to occur together, and what they are likely to connect with. Thus a circuit can be defined and looked at physiologically. Immunopeptide cytochemistry however does not tell us what makes cells produce what they do and why they release a particular substance in a particular place at a specific time in the life cycle of the animal. It is nevertheless an important tool in neurobiology when reservations are used in the interpretations of observations.

If we are to make sense of these molecules it is important to find out if they have some value. The ultimate test of the relevance of a molecule being if it has any biological activity. It does not mean that the molecule has to be discarded if it does not have biological activity but it does mean that it is serving some purpose if it possesses activity.

Molecules which do not reflect biological activity but which are nonetheless important are DNA and RNA. DNA (gene) expression is very important for all these molecules, and molecular biology is an important aspect of neurobiology. This field is however very limited as far as invertebrates are concerned as material and scientific rewards are much greater in the mammalian field. This limitation is clearly borne out by the absence of a molecular biologist at this meeting despite attempts to invite one. In looking at the molecular DNA level we are looking at specific explanations and must ask the right questions. Cells can be identified by their DNA expression and neurones are the same in that we can understand more about their layout if we understand the gene which is actually making it. However, we will not be able to understand integration of the nervous system by looking at DNA expression alone. We cannot understand how the nervous system works by understanding the things which are expressed in neurones because there is the property of the system which is not expressed at the molecular leval. Thus all levels (molecular, cellular, tissue, organ) of the nervous system must be considered - though not necessarily in a single work, but they must be kept in mind.

Why does an animal live where it does? How does it behave? What does it respond to in its environment? These are questions which interest a sensory physiologist, as "Life is like a sewer, what you get out of it tends to form what you put into it". All living things are bombarded at all times with an input. There is something which is continually changing in the environment and the nature of the input and the way the animal responds to it is an interesting feature in the way the subsequent nervous system behaves. There is tremendous input going into the nervous system and the number of neurones involved is huge. There is a great deal more to the nervous system than the electrical patterns generated. Animals are not mechanically put together, they actually do receive information from outside themselves and are just as sensitive to these signals as internal ones. Inputs of external signals can change and bias the nervous systems such that the central pattern generated can be turned ON and OFF only by something which comes from outside. In other words sense organs are cues and triggers which should not be ingored.

The peripheral nervous system consists of sense organs (which feed information in) and perhaps the neuromuscular system (which is the final output of the nervous system). There is therefore the consensus that the sensory cells are located peripherally but this is not generally true. For example in the leech there are sensory cells located in the ganglia (and we do tend to think of the ganglia as part of the central nervous system) yet their function is peripheral. Thus there are some nerve cells which are located centrally but which react peripherally. Therefore it is difficult to separate what we define as the central and the peripheral nervous systems - this cannot be done solely from a positional basis.

ANATOMY STRUCTURE/PHYSIOLOGY/BEHAVIOUR

Anatomical stains give some idea of the structure of the cell involved - ideas on its size and shape. In addition, stains give some hint of the activity of the cell. Thus we can ask: 1) can structure predict physiology or to what extent does it predict physiology? 2) coupled to present recording techniques, to what degree does physiology predict behaviour?

Every structure has a function but there are limits to determining structure. Difficulties in preparation arise from the penetration of materials into neurones - it is especially difficult to get materials into junctions between cells, and cells and muscles. Anatomy tells us what the possibilities are. It does not, for example, tell us what type of contact is made, whether the contact is functional and under what conditions it is functional, whether it is excitatory or inhibitory, etc. but it does tell us that the contact is a good place to look at if we are interested in activity. Anatomy also indicates to us what we cannot do in one physiological experiment. It can tell us the number of inputs a neurone has, and where they are coming from, so that we can actually test under different conditions the sort of things the neurone we are recording from is concerned. Taken in this way anatomy is invaluable and can be used to predict where to apply physiological techniques. Furthermore, in some cases, anatomy may be the only means of study. It is impossible, for example, to obtain readings from some nematodes because of their size; but they are important organisms in neurobiological research. Painstaking serial reconstruction of the 302 neurones in the nematode, Caenorhabditis elegans showed how they are interconnected. Thus by comparing the anatomy of a normal organism and behavioural mutants (where the cell got into the wrong position) anatomy can be used to predict behaviour.

In physiological studies the animal is taken and dissected, cells taken out and electrodes placed into them to obtain recordings. These electrical records will correlate with movements of this remanent animal, but what does it tell of the behaviour of the animal? Furthermore, is it possible to tell how an animal relates to the whole world by recording from a few neurones simultaneously? Behaviour is the reaction of the animal in time and space, its continual interactions with the environment in surviving and making its way through its whole life. If we remove the nervous system from the animal we are looking at the nervous system and not at the behaviour of the animal. We can understand certainly the neuronal interactions which generate a particular electrical input but this is not behaviour. One cell may respond in one way, while several cells may respond in one or different ways. How do we explain these observations? It is important to have the concept at the cellular level but we have to continually relate back to the animal. Furthermore it is important to recognise that the integration process of going from one level (cellular) to another (behavioural) is much easier in some animals than others. It is not a single problem but a very different problem in each phylum. It should also be remembered that with the basic modern complex we cannot ask any physiological questions if we do not have a quantified behaviour. We really have to measure in time all aspects of the cell which are necessary for that behaviour. Or on the contrary, we can start with the behaviour and then go to the neurones only if we have described all the parameters of the behaviour.

There is no behaviour without the brain and if the brain is removed the animal will not react. However in this group (a neurobiological group) too much emphasis is placed on the brain. But the total environment, both internal and external, should be considered because some other physiological factors may play a role either in themselves or indirectly via the brain in modifying behaviour. An analogous situation existed in endocrine studies when for a long time all attention was concentrated on the influence of a single hormone on a particular system.

Mathematical models and computer simulations allow us to deal with the various parameters. There is a limit to the number of variables which we can sieve out and deal simultaneously - a limitation the computer does not have. Thus there is nothing good or bad about the computer - it is just a means of permitting us to arrive at the complex models which we ourselves, unaided, would have difficulty in conceiving.

Studies likely to gain prominence in the future include the intercelular messengers. Studies on what happens after these molecules attach themselves to the receptors and what actually happens in the cell. It is likely that we will find a very great range of very subtle changes occurring under different conditions according to the different molecules attached to the receptors and according to the pathways they use. We are at present only scratching the surface of this field of research, but the first promising results should be appearing in the near future. Neurones are not just cytoplasm. They are actually, in themselves, intracellularly very variable such that the biochemical interactions within neurones are very complex, besides in time they vary their responsiveness. This is already evident in preparations in which we know that in response to serotonin channels are modulated in a such a way so as to change the activity of the neurones in a circuit. Furthermore, the nervous system does not consist of neurones alone, glial cells also play an important role, the importance of which we are only just beginning to understand.

We have to be very open minded in the influences we are looking at when we are analysing neuronal functions and we must take into account all sorts of levels of actions - from the molecular level to the whole animal and its environment to behavioural ecology. Obviously we cannot do this at all levels at once but in order to ask the right questions we must be aware on which level we are working on and how that level interacts with all the other levels of explanation.

Another aspect of invertebrate neurobiology which has received very little attention is the larvae of lower groups (e.g. sponges, cnidarians, etc.). These animals are very easy to get and keep in captivity; in addition they develop fairly rapidly. These characteristics make them highly suitable material for studying the things mentioned at this ASI.

SPECIES INDEX

Abylopsis 240
Acetes 328, 343
Achatina 50-52
Acheta 160, 194
-- *domesticus* 606
Acholoe 581, 583
Acrographinotus 312
Actinia 11
Aequorea 12, 13, 228, 230, 574
-- *aequorea* 234, 235
Agalma 240
Aglantha 9, 11-13, 229, 235, 242
-- *digitale* 230, 235, 236
Anodonta 10
Anthopleura 121, 122, 243
-- *elegantissima* 110, 119, 120
Antromysis 344
Aplysia 1, 8, 16, 17, 29-35, 42, 50-52, 82, 88, 109, 115, 116
-- *brasiliana* 110
-- *californica* 108, 109, 111, 116
Araneus 314
-- *diadematus* 313
Argiope aurantia 304
Ariolimax californica 52
Artemia 328
Ascaris 15
Asellus 344
Astacus 16, 342
Asterias 497, 501, 505
-- *amurensis* 501
Astrangia 12
Atya 339
Aurelia 9, 12
Austropotamobius 328, 343

Balanus 16
Bathyporeia 340-344
Beroe 246, 250
Biomphalaria 34
Blabera 160
Bolinopsis 579
Bombina bombina 107
Bombyx 135-140, 142, 143, 152, 159, 160, 162
-- *mori* 135-137, 139, 143, 147, 154, 158, 161, 163
Bulla 50

Caenorhabditis elegans 649
Calliactis 123, 221, 222
-- *parasitica* 119, 123, 221
Calliphora 80, 138, 140, 142, 146, 147, 149, 150, 152-156, 174-176, 181-187, 189, 190, 192, 194-197, 199, 201, 353, 354, 357, 359, 360, 362, 366, 367, 371, 373-375, 378, 381-385
-- *vomitaria* 134, 137, 138, 140, 146, 148, 151, 154
Cancer 328
-- *pagurus* 342
Carausius morosus 114, 117
Carcinus 328, 344
-- *maenas* 112, 116, 333, 342
Carybdea 225, 226
Carybdea rastonii 223, 224, 226
Caryophyllia smithii 222
Cataglyphis bicolor 193

Ceriantheopsis 11, 12
Chaetopterus 583-586
Chelophyes 229, 240
Cherax destructor 329
Chirocephalus 328
Chrysaora 9, 12
Ciona 527, 528, 530, 532, 551, 554
Clavularia 11
Clitumnus 160
Conchoecia 344
Cucamaria 498
Cyanea 6-8, 11-13, 224, 225, 227, 228, 594
-- *capillata* 223-225
Cylindroliberis 344

Daphnia 343
Dendraster 504
Dendrodoa 535, 536, 540, 549, 550, 552
Dolioletta 535, 549, 552
Doliolum 528, 550-552
Drosophila 173, 181, 182, 193, 194, 199, 362, 366, 367, 375

Echinus 486, 500
Emerita 333, 334, 341
-- *analoga* 343
Ephydatia 216
Ephysa 574
Eristalis 152
-- *aeneus* 143, 146, 147
Eucidaris 500
Euphysa 537, 538, 574
Eurhamphea 578
Euspongia 216

Formica 312
Forskalia 240
-- *edwardsii* 241
Fritillaria 532, 534
Funchalia 328

Galathea 334, 341, 344
-- *strigosa* 333
Galeodes 313
Galleria 80
Gammarus 341
Gastrocotyle 7
-- *trachuri* 14
Geogarypus 310
Glomeridia 308

Glycimeris 17
Gomphocerripus rufus 46
Gonionemus 11, 12
Goniopora 218
-- *lobata* 218, 219
Grapsis 15
Gryllus 160
-- *assimilus* 47
-- *bimaculatus* 47, 606, 608
-- *campestris* 604, 608

Haementeria 266, 272
-- *ghilianii* 272
Haemopis 89, 278, 283, 287, 289
-- *marmorata* 288
-- *sanguisuga* 70, 79, 91
Haliclystus 9, 11, 12
Halistaura 574
Halistemma 240
Harmothoë 15
-- *imbricata* 581-583
-- *lunulata* 581-583
Heliothis 117
-- *zea* 113, 117
Helisoma 34, 52
-- *trivolvis* 33
Helix 50, 52, 109
-- *aspersa* 52, 109, 110
-- *pomatia* 52, 67
Helobdella 287
-- *triserialis* 288, 290
Hermissenda 50-52
Hesperonoe 583
-- *complanata* 581
Heterometrus 313
Hippopodius 229, 240, 575, 576
Hippospongia 216
Hirudo 14, 265, 266, 272, 274, 276-279, 281, 283, 284, 287
-- *medicinalis* 14, 30, 265, 285
Homarus 16, 324, 328, 333, 334, 341, 342-344
-- *americanus* 112, 326, 343, 344, 346
-- *gammarus* 324, 326, 333-335
Hydra 5, 7, 9, 11-13, 19, 118, 228, 231
Hydractinia 11, 228
Hyla caerulea 107
Hymenolepis microstoma 14

Idotea 344
Illex 17
Irogulus nepaeformis 308

Leptinotarsa decemlineata 189
Leptodora 466
Leucophaea 158
-- *maderae* 114, 154, 158
Ligidium 342, 343
Limulus 16, 312
Littorina 50, 53
Locusta 117, 144, 145, 158, 160
-- *migratoria* 112, 113, 117, 145
Loligo 17, 445, 448, 452-455
-- *pealei* 454
-- *vulgaris* 443, 445-447, 450, 451, 454-456
Lophelia pertusa 218
Luciola 588, 589
Luidia 504
Lymnaea 17, 50-52, 88
Lymnea 34, 116
-- *stagnalis* 33, 109-111, 116
Leucophaea 158
-- *maderae* 114, 154, 158
Ligidium 342, 343
Limulus 16, 312
Littorina 50, 53
Locusta 117, 144, 145, 158, 160
-- *migratoria* 112, 113, 117, 145
Loligo 17, 445, 448, 452-455
-- *pealei* 54
-- *vulgaris* 443, 445-447, 450, 451, 454-456
Lophelia pertusa 218
Luciola 588, 589
Luidia 504
Lymnaea 17, 50-52, 88
Lymnea 34, 116
-- *stagnalis* 33, 109-111, 116

Macrobdella 278
Macrocallista nimbosa 110
Manduca 16, 117, 135, 137, 138, 144-146, 152, 638
-- *sextra* 15, 78, 113, 117, 137, 138, 147, 189, 638
Marinogammarus 343
Marthasterias 504
Melanopus femmurrubrum 46
-- *sp.* 46
Metridium 9, 11, 12
Microstomum lineare 14
Mimosa 216
Mnemiopsis 251, 578-580, 594
-- *leidyi* 579
Mola 548
Muricea 12
-- *californica* 220

Musca 80, 181, 182, 360, 374
-- *domestica* 15, 372
Myxicola 14
-- *infundibulum* 14

Nanomia 12, 240, 241
Nauphoeta cinera 113, 117
Navanax 50-52
Necturus 82, 92
Neomysis 34-344
-- *integer* 343
Neophrynus 312
Nepanthia 501
Nereis 7, 14, 71
-- *virens* 265
Notodromas 344
Notoplana 14

Obelia 595
-- *geniculata* 575
Octopus 445, 447-448, 452
-- *vulgaris* 17, 443, 445-452
Oikopleura 528, 532-534, 536, 537, 540, 552
-- *albicans* 535
-- *dioica* 535, 538, 539
-- *labradoriensis* 539
-- *longicauda* 538
Onchidium verruculatum 33
Ophiura 501
-- *ophiura* 488, 499, 506, 507, 512
-- *texturata* 500
Orchestia 343
Orconectes 328
Ostrinia nubilalis 143

Pachylus 303, 305, 306, 311, 312
-- *quinamavidensis* 303, 305, 307, 311
Pacifastacus 328
-- *leniusculus* 333
Pagurus 344
Palinurus 328
-- *vulgaris* 333
Pandalus 116
-- *borealis* 112, 116
Panulirus 324, 328
-- *argus* 333, 344, 346
-- *interruptus* 344, 346
Paragrapsus 344
Pegea confederata 544
Periplaneta 145, 157-160, 188, 193
-- *americana* 112, 113, 117, 143,

144, 147, 157, 188
Phallusia 528
Phialidium 230
Philosamia 137
-- *cynthia ricini* 137
Photuris 587, 589, 590
Phrynichus 313
Planorbis 17, 51, 52, 80, 89
-- *corneus* 77, 81, 85
Pleurobrachia 246, 248-250, 580
-- *pileus* 244-246
Pleurobranchaea 50-53
Podocoryne 11, 12
Pollistes 160
Polyorchis 11-13, 228, 229, 231, 233, 236, 238-240, 242-243
-- *penicillatus* 230, 231-234, 237, 238, 243
Porcellio 312, 342
Proboscidactyla 238
Procambarus 16, 326, 342
Pugettia producta 341, 342, 343
Pyrosoma 528, 529, 534

Renilla 122, 229, 576-579, 585, 594, 595
-- *köllikeri* 110, 122, 577, 578
Rhabdocalyptus 214, 216
-- *dawsoni* 214, 215
Rhodnius 199, 200
-- *prolixus* 199
Rhodnius prolixus 199

Saccharomyces cerevisiae 161
Sagitta bipunctata 470
-- *crassa* 462, 465, 470, 472
-- *hexaptera* 468, 469
-- *setosa* 469, 480
Salpa 528, 532, 536
-- *fusiformis* 541, 542, 544-549
-- *maxima* 532, 553

Sarcophaga bullata 192
Sarsia 8, 11, 238
Sclerasterias 504
Scotolemon 308
Sepia 17, 74, 454
Sergestes 328
Shistocerca alutecea 46
-- *gregaria* 46, 113, 117
Siro rubens 308
Spadella schizoptera 470
Spirocodon 11, 238, 239
Spadella schizoptera 470
Spirocodon saltatrix 239
Srybula fuscouittata 46
Stichopus 497
Stomotaca 11, 238
Stomoxys calcitrans 158
Stomphia 222
Strongylocentrotus intermedius 563
Sulculeolaria 240
Sympetrum 193

Tabanus 158
Tamoya 11
Telyphonus 313
Tethya 216
Tetragonurus 548
Thaumetopoea pityocampa 154

Uca 116
-- *pugilata* 112
-- *pugilator* 116

Veretillum 9, 11, 13, 576
Verongia 216
Virgularia 11

Zygilla 16

SUBJECT INDEX

Abdominal, muscle receptor organ 336, 342
Aboral nervous system 566
-- poles 244, 247, 560
-- ring system 559, 564, 567, 570
-- sinus 565
Acarina 304, 312
Acetylcholine 32, 33, 69, 86, 118, 172, 276, 278, 401, 478, 491, 495, 498, 499, 544, 550, 559, 564, 568, 570, 586, 636
-- receptor 31, 402, 403
Acetylcholinesterase 498, 551, 583, 586
Acochilidiacea 50
Acoustic interneurons 378, 379
Across fibre pattern 624-631, 637, 638
Actin 88
Actiniara 122, 221
Action potential 30, 35, 41, 106, 225, 227, 231, 241, 251, 279, 413, 551, 579, 588
-- -- antidromic 400
-- -- calcium dependent 399
-- -- depression 30
-- -- epithelial system 534
-- -- in Hydrozoa 228-230
-- -- -- nociceptive cell 270
-- -- -- Siphonophora 240
-- -- -- sponge 216
-- -- regenerative 281
-- -- tunicate epithelia 535-550
Adenosin tri-phosphate (ATP) 83, 92
Adenylate cyclase 108
-- -- activation 92
Adrenal medulla 106
Adrenaline 497, 498, 578, 580, 586, 590

Adrenergic antagonist 590
-- system, Beta-like 580
Aesthete 324, 339, 341-345
-- seta 325
Agalenidentypus 316
Agelenidae 316
Alpha bungarotoxin 402, 403
Amacrine neurons 191
Amaurobiidae 316
Amblypygi 303, 310-313
Ameobocyte 578
Aminergic transmitter action 590, 594
Amino acids 2, 88, 106, 108, 114, 115, 121, 122, 395, 503, 506, 647
-- -- sequences 133-162
-- -- tryptophan lines 146
Aminopeptidases 106, 121
Aminopyridine 537
Amphioxus 555
Amphipoda 330, 340, 342-344
-- setae 323, 339
Ampullae 487
-- tubefoot system 492
Anamorphose 310
Anaspidea 50
Anechoic chamber 604
Anemones *See also* sea anemone 221-223
Aneurogenic character 586
Angiotensin 161
Angular velocity in cricket 609, 610, 613
Anisomycin 34
Annelida 4, 7, 10, 14-18, 45, 62-65, 71, 75, 89, 265-294, 303, 309, 311, 312, 327, 574, 592
-- luminescent system 580

Annuli 267, 271, 274, 281, 289, 340
-- erection 275
Anomura 337
Antenna 323, 327, 328, 331, 332, 340, 342, 355, 357, 358, 367
-- in Insecta 629, 634
-- mechanoreceptors 329
Antennal glomeruli 175, 177, 178, 356, 357, 366, 367, 371, 379
-- -- chemosensory 176
-- -- immunocytochemical mapping 173-191
-- interneuron 384
-- lobe 176, 180
-- -- GABAi neurons 182
-- -- of the moth 15
-- mechanosensory afferents 371, 374, 376
-- pedicellar 375
-- projections 353
-- receptors 356
-- -- central projections 366, 367
-- segment 355, 366
-- -- sensory pathways 367
-- seta 340, 343-345
-- scolopidia 342
-- system 354
Antenno-glomerular tract 183, 371
Antennular annuli 325
-- organs 341
-- proprioceptor 342
Antennule 327, 332, 346
-- chemoreceptors 329
-- muscle receptor organ 342
-- setae 342
Anterior optic tubercle 378
-- sucker in medicinal leech 267
Anteriormost organs 329
Antho-RF amide 110, 123
Anthomedusa, electrical coupling 232
Anthozoa 9, 11-13, 122, 123, 217-223, 574, 576-578, 592
Antibodies, anti-bovine insulin 143
-- -- ARG-PHE-amide 578
-- -- corticotropin releasing factor 194
Antidromic stimulation 404
Antigens, surface 287
Antimotilin 194
Anti-myosin 276
Antiserum, RF amide 118-120
Apical organ in ctenophore 244, 247
Arachnida 303- 305, 310-314, 317, 646
Araneida 314
Araneomorpha 313
Archeogastropoda 50
Arginine 108, 119
-- vasopressin 120
Arista 366
Arrowworms *See* Chaetognatha
Arthropoda 6-8, 10-18, 45-48, 62-65, 71, 72, 75, 80, 86, 89, 117, 273, 303, 305, 309, 311, 312, 330, 354, 574, 593
Ascending neuron in Insecta 374, 606, 607, 613, 615, 631, 633, 635
Ashelmintha 62, 63
Ascidiacea 527-530, 532, 534-536, 549-554
-- excitable epithelia 549
Asteroidea 485-487, 493, 503, 504, 593
Atropine 586
Attraction-repulsion, phenomenon 34
Auditory interneuron in cricket 605, 612
-- receptors 634
Autozooid 576-578
Axon *See also* giant axon, giant fibre, giant nerve fibre, nerve fibre,
-- antibody binding 288
-- giant in cockroach 395, 396, 423
-- -- -- Crustacea 4, 6, 12, 16, 69, 327
-- -- -- earthworm 4
-- -- -- squid 1, 67, 69, 75, 88, 398
-- glial-less 499
-- hyponeural motor 487, 488
-- Langerhans 533, 534, 537, 539
-- mechanoreceptor 346
-- medial giant in crayfish 67
-- motor 488, 495, 508, 509
-- -- giant 229, 242, 456
-- outgrowth 288
-- ring-giant 235, 242
-- sensitive cell 285, 287
-- sensory ectoneural 489
-- subset, surface antigens 287
-- syncytial motor giant 235
-- wrapping 74-78
Axoplasmic resistivity 397

Bag cells 111, 115
Barnacle 16, 326, 330
Basic cellular mechanism 514
Basiepithelial plexus neurones 489
Basommatophora 50, 52
Bee 15, 172, 173, 177, 181, 182, 201, 382, 632
-- antennal innervation 181, 182
-- brain, catecholamine-containing neuron 190
-- -- GABAi neurons 181, 182
-- honey 83, 84, 172, 173, 177, 181, 360, 378, 379, 381
-- -- neuroactive substance 171-202
Behavior, circuitry underlying 413-423
-- control in Cnidaria 217-243
-- -- in Insecta 393-425
-- egg laying 109, 115, 116

-- feeding, gastropod 49-53, 115
-- neural control 621-639
-- phonotactic in cricket 603-617
Beroida 244, 245, 248
Bicuculline 583
Bimodal receptors 339, 342-345
Binocular pathways 364
Biogenic amine 583
Bioluminescence, control in Ctenophora 244
-- neural control mechanisms *See also* luminescent systems, 573-595
Biphasic pulse 577
Bipolar cell 507, 508
Bivalvia 5, 10, 17, 591, 592
Blast cell 290
Blastema 505
Blastomeres 289, 292
Blastozooid 541-549
-- attachment plaques 548
-- chains 547
-- -- separation 548
-- -- swimming behaviour 546
-- epithelial conduction system 549
Blood-borne modulator substances 625
--brain barrier in Crustacea 73
-- -- -- in cuttlefish 74
-- -- -- in Insecta 62, 69-74, 394-396, 406, 410, 423, 424
Blowfly 134, 146, 148, 150, 153, 156, 172, 173, 190, 191, 195, 197, 355, 359, 360, 363, 366
-- brain, catecholamine-containing neuron 190
-- neuroactive substance 171-202
-- serotonin-immunoreactive neurons in brain 173-177
-- -- -- -- in ventral ganglia 195-197
Body-coxopodite joint 331
-- limb joint 332
Bombesin 107, 239
-- like 239
Brain catecholamines in Insecta 189, 190
-- formation in Cephalopoda 445-452
-- GABA-like immunoreactive neurons in insect 181, 182
-- gastrin/CCK-like immunoreactive neurons in Insecta 184-188
-- hemispheres interconnections 173-202
-- in Araneida 314
-- -- Ascidiacea 528
-- octopaminergic neurons 190, 191
-- variation in Chelicerata 312, 313
Branchial cavity 541
-- chamber 527, 528
Branchiopoda 328
Briefly 638
Brittlestar 483, 484, 487, 488, 495, 497, 500-502, 507, 512
Buccal tentacles in Holothuria 487
Burfield's terminology 462
Bursicon 184
Burst generation 418, 419
Bursting system 236, 237, 239
-- -- synapses 242

Cadophore organ 531
Caerulein 107
Calanoid copepod 324, 329
Calcarea 214, 216
-- "nerve-like" cells 215
Calcitonin 108, 184
Calcium 31, 403, 494, 537, 551
-- channel 575
-- concentration 216, 225
-- current 35, 250
-- dependent action 583, 590, 595
-- spike 506, 507
Calmodulin 647
Calycophora 240
Calyx 177, 370, 371, 378-382
Capitate junction 64
-- projection 65, 88, 90
Carapace setae 342
Carboxy-peptidases 106, 108
Carbohydrate metabolism 91
Cartridges 360, 362
Catabolites, removing 87
Catch mechanism 494, 495
Catecholamines 108, 118, 172, 362, 363, 382, 384, 486, 492, 499, 504, 580, 636
-- containing neurons 175-177
-- -- -- mapping 189, 190
-- -- -- patterns of distribution 200, 201
-- function 189
-- like mechanisms 578
Catecholaminergic pathways 382
Caudodorsal cells 111, 116
Cellularia 214, 216
Central body complex 353, 356, 378, 382
-- -- -- organization 383-386
-- -- in Chelicerata 312-316
-- -- -- Insecta 174-177, 357, 378, 380, 631, 638
-- monoaminergic neurons 188
-- nervous system, comparative aspect 309-311
-- -- -- in Coelenterate 589, 591
-- -- -- -- Echinodermata 503
-- -- -- -- Gastropoda 31, 33, 36
-- -- -- -- mollusc 17
-- -- -- -- tunicates 552-554
Cephalaspidea 50
Cephalisation 327
Cephalic system 310

Cephalopoda 4, 7, 17, 18, 62, 74, 591, 592, 594
Cercal afferent, connection 402
-- mechanosensory hair 403
Cerebral invaginations in polychaetes 309
Cestid 248
Cestoda 14
Chaetognatha 461-480
Chaetopteridae 580, 583-586
Chelae 45, 326, 346
Chelicerata 303-317
-- slit sensillae 330
Chemical communication 2, 316
-- perception 329
Chemiluminescent reactant 594
Chemo-receptive channels in Chelicerata 304
Chemoreceptor in Annelida 280, 281, 284, 285
-- -- Crustacea 323, 338, 339-346, 366
-- -- Echinodermata 499, 502, 506
Chemosensory cells in tunicate 532
-- neurons 265
Chiefly 629
Chilipods 303, 311
Chinese silkworm 135
Chloride 67, 83, 225, 403
-- conductance 30
Chlorimipramine 31
Choanoderm flagella 214
Choanosome membrane 216
Cholecystokinin 107, 120, 134-135, 150-153, 239
Choline 537
-- acetyltransferase 194, 362, 363
Cholinergic activation system 69
-- excitation 586
-- longitudinal muscle 276
-- mechanism 591
-- nerve in Echinodermata 491, 493
-- plexus 491
-- receptors 69
-- skeletal muscle in Echinodermata 487
-- system 580, 586
Cholesterol 78
Chordotonal organ 323, 326, 330-332, 336, 341, 342
-- receptor 326, 331
Chromatophores 116, 500
Cilia 243, 245, 491, 496, 501, 502
Ciliary band 504
-- beating 497
-- in Ctenophora 248
-- rootlet area 531
-- tufts of Chaetognatha 461, 462, 464, 478
Ciliated cells 281, 284, 289, 462, 534, 553

-- chemosensory neurons 266
-- epithelial cell 502
-- grooves 247, 248
-- receptor cell 500-502
-- sensory cells 531
Cirri 15, 281, 330, 490, 491
Cladocera 343
Clam 568
Claw 325, 326
Clitellate annelids 289
Clubionidae 313, 316
Cnidaria 5-13, 18, 19, 109, 110, 213-252, 574-578, 585, 592, 647, 650
Cnidoblasts 217
Cockroach 15, 80, 113, 117, 143, 145, 147, 157, 158, 172-202, 378, 380, 395, 396, 402, 410-413, 423
Coelenterata 12, 18, 62, 63, 75, 118, 123, 573-580, 591, 594
Coelomic cavity 501
-- factors 496
-- fluid 495, 568
Coenasarc 218
Collagen 484, 492, 496
Collagenous connective tissue 494, 496
Colloblasts 245, 249
Colorado potato beetle 152, 172, 189, 194, 197
Column 221, 222
Columnar centrifugal neurons 362
-- descending neuron 386
-- laminal organization 360
-- lobula neurons 371, 374
-- neurons 355
Comb plates 243, 248, 249
-- rows 244, 245, 247, 248
Commissure, brachial supraoesophageal in Cephalopoda 446
-- in Chaetognatha 462
-- -- insect ganglia 423
-- interbrachial ring in Cephalopoda 452
-- oesophageal of Chaetognatha 463, 470-472, 474
-- posterior visual in Araneida 315, 316
-- subcerebral 51
-- vestibular of Chaetognatha 472
Comparative embryogeny 303
Compound potential 506
Conditioning factors 34
Conduction system, ectodermal 123, 540
-- longitudinal 487
-- velocity 74, 78, 539, 542
-- -- of nerve impulse 76
Connective, brachial in Cephalopoda 446
-- brachio-palliovisceral in Cephalopoda 448
-- buccal superior-brachial in

Cephalopoda 446
-- cerebral posterior in Cephalopoda 446
-- cerebrobrachial in Cephalopoda 446
-- cerebro-buccal in Gastropoda 32, 49-52
-- cerebro-subradular in Cephalopoda 446
-- cervical in Insecta 176, 177, 356, 357, 372, 609, 633, 634
-- circumoesophageal in Insecta 188, 356, 635
-- frontal in Insecta 356
-- -- -- Chaetognatha 462, 463, 470, 471
-- in Annelida 267, 271, 275, 278, 285, 287, 289, 292
-- in Chaetognatha 473
-- in Gastropoda 42
-- in Insecta 395, 396, 400
-- interbuccal in Cephalopoda 446
-- interganglionic in Crustacea 112
-- -- -- Echinodermata 486
-- main in Chaetognatha 462, 463, 466, 475, 478
-- mantle 445
-- pallial 446
-- tissue innervation 494-496
Contraction potential 575
Convergent evolution 44, 45
Co-ordinate locomotion 503
-- contractures 491
-- motor input 511
Copepoda 328, 329
Coral 12, 217, 218
-- gorgonian 220
-- scleractinian 219
Cord stretch receptor 336
Corn borer 117
Corona (ciliary loop) of Chaetognatha 461-466
Corpora allata 144
-- cardiaca 112, 113, 117, 144, 160, 189, 201, 371
-- -- corpora allata complex 134
-- pedunculata 303, 304, 308, 312-317
Corpus allatus 143, 152
-- cardiacum 140, 146, 148, 150, 152, 154
-- -- in hoverfly 143
Cortex ganglionic in Araneida 314-316
-- visual in Araneida 314-315
-- cell in Insecta 589
Corticotropin 108, 184
-- releasing factor 184
Co-transmitter in motoneuron 401
Coxa 327, 332
-- basipodite joint 330
-- body joint 331
-- muscle receptor organ 336

Coxo-thoracic sensors 326
Crab 15, 116, 117, 327, 328, 337
-- fiddler 116
-- horseshoe 16
Crayfish 4, 6-8, 11, 12, 16-19, 47, 48, 65-69, 80, 88, 325, 328, 336, 337, 341, 413
-- australian 329
-- claw, branching pattern 325
Cricket 46, 65, 394, 397, 399, 407, 412, 413, 603-617, 622, 632, 638
Crinoidea 485, 490, 502
Crustacea 8-12, 15, 45, 62, 73-76, 86, 88, 90, 112, 117, 158, 303, 308, 311, 312, 323-346, 466, 593, 594
Ctenizidae 316
Ctenophora 6, 9, 13, 14, 18, 213-252, 574, 578-580, 591, 592
Ctenophore, cydippid 244, 246
Cuboidal cell in Annelida 581
Cubomedusae 11, 224, 226
Cubozoa 223
Cumacea 328, 339
Cupular organs 530, 532
Curare 537, 580
Current-voltage response, curves 41
Cuticle 327, 338-341
Cuticular hair 338
Cuttlefish 74
Cyclic AMP 69, 108
Cydippid 248
Cysteines 142
Cytoskeletal role 491

Dactylopodites 339
Decapoda 45, 308, 324-330, 340-345, 443, 444, 448, 454
-- limbs 346
-- megalopa 342
-- zoeae 342
Deep sea shrimps 328
Delayed initiation system (Dis) 221, 222
Demospongiae 214, 216
-- "nerve-like" cells 215
Dermal light sense 503
-- papulae in asteroid 493
Descending contralateral movement detector (Dcmd) 367, 369, 407, 410-412, 416
-- interneuron in Insecta 368, 372, 623, 625, 629-631, 633-635
-- neurons, columnar 375, 386
-- -- deutocerebral 371, 372
-- -- giant 372, 374, 375
-- -- -- in fly 629
-- -- in cricket 606, 607
-- -- postoral 371

– – preoral 371, 372
– – protocerebral 371
– – sensory inputs 358
– pathways, organization 367-378
Desert ants 366
Desmosomes 62, 64, 65, 76
Deuterostomes 480
Deutocerebrum 174, 178, 310, 312, 355-357, 363, 366, 371, 374
– immunocytochemical mapping 173-191
– organ 308
Diapause, control "factors" 134
Diffuse nerve net 226, 227, 244, 245
Diphyids 241
Diplopoda 303, 308, 310, 311, 312
Diptera 80, 355, 409, 413
Disinhibitory connection 409
Distal fan 280
Disynaptic disinhibition 419
DNA 115, 647, 648
– transcript 108
Doliolida 529, 530, 531, 535, 550, 551, 554
– excitable epithelia 549
– muscle potentials 552
Dopa/5-hydroxytryptophan decarboxylase 173
Dopamine 86, 191, 497-499, 554, 568, 578
– Beta-hydroxylase 384, 385
– cellular localization 189
– noradrenalin 384
– sequestering (DAs) neurons 200
Dorsal organ 326, 339, 342, 345
– rim area 363
– unpaired median (DUM) 399, 423, 589-591
Dragonfly 193, 355
Drassidae 316
Drone 65, 67, 68, 83, 84, 92
Dyadic type, synapse 3-7, 14-17
Dysderidae 316

Earthworm 4-7, 14, 15, 19, 74, 75, 78, 265, 274, 277, 279, 281
Echinodermata 5, 6, 17, 18, 62, 63, 75, 478, 480, 483-515, 591, 593, 639, 646
Echinoidea 5, 485-487, 503
– lantern 487, 490, 493
– spawning mechanisms 559-571
– sphaeridia 500
Ectodermal conduction system 123, 540
– nerve net 240
– precurcor bandlets 291
Ectoneural/hyponeural connection 487-489
– innervation in skeletal muscle system 493

– nerve plexus 491, 492
– nervous system 485-515
– neuron, longitudinal 509, 512
– single cell function 509-511
– – – morphology 508, 509
Electromyogram 417
Eleutherozoa 514
Ellipsoid boides 190
Elytral luminescent system 583
Embryonic development of peripheral nervous system 289-293
– nervous system in Cephalopoda 443-457
– organ in adult Chelicerata 308
– segmentation 291
Embryogeny comparative 303
Endocrine cells 107, 116
Endocytosis 90
Endodermal conduction system 123, 540
Endorphin, Alpha- 154
– – antiserum 154
– Beta- 108
Endostyle 528, 541
Endothelial cell in Echinodermata 494, 497
Enkephalins 184
– endorphin 133, 134
– receptors 154
Entoneural nervous system 485
Enzymes *See* specific name
Epithelial action potentials, effects 542
– conducting, electrical properties 538
– – systems 216, 217, 223, 228, 241, 252, 530, 536
– propagation of excitability 574
Epitheliomuscular cell 13, 217, 586, 591
Erythrophore 112, 116
Escape system in tunicate 539
– swimming control in Trachymedusae 236
Eserine 565, 570
Ethanolic phosphotungstic acid (EPTA) 17
Eumetazoa 213
European cornborer 143
– pennatulacea 576
Evisceration factor 495
Evolutionary theory 44
Excitable epithelia in tunicates 535-550
Excitatory post-synaptic potential (EPSP) 236, 237, 239, 242, 249, 398, 402-404, 411-413, 510-513
– connection 415-419
Exocrine goblet cell 586
– secretion 576
Exoskeleton in Cnidaria 217
External mechanical signals 328
Exteroreceptor cells in Larvacea 534

Extracellular potential 85
-- recording techniques 505-507
Extra-ocellar light 501
Eye in Annelida 281
-- -- Arachnida 312
-- -- Cephalopoda 450
-- -- Chaetognatha 461-463
-- -- Crustacea 15, 327
-- -- Insecta 7, 15, 16, 61, 65, 68, 80, 83, 84, 86, 92, 355, 357, 358, 360-367, 403, 407, 409, 411, 423, 629, 634
-- lensed in Scyphomedusa 226
-- stalk (x-organ) 116

Fasicular glia 71
Fast extensor tibiae motor neuron (Feti) 46, 47, 399-401
Femur-tibia joint, control 627
Fibre, see axon, nervous fibre, giant
Filaments fine 14, 16
Filistatidae 316
Firefly 586-591, 594
-- european 589
-- lantern 573, 587, 589
Flagella in sponge 214-216
Flagellar segments 375
Flagellated collar bodies 214
Flagellum 366
Flashes, bioluminescent 575, 578-580, 588-590
Flatworm 7, 14, 18, 19
Fleshfly 192
Flight motor pattern 394, 415, 416, 423
-- muscle in locust 626
-- neuron in locust 419
-- oscillator 417
-- phasic sensory input 416, 420-422, 424, 425
-- system 413-423
-- -- in Diptera 409
-- -- -- Orthoptera 407, 409, 417, 418, 424
-- -- reciprocal inhibition 420
Fluorescent orthochromatic cell 586
Fly See also blowfly, briefly, chiefly, dragonfly, firefly, fleshfly, fruitfly, horsefly, hoverfly, housefly, mayfly 7, 15, 16, 86, 88, 141, 142, 150, 158, 173, 181, 182, 201, 354, 356, 360, 363, 382, 632
FMKF amide 120
FMRF amide 33, 184, 194, 198, 201, 239, 243, 277, 384
-- role in leech 278
-- like immunoreactivity 363
Fodrin-like polypeptide 88
Follicles 303
-- cell in Echinodermata 497

Food rejection, reflexes 543
Foregut 452
Foveal area 363
Freeze behaviour 506, 509, 510, 512, 513
Frenzies reminiscent 576
Frenzy 575
Fringed setae 339, 343, 345, 346
Fritillariid larvaceans 534, 539
Fruitfly 48
Funnel 450, 455
-- canals 339, 344-346
-- muscle 454
-- valves 445
Furcilia 328

GABA (Gamma-aminobutyric acid) 86, 172, 194, 277, 360, 362, 384, 385, 498, 554, 559-563, 568, 570, 586
-- ergic system 586
-- -- inhibitory motoneurons 277
-- -- transmission 401
-- immunoreactive fibers 360
-- -- neurons 382
-- like immunoreactive neurons 171-202, 363,
Galactocerebrosides 78
Gamete-shedding substance (GSS) 497, 498
Ganglion abdominal in Crustacea 16, 65, 334, 335
-- -- -- Gastropoda 32, 33, 42, 82, 111, 115, 116
-- -- -- Insecta 147, 150, 151, 197, 199, 201, 202, 357, 587-590
-- arm in Cephalopoda 450, 452
-- basal in Cephalopoda 443
-- brachial in Cephalopoda 443, 448, 450, 452
-- brain in Chaetognatha 470
-- buccal, contralateral projections 52
-- -- in Cephalopoda 443, 452
-- -- -- Gastropoda 32, 34, 49-53, 65, 66, 77, 81, 91, 110, 111, 115
-- caudal in Annelida 292, 293
-- -- -- tunicates 528, 533, 534, 537, 539, 554
-- cell, tadpole retinal 35
-- -- in Polychaeta 309
-- cerebral, contralateral projection 52
-- -- in Cephalopoda 443-447, 450, 454
-- -- -- Chaetognatha 461-480
-- -- -- Chelicerata 303, 311
-- -- -- Gastropoda 32, 33, 49-53, 109, 111, 115, 116

-- -- serotonergic projection 53
-- cheliceral 311
-- circumoesophageal in Insecta 356
-- development in Chelicerata 305-308
-- ectoneural ring in Echinodermata 493
-- elytral in Annelida 580, 581
-- frontal in Chaetognatha 462, 470
-- gastric in Cephalopoda 443, 447, 453
-- head in Cephalopoda 443
-- -- -- Gastropoda 111, 116
-- -- -- leech 266
-- hypocerebral in Insecta 147, 148, 152
-- hyponeural in Echinodermata 486-488, 507, 508
-- inferior buccal in Cephalopoda 446
-- labial in Insecta 356
-- mandibular in Insecta 355
-- marginal in Cnidaria 7
-- mass, suboesophageal in Cephalopoda 443, 445
-- maxillary in Insecta 355
-- meso-metathoracic in Insecta 199
-- mesothoracic in Insecta 186, 197, 357, 360, 407, 419
-- metathoracic in Insecta 15, 150, 195, 197, 199, 357, 368, 400, 403-407, 411, 417, 419, 625
-- occipital in Chelicerata 311
-- oesophageal in Chaetognatha 461-463, 470, 471, 473
-- -- -- Crustacea 324
-- olfactory in Cephalopoda 443
-- opisthozomial in Chelicerata 310
-- optic in Cephalopoda 443, 444, 445, 447, 450, 454
-- -- -- Chelicerata 311
-- -- -- Crustacea 112, 116
-- oral in Echinodermata 504
-- pallioviscerai in Cephalopoda 453, 454
-- parapodial in Annelida 581
-- parietal in Chelicerata 311
-- pedal in Cephalopoda 443, 444, 446, 448, 450, 452, 453, 454, 455
-- -- -- Gastropoda 115
-- peduncular in Cephalopoda 443
-- precheliceral in Chelicerata 311
-- prosomal in Chelicerata 310
-- prothoracic in Insecta 197, 199, 357, 376, 605, 610
-- protocerebral in Insecta 201
-- segmental, embryonic origin 292
-- -- hyponeural in Echinodermata 495, 508
-- -- in Annelida 42, 48, 61, 70, 78, 87, 88, 91, 266, 267, 272-279, 284, 292
-- -- -- Insecta 196, 632, 634, 636

-- -- of radial cord 483, 499, 515
-- stellate, development 452
-- -- in Cephalopoda 443, 445, 447, 450, 456
-- stomatogastric in Crustacea 16, 324, 637
-- subacetabular in Cephalopoda 447, 452
-- suboesophageal, immunocytochemical mapping 173-191
-- -- in Annelida 291, 193
-- -- -- Chaetognatha 461-463, 472, 474
-- -- innervation 173-191, 355
-- -- in Insecta 147, 148, 150, 154, 160, 173-176, 178, 201, 202, 310, 312, 355-357, 359, 366, 367, 371, 629, 632-637
-- -- origine 272
-- subradular in Cephalopoda 443, 452, 453
-- supraoesophageal, embryonic origin 292
-- -- in Insecta 355
-- tail in medicinal leech 267
-- terminal in insect 195, 197
-- -- abdominal in Insecta 402
-- thoracic in Insecta 146-152, 160, 177, 187, 197, 199-202, 355, 356, 359, 368, 369, 372, 375, 376, 378, 394, 411, 416, 420, 625, 629, 633
-- thoracico-abdominal in Insecta 176, 187, 198, 199, 355
-- ventral in Annelida 30
-- -- -- Chaetognatha 461-463, 465, 472-479
-- -- -- Chelicerata 310
-- -- -- Insecta 171, 172, 180, 188, 190, 198-202, 367, 379
-- -- neuroactive substances 194-200
-- vestibular of Chaetognatha 461-463, 465, 470, 471, 473
-- visceral in Cephalopoda 443-450
-- -- -- Gastropoda 85
Ganglionic mass, supraoesophageal 632
-- pits 309
Gap junction *See also* electrical synapse 3-6, 10-16, 19, 62-68, 71, 73, 85, 90, 223, 227-229, 231, 242, 249, 252, 395, 492, 531-537, 543, 546, 550-552, 576, 579, 581
-- like junction 88
Gastrin 107
-- cholecystokinin 119, 133, 150-153, 172, 184, 194, 384
-- -- function 187
-- -- like immunoreactive neurons 171-202, 362, 363, 381, 382
Gastrodermal cell 591

Gastropoda 8, 17, 18, 42, 45, 49-53, 65, 66, 75, 80, 86, 88, 637
Gastrozooid 240, 241
Gastrulation 291
Geotactic responses by ctenophores 247
Germinal band 290, 291
-- layers 28
-- plate 272, 289-291
Giant cell in Cephalopoda 446
-- cerebral neuron 49
-- cholinergic cell (cell R2) 32
-- fiber nerve net 223
-- -- in Annelida 15, 19, 74, 334
-- -- -- Cephalopoda 17, 76
-- -- -- Crustacea 16
-- -- -- Echinodermata 486, 490, 505
-- glial cell 67, 70
-- interneuron 278, 398, 399, 402, 410, 411
-- nerve fibre system 445, 453-457
-- neuron in Echinodermata 489, 507, 511
-- smooth muscle cell of Ctenophora 250, 251
-- suboesophageal efferent neurons 176
Gland, atrial 111, 115, 116
-- mucus 277, 294
-- photogenic 584, 586
-- prothoracic 134, 137
-- salivary 173, 277
-- sinus 112, 116
Glia mesaxonal 395
-- role of 70-93, 396, 397
-- sub-perineurial 396
Glial cell 1, 61-93, 285, 287, 336, 355, 394
-- fasiculation 285
-- processes 398
-- trophospongial processes 397
Gliointerstitial granules 86
-- process 585
Glio-vascular system 74
Globuli 303-308, 312, 315, 316
-- cells 190, 191, 378
Glomeruli 304, 316
-- chemosensory 190
Glomerular 353, 356, 371, 378
Glucagon 133, 143-146, 184
-- amino acid sequences 145
Glucose 91
-- 6-phosphatase 89
Glutamate 86, 121, 172, 401, 554
Glutamatergic transmission 401
Glutamic acid 559, 568
-- -- decarboxylase 570, 571
Glutamine 86, 106, 108, 121
Glutaraldehyde 487, 488
Gly-Arg-Phe-amide 122
Glycine 108
Glycogen metabolism in glia cell 87-93

Glycogenolysis 91
Glycosylated antigens 288
Gonad 559, 560, 562, 566, 570
-- arrhythmic contraction 559, 563, 564, 568
-- maturation in Echinodermata 497
-- rhythmic contractions 511, 559, 563-571
Gonadal contraction in Echinoida 561
Glycocalyx 496, 537
Gonoduct 559, 561, 562, 565, 567, 570
Gonyleptidae opilion 303-305, 311, 312, 317
Granular amorphous substance 489
Gravity detection function 500, 501
-- receptors 534
Growth cone 34, 35
-- of nervous system in Crustacea 324-326
Gut endocrine cells 147
Gymnosomata 50

H-shaped tube 610
Haemal system 498
Haltere afferents 377
-- interneuron 376
Heart accessory (Ha) modulatory neurons 277
-- excitor (He) motoneurons 274, 277
Helically-coiled process 249
Head segment in medicinal leech 267
Helical arrangement of smooth muscles 490
Helicidae 52
Hemichordata 591
Hemi-desmosomes 64, 70, 71
Hemipteran, catecholamine-containing in ventral ganglia 199
Heterocellular glial-neuron 62
-- junction 65
Hexacorals 217, 218
Hexactinellida 214-216
-- flagellar arrest 215
Hexamethonium 564, 565, 570
Hirudinea 265
Histamine 86, 191, 194, 362, 384, 385, 498
Histidine 148
Holothuroidea 485, 487, 490, 493-498, 501, 502
Hoover neuron 278, 279
Homeostatic mechanism 396
Homocellular glial-glial junctions 62
Homologies cellular 1, 41-54
-- cladistic cellular 45
-- serial 45
Homology 43-46
Homoplasty 44, 45
Horizon detection 358

Horizontal motion sensitive neurons 375-377
Hormone *see also* neuropeptide
-- adipokinetic 112, 113, 117, 143, 145, 184
-- Alfa-melanocyte stimulating 184
-- anti-growth 194
-- caudodorsal cell 111
-- diuretic in locust 160
-- eclosion of silkworm 158, 159
-- egg-laying (ELH) 111, 115, 116
-- juvenile (JH) 134
-- light adapting 112, 116
-- melanocyte-stimulating (MSH) 156
-- pigment dispersing 112
-- prothoracicotropic (PTTH) 133-143, 161
-- red pigment concentrating (RPCH) 112, 116, 117
Horsefly 158
Housefly 15, 65, 355, 363
Hoverfly 143, 146, 147
Hyaluronic acid 76
Hydroid calyptoblastoid 575
Hydromedusae 5, 8, 9, 11-113, 227, 230, 235, 240, 574
-- motor system 229, 230
Hydropolyps 228, 240
Hydrostatic skeleton in Cnidaria 217
Hydrozoa 5, 9-12, 118, 223, 537, 574-576, 591, 592
-- action potentials 228, 229
Hyperglycaemic factors 113, 117, 143-146
Hypertrehalocaemic factors 144
Hypoglycaemic factors 140, 143, 146
Hyponeural innervation 493, 495
-- motor neuron 18
-- nervous system 485-515
-- single cell 507-509

Identified interneuron, references 414
-- neurons 1, 41-54, 355, 646
-- -- chemically 171-202
-- -- in cricket 610
Immunicytochemical mapping 171-202
Immunohistochemically identified systems in Hydrozoa 139, 240
Indian silkworm 137
Information collector 623
-- integrator 623
Infra-cerebral organ in peripates 309
Inhibitory connection 406, 415, 418, 419, 589
-- neuron 86, 408
-- postsynaptic potential (IPSP) 32, 403, 404, 406, 407, 416, 420, 510-513
-- transmitter 401
Inner nerve-rings 230, 231

Insecta 7, 15, 16, 61, 68-76, 79, 83, 86, 112-114, 116, 117, 184, 188, 189, 304, 305, 311, 312, 330, 336, 425, 573, 593, 646
-- catecholamines in brain 189, 190
-- indian stick 117
-- metamorphosis 62
Insulin 133-143, 184, 646, 647
Intercellular junction 396
Interganglionic fibres 292
Internal proprioceptors 330
Interneuron, flight 418, 421
-- giant 398, 399, 402, 410, 411
-- interganglionic 397, 399, 413
-- mechanoreceptor 378
-- non-spiking 397, 403, 409, 416
-- optic 384
-- premotor 417, 418, 420
-- S 49
-- sensory 413, 416
-- thoracic in locust 415
Interneuronal, gap junction 13
Interoreceptor 489, 501, 502, 506
Intersegmental fibres 324
-- interneuron in Insecta 399, 623, 625, 629-631, 633, 635
-- thoracic interneurons 375
Intracellurar techniques 507-514
Intracleft filament 9
Inulin 74
Ionic basis of action potential 507
-- conductance 398
-- diffusion 396
-- exchange 90
-- regulation in glial cells 82-86
Ischyropsalidae 308
Isopoda 330, 339-344
Isoproterenol 578
Iulidae 308

Jellyfish 4-7, 9, 11, 18, 224, 537
Jet-propulsion driven 529
Johnston's organ 366, 374
Jonction of glial cells 64
Jump trigger neuron (M) 416
Jumping system 413-423
-- -- in Diptera 409
Junctional "plugs" 214
Juxtaligamental cell 18, 484, 488, 491, 495, 508
-- tissue 489, 495, 508

Kenyon cells 355, 378, 382
Kolossale ganglienzellen 42

Lamella neural 394, 395
Lamina 15, 175, 177, 191-193, 360-363

-- intracellular 501
-- basal 18, 70
-- columnar organization 360
-- extracellular 5
-- neural 73
Laminae of glial cells 75-77
Laminar interneuron 74
Lampirid fireflies 586-591
Langerhans axon 533, 534, 537, 539
-- bristle receptors 532, 537, 539
-- receptor 533
Lantern muscles 498
-- nerve roots 587
Large bilateral optic lobe 5-HTi neurons (LBO5HT) 175
Larvacea 527, 532, 534, 536, 537, 550-554
Larval ciliary 501
-- fireflies 587, 589, 590
-- hoverfly 147
-- light organ 590
-- mussel 5
-- nervous systems in Echinodermata 497, 504
-- ocelli 501
-- silkworm 143, 147
-- zoea of Decapoda 324
Lateral horn 183, 188, 371, 381, 382
Lateralis receptors 530
LE mechanosensory cell 33
Learning, echinoderm 504
-- sites 382
Leech 14, 29-31, 36, 42, 45, 49, 54, 61, 65-71, 73, 78-83, 87-91, 199, 265-267, 281
-- chemosensory organs 284, 285
-- embryos 288
-- glossiphoniid, development 289
-- musculature 275
Lensed eyes in Scyphomedusa 226
Lepidoptera 154
Leptomedusae 230
-- control of swimming 234, 235
Leucokinin 114, 117
Light organ (lantern) 587
Limbs (body-coxal joint) 323, 328-333, 346
-- abdominal 323
-- bud 327
Limulus 312, 317
Lip sensilla 284, 285
Lipofuscin 80, 81
Lipotropin, Beta- 108, 154
Lobate 578
Lobe, Alpha 174, 175, 177, 179, 357, 378, 379, 381, 382
-- antennal 358, 366
-- -- input 384
-- anterior basal 444, 446, 448, 453
-- Beta 174, 175, 177, 179, 357, 371, 378-382

-- brachial 444, 446, 448, 452
-- buccal 444
-- chromatophore 445
-- gonadal 569
-- inferior frontal 446, 453
-- magnocellular 445, 454, 455
-- optic 353-367, 382, 384, 445, 450
-- palliovisceral 446, 454, 455
-- pedal 445, 446, 454
-- posterior basal 444, 446, 448, 453
-- -- buccal 446
-- sub-frontal 446
-- sub-oesophageal 444, 448
-- sub-vertical 446
-- superior buccal 446, 448
-- -- frontal 446, 453
-- vertical 444, 446, 448, 453
Lobster 6, 10, 11, 16, 18, 86, 116, 324-328, 637
-- optic lamina 15
Lobula 175, 177, 185, 189, 191, 192, 194, 201, 355, 357, 359-362, 366, 370, 371, 381
-- giant movement detector (LGMD) 367, 369, 401, 407
-- neurons 374
-- plate 175, 353, 360-362, 370, 371, 373, 375, 376
-- -- columnar organization 363
Local conducting system 221, 222
Locomotion co-ordinate 503
-- in Ctenophora 247-249
Locomotory integration 504
Locust 15, 46, 65, 145, 160, 172, 173, 177, 190, 191, 194, 195, 197, 198, 201, 202, 336, 360, 367, 393-425, 625-627, 632, 635, 637
-- desert 117
-- embryos, identified neuron 48
-- migratory 117
-- neuroactive substance 171-202
-- serotonin-immunoreactive neurons in ventral ganglia 195-197
LPLRF amide 119, 120
Luminescence *See also* photocyte 573-595
-- control, neuropharmacological basis 578
-- extracellular 584
-- intersegmental propagation 585
-- intracellular 574, 580
-- stereotyped defensive reaction 579
Luminescent clouds 586, 592, 593
-- coordination 580
-- flashes 575, 578-580, 588-590
-- frenzy 577
-- glows 585
-- mucus 584, 586
-- organ, photophore 591

-- potential (LP) 575
-- secretion 578, 585
-- systems 592, 593
-- wave 576, 577
Luq cell 32
Lycosidae 304, 313, 315, 317
Lysine 108, 119, 148
Lyssomanidae 313

Macromere 289, 290
-- Yolky vegetal 289
Macromolecular exchange 90
Macrura 324, 337
Macula 453
Mantle 445, 450 452, 454, 455
Maturation inducing substance (MIS) 497, 498
Mayflies 366
Mechanisms of spawning in Echinoidea 559-571
Mechano-receptive channels in Chelicerata 304
Mechanoreceptor afferents in Insecta 375,
-- in Crustacea 323, 334, 338-346
-- -- Ctenophora 247
-- -- Echinodermata 502, 506
-- -- Insecta 632
-- -- Tunicates 530-534, 539
Mechanosensory cells in Annelida 265, 279, 281, 285
-- neurons in Insecta 361, 371, 378, 379
-- -- peripheral fields and sensory terminals 266-272
-- organs in Crustacea 341
-- systems in Insecta 357, 358
Medulla 175, 177, 186, 191-194, 201, 357, 359-363, 366, 370, 371, 375, 450
-- layered organization 362
-- neurons 188
Medullar cord 452
Medusa-like nectophore 576
Medusae 119, 223, 239
Megalopa 328, 342
Melanophores 112
Melatonin 384, 385
Melanotropin 108
Membrane apical 395, 396
-- basolateral 395, 396
-- capacitance 397
-- conductance 401
-- permeabilities of glial cells 67
-- postcytoplasmic densification 17
-- postjunctional specialization 15
-- postsynaptic 13, 14, 118, 401, 403, 404, 409, 411
-- -- polarization 403
-- potential 31, 67, 68, 83, 418, 420

-- presynaptic 106, 118
-- -- potential 403
-- receptoral 501
-- resistivity 397
Memory sites in Insecta 382
Mesocerebrum 511
Mesogastropoda 50, 53
Mesogleal nerve net 578, 579, 580
-- neurons in Ctenophora 246
Mesenteries 221, 222
Mesodermal precursor bandlet 291
Mesoskeleton in Skeleton 217
Mesozoa 63
Metabolic syncytium 68
Metabolite exchanging 87
Metacerebral cell (MCC) in Aplysia 32, 33
-- -- in Insecta 637
-- giant cell (MCG) in Gastropoda 49-53
Metachronal coordination 247, 248
Metameric nervous system 172
Metazoa 3, 4, 213, 216, 484, 497
Methionine 33
-- enkephaline 461, 478, 480
Methyladenine 498
Microglia 79
Microglial cell 78, 79, 82
Micromeres 289
Microtubule function 71
-- synaptic terminal 3, 4
Mollusca 5-10, 16-18, 31, 33, 49, 50, 62, 63, 71, 74, 75, 80, 86, 89, 109-111, 116, 118, 243, 289, 405, 499, 591, 592, 637
Monoamines 2, 30, 92, 491, 496, 497, 504, 505, 583, 625
-- oxidase 173, 499
-- transmitters 91
Monoaminergic mechanism 591
-- receptor 583
Monogenea 7
Monosynaptic connection 401, 405, 406, 412
-- reflex arc 118
Monstrillidae 324
Morphogenesis, control factors 134
Mosquito 147
Moth 6, 10, 15, 16, 78, 135, 192, 420
Motor, nerve net 223-227
-- system in Echinodermata 490-497
-- -- -- Hydromedusae 229-230
Motorneurone 30, 397, 399, 401, 403-405, 407, 410
-- annuli erection 274, 275, 277
-- anterior tergocoxal in locust 422
-- B1 and B2 32
-- C3 109
-- L2, L6, L10, L11 32
-- cholinergic 277

-- co-transmitter 401
-- depressor 419
-- elevator 419, 421
-- extensor 404, 415, 627
-- flexor 404, 415, 416, 627
-- flight 417, 418
-- gill 33
-- heart excitor (HE) 274, 277
-- in Insecta 625, 626, 628
-- large cervical nerve 376
-- -- longitudinal (L cell) 276, 277
-- properties and peripheral fields in Annelida 274-277
-- tergosternal in locust 422
-- thoracic in Insecta 635
Moult 134, 324-327, 345
Mouth part receptor (MPR) systems 326
mRNA 115
-- formation 108
Mucus 277, 497, 585, 591,
Multiciliated cells 281, 283, 284,
Muscarinic receptor 498, 583, 586
Muscle action potentials in Hydrozoa 234
-- anterior byssus retractor 109
-- cardiac in Mollusca 110
-- caudal in tunicate 540, 551
-- circular, embryonic origin 292
-- -- in Annelida 274, 275, 277, 279
-- -- subumbrellar 223
-- contraction, control 625-627
-- dorsal longitudinal in locust 420, 421
-- dorsoventral in Annelida 275
-- ectodermal longitudinal in Hydrozoa 240
-- -- radial in Hydrozoa 230
-- -- striated in Scyphozoa 225
-- endodermal in Cnidaria 110, 123
-- extensor tibia in Insecta 158
-- flight 376
-- hindgut in Insecta 188
-- hyperneural in Insecta 158
-- innervation in Echinodermata 490-494
-- -- in tunicate 550
-- intervertebral in Echinodermata 487, 493, 495, 507, 508
-- lantern retractor 493, 495
-- locomotor in Chaetognatha 478
-- longitudinal, embryonic origin 292
-- -- in Annelida 274-276, 279
-- -- -- gastrozooid 241
-- -- response to stretch 282
-- -- retractor in Echinoidea 564
-- mantle 455
-- mesodermal in Echinodermata 487, 489
-- metathoracic tarsal levator in locust 410
-- neck in Insecta 376, 377

-- oblique, embryonic origin 292
-- -- in Annelida 275, 279
-- papillar in Annelida 281
-- peripheral modulation 277, 278
-- procteal in Insecta 112, 114, 117
-- proctodeal in Insecta 158
-- pyloric in Insecta 637
-- radial in Echinodermata 498
-- -- of the mesenteries 119
-- radula protractor 109
-- receptor organ 323, 330, 332, 335, 336
-- retractor of the mesenteries 119
-- skeletal in Insecta 188
-- sphincter in Anthozoa 221
-- striated subumbrellar 226, 230
-- subepidermal in Ctenophora 244
-- tail in Echinodermata 487
-- tentacle in Gastropoda 109
-- tentacular longitudinal in Ctenophora 249
-- tergotrochanter 375
-- visceral in Echinoidea 491, 492
-- -- -- Insecta 173
-- walking leg in Crustacea 112
Mushroom body 201, 202, 353-358, 360, 267, 271, 278-284, 386
-- -- calyx 175-178
-- -- GABAi fibers 182
-- -- general organization 378-382
-- -- immunocytochemical mapping 173-191
-- -- in Insecta 631, 638
-- -- peduncle 175-178
Mussel 5
-- fresh water 10
Mutable connective tissue 495
Myelin 76, 78
Mygalomorpha 313
Myoactive substance 117, 184
Myochordotonal organ 332
Myoepithelium 234, 235, 492, 494
Myofilaments 487, 488, 492
Myogenic movements 529
Myoneural junction in Annelida 15
Myriapods 303, 305, 309, 311
Mysidacea 342-344

Nauplius larvae 326, 328, 329
Near field 328
Nectophore 240, 241, 576
Nematocyst 217
Nematoda 15, 649
Neogastropoda 50
Nephridia 293
-- neurones associated 283, 284
Nephridial function 284
-- neurones, afferents 280
Nephridiopore 270, 280
Nernst equition 67

Nerve, *see also* connective, giant fibres and axon
-- antennal in Insecta 178, 189
-- arm 445
-- anterior occulomotor 447
-- -- vena cava 447
-- caudal in Chaetognatha 463, 472, 475, 477
-- circular in Echinoidea 559
-- collar 445, 446
-- commissural in Chaetognatha 462
-- cord, abdominal in Insecta 588
-- -- circumanal in Echinoidea 565, 566, 567
-- -- crush in Annelida 79
-- -- dorsal in tunicate 528, 531, 539
-- -- ectoneural 486
-- -- in crayfish 337
-- -- -- Echinoidea 565
-- -- mechanoreceptors 334
-- -- radial 483, 485, 486, 488, 489, 505, 508, 512, 559, 560, 567
-- -- ventral, development 291
-- -- -- in Annelida 15, 266, 267
-- -- -- -- Insecta 158, 160, 395, 396, 407, 410, 629, 631, 633-635
-- coronal of Chaetognatha 462-464, 466, 471, 472
-- dorsal of Chaetognatha 463, 470, 471
-- ectoneural in Echinodermata 18, 487, 492, 498, 499, 513
-- epithelio-motor 543
-- fibers, cholinergic in Echinoida 564
-- -- in tunicate 529
-- frontal in Insecta 189
-- hyponeural in Echinodermata 484, 489, 496, 498, 508
-- infundibuli 446
-- labial 447
-- lateral oesophageal of Chaetognatha 463, 471
-- like juxtaligamental cell 496
-- mandibular of Chaetognatha 463, 470, 471, 473
-- marginal in Echinodermata 485
-- maxillary-labellar in Insecta 176
-- musculi retractoris pallii mediani 447
-- neck motorneurons 376
-- net 213-252
-- -- mesogleal 578-580, 592
-- -- in Cnidaria 6, 118
-- -- -- colonial Anthozoa 218-221, 576, 577
-- -- -- Ctenophora 246
-- -- pulse 576, 577
-- -- zooid 577
-- optic of Chaetognatha 462-467, 471, 472
-- pallial 445, 453
-- peripheral of Chaetognatha 461, 466
-- plexus, ectoneural 491, 492
-- -- subepidermal 592
-- posterior lip in Gastropoda 32
-- post-orbital in Cephalopoda 453
-- radial in Anthomedusae 239
-- -- -- Chaetognatha 463, 472, 477
-- -- -- Limnomedusae 230
-- ring 63
-- -- aboral 559, 564, 570
-- -- circumoral in Echinodermata 485-487, 503, 505, 514, 515
-- -- in Anthomedusae 230, 231
-- -- -- Limmomedusae 230
-- roots 42
-- -- in Annelida 278
-- -- -- Crustacea 346
-- staticus in Cephalopoda 447
-- sympathique in Cephalopoda 447
-- trunks, circumferential 266
-- ventral oesophageal of Chaetognatha 463, 472, 474
-- vestibular of Chaetognatha 463, 470, 471, 473
-- visceral in Cephalopoda 445, 447
Nervi corporis cardiaca 147
Nervous chain, metamerisation 306
-- mass, sub-oesophagic 306, 308
-- system, occurence of glia 63
-- -- primitive 574
-- -- segmental architecture 293, 294
-- tissue, regulation of energy 92
Neural lamellae 71, 74
Neuroactive substances *see also* neurotransmitter, transmitter, peptide, neuropeptide, 171-202, 353, 355, 360, 378, 401, 469, 470, 477, 478, 480
-- -- in ventral ganglia 194-200
-- -- patterns of distribution 200-202
Neuroblast processus 305, 309
Neuroectodermal organ 583
Neuroeffector junction in ctenophore 13
Neuroendocrine organ 6
Neuroepithelial transmitter 544
Neuroglia 62, 73, 88, 395
-- nerve glue 70
Neuroglial cell 61-93
Neurohaemal complex 134
-- organs 145
Neurohormone D 145
Neurokinin A 108
Neurolemma 306, 308
Neuromeres 303, 311
Neuromodulation (-tor) 401, 420, 424, 480, 498, 499
Neuromuscular junction, *see also* synapes neuromuscular, 3, 5, 13,

86, 116, 118, 190
-- -- in Cephalopoda 462, 478, 479
-- -- -- Ctenophora 250
-- -- -- Echinodermata 6, 560
-- -- -- Gastropoda 17
-- -- -- Hydrozoa 242
-- -- -- Scyphozoa 223
-- -- -- tunicates 539, 544
-- system in Echinodermata 485, 499
-- -- -- Insecta 401, 648
-- -- -- tunicate 550-552
Neuro-neuronal junctions in Hydrozoa 242
Neuron *see also* specific names
-- central mechanosensory 49
-- columnar 355
-- descending *see also* descending, 386
-- development 415
-- identified *see also* identified, 1, 41-54, 355, 413-415, 646
-- insulation 74-78
-- L 397, 406
-- lobula 371, 374
-- mapping 171-202
-- R3, R14 116
-- removal 78-80
Neuronal circuit, modulation 636
-- conduction system SS-1 and SS-2 123, 221, 222
-- exchange, glial cells 87-93
-- growth 78-82
-- integration 2
Neuropeptide *see also* peptide, 2, 91
-- evolutionary conservation 133-162
-- functions 184
-- immunocytochemistry 182-189
-- list of 110-114
-- synthesis 107-109
Neuro-photocyte transmission 574, 578, 580, 583, 590, 591
Neurophysins 184
Neurosecretory cell in crayfish 8
-- -- -- Echinodermata 486, 491, 493, 495, 496
-- -- -- Insecta 397, 399
-- ending 13-18
-- motor junction 3
-- neuron in Aplysia 8
Neurotensin (-like) 108, 239, 461, 478, 480
Neurotransmitter *see also* transmitter 31, 32, 34, 461, 462, 480, 498, 499
-- mechanism, identification 574
-- neuropeptides 105-123
Nicotinic receptors 86, 498
Nodula 186, 190
Non-glomerular neuropil 353, 356, 363, 367, 378, 379, 381, 383, 384
Non-muscular contraction system in Echinodermata 490, 491

-- tissues innervation in Echinodermata 496, 497
Non-spiking neuron 399, 412
Noradrenaline 189, 191, 384, 497-499, 554, 578, 589, 590
Notaspidea 50
Notochord 550
Notopod 584, 585
Noxious sensitive cell (N cell) 30, 266, 267, 272, 287
-- -- -- receptive fields 269
Nudibranchia 50, 52

Ocellar foci 356, 360
-- interneurons in Insecta 371, 375, 366, 382
-- system in Insecta 181, 353, 354, 358-360, 406
Ocelli in Cubomedusae 11
-- -- Echinodermata 500, 501
-- -- Hydrozoa 236, 238, 239
-- -- Insecta 15, 178, 355-360, 367, 370, 403, 629
-- -- Scypomedusae 226
Octocoral 12, 217, 222
Octopamine 69, 73, 172, 190, 199, 401, 420, 590, 591, 637
-- containing neurons, patterns of distribution 201
-- functional role 190
-- transperineurial potassium permeability 73
Octopaminergic neurons in brain 190, 191
Octopoda 443, 444, 448, 452, 454
Octopus 7, 17
Oculo-Ganglionar complex 451
Oesophageal foramen 174, 178, 185, 383
-- pores 343, 345
-- receptors 341, 345
-- sensors 323
Oikoplastic epithelium 533
Oikopleurid larvacea 539
-- -- excitable epithelia 537-540
Olfactory detection in Crustacea 341, 345
-- relay neurons in Insecta 374
-- system in Insecta 358
Oligochaeta 62, 265, 266, 279, 289, 292, 293, 580
Oligodendrocytes 78
Oligopeptide 560
Omega neurons 605, 610, 611, 615
-- -- auditory responses 612
-- -- morphology and physiology 616
Ommatidium 84, 116, 325, 366
-- fiber pathways 364
Onychophora 303, 308, 309, 311, 312

669

Oozooid 528, 531, 542-545, 548
-- epithelial conduction system 549
Ophiuroidea 485, 486, 487, 489, 493, 495, 496, 503, 505-507, 593
Opilions 303, 304, 305, 308, 311, 312, 317
Opioids 133, 153-156
Opisthobranchia 50, 52, 53
Opisthosoma 306, 310
Optic channel in Chelicerates 313, 314
-- foci 182, 358, 361, 363
-- interneuron 384
-- lamina in fly 88
-- lobes 171-180, 588, 589
-- -- immunocytochemical mapping 173-194
-- mass 304, 315
-- -- of lateral eyes in Araneide 314-316
-- nerve of Chaetognathe 462-467, 471, 472
-- tract 371
-- tubercle 178
-- vesicle 450
Optomotor interneuron 631
Oral arms 227
-- disc 221
-- poles 244, 560
Orthochromatic cell 584, 585
Orthoptera 46, 47, 393, 413, 415
Oscillator system ("O") 238
-- -- neurons 228
Oscular membrane 216
Ostracod 344
Outer nerve ring systems in Hydrozoa 235-239
-- skin pulse (OSP), conduction 546
-- -- -- system 543, 546, 548, 553
Oval organ 334, 337
Oxyopidae 313
Oxytocin 105, 106, 160, 184, 239
-- vasopressin - reactive 637
Oyster 568

Paracrine, action 106
Pagoda cell 31
Palinurid decapod 324
Palpons 241
Pancreatic polypeptide-like immunoreactivity 198, 199
Papillae 461-464, 470, 500
Paracrine action 115
Paragnath 330
-- receptors 331-334
Paramembranous densities 3-5, 9, 12, 17
Paramyosin-like filaments 490
Pars intercerebralis 153, 178, 184, 185, 189, 201
Parasitic copepods 324

Path finding of axons, mechanisms 287
Pedal disc 222
Pedicellar giant campaniform sensillum 366, 375
Pedicellariae 490, 493, 500, 502
Pedicellus 366
Pennatulid 219
Pennatulacea 122, 576, 577, 579
Peptidases 108
Peptide *see also* neuropeptide, polypeptide and specifics names, 69, 92, 105-123, 133-164, 243, 265, 401, 497, 625, 636, 637
-- brain-gut 134
-- cardioactive 108-120, 499
-- -- amino acid sequences 145
-- diuretic in insect 159
-- effect in glial cells 69
-- FMRF- like 276
-- hyperglycaemic, purification of 144-146
-- myoactive 113, 117, 182
-- vasoactive intestinal (VIP) 146, 184
Peptidergic neurones 105, 478
-- -- patterns of distribution 201
-- pathways 382
Pericardial organ 112, 116, 117
Perineurial cell 71, 74, 80, 396
-- glia 73, 395, 396
-- junctions 73
Perineurium 61, 64, 71, 73
Peri-pharyngeal band 333
-- system 330
Peripheral nerves, organisation of axons 285-289
-- nervous system embryonic origin 289-293
-- -- -- in Echinodermata 503
-- -- -- -- Insecta 355, 648
-- -- -- -- tunicate 529
Peripherally located neurones in Annelida 278-285
Peripheryngeal ciliated bands 541
Periplanetin 113, 145
Perisympathetic organs 355
Perivascular glial cell 74
Phagocytic activity of glial cell 78-80
Phenolamine octopamine 190
Phenylalanine 148
Pheromon 313
-- detectors 341
-- like, role of 498
Phonotaxis in cricket 413, 603-617, 614
Phosphatases 80
Phospatidyl inositol metabolic cycle 69
Phosphoglycerolipids 78
Phospholipids 78
Photic inhibition 589
-- modulation 589

-- response in Echinodermata 506
-- stimulation 588, 589
Photocyte *see also* luminescent system, luminescence 573-595
-- in Ctenophora 245
-- -- lantern 587
-- intracellular control mechanism 594
-- location 577
-- memrane resistance 590
-- neurotransmitter 578, 580
-- postsynaptic potential (PSPs) 581
Photocytic excitation -luminescence 590
Photogenic tissues 578
Photomotor neuron 588, 589, 590, 594
Photophore 591
Photoprotein 583, 586
-- polynoidin 580
Photoreceptors 74, 80, 85, 92, 594
-- axonal organization 366
-- in barnacle 16
-- -- Cephalopoda 17, 452
-- -- Echinodermata 500, 501, 503
-- -- Hydrozoa 239
-- -- Insecta 357, 360, 363
-- -- tunicate 534
-- photostimulation of 84
-- small diameter afferents 280
-- transmitter 194, 362
Photosensory neurons 265
Photosomes 580, 583
Phototactic movements 461, 462
Phrynes 311
Phyllopodous 328
Phylogenetic homology 44
-- proximity in Gastropoda 49-53
Phylogeny 44, 45
Physonectid nervous system 240, 241
Picrotoxin 403, 419, 568, 570
Pigmented cell 360, 582
Pisauridae 313
Plate setae 342, 343
Platyhelmintha 2, 7, 14, 62, 63
Pleopods 324, 330, 331, 332
-- receptor 331
Plexus, basi-epithelial 485, 486, 503
-- endo-epithelial 499
-- visceral 497
Pneumatophores 240
Podial pit 500
-- sensory receptor 504
Polarized light perception 366, 376
Polster cells 248, 249
Polychaeta 4, 7, 14, 15, 18, 266, 279, 289, 309, 311, 580, 585, 591, 592
Polyclonal antisera 118
Polynoid 573, 580-583, 586, 591, 594
-- system 584

Polyp 10-12, 218
-- colonial conducting system 218-221
-- fresh water 11
Polypeptides 33
-- in glial cells 88
-- pancreatic (PP) 133, 147-149, 184, 194
-- vasoactive -intestinal 107
Polypoid 217
Porifera 2, 63, 213-252
-- conducting system 213-216
Postembryonic period in chelicerates 306-308
Posterior mechanosensory antennal centers 356
Postsynaptic densities 4, 5, 6, 18
-- dyad 14
-- membrane 13, 14, 17, 401, 403
-- potentials 31, 106, 405, 409, 411
-- spine 16
-- -- in octopus statocyst 7
Potassium accumulation 393, 395, 397, 410
-- channel 82, 537
-- concentration 68, 71, 73, 84, 85, 225, 229
-- conductance 69, 86, 399
-- currents 85, 229, 232, 250
-- extracellular 68, 82, 410
-- permeability of glial cell 67, 69, 73
-- pump 74, 85
-- regulation of 73
Preantennal segment 355
Premandibular segment 355
Pressure mechanosensory neurons, axon outgrowth 272-274
-- sensitive cell (P cell) 30, 31, 266, 267, 271, 272, 285, 287
-- -- -- receptive fields 269
Presynaptic, density 3, 5, 7, 16-18
-- inhibition 393, 410-413
-- membrane, v-shaped depression 16
-- spike 407
-- triad 6, 9, 13
Prey capture and ingestion in ctenophore 249, 250
Procerebrum 51
Processing network, definitions 623
Proctolin 112, 116, 117, 172, 184, 194, 637
-- in Insecta 157-159, 363, 384, 401
-- like immunoreactive neurons in Insecta 188, 189, 197, 198, 201
Prolactin 106
-- anti-rat 194
Pronase 537
Pro-opiomelanocortin 108, 153, 156
Proprioceptive feedback 416, 631
-- information 416, 420
-- innervation in Crustacea 332
Proprioceptor chordotonal 326

-- in Annelida 278
-- -- Crustacea 327, 330-338, 341
-- -- Echinodermata 501, 502
-- -- Insecta 634
-- -- tunicate 530
Prosobranchia 50, 53
Prosopyles 214
Prosternal organ 377
Proteases 107, 288
-- chymotrypsin 34
-- trysin 34
Proteoglycon matrix 496
Prothoracic auditory neurons 612
-- motorneurons 377
Protocerebral bridge 174, 175, 177, 315, 356, 384
-- centers 353, 357
-- commissure 378
-- hemispheres 147
-- mediocaudal cluster 175
-- structure, classification of spiders 304
Protocerebrum, immunocytochemical mapping 173-191
-- in Chelicerata 304, 308, 310, 312, 316, 317
-- -- Insecta 148, 152, 174, 175, 178, 355-360, 363, 366, 367, 371, 374, 378, 380, 588
-- organization 173-191, 383-386
Protochordate 593
Protolin-like immunoreactive neurons 171-202
Proventriculus 154
Proximal fan 280
Pseudoscorpions 303, 304, 310, 311, 312
Puerulus 325, 328
Pulmonata 49, 50, 52, 53
Pycnogonids 303, 305, 309, 311
Pyroglutamate 106, 108, 121, 146
Pyrosomida 529, 550
Pyruvate kinase 92

Rachidial contraction 577
Radial attachment zone 76
-- digestive canal 235
Radii 514
Radioimmunoassay, RF amide 120
Radula 49, 452
Ragworm 265
Ramus 327
Receptor acceptor 647
-- acetylcholine (ACh) 31, 402, 403
-- Beta-adrenergic type 578
-- enkephalin 154
-- glial cell 69
-- insulin 142
-- lateralis 530
-- monoaminergic 583

-- muscarinic 498, 583, 586
-- non-ciliated in Echinodermata 500
-- prothoracicotropic hormone 139
-- stretch afferent 278-285
Reflex control in Echinodermata 503
-- republic of ganglion 559
Regeneration in Crustacea 327
-- of the nervous system in Echinodermata 505
Relaxin 161, 163
Repaird mecanism of glial cell 80
Reproduction, control "factors" 134
Reptantia 308
Retina see also photoreceptor
-- in Cephalopoda 447
-- -- Insecta 65, 67, 68, 83, 84, 360, 363, 366
-- lamina complex 589
-- stratified in Chelicerata 314
Retinal pigment cells 112
-- s-antigen 384
Retinotopic columnar elements 191, 361
Retrocerebral complex 147, 150, 187
-- organ of Chaetognatha 465-467, 469
Retzius (R) cell 30, 31, 33, 48, 277, 294
RF amide 109, 119-122, 239
-- immunoreactive neurons 240
-- like 121
Rhabdite-secreting cell 14
Rhabdom 84, 360
Rhabdomeres 16
Rhizocephala cirripedes 324
Rhodamine 239
Rhodopsin-like substance 462
Rhopalia 226, 277, 252
Rhopdial pacemakers 226
RNase treatment 34
RNA 647
Rohde's fibre 278
Roots in medicinal leech 267
RY amide 119

Sacoglossa 50
Salp, excitable epithelia 540-549
-- mechanoreceptor cell 531, 532
-- brain, organisation 553
Salpida 529, 536
Salticidae 304, 313-317
Sarconeural junction 14
Sarcoplasm, granular 16
Scalariform junction 62, 64
Scaphognathite 323, 337
-- oval organ 334
Scavenger cell 78, 79
Schwann cell 67, 69, 78, 86
-- like cell 65, 88
Sclerites 333

Sclerospongiae 214
Scolopale 341
Scolopendre 309
Scolopideal sensillae 366
Scolopidia 336, 342
Scorpions 303, 310-313
Scyphistoma 12
Scyphomedusae 12, 13, 223, 227, 252
Scyphozoa 6, 9, 11, 12
-- nervous system 223-227
Sea anemone 9, 11, 12, 119-123, 217, 218, 221
-- cucumber 568
-- hare 16
-- pansy 576
-- pen 11, 13
-- slug 30, 34
-- urchin 2, 5, 10, 489, 492, 493, 495, 498, 501, 502, 505, 560, 564, 568
Searching behaviour in Echinodermata 506
Seastars 568
Second messenger system 69, 92, 108
Secretin 146, 184
-- amino acid sequences 145
-- antisera 146
Segmental architecture of nervous system 293, 294
Sensillae chemosensory 345
-- cuticular articulate PEG (CAP) 324
-- in Annelida 281, 284, 285, 288, 289
-- -- Crustacea 324, 325, 327, 336, 338, 342-346
-- -- Insecta 366
Sensory cell, see also pressure, touch and noxious sensitive cell, 278
-- -- in Chaetognatha 464
-- fields, development in Annelida 272-274
-- interneurone-motor reflex 486
-- papilla 582
-- receptors in tunicates 530-534
-- system in Crustacea 323-346
-- -- -- Echinodermata 499-503
-- -- -- Insecta 357-367
Sepioidae 454
Septal junction 19
Septate junction 62, 64, 73, 77, 395, 579
Serotonergic cell 49
-- neuron 53
-- like neuron 594
-- pathways 382
Serotonin (5-HT) 30-35, 49, 50, 67, 69, 83, 86, 88, 91, 106, 108, 118, 172, 173, 191, 194, 199, 201, 277, 360, 379, 382-385, 461, 469, 477, 480, 568, 582, 583, 637

-- containing retzius cells 277
-- immunoreactive neuron 578
-- -- -- in insect brain 173-181, 195, 197, 361, 380
-- -- -- patterns of distribution 200
Setae 325-328, 338-346
Sheath of glial cells 75, 76, 88
Shrimp 78, 116
Sicariidae 316
Silkworm 143, 147, 161
Sinephrine 590
Siphonophora 12, 227, 241, 575
-- nervous system 240-242
Siphonozooid 576, 577
Skeleton, hydroelastic 490
Skeletal muscle system in Echinodermata 485, 490-, 491, 493, 494
Skin mechanosensory neurons 266
-- reinnervation in Annelida 272
Slow extensor tibiae motor neuron (Seti) 46, 47
-- system 1 and 2 221, 222, 240, 241
Slug 52
Snail 17, 29, 30, 33-36, 52, 65, 67, 81, 83, 85, 89, 91, 108, 115
Sodium channels 399
-- conductance 399
-- current 229, 250
-- potassium pumps 73
-- pump 74
Solifuge 308, 310, 311, 312
Somatostatin 106, 107, 184, 194
-- like immunoreactive neuron 362
Song motor pattern 394
-- recognition in cricket 603-605
Sound localization in cricket 609-613
Spatial buffer theory 82-86
Spawning, mechanisms in echinoid 559-571
Sphaeridia 501, 506
Spherical organ in Chelicerata 303
Sphingomyelin 78
Spiders 303, 304, 311, 313, 316, 317, 336
-- hunting 304, 315
-- phylogeny 313
-- primitive 313
-- sedentary 312, 316
-- spinner 314
-- tetrapneumones 310
-- vagabond 312, 314
-- whip 16, 303
Spike initiation zone 398-401
-- tetrodotoxin insensitive 507
Spiking interneuron list in locust 408
Spine synapse 3, 11, 16, 17
-- -- in cephalopods 4
Spiny lobsters 328
Spirocyst 217
Sponges 214, 650
-- nerve cells 214

Squid 1, 17, 67, 69, 75-78, 86, 88, 398, 399, 443, 444, 450, 454-456
Stages, interconnexions 631
Starfish 485, 500-502, 505
Station, definitions 623
Statocyst, bearing structure 244
-- in Ctenophora 247, 248
-- -- Cephalopoda 7, 17, 445-450, 453, 454
-- -- Larvacea 534
-- -- Scyphomedusa 226
-- seta in Crustacea 342
Statolith 247
Stauromedusae 11, 223
Stem 240, 241
-- nervous system 241
Stomatogastric nervous system 172, 312, 355, 356
Stomatopoda 328, 330
Stomodeum 447, 452
Stretch receptor afferents 278-285
-- -- in Annelida 286
-- -- -- Crustacea 337
-- release, depolarising responses 283
Stylommatophora 50, 52
Subgenual organ 617
Sub-oesophageal brain mass 448, 452
-- complex 444
Subperineurial glia 73, 395
Subspherical organs in Chelicerata 303
Substance P 108, 184, 239, 384
Subsynaptic cisterna 6, 14, 17
-- -- of endoplasmic reticulum 9, 12
Sucrose 74
Sulphatides 78
Supra-oesophageal mass in Cephalopoda 444
Surface antigens 287
-- -- in embryonic peripheral nervous system 287-289
-- sensory system in Crustacea 338-346
Swimming motor neuron, electrical coupling 232
-- -- systems in Hydrozoa 230-235
Symphyla 303, 309, 311
Symplasma 214
Synapses, bursting system 242
-- chemical 3-6, 8, 10, 11, 18, 19, 31, 32, 36, 118, 2116, 228, 401, 403, 409, 469
-- -- bidirectional 8, 227
-- -- inhibitory 16, 32, 413, 418, 510, 511
-- electrical *see also* gap junction, 3-7, 13, 15, 16, 19, 29-36, 216, 227, 242, 407, 415, 490, 494
-- epithelio-neural 548
-- giant of squid 455
-- in Annelida 14

-- -- Arthropoda 15, 16
-- -- Chaetognatha 468, 478
-- -- Cnidaria 7-13
-- -- Ctenophora 13, 14
-- -- Echinodermata 17, 18, 487, 489, 490, 494-495
-- -- Mollusca 16
-- interneuronal 5, 11-14, 18, 393, 435
-- neuromuscular *see also* neuromuscular jonction, 5, 6, 9, 10-12, 16, 18, 109, 234, 235
-- -- in Crustacea 16
-- -- -- Ctenophora 9
-- -- -- Echinodermata 489
-- -- -- Hydrozoa 237
-- photoreceptor 358
-- symmetrical 7, 11, 14, 223
-- ultrastructure 3-19, 489, 490
Synaptic connectivity 44
-- vesicles 3-19, 118, 277, 469, 470, 487, 491-493, 495, 498, 499, 580
Syncytium 82

Tagmatisation 327
Tanaids 328
Taste in Crustacea 345
Tectum 35
Teloblast 289, 290
-- ectodermal 292
-- -- descendant pattern 293
-- mesodermal 292
Telson flexor insertions 333
-- setae in Crustacea 342
Tentacle 217, 221, 222, 227, 230, 237, 245 249
-- activation system 222
-- contraction control 238
Tentacular filaments 245, 249
Terminal pore 567
Test in Echinoidea 560, 562, 563
-- -- tunicate 529, 530, 531, 543, 548
Tetraethylammonium (TEA) 399, 423
Tetrodotoxin 229, 507, 537
Teuthoidea 454
Thaliacea 527, 529, 530
Thecosomata 50
Theraphosidae 313
Theridiidae 316
Thin-walled setae 345
Thomidisae 313
Thoracic neurohaemal areas 201
Thoracico-coxal, muscle receptor organ 337, 342
Thorax 375
Through conducting fiber in cricket 606
-- -- nerve net (TCNN) 218, 221, 222

674

Tight junction 62, 64, 71, 73, 77, 395, 396
Tobacco hornworn 117, 135
Tonofibrils 71, 500
Touch, sensitive cell (T cell) 30, 266, 267, 271, 272, 282, 285
— — — distribution 268
— — — receptive fields 269-271
Trabecular syncytium 214, 216
Tracheal end-organ 589, 590
— system 610
Trachymedusae 235, 236
Tractus olfactorio-globularis 190
— opticus 447
Transglial channels 88, 90
Transmembrane potential in leech 279
— — — locust 409
Transmitter in Echinodermata 498, 499
— — Hydrozoa 242, 243
— — Insecta 401-403
— release, threshold 403, 409
Triadic densifications 4
Tritocerebral commissure giant neuron 629
Tritocerebrum in Chelicerata 312
— — Insecta 174, 177, 178, 355-360, 371
— immunocytochemical mapping 173-191
Trocophore larva 289, 309
Trophospongia 64, 65, 87-89
Tryptophan 142, 148, 173
— hydroxylase 173
Tube dwellers 265
Tubocurarine 544
Tubular lattices 90
Tunicate 2, 527-555
Two-way conduction junction 227
Tympanal membrane 610
— organ 605, 617, 635

Uniciliated cells 283
Uridine 88
Urocteidae 316
Uropods 323, 327, 330-337
— coxa 344
— muscle receptor organ 333-337, 343, 344
— receptors 331
Uropygids 312, 313

Vasopressin 105, 106, 119, 133, 160
Vasotocin 184
Ventral bodies 356
— cerebral organ 305
— chain, *see also* nerve cord
— — differentiation in Chelicerata 306
— — in chelicerata 311
— — — Crustacea 324
— organs, embryonic 303-317
— — posterior 306
— thoracic neurons 199
Vertical lobe system 444
— motion sensitive neuron 373-377
Visceral innervation in Echinodermata 486
Visual channels 304
— cortex in Araneida 314, 315
— input neuron 367, 378
— interneuron 374
— system in insect 354, 358, 360-366

Water-movement detecting neurons 265, 266
— — detection in leech 266, 289
— — detectors, afferent 280
— vascular system, muscles of 490, 492-494
Wave-like response 576
Worm *see also* arroworm, chinese silkworm, flatworm, indian silkworm, ragworm
— chaetopterid 573, 580, 583-586
— fan 265
— nereid 281
— peacock 265
— polynoid 573, 580-583, 586, 591, 594

Xiphosura 303, 310, 311, 312

Zoea 328, 342
Zooids 545-549
— locomotor activites 545

Tight junction 82, 64, 71, 75, 77, 395, 556
Tobacco hornworm 172, 155
Torotrix iris 71, 500
Touch, sensitive cell (T cell) 30, 266, 261, 271, 272, 282, 285
— distribution 268
— receptive fields 269–271
Trabecular synaptism 214, 216
Tracheal end-organ 583, 590
— system 610
Tract-neurones 255, 256
Tractus olfactorio-globularis 100
— opticus 442
Transjoint channels 89, 90
Transmembrane potential in toad 379
— toad 405
Transmitter in Echinodermata 583, 599
Trehalose 242, 243
— Insect 401–403
— release threshold 402, 405
Trieste densifications
Triloarchetal nommissural giant neuron 629
Tritocerebrum in Chelicerata 312
Insects 174, 177, 178, 385–360, 371
— immunocytochemistry (reaction) 177, 181
Trochantera femur 280, 705
Tropomyosin c1, 60, 71–82
Trypsinolysin 140, 142, 175
Turbidylmea 76
Ultraviolet 256
Urochordates 514
Urodele reflexes 90
Use sets 2, 527–557
Wet-dry proprioceptor (muscle) 707
— Urodeles reflexes 670
— USE sets 514, 455

Utricula 349
Vagine
— of brain 85
— of the eye
Vasinoma 621, 672

Vasopressin 108, 105, 179, 182, 180
Yasotocin 184
Ventral bodies 388
— caruncular organ 205
— chain, see also nerve cord
— — differentiation in Chelicerata 308
— — in Chelicerata 311
— — Crustacea 324
— — organs, embryonic 305–312
— — posterior 308
— — thoracic neurones 193
Ventral lobe system 644
— motion sensitive neuron 374–377
Visceral innervation in Echinodermata 431
Visual channels 306
— cortex in Anura 68 314, 315
— input neuron 307, 318
— Interneuron 314
— system in insect 577, 459, 560–599

Water-movement detecting neurons 255, 256
— — detection in leech 265, 266
— — detectors, afferent 260
— — vascular system, muscles of 490, 492–494
Wave-like responses 278
Worm, see also Annelida, Chacta
shanna, Flatworm, Rotam
Nemera, Nemertini
— shaenorbida 572, 590, 595, 596
— fan 256
— nereid 281
— peanut 156
— polynold 575, 580–585, 586, 701, 734

Ytokaine 709, 710, 711, 512

MIX
Papier aus verantwortungsvollen Quellen
Paper from responsible sources
FSC® C105338

If you have any concerns about our products,
you can contact us on
ProductSafety@springernature.com

In case Publisher is established outside the EU,
the EU authorized representative is:
**Springer Nature Customer Service Center GmbH
Europaplatz 3, 69115 Heidelberg, Germany**

Printed by Libri Plureos GmbH
in Hamburg, Germany